HANDBOOK OF ALGEBRA

VOLUME 4

HANDBOOK OF ALGEBRA
Volume 4

edited by
M. HAZEWINKEL
CWI, Amsterdam

AMSTERDAM • BOSTON • HEIDELBERG • LONDON • NEW YORK • OXFORD
PARIS • SAN DIEGO • SAN FRANCISCO • SINGAPORE • SYDNEY • TOKYO

North-Holland is an imprint of Elsevier

North-Holland is an imprint of Elsevier
Radarweg 29, PO Box 211, 1000 AE Amsterdam, The Netherlands
The Boulevard, Langford Lane, Kidlington, Oxford OX5 1GB, UK

First edition 2006

Library of Congress Cataloging-in-Publication Data
A catalog record for this book is available from the Library of Congress

British Library Cataloguing in Publication Data
A catalogue record for this book is available from the British Library

ISBN-13: 978-0-444-52213-9
ISBN-10: 0-444-52213-1
ISSN: 1570-7954

For information on all North-Holland publications
visit our website at books.elsevier.com

Printed and bound in The Netherlands

06 07 08 09 10 10 9 8 7 6 5 4 3 2 1

Preface

Basic philosophy

Algebra, as we know it today (2005), consists of a great many ideas, concepts and results. A reasonable estimate of the number of these different "items" would be somewhere between 50 000 and 200 000. Many of these have been named and many more could (and perhaps should) have a "name", or other convenient designation. Even a nonspecialist is quite likely to encounter most of these, either somewhere in the published literature in the form of an idea, definition, theorem, algorithm, ... somewhere, or to hear about them, often in somewhat vague terms, and to feel the need for more information. In such a case, if the concept relates to algebra, then one should be able to find something in this Handbook; at least enough to judge whether it is worth the trouble to try to find out more. In addition to the primary information the numerous references to important articles, books, or lecture notes should help the reader find out more.

As a further tool the index is perhaps more extensive than usual, and is definitely not limited to definitions, (famous) named theorems and the like.

For the purposes of this Handbook, "algebra" is more or less defined as the union of the following areas of the Mathematics Subject Classification Scheme:

- 20 (Group theory)
- 19 (K-theory; this will be treated at an intermediate level; a separate Handbook of K-theory which goes into far more detail than the section planned for this Handbook of Algebra is under consideration)
- 18 (Category theory and homological algebra; including some of the uses of category in computer science, often classified somewhere in section 68)
- 17 (Nonassociative rings and algebras; especially Lie algebras)
- 16 (Associative rings and algebras)
- 15 (Linear and multilinear algebra, Matrix theory)
- 13 (Commutative rings and algebras; here there is a fine line to tread between commutative algebras and algebraic geometry; algebraic geometry is definitely not a topic that will be dealt with in this Handbook; there will, hopefully, one day be a separate Handbook on that topic)
- 12 (Field theory and polynomials)
- 11 The part of that also used to be classified under 12 (Algebraic number theory)
- 08 (General algebraic systems)
- 06 (Certain parts; but not topics specific to Boolean algebras as there is a separate three-volume Handbook of Boolean Algebras)

Planning

Originally (1992), we expected to cover the whole field in a systematic way. Volume 1 would be devoted to what is now called Section 1 (see below), Volume 2 to Section 2, and so on. A quite detailed and comprehensive plan was made in terms of topics that needed to be covered and authors to be invited. That turned out to be an inefficient approach. Different authors have different priorities and to wait for the last contribution to a volume, as planned originally, would have resulted in long delays. Instead there is now a dynamic evolving plan. This also permits to take new developments into account.

Chapters are still by invitation only according to the then current version of the plan, but the various chapters are published as they arrive, allowing for faster publication. Thus in this Volume 4 of the Handbook of Algebra the reader will find contributions from 5 sections.

As the plan is dynamic suggestions from users, both as to topics that could or should be covered, and authors, are most welcome and will be given serious consideration by the board and editor.

The list of sections looks as follows:

Section 1: Linear algebra. Fields. Algebraic number theory
Section 2: Category theory. Homological and homotopical algebra. Methods from logic (algebraic model theory)
Section 3: Commutative and associative rings and algebras
Section 4: Other algebraic structures. Nonassociative rings and algebras. Commutative and associative rings and algebras with extra structure
Section 5: Groups and semigroups
Section 6: Representations and invariant theory
Section 7: Machine computation. Algorithms. Tables
Section 8: Applied algebra
Section 9: History of algebra

For the detailed plan (2005 version), the reader is referred to the Outline of the Series following this preface.

The individual chapters

It is not the intention that the handbook as a whole can also be a substitute undergraduate or even graduate, textbook. Indeed, the treatments of the various topics will be much too dense and professional for that. Basically, the level should be graduate and up, and such material as can be found in P.M. Cohn's three volume textbook 'Algebra' (Wiley) should, as a rule, be assumed known. The most important function of the articles in this Handbook is to provide professional mathematicians working in a different area with a sufficiency of information on the topic in question if and when it is needed.

Each of the chapters combines some of the features of both a graduate level textbook and a research-level survey. Not all of the ingredients mentioned below will be appropriate in each case, but authors have been asked to include the following:

- Introduction (including motivation and historical remarks)
- Outline of the chapter
- Basic concepts, definitions, and results. (These may be accompanied by proofs or (usually better) ideas/sketches of the proofs when space permits)
- Comments on the relevance of the results, relations to other results, and applications
- Review of the relevant literature; possibly complete with the opinions of the author on recent developments and future directions
- Extensive bibliography (several hundred items will not be exceptional)

The present

Volume 1 appeared in 1995 (copyright 1996), Volume 2 in 2000, Volume 3 in 2003. Volume 5 is planned for 2006. Thereafter, we aim at one volume every two years (or better).

The future

Of course, ideally, a comprehensive series of books like this should be interactive and have a hypertext structure to make finding material and navigation through it immediate and intuitive. It should also incorporate the various algorithms in implemented form as well as permit a certain amount of dialogue with the reader. Plans for such an interactive, hypertext, CDROM (DVD)-based version certainly exist but the realization is still a nontrivial number of years in the future.

Kvoseliai, July 2005 Michiel Hazewinkel

Kaum nennt man die Dinge beim richtigen Namen
so verlieren sie ihren gefährlichen Zauber

(You have but to know an object by its proper name
for it to lose its dangerous magic)

Elias Canetti

Outline of the Series

(as of July 2005)

Philosophy and principles of the Handbook of Algebra

Compared to the outline in Volume 1 this version differs in several aspects.

First, there is a major shift in emphasis away from completeness as far as the more elementary material is concerned and towards more emphasis on recent developments and active areas. Second, the plan is now more dynamic in that there is no longer a fixed list of topics to be covered, determined long in advance. Instead there is a more flexible nonrigid list that can and does change in response to new developments and availability of authors.

The new policy, starting with Volume 2, is to work with a dynamic list of topics that should be covered, to arrange these in sections and larger groups according to the major divisions into which algebra falls, and to publish collections of contributions (i.e. chapters) as they become available from the invited authors.

The coding below is by style and is as follows.

- **Author(s) in bold**, followed by chapter title: articles (chapters) that have been received and are published or are being published in this volume.
- Chapter title in *italic*: chapters that are being written.
- Chapter title in plain text: topics that should be covered but for which no author has yet been definitely contracted.

Chapters that are included in Volumes 1–4 have a (x; yy pp.) after them, where ‘x’ is the volume number and ‘yy’ is the number of pages.

Compared to the plan that appeared in Volume 1 the section on “Representation and invariant theory” has been thoroughly revised. The changes of this current version compared to the one in Volume 2 (2000) and Volume 3 (2003) are relatively minor: mostly the addition of quite a few topics.

Editorial set-up

Managing editor: M. Hazewinkel.

Editorial board: M. Artin, M. Nagata, C. Procesi, O. Tausky-Todd,† R.G. Swan, P.M. Cohn, A. Dress, J. Tits, N.J.A. Sloane, C. Faith, S.I. Ad’yan, Y. Ihara, L. Small, E. Manes, I.G. Macdonald, M. Marcus, L.A. Bokut’, Eliezer (Louis Halle) Rowen, John S. Wilson, Vlastimil Dlab. Note that three editors have been added startingwith Volume 5.

Planned publishing schedule (as of July 2005)

1996: Volume 1 (published)
2001: Volume 2 (published)
2003: Volume 3 (published)

2005: Volume 4 (last quarter)
Further volumes at the rate of one every year.

Section 1. Linear algebra. Fields. Algebraic number theory

A. *Linear Algebra*

G.P. Egorychev, Van der Waerden conjecture and applications (1; 22 pp.)
V.L. Girko, Random matrices (1; 52 pp.)
A.N. Malyshev, Matrix equations. Factorization of matrices (1; 38 pp.)
L. Rodman, Matrix functions (1; 38 pp.)
Correction to the chapter by L. Rodman, Matrix functions (3; 1 p.)
J.A. Hermida-Alonso, Linear algebra over commutative rings (3, 59 pp.)
Linear inequalities (also involving matrices)
Orderings (partial and total) on vectors and matrices
Positive matrices
Structured matrices such as Toeplitz and Hankel
Integral matrices. Matrices over other rings and fields
Quasideterminants, and determinants over noncommutative fields
Nonnegative matrices, positive definite matrices, and doubly nonnegative matrices
Linear algebra over skew fields

B. *Linear (In)dependence*

J.P.S. Kung, Matroids (1; 28 pp.)

C. *Algebras Arising from Vector Spaces*

Clifford algebras, related algebras, and applications

D. *Fields, Galois Theory, and Algebraic Number Theory*

(There is also an article on ordered fields in Section 4)
J.K. Deveney, **J.N. Mordeson**, Higher derivation Galois theory of inseparable field extensions (1; 34 pp.)
I. Fesenko, Complete discrete valuation fields. Abelian local class field theories (1; 48 pp.)
M. Jarden, Infinite Galois theory (1; 52 pp.)
R. Lidl, **H. Niederreiter**, Finite fields and their applications (1; 44 pp.)
W. Narkiewicz, Global class field theory (1; 30 pp.)
H. van Tilborg, Finite fields and error correcting codes (1; 28 pp.)
Skew fields and division rings. Brauer group
Topological and valued fields. Valuation theory
Zeta and L-functions of fields and related topics
Structure of Galois modules
Constructive Galois theory (realizations of groups as Galois groups)
Dessins d'enfants
Hopf Galois theory

E. Nonabelian Class Field Theory and the Langlands Program

(To be arranged in several chapters by Y. Ihara)

F. Generalizations of Fields and Related Objects

U. Hebisch, **H.J. Weinert**, Semi-rings and semi-fields (1; 38 pp.)
G. Pilz, Near rings and near fields (1; 36 pp.)

Section 2. Category theory. Homological and homotopical algebra. Methods from logic

A. Category Theory

S. MacLane, **I. Moerdijk**, Topos theory (1; 28 pp.)
R. Street, Categorical structures (1; 50 pp.)
B.I. Plotkin, Algebra, categories and databases (2; 68 pp.)
P.S. Scott, Some aspects of categories in computer science (2; 73 pp.)
E. Manes, Monads of sets (3; 87 pp.)
Operads

B. Homological Algebra. Cohomology. Cohomological Methods in Algebra. Homotopical Algebra

J.F. Carlson, The cohomology of groups (1; 30 pp.)
A. Generalov, Relative homological algebra. Cohomology of categories, posets, and coalgebras (1; 28 pp.)
J.F. Jardine, Homotopy and homotopical algebra (1; 32 pp.)
B. Keller, Derived categories and their uses (1; 32 pp.)
A.Ya. Helemskii, Homology for the algebras of analysis (2; 122 pp.)
Galois cohomology
Cohomology of commutative and associative algebras
Cohomology of Lie algebras
Cohomology of group schemes

C. Algebraic K-theory

A. Kuku, Classical algebraic K-theory: the functors K_0, K_1, K_2 (3; 40 pp.)
A. Kuku, Algebraic K-theory: the higher K-functors (4; 72 pp.)
Grothendieck groups
K_2 and symbols
KK-theory and EXT
Hilbert C^*-modules
Index theory for elliptic operators over C^* algebras
Simplicial algebraic K-theory
Chern character in algebraic K-theory
Noncommutative differential geometry
K-theory of noncommutative rings
Algebraic L-theory

Cyclic cohomology
Asymptotic morphisms and E-theory
Hirzebruch formulae

D. Model Theoretic Algebra

(See also P.C. Eklof, Whitehead modules, in Section 3B)
M. Prest, Model theory for algebra (3; 28 pp.)
M. Prest, Model theory and modules (3; 27 pp.)
Logical properties of fields and applications
Recursive algebras
Logical properties of Boolean algebras
F.O. Wagner, Stable groups (2; 40 pp.)
The Ax–Ershov–Kochen theorem and its relatives and applications

E. Rings up to Homotopy

Rings up to homotopy
Simplicial algebras

Section 3. Commutative and associative rings and algebras

A. Commutative Rings and Algebras

(See also C. Faith, Coherent rings and annihilator conditions in matrix and polynomial rings, in Section 3B)
J.P. Lafon, Ideals and modules (1; 24 pp.)
General theory. Radicals, prime ideals etc. Local rings (general). Finiteness and chain conditions
Extensions. Galois theory of rings
Modules with quadratic form
Homological algebra and commutative rings. Ext, Tor, etc. Special properties (p.i.d., factorial, Gorenstein, Cohen–Macauley, Bezout, Fatou, Japanese, excellent, Ore, Prüfer, Dedekind, ... and their interrelations)
D. Popescu, Artin approximation (2; 34 pp.)
Finite commutative rings and algebras (see also Section 3B)
Localization. Local–global theory
Rings associated to combinatorial and partial order structures (straightening laws, Hodge algebras, shellability, ...)
Witt rings, real spectra
R.H. Villareal, Monomial algebras and polyhedral geometry (3; 58 pp.)

B. Associative Rings and Algebras

P.M. Cohn, Polynomial and power series rings. Free algebras, firs and semifirs (1; 30 pp.)
Classification of Artinian algebras and rings
V.K. Kharchenko, Simple, prime, and semi-prime rings (1; 52 pp.)

A. van den Essen, Algebraic microlocalization and modules with regular singularities over filtered rings (1; 28 pp.)
F. Van Oystaeyen, Separable algebras (2; 43 pp.)
K. Yamagata, Frobenius rings (1; 48 pp.)
V.K. Kharchenko, Fixed rings and noncommutative invariant theory (2; 38 pp.)
General theory of associative rings and algebras
Rings of quotients. Noncommutative localization. Torsion theories
von Neumann regular rings
Semi-regular and pi-regular rings
Lattices of submodules
A.A. Tuganbaev, Modules with distributive submodule lattice (2; 16 pp.)
A.A. Tuganbaev, Serial and distributive modules and rings (2; 19 pp.)
PI rings
Generalized identities
Endomorphism rings, rings of linear transformations, matrix rings
Homological classification of (noncommutative) rings
S.K. Sehgal, Group rings and algebras (3; 87 pp.)
Dimension theory
V. Bavula, Filter dimension (4; 29 pp.)
A. Facchini, The Krull–Schmidt theorem (3; 41 pp.)
Duality. Morita-duality
Commutants of differential operators
E.E. Enochs, Flat covers (3; 14 pp.)
C. Faith, Coherent rings and annihilator conditions in matrix and polynomial rings (3; 30 pp.)
Rings of differential operators
Graded and filtered rings and modules (also commutative)
P.C. Eklof, Whitehead modules (3; 25 pp.)
Goldie's theorem, Noetherian rings and related rings
Sheaves in ring theory
A.A. Tuganbaev, Modules with the exchange property and exchange rings (2; 19 pp.)
Finite associative rings (see also Section 3A)
Finite rings and modules
T.Y. Lam, Hamilton's quaternions (3; 26 pp.)
A.A. Tuganbaev, Semiregular, weakly regular, and π-regular rings (3; 22 pp.)
Hamiltonian algebras
A.A. Tuganbaev, Max rings and V-rings (3; 20 pp.)
Algebraic asymptotics
(See also "Freeness theorems in groups and rings and Lie algebras" in Section 5A)

C. *Coalgebras*

W. Michaelis, Coassociative coalgebras (3; 202 pp.)
Co-Lie-algebras

D. Deformation Theory of Rings and Algebras (Including Lie Algebras)

Deformation theory of rings and algebras (general)
Yu. Khakimdzanov, Varieties of Lie algebras (2; 31 pp.)
Deformation theoretic quantization

Section 4. Other algebraic structures. Nonassociative rings and algebras. Commutative and associative algebras with extra structure

A. Lattices and Partially Ordered Sets

Lattices and partially ordered sets
A. Pultr, Frames (3; 67 pp.)
Quantales

B. Boolean Algebras

C. Universal Algebra

Universal algebra

D. Varieties of Algebras, Groups, ...

(See also Yu. Khakimdzanov, Varieties of Lie algebras, in Section 3D)
V.A. Artamonov, Varieties of algebras (2; 29 pp.)
Varieties of groups
V.A. Artamonov, Quasivarieties (3; 23 pp.)
Varieties of semigroups

E. Lie Algebras

Yu.A. Bahturin, M.V. Zaitsev, A.A. Mikhailov, Infinite-dimensional Lie superalgebras (2; 34 pp.)
General structure theory
Ch. Reutenauer, Free Lie algebras (3; 17 pp.)
Classification theory of semisimple Lie algebras over **R** and **C**
The exceptional Lie algebras
M. Goze, Y. Khakimdjanov, Nilpotent and solvable Lie algebras (2; 47 pp.)
Universal enveloping algebras
Modular (ss) Lie algebras (including classification)
Infinite-dimensional Lie algebras (general)
Kac–Moody Lie algebras
Affine Lie algebras and Lie super algebras and their representations
Finitary Lie algebras
Standard bases
A.I. Molev, Gelfand–Tsetlin bases for classical Lie algebras (4; 62 pp.)
Kostka polynomials

F. Jordan Algebras (finite and infinite dimensional and including their cohomology theory)

G. Other Nonassociative Algebras (Malcev, alternative, Lie admissable, ...)

Mal'tsev algebras
Alternative algebras

H. Rings and Algebras with Additional Structure

Graded and super algebras (commutative, associative; for Lie superalgebras, see Section 4E)
Topological rings
M. Cohen, **S. Gelaki**, **S. Westreich**, Hopf algebras (4; 67 pp.)
Classification of pointed Hopf algebras
Recursive sequences from the Hopf algebra and coalgebra points of view
Quantum groups (general)
A.I. Molev, Yangians and their applications (3; 53 pp.)
Formal groups
p-divisible groups
F. Patras, Lambda-rings (3; 26 pp.)
Ordered and lattice-ordered groups, rings and algebras
Rings and algebras with involution. C^*-algebras
A.B. Levin, Difference algebra (4; 94 pp.)
Differential algebra
Ordered fields
Hypergroups
Stratified algebras
Combinatorial Hopf algebras
Symmetric functions
Special functions and q-special functions, one and two variable case
Quantum groups and multiparameter q-special functions
Hopf algebras of trees and renormalization theory
Noncommutative geometry à la Connes
Noncommutative geometry from the algebraic point of view
Noncommutative geometry from the categorical point of view
Solomon descent algebras

I. Witt Vectors

Witt vectors and symmetric functions. Leibniz Hopf algebra and quasi-symmetric functions

Section 5. Groups and semigroups

A. Groups

A.V. Mikhalev, **A.P. Mishina**, Infinite Abelian groups: methods and results (2; 36 pp.)
Simple groups, sporadic groups
Representations of the finite simple groups

Diagram methods in group theory
Abstract (finite) groups. Structure theory. Special subgroups. Extensions and decompositions
Solvable groups, nilpotent groups, p-groups
Infinite soluble groups
Word problems
Burnside problem
Combinatorial group theory
Free groups (including actions on trees)
Formations
Infinite groups. Local properties
Algebraic groups. The classical groups. Chevalley groups
Chevalley groups over rings
The infinite dimensional classical groups
Other groups of matrices. Discrete subgroups
M. Geck, **G. Malle**, Reflection groups (4; 47 pp.)
M.C. Tamburini, **M. Vsemirnov**, Hurwitz groups and Hurwitz generation (4; 42 pp.)
Groups with BN-pair, Tits buildings, ...
Groups and (finite combinatorial) geometry
"Additive" group theory
Probabilistic techniques and results in group theory
V.V. Vershinin, Braids, their properties and generalizations (4; 39 pp.)
L. Bartholdi, **R.I. Grigorchuk**, **Z. Šuniḱ**, Branch groups (3; 124 pp.)
Frobenius groups
Just infinite groups
V.I. Senashov, Groups with finiteness conditions (4; 27 pp.)
Automorphism groups of groups
Automorphism groups of algebras and rings
Freeness theorems in groups and rings and Lie algebras
Groups with prescribed systems of subgroups
(see also "Groups and semigroups of automata transformations" in Section 5B)
Automatic groups
Groups with minimality and maximality conditions (school of Chernikov)
Lattice-ordered groups
Linearly and totally ordered groups
Finitary groups
Random groups
Hyperbolic groups

B. Semigroups

Semigroup theory. Ideals, radicals, structure theory
Semigroups and automata theory and linguistics
Groups and semigroups of automata transformations
Cohomology of semigroups

C. Algebraic Formal Language Theory. Combinatorics of Words

D. Loops, Quasigroups, Heaps, ...

Quasigroups in combinatorics

E. Combinatorial Group Theory and Topology

(See also "Diagram methods in group theory" in Section 5A)

Section 6. Representation and invariant theory

A. Representation Theory. General

Representation theory of rings, groups, algebras (general)
Modular representation theory (general)
Representations of Lie groups and Lie algebras. General

B. Representation Theory of Finite and Discrete Groups and Algebras

Representation theory of finite groups in characteristic zero
Modular representation theory of finite groups. Blocks
Representation theory of the symmetric groups (both in characteristic zero and modular)
Representation theory of the finite Chevalley groups (both in characteristic zero and modular)
Modular representation theory of Lie algebras

C. Representation Theory of 'Continuous Groups' (linear algebraic groups, Lie groups, loop groups, ...) and the Corresponding Algebras

Representation theory of compact topological groups
Representation theory of locally compact topological groups
Representation theory of $SL_2(\mathbf{R})$, ...
Representation theory of the classical groups. Classical invariant theory
Classical and transcendental invariant theory
Reductive groups and their representation theory
Unitary representation theory of Lie groups
Finite dimensional representation theory of the ss Lie algebras (in characteristic zero); structure theory of semi-simple Lie algebras
Infinite dimensional representation theory of ss Lie algebras. Verma modules
Representation of Lie algebras. Analytic methods
Representations of solvable and nilpotent Lie algebras. The Kirillov orbit method
Orbit method, Dixmier map, ... for ss Lie algebras
Representation theory of the exceptional Lie groups and Lie algebras
(See also A.I. Molev, Gelfand–Tsetlin bases for classical Lie algebras, in Section 4E)
Representation theory of 'classical' quantum groups
A.U. Klimyk, Infinite dimensional representations of quantum algebras (2; 27 pp.)

Duality in representation theory
Representation theory of loop groups and higher dimensional analogues, gauge groups, and current algebras
Representation theory of Kac–Moody algebras
Invariants of nonlinear representations of Lie groups
Representation theory of infinite-dimensional groups like GL_∞
Metaplectic representation theory

D. Representation Theory of Algebras

Representations of rings and algebras by sections of sheafs
Representation theory of algebras (Quivers, Auslander–Reiten sequences, almost split sequences, ...)
Quivers and their representations
Tame algebras
Ringel–Hall algebras

E. Abstract and Functorial Representation Theory

Abstract representation theory
S. Bouc, Burnside rings (2; 64 pp.)
P. Webb, A guide to Mackey functors (2; 30 pp.)

F. Representation Theory and Combinatorics

G. Representations of Semigroups

Representation of discrete semigroups
Representations of Lie semigroups

H. Hecke Algebras

Hecke–Iwahori algebras

I. Invariant Theory

Section 7. Machine computation. Algorithms. Tables

Some notes on this volume: Besides some general article(s) on machine computation in algebra, this volume should contain specific articles on the computational aspects of the various larger topics occurring in the main volume, as well as the basic corresponding tables. There should also be a general survey on the various available symbolic algebra computation packages.

The CoCoA computer algebra system
Combinatorial sums and counting algebraic structures
Groebner bases and their applications

Section 8. Applied algebra

Section 9. History of algebra

(See also K.T. Lam, Hamilton's quaternions, in Section 3B)
History of coalgebras and Hopf algebras
Development of algebra in the 19-th century

Contents

List of Contributors

V. Bavula, *University of Sheffield, Sheffield, e-mail: v.bavula@sheffield.ac.uk*
M. Cohen, *Ben Gurion University of the Negev, Beer Sheva, e-mail: mia@cs.bgu.ac.il*
M. Geck, *King's College, Aberdeen University, Aberdeen, e-mail: geck@maths.abdn.ac.uk*
S. Gelaki, *Technion, Haifa, e-mail: gelaki@tx.technion.ac.il*
A. Kuku, *Institute for Advanced Study, Princeton, NJ, e-mail: kukuao@muohio.edu*
A.B. Levin, *The Catholic University of America, Washington, DC, e-mail: Levin@cua.edu*
G. Malle, *Universität Kaiserslautern, Kaiserslautern, e-mail: malle@mathematik.uni-kl.de*
A.I. Molev, *University of Sydney, Sydney, e-mail: alexm@maths.usyd.edu.au*
V.I. Senashov, *Institute of Computational Modelling of Siberian Division of Russian Academy of Sciences, Krasnoyarsk, e-mail: sen@icm.krasn.ru*
M.C. Tamburini, *Universitá Cattolica del Sacro Cuore, Brescia, e-mail: c.tamburini@dmf.unicatt.it*
V.V. Vershinin, *Université Montpellier II, Montpellier, e-mail: vershini@math.univ-montp2.fr*
Sobolev Institute of Mathematics, Novosibirsk, e-mail: versh@math.nsc.ru
M. Vsemirnov, *St. Petersburg Division of Steklov Institute of Mathematics, St. Petersburg, e-mail: vsemir@pdmi.ras.ru*
S. Westreich, *Bar-Ilan University, Ramat-Gan, e-mail: swestric@mail.biu.ac.il*

List of Contributors

Section 2C
Algebraic K-theory

Higher Algebraic K-Theory

Aderemi Kuku

Institute for Advanced Study, Princeton, NJ 08540, USA
E-mail: *kukuao@muohio.edu*

Contents

HANDBOOK OF ALGEBRA, VOL. 4
Edited by M. Hazewinkel

Introduction

This chapter is a sequel to "Classical Algebraic K-Theory: The Functors K_0, K_1, K_2" published in Volume 3 of Handbook of Algebra [79]. The unexplained notions in this chapter are those of [79]. Here, we shall provide higher-dimensional analogues to quite a number of results in [79].

As already observed in [79], the functor K_0 was defined by A. Grothendieck, K_1 by H. Bass and K_2 by J. Milnor. The definition of K_2 by Milnor in 1967 inspired many mathematicians to search for higher K-groups and the next five years (1967–1972) witnessed a lot of vigorous activity in this respect. During this period, several higher K-theories were proposed; notably by D. Quillen, [114,111], S. Gersten, [36], R.G. Swan, [138], I. Volodin, [151], J. Milnor, [105], and F. Keune, [64]. These theories are briefly reviewed in Section 2 with connections between them highlighted. By far the most successful among the theories are those of D. Quillen. Hence, a substantial part of this chapter is devoted to developments of the subject arising from Quillen's work.

We now review the contents of this chapter. Section 1 is a brief discussion of some of the central notions in most constructions of higher K-theories – simplicial objects, classifying spaces and spectra (see 1.1, 1.2).

In Section 2, we define Quillen K-theory, K_n^Q (2.1); K-theory of Gersten and Swan, K_n^S (2.2); K-theory of Karoubi and Villamayor K_n^{k-v} (2.3); Volodin K-theory, K_n^V (2.4); and Milnor K-theory, K_n^M (2.5) – also highlighting some connections between them, e.g., that $K_n^Q(A)$ coincides with $K_n^V(A)$ and $K_n^S(A)$ while $K_n^Q(A)$ coincides with $K_n^{k-v}(A)$ when A is regular.

In Section 3, we define higher K-theory of exact, symmetric monoidal and Waldhausen categories, providing copious examples in each situation (see 3.1, 3.2, 3.3). Thus we discuss for exact categories, higher K-theory of rings and schemes; mod-m and profinite higher K-theory; equivariant higher K-theory, etc. For Waldhausen categories, for instance, we discuss K-theory of perfect complexes and stable derived categories (see 3.3.10).

In Section 4, we highlight, with copious examples, some fundamental results in higher K-theory, most of which have classical analogues at the zero-dimensional level. The topics covered include the resolution theorem for exact categories (4.1); the additivity theorem for exact and Waldhausen categories (4.2), the devissage theorem (4.3); localization sequences (4.4) leading to the Gersten conjecture and fundamental theorems for higher K- and G-theories (4.4.3 and 4.5). We also discuss Waldhausen's fibration sequence, localization sequence for Waldhausen's K-theory and a long exact sequence which realizes the cofibre of the Cartan maps as K-theory of a Waldhausen category. Finally, we discuss excision, Mayer–Vietoris sequences and long exact sequence associated to an ideal.

In Section 5, we define Galois, étale and motivic cohomologies and discuss their interconnections as well as connections with K-theory. We discuss in 5.2 the Bloch–Kato conjecture including parts of it earlier proved – the Milnor conjecture and the Merkurjev–Suslin theorem. In 5.3 we discuss Zariski and étale cohomology as well as connections between them (see 5.3.6). Next we define motivic cohomology which we identify with Lichtenbaum (étale) cohomology groups for smooth k-varieties as well as the connection of this to a special case of Bloch–Kato conjecture 5.4.8. Next we discuss Bloch's higher Chow groups and their connections with K-theory and motivic cohomology.

In Section 6, we discuss higher K-theory of rings of integers in local and global fields. In 6.2 we define étale Chern characters of Soulé (6.2.3) with the observation that there are alternative approaches through "anti-Chern" characters defined by B. Kahn, [56], and maps from étale K-theory to étale cohomology due to Dwyer and Friedlander, [26]. We discuss the Quillen–Lichtenbaum conjecture and record for all $n \geqslant 2$ computations of $K_n(O_S)(2)$, $K_n(O_S)(l)$ in terms of étale cohomology groups where O_S is a ring of integers in number field F as well as $K_n(O_S)$ when F is totally imaginary. We also briefly discuss the motivic Chern characters of Pushin used to identify K-groups of number fields and their integers with motivic cohomology groups (see 6.2.18). In 6.3, we treat higher K-theory and zeta functions including the Lichtenbaum conjecture, Wiles theorem and their consequences. Finally we review in 6.4 some more explicit computations of $K_n(\mathbb{Z})$.

Section 7 deals with the higher K-theory of orders, group rings and modules over EI categories. In 7.1, we review some finiteness results due to Kuku, e.g.: If F is a number field with integers O_F and Λ any O_F-order in a semi-simple F-algebra, then for all $n \geqslant 1$, $K_n(\Lambda)$, $G_n(\Lambda)$ are finitely generated, $SK_n(\Lambda)$, $SG_n(\Lambda)$ are finite and $\operatorname{rank} K_n(\Lambda) = \operatorname{rank} G_n(\Lambda)$ (see 7.1.4, 7.1.6 and 7.1.11). We also discuss the result due to R. Laubenbacher and D. Webb that $SG_n(O_F G) = 0$ for all $n \geqslant 1$ (see 7.1.8) as well as the result of Kuku and Tang that for all $n \geqslant 1$, $G_n(O_F V)$ is finitely generated where V is virtually infinite cyclic group (see 7.1.10). We also exhibit D. Webb's decomposition of $G_n(RG)$, R a Noetherian ring and G finite Abelian group as well as extensions of the decomposition to some non-Abelian groups, e.g., quaternion and dihedral groups.

Next we review in 7.2 results on higher class groups $Cl_n(\Lambda)$, $n \geqslant 0$, for orders. First we observe Kuku's result that $Cl_n(\Lambda)$ is finite for all $n \geqslant 1$, as well as a result due to Laubenbacher and Kolster that the only p-torsion possible in odd-dimensional class groups $Cl_{2n-1}(\Lambda)$ are for primes p lying below prime ideals $\underline{q}$ for which $\hat{\Lambda}_{\underline{q}}$ are not maximal. An analogous result due to Guo and Kuku for even-dimensional class groups $Cl_{2n}(\Lambda)$ is given in 7.2.9 for Eichler orders in quaternion F-algebras and hereditary orders in semi-simple F-algebras. In 7.3, we discuss Kuku's results on profinite K-theory of orders and group-rings providing several l-completeness results for orders in algebras over number fields and p-adic fields as well as showing that for p-adic orders Λ, $G_n(\Lambda)_l$, $K_n(\Sigma)_l$ are finite groups if $l \neq p$ and that $K_n(O_F G)_l$ is also finite for any finite group G. In 7.4, we exhibit several finiteness results on higher K-theory of modules over 'EI' categories.

The last Section 8 deals with equivariant higher algebraic K-theory together with relative generalizations – for finite group action – due to Dress and Kuku with the observation that there are analogous theories for profinite groups and compact Lie group actions due to Kuku, [70,77]. Time and space prevented us from including the latter two cases. We also remark that K. Shimakawa, [127], provided a G-spectrum formulation of the absolute part of the theory discussed in Section 8, but again time and space has prevented us from going into this.

1. Simplicial objects, classifying spaces, and spectra

In this opening section, we briefly review some of the central notions in the construction of higher K-theories.

1.1. *Simplicial objects and classifying spaces*

1.1.1. DEFINITION. Let Δ be the category defined as follows: $\mathrm{ob}(\Delta) =$ ordered sets $\underline{n} = \{0 < 1 < \cdots < n\}$. The set $\mathrm{Hom}_\Delta(\underline{m}, \underline{n})$ of morphisms from $\underline{m}$ to $\underline{n}$ consists of maps $f : \underline{m} \to \underline{n}$ such that $f(i) \leqslant f(j)$ for $i < j$.

Let $\mathcal{A}$ be any category. A simplicial object in $\mathcal{A}$ is a contravariant functor $X. : \Delta \to \mathcal{A}$ where we write X_n for $X(\underline{n})$. Thus, a simplicial set (resp. group; resp. ring; resp. space, etc.) is a simplicial object in the category of sets (resp. groups; resp. rings; resp. spaces, etc.). A co-simplicial object is a covariant functor $X : \Delta \to \mathcal{A}$.

Equivalently one could define a simplicial object in a category $\mathcal{A}$ as a set of objects X_n $(n \geqslant 0)$ in $\mathcal{A}$ and a set of morphisms $\delta_i : X_n \to X_{n-1}$ $(0 \leqslant i \leqslant n)$ called face maps as well as a set of morphisms $s_j : X_n \to X_{n+1}$ $(0 \leqslant j \leqslant n)$ called degeneracy maps satisfying certain "simplicial identities" – [165, p. 256]. We shall denote the category of simplicial sets by $\mathcal{S}$. sets.

1.1.2. DEFINITION. The geometric n-simplex is the topological space

$$\hat{\Delta}^n = \bigl\{(x_0, x_1, \ldots, x_n) \in \mathbb{R}^{n+1} \mid 0 \leqslant x_i \leqslant 1\ \forall i \text{ and } \Sigma x_i = 1\bigr\}.$$

The functor $\hat{\Delta} : \Delta \to$ spaces given by $\underline{n} \to \hat{\Delta}^n$ is a co-simplicial space.

1.1.3. DEFINITION. Let X_* be a simplicial set. The geometric realization of X_* written $|X_*|$ is defined by $|X_*| := X \times_\Delta \hat{\Delta} = \bigcup_{n \geqslant 0} (X_n \times \hat{\Delta}^n)/\approx$ where the equivalence relation '$\approx$' is generated by $(x, \varphi_*(y)) \approx (\varphi^*(x), y)$ for any $x \in X_n$, $y \in Y_m$, and $\varphi : \underline{m} \to \underline{n}$ in Δ and where $X_n \times \hat{\Delta}^n$ is given the product topology and X_n is considered as a discrete space.

1.1.4. EXAMPLES/REMARKS.

(i) Let T be a topological space, $\mathrm{Sing}_* T = \{\mathrm{Sing}_n T\}$ where $\mathrm{Sing}_n T = \{$continuous maps $\hat{\Delta}^n \to T\}$. A map $f : \underline{n} \to \underline{m}$ determines a linear map $\hat{\Delta}^n \to \hat{\Delta}^m$ and hence induces a map $\hat{f} : \mathrm{Sing}_m T \to \mathrm{Sing}_n T$. So, $\mathrm{Sing}_* T : \Delta \to$ sets is a simplicial set. $\mathrm{Sing}_* T$ is called a Kan complex.

(ii) For any simplicial set X_*, $|X_*|$ is a CW-complex with X_n in one-one correspondence with n-cells in $|X_*|$.

(iii) For any simplicial sets X_*, Y_*, $|X_*| \times |Y_*| \cong |X_* \times Y_*|$ where the product is such that $(X_* \times Y_*)_n = X_n \times Y_n$.

1.1.5. DEFINITION. Let $\mathcal{A}$ be a small category. The nerve of $\mathcal{A}$ written $N\mathcal{A}$ is the simplicial set whose n-simplices are diagrams

$$\mathcal{A}_n = \bigl\{A_0 \xrightarrow{f_1} A_2 \longrightarrow \cdots \xrightarrow{f_n} A_n\bigr\},$$

where the A_i are $\mathcal{A}$-objects and the f_i are $\mathcal{A}$-morphisms. The classifying space of $\mathcal{A}$ is defined as $|N\mathcal{A}|$ and is denoted by $B\mathcal{A}$.

1.1.6. PROPERTIES OF $B\mathcal{A}$.

(i) $B\mathcal{A}$ is a CW-complex whose n-cells are in one-one correspondence with the diagrams $\mathcal{A}_n$ above (see 1.1.4(ii)).

(ii) From 1.1.4(iii), we have, for small categories $\mathcal{C}$, $\mathcal{D}$ (I) $B(\mathcal{C} \times \mathcal{D}) \approx B\mathcal{C} \times B\mathcal{D}$ where $B\mathcal{C} \times B\mathcal{D}$ is given the compactly generated topology (see [128]). In particular we have the homeomorphism (I) if either $B\mathcal{C}$ or $B\mathcal{D}$ is locally compact (see [128]).

(iii) Let F, G be functors, $\mathcal{C} \to \mathcal{D}$ (where $\mathcal{C}$, $\mathcal{D}$ are small categories). A natural transformation of functors $\eta : F \to \mathcal{C}$ induces a homotopy $B\mathcal{C} \times I \to B\mathcal{D}$ from $B\mathcal{C}$ to $B\mathcal{D}$.

(iv) If $F : \mathcal{C} \to \mathcal{D}$ has a left or right adjoint, then F induces a homotopy equivalence.

(v) If $\mathcal{C}$ is a category with initial or final object, then $B\mathcal{C}$ is contractable.

1.1.7. EXAMPLES.

(i) A discrete group G can be regarded as a category with one object G whose morphisms can be identified with the elements of G.

The nerve of G, written N_*G is defined as follows: $N_n(G) = G^n$, with face maps δ_i given by

$$\delta_i(g_1 \dots g_n) = \begin{cases} (g_2, \dots, g_n), & i = 0, \\ (g_1, \dots, g_i g_{i+1}, \dots, g_n), & 1 \leqslant i < n-1, \\ (g_1, \dots, g_{n1}), & i = n-1, \end{cases}$$

and degeneracies s_i given by

$$s_i(g_1, g_2, \dots, g_n) = (g_1, g_i, 1, g_{i+1}, g_n).$$

The classifying space BG of G is defined as $|N_*(G)|$ and it is a connected CW-complex characterized up to homotopy type by the property that $\pi_1(BG, *) = G$ and $\pi_n(BG, *) = 0$ for all $n > 0$ where $*$ is some basepoint of BG. Note that BG has a universal covering space usually denoted by EG (see [165]).

Note that the term classifying space of G comes from the theory of fibre bundles. So, if X is a finite cell complex, the set $[X, BG]$ of homotopy classes of maps $X \to BG$ gives a complete classification of the fibre bundles over X with structure group G.

(ii) Let G be a topological group (possibly discrete) and X a topological G-space. The translation category $\underline{X}$ of X is defined as follows: $\mathrm{ob}(\underline{X})$ = elements of X; $\mathrm{Hom}_{\Delta}(x, x') = \{g \in G \mid gx = x'\}$. Then the nerve of $\underline{X}$ is the simplicial space equal to $G^n \times X$ in dimension n. $B\underline{X} = |\text{nerve of } X|$ is the Borel space $EG \underset{G}{\times} X$ (see [94]).

(iii) Let $\mathcal{C}$ be a small category, $F : \mathcal{C} \to \underline{Sets}$ a functor, then $\mathcal{C}_F$ is the category defined as follows

$$\mathrm{ob}\,\mathcal{C}_F = \{(C, x) \mid C \in \mathrm{ob}\,\mathcal{C}, x \in F(C)\}.$$

A morphism from (C, x) to (C', x') is a morphism $f: C \to C^1$ in $\mathcal{C}$ such that $f_*(x) = x'$.

The homotopy colimit of F is defined as hocolim $F := B\mathcal{C}_F$. This construction is also called the Bonsfield–Kan construction. If the functor F is trivial, we have $B\mathcal{C}_F = B\mathcal{C}$.

1.1.8. Let $\mathcal{C} = \mathcal{C}^{\text{top}}$ be a topological category (i.e. the objects in $\mathcal{C}$ as well as $\text{Hom}_{\mathcal{C}}(X, Y)$ $(X, Y \in \mathcal{C})$ are topological spaces). Then $N\mathcal{C}^{\text{top}}$ is a simplicial topological space and $B\mathcal{C}^{\text{top}} = |N\mathcal{C}^{\text{top}}|$ the geometric realization of $N\mathcal{C}^{\text{top}}$. We could regard the identity map as a continuous function $\mathcal{C}^\delta \to \mathcal{C}^{\text{top}}$ between topological categories and get an induced continuous maps $B\mathcal{C}^\delta \to B\mathcal{C}^{\text{top}}$. (Here $\mathcal{C}^\delta$ is a discrete category, i.e. $\mathcal{C}$ with discrete topology on objects.)

1.1.9. EXAMPLES.

(i) Any topological group $G = G^{\text{top}}$ is a topological category: $\pi_1(BG^\delta) = G^\delta$, $\pi_j(BG^\delta) = 0$ if $j \neq 1$, $\Omega BG^{\text{top}} = G^{\text{top}}$. Hence $\pi_i(BG^{\text{top}}) = \pi_{i-1}G^{\text{top}}$ for $i > 0$.

(ii) If A is a C^*-algebra with identity, then put G in (i) as $G = \text{GL}(A) = \bigcup_n \text{GL}_n(A)$, and $\pi_i(B\text{GL}(A)) = \pi_{i-1}(\text{GL}(A))$ which is by definition $K_i^{\text{top}}(A)$ (higher 'topological' K-theory of A). $K_0^{\text{top}}(A) = \pi_0(\text{GL}(A)) = K_0(\mathcal{P}(A)^\delta)$, the usual Grothendieck group of A and $K_1(A) = \text{GL}_\infty(A)/\text{GL}_0(A)$ where $\text{GL}_0(A)$ is the connected component of the identity in $\text{GL}(A)$. In fact, Bott periodicity is satisfied, i.e. $K_n(A) \cong K_{n+2}(A)$ for all $n \geqslant 0$ (see [18]) and so, this theory is $\mathbb{Z}_2$-graded, having only $K_0^{\text{top}}(A) = K_0(A)$ and $K_1^{\text{top}}(A)$.

(iii) If $A = \mathbb{C}$ in (ii) and we denote by U_n the unitary groups, then BU_n is homotopy equivalent to $B\text{GL}_n(\mathbb{C})^{\text{top}}$ (because U_n is a deformation retract of $\text{GL}_n(\mathbb{C})^{\text{top}}$). Since $\text{GL}_n(\mathbb{C})$ is connected, we have $K_1^{\text{top}}(\mathbb{C}) = 0$, and $K_0(\mathbb{C}) = K_0^{\text{top}}(\mathbb{C}) \approx \mathbb{Z}$.

1.1.10. REMARKS.

(i) Given a simplicial object $A = \{A_n\}$ in an Abelian category, there exists a chain complex $(C(A), d)$, i.e.

$$C(A): \cdots \to C_n \to C_{n-1} \to C_{n-2} \to \cdots,$$

where $C_n = A_n$ and $d_n: C_n \to C_{n-1}$ is given by $d_n = \delta_0 - \delta_1 + \cdots + (-1)^n \delta_n$.

(ii) If R is a ring, then there exists a functor

$$\textit{Sets} \to R\textit{-Mod}: X \to R[X] = \textit{free } R\text{-module on } X.$$

If $X = \{X_n\}$ is a simplicial set, then $R[X] = \{R[X_n]\}$ is a simplicial R-module and $H_*(X, R) :=$ homology of the chain complex associated to $R[X]$ (see (i) above).
Also $H_*(X, R) = H_*(|X|, R)$, the singular homology of X.

1.1.11. Let $G = \{G_n\}$ be a simplicial group with face maps $\delta_i : G_n \to G_{n-1}$ and degeneracies $s_i : G_n \to G_{n+1}$ $(0 \leqslant i \leqslant n)$. Define $\pi_n G = H_n / d_{n+1} K_n$ where $H_n \subset K_n \subset G_n$ are defined by

$$K_n := \ker(\delta_0) \cap \cdots \cap \ker(\delta_{n-1})$$

and

$$H_n = K_n \cap \big(\ker(\delta_n)\big).$$

Say that G is acyclic if $\pi_n(G) = 0\ \forall n$.

We can regard a simplicial ring as a simplicial group using its additive structure and we say that a simplicial ring is acyclic if $\pi_n R = 0$ for all n.

1.1.12. A simplicial ring R is said to be free if there exists a basis B_n of R_n as a free ring for all n and $s_i(B_n) \subset B_{n+1}$ for all i and all n.

A simplicial ring $R = \{R_i\}$ is said to have a unit if each R_i has a unit and all δ_i and s_i are unit preserving.

1.2. *Spectra – brief introduction*

1.2.1. REMARKS. The importance of spectra for this chapter has to do with the fact that higher K-groups are often expressed as homotopy groups of spectra $\underline{E} = \{E_i\}$ whose spaces $E_i \approx \Omega^k E_{i+k}$ (for k large) are infinite loop spaces. (It is usual to take $i = 0$ and consider E_0 as an infinite loop space.) Also to each spectrum can be associated generalized cohomology theory and vice-versa. Hence algebraic K-theory can always be endowed with the structure of a generalized cohomology theory. We shall come across these notions copiously in later sections.

1.2.2. DEFINITION. A spectrum $\underline{E} = \{E_i\}$, $i \in \mathbb{Z}$, is a sequence of based spaces E_n and based homeomorphisms $E_i \approx \Omega E_{i+1}(I)$. If we regard $E_i = 0$ for negative i, call $\underline{E}$ a connective spectrum.

A map $f : \underline{E} = \{E_i\} \to \{F_i\} = \underline{F}$ of spectra is a sequence of based continuous maps strictly compatible with the given homeomorphism (I). The spectra form a category which we shall denote by $\mathcal{S}$pectra.

1.2.3. From the adjunction isomorphism $[\Sigma X, Y] = [X, \Omega Y]$ for spaces X, Y, we have $\pi_n(\Omega E_i) \cong \pi_{n+1}(E_1)$, and so, we can define the homotopy groups of a connective spectrum $\underline{E}$ as $\pi_n(\underline{E}) = \pi_n(E) = \pi_{n+1}(E_1) = \cdots = \pi_{n+i}(E_i)$.

1.2.4. Each spectrum $\underline{E} = \{E_n\}$ gives rise to an extraordinary cohomology theory E^n in such a way that if X_+ is the space obtained from X by adjoining a base point, $E^n(X) = [X_+, E_n]$ and conversely.

One can also associate to $\underline{E}$ a homology theory defined by

$$E_n(X) = \lim_{k\to\infty} \pi_{n+k}(E_k \wedge X_+).$$

1.2.5. EXAMPLES.

(i) *Eilenberg–MacLane spectrum.*

Let $E_s = K(A, s)$ where each $K(A, s)$ is an Eilenberg–MacLane space where A is an Abelian group and $\pi_n(K(A, s)) = \delta_{is}(A)$. By adjunction isomorphism, we have $K(A, n) \approx \Omega K(A, n+1)$, and get the Eilenberg–MacLane spectrum whose associated cohomology theory is ordinary cohomology with coefficients in A, otherwise defined by means of singular chain complexes.

(ii) *The suspension spectrum.*

Let X be a based space. The n-th space of the suspension spectrum $\Sigma^\infty X$ is $\Omega^\infty \Sigma^\infty(\Sigma^n X)$ and the homotopy groups are $\pi_n(\Sigma^\infty X) = \lim_{k\to\infty} \pi_{n+k}(\Sigma^k X)$. When $X = S^0$, we obtain the sphere spectrum $\Sigma^\infty(S^0)$ and $\pi_n(\Sigma^\infty(S^0)) = \lim_{k\to\infty} \pi_{n+k}(S^k)$ is called the stable n-stem and denote by π_n^S.

Note that there is an adjoint pair $(\Sigma^\infty, \Omega^\infty)$ of functors between spaces and spectra and we can write $\Sigma^\infty X = \{X, \Sigma X, \Sigma^2 X, \ldots\}$. Also if $\underline{E}$ is an Ω-spectrum, $\Omega^\infty \underline{E}$ is an infinite loop space. (Indeed, every infinite loop space is the initial space of an Ω-spectrum and $\pi_n(\underline{E}) = [\Sigma^\infty S^n, \underline{E}] = \pi_n(\Omega^\infty \underline{E})$.)

2. Definitions of and relations between several higher algebraic K-theories (for rings)

In this section, we define the higher K-functors K_n^Q (Quillen K-theory), K_n^S (K-theory of Swan), K_n^{k-v} (Karoubi–Villamayor K-theory), K_n^M (Milnor K-theory) and K_n^V (Volodin K-theory) for arbitrary rings with identity and discuss connections between the theories. Because K_n^Q has been the most successful and has been most often used, we shall eventually write K_n for K_n^Q.

2.1. K_n^Q – *the K-theory of Quillen*

The definition of $K_n^Q(A)$, A any ring with identity, will make use of the following result.

2.1.1. THEOREM [94,111]. *Let X be a connected CW-complex, N a perfect normal subgroup of $\pi_1(X)$. Then there exists a CW-complex X^+ (depending on N) and a map $i : X \to X^+$ such that*

(i) $i_* : \pi_1(X) \to \pi_1(X^+)$ *is the quotient map* $\pi_1(X) \to \pi_1(X^+)/N$.

(ii) *For any $\pi_1(X^+)/N$-module L, there is an isomorphism $i_* : H_*(X, i^*L) \to H_*(X^+, L)$ where i^*L is L considered as a $\pi_1(X)$-module.*

(iii) *The space X^+ is universal in the sense that if Y is any CW-complex and $f : X \to Y$ is a map such that $f_* : \pi_1(X) \to \pi_1(Y)$ satisfies $f_*(N) = 0$, then there exists a unique map $f^+ : X^+ \to Y$ such that $f^+ i = f$.*

2.1.2. DEFINITION. Let A be a ring and take $X = B\mathrm{GL}(A)$ in Theorem 2.1.1. Then $\pi_1 B\mathrm{GL}(A) = \mathrm{GL}(A)$ contains $E(A)$ as a perfect normal subgroup. Hence by Theorem 2.1.1, there exists a space $B\mathrm{GL}(A)^+$. Define $K_n(A) = \pi_n(B\mathrm{GL}(A)^+)$.

2.1.3. *Hurewitz map.* For any ring A with identity, there exist Hurewitz maps:

(i) $h_n : K_n(A) = \pi_n(B\mathrm{GL}(A)^+) \to H_n(B\mathrm{GL}(A)^+, \mathbb{Z}) \approx H_n(\mathrm{GL}(A), \mathbb{Z})\ \forall n \geqslant 1$,

(ii) $h_n : K_n(A) = \pi_n(BE(A)^+) \to H_n(BE(A)^+, \mathbb{Z}) \approx H_n(E(A), \mathbb{Z})\ \forall n \geqslant 2$,

(iii) $h_n : K_n(A) = \pi_n(B\mathrm{St}(A)^+) \to H_n(B\mathrm{St}(A)^+, \mathbb{Z}) \approx H_n(\mathrm{St}(A), \mathbb{Z})\ \forall n \geqslant 3$.

Note that $B\mathrm{GL}(A)^+$ is connected, $BE(A)^+$ is simply connected (i.e. one-connected) and $B\mathrm{St}(A)^+$ is 2-connected.

For a comprehensive discussion of Hurewitz maps, see [6].

2.1.4. EXAMPLES/REMARKS. For $n = 0, 1, 2$, $K_n(A)$ as defined in Section 2.1.2 can be identified respectively with the classical $K_n(A)$.

(i) $\pi_1(B\mathrm{GL}(A)^+) = \mathrm{GL}(A)/E(A) = K_1(A)$.

(ii) Note that $BE(A)^+$ is the universal covering space of $B\mathrm{GL}(A)^+$ and so, we have

$$\pi_2\big(B\mathrm{GL}(A)^+\big) \approx \pi_2\big(BE(A)^+\big) \approx H_2\big(BE(A)^+\big) \cong H_2\big(BE(A)\big) \\ \cong H_2\big(E(A)\big) \approx K_2(A).$$

(iii) $K_3(A) = H_3(\mathrm{St}(A))$. For a proof, see [38].

(iv) If A is a finite ring, then $K_n(A)$ is finite (see [73] for a proof).

(v) For a finite field $\mathbb{F}_q$, $K_{2n}(\mathbb{F}_q) = 0$, $K_{2n-1}(\mathbb{F}_q) = \mathbb{Z}/(q^n - 1)$ (see [112]). In later Sections 3–8, we shall come across many computations of $K_n(A)$, for various rings, fields, etc.

2.2. K_n^S – *the K-theory of Swan and Gersten*

2.2.1. In [138], R.G. Swan defined higher K-functors by resolving the functor GL in the category of functors and S.M. Gersten in [36] defined higher K-functors by introducing a cotriple construction in the category $\mathcal{R}$ing of rings. Swan showed in [142] that Gersten's resolution applied to GL gives Swan's groups. As has been the tradition, we denote this theory by $K_n^S(A)$.

2.2.2. *Cotriples.* A cotriple (T, ε, δ) in a category $\mathcal{A}$ is an endofunctor $T : \mathcal{A} \to \mathcal{A}$ together with natural transformations $\varepsilon : T \to \mathrm{id}_{\mathcal{A}}$ and $\delta : T \to T^2$ such that the following diagrams commute for every object A.

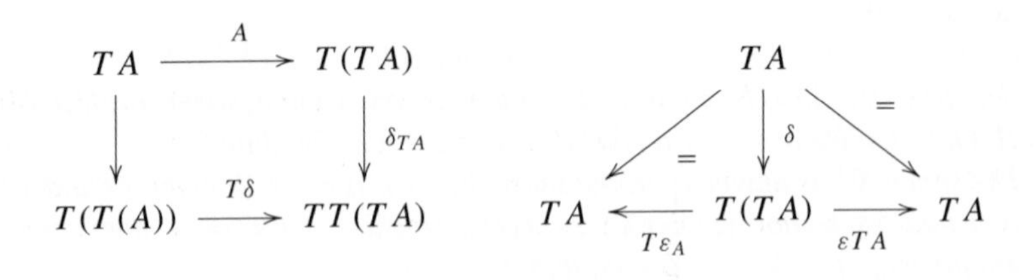

2.2.3. REMARKS.

(i) If $\mathcal{A} \underset{V}{\overset{L}{\rightleftarrows}} \mathcal{B}$ is an adjoint situation where L is left adjoint to V, then $T = LV : \mathcal{B} \to \mathcal{B}$ is part of a cotriple (T, ε, δ), where $\varepsilon : LV \to \mathrm{id}_{\mathcal{B}}$ is the counit of the adjunction.

(ii) Given a cotriple T on $\mathcal{A}$, and $A \in \mathrm{ob}\,\mathcal{A}$, we have a simplicial object $T^*A = \{T^n A\}$ of $\mathcal{A}$ with face maps $\delta_i = T^i \in T^{n-i} : T^{n+1}A \to T^n A$ and degeneracy maps $s_i = T^i \delta T^{n-1}$, $T^{n+1}A \to T^{n+2}A$.

2.2.4. Let $\mathcal{R}ing$ be the category of rings and for any ring A, let FA be the free ring on the underlying set of A. Then F is a functor $\mathcal{S}et \to \mathcal{R}ing$ adjoint to the forgetful functor and the adjointness yields a morphism $\varepsilon : FA \to A$ and a morphism $\delta : FA \to F^2 A$ such that (F, ε, δ) is a cotriple in $\mathcal{R}ing$.

Let $|r|$ be the free generator of FA corresponding to $r \in A$. Then $\varepsilon(|r|) = r$ and $\delta(|r|) = \|r\|$. So, we obtain the augmented simplicial ring:

$$F^*A : R \longleftarrow FA \;\substack{\longrightarrow \\ \longleftarrow \\ \longrightarrow}\; F^2A \;\substack{\longrightarrow \\ \longleftarrow \\ \longrightarrow \\ \longleftarrow \\ \longrightarrow}\; F^3A \cdots .$$

2.2.5. Define $K_n^S(A) = \hat{\pi}_n(\mathrm{GL}(F^*A))$ where $\hat{\pi}_n = \pi_n$, $n \geqslant 1$, $\hat{\pi}_0(\mathrm{GL}(F^*A)) = \ker(\pi_0(\mathrm{GL}(F^*A)) \xrightarrow{\varepsilon} \mathrm{GL}(A))$ and $\hat{\pi}_{-1}(\mathrm{GL}(F^*(A))) = \mathrm{Coker}(\mathrm{GL}(F^*A) \to \mathrm{GL}(A))$.

2.2.6. THEOREM [138]. $K_n^S(FA) = 0$.

2.2.7. THEOREM [38]. $K_n^Q(FA) = 0$.

2.2.8. THEOREM [3]. *For any ring A, there exists an exact sequence*

$$\to K_{n+1}(A) \to K_{n+1}^S(A) \to K_n(FA) \to K_n(A) \to K_n^S(A) \to .$$

2.2.9. COROLLARY (Connection with Quillen K-theory). $K_n^S(A) = K_n^Q(A)$ *for any ring A.*

PROOF. This follows from 2.2.6, 2.2.7 and 2.2.8. □

2.3. K_n^{k-v} – *the K-theory of Karoubi and Villamayor*

2.3.1. Let $R(\Delta^n) = R[t_0, t_1, t_n]/(\Sigma t_i - 1) \simeq R[t_1, \ldots, t_n]$. Applying the functor GL to $R(\Delta^n)$ yields a simplicial group $\mathrm{GL}(R(\Delta^*))$.

2.3.2. DEFINITION. Let R be a ring with identity. Define the Karoubi–Villamayor K-groups by $K_n^{k-v}(R) = \pi_{n-1}(\mathrm{GL}(R[\Delta^*])) = \pi_n(B\mathrm{GL}(R[\Delta^*]))$ for all $n \geqslant 1$. Note that $\pi_0(\mathrm{GL}(R[\Delta^*]))$ is the quotient $\mathrm{GL}(R)/\mathrm{uni}(R)$ of $K_1(R)$ where $\mathrm{uni}(R)$ is the subgroup of $\mathrm{GL}(R)$ generated by unipotent matrices, i.e. matrices of the form $1 + N$ for some nilpotent matrix N.

2.3.3. THEOREM [151].

(i) *For* $p \geqslant 1, q \geqslant 0$, *there is a spectral sequence* $E^1_{pq} = K_p(R[\Delta^q]) \Longrightarrow K^{k-v}_{p+q}(R)$.

(ii) *If* R *is regular, then the spectral sequence in* (i) *above degenerates and* $K_n(R) = K^{k-v}_n(R)$ *for all* $n \geqslant 1$.

2.3.4. DEFINITION. A functor F : Rings $\to$ $\mathbb{Z}$-mod (Chain complexes etc.) is said to be homotopy invariant if for any ring R, the natural map $R \to R[t]$ induces an isomorphism $F(R) \approx F(R[t])$. Note that if F is homotopy invariant, then the simplicial object $F(R[\Delta^*])$ is constant.

2.3.5. THEOREM [38]. *The functors* K^{k-v}_n : *Rings* $\to$ $\mathbb{Z}$-mod *are homotopy invariant, i.e.* $K^{k-v}_n(R) \cong K^{k-v}_n(R[t])$ *for all* $n \geqslant 1$.

2.4. K^V_n – *Volodin K-theory*

2.4.1. Let A be a ring with identity, γ a partial ordering of $\{1, 2, \ldots, n\}$ and $T^\gamma(A) := \{1 + (a'_{ij}) \in \mathrm{GL}_n(A) \mid a_{ij} = 0\, \forall i \overset{\gamma}{\not<} j\}$. Note that if γ is the standard ordering $\{1 < \cdots < n\}$, then $T^\gamma(A)$ is the subgroup of upper triangular matrices. The inclusion $T^\gamma(A) \hookrightarrow \mathrm{GL}(A)$ induces a cofibration on the classifying space $BT^\gamma(A) \hookrightarrow B\mathrm{GL}(A)$.

2.4.2. Define the Volodin space $X(A)$ by $X(A) := \bigcup_\gamma BT^\gamma(A)$.

2.4.3. THEOREM [151]. *For any ring A with identity, the connected space* $X(A)$ *is acyclic* ($\widetilde{H}_n(X(A)) = 0$) *and is simple in dimension* $\geqslant 2$.

2.4.4. DEFINITION. Define $K^V_n(A) := \pi_{n-1}(X(A))$.

2.4.5. *Connections with Quillen K-theory.*

THEOREM [94]. *There exists a natural homotopy fibration* $X(A) \to B\mathrm{GL}(A) \to B\mathrm{GL}(A)^+$ *and hence* $\pi_1(X(A)) = \mathrm{St}(A)$, $\pi_n(X(A)) = K_{n+1}(A)$ *for all* $n \geqslant 2$, *i.e.*

$$K^V_n(A) = \pi_n\big(X(A)\big) = K_{n+1}(A) \quad \forall n \geqslant 2.$$

2.5. K^M_n – *Milnor K-theory*

2.5.1. Let A be a commutative ring with identity and $T(A^*)$ the tensor algebra over $\mathbb{Z}$ where A^* is the Abelian group of invertible elements of A. For any $x \in A^* - \{1\}$, the elements $x \otimes (1 - x)$ and $x \otimes (-x)$ generate a 2-sided ideal I of $T(A^*)$. The quotient $T(A^*)/I$ is a graded Abelian group whose component in degree 0, 1, 2 are respectively $\mathbb{Z}$, A^* and $K^M_2(A)$ where $K^M_2(A)$ is the classical K_2-group, see [105,79].

2.5.2. *Connections with Quillen K-theory.*

(i) As remarked above $K_n^M(A) = K_n^Q(A)$ for $n \leqslant 2$.

(ii) First observe that there is a well defined product $K_m^Q(A) \times K_n^Q(A) \to K_{m+n}^Q(A)$, due to J.L. Loday (see [95]). Now, there exists a map $\varphi: K_n^M(A) \to K_n^Q(A)$ constructed as follows: We use the isomorphism $K_1(A) \simeq A^*$ to embed A^* in $K_1(A)$ and use the product in Quillen K-theory to define inductively a map $(A^*)^n \to K_1(A)^n \to K_n(A)$, which factors through the exterior power $\Lambda^n A^*$ over $\mathbb{Z}$, and hence through the Milnor K-groups $K_n^M(A)$ yielding the map $\varphi: K_n^M(A) \to K_n(A)$.

If F is a field, we have the following more precise result due to A. Suslin.

2.5.3. THEOREM [132]. *The kernel of* $\varphi: K_n^M(F) \to K_n(F)$ *is annihilated by* $(n-1)!$.

We shall discuss more connections between Milnor and Quillen K-theories (especially for fields) in Section 5.

3. Higher K-theory of exact, symmetric monoidal and Waldhausen categories

3.1. *Higher K-theory of exact categories – definitions and examples*

In [79, Section 3], we discussed K_0 of exact categories $\mathcal{C}$, providing copious examples. In this section, we define $K_n(\mathcal{C})$ for all $n \geqslant 0$ with the observation that this definition generalizes to higher dimensions the earlier ones at the zero-dimensional level.

3.1.1. DEFINITION. Recall [108], [79, 3.1], that an exact category is a small additive category $\mathcal{C}$ (which is embeddable as a full subcategory of an Abelian category $\mathcal{A}$) together with a family $\mathcal{E}$ of short exact sequences $0 \to C' \xrightarrow{i} C \xrightarrow{j} C'' \to 0$ (I) such that $\mathcal{E}$ is the class of sequences (I) in $\mathcal{C}$ that are exact in $\mathcal{A}$ and $\mathcal{C}$ is closed under extensions (i.e. for any exact sequence $0 \to C' \xrightarrow{i} C \xrightarrow{j} C'' \to 0$ in $\mathcal{A}$ with C', C'' in $\mathcal{C}$, we have $C \in \mathcal{C}$).

In the exact sequence (I) above, we shall refer to i as an inflation or admissible monomorphism, j as a deflation or admissible epimorphism; and to the pair (i, j) as a conflation.

Let $\mathcal{C}$ be an exact category. We form a new category $Q\mathcal{C}$ whose objects are the same as objects of $\mathcal{C}$ such that for any two objects $M, P \in \mathrm{ob}(Q\mathcal{C})$, a morphism from M to P is an isomorphism class of diagrams $M \overset{j}{\twoheadleftarrow} N \overset{i}{\rightarrowtail} P$ where i is admissible monomorphism and j is an admissible epimorphism in $\mathcal{C}$, i.e. i and j are part of some exact sequences $0 \to N \overset{i}{\rightarrowtail} P \overset{j}{\twoheadrightarrow} P' \to 0$ and $0 \to N' \to N \twoheadrightarrow M \to 0$ respectively.

Composition of arrows $M \twoheadleftarrow N \rightarrowtail P$ and $P \twoheadleftarrow R \rightarrowtail T$ is defined by the following diagram which yields an arrow

$$M \twoheadleftarrow N \times_P R \rightarrowtail T$$

in $Q\mathcal{C}$

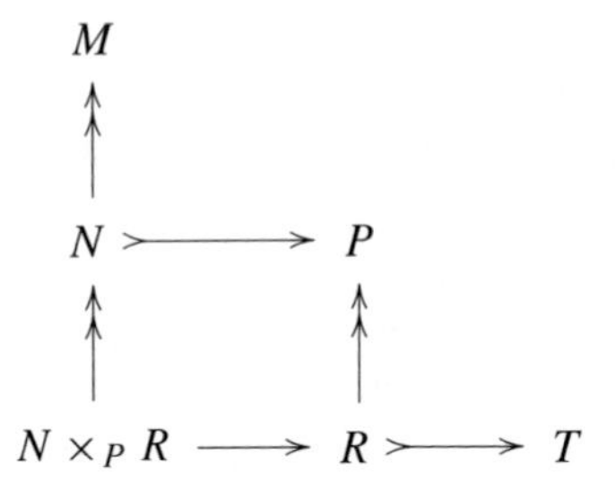

3.1.2. DEFINITION. For all $n \geqslant 0$, define

$$K_n(\mathcal{C}) := \pi_{n+1}(BQ\mathcal{C}, o)$$

(see [114]).

3.1.3. We could also obtain $K_n(\mathcal{C})$ via spectra. For example, we could take the Ω-spectrum (see 1.2) $\underline{BQ\mathcal{C}} = \{\Omega BQ\mathcal{C}, BQ\mathcal{C}, BQ^2\mathcal{C}, \ldots\}$ where $Q^i\mathcal{C}$ is the multicategory defined in [154] and $\pi_n(\underline{BQ\mathcal{C}}) = K_n(\mathcal{C})$.

3.1.4. EXAMPLES.

(i) For any ring A with identity, the category $\mathcal{P}(A)$ of finitely generated projective modules over A is exact and we shall write $K_n(A)$ for $K_n(\mathcal{P}(A))$.

Note that for all $n \geqslant 1$, $K_n(A)$ coincides with the groups $\pi_n(B\mathrm{GL}(A)^+)$ defined in 2.1.2.

(ii) Let A be a left Noetherian ring. Then $\mathcal{M}(A)$, the category of finitely generated (left) A-modules is an exact category and we denote $K_n(\mathcal{M}(A))$ by $G_n(A)$. The inclusion functor $\mathcal{P}(A) \to \mathcal{M}(A)$ induces a homomorphism $K_n(A) \to G_n(A)$.

If A is regular, then $K_n(A) \approx G_n(A)$ (see 4.1.2).

(iii) Let X be a scheme, see [128], $\mathcal{P}(X)$ the category of locally free sheaves of O_X-modules of finite rank (or equivalently category of finite-dimensional (algebraic) vector bundles on X). Then $\mathcal{P}(X)$ is an exact category and we write $K_n(X)$ for $K_n(\mathcal{P}(X))$, see [114].

If $X = \mathrm{Spec}(A)$ for some commutative ring A, then we have an equivalence of categories:

$$\mathcal{P}(X) \to \mathcal{P}(A) : E \to \Gamma(X, E) = \{A\text{-modules of global sections}\}$$

with an inverse equivalence $\mathcal{P}(A) \to \mathcal{P}(X)$ given by

$$P \to \widetilde{P} : U \to O_X(U) \otimes_A P.$$

So,

$$K_n(A) \approx K_n(X).$$

(iv) If X is a Noetherian scheme, then the category $\mathcal{M}(X)$ of coherent sheaves of O_X-modules is exact. We write $G_n(X)$ for $K_n(\mathcal{M}(X))$. If $X = \operatorname{Spec}(A)$, then we have an equivalence of categories $\mathcal{M}(X) \approx \mathcal{M}(A)$ and $G_n(X) \approx G_n(A)$.

(v) Let R be a commutative ring with identity, Λ an R-algebra that is finitely generated as an R-module, $\mathcal{P}_R(\Lambda)$ the category of left Λ-lattices. Then $\mathcal{P}_R(\Lambda)$ is an exact category and we write $G_n(R,\Lambda)$ for $K_n(\mathcal{P}_R(\Lambda))$. If $\Lambda = RG$, G finite group, write $G_n(R,G)$ for $G_n(R,RG)$. If R is regular, then $G_n(R,\Lambda) \approx G_n(\Lambda)$, see [67].

(vi) Let G be a finite group, S a G-set, $\underline{S}$ the translation category of S (or category associated to S) see [72] or [25], or 1.1.7(ii). Then, the category $[\underline{S},\mathcal{C}]$ of functors from $\underline{S}$ to an exact category $\mathcal{C}$ is also an exact category. We denote by $K_n^G(S,\mathcal{C})$ the Abelian group $K_n([\underline{S},\mathcal{C}])$. As we shall see later $K_n^G(-,\mathcal{C}) : \underline{GSet} \to \underline{Ab}$ is a 'Mackey' functor see [24] or [25] or [80].

If $S = G/G$, and $\mathcal{C}_G$ denotes the category of representations of G in $\mathcal{C}$, then $[\underline{G/G},\mathcal{C}] \approx \mathcal{C}_G$. In particular, $[\underline{G/G},\mathcal{P}(R)] \approx \mathcal{P}(R)_G \approx \mathcal{P}_R(RG)$ and so $K_n^G[\underline{G/G},\mathcal{P}(R)] \approx K_n(\mathcal{P}(R)_G) \approx G_n(R,G) \approx G_n(RG)$ if R is regular. As explained in [79,72], when $R = \mathbb{C}$, $K_0(\mathcal{P}(\mathbb{C})_G) \approx G_0(\mathbb{C},G) \approx G_0(\mathbb{C}G) =$ Abelian group of characters $\chi : G \to G$.

We shall discuss relative generalizations of this in Section 8.

(vii) Let X be a compact topological space, $F = \mathbb{R}$ or $\mathbb{C}$. Then the category $VB_F(X)$ of vector bundles on X is an exact category and we can write $K_n(VB_F(X))$ as $K_n^F(X)$.

(viii) Let X be an H-space, m, n positive integers, M_m^n an n-dimensional mod-m Moore space, i.e. the space obtained from S^{n-1} by attaching an n-cell via a map of degree m (see [16] or [107]). Write $\pi_n(X,\mathbb{Z}/m)$ for $[M_m^n, X]$ the set of homotopy classes of maps form M_m^n to X. If $X = BQ\mathcal{C}$ where $\mathcal{C}$ is an exact category, write $K_n(\mathcal{C},\mathbb{Z}/m)$ for $\pi_{n+1}(BQ\mathcal{C},\mathbb{Z}/m)$, $n \geqslant 1$, and call this group the mod-m higher K-theory of $\mathcal{C}$. This theory is well defined for $\mathcal{C} = \mathcal{P}(A)$ where A is any ring with identity and we write $K_n(A,\mathbb{Z}(m))$ for $K_n(\mathcal{P}(A),\mathbb{Z}/m)$. If X is a scheme write $K_n(X,\mathbb{Z}/m)$ for $K_n(\mathcal{P}(X),\mathbb{Z}/m)$. For a Noetherian ring A, write $G_n(A,\mathbb{Z}/m)$ for $K_n(\mathcal{M}(A),\mathbb{Z}/m)$ while for a Noetherian scheme X, we shall write $G_n(X,\mathbb{Z}/m)$ for $K_n(\mathcal{M}(X),\mathbb{Z}/m)$. For the applications, it is usual to consider $m = \ell^s$ where ℓ is a prime and s a positive integer (see [16] or [78]).

(ix) Let G be a discrete Abelian group, $M^n(G)$ the space with only one non-zero reduced integral cohomology group $\widetilde{H}^n(M^n(G))$. Suppose that $\widetilde{H}^n(M^n(G)) = G$. If we write $\pi_n(X,G)$ for $[M^n(G), X]$, and we put $G = \mathbb{Z}/m$, we recover (viii) above since $M_m^n = M^n(\mathbb{Z}/m)$. If $G = \mathbb{Z}$, $M^n(\mathbb{Z}) = S^n$ and so, $\pi_n(X,\mathbb{Z}) = [S^n, X] = \pi_n(X)$.

(x) With notations as in (ix), let $M_{\ell^\infty}^{n+1} = \varinjlim_s M_{\ell^s}^{n+1}$. For all $n \geqslant 0$, we shall denote $[M_{\ell^\infty}^{n+1}, B\mathcal{C}]$ ($\mathcal{C}$ an exact category) by $K_n^{\mathrm{pr}}(\mathcal{C},\widehat{\mathbb{Z}}_\ell)$ and call this group the profinite (higher) K-theory of $\mathcal{C}$. By way of notation, we shall write $K_n^{\mathrm{pr}}(A,\widehat{\mathbb{Z}}_l)$ if $\mathcal{P} = \mathcal{M}(A)$, A any ring with identity: $G_n^{\mathrm{pr}}(A,\widehat{\mathbb{Z}}_l)$ if $\mathcal{C} = \mathcal{M}(A)$, A Noetherian; $K_n^{\mathrm{pr}}(X,\widehat{\mathbb{Z}}_i)$ if $\mathcal{C} = \mathcal{P}(X)$, X any scheme and $G_n^{\mathrm{pr}}(X,\widehat{\mathbb{Z}}_i)$ if $\mathcal{C} = \mathcal{M}(X)$, X a Noetherian scheme. For a comprehensive study of these constructions and applications especially to orders and grouprings, see [78] or 7.3.

3.2. *Higher K-theory of symmetric monoidal categories – definitions and examples*

3.2.1. A symmetric monoidal category is a category $\mathcal{S}$ equipped with a functor $\perp$: $\mathcal{S}\times\mathcal{S}\to\mathcal{S}$ and a distinguished object '0' such that '$\perp$' is 'coherently' associative and commutative in the sense of MacLane (i.e. satisfying properties and diagrams in [79, 1.4.1]). Note that $B\mathcal{S}$ is an H-space (see [39]).

3.2.2. EXAMPLES.

(i) Let $(\mathrm{Iso}\,\mathcal{S})$ denotes the subcategory of isomorphisms in $\mathcal{S}$, i.e. $\mathrm{ob}(\mathrm{Iso}\,\mathcal{S})=\mathrm{ob}\,\mathcal{S}$; morphisms are isomorphisms in $\mathcal{S}$. $\pi_0(\mathrm{Iso}\,\mathcal{S})=$ set of isomorphism classes of objects of $\mathcal{S}$. Then $\mathcal{S}^{\mathrm{iso}}:=\pi_0(\mathrm{Iso}\,\mathcal{S})$ is a monoid.

$\mathrm{Iso}(\mathcal{S})$ is equivalent to the disjoint union $\coprod \mathrm{Aut}_{\mathcal{S}}(S)$ and $B(\mathrm{Iso}\,\mathcal{S})$ is homotopy equivalent to $\coprod B(\mathrm{Aut}_{\mathcal{S}}(S)$, $S\in\mathcal{S}^{\mathrm{iso}}$.

(ii) If $\mathcal{S}=\underline{FSet}$ in (1), $\mathrm{Aut}_{\underline{FSet}}(S)\simeq\Sigma_n$ (symmetric group on n letters) $\mathrm{Iso}(\underline{FSet})$ is equivalent to the disjoint union $\coprod\Sigma_n$. $B(\mathrm{Iso}(\underline{FSet}))$ is homotopy equivalent to $\coprod B\Sigma_n$.

(iii) $B(\mathrm{Iso}\,\mathcal{P}(R))$ is equivalent to the disjoint union $\coprod B\mathrm{Aut}(P)$, $P\in\mathcal{P}(R)$.

(iv) Let $\mathcal{F}(R)=$ category of free R-modules $(\mathrm{Iso}\,\mathcal{F}(R))=\coprod \mathrm{GL}_n(R)$ and $B(\mathrm{Iso}(\mathcal{F}(R))$ is equivalent to the disjoint union $\coprod B\mathrm{GL}_n(R)$. If R satisfies the invariant basis property, then $\mathrm{Iso}(\mathcal{F}(R))$ is a full subcategory of $\mathrm{Iso}(\mathcal{P}(R))$ and $\mathrm{Iso}(\mathcal{F}(R))$ is cofinal in $\mathrm{Iso}\,\mathcal{P}(R)$.

3.2.3. Suppose that every map in $\mathcal{S}$ is an isomorphism and every translation $S\perp$: $\mathrm{Aut}_{\mathcal{S}}(T)\to\mathrm{Aut}_{\mathcal{S}}(S\perp T)$ is an injection. We now define a category $\mathcal{S}^{-1}\mathcal{S}$ such that $K(\mathcal{S})=B(\mathcal{S}^{-1}\mathcal{S})$ is a 'group completion' of $B\mathcal{S}$.

Recall that a group completion of a homotopy commutative and homotopy associative H-space X is an H-space Y together with an H-space map $X\to Y$ such that $\pi_0(Y)$ is the group completion of (i.e. the Grothendieck group associated to) the monoid $\pi_0(X)$ (see [79, 1.1]) and the homology ring $H_*(Y,R)$ is isomorphic to the localization $\pi_0(X)^{-1}H_*(X,R)$ of $H_*(X,R)$.

3.2.4. DEFINITION. Define $\mathcal{S}^{-1}\mathcal{S}$ as follows:

$$\mathrm{ob}(\mathcal{S}^{-1}\mathcal{S})=\{(S,T)\mid S,T\in\mathrm{ob}\,\mathcal{S}\},$$

$$\mathrm{mor}_{\mathcal{S}^{-1}\mathcal{S}}\big((S_1,T_1),(S_1^1,T_1^1)\big)=\left\{\begin{matrix}\text{equivalence class of composites}\\ (S_1,T_1)\xrightarrow{S\perp}(S\perp S_1,S\perp T_1)\xrightarrow{(f,g)}(S_1^1,T_1^1)\end{matrix}\right\}$$

NOTES.

(i) The composite $(S_1,T_1)\xrightarrow{S\perp}(S\perp S_1,S\perp T_1)\overset{(f,g)}{\twoheadrightarrow}(S_1',T_1')$ is said to be equivalent to $(S_1,T_1)\xrightarrow{T\perp}(T\perp S_1,T\perp T_1)\xrightarrow{(f^1,g^1)}(S_1',T_1')$ if there exists an isomorphism $\alpha: S\approx T$ in $\mathcal{S}$ such that composition with $\alpha\perp S_1,\alpha\perp T_1$ sends f' and g' to f.

(ii) Since we have assumed that every translation is an injection in 3.2.3, it means that $\mathcal{S}^{-1}\mathcal{S}$ determines its objects up to unique isomorphism.

(iii) $\mathcal{S}^{-1}\mathcal{S}$ is a symmetric monoidal category with $(S,T)\perp(S',T')=(S\perp S',T\perp T')$ and the functor $\mathcal{S}\to\mathcal{S}^{-1}\mathcal{S}: S\to(o,S)$ is monoidal. Hence $B(\mathcal{S}^{-1}\mathcal{S})$ is an H-space (see [39]).

(iv) $B\mathcal{S}\to B(\mathcal{S}^{-1}\mathcal{S})$ is an H-space map and $\pi_0(\mathcal{S})\to\pi_0(\mathcal{S}^{-1}\mathcal{S})$ is a map of Abelian monoids.

(v) $\pi_0(\mathcal{S}^{-1}\mathcal{S})$ is an Abelian group.

3.2.5. EXAMPLES.

(i) If $\mathcal{S}=\coprod \mathrm{GL}_n(R)=\mathrm{Iso}\,\mathcal{F}(R)$, then $B(\mathcal{S}^{-1}\mathcal{S})$ is a group completion of $B\mathcal{S}$ and $B(\mathcal{S}^{-1}\mathcal{S})$ is homotopy equivalent to $\mathbb{Z}\times B\mathrm{GL}(R)^+$, see [39] or [167], for a proof. See theorem 3.2.8 below for a more general formulation of this.

(ii) For $\mathcal{S}=\mathrm{Iso}(\mathcal{F}\mathcal{S}et)$, $B(\mathcal{S}^{-1}\mathcal{S})$ is homotopy equivalent to $\mathbb{Z}\times B\Sigma^+$ where Σ is the infinite symmetric group (see [167]).

3.2.6. DEFINITION. Let $\mathcal{S}$ be a symmetric monoidal category in which every morphism is an isomorphism.

Define

$$K_n^{\perp}(\mathcal{S}):=\pi_n\big(B(\mathcal{S}^{-1}\mathcal{S})\big).$$

NOTE. $K_0^{\perp}(\mathcal{S})$ as defined above coincides with $K_0^{\perp}(\mathcal{S})$ as defined in [79, 1.4]. This is because $K_0^{\perp}(\mathcal{S})=\pi_0(B(\mathcal{S}^{-1}\mathcal{S}))$ is the group completion of the Abelian monoid $\pi_0(\mathcal{S})=\mathcal{S}^{\mathrm{iso}}$. For a proof, see [167].

3.2.7. REMARKS. Suppose that $\mathcal{S}$ is a symmetric monoidal category which has a countable sequence of objects $S_1,S_2,\ldots$ such that $S_{n+1}=S_n\perp T_n$ for some $T_n\in\mathcal{S}$ and satisfying the cofinality condition, i.e. for every $S\in\mathcal{S}$, there exist an S' and an n such that $S\perp S'\approx S_n$. If this situation obtains, then we can form $\mathrm{Aut}(\mathcal{S})=\mathrm{colim}_{n\to\infty}\mathrm{Aut}_{\mathcal{S}}(S_n)$.

3.2.8. THEOREM [167]. *Suppose that $\mathcal{S}=\mathrm{Iso}(\mathcal{S})$ is a symmetric monoidal category whose translations are injections, and that the conditions of* 3.2.7 *are satisfied so that the group* $\mathrm{Aut}(\mathcal{S})$ *exists. Then the commutator subgroup E of* $\mathrm{Aut}(\mathcal{S})$ *is a perfect normal subgroup*; $K_1(S)=\mathrm{Aut}(S)/E$ *and* $B\mathrm{Aut}(S)^+$ *is the connected component of the identity in the group completion of* $B(\mathcal{S}^{-1}\mathcal{S})$.

Hence $B(\mathcal{S}^{-1}\mathcal{S})\cong K_0(\mathcal{S})\times B\mathrm{Aut}(\mathcal{S})^+$.

3.2.9. EXAMPLE. Let R be a commutative ring with identity. We saw in [79, 1.43] that $(\mathcal{S}=\underline{\mathrm{Pic}}(R),\otimes)$ is a symmetric monoidal category. Since $\pi_0(\mathcal{S})$ is a group, $\mathcal{S}$ and $\mathcal{S}^{-1}\mathcal{S}$ are homotopy equivalent (see [167]). Hence we get $K_0\,\underline{\mathrm{Pic}}(R)=\mathrm{Pic}(R)$, $K_1(\underline{\mathrm{Pic}}(R))=U(R)$ (units of R), and $K_n(\underline{\mathrm{Pic}}(R))=0$ for all $n\geqslant 2$.

3.3. *Higher K-theory of Waldhausen categories – definitions and examples*

3.3.1. DEFINITION. A category with cofibrations is a category $\mathcal{C}$ with zero object together with a subcategory $\mathrm{co}(\mathcal{C})$ whose morphisms are called cofibrations written $A \rightarrowtail B$ and satisfying the axioms

(C1) Every isomorphism in $\mathcal{C}$ is a cofibration.

(C2) If $A \rightarrowtail B$ is a cofibration and $A \to C$ any $\mathcal{C}$-map, then the pushout $B \cup_A C$ exists in $\mathcal{C}$

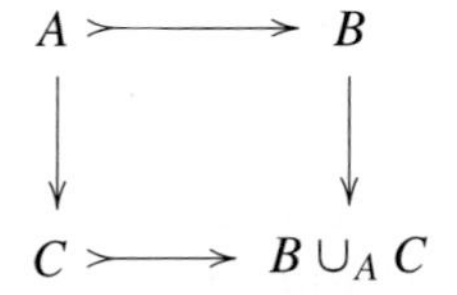

- Hence coproducts exists in $\mathcal{C}$ and each cofibration $A \rightarrowtail B$ has a cokernel $C = B/A$.
- Call $A \rightarrowtail B \twoheadrightarrow B/A$ a cofibration sequence.

(C3) The unique map $0 \to B$ is a cofibration $\forall$ $\mathcal{C}$-objects B.

3.3.2. DEFINITION. A Waldhausen category (or W-category for short) $\mathcal{C}$ is a category with cofibrations together with a subcategory $w(\mathcal{C})$ of weak equivalences (w.e. for short) containing all isomorphisms and satisfying.

GLUING AXIOM FOR WEAK EQUIVALENCES (W1). For any commutative diagram

$$\begin{array}{ccccc} C & \longleftarrow & A & \rightarrowtail & B \\ \downarrow\wr & & \downarrow\wr & & \downarrow\wr \\ C' & \longleftarrow & A' & \rightarrowtail & B' \end{array}$$

in which the vertical maps are weak equivalences and the two right horizontal maps are cofibrations, the induced map $B \cup_A C \to B' \cup_{A'} C'$ is also a weak equivalence.

We shall sometimes denote $\mathcal{C}$ by $(\mathcal{C}, w)$.

3.3.3. DEFINITION. A Waldhausen subcategory $\mathcal{A}$ of a W-category $\mathcal{C}$ is a subcategory which is also W-category such that (a) the inclusion $\mathcal{A} \subseteq \mathcal{C}$ is an exact functor, (b) the cofibrations in $\mathcal{A}$ are the maps in $\mathcal{A}$ which are cofibrations in $\mathcal{C}$ and whose cokernel lies in $\mathcal{A}$ and (c) the weak equivalences in $\mathcal{A}$ are the weak equivalences of $\mathcal{C}$ which lie in $\mathcal{A}$.

3.3.4. DEFINITION. A W-category $\mathcal{C}$ is said to be saturated if whenever (f, g) are composable maps and fg is a w.e., then f is a w.e., iff g is.

- The cofibrations sequences in a W-category $\mathcal{C}$ form a category $\mathcal{E}$. Note that $\mathrm{ob}(\mathcal{E})$ consists of cofibrations sequences $E : A \rightarrowtail B \twoheadrightarrow C$ in $\mathcal{C}$. A morphism $E \to E' : A' \rightarrowtail B' \twoheadrightarrow C'$ in $\mathcal{E}$ is a commutative diagram

$$\begin{array}{ccccccc} E & & A & \rightarrowtail & B & \twoheadrightarrow & C \\ \downarrow & & \downarrow & & \downarrow & & \downarrow \\ E' & & A' & \rightarrowtail & B' & \twoheadrightarrow & C' \end{array} \qquad \text{(I)}$$

To make $\mathcal{E}$ a W-category, we define a morphism $E \to E'$ in $\mathcal{E}$ to be a cofibration if $A \to A', C \to C'$ and $A' \cup_A B \to B'$ are cofibrations in $\mathcal{C}$ while $E \to E'$ is a w.e. if its component maps $A \to A', B \to B', C \to C'$ are w.e. in $\mathcal{C}$.

3.3.5. EXTENSION AXIOM. A W-category $\mathcal{C}$ is said to satisfy the extension axiom if for any morphism $f : E \to E'$ as in 3.3.4, maps $A \to A', C \to C'$ being w.e. implies that $B \to B'$ is also a w.e.

3.3.6. EXAMPLES.

(i) Any exact category $\mathcal{C}$ is a W-category where the cofibrations are the admissible monomorphisms and the w.e. are isomorphisms.

(ii) If $\mathcal{C}$ is any exact category, then the category $\mathrm{Ch}_b(\mathcal{C})$ of bounded chain complexes in $\mathcal{C}$ is a W-category where the w.e. are quasi-isomorphisms (i.e. isomorphisms on homology) and a chain map $\underline{A}. \to \underline{B}.$ is a cofibration if each $A_i \to B_i$ is a cofibration (admissible monomorphism) in $\mathcal{C}$.

(iii) Let $\mathcal{C}$ = category of finite based CW-complexes. Then $\mathcal{C}$ is a W-category where the cofibrations are cellular inclusion and the w.e. are homotopy equivalences.

(iv) If $\mathcal{C}$ is a W-category, define $K_0(\mathcal{C})$ as the Abelian group generated by objects of $\mathcal{C}$ with relations

(i) $A \xrightarrow{\sim} B \Rightarrow [A] = [B]$.

(ii) $A \rightarrowtail B \twoheadrightarrow C \Rightarrow [B] = [A] + [C]$.

Note that this definition agrees with the earlier $K_0(\mathcal{C})$ given in [79, 3.1] for an exact category.

3.3.7. In order to define the K-theory space $K(\mathcal{C})$ such that

$$\pi_n\big(K(\mathcal{C})\big) = K_n(\mathcal{C})$$

for a W-category $\mathcal{C}$, we construct a simplicial W-category $S_*\mathcal{C}$, where $S_n\mathcal{C}$ is the category whose objects $A.$ are sequences of n cofibrations in $\mathcal{C}$, i.e.

$$A_\bullet : 0 = A_0 \rightarrowtail A_1 \rightarrowtail A_2 \rightarrowtail \cdots \rightarrowtail A_n$$

together with a choice of every subquotient $A_{ij} = A_j/A_i$ in such a way that we have a commutative diagram

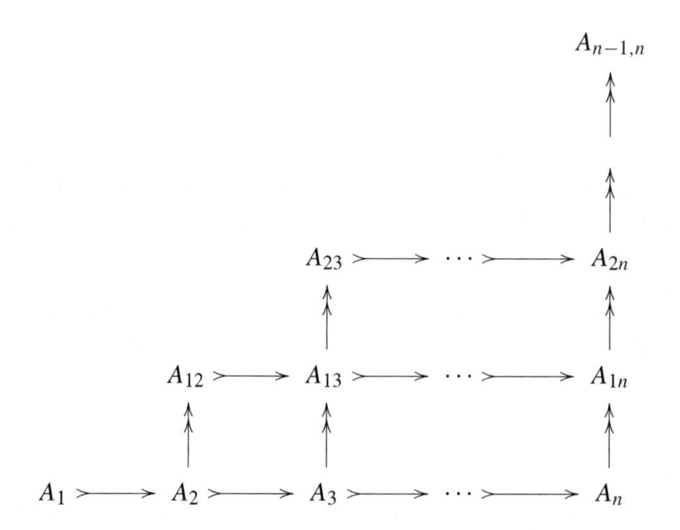

By convention put $A_{jj} = 0$ and $A_{0j} = A_j$.

A morphism $A_\bullet \to B.$ is a natural transformation of sequences.

A weak equivalence in $S_n(\mathcal{C})$ is a map $A_\bullet \to B_\bullet$ such that each $A_i \to B_i$ (and hence each $A_{ij} \to B_{ij}$) is a w.e. in $\mathcal{C}$. A map $A_\bullet \to B_\bullet$ is a cofibration if for every $0 \leqslant i < j < k \leqslant n$ the map of cofibration sequences is a cofibration in $\mathcal{E}(\mathcal{C})$.

$$\begin{array}{ccccc} A_{ij} & \rightarrowtail & A_{ik} & \twoheadrightarrow & A_{jk} \\ \downarrow & & \downarrow & & \downarrow \\ B_{ij} & \rightarrowtail & B_{ik} & \twoheadrightarrow & B_{jk} \end{array}$$

For $0 < i \leqslant n$, define exact functors $\delta_i : S_n(\mathcal{C}) \to S_{n+1}(\mathcal{C})$ by omitting A_i from the notation and re-indexing the A_{jk} as needed. Define $\delta_0 : S_n(\mathcal{C}) \to S_{n+1}(\mathcal{C})$ where δ_0 omits the bottom arrow. We also define $s_i : S_n(\mathcal{C}) \to S_{n+1}(\mathcal{C})$ by duplicating A_i and re-indexing (see [154]).

We now have a simplicial category $n \to wS_n\mathcal{C}$ with degree-wise realization $n \to B(wS_n\mathcal{C})$, and denote the total space by $|wS.\mathcal{C}|$ (see [154]).

3.3.8. DEFINITION. The K-theory space of a W-category $\mathcal{C}$ is $K(\mathcal{C}) = \Omega|wS.\mathcal{C}|$. For each $n \geqslant 0$, the K-groups are defined as $K_n(\mathcal{C}) = \pi_n(K\mathcal{C})$.

3.3.9. By iterating the $S_\bullet$ construction, one can show (see [154]) that the sequence

$$\{\Omega|wS_\bullet\mathcal{C}|, \Omega|wS_\bullet S_\bullet\mathcal{C}|, \ldots, \Omega|wS_\bullet^n\mathcal{C}|\ldots\}$$

forms a connective spectrum $\mathbb{K}(\mathcal{C})$ called the K-theory spectrum of $\mathcal{C}$. Hence $K(\mathcal{C})$ is an infinite loop space, see 1.2.2.

3.3.10. EXAMPLES.

(i) Let $\mathcal{C}$ be an exact category, $\mathrm{Ch}_b(\mathcal{C})$ the category of bounded chain complexes over $\mathcal{C}$. It is a theorem of Gillet and Waldhausen that $K(\mathcal{C}) \cong K(\mathrm{Ch}_b(\mathcal{C}))$ and so, $K_n(\mathcal{C}) \simeq K_n(\mathrm{Ch}_b(\mathcal{C}))$ for every $n \geqslant 0$ (see [32]).

(ii) *Perfect complexes.* Let R be any ring with identity and $\mathcal{M}'(R)$ the exact category of finitely presented R-modules. (Note that $\mathcal{M}'(R) = \mathcal{M}(R)$ if R is Noetherian.) An object $M_\bullet$ of $\mathrm{CH}_b(\mathcal{M}'(R))$ is called a perfect complex if $M_\bullet$ is quasi isomorphic to a complex in $\mathrm{Ch}_b(\mathcal{P}(R))$. The perfect complexes form a Waldhausen subcategory $\mathrm{Perf}(R)$ of $\mathrm{Ch}_b(\mathcal{M}'(R))$. So, we have

$$K(R) \simeq K\big(\mathrm{Ch}_b\big(\mathcal{P}(R)\big)\big) \cong K\big(\mathrm{Perf}(R)\big).$$

(iii) *Derived categories.* Let $\mathcal{C}$ be an exact category and $H^b(\mathcal{C})$ the (bounded) homotopy category of $\mathcal{C}$, i.e. the stable category of $\mathrm{Ch}_b(\mathcal{C})$ (see [63]). So, $\mathrm{ob}(H^b(\mathcal{C})) = \mathrm{Ch}_b(\mathcal{C})$ and morphisms are homotopy classes of bounded complexes. Let $A(\mathcal{C})$ be the full subcategory of $H^b(\mathcal{C})$ consisting of acyclic complexes (see [63]). The derived category $D^b(\mathcal{C})$ of $\mathcal{E}$ is defined by $D^b(\mathcal{C}) = H^b(\mathcal{C})/A(\mathcal{C})$. A morphism of complexes in $\mathrm{Ch}_b(\mathcal{C})$ is called a quasi-isomorphism if its image in $D^b(\mathcal{C})$ is an isomorphism. We could also define the unbounded derived category $D(\mathcal{C})$ from unbounded complexes $\mathrm{Ch}(\mathcal{C})$.

Note that there exists a faithful embedding of $\mathcal{C}$ in an Abelian category $\mathcal{A}$ such that $\mathcal{C} \subset \mathcal{A}$ is closed under extensions and the exact functor $\mathcal{C} \to \mathcal{A}$ reflects exact sequences. So, a complex in $\mathrm{Ch}(\mathcal{C})$ is acyclic iff its image in $\mathrm{Ch}(\mathcal{A})$ is acyclic. In particular, a morphism in $\mathrm{Ch}(\mathcal{C})$ is a quasi-isomorphism iff its image in $\mathrm{Ch}(\mathcal{A})$ is a quasi-isomorphism. Hence, the derived category $D(\mathcal{C})$ is the category obtained from $\mathrm{Ch}(\mathcal{C})$ formally inverting quasi-isomorphisms.

(iv) *Stable derived categories and Waldhausen categories.* Now let $\mathcal{C} = \mathcal{M}'(R)$. A complex $M.$ in $\mathcal{M}'(R)$ is said to be compact if the functor $\mathrm{Hom}(M., -)$ commutes with arbitrary set-valued coproducts. Let $\underline{\mathrm{Comp}(R)}$ denote the full subcategory of $D(\mathcal{M}'(R))$ consisting of compact objects. Then we have $\underline{\mathrm{Comp}(R)} \subset D^b(\mathcal{M}'(R)) \subset D(\mathcal{M}'(R))$.

Define the stable derived category of bounded complexes $\underline{D}^b(\mathcal{M}'(R))$ as the quotient category of $D^b(\mathcal{M}'(R))$ with respect to $\underline{\mathrm{Comp}(R)}$. A morphism of complexes in $\mathrm{Ch}_b(\mathcal{M}'(R))$ is called a stable quasi-isomorphism if its image in $\underline{D}^b(\mathcal{M}'(R))$ is an isomorphism. The family of stable quasi-isomorphisms in $\mathcal{A} = \mathrm{Ch}_b(\mathcal{M}'(R))$ is denoted $\omega\mathcal{A}$.

(v) THEOREM.

(1) $w(\mathrm{Ch}_b(\mathcal{M}'(R))$ *forms a set of weak equivalences and satisfies the saturation and extension axioms.*

(2) $\mathrm{Ch}_b(\mathcal{M}'(R))$ *together with the family of stable quasi-isomorphisms is a Waldhausen category.*

4. Some fundamental results and exact sequences in higher K-theory

4.1. *Resolution theorem*

4.1.1. RESOLUTION THEOREM FOR EXACT CATEGORIES [114]. *Let $\mathcal{P} \subset \mathcal{H}$ be full exact subcategories of an Abelian category $\mathcal{A}$, both closed under extensions and inheriting their exact structure from $\mathcal{A}$. Suppose that* (1) *every object M of $\mathcal{H}$ has a finite $\mathcal{P}$-resolution and* (2) *$\mathcal{P}$ is closed under kernels in $\mathcal{H}$, i.e. if $L \to M \twoheadrightarrow N$ is an exact sequence in $\mathcal{H}$ with $M, N \in \mathcal{P}$, then L is also in $\mathcal{P}$. Then $K_n\mathcal{P} \cong K_n\mathcal{H}$ for all $n \geqslant 0$.*

4.1.2. REMARKS AND EXAMPLES.

(i) Let R be a regular Noetherian ring. Then by taking $\mathcal{H} = \mathcal{M}(R)$, $\mathcal{P} = \mathcal{P}(R)$ in 4.1.1, we have $K_n(R) \simeq G_n(R)$ for all $n \geqslant 0$.

(ii) Let R be any ring with identity and $\mathcal{H}(R)$ the category of all R-modules having finite homological dimension (i.e. having a finite resolution by finitely generated projective R-modules), $\underline{H}_s(R)$ the subcategory of modules in $\mathcal{H}(R)$ having resolutions of length $\leqslant s$. Then by 4.1.1 applied to $\mathcal{P}(R) \subseteq \underline{H}_s(R) \subseteq \underline{H}(R)$ we have $K_n(R) \cong K_n(\underline{H}(R)) \cong K_n(\underline{H}_s(R))$ for all $s \geqslant 1$.

(iii) Let $T = \{T_i\}$ be an exact connected sequence of functors from an exact category $\mathcal{C}$ to an Abelian category, i.e. given an exact sequence $0 \to M' \to M \to M'' \to 0$ in $\mathcal{C}$ there exists a long exact sequence $\cdots \to T_2M'' \to T_1M' \to T_1M \to$. Let $\mathcal{P}$ be the full subcategory of T-acyclic objects (i.e. objects M such that $T_n(M) = 0$ for all $n \geqslant 1$), and assume that for each $M \in \mathcal{C}$, there is a map $P \twoheadrightarrow M$ such that $P \in \mathcal{P}$ and that $T_nM = 0$ for n sufficiently large. Then $K_nP \cong K_n\mathcal{C}$ $\forall n \geqslant 0$ (see [114]).

(iv) As an example of (iii) let A, B be a Noetherian rings, $f: A \to B$ a homomorphism, B a flat A-module, then we have a homomorphism of K-groups: $(B \otimes_A\ ?)_*: G_n(A) \to G_n(B)$ (since $B \otimes_A\ ?$ is exact). Let B be of finite tor-dimension as a right A-module. Then by applying (iii) above, to $\mathcal{C} = \mathcal{M}(A)$, $T_i(M) = \mathrm{Tor}_i^A(B, M)$ and taking $\mathcal{P}$ as the full subcategory of $\mathcal{M}(A)$ consisting of M such the $T_iM = 0$ for $i > 0$, we have $K_n(\mathcal{P}) \approx G_n(A)$.

(v) Let $\mathcal{C}$ be an exact category and $\mathrm{Nil}(\mathcal{C})$ the category whose objects are pairs (M, ν) with $M \in \mathcal{C}$ and ν is a nilpotent endomorphism of M. Let $\mathcal{C}_0 \subset \mathcal{C}$ be an exact subcategory of $\mathcal{C}$ such that every object of $\mathcal{C}$ has a finite $\mathcal{C}_0$-resolution. Then every object of $\mathrm{Nil}(\mathcal{C})$ has a finite $\mathrm{Nil}(\mathcal{C}_0)$ resolution and so, by 4.1.1,

$$K_n(\mathrm{Nil}(\mathcal{C}_0)) \approx K_n(\mathrm{Nil}(\mathcal{C})).$$

4.2. *Additivity theorem (for exact and Waldhausen categories)*

4.2.1. Let $\mathcal{A}, \mathcal{B}$ be exact categories. A sequence of functors $F' \to F \twoheadrightarrow F''$ from $\mathcal{A}$ to $\mathcal{B}$ is called an exact sequence of exact functors if $0 \to F'(A) \to F(A) \to F''(A) \to 0$ is an exact sequence in $\mathcal{B}$ for every $A \in \mathcal{A}$.

Let $\mathcal{A}$, $\mathcal{B}$ be Waldhausen categories. If $F'(A) \rightarrowtail F(A) \twoheadrightarrow F''(A)$ is a cofibration sequence in $\mathcal{B}$ and for every cofibration $A \rightarrowtail A'$ in $\mathcal{A}$, $F(A) \cup_{F'(A)} F'(A') \to F(A')$ is a cofibration in $\mathcal{B}$ say that $F' \rightarrowtail F \twoheadrightarrow F''$ a short exact sequence or a cofibration sequence of exact functors.

4.2.2. ADDITIVITY THEOREM. *Let $F' \rightarrowtail F \twoheadrightarrow F''$ be a short exact sequence of exact functors from $\mathcal{A}$ to $\mathcal{B}$ where both $\mathcal{A}$ and $\mathcal{B}$ are either exact categories or Waldhausen categories. Then $F_* \simeq F'_* + F''_* : K_n(\mathcal{A}) \to K_n(\mathcal{B})$.*

4.2.3. REMARKS AND EXAMPLES.

(i) It follows from 4.2.2 that if $0 \to F_1 \to F_2 \to \cdots \to F_s \to 0$ is an exact sequence of exact functors $\mathcal{A} \to \mathcal{B}$ then

$$\sum_{k=0}^{s} (-1)^k F_k = 0 : K_n(\mathcal{A}) \to K_n(\mathcal{B})$$

for all $n \geqslant 0$ (see [167]).

(ii) Let X be a scheme, $E \in \mathcal{P}(X)$ (see 3.1.4(iii)). Then we have an exact functor $(E \otimes ?) : \mathcal{P}(X) \to \mathcal{P}(X)$ which induces homomorphisms $K_n(X) \to K_n(X)$.

If $0 \to E' \to E \to E'' \to 0$ is an exact sequence in $\mathcal{P}(X)$, then by 4.2.2 $(E \otimes ?)_* = (E' \otimes ?)_* + (E'' \otimes ?)_* : K_n(X) \to K_n(X)$. Hence we obtain a homomorphism

$$K_0(X) \otimes K_n(X) \to K_n(X) : (E) \otimes y \to (E \otimes ?)_* y, \quad y \in K_n(X),$$

making each $K_n(X)$ a $K_0(X)$-module.

(iii) *Flasque categories.* An exact (or Waldhausen) category is called flasque if there is an exact functor $\infty : \mathcal{A} \to \mathcal{A}$ and a natural isomorphism $\infty(A) \cong A \amalg \infty(A)$; i.e. $\infty \cong 1 \amalg \infty$ where 1 is the identity functor. By 4.2.2, $\infty_* = 1_* \amalg \infty_*$ and hence the identity map $1_* : K(\mathcal{A}) \to K(\mathcal{A})$ is null homotopic. Hence $K(\mathcal{A})$ is contractible and so $\pi_n(K(\mathcal{A})) = K_n(\mathcal{A}) = 0$ for all n.

4.3. *Devissage*

4.3.1. DEVISSAGE THEOREM [114]. *Let $\mathcal{A}$ be an Abelian category, $\mathcal{B}$ a non-empty full subcategory closed under subobjects, quotient objects and finite products in $\mathcal{A}$. Suppose that every object M of $\mathcal{A}$ has a finite filtration $0 = M_0 \subset M_1 \subset \cdots \subset M_n = M$ such that $M_i / M_{i-1} \in \mathcal{B}$ for each i, then the inclusion $Q\mathcal{B} \to Q\mathcal{A}$ is a homotopy equivalence. Hence $K_i(\mathcal{B}) \cong K_i(\mathcal{A})$.*

4.3.2. COROLLARY [114]. *Let $\underline{a}$ be a nilpotent two-sided ideal of a Noetherian ring R. Then for all $n \geqslant 0$, $G_n(R/\underline{a}) \cong G_n(R)$.*

4.3.3. EXAMPLES.

(i) Let R be an Artinian ring with maximal ideal $\underline{m}$ such that $\underline{m}^r = 0$ for some r. Let $k = R/\underline{m}$ (e.g., $R \equiv \mathbb{Z}/p^r$, $k \equiv \mathbb{F}_p$). In 4.3.1 put $\mathcal{B} =$ category of finite-dimensional k-vector spaces and $\mathcal{A} = \mathcal{M}(R)$. Then we have a filtration $0 = \underline{m}^r M \subset \underline{m}^{r-1} M \subset \cdots \subset \underline{m} M \subset M$ for any $M \in \mathcal{M}(R)$. Hence by 4.3.1, $G_n(R) \approx K_n(k)$.

(ii) Let X be a Noetherian scheme, $i : Z \subset X$ the inclusion of a closed subscheme. Then $\mathcal{M}(Z)$ is an Abelian subcategory of $\mathcal{M}(X)$ via the direct image i: $\mathcal{M}(Z) \subset \mathcal{M}(X)$. Let $\mathcal{M}_Z(X)$ be the Abelian category of O_X-modules supported on Z, $\underline{a}$ an ideal sheaf in O_X such that $O_X/\underline{a} = O_Z$. Then every $M \in M_Z(X)$ has a finite filtration $M \supset M\underline{a} \supset M\underline{a}^2 \supset \cdots$ and so, by devissage, $K_n(\mathcal{M}_Z(X)) \approx K_n(\mathcal{M}(Z)) \approx G_n(Z)$.

4.4. *Localization*

4.4.1. A full subcategory $\mathcal{B}$ of an Abelian category $\mathcal{A}$ called a Serre subcategory if whenever:

$$0 \to M' \to M \to M'' \to 0$$

is an exact sequence in $\mathcal{A}$, then $M \in \mathcal{B}$ if and only if $M', M'' \in \mathcal{B}$. Given such a $\mathcal{B}$, construct a quotient Abelian category $\mathcal{A}/\mathcal{B}$ as follows:

$$\mathrm{ob}(\mathcal{A}/\mathcal{B}) = \mathrm{ob}\,\mathcal{A}.$$

$\mathcal{A}/\mathcal{B}(M, N)$ is defined as follows: If $M' \subseteq M$, $N' \subseteq N$ are subobjects such that $M/M' \in \mathrm{ob}(\mathcal{B})$, $N' \in \mathrm{ob}(\mathcal{B})$, then there exists a natural isomorphism $\mathcal{B}(M, N) \to \mathcal{B}(M', N/N')$.

As M', N' range over such pairs of objects, the group $\mathcal{B}(M', N/N')$ forms a direct system of Abelian groups and we define

$$\mathcal{A}/\mathcal{B}(M, N) = \varinjlim_{(M',N')} \mathcal{B}(M', N/N').$$

NOTE. Let $T : \mathcal{A} \to \mathcal{A}/\mathcal{B}$ be the quotient functor: $M \mapsto T(M) = M$.

(i) $T : \mathcal{A} \to \mathcal{A}/\mathcal{B}$ is additive functor.

(ii) If $\mu \in \mathcal{A}(M, N)$ then $T(\mu)$ is null if and only if $\operatorname{Ker} \mu \in \mathrm{ob}(\mathcal{B})$ and $T(\mu)$ is an epimorphism if and only if $\operatorname{Coker} \mu \in \mathrm{ob}(\mathcal{B})$.

(iii) $\mathcal{A}/\mathcal{B}$ is an additive category such that $T : \mathcal{A} \to \mathcal{A}/\mathcal{B}$ is an additive functor.

4.4.2. LOCALIZATION THEOREM [114]. *If $\mathcal{B}$ is a Serre subcategory of an Abelian category $\mathcal{A}$, then there exists a long exact sequence*:

$$\begin{aligned} - - - \to K_n(\mathcal{B}) \to K_n(\mathcal{A}) \to K_n(\mathcal{A}/\mathcal{B}) \to K_{n-1}(\mathcal{B}) \to - - - \\ - - - \to K_0(\mathcal{B}) \to K_0(\mathcal{A}) \to K_0(/\mathcal{B}) \to 0. \end{aligned} \tag{I}$$

4.4.3. EXAMPLES.

(i) Let A be a Noetherian ring, $S \subset A$ a central multiplicative system; $\mathcal{A} = \mathcal{M}(A)$, $\mathcal{B} = \mathcal{M}_S(A)$, the category of finitely generated S-torsion A-modules, $\mathcal{A}/\mathcal{B} \simeq \mathcal{M}(A_S) =$ category of finitely generated A_S-modules.

Let T be the quotient functor $\mathcal{M}(A) \to \mathcal{M}(A)/\mathcal{M}_S(A)$, $u : \mathcal{M}(A)/\mathcal{M}_S(A) \to \mathcal{M}(A_S)$ is an equivalence of categories such the $u.T \simeq L$, where $L : \mathcal{M}(A) \to \mathcal{M}(A_S)$. We thus have an exact sequence $K_{n+1}(\mathcal{M}(A_S)) \to K_n(\mathcal{M}_S(A)) \to K_n(\mathcal{M}(A)) \to K_n(\mathcal{M}(A_S)) \to K_{n-1}(\mathcal{M}_S(A))$, that is:

$$\cdots \to K_n(\mathcal{M}_S(A)) \to G_n(A) \to G_n(A_S) \to K_{n-1}(\mathcal{M}_S(A)) \to \cdots.$$

(ii) Let $A = R$ in (i) be a Dedekind domain with quotient field F, $S = R \backslash \{0\}$. Then, one can show that

$$\mathcal{M}_S(R) = \bigcup_{\underline{m}} \mathcal{M}(R/\underline{m}^k)$$

as $\underline{m}$ runs through all maximal ideals of R.

So,

$$\begin{aligned} K_n(\mathcal{M}_S(R)) &\simeq \bigoplus_{\underline{m}} \lim_{k \to \infty} G_n(R/\underline{m}^k) \\ &= \bigoplus_{\underline{m}} G_n(R/\underline{m}) = \bigoplus_{\underline{m}} K_n(R/\underline{m}). \end{aligned}$$

So, (I) gives

$$\begin{aligned} &\to K_{n+1}(F) \to \bigoplus_{\underline{m}} K_n(R/\underline{m}) \to K_n(R) \to K_n(F) \to \bigoplus_{\underline{m}} K_{n-1}(R/\underline{m}) \\ &\to \cdots \bigoplus_{\underline{m}} \to K_2(R/\underline{m}) \to K_2(R) \to K_2(F) \to \bigoplus K_1(R/\underline{m}) \\ &\to K_1(R) \to K_1(F) \to \bigoplus K_0(R/\underline{m}) \to K_0(R) \to K_0(F), \end{aligned}$$

that is

$$\begin{aligned} &\cdots \to \cdots \to \bigoplus K_2(R/m) \to K_2(R) \to K_2(F) \to \bigoplus (R/m)^* \\ &\to R^* \to F^* \to \bigoplus(\mathbb{Z}) \to \mathbb{Z} \oplus Cl(R) \to \mathbb{Z} \to 0. \end{aligned}$$

(iii) Let R in (i) be a discrete valuation ring (e.g., the ring of integers in a p-adic field) with unique maximal ideal $\underline{m} = sR$. Let $F =$ quotient field of R. Then $F = R[\frac{1}{s}]$, with residue field $= R/m = k$. Hence, we obtain the following exact sequence

$$\begin{aligned} &\to K_n(k) \to K_n(R) \to K_n(F) \to K_{n-1}(k) \to \cdots \to K_2(k) \to K_2(R) \\ &\qquad \to K_2(F) \to K_1(k) \to \cdots \to K_0(F) \to 0. \end{aligned} \tag{II}$$

Gersten's conjecture says that the sequence (II) breaks up into split short exact sequences

$$0 \to K_n(R) \xrightarrow{\alpha_n} K_n(F) \xrightarrow{\beta_n} K_{n-1}(k) \to 0.$$

For this to happen, one must have that for all $n \geqslant 1$, $K_n(k) \to K_n(R)$ is the zero map and that there exists a map $K_{n-1}(k) \xrightarrow{\eta_n} K_n(F)$ such that $K_n(F) \simeq K_n(R) \oplus K_{n-1}(k)$, i.e. $\beta_n\eta_n = 1_{K_{n-1}}(k)$.

True for $n = 0$, $K_0(R) \simeq K_0(F) \simeq \mathbb{Z}$.

True for $n = 1$, $K_1F \simeq F^*$, $K_1(R) = R^*$, $F^* = R^* \times \{s^n\}$.

True for $n = 2$.

$$0 \to K_2(R) \xrightarrow{\alpha_2} K_2(F) \xrightarrow{\beta_2} K_1(k) \to 0.$$

Here β_2 is the tame symbol. If characteristic of F = characteristic of k, then Gersten's conjecture is also known to be true. When k is algebraic over F_p, then Gersten's conjecture is also true. It is not known in the case when $\operatorname{char}(F) = 0$ or $\operatorname{char}(k) = p$.

(iv) Let R be a Noetherian ring, $S = \{s^n\}$ a central multiplicative system $\mathcal{B} = \mathcal{M}_S(R)$, $\mathcal{A} = \mathcal{M}(R)$.

$$\mathcal{A}/\mathcal{B} = \mathcal{M}(R_S) = \bigcup_{n=1}^{\infty} \mathcal{M}\big(R/s^nR\big).$$

Then (I) gives

$$\cdots \to G_{n+1}(R_S) \to K_n\big(\mathcal{M}_S(R)\big) \to G_n(R) \to G_n(R_S)$$
$$\to K_{n-1}\big(\mathcal{M}_S(R)\big).$$

Note that $K_n(\mathcal{M}_S(R)) = K_n(\bigcup_{n=1}^{\infty} \mathcal{M}(R/s^nR))$.

Now, by devissage $G_n(R/s^nR) \simeq G_n(R/sR)$.

Hence $K_n(\bigcup_{n=1}^{\infty} \mathcal{M}(R/s^nR)) = \lim_{n\to\infty} G_n(R/s^nR) = G_n(R/sR)$. So, we have

$$\ldots G_{n+1}\left(R\left(\frac{1}{s}\right)\right) \to G_n(R/sR) \to G_n(R) \to G_n\left(R\left(\frac{1}{s}\right)\right)$$
$$\to G_{n-1}(R/sR) \to \cdots.$$

(v) Let R be the ring of integers in a p-adic field F, Γ a maximal R-order in a semi-simple F-algebra Σ, if $S = R\backslash\{0\}$, then $F = R_S$

$$\mathcal{B} = \mathcal{M}_S(\Gamma), \qquad \mathcal{A} = \mathcal{M}(\Gamma), \qquad \mathcal{A}/\mathcal{B} = \mathcal{M}(\Sigma).$$

Then sequence (I) yields an exact sequence

$$\cdots \to K_n(\Gamma) \to K_n(\Sigma) \to K_{n-1}(\mathcal{M}_S(\Gamma)) \to K_{n-1}(\Gamma) \\ \to K_{n-1}(\Sigma). \qquad \text{(I)}$$

One can see from (iv) that if $\underline{m} = \pi R$ is the unique maximal ideal of R, then $K_n(\mathcal{M}_S(\Gamma)) = \lim_{n\to\infty} G_n(\Gamma/\pi^n\Gamma) = G_n(\Gamma/\pi\Gamma) \simeq K_n(\Gamma/\operatorname{rad}\Gamma)$ (see [25]). Here $\Sigma = \Gamma_S$ where $S = \{\pi^i\}$. We have also used above the corollary to devissage which says that if $\underline{a}$ is a nilpotent ideal in a Noetherian ring R, then $G_n(R) \simeq G_n(R/\underline{a})$ (see 4.3.2).

(vi) Let R be the ring of integers in an algebraic number field F, Λ any R-order in a semi-simple F-algebra Σ. Let $S = R = 0$. Then we have the following exact sequence

$$\cdots \to K_n(\mathcal{M}_S(\Lambda)) \to G_n(\Lambda) \to G_n(\Sigma) \to K_{n-1}(\mathcal{M}_S(\Lambda)) \to \cdots.$$

One can show that $K_n(\mathcal{M}_S(\Lambda)) \simeq \bigoplus G_n(\Lambda/\underline{p}\Lambda)$ where $\underline{p}$ runs through all the prime ideals of R. See [69] for further details about how to use this sequence to obtain finite generation of $G_n(\Lambda)$, and the fact that $SG_n(\Lambda)$ is finite (see [69,71]).

(vii) Let X be a Noetherian scheme, U an open subscheme of X, $Z = X \backslash U$, the closed complement of U in X. Put $\mathcal{A} = \mathcal{M}(X) =$ category of coherent 0_X-modules, and let $\mathcal{B}$ be the category of coherent O_X-modules whose restriction to U is zero (i.e. the category of coherent modules supported by Z). Let $\mathcal{A}/\mathcal{B}$ be the category of coherent O_U-modules. Then we have the following exact sequence

$$\ldots G_n(Z) \to G_n(X) \to G_n(U) \to G_{n-1}(Z) \to \cdots \to G_0(Z) \\ \to G_0(X) \to G_0(U) \to 0.$$

So far, our localization results have involved mainly the G_n-theory which translates into K_n-theory when the rings involved are regular. We now obtain localization for K_n-theory.

4.4.4. THEOREM. *Let S be a central multiplicative system for a ring R, $\mathcal{H}_S(R)$ the category of S-torsion finitely generated R-modules of finite projective dimension. If S consists of non-zero divisors, then there exists an exact sequence*

$$\cdots \to K_{n+1}(R_S) \to K_n(\mathcal{H}_S(R)) \xrightarrow{\eta} K_n(R) \xrightarrow{\alpha} K_n(R_S) \to \cdots.$$

For a proof see [39].

4.4.5. REMARKS. It is still an open problem to understand $K_n(\mathcal{H}_S(R))$ for various rings R.

If R is regular (e.g., $R = \mathbb{Z}$, the integers in a number field, a Dedekind domain, a maximal order), then $\mathcal{M}_S(R) = \mathcal{H}_S(R)$ and

$$K_n(\mathcal{H}_S(R)) = K_n(\mathcal{M}_S(R)); \qquad G_n(R) = K_n(R).$$

So we recover G-theory. If R is not regular, then $K_n(\mathcal{H}_S(R))$ is not known in general.

4.4.6. DEFINITION. Let $\alpha : A \to B$ be a homomorphism of rings A, B. Suppose that s is a central non-zero divisor in $\mathcal{B}$. Call α an analytic isomorphism along s if $A/sA \simeq B/\alpha(s)B$.

4.4.7. THEOREM. *If $\alpha : A \to B$ is an analytic isomorphism along $s \in S = \{s^i\}$ where s is a central non-zero divisor, then $H_S(A) = H_S(B)$.*

PF follows by comparing the localization sequences for $A \to A[\frac{1}{s}]$ and $B \to B[\frac{1}{s}]$.

4.5. *Fundamental theorem for higher K-theory*

4.5.1. Let $\mathcal{C}$ be an exact category, $\mathrm{Nil}(\mathcal{C})$ the category of nilpotent endomorphisms in $\mathcal{C}$, i.e. $\mathrm{Nil}(\mathcal{C}) = \{(M, \nu) \mid M \in \mathcal{C},\ \nu$ a nilpotent endomorphism of $M\}$. Then we have two functors $Z : \mathcal{C} \to \mathrm{Nil}(\mathcal{C}) = Z(M) = (M, 0)$ (where '0' = zero endomorphism) and $F : \mathrm{Nil}(\mathcal{C}) \to \mathcal{C}$: $F(M, \nu) = M$ satisfying $FZ = 1_{\mathcal{C}}$. Hence we have a split exact sequence $0 \to K_n(\mathcal{C}) \xrightarrow{Z} K_n(\mathrm{Nil}(\mathcal{C})) \to \mathrm{Nil}_n(\mathcal{C}) \to 0$ which defines $\mathrm{Nil}_n(\mathcal{C})$ as the cokernel of Z.

Hence $K_n(\mathrm{Nil}(\mathcal{C})) \simeq K_n(\mathcal{C}) \oplus \mathrm{Nil}_n(\mathcal{C})$.

4.5.2. Let R be a ring with identity, $\underline{H}(R)$ the category of R-modules of finite homological dimension, $\underline{H}_S(R)$ the category of S-torsion objects of $\underline{H}(R)$, $\mathcal{M}_S(R)$ the category of finitely generated S-torsion R-modules. One can show (see [167]) that if $S = T_+ = \{t^i\}$, the free Abelian monoid on one generator t, then there exist isomorphisms $\mathcal{M}_{T_+}(R[t]) \simeq \mathrm{Nil}(\mathcal{M}(R))$, $\underline{H}_{T_+}(R[t]) \simeq \mathrm{Nil}(\underline{H}(R))$ and $K_n(\underline{H}_{T_+}(R[t])) \simeq K_n(R) \oplus \mathrm{Nil}_n(R)$ where we write $\mathrm{Nil}_n(R)$ for $\mathrm{Nil}_n(\mathcal{P}(R))$.

Moreover, the localization sequence 4.4.4 breaks up into short exact sequences

$$0 \to K_n(R[t]) \to K_n(R[t, t^{-1}]) \xrightarrow{\partial} K_{n-1}(\mathrm{Nil}(R)) \to 0.$$

4.5.3. FUNDAMENTAL THEOREM OF HIGHER K-THEORY [114]. *Let R be a ring with identity. Define for all $n \geqslant 0$ $NK_n(R) := \mathrm{Ker}(K_n(R[t]) \xrightarrow{\tilde{i}_+} K_n(R))$ where $\tilde{i}_+$ is induced by the augmentation $t \mapsto 1$.*

Then there are canonical decompositions for all $n \geqslant 0$

(i) $K_n(R[t]) \simeq K_n(R) \oplus NK_n(R)$.

(ii) $K_n(R[t, t^{-1}]) \cong K_n(R) \oplus NK_n(R) \oplus NK_n(R) \oplus K_{n-1}(R)$.

(iii) $K_n(\mathrm{Nil}(R)) \cong K_n(R) \oplus NK_{n+1}(R)$.

The above decompositions are compatible with a split exact sequence

$$0 \to K_n(R) \to K_n(R[t]) \oplus K_n(R[t^{-1}]) \to K_n(R[t, t^{-1}]) \to K_{n-1}(R) \to 0.$$

We close this subsection with the fundamental theorem for G-theory.

4.5.4. THEOREM. *Let R be a Noetherian ring. Then*
(i) $G_n(R[t]) \simeq G_n(R)$.
(ii) $G_n(R[t,t^{-1}]) \simeq G_n(R) \oplus G_{n-1}(R)$.

4.6. *Some exact sequences in the K-theory of Waldhausen categories*

4.6.1. *Cylinder functors.* A Waldhausen category has a cylinder functor if there exists a functor $T : \mathrm{Ar}\,\mathcal{A} \to \mathcal{A}$ together with three natural transformations p, j_1, j_2 such that to each morphism $f : A \to B$, T assigns an object Tf of $\mathcal{A}$ and $j_1 : A \to Tf$, $j_2 : B \to Tf$, $p : Tf \to B$ satisfying certain properties (see [154]).

CYLINDER AXIOM. For all f, $p : Tf \to B$ is in $w(\mathcal{A})$.

4.6.2. Let $\mathcal{A}$ be a Waldhausen category. Suppose that $\mathcal{A}$ has two classes of weak equivalences $\nu(\mathcal{A})$, $w(\mathcal{A})$ such that $\nu(\mathcal{A}) \subset w(\mathcal{A})$. Assume that $w(\mathcal{A})$ satisfies the saturation and extension axioms and has a cylinder functor T which satisfies the cylinder axiom. Let $\mathcal{A}^w$ be the full subcategory of $\mathcal{A}$ whose objects are those $A \in \mathcal{A}$ such that $0 \to A$ is in $w(\mathcal{A})$. Then $\mathcal{A}^\omega$ becomes a Waldhausen category with $\mathrm{co}(\mathcal{A}^\omega) = \mathrm{co}(\mathcal{A}) \cap \mathcal{A}^\omega$ and $\nu(\mathcal{A}^\omega) = \nu(\mathcal{A}) \cap \mathcal{A}^\omega$.

4.6.3. THEOREM (Waldhausen fibration sequence, [154]). *With the notations and hypothesis of* 4.6.2, *suppose that $\mathcal{A}$ has a cylinder functor T which is a cylinder functor for both $\nu(\mathcal{A})$ and $\omega(\mathcal{A})$. Then the exact inclusion functors $(\mathcal{A}^\omega, \nu) \to (\mathcal{A}, \omega)$ induce a homotopy fibre sequence of spectra*

$$K(\mathcal{A}^\omega, \nu) \to K(\mathcal{A}, \nu) \to K(\mathcal{A}, \omega)$$

and hence a long exact sequence

$$K_{n+1}(\mathcal{A}, \omega) \to K_n(\mathcal{A}^\omega) \to K_n(\mathcal{A}, \nu) \to K_n(\mathcal{A}, \omega) \to .$$

The next result is a long exact sequence realizing the cofibre of the Cartan map as K-theory of a Waldhausen category, see [33].

4.6.4. THEOREM [33]. *Let R be a commutative ring with identity. The natural map $K(\mathcal{P}(R)) \to K(\mathcal{M}'(R))$ induced by $\mathcal{P}(R) \hookrightarrow \mathcal{M}'(R)$ fits into a cofibre sequence of spectra $K(R) \to K(\mathcal{M}'(R)) \to K(\mathcal{A}, \omega)$ where $(\mathcal{A}, \omega)$ is the Waldhausen category of bounded chain complexes over $\mathcal{M}'(R)$ with weak equivalences being quasi-isomorphisms. In particular, we have a long exact sequence*

$$\cdots \to K_{n+1}(\mathcal{A}, \omega) \to K_n(R) \to G'_n(R) \to K_{n-1}(\mathcal{A}, \omega) \to \cdots,$$

where

$$G_n'(R) = K_n(\mathcal{M}'(R)).$$

(See 7.1.16 for applications to orders.)

We close this subsection with a generalization of the localization sequence 4.4.3. In 4.6.5 below, the requirement that S contains no zero divisors is removed.

4.6.5. THEOREM [144]. *Let S be a central multiplicatively closed subset of a ring R with identity,* $\mathrm{Perf}(R, S)$ *the Waldhausen subcategory of* $\mathrm{Perf}(R)$ *consisting of perfect complexes $M_\bullet$ such that $S^{-1}M$ is an exact complex. Then* $K(\mathrm{Perf}(R, S)) \to K(R) \to K(S^{-1}R)$ *is a homotopy fibration. Hence there is a long exact sequence.*

$$\dots K_{n+1}(S^{-1}R) \xrightarrow{\delta} K_n(\mathrm{Perf}(R, S)) \to K_n(R) \to K_n(S^{-1}R) \to \cdots.$$

4.7. *Excision; relative and Mayer–Vietoris sequences*

4.7.1. Let Λ be a ring with identity, $\underline{a}$ a 2-sided ideal of Λ. Define $F_{\Lambda,\underline{a}}$ as the homotopy fibre of $B\mathrm{GL}(\Lambda)^+ \to B\overline{\mathrm{GL}}(\Lambda/\underline{a})^+$ where $\overline{\mathrm{GL}}(\Lambda/\underline{a}) = \mathrm{image}(\mathrm{GL}(\Lambda) \to \mathrm{GL}(\Lambda/\underline{a}))$. Then $F_{\Lambda,\underline{a}}$ depends not only on $\underline{a}$ but also on Λ.

If we denote $\pi_n(F_{\Lambda,\underline{a}})$ by $K_n(\Lambda, \underline{a})$, then we have a long exact sequence

$$\to K_n(\Lambda, \underline{a}) \to K_n(\Lambda) \to K_n(\Lambda/\underline{a}) \to K_{n-1}(\Lambda, \underline{a}) \to \tag{I}$$

from the fibration $F_{\Lambda,\underline{a}} \to B\mathrm{GL}(\Lambda)^+ \to \overline{B\mathrm{GL}}(\Lambda/\underline{a})^+$.

4.7.2. DEFINITION. Let B be any ring without unit and $\widetilde{B}$ the ring with unit obtained by formally adjoining a unit to B, i.e. $\widetilde{B}$ = set of all $(b, s) \in B \times \mathbb{Z}$ with multiplication defined by $(b, s)(b', s') = (bb' + sb' + s'b, ss')$.

Define $K_n(B)$ as $K_n(\widetilde{B}, B)$. If Λ is an arbitrary ring with identity containing B as a 2-sided ideal, then B is said to satisfy excision for K_n if the canonical map. $K_n(B) := K_n(\widetilde{B}, B) \to K_n(\Lambda, B)$ is an isomorphism for any ring Λ containing B. Hence, if in 4.7.1 $\underline{a}$ satisfies excision, then we can replace $K_n(\Lambda, \underline{a})$ by $K_n(\underline{a})$ in the long exact sequence (I). We denote $F_{\tilde{\underline{a}},\underline{a}}$ by $F_{\underline{a}}$.

4.7.3. We now present another way to understand $F_{\underline{a}}$ (see [17]). Let $\Gamma_n(\underline{a}) := \mathrm{Ker}(\mathrm{GL}_n(\underline{a} \oplus \mathbb{Z})) \to \mathrm{GL}_n(\mathbb{Z})$ and write $\Gamma(\underline{a}) = \varinjlim \Gamma_n(\underline{a})$. Let Σ_n denote the $n \times n$ permutation matrices. Then Σ_n can be identified with the n-th symmetric group. Put $\Sigma = \varinjlim \Sigma_n$. Then Σ acts on $\Gamma(\underline{a})$ by conjugation and so, we can form $\widetilde{\Gamma}(\underline{a}) = \Gamma(\underline{a}) \rtimes \Sigma$. One could think of $\widetilde{\Gamma}(a)$ as the group of matrices in $\mathrm{GL}_n(\underline{a} \oplus \mathbb{Z})$ whose image in $\mathrm{GL}_n(\mathbb{Z})$ is a permutation matrix. Consider the fibration $B\Gamma\underline{a} \to B\widetilde{\Gamma}(\underline{a}) \to B(\Sigma)$. Note that $B(\Sigma)$, $B\widetilde{\Gamma}(\underline{a})$ has an associated $+$-construction which are infinite loop spaces. Define $F_{\underline{a}}$ as the homotopy fibre

$$F_{\underline{a}} \to B\widetilde{\Gamma}(\underline{a})^+ \to B\Sigma^+.$$

Then, for any ring Λ (with identity) containing $\underline{a}$ as a two-sided ideal, we have a map of fibrations

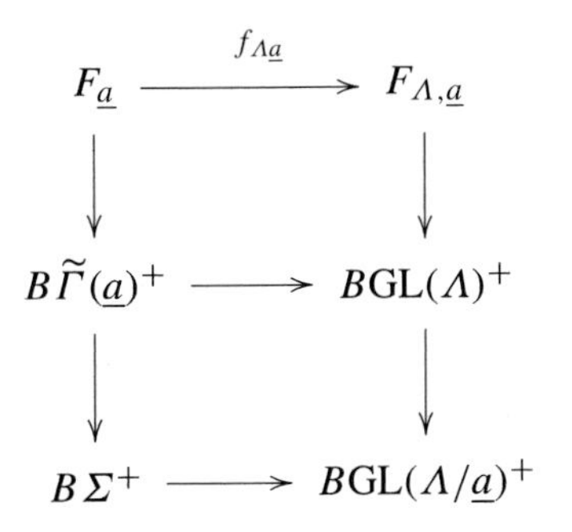

4.7.4. DEFINITION. Let $\underline{a}$ be a ring without unit, $S \subseteq \mathbb{Z}$ a multiplicative subset. Say that $\underline{a}$ is an S-excision ideal if for any ring Λ with unit containing $\underline{a}$ as a 2-sided ideal, $f_{\Lambda,\underline{a}}$ induces an isomorphism $\pi_*(F_{\underline{a}}) \otimes S^{-1}\mathbb{Z} \approx \pi_*(F_{\Lambda,\underline{a}}) \otimes S^{-1}\mathbb{Z}$.

4.7.5. THEOREM [17]. *Let $\underline{a}$ be a ring without unit and $S \subseteq \mathbb{Z}$ a multiplicative set such that $\underline{a} \otimes S^{-1}\mathbb{Z} = 0$ or $\underline{a} \otimes S^{-1}\mathbb{Z}$ has a unit. Then $\underline{a}$ is an S-excision ideal and*

$$H_*\big(F_{\underline{a}}, S^{-1}\mathbb{Z}\big) \cong H_n\big(\Gamma(\underline{a}); S^{-1}\mathbb{Z}\big).$$

4.7.6. EXAMPLES/APPLICATIONS.

(i) If $\underline{a}$ is a 2-sided ideal in a ring Λ with identity such that $\Lambda/\underline{a}$ is annihilated by some $s \in \mathbb{Z}$, then the hypothesis of 4.7.5 is satisfied by $S = \{s^i\}$ and $\underline{a}$ is an S-excision ideal.

(ii) Let R be the ring of integers in a number field F, Λ an R-order in a semi-simple F-algebra Σ, Γ a maximal R-order containing Λ. Then there exists an $s \in \mathbb{Z}$, $s > 0$, such that $s\Gamma \subset \Lambda$ and so $\underline{a} = s\Gamma$ is a 2-sided ideal in both Λ and Γ. Since s annihilates $\Lambda/\underline{a}$ (also $\Gamma/\underline{a}$), $\underline{a}$ is an S-excision ideal and so, we have a long exact Mayer–Vietoris sequence

$$\to K_{n+1}(\Gamma/\underline{a})\left(\frac{1}{s}\right) \to K_n(\Lambda)\left(\frac{1}{s}\right) \to K_n(\Lambda/\underline{a})\left(\frac{1}{s}\right) \oplus K_n(\Gamma)\left(\frac{1}{s}\right)$$
$$\to K_n(\Gamma/\underline{a})\left(\frac{1}{s}\right) \to,$$

where we have written $A(\frac{1}{s})$ for $A \otimes \mathbb{Z}(\frac{1}{s})$ for any Abelian group A.

(iii) Let Λ be a ring with unit and $K_n(\Lambda, \mathbb{Z}/r)$ the K-theory with mod-r coefficients (see [16]). Let $S = \{s \in \mathbb{Z} \mid (r, s) = 1\}$. Then multiplication by $s \in S$ is invertible on $K_n(\Lambda, \mathbb{Z}/r)$. Hence for an S-excision ideal $\underline{a} \subset \Lambda$, $\pi_*(F_{\Lambda,\underline{a}}) \otimes S^{-1}\mathbb{Z} \simeq \pi_*(F_{\underline{a}}) \otimes S^{-1}\mathbb{Z}$ implies that $\pi_*(F_{\Lambda,\underline{a}}; \mathbb{Z}/r) \cong \pi_*(F_{\underline{a}}, \mathbb{Z}/r)$.

If we write $\mathbb{Z}_{(r)}$ for $S^{-1}\mathbb{Z}$ in this situation, we have that $K_n(\Lambda, \mathbb{Z}/r)$ satisfies excision on the class of ideals $\underline{a}$ such that $\underline{a} \otimes \mathbb{Z}_{(r)} = 0$ or $\underline{a} \otimes \mathbb{Z}_{(r)}$ has a unit.

5. Higher K-theory and connections to Galois, étale and motivic cohomology theories

5.1. *Higher K-theory of fields*

The importance of K-theory of fields lies not only in its connections with such areas as Brauer groups; Galois, étale, motivic cohomologies; symbols in arithmetic; zeta functions; Bernoulli numbers, etc., but also because understanding K-theory of fields also helps to understand K theory of other rings, e.g., K-theory of rings of integers in number fields, p-adic fields, algebras over such rings (e.g., orders) as well as K-theory of varieties and schemes.

First, we present some well known calculations of the K-theory of some special fields. Interested readers may see [41,132] for a comprehensive survey of this topic.

5.1.1. THEOREM [112].

(a) *Let F_q be a finite field of order q. Then $K_{2n}(F_q) = 0$ and $K_{2n-1}(F_q)$ is a cyclic group of order $q^n - 1$ for all $n \geqslant 1$.*

(b) *If E is finite extension of F_q, then the natural map $K_{2n-1}(F_q) \to K_{2n-1}(E)$ is injective. Moreover the automorphisms of E over F_q act on $K_{2n-1}(E)$ by multiplication by q^n. If E/F_q is Galois then the natural map $K_{2n-1}(F_q) \to K_{2n-1}(E)$ is an isomorphism.*

(c) *If E is the algebraic closure of F_q, then for all $n \geqslant 1$, $K_{2n}(E) = 0$ and $K_{2n-1}(E) \cong (Q/\mathbb{Z}) := \bigoplus(Q_\ell/\mathbb{Z}_\ell)$.*

On $K_{2n}(F)$, the Frobenius automorphism of F_q acts through multiplication by q^n.

For more general fields we have

5.1.2. THEOREM [132]. *Let F be an algebraically closed field. Then for $n \geqslant 1$, $K_{2n}(F)$ is uniquely divisible and $K_{2n-1}(F)$ is the direct sum of a uniquely divisible group and a group isomorphic to $Q/\mathbb{Z}$.*

5.1.3. THEOREM [34]. *Let F be a field of positive characteristic p. Then $K_n(F)$ has no p-torsion.*

5.1.4. For any commutative ring A, we have a composition $\otimes : \mathrm{GL}_n(A) \times \mathrm{GL}_p(A) \to \mathrm{GL}_{np}(A)$ which induces $\gamma_{n,p} : B\mathrm{GL}_n(A)^+ \times B\mathrm{GL}_p(A)^+ \to B\mathrm{GL}_{n,p}(A)^+ \to B\mathrm{GL}(A)^+$. Also $\gamma_{n,p}$ induces $B\mathrm{GL}(A)^+ \times B\mathrm{GL}(A)^+ \to B\mathrm{GL}(A)^+$ which induces a product: $K_n(A) \times K_m(A) \to K_{n+m}(A)$. The product $*$ endows $\bigoplus_{n=1}^{\infty} K_n(A)$ with the structure of a skew-commutative graded ring. If $m = n = 1$, then the product coincides with the Steinberg symbol (up to sign) discussed in [79]. This is in particular true for fields.

5.1.5. THEOREM [144]. *Let F be a number field with r_1 real places and r_2 complex places. Then*

$$\operatorname{rank} K_n(F) = \begin{cases} 1, & n = 0, \\ \infty, & n = 1, \\ 0, & n = 2k,\ k > 0, \\ r_1 + r_2, & n = 4k+1,\ k > 0, \\ r_2 & n = 4k+3. \end{cases}$$

5.1.6. REMARKS. The above is also true for $K_n(R)$ if R is the ring of integers in F for $n \geqslant 2$ and $n = 0$. It is classical that rank $K_1(R) = r_1 + r_2 - 1$ (see [9]). It is a result of Quillen that $K_n(R)$ is finitely generated for all $n \geqslant 1$ (see [115]).

5.1.7. THEOREM [41]. $K_2\mathbb{Q} \simeq \mathbb{Z}/2 \oplus (\bigoplus_p U(\mathbb{Z}/p))$.

5.1.8. THEOREM [89]. $K_3\mathbb{Q} \simeq K_3(\mathbb{Z}) \simeq \mathbb{Z}/48$.

We now recall the definition of Milnor K-theory in the present context.

5.1.9. DEFINITION. Let F be any field, F^* the group of non-zero elements of F. Define $K_*^M(F) = T(F^*)/J$ where $T(F)$ is the tensor algebra over F^* and J the ideal generated by all $x \otimes (1-x)$. Thus $K_n^M(F) = (F^* \otimes F^* \otimes \cdots \otimes F^*)/J_n$ where J_n is the subgroup of $F^* \otimes F^* \otimes \cdots \otimes F^*$ generated by all $a_1 \otimes a_2 \otimes \cdots \otimes a_n$ such that $a_i + a_j = 1$ for some $i \neq j$. The image of $x_1 \otimes x_2 \otimes \cdots \otimes x_n$ in $K_*^M(F)$ is denoted by $\{x_1, x_2, \ldots, x_n\}$. Note that by the Matsumoto theorem, $K_2(F) = K_2^M(F) = F^* \times F^*/\langle x \otimes (1-x)\rangle$. Also, $K_1(F) = K_1^M(F) = F^*$.

We also have a map $\varphi : K_n^M(F) \to K_n(F)$ defined by: $\{x_1, x_2, \ldots, x_n\} \to x_1 * x_2 \cdots * x_n$ where $*$ is the product defined in 5.1.4.

5.1.10. THEOREM [132]. *The kernel of $\varphi : K_n^M(F) \to K_n(F)$ is annihilated by $(n-1)!$.*

5.1.11. THEOREM [134]. *Let F be an infinite field. Then, there exists an isomorphism $H_n(\mathrm{GL}_n(F)) \simeq H_n(\mathrm{GL}_{n+1}(F), \mathbb{Z}) \simeq \cdots \simeq H_n(\mathrm{GL}(F), \mathbb{Z})$ and an exact sequence*

$$H_n\big(\mathrm{GL}_{n-1}(F), \mathbb{Z}\big) \to H_n\big(\mathrm{GL}_n(F), \mathbb{Z}\big) \to K_n^M(F) \to 0.$$

We also have a map (cup product)

$$H_1\big(\mathrm{GL}_1(F), \mathbb{Z}\big) \otimes \cdots \otimes H_1\big(\mathrm{GL}_1(F), \mathbb{Z}\big) \to H_n\big(\mathrm{GL}_n(F), \mathbb{Z}\big),$$

that is

$$F^* \otimes \cdots \otimes F^* \to H_n\big(\mathrm{GL}_n(F), \mathbb{Z}\big)$$

as well as a map $\delta : K_n^M(F) \to H_n(\mathrm{GL}_n(F), \mathbb{Z})$ *which makes the following diagram commutative*

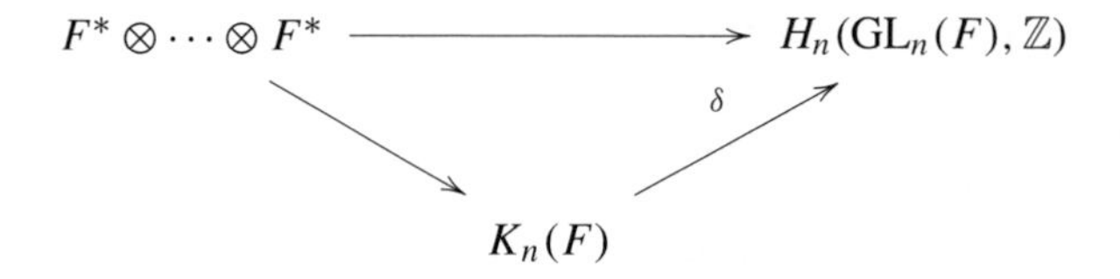

5.1.12. THEOREM [134]. *There exists a map* $\psi : K_n F \to K_n^M(F)$ *defined by*

$$K_n(F) = \pi_n\bigl(B\mathrm{GL}(F)^+\bigr) \to H_n\bigl(\mathrm{GL}(F), \mathbb{Z}\bigr) \simeq H_n\bigl(\mathrm{GL}_n(F), \mathbb{Z}\bigr)$$
$$\to H_n\bigl(\mathrm{GL}_n(F), \mathbb{Z}\bigr)/Im\bigl(H_n\bigl(\mathrm{GL}_{n-1}(F), \mathbb{Z}\bigr)\bigr) \simeq K_n^M(F)$$

such that

(i) $\varphi \circ \psi = c_*$,

(ii) $\psi \circ \varphi =$ *multiplication by* $(-1)^{n-1}(n-1)!$.

5.1.13. THEOREM [132]. *Let* F *be a number field with* r_1 *real places. Then* $K_n^M(F) \cong (\mathbb{Z}/2)^{r_1}$ *for* $n \geqslant 3$.

5.1.14. THEOREM [132]. *Let* F *be a real number field. Then the map* $K_4^M(F) \to K_4(F)$ *is not injective and the map* $K_n^M(F) \to K_n(F)$ *is zero for* $n \geqslant 5$.

5.1.15. DEFINITION.

$$K_n^{\mathrm{ind}}(F) := \mathrm{Coker}\bigl(K_n^M(F) \xrightarrow{\varphi} K_n(F)\bigr).$$

NOTE. It is well know that $K_3^{\mathrm{ind}}(F) \neq 0$ for all fields, i.e. φ is never surjective. Also the map is not injective in general.

5.1.16. THEOREM [132]. *Let* L/F *be a field extension such that* F *is algebraically closed in* L. *Then the map:* $K_3^{\mathrm{ind}} F \to K_3^{\mathrm{ind}} L$ *in an isomorphism on the torsion and cotorsion. Hence* $K_3^{\mathrm{ind}} L / K_3^{\mathrm{ind}} F$ *is uniquely divisible.*

5.2. *Galois cohomology*

5.2.1. Let G be a profinite group. A discrete G-module is a G-module A such that if A is given the discrete topology the multiplication map $G \times A \to A$ is continuous. If A is a discrete G-module then for every $a \in A$, the stabilizer U of a is an open subgroup of G and

$$a \in A^U = \{a \in A \mid gA = A \text{ for all } g \in U\}.$$

Note that A is a discrete G-module iff $\bigcup A^U = A$. Note also that if G is a finite group, then every G-module is discrete.

The category $\mathcal{C}_G$ of discrete G-modules is an Abelian subcategory of $\underline{G\text{-mod}}$ with enough injectives. The right derived functor of the left exact functor $\mathcal{C}_G \to Ab$; $A \to A^G$ where $A^G = \{a \in A \mid ga = a\ \forall g \in G\}$ are the cohomology groups of G with coefficients in A and denoted by $H^*(G, A)$, see [165].

5.2.2. Let A be a discrete G-module (G a profinite group), $C^n(G, A)$ the Abelian group of continuous maps $G^n \to A$ and $C^n(G, A) = \varinjlim(C^n(G/U, A^U))$ where U runs through all open normal subgroups of G.

5.2.3. THEOREM. *If G is a profinite group and A a discrete G-module*

$$\begin{aligned} H^n(G, A) &\cong H^n\big(C^*(G, A)\big) \\ &\cong \varinjlim_U H^n\big(G/U, A^U\big), \end{aligned}$$

where U runs through all open normal subgroups of G.

5.2.4. EXAMPLE. Let F_s be the separable closure of a field F, i.e. F_s is the subfield of the algebraic closure $\overline{F}$ consisting of all elements separable over F and $F_s = \overline{F}$ if $\text{char}(F) = 0$.

Krull's theorem (see [165]) says that the group $\text{Gal}(F_s/F) \cong \varprojlim \text{Gal}(F_i/F)$ where F_i runs through all finite Galois extensions of F. As such $\text{Gal}(F_s/F)$ is a profinite group and F_s is a discrete $\text{Gal}(F_s/F)$-module.

We shall denote $H^n(\text{Gal}(F_s/F), F_s)$ by $H^n(F, F_s)$.

5.2.5. Let F be field and $B_r(F)$ the Brauer group of F, i.e. the group of stable isomorphism classes of central simple F-algebras with multiplication given by the tensor product of algebras. See [119].

A central simple F-algebra A is said to be split by an extension E of F if $E \otimes A$ is E-isomorphic to $M_r(E)$, the algebra of $r \times r$ matrices over E, for some positive integer r. It is well known (see [60]) that such an E can be taken as some finite Galois extension of F. Let $B_r(F, E)$ be the group of stable isomorphism classes of E split central simple algebras. Then $B_r(F) := B_r(F, F_s)$ where F_s is the separable closure of F.

5.2.6. THEOREM [119]. *Let E be a Galois extension of a field F, $G = \text{Gal}(E/F)$. Then there exists an isomorphism $H^2(G, E^*) \cong B_r(F, E)$. In particular $B_r(F) \cong H^2(G, F_s^*)$ where $G = \text{Gal}(F_s/F) = \varinjlim \text{Gal}(E_i/F)$, where E_i runs through the finite Galois extensions of F.*

5.2.7. Now, for any $m > 0$, let μ_m be the group of m-th roots of 1, $G = \text{Gal}(F_s/F)$, then we have the Kummer sequence of G-modules

$$0 \to \mu_m \to F_s^* \to F_s^* \to 0$$

from which we obtain an exact sequence of Galois cohomology groups

$$F^* \xrightarrow{m} F^* \to H^1(F, \mu_m) \to H^1(F, F_s^*) \to \cdots,$$

where $H^1(F, F_s^*) = 0$ by the Hilbert theorem 90. So we obtain an isomorphism $\chi_m : F^*/mF^* \cong F^* \otimes \mathbb{Z}/m \to H^1(F, \mu_m)$.

Now, the composite

$$\begin{aligned} F^* \otimes_{\mathbb{Z}} F^* &\to (F^* \otimes_{\mathbb{Z}} F^*) \otimes \mathbb{Z}/m \to H^1(F, \mu_m) \otimes H^1(F, \mu_m) \\ &\to H^2(F, \mu_m^{\otimes 2}) \end{aligned}$$

is given by $a \otimes b \to \chi_m(a) \cup \chi_m(b)$ (where $\cup$ is the cup product) which can be shown to be a Steinberg symbol inducing a homomorphism $g_{2,m} : K_2(F) \otimes \mathbb{Z}/m\mathbb{Z} \to H^2(F, \mu_m^{\otimes 2})$.

We then have the following result due to A.S. Merkurjev and A.A. Suslin, see [100].

5.2.8. THEOREM [100]. *Let F be a field, m an integer > 0 such that the characteristic of F is prime to m. The map*

$$g_{2,m} : K_2(F)/mK_2(F) \to H^2(F, \mu_m^{\otimes 2})$$

is an isomorphism where $H^2(F, \mu_m^{\otimes 2})$ can be identified with the m-torsion subgroup of $B_r(F)$.

5.2.9. REMARKS. By generalizing the process outlined in 5.2.7 above, we obtain a map

$$g_{n,m} : K_n^M(F)/mK_n^M(F) \to H^n(F, \mu_m^{\otimes n}). \tag{I}$$

It is a conjecture of Bloch and Kato that $g_{n,m}$ is an isomorphism for all F, m, n. So, 5.2.8 is the $g_{2,m}$ case of the Bloch–Kato conjecture when m is prime to the characteristic of F. Furthermore, A. Merkurjev proved that 5.2.8 holds without any restriction of F with respect to m.

It is also a conjecture of Milnor that $g_{n,2}$ is an isomorphism. In 1996, V. Voevodsky proved that $g_{n,2^r}$ is an isomorphism for any r, see [106].

5.3. *Zariski and étale cohomologies*

5.3.1. A site is a small category $\mathcal{S}$. For $T \in \mathcal{S}$, let $\mathrm{Cov}(T)$ denote a family of coverings of T where a covering of T is a family of maps $\{p_i : U_i \to T\}$ satisfying the following properties

(1) If $f : S \to T$ is an isomorphism, then $\{f\} \in \mathrm{Cov}(T)$.
(2) If $\{p_\alpha : T_\alpha \to T\}_\alpha \in \mathrm{Cov}(T)$ and for each α we are given a cover $\{q_{\beta\alpha} : U_{\beta\alpha} \to T_\alpha\}_\beta$, then $\{p_\alpha \cdot q_{\beta\alpha} : U_{\beta\alpha} \to T\}_{\beta\alpha}$ is a cover.

(3) If $\{T_\alpha \to T\}_\alpha \in \mathrm{Cov}(T)$ and $g: V \to T$ is an arbitrary morphism, then for all α, the fibre product $T_\alpha \times_T V \to T$ exists and $\{T_\alpha \times_T T \to V\}$ is a cover.

A sheaf F on a site $\mathcal{S}$ with values in a suitable category $\mathcal{A}$ is a contravariant functor (presheaf) $F: \mathcal{S} \to \mathcal{A}$ such that for every covering $\{T_\alpha \to T\}_\alpha$, the sequence

$$F(T) \to \prod_\alpha F(T\alpha) \rightrightarrows \prod_{\beta\alpha} F\big(T_\alpha \times_T T_\beta\big)$$

is a difference kernel or equalizer diagram in $\mathcal{A}$.

If, for example, the category $\mathcal{A}$ is Abelian, then the category of sheaves on the site $\mathcal{S}$ has enough injectives; i.e. every object A or complex A^* has an injective resolution $A \to I$ or $A^* \to I^*$. The cohomology (resp. hypercohomology) of X is then the cohomology of the complex $I(X)$ (resp. $I^*(X)$).

5.3.2. *Zariski cohomology.* Let $\mathcal{S}ch$ be a suitable category of schemes, $S \in \mathcal{S}ch$. The big Zariski site S_{Zar} (on S) is the category such that $\mathrm{ob}(S_{\mathrm{Zar}})$ = schemes over S and for $X \in \mathcal{S}ch$, a covering family is a Zariski open cover by open subschemes of X. If $\mathcal{A} = \mathcal{S}ets$ or $\mathbb{Z}$-mod, etc., then a sheaf on $\mathcal{S}ch$ is a presheaf which when restricted to any scheme $X \in \mathcal{S}ch$ is actually a sheaf in the usual Zariski topology of X.

A cohomology theory on $\mathcal{S}ch$ consists of a bounded below complex of sheaves Γ^* (which we could assume to be a complex of injectives) on S_{Zar}. If X is a scheme over S, then the cohomology of X with coefficient in Γ is defined as the hypercohomology of X with coefficient in the restriction of Γ^* to X and is denoted by $H^*(X, T^*)$.

5.3.3. *Étale cohomology.* Let $\mathcal{S}ch$ be a suitable category of schemes, $S \in \mathcal{S}ch$. The big étale site $\mathrm{S}_{\text{ét}}$ of S is defined as follows: $\mathrm{ob}(\mathrm{S}_{\text{ét}})$ = schemes $X \in \mathcal{S}ch$ over S. Coverings are étale coverings; i.e. families $\{U_i \xrightarrow{p_i} X\}$ of maps where each p_i is an étale mapping, i.e. each p_i is flat and unramified (see [101]).

For $X \in \mathcal{S}ch$, we shall write $H^n_{\mathrm{et}}(X, M)$ for the étale hypercohomology of X where M is a complex of sheaves of the étale topology.

If R is a commutative ring with identity, we shall write $H^n_{\mathrm{et}}(R, M)$ for $H^n_{\mathrm{et}}(\mathrm{Spec}(R), M)$. (Here M is a complex of sheaves on étale site of $\mathrm{Spec}(R)$.)

5.3.4. REMARKS AND NOTATIONS. In 5.3.3, M could be finite, torsion or profinite and could take any of the following forms.

In what follows ℓ is a rational prime, i a positive integer

(i) $G_m = G_m(R) = R^*$ = units of R.
(ii) $\mu(R)$ = torsion subgroup of R^* = roots of unity in R.
(iii) $\mu_{\ell^\nu}(R) = \ell^\nu$-th roots of unity in R.
(iv) $\mu_{\ell^\infty}(R)$ = all ℓ-th power roots of unity in R.
(v) $\mu_{\ell^\nu} = \mathbb{Z}/\ell^\nu$ (in additive notation).
(vi) $\mu_{\ell^\nu}(i) = \mathbb{Z}/\ell^\nu(i)$ where $\mu_{\ell^\nu}(i) = \mu_{\ell^\nu}^{\otimes i} = \underbrace{\mu_{\ell^\nu} \otimes \mu_{\ell^\nu} \otimes \cdots \otimes \mu_{\ell^\nu}}_{i \text{ times}}$.

(vii) $W(i) = \varinjlim \mathbb{Z}/_{\ell^\nu}(i)$, the direct limit taken over the injections $\mathbb{Z}/_{\ell^\nu}(i) \to \mathbb{Z}/\ell^{\nu+1}(i)$. Note that $W(i)$ is a discrete torsion group.

(viii) $\mathbb{Z}_\ell(i) = \varprojlim \mathbb{Z}/\ell^\nu(i)$, the inverse limit taken over projections $\pi : \mathbb{Z}/\ell^{\nu+1}(i) \to \mathbb{Z}/\ell^\nu(i)$ is a profinite group.

5.3.5. EXAMPLES.

(i) For any commutative ring R with identity, $\mathrm{Pic}(R) \simeq H^1_{\mathrm{et}}(R, G_m)$, $\mathrm{Br}(R) \simeq H^2_{\mathrm{et}}(R, G_m)$. If R is a field, $\mathrm{Pic}(F) \simeq H^1(F, G_m) = 0$ (see [166]).

(ii) Let F be a number field with ring of integers O_F, ℓ a prime, $O'_F = O_F(\frac{1}{\ell})$. Then $H^2_{\mathrm{et}}(O'_F, \mathbb{Z}_\ell(n))$ is finite and $= 0$ for almost all primes ℓ. Also

$$rk_{\mathbb{Z}} H^1_{\mathrm{et}}\big(O'_F, \mathbb{Z}_\ell(n)\big) = \begin{cases} r_1 + r_2 & \text{if } n \text{ is odd} > 1, \\ r_2 & \text{if } n \text{ is even.} \end{cases}$$

(iii) Let E be a p-adic field, then each $H^n_{\mathrm{et}}(E, W(i))$ is a discrete torsion group, $H^n_{\mathrm{et}}(E, \mathbb{Z}_\ell(i))$ is a profinite group.

In particular for $n \neq 0, 1, 2$, $H^n_{\mathrm{et}}(E, W(i)) = 0$ and $H^n_{\mathrm{et}}(E, \mathbb{Z}_\ell(i)) = 0$, see [166]. If E has residue field $\mathbb{F}_q$ of characteristic $p \neq \ell$, then for all $i > 1$,

$$H^n_{\mathrm{et}}\big(E, W(i)\big) \simeq \begin{cases} \mathbb{Z}/w_i(\mathbb{F}_q), & n = 0, \\ \mathbb{Z}/w_{i+1}(\mathbb{F}_q), & n = 1, \\ 0, & \text{otherwise} \end{cases}$$

(see [166]). Also see 6.2.9 for the definition of $\omega_i(F)$.

(iv) Let F be a number field with separable closure F_s, ℓ a prime. Then $H^n(\mathrm{Gal}(F_s/F), \mu_\ell^{\otimes n}) \simeq H^n_{\mathrm{et}}(\mathrm{Spec}(F), \mu_\ell^{\otimes n})$, see [166].

5.3.6. Let ℓ be a rational prime, F a number field with ring of integers O_F. An extension E of F is said to be unramified outside ℓ when given any prime ideal $\underline{p}$ of O_F such that ℓ does not divide $\underline{p}$, we have $\underline{p}O_F = \underline{p}_1\underline{p}_2\cdots\underline{p}_n$ where $\underline{p}_1, \underline{p}_2, \ldots, \underline{p}_n$ are distinct prime ideals in O_E.

Let $\overline{F}$ be the algebraic closure of F and Γ the union of all finite extensions E of F which are unramified outside ℓ.

Let $F(\mu_{\ell^\infty}) = \bigcup_{\nu \geqslant 1} F(\mu_{\ell^\nu})$ be the maximal ℓ-cyclotomic extension of F. Then $F(\mu_{\ell^\infty}) \subset \Gamma$.

For any $i \in \mathbb{Z}$, the $\mathrm{Gal}(F(\mu_{\ell^\infty})/F)$-module $\mathbb{Z}_\ell(i)$ is defined as the additive group $\mathbb{Z}_\ell := \varprojlim \mathbb{Z}/\ell^\nu$ equipped with Galois action $g(x) = \varepsilon(g)^i x$ where $\varepsilon : \mathrm{Gal}(F(\mu_{\ell^\infty})/F) \to \mathbb{Z}_\ell^*$ is defined by $g(\xi) = \xi^{\varepsilon(g)}$ for all $\xi \in \overline{F}$ such that $\xi^{\ell^\nu} = 1$ for some $\nu \geqslant 1$.

One can regard $\mathbb{Z}_\ell(i)$ as a continuous $\mathrm{Gal}(\Gamma/F)$-module via the projection $\mathrm{Gal}(\Gamma/F) \to \mathrm{Gal}(F(\mu_{\ell^\infty})/F)$. One can identify the Galois cohomology groups $H(\mathrm{Gal}(\Gamma/F), \mathbb{Z}_\ell(i))$ with $H^k_{\mathrm{et}}(O'_F, \mathbb{Z}_\ell(i))$; i.e. $H^k_{\mathrm{et}}(O'_F, \mathbb{Z}_\ell(i)) \simeq H^k(\mathrm{Gal}(\Gamma/F), \mathbb{Z}_\ell(i))$ where $O'_F = O_F(\frac{1}{\ell})$. Some authors use this Galois cohomology group to define $H^k_{\mathrm{et}}(O'_F, \mathbb{Z}_\ell(i))$.

5.4. *Motivic cohomology*

5.4.1. Let $Sm(k)$ be the category of smooth quasi-projective varieties over a field k (usually of char 0). For $X, Y \in Sm(k)$, let $c(X, Y) :=$ free Abelian group on the set of closed irreducible subvarieties $Z \subset X \times_k Y$ for which the projection $Z \to X$ is finite and surjective onto an irreducible component of X. Call $c(X, Y)$ the group of finite correspondences from X to Y.

Define a category $Sm\,\mathrm{Cor}(k)$ as follows

$$\mathrm{ob}\big(Sm\,\mathrm{Cor}(k)\big) = \mathrm{ob}\big(Sm(k)\big),$$

$$\mathrm{Hom}_{Sm\,\mathrm{Cor}(k)}(X, Y) := c(X, Y).$$

For any smooth k-variety X, define a functor $L[X] : (Sm\,\mathrm{Cor}(k))^{\mathrm{op}} \to \underline{Ab} : Y \mapsto c(Y, X)$.

5.4.2. Let $F : (Sm(k))^{\mathrm{op}} \to Ab$ be a presheaf of Abelian groups. Define a singular chain complex $\underline{C}_\bullet(F)$ as the presheaf of complexes $\underline{C}_\bullet(F) : Sm(k)^{\mathrm{op}} \to \underline{\text{chain complexes}} : X \to \underline{C}_\bullet(F)(X)$ where $\underline{C}_*(F)X$ is the chain complex associated to the $H_*^S(X)$, the singular homology groups of $X :=$ homology groups of the chain complex $\underline{C}_*(X)(\mathrm{spec}(k))$. Now define a presheaf of chain complexes

$$\underline{C}_* L[X] : \big(Sm\,\mathrm{Cor}(k)\big)^{\mathrm{op}} \to \underline{\text{chain complexes}} : Y \mapsto \underline{C}_* L(X)(Y).$$

Now let $G_m = \mathbb{A}^1 \backslash \{0\}$ and for $i \geqslant 0$, let $L_i =$ cokernel of the morphism $\bigoplus L((G_m)^{i-1}) \to L((G_m)^i)$ induced by $(G_m)^{i-1} \subset (G_m)^i$ which puts the j-th co-ordinate equal to 1.

Now let $Z(i)$ be the presheaf of complexes on $Sm(k)$ given by $Z(i) = \underline{C}_*(L_i)[-1]$, the i-th desuspension of $\underline{C}_*(L_i)$. (See [106].)

5.4.3. Definition. Let X be a smooth k-variety, $i \geqslant 0$. The Zariski hypercohomology of X with coefficients in $Z(i)$ is denoted by $H_B^*(X, Z(i))$. The bigraded Abelian group $H_B^*(X, Z(-))$ is called the Beilinson motivic cohomology (or just motivic cohomology) with coefficients in $\mathbb{Z}$. It is usual to denote $H_B^*(X, \mathbb{Z}(*))$ by $H_{\mathcal{M}}^*(X, \mathbb{Z}(*))$.

For any Abelian group A, $X \in Sm(k)$, define the motivic cohomology of X with coefficients in A by $H_{\mathcal{M}}^*(X, A(*))$ where $A(i) = \mathbb{Z}(i) \otimes A$.

5.4.4. Examples.

(i) For any

$$X \in Sm(k), \quad H_{\mathcal{M}}^n\big(X, Z(0)\big) = \begin{cases} 0 & \text{if } n \neq 0, \\ \mathbb{Z}^{\pi_0(X)} & \text{if } n = 0, \end{cases}$$

where $\pi_0(X) =$ set of connected components of X.

(ii)

$$H_{\mathcal{M}}^n\big(X, Z(i)\big) \cong \begin{cases} 0 & \text{if } n \neq 1, 2, \\ O_X^* & \text{for } n = 1, \\ \mathrm{Pic}(X) & \text{for } n = 2. \end{cases}$$

(iii) $k^* \equiv H^1_{\mathcal{M}}(\mathrm{Spec}(k), Z(1))$.

(iv) Let $H^*_{\mathcal{M}}(k) :=$ graded commutative ring equal in degree n to $H^n_{\mathcal{M}}(\mathrm{Spec}(k), Z(n))$. Then for any field k, $K^M_*(k) \simeq H^*_{\mathcal{M}}(k)$. (See [106].)

(v) For any field k and any integers $n \geqslant 0$, $m > 0$, we have

$$K^M_n(k)/m \simeq H^n_{\mathcal{M}}\big(\mathrm{Spec}(k), \mathbb{Z}/m(n)\big)$$

(see [106]).

(vi) Let ℓ be a rational prime, F a global field with ring of integers O_F, $O'_F = O_F(\frac{1}{\ell})$. Then there are isomorphisms $H^i_{\mathcal{M}}(O_F, \mathbb{Z}(n)) \otimes \mathbb{Z}_\ell \simeq H^i_{\mathrm{et}}(O'_F, \mathbb{Z}_\ell(n))$ for all primes ℓ, $n \geqslant Z$ and $i = 1, 2$ (see [106]).

(vii) Let F be a field of characteristic $\neq p$. If the Bloch–Kato conjecture holds, then there is an isomorphism

$$H^i_{\mathcal{M}}\big(F, \mathbb{Z}/\ell^\nu(n)\big) \simeq \begin{cases} H^i_{\mathrm{et}}(F, \mathbb{Z}/\ell^\nu(n)), & 0 \leqslant i \leqslant n, \\ 0, & \text{otherwise} \end{cases}$$

(see [106]).

5.4.5. Let A be an Abelian group and X any smooth k-variety. Let $H^*_L(X, A(i))$ be the hypercohomology groups of X of the presheaves $A(i)$ in the étale topology. Call $H_L(X, A(i))$ the Lichtenbaum (or étale) cohomology groups of X with coefficients in A.

5.4.6. THEOREM [106]. *Let k be a field, X a smooth k-variety, A a Q-vector space. Then*

$$H^*_{\mathcal{M}}\big(X, A(*)\big) \xrightarrow{\sim} H^*_L\big(X, A(*)\big).$$

5.4.7. THEOREM [106]. *Let A be a torsion Abelian group of torsion prime to* $\mathrm{char}(k)$. *Let $A_{\bar{\mathrm{et}}}$ be the constant étale sheaf associated to A. For $i \geqslant 0$, let $A_{\bar{\mathrm{et}}}(i) = i$-th Tate twist. Then, for all $i \geqslant 0$ and any smooth k-variety X, we have*

$$H^n_L\big(X, A(i)\big) \simeq H^n_{\mathrm{et}}\big(X, A_{\bar{\mathrm{et}}}(i)\big).$$

We close this subsection with the following result

5.4.8. THEOREM [106]. *Let k be a field which admits resolution of singularities (e.g.,* $\mathrm{char}(k) = 0$*), ℓ a prime number not equal to* $\mathrm{char}(k)$. *Then for any integer $n \geqslant 0$, the following are equivalent*

(i) *For any field extension $k \subset F$ of finite type,*

$$g_{n,\ell}\colon K^M_n(F)/\ell K^M_n(F) \simeq H^n\big(\mathrm{Gal}(F_s/F), \mu_\ell^{\otimes n}\big).$$

(ii) *For any smooth K-variety X, and any integer $m \in \{0, 1, \ldots, n+1\}$, we have an isomorphism*

$$H^m_{\mathcal{M}}\big(X, \mathbb{Z}_\ell(n)\big) \simeq H^m_L\big(X : \mathbb{Z}_\ell(n)\big).$$

(iii) *For any field extension $k \subset F$ of finite type we have*

$$H_L^{n+1}\big(\mathrm{Spec}(F); \mathbb{Z}_\ell(n)\big) = 0.$$

5.4.9. REMARKS. In 5.4.8, (i) is a special case of the Bloch–Kato conjecture which is believed to have now been proved by Rost and Voevodsky. The case $g_{2,\ell}$ was proved by Voevodsky in 1997 (see [106]). (ii) was a conjecture of S. Lichtenbaum.

5.5. *Connections to Bloch's higher Chow groups*

5.5.1. Let k be a field and $\Delta^n = \mathrm{Spec}(R)$ where $R = k[t_0 \ldots t_n]/(\Sigma t_j - 1)$. Let X be a quasi-projective variety, $z^i(X, n) :=$ free Abelian group on the set of codim i subvarieties Z of $X \times_k \Delta^n$ which meet all faces properly (see [106]). Call $z^i(X, *)$ Bloch's cycle group. $z^i(X, *)$ is a simplicial Abelian group.

Define $CH^i_j(X) = (\pi_j z^i(X, *)) = H_j$ of the associated chain complex.

5.5.2. REMARKS/EXAMPLES.

(i) For $j = 0$, $CH^i_0(X)$ is the classical Chow group of codim i cycles in X modulo rational equivalence.

(ii) $CH^i_j(X) = 0$ for $i > j + \dim(X)$. See [106].

(iii) $CH^i_j(X) \simeq CH^i_j(X[t])$ where $X(t) = X \times_k \mathrm{Spec}(k[t])$, i.e. the higher Chow groups are homotopy invariant.

(iv) If X is smooth

$$CH^1_j(X) = \begin{cases} \mathrm{Pic}(X), & j = 0, \\ H^0(X, O_X), & j = 1, \\ 0, & \text{otherwise}. \end{cases}$$

5.5.3. THEOREM [106]. *For every field k and all n,*

$$K_n^M(k) \cong CH^n_n(k).$$

Here $X = \mathrm{Spec}(k)$.

5.5.4. THEOREM [106]. *Let X be a smooth scheme of finite type over a field k. Then for $n \geqslant 0$, there is an isomorphism*

$$K_n(X)_Q \simeq \bigoplus_{d \geqslant 0} CH^d_n(X)_Q,$$

where $A_Q := A \otimes Q$ for any Abelian group A.

5.5.5. THEOREM [106]. *Let X be a smooth algebraic variety. There exists a spectral sequence*

$$E_2^{p,q} = CH^{-q}_{-q-q}(X) \Rightarrow K_{-p-q}(X).$$

5.5.6. THEOREM [106]. *Let k be a field of characteristic zero (more generally a field which admits the resolution of singularities). Then, for each smooth scheme X of finite type over k, there are natural isomorphisms*

$$H^p_{\mathcal{M}}\big(X, \mathbb{Z}(q)\big) \simeq CH^p_{2q-p}(X).$$

6. Higher K-theory of rings of integers in local and global fields

6.1. *Some earlier general results on the higher K-theory of ring of integers in global fields*

The first result is due to D. Quillen.

6.1.1. THEOREM [40,115]. *Let O_F be the ring of integers in a global field F. Then for all $n \geqslant 1$, the Abelian groups $K_n(O_F)$ are finitely generated.*

The next result, due to A. Borel, computes the rank of $K_n(O_F)$, $K_n(F)$ for $n \geqslant 2$.

6.1.2. THEOREM [13]. *Let F be a number field and let us write $[F : \mathbb{Q}] = r_1 + 2r_2$, where r_1 is the number of distinct embeddings of F into $\mathbb{R}$ and r_2 the number of distinct conjugate pairs of embeddings of F into $\mathbb{C}$ with image not contained in $\mathbb{R}$.*

(i) *If R denotes either the number field F or its ring of algebraic integers O_F, then the rational cohomology of the special linear group $\mathrm{SL}(R)$ is given by*

$$H^*\big(\mathrm{SL}(R); \mathbb{Q}\big) \simeq \Big(\bigotimes_{1 \leqslant j \leqslant r_1} A_j\Big) \otimes \Big(\bigotimes_{1 \leqslant k \leqslant r_2} B_k\Big),$$

where j runs over all distinct embeddings of F into $\mathbb{R}$, k over all distinct conjugate pairs of embeddings of F into $\mathbb{C}$ with image not contained in $\mathbb{R}$, and where A_j and B_k are the following exterior algebras:

$$A_j = \Lambda_{\mathbb{Q}}(x_5, x_9, x_{13}, \ldots, x_{4l+1}, \ldots) \quad \textit{and}$$
$$B_k = \Lambda_{\mathbb{Q}}(x_3, x_5, x_7, \ldots, x_{2l+1}, \ldots),$$

where $\deg(x_j) = j$.

(ii) *If R denotes either the number field F or its ring of algebraic integers O_F, then for any integer $i \geqslant 2$,*

$$K_i(R) \otimes \mathbb{Q} \simeq \begin{cases} 0 & \textit{if } i \textit{ is even}, \\ \mathbb{Q}^{r_1+r_2} & \textit{if } i \equiv 1 \bmod 4, \\ \mathbb{Q}^{r_2} & \textit{if } i \equiv 3 \bmod 4. \end{cases}$$

6.1.3. REMARKS.

(i) As consequences of 6.1.2, we have

(a) For all $n \geqslant 1$, the $K_{2n}(O_F)$ are finite groups and

$$\operatorname{rank} K_{2n-1}(O_F) = \begin{cases} r_1 + r_2 & \text{if } n \text{ is odd,} \\ r_2 & \text{if } n \text{ is even.} \end{cases}$$

(b) For all $n \geqslant 1$, $K_{2n-1}(F)$ is finitely generated and $K_{2n}(F)$ is torsion.

(ii) Let R be a Dedekind domain with quotient field F and finite residue field $R/\underline{p}$ for each prime ideal $\underline{p}$ of R. Then it follows from the localization sequence 4.4.3(ii) that we have an exact sequence (for all $n \geqslant 1$)

$$0 \to K_{2n}(R) \to K_{2n}(F) \to \bigoplus_{\underline{p}} K_{2n-1}(R/\underline{p}) \to K_{2n-1}(R)$$
$$\to K_{2n-1}(F) \to 0. \qquad \text{(I)}$$

The next result, due to C. Soulé shows that we do get short exact sequences out of the sequence (I).

6.1.4. THEOREM [129]. *Let O_F be the ring of integers in a number field F. Then*

(i) $K_{2n-1}(O_F) \approx K_{2n-1}(F)$ *for all* $n \geqslant 1$.

(ii) *For all $n \geqslant 1$, $K_{2n}(F)$ is an infinite torsion group and we have an exact sequence*

$$0 \to K_{2n}(O_F) \to K_{2n}(F) \to \bigoplus_{\underline{p}} K_{2n-1}(O_F/\underline{p}),$$

where $\underline{p}$ runs through the prime ideals of O_F.

6.1.5. EXAMPLES.

(i) If $F = \mathbb{Q}$, $O_F = \mathbb{Z}$ then $K_3(\mathbb{Z}) \simeq K_3(\mathbb{Q}) \simeq \mathbb{Z}/48$. That $K_3(\mathbb{Z}) \simeq \mathbb{Z}/48$ is a result of Lee and Szcarba, see [89].

(ii) $F = \mathbb{Q}(i)$, then $K_3(Q(i)) \simeq \mathbb{Z} \oplus \mathbb{Z}/24$.

6.1.6. Let F be a global field of characteristic p, i.e. F is finitely generated of finite transcendence degree over $\mathbb{F}_q$. It is well known (see [128]) that there is a unique smooth projective curve X over $\mathbb{F}_q$ whose function field is F, i.e. $F = \mathbb{F}_q(X)$. If S is a non-empty set of closed points of X, then $X \backslash S$ is affine and we shall refer to the coordinate ring of $X \backslash S$ as the ring of S-integers in F.

6.1.7. THEOREM [166]. *Let X be a smooth projective curve over a finite field of characteristic p, then*

(i) *The groups $K_n(X)$ are finite groups of order prime to p.*

(ii) *If R is the ring of S-integers in $F = \mathbb{F}_q(X)$ ($S \neq \emptyset$ as in 6.1.6) then*

(a) $K_i(R) \cong R^* \simeq \mathbb{F}_q^* \times \mathbb{Z}^s$, $|S| = s + 1$.

(b) *For $n \geqslant 2$, $K_n(R)$ is a finite group of order prime to p.*

For further information on $K_n(X)$, see [166].

Next, we provide some general results on the higher K-theory of integers in local fields.

6.1.8. THEOREM [166]. *Let E be a local field with discrete valuation ring A and residue class field $\mathbb{F}_q$ of characteristic p. Assume that* $\mathrm{char}(E) = p$, *then* $E = \mathbb{F}_q((\pi))$, $A = \mathbb{F}_q[[\pi]]$ *where π is a uniformizing parameter of A and we have for $n \geqslant 2$*

(i) $K_n(E) \simeq K_n(A) \oplus K_{n-1}(\mathbb{F}_q)$,

(ii) $K_n(A) \simeq K_n(\mathbb{F}_q) \oplus U_n$ *where the U_n are uncountable uniquely divisible Abelian groups.*

REMARKS. 6.1.8 is a generalization of Moore's theorem which states that $K_2(E) \simeq \mu(E) \times U_2$ and $K_2(A) \cong \mu_{p^\infty}(E) \times U_2$ (see [166]).

6.2. *Étale and motivic Chern characters*

6.2.1. A lot of the computations of K-theory of integers in number fields and p-adic fields have been through mapping K-theory into "étale cohomology via étale Chern characters" initially defined by C. Soulé, see [129]. We now briefly review this construction with the observation that since Soulé's definition, there have been other approaches, e.g., maps from étale K-theory to étale cohomology due to Dwyer and Friedlander, [26], and "anti-Chern" characters, i.e. maps from étale cohomology to K-theory due to B. Kahn, [55]. Moreover the various interconnections already outlined in Section 5 between K-theory, Galois, étale and motivic cohomologies have also made computations of the K-theory of $\mathbb{Z}$ more accessible.

6.2.2. Let X be an H-space, M_m^n the n-dimensional mod-m Moore space. We write $\pi_n(X, \mathbb{Z}/m)$ for $[M_m^n, X]$ (see 3.1.4(viii)). Note then $\pi_n(X, \mathbb{Z}/m)$ is a group for $n \geqslant 2$.

Let $h_n : \pi_n(X, \mathbb{Z}/m) \to H_n(X, \mathbb{Z}/m)$ be the mod-m Hurewitz map defined by $\alpha \in [M_m^n, X] \mapsto \alpha_*(\varepsilon_n)$ where ε_n is the generator of $\mathbb{Z}/m \simeq H_n(S^n, \mathbb{Z}/m) \simeq H_n(M_m^n, \mathbb{Z}/m)$ and α_* is the homomorphism $H_n(M_m^n, \mathbb{Z}/m) \to H_n(X, \mathbb{Z}/m) : \varepsilon_n \to \alpha_*(\varepsilon_n)$.

Then, h_n is a group homomorphism for $n \geqslant 3$. If $X = B\mathrm{GL}(A)^+$, then we have a mod-m Hurewitz map $K_n(A, \mathbb{Z}/m) \to H_n(\mathrm{GL}(A), \mathbb{Z}/m)$ (I) which is a group homomorphism.

6.2.3. *Soulé's étale Chern character.* Let l be a rational prime, A a commutative ring with identity such that $1/\ell \in A$. Let G be a group acting on $\mathrm{Spec}(A)$ and $\rho : G \to \mathrm{GL}_s(A)$ (s an integer > 1) a representation of G, $c_i(\rho) \in H_{\mathrm{et}}^{2i}(A, G, \mu_{\ell^\nu}^{\otimes i})$ where $H_{\mathrm{et}}^{2i}(A, G, \mu_{\ell^\nu}^{\otimes i})$ are étale cohomology of G-sheaves on the étale site of $\mathrm{Spec}(A)$. (See [101] and 5.3.3.) Assume that G acts trivially on $\mathrm{Spec}(A)$ then there exists a homomorphism

$$\phi : H_{\mathrm{et}}^{2i}(A, G, \mu_{\ell^\nu}^{\otimes i}) \to \coprod_{k=0}^{2i} \mathrm{Hom}\big(H_{2i-k}(G, \mathbb{Z}/(\ell^\nu)), H_{\mathrm{et}}^k(A, \mu_{\ell^\nu}^{\otimes i})\big)$$

mapping $c_i(\rho)$ to $c_{ik}(\rho): H_{2i-k}(G, \mathbb{Z}/\ell^\nu) \to H^k_{\text{et}}(A, \mu_{\ell^\nu}^{\otimes i})$. Put $G = \text{GL}_s(A)$, we obtain a map

$$c_{ik}(\text{id}): H_n(\text{GL}(A), \mathbb{Z}/\ell^\nu) \to H^k_{\text{et}}(A, \mu_{\ell^\nu}^{\otimes i}), \tag{II}$$

where $n + k = 2i$, $n \geqslant 2$.

Composing (I) in 6.2.2 with (II) above, yields the Soulé Chern characters

$$c_{ik}: K_n(A, \mathbb{Z}/\ell^\nu) \to H^k_{\text{et}}(A, \mu_{\ell^\nu}^{\otimes i}), \quad n + k = 2i, n \geqslant 2. \tag{III}$$

6.2.4. Recall that if X is an H-space, then the quotient map $M^n_{\ell^\nu} \to S^n$ (see 3.1.4(ix)) induces a map

$$[S^n, X] \to [M^n_{\ell^\nu} X], \quad \text{i.e.} \quad \pi_n(X) \to \pi_n(X, \mathbb{Z}/\ell^\nu), \tag{IV}$$

where $n + k = 2i$. If $X = B\text{GL}(A)^+$, A as in 6.2.3, then we have a homomorphism $K_n(A) \to K_n(A, \mathbb{Z}/\ell^\nu)$, and by composing this map with (III) of 6.2.3, we obtain a homomorphism

$$c_{ik}: K_n(A) \to K_n(A, \mathbb{Z}/\ell^\nu) \to H^k_{\text{et}}(A, \mu_{\ell^\nu}^{\otimes i}), \quad n + k = 2i. \tag{V}$$

6.2.5. REMARKS/EXAMPLES.

(i) Let A be a Dedekind domain with field of fractions F, ℓ a prime. Assume that A contains $1/\ell$. Then we have localization sequences

$$\cdots \to K_n(A, \mathbb{Z}/\ell^\nu) \to K_n(F, \mathbb{Z}/\ell^\nu) \to \bigoplus_v K_{n-1}(k_v, \mathbb{Z}/\ell^\nu)$$
$$\to K_{n-1}(A, \mathbb{Z}/\ell^\nu) \to \cdots$$

and

$$0 \to H^1_{\text{et}}(A, \mu_{\ell^\nu}^{\otimes i}) \to H^1_{\text{et}}(F, \mu_\ell^{\otimes i}) \to \bigoplus_v H^0_{\text{et}}(k_v, \mu_{\ell^\nu}^{\otimes i-1})$$
$$\to H^2_{\text{et}}(A, \mu_{\ell^\nu}^{\otimes i}) \to \cdots.$$

(ii) If in (i) A contains $\frac{1}{\ell}$ and ξ_m, $m = \ell^\nu$, $m \not\equiv 2$ (4), and if all residue fields k_v are finite, then there is a map of localization sequences.

$$\begin{array}{ccccc}\cdots \longrightarrow & K_{n+1}(F,\mathbb{Z}/m) & \xrightarrow{\delta} & \bigoplus K_n(k_v,\mathbb{Z}/m) \\ & \downarrow & & \downarrow{\scriptstyle (i-1)c_{i-1,k-2}} \\ \cdots \longrightarrow & H^{k-1}_{\mathrm{et}}(F,\mu_m^{\otimes i}) & \xrightarrow{\delta} & \bigoplus H^{k-2}_{\mathrm{et}}(k_v,\mu_m^{\otimes i})\end{array}$$

$$\begin{array}{ccccc}\xrightarrow{\tilde{i}} & K_n(A,\mathbb{Z}/m) & \longrightarrow & K_n(F,\mathbb{Z}/m) & \longrightarrow \cdots \\ & \downarrow{\scriptstyle c_{ik}} & & \downarrow{\scriptstyle c_{ik}} & \\ \longrightarrow & H^k_{\mathrm{et}}(A,\mu_m^{\otimes i}) & \longrightarrow & H^i_{\mathrm{et}}(F,\mu_m^{\otimes i}) & \end{array}$$

(iii) If A is the ring of integers in a p-adic field F with residue field k, and $1/\ell \in A$ (ℓ a rational prime, $\ell \neq p$), then $K_n(A,\mathbb{Z}/\ell^\nu) \simeq K_n(k,\mathbb{Z}/\ell^\nu)$ and $H^*_{\mathrm{et}}(A,\mu_{\ell^\nu}^{\otimes i}) \cong H^*_{\mathrm{et}}(k,\mu_{\ell^\nu}^{\otimes i})$ and c_{ij} can be identified with multiplication by $\pm(i-1)!$ (see [166]).

6.2.6. Let F be a global field with ring of integers O_F, ℓ a prime, $\ell \neq \mathrm{char}(F)$, $O'_F = O_F(\frac{1}{\ell})$, then by tensoring (V) ($A = O'_F$) with $\mathbb{Z}_\ell$ we obtain map

$$\mathrm{ch}_{n,k} : K_{2n-k}(O_F) \otimes \mathbb{Z}_\ell \to H^k_{\mathrm{\bar{e}t}}(O'_F, \mathbb{Z}_\ell(n)).$$

6.2.7. QUILLEN–LICHTENBAUM (Q–L) CONJECTURE. *Let F be a global field with ring of integers O_F, ℓ a rational prime, $O'_F = O_F(\frac{1}{\ell})$. Then the Chern characters*

$$\mathrm{ch}^\ell_{i,n} : K_{2n-k}(O_F) \otimes \mathbb{Z}_\ell \to H^k_{\mathrm{er}}(O'_F, \mathbb{Z}_\ell(n))$$

are isomorphisms for $n \geqslant 2$, $i = 1, 2$ unless $\ell = 2$ and F is a number field with a real embedding.

The following result 6.2.8 establishes the Q–L conjecture in a special case $k \leqslant 2(n-1)$.

6.2.8. THEOREM. *In the notation of* 6.2.6, *when $k = 1$ or* 2, *the mapping*

$$\mathrm{ch}_{n,k} : K_{2n-k}(O_F) \otimes \mathbb{Z}_\ell \to H^k_{\mathrm{et}}(O'_F, \mathbb{Z}_\ell(n))$$

satisfies

(i) *If $2n - k \geqslant 2$ and $\ell \neq 2$, then $\mathrm{ch}_{n,k}$ is split surjective.*
(ii) *If $2n - k = 2$, $2n - k = 3$, $\ell \neq 2$ or when $\sqrt{-1} \in F$, $2n - k \geqslant 2$ and $\ell = 2$, then $\mathrm{ch}_{n,k}$ is an isomorphism.*

6.2.9. DEFINITION. Let ℓ be a fixed rational prime. For any field F, define integers $\omega^\ell_i(F)$ by $\omega^{(\ell)}_i(F) := \max\{\ell^\nu \mid \mathrm{Gal}(F(\mu_{\ell^\nu})/F)$ has exponent dividing $i\}$.

If there is no maximum ν, put $\omega_i^{(\ell)}(F)=\ell^\infty$. Suppose that $\omega_i^{(\ell)}(F)=1$ for almost all ℓ and that $\omega_i^{(\ell)}(F)$ is finite otherwise, write $\omega_i(F)=\prod\omega_i^\ell(F)$.

Note that if $F(\mu_\ell)$ has only finitely many ℓ-primary roots of unity for all primes ℓ and $[F(\mu_\ell):F]\to\infty$ as $\ell\to\infty$, then the $\omega_i(F)$ are finite for all $i\neq 0$. This is true for all local and global fields. See [166].

Also note that $\omega_i(F)=|H^0(F,Q/\mathbb{Z}(n))|$.

6.2.10. EXAMPLES.

(i) $w_i(\mathbb{F}_q)=q^i-1$ $\forall$ positive integers i. Hence $K_{2n-1}(\mathbb{F}_q)=q^i-1$ unless $(p-1)|i$.

(ii) $\omega_i(\widehat{Q}_p)=q^i-1$ unless $(p-1)|i$. If $i=(p-1)p^b m$, $(p\nmid m)$, then $\omega_i(\widehat{Q}_p)=(q^i-1)p^{1+b}$.

For further information and results on the $\omega_i(F)$, see [166].

6.2.11. Let F be a number field. Then each real embedding $\delta_i: F\to\mathbb{R}$ induces a map $F^*\to R^*\to\mathbb{Z}/2$ which detects the sign of F^* under δ_i. The sign map $\delta: F^*\to(\mathbb{Z}/2)^{r_1}$ is the sum of the δ_i and δ is surjective. $\operatorname{Ker}\delta$ is called the group of totally positive units in F and denoted by F_+^*.

Now let S be a finite set of places of F, O_S the ring of S-integers of F. Then the kernel of $\delta|_{O_S}: O_S\to F^*\to(\mathbb{Z}/2)^{r_i}$ is a subgroup of O_S^* called the subgroup of totally positive units in O_S and denoted by O_{S+}^*. The sign map $\delta|_{O_S}$ factors through $F^*/2=H^1(F,\mathbb{Z}/2)$ and hence factors through $\delta^1: H^1(O_S,\mathbb{Z}/2)\to(\mathbb{Z}/2)^{r_i}$. The signature defect $j(O_S)$ of O_F is defined as the dimension of the cokernel of δ^1. Note that $j(O_S)\leqslant j(O_F)$ and that $j(F)=0$ if $0\leqslant j(O_S)<r_1$.

The narrow Picard group, $\operatorname{Pic}_+(O_S)$ is the cokernel of the restricted divisor map $F_+^*\to\bigoplus_{\underline{p}\notin S}\mathbb{Z}$.

6.2.12. THEOREM [118]. *Let ℓ be an odd prime, F a number field, O_S a ring of S-integers in F containing $\frac{1}{2}$, j the signature defect of O_S. Denote $\omega_i^{(2)}(F)$ by ω_i. Then, there exists an integer m, $j\leqslant m<r$, such that for all $n\geqslant 2$, the 2-primary subgroup $K_n(O_S)(2)$ of $K_n(O_S)$ is given by*

$$K_n(O)_S(2)\cong\begin{cases} H^2_{\text{et}}(O_S,\widehat{\mathbb{Z}}_2(4a+1)) & \textit{for } n=8a,\\ \mathbb{Z}/2 & \textit{for } n=8a+1,\\ H^2_{\text{et}}(O_S,\widehat{\mathbb{Z}}_2(4a+2)) & \textit{for } n=8a+2,\\ (\mathbb{Z}/2)^{r_1-1}\oplus\mathbb{Z}/2\omega_{4a+2} & \textit{for } n=8a+3,\\ (\mathbb{Z}/2)^m\rtimes H^2_{\text{et}}(O_S,\mathbb{Z}_2(4a+3)) & \textit{for } n=8a+4,\\ 0 & \textit{for } n=8a+6,\\ H^2_{\text{et}}(O_S;\widehat{\mathbb{Z}}_2(4a+4)) & \textit{for } n=8a+6,\\ \mathbb{Z}/\omega_{4a+4} & \textit{for } n=8a+7. \end{cases}$$

The next result is on the odd prime torsion subgroup of $K_n(O_S)$.

6.2.13. THEOREM [166]. *Let ℓ be an odd prime, F a number field, O_S a ring of S-integers of F, $O'_S = O_S(\frac{1}{\ell})$. Then for all $n \geqslant 2$, we have*

$$K_n(O_S)(\ell) \cong \begin{cases} H^2_{\text{et}}(O'_S, \mathbb{Z}_\ell(i+1)) & \text{for } n = 2i > 0, \\ \mathbb{Z}_\ell^{r_2} \oplus \mathbb{Z}/\omega_i^\ell(F) & \text{for } n = 2i-1, i \text{ even}, \\ \mathbb{Z}_\ell^{r_1+r_2} \oplus \mathbb{Z}/\omega_i^{(\ell)}(F) & \text{for } n = 2i-1, i \text{ odd}. \end{cases}$$

The following result gives a picture for $K_n(O_S)$ when F is totally imaginary.

6.2.14. THEOREM [166]. *In the notation of* 6.2.13, *let F be totally imaginary. Then for all $n \geqslant 2$*

$$K_n(O_S) \cong \begin{cases} \mathbb{Z} \oplus \text{Pic}(O_S) & \text{for } n = 0, \\ \mathbb{Z}^{r_2+|S|-1} \oplus \mathbb{Z}/\omega_i & \text{for } n = 1, \\ \bigoplus_i H^2_{\text{et}}(O'_S, \mathbb{Z}_\ell(i+1)) & \text{for } n = 2i \geqslant 2, \\ \mathbb{Z}^{r_2} \oplus \mathbb{Z}/\omega_i & \text{for } n = 2i-1 \geqslant 3. \end{cases}$$

The next result which gives the picture of $K_n(O_F)$ for each odd $n \geqslant 3$ is a consequence of 6.2.13 and 6.2.14 and some other results which can be found in [166]. This reference also contains details of the proof.

6.2.15. THEOREM [166]. *Let F be a number field, O_S a ring of S-integers. Then*

(i) *$K_n(O_S) \simeq K_n(F)$ for each odd $n \geqslant 3$.*

(ii) *If F is totally imaginary then*

$$K_n(F) \cong \mathbb{Z}^{r_2} \oplus \mathbb{Z}/\omega_i(F).$$

(iii) *If $i = \frac{n+1}{2}$, then*

$$K_n(O_S) \simeq K_n(F) \cong \begin{cases} \mathbb{Z}^{r_1+r_2} \oplus \mathbb{Z}/\omega_i(F), & n \equiv 1 \pmod 8, \\ \mathbb{Z}^{r_1} \oplus \mathbb{Z}/2\omega_i(F) \oplus (\mathbb{Z}/2)^{r_1-1}, & n \equiv 3 \pmod 8, \\ \mathbb{Z}^{r_1+r_2} \oplus \mathbb{Z}/\frac{1}{2}\omega_i(F), & n \equiv 5 \pmod 8, \\ \mathbb{Z}^{r_2} \oplus \mathbb{Z}/\omega_i(F), & n \equiv 7 \pmod 8. \end{cases}$$

6.2.16. It follows from 6.2.15 that $K_n(Q) \simeq \mathbb{Z}$ for all $n \equiv 5 \pmod 8$ (since $w_i(Q) = 2$); more generally, if F has a real embedding, and $n \equiv 5 \pmod 8$ then $K_n(F)$ has no 2-primary torsion. (Since $\frac{1}{2}\omega_i(F)$ is an odd integer.)

6.2.17. THEOREM [166]. *Let E be a p-adic field of degree d over $\widehat{Q}_{p'}$, with ring of integers A. Then for all $n \geqslant 2$, we have*

(i)
$$K_n(A, \widehat{\mathbb{Z}}_p) \cong K_n(E, \widehat{\mathbb{Z}}_p) \cong \begin{cases} \mathbb{Z}/\omega_i^{(p)} E, & n = 2i, \\ (\widehat{\mathbb{Z}}_p)^d \oplus \mathbb{Z}/\omega_i^{(p)}(E), & n = 2i-1. \end{cases}$$

(ii) $K_{2i-1}(E, \mathbb{Z}/p^\nu) \cong H^1(E, \mu_{p\nu}^{\otimes i})$ *for all* i *and* ν *and for all* i *and* $\ell^\nu > \omega_i$, $H^1(E, \mu_{p\nu}^{\otimes i}) \cong (\mathbb{Z}/p^\nu)^d \oplus \mathbb{Z}/\omega_{i-1}$.

(iii) $K_n(E, \widehat{\mathbb{Z}}_p)$ *are finitely generated* $\widehat{\mathbb{Z}}_p$*-modules.*

6.2.18. REMARKS.

(i) Pushin, [110], has constructed motivic Chern characters $\mathrm{ch}^M_{i,n} : K_{2n-i}(F) \to H^i_{\mathcal{M}}(F, \mathbb{Z}(n))$ for $n \geqslant 2$, $i = 1, 2$, which induce étale Chern characters after tensoring by $\mathbb{Z}_\ell$.

(ii) For all $n \geqslant 2$, there exist isomorphisms $K_{2n-2}(F) \cong H^2_{\mathcal{M}}(F, \mathbb{Z}(n))$ and $K_{2n-1}(F) \cong H^1_{\mathcal{M}}(F, \mathbb{Z}(n))$ up to 2-torsion, see [166].

(iii) In view of 5.4.4(v), we have up to 2-torsion isomorphisms $K_{2n-2}(O_F) \simeq H^2_{\mathcal{M}}(O_F, \mathbb{Z}(n))$ and $K_{2n-1}(O_F) \cong H^1_{\mathcal{M}}(O_F, \mathbb{Z}(n))$.

6.3. *Higher K-theory and zeta functions*

6.3.1. DEFINITION. Let s be a complex number with $\mathrm{Re}(s) > 1$. The Riemann zeta function is defined as the convergent series $\zeta(s) = \sum_{n \geqslant 1} \frac{1}{n^s}$.

Note that $\zeta(s)$ admits a meromorphic continuation to the whole complex plane with a simple pole at $s = 1$ and no pole anywhere else.

For connections with K-theory we shall be interested in the values of $\zeta(s)$ at negative integers.

6.3.2. The Bernoulli numbers B_n, $n \geqslant 0$, are the rational numbers which arise in the power series expansion of $\frac{x}{e^x - 1} = \sum_{n=0}^{\infty} B_n \frac{x^n}{n!}$.

6.3.3. THEOREM.

(i) *For* n *odd* $\geqslant 3$, $\zeta(s)$ *has a simple zero at* $s = 1 - n$. *For* n *even* > 0, $\zeta(1-n) \neq 0$.

(ii) *For* $n \geqslant 1$, $\zeta(1-n) = (-1)^{n-1} \frac{B_n}{n}$.

6.3.4. Let F be a number field with ring of integers O_F, $\underline{a}$, an ideal of O_F, $N(\underline{a}) = |O_F/\underline{a}|$.

Then the Dedekind generalization of Riemann zeta function is defined by $\zeta_F(s) := \sum_{0 \neq \underline{a} \subset O_F} \frac{1}{N(\underline{a})^s}$. Note that $\zeta_F(s)$ is convergent for $\mathrm{Re}(s) > 1$ and can be extended to a meromorphic function on $\mathbb{C}$ with a single pole at $s = 1$.

6.3.5. THEOREM. *Let* F *be a number field and let* r_1 = *number of real embeddings of* F, r_2 = *number of pairs of complex embeddings of* F. *Let* d_n *be the order of vanishing of* $\zeta(s)$ *at* $s = 1 - n$.

Then

$$d_n = \begin{cases} r_1 + r_2 - 1 & \text{if } n = 1, \\ r_1 + r_2 & \text{if } n \geqslant 3 \text{ is odd}, \\ r_2 & \text{if } n = \text{is even}. \end{cases}$$

In view of 6.3.5, we now have a reformulation of 6.1.3(i)(a) in the following.

6.3.6. THEOREM. *Let F be a number field and O_F the ring of integers of F. Then for all $n \geqslant 2$*

$$K_{2n-1}(O_F) \simeq \mathbb{Z}^{d_n} \oplus \textit{finite},$$

i.e. the rank of $K_{2n-1}(O_F)$ is the order of vanishing of ξ_F at $s = 1 - n$.

6.3.7. DEFINITION. Define $\xi^*(-2m)$ as the first non-vanishing coefficient in the Taylor expansion around $s = -2m$ and call this the special value of the zeta function at $s = -2m$.

6.3.8. Let F be a number field. Recall from [14] that there exist higher regulator maps $X_n^B(F): K_{2n-1}(O_F) \to \mathbb{R}^{d_n}$ with finite kernel and image a lattice of rank d_n. The covolume of the lattice is called the Borel regulator and denoted by $R_n^B(F)$. Borel proved that $\xi_F^*(1-n) = q_n \cdot R_n^B(F)$ where q_n is a rational number.

6.3.9. LICHTENBAUM CONJECTURE [91,92]. *For all $n \geqslant 2$*

$$\xi_F^*(1-n) = \pm \frac{|K_{2n-2}(O_F)|}{|K_{2n-1}(O_F)_{\text{tors}}|} R_n^B(F)$$

up to powers of 2.

6.3.10. REMARKS.

(i) Birch and Tate had earlier conjectured a special case of 6.3.10 above viz. for a totally real number field F, $\xi_F(-1) = \pm\frac{|K_2(O_F|}{\omega_2(F)}$.

(ii) The next result is due to A. Wiles, [170].

6.3.11. THEOREM [170]. *Let F be a totally real number field. If ℓ is odd, and $O'_F = O_F(\frac{1}{\ell})$, then for all even $n = 2i$*

$$\xi_F(1-n) = \frac{|H^2_{\text{et}}(O'_F, \mathbb{Z}_\ell(n))|}{|H^1_{\text{et}}(O'_F, \mathbb{Z}_\ell(n))|} u_n,$$

where u_i is a rational number prime to ℓ.

6.3.12. THEOREM [166]. *Let F be a totally real number field. Then*

$$\zeta_F(1-2k) = (-1)^{kr_1} \frac{|K_{4k-2}(O_F)|}{|K_{4k-1}(O_F)|}$$

up to a factor of 2.

6.3.13. THEOREM [166]. *Let F be a totally real number field, O_F the ring of integers of F, $O'_F = O_F(\frac{1}{\ell})$. Then for all even $i > 0$*

$$2^{r_1} \frac{|K_{2i-2}(O_F)|}{|K_{2i-1}(O_F)|} = \frac{\prod_\ell (H^2_{\text{et}}(O'_F, F_2(\ell)))}{\prod_\ell |H^1_{\text{et}}(O'_F, \mathbb{Z}_2(o))|}.$$

6.3.14. DEFINITION. Let F be a global field of characteristic p and X the associated smooth projective curve over $\mathbb{F}_q$, i.e. $F = \mathbb{F}_q(X)$, O_F the integral closure of $\mathbb{F}_q[t]$ in F. The maximal ideals of O_F are the finite primes in F and the associated zeta function is defined by

$$\zeta_F(s) = \prod_{\underline{p} \text{ finite}} \frac{1}{(1 - N(\underline{p})^{-s})}.$$

6.3.15. THEOREM. *Let F be a global field of characteristic $p > 0$. Then for all $n \geqslant 2$, we have*

$$\zeta_F(1-n) = \pm \frac{H^2(O_F, \mathbb{Z}(n))}{\omega_n(F)}.$$

6.4. *Higher K-theory of $\mathbb{Z}$*

In this subsection, we briefly survey the current situation with the computations of $K_n(\mathbb{Z})$, $n \geqslant 0$. Again, details of arguments leading to the computations can be found in [166]. First we list specific information for $K_n(\mathbb{Z})$, $n = 0, 1, 2, 3, 4, 5, 6, 7, 9, 10$.

6.4.1. THEOREM. *$K_0(\mathbb{Z}) \simeq \mathbb{Z}$, $K_1(\mathbb{Z}) \simeq \mathbb{Z}/2$, $K_2(\mathbb{Z}) \simeq \mathbb{Z}/2$, $K_3(\mathbb{Z}) \simeq \mathbb{Z}/48$, see* [89]. *$K_4(\mathbb{Z}) = 0$, see* [117], *$K_5(\mathbb{Z}) = \mathbb{Z}$, see* [166], *$K_6(\mathbb{Z}) = 0$, see* [166], *$K_7(\mathbb{Z}) = \mathbb{Z}/240$, see* [166], *$K_9(\mathbb{Z}) = \mathbb{Z} \oplus \mathbb{Z}/2$, $K_{10}(\mathbb{Z}) = \mathbb{Z}/2$,* [166].

More generally we have.

6.4.2. THEOREM.

$$K_n(\mathbb{Z}) = \begin{cases} \text{finite } \forall n > 0, & n \not\equiv 1 \ (4), \\ \mathbb{Z} + \text{finite} & \text{if } n \equiv 1 \ (4). \end{cases}$$

Using 6.2.10 and 6.2.12, one can prove.

6.4.3. THEOREM [166]. *Let ℓ be an odd rational prime*

(i) *$K_{2n-1}(\mathbb{Z})$ has ℓ-torsion exactly when $n \equiv 0 \bmod(\ell - 1)$.*
(ii) *The ℓ-primary subgroups of $K_{2n}(\mathbb{Z})$ are $H^2_{\text{et}}(\mathbb{Z}(\frac{1}{\ell}); \widehat{\mathbb{Z}}_\ell(i+1))$.*

The following result is a consequence of 6.3.13 and shows that the Lichtenbaum conjecture holds up to a factor of 2.

6.4.4. THEOREM [166]. *For all $n \geqslant 1$, we have*

$$\frac{|K_{4n-2}(\mathbb{Z})|}{|K_{4n-1}(\mathbb{Z})|} = \frac{b_n}{4n} = \frac{-1^n}{2}\zeta(1-2n).$$

Hence if c_n denotes the numerator of $\frac{b_n}{4n}$, then $|K_{4n-2}| = \begin{cases} c_n, & n \text{ even}, \\ 2c_n, & n \text{ odd}. \end{cases}$

6.4.5. DEFINITION. A prime p is said to be regular if $\mathrm{Pic}(\mathbb{Z}[\mu_p])$ has no element of exponent p, i.e. p does not divide the order h_p of $\mathrm{Pic}(\mathbb{Z}[\mu_p])$. Note that p is regular iff $\mathrm{Pic}(\mathbb{Z}[\mu_p \nu])$ has no p-torsion for all ν.

6.4.6. THEOREM [166]. *If ℓ is an odd regular prime, then $K_{2n}(\mathbb{Z})$ has no ℓ-torsion.*

The following result shows that the 2-primary subgroups of $K_n(\mathbb{Z})$ are essentially periodic of period 8.

6.4.7. THEOREM [166].

$$K_n(\mathbb{Z})(2) = \begin{cases} \mathbb{Z}/2 & \text{if } n \equiv 1 \pmod 8, \\ \mathbb{Z}/2 & \text{if } n \equiv 2 \pmod 8, \\ \mathbb{Z}/16 & \text{if } n \equiv 3 \pmod 8, \\ 0 & \text{if } n \equiv 4 \pmod 8, \\ 0 & \text{if } n \equiv 5 \pmod 8, \\ 0 & \text{if } n \equiv 6 \pmod 8, \\ \mathbb{Z}/6a & \text{if } n \equiv 7 \pmod 8, \\ 0 & \text{if } n \equiv 8 \pmod 8. \end{cases}$$

6.4.8. VANDIVER'S CONJECTURE. *If ℓ is an irregular prime, then* $\mathrm{Pic}(\mathbb{Z}[\xi_\ell + \xi_\ell^{-1}])$ *has no ℓ-torsion.*

This conjecture is equivalent to the expression of the natural representation of $G = \mathrm{Gal}(Q(\xi_\ell)/Q)$ on $\mathrm{Pic}(\mathbb{Z}[\xi_\ell])/\ell$ as a sum of G-modules $\mu_\ell^{\otimes i}$ when i is odd.

Note that Vandiver's conjecture for ℓ is equivalent to the assertion that $K_{4n}(\mathbb{Z})$ has no ℓ-torsion for all $n < \frac{\ell-2}{2}$, see [166].

The following result is due independently to S. Mitchel and M. Kurihara.

6.4.9. THEOREM [84]. *If Vandiver's conjecture holds, then*

$$K_n(\mathbb{Z}) = \begin{cases} \mathbb{Z} \oplus \mathbb{Z}/2 & \textit{if } n \equiv 1 \pmod 8, \\ \mathbb{Z}/2c_k & \textit{if } n \equiv 2 \pmod 8, \\ \mathbb{Z}/2\omega_{2k} & \textit{if } n \equiv 3 \pmod 8, \\ 0 & \textit{if } n \equiv 4 \pmod 8, \\ \mathbb{Z} & \textit{if } n \equiv 5 \pmod 8, \\ \mathbb{Z}/c_k & \textit{if } n \equiv 6 \pmod 8, \\ \mathbb{Z}/\omega_{2k} & \textit{if } n \equiv 7 \pmod 8, \\ 0 & \textit{if } n \equiv 8 \pmod 8. \end{cases}$$

7. Higher *K*-theory of orders, group-rings and modules over 'EI'-categories

7.1. *Higher K-theory of orders and group-rings*

7.1.1. We recall that if R is a Dedekind domain with quotient field F, and Λ is any R-order in a semi-simple F-algebra Σ, then $SK_n(\Lambda) := \operatorname{Ker}(K_n(\Lambda) \to K_n(\Sigma))$ and $SG_n(\Lambda) := \operatorname{Ker}(G_n(\Lambda) \to G_n(\Sigma))$. Also, any R-order in a semi-simple F-algebra Σ can be embedded in a maximal R-order Γ which has well-understood arithmetic properties relative to Σ. More precisely if $\Sigma = \prod_{i=1}^{r} M_{n_i}(D_i)$ then Γ is Morita equivalent to $\prod_{i=1}^{r} M_{n_i}(\Gamma_i)$ where the Γ_i are maximal orders in the division algebra D_i and so $K_n(\Gamma) \approx \bigoplus K_n(\Gamma_i)$ while $K_n(\Sigma) \approx \bigoplus K_n(D_i)$. So, the study of K-theory of maximal orders in semi-simple algebras can be reduced to the K-theory of maximal orders in division algebras.

As remarked in [79] the study of $SK_n(\Lambda)$ facilitates the understanding of $K_n(\Lambda)$ apart from the various known topological applications known for $n = 0, 1, 2$, where $\Lambda = ZG$ for some groups G that are usually fundamental groups of some spaces (see [163,150,103]). Also $SK_n(\Lambda)$ is connected to the definition of higher class groups which generalizes to higher K-groups the notion of class groups of orders and group-rings. (See [48,65].)

First, we have the following result 7.1.2 on SK_n of maximal orders in p-adic semi-simple algebras. This result as well as the succeeding ones, 7.1.4 to 7.1.6, are due to Kuku.

7.1.2. THEOREM [68]. *Let R be the ring of integers in a p-adic field F, Γ a maximal R-order in a semi-simple F-algebra Σ. Then for all $n \geqslant 1$,*

(i) $SK_{2n}(\Gamma) = 0$.
(ii) $SK_{2n-1}(\Gamma) = 0$ *if and only if Σ is unramified over its centre (i.e. Σ is a direct product of matrix algebras over fields).*

7.1.3. REMARK.

(i) The result above is a generalization to higher K-groups of an earlier result for SK_1.
(ii) Before discussing arbitrary orders, we record the following consequences of 7.1.2
  (a) 7.1.2 holds for group rings $\Gamma = RG$ where G is a finite group of order relatively prime to p and FG splits, see [19].
  (b) If $\underline{m} = \operatorname{rad}\Gamma$, then for all $n \geqslant 1$, the transfer map: $K_{2n-1}(\Gamma/\underline{m}) \to K_{2n-1}(\Gamma)$ is non-zero unless Σ is a product of matrix algebras over fields. Hence, the Gersten conjecture does not hold in the non-commutative case.

(c) For all $n > 0$, we have an exact sequence $0 \to K_{2n}(\Gamma) \to K_{2n-1}(\Sigma) \to K_{2n-1}(\Gamma/\underline{m}) \to 0$ if and only if Σ is a direct product of matrix algebras over fields.

We now discuss arbitrary orders and the global situations.

7.1.4. THEOREM [74]. *Let R be the ring of integers in a number field F, Λ any R-order in a semi-simple F-algebra Σ. Then for all $n \geqslant 1$*

(i) *$K_n(\Lambda)$ is a finitely generated Abelian group.*
(ii) *$SK_n(\Lambda)$ is a finite group.*
(iii) *If $\hat{\Lambda}_{\underline{p}}$ denotes the completion of Λ at a prime $\underline{p}$ of R, then $SK_n(\hat{\Lambda}_{\underline{p}})$ is a finite group.*

7.1.5. REMARKS.

(i) The above results 7.1.4 hold for $\Lambda = RG$, $\hat{\Lambda}_{\underline{p}} = \widehat{R}_{\underline{p}}G$ where G is a finite group and $\widehat{R}_p$ is the completion of R at a prime $\underline{p}$ of R.
(ii) 7.1.4(i) was proved in [73]. Theorem 2.1.1, 7.1.4 (ii) and (iii) were proved for group-rings RG, $\widehat{R}_{\underline{p}}G$ in [73], Theorem 3.2 and later for R-orders in [74], Theorem 1.1 (ii) and (iii).

We next discuss similar finiteness results for G-theory.

7.1.6. THEOREM [69,74]. *Let R be the ring of integers in a number field F, Λ any R-order in a semi-simple F-algebra Σ, $\underline{p}$ any prime ideal of R. Then for all $n \geqslant 1$*

(i) *$G_n(\Lambda)$ is a finitely generated Abelian group.*
(ii) *$G_{2n-1}(\Lambda_{\underline{p}})$ is a finitely generated Abelian group.*
(iii) *$SG_{2n}(\Lambda) = SG_{2n}(\Lambda_{\underline{p}}) = SG_{2n}(\hat{\Lambda}_{\underline{p}}) = 0$.*
(iv) *$SG_{2n-1}(\Lambda)$ is a finite group.*
(v) *$SG_{2n-1}(\hat{\Lambda}_{p})$, $SG_{2n-1}(\hat{\Lambda}_{\underline{p}})$ are finite groups of order relatively prime to the rational primes lying below $\underline{p}$.*

7.1.7. REMARKS.

(i) The results above hold for $\Lambda = RG$, $\hat{\Lambda}_{\underline{p}} = R_pG$, $\hat{\Lambda}_{\underline{p}} = \widehat{R}_pG$.
(ii) One can show from 7.1.6(i) that if A is any R-algebra finitely generated as an R-module, then $G_n(A)$ is finitely generated (see [73, Theorem 2.3]).

We also have the following result on the vanishing of $SG_n(\Lambda)$ (see [87]) due to Laubenbacher and Webb.

7.1.8. THEOREM [87]. *Let R be a Dedekind domain with quotient field F, Λ any R-order in a semi-simple F-algebra. Assume that*

(i) *$SG_1(\Lambda) = 0$.*
(ii) *$G_n(\Lambda)$ is finitely generated for all $n \geqslant 1$.*
(iii) *$R/\underline{p}$ is finite for all primes $\underline{p}$ or R.*

(iv) *If ξ is an ℓ^s-th root of unit for any rational prime ℓ and positive integer s, R the integral closure of R in $F(\xi)$, then $SG_1(\widetilde{R} \otimes_R \Lambda) = 0$.*

Then $SG_n(\Lambda) = 0$ for all $n \geqslant 1$.

7.1.9. REMARKS. It follows from 7.1.8 that if R is the ring of integers in a number field F then (i) $SG_n(RG) = 0$, (ii) $SG_n(\hat{\Lambda}_{\underline{p}}) = 0$ for all prime ideals $\underline{p}$ of R and all $n \geqslant 1$ (see [78, 1.9]), if Λ satisfies the hypotheses of 7.1.8.

The following extension of finiteness results for G-theory to group-rings of infinite groups in significant. This result is due to Kuku and Tang, see [82].

7.1.10. THEOREM [82]. *Let $V = G \rtimes_\alpha T$ be the semi-direct product of a finite group G of order p with an infinite cyclic group $T = \langle t \rangle$ with respect to the automorphism $\alpha : G \to G : g \to tgt^{-1}$. Then for all $n \geqslant 0$, $G_n(RV)$ is a finitely generated Abelian group, where R is the ring of integers in a number field.*

We also have the following result on the ranks of $K_n(\Lambda)$, $G_n(\Lambda)$, due to Kuku, see [76].

7.1.11. THEOREM [76]. *Let R be the ring of integers in a number field F, Λ an R-order in a semi-simple F-algebra Σ, Γ a maximal R-order containing Λ.*

Then for all $n \geqslant 2$

$$\operatorname{rank} K_n(\Lambda) = \operatorname{rank} G_n(\Lambda) = \operatorname{rank} K_n(\Sigma) = \operatorname{rank} K_n(\Gamma).$$

7.1.12. REMARKS.

(i) The above results 7.1.11 hold for $\Lambda = RG$ where G is any finite group.

(ii) The ranks of $K_n(R)$ and $K_n(F)$ have been discussed in Section 6.

It then means that if Σ is a direct product of matrix algebras over fields and Γ is a maximal order in Γ, then $\operatorname{rank} K_n(\Gamma) = \operatorname{rank} K_n(\Sigma)$ is completely determined since $\Sigma = \prod M_{n_i}(F_i)$ and $\Gamma = \prod M_{n_i}(R_i)$ where R_i is the ring of integers in F_i. Also by theorem 7.1.11 this is equal to $\operatorname{rank} G_n(\Lambda)$ as well as $\operatorname{rank} K_n(\Lambda)$ if Λ is any R-order contained in Γ.

However, if Σ does not split, there exists a Galois extension E of F which splits Σ, in which case we can reduce the problem to that computation of the ranks of K_n of fields.

We next obtain some results connecting K_n and G_n through the Cartan map. Recall that if A is any Noetherian ring, the inclusions $\underline{P}(A) \to \underline{M}(A)$ induces a homomorphism $K_n(A) \to G_n(A)$ for all $n \geqslant 0$. First we have the following result, whose proof uses methods of equivariant higher K-theory, is due to Dress and Kuku (see [25]).

7.1.13. THEOREM [25]. *Let k be a field of characteristic p, G a finite group. Then the inclusion functor $\underline{P}(kG) \mapsto \underline{M}(kG)$ induces an isomorphism $\mathbb{Z}(\frac{1}{p}) \otimes K_n(kG) \mapsto Z(\frac{1}{p}) \otimes G_n(kG)$ for all $n \geqslant 0$.*

As a consequence of 7.1.13 one can prove the following result 7.1.14 and 7.1.15 due to Kuku, [73].

7.1.14. COROLLARY [73]. *Let G be a finite group, k a finite field of characteristic $p \neq 0$. Then for all $n \geqslant 1$,*

(1) *$K_{2n}(kG)$ is a finite p-group.*
(2) *The Cartan map $K_{2n-1}(kG) \to G_{2n-1}(kG)$ is surjective and $\operatorname{Ker} \varphi_{2n-1}$ is the Sylow-p-subgroup of $K_{2n-1}(kG)$.*

Finally we have the following:

7.1.15. THEOREM [74]. *Let R be the ring of integers in a number field F, G a finite group, $\underline{p}$ a prime ideal of R. Then for all $n \geqslant 1$.*

(i) *The Cartan homomorphisms $K_n((R/\underline{p})G) \to G_n((R/\underline{p})G)$ are surjective.*
(ii) *The Cartan homomorphism $K_n(RG) \to G_n(RG)$ induces a surjection $SK_n(RG) \to SG_n(RG)$.*
(iii) *$SK_{2n}(RG) = \operatorname{Ker}(K_{2n}(RG) \to G_{2n}(RG))$.*

7.1.16. REMARKS. Recall from 4.6.4 that if R is the ring of integers in a number field F and Λ is any R-order in a semi-simple F-algebra Σ, then we have a long exact sequence

$$\cdots \to K_{n+1}(\mathcal{A}, \omega) \to K_n(\Lambda) \to G_n(\Lambda) \to K_{n-1}(\mathcal{A}, \omega) \to \cdots,$$

where $(\mathcal{A}, \omega)$ is the Waldhausen category of bounded chain complexes over $\mathcal{M}(\Lambda)$ with weak equivalences as quasi-isomorphisms.

It follows from 7.1.4 and 7.1.6 that for all $n \geqslant 1$, $K_{n+1}(\mathcal{A}, \omega)$ is a finitely generated Abelian group.

7.1.17. For the rest of this subsection, we shall focus briefly on the higher G-theory of Abelian group-rings and some ramifications of this theory to non-commutative cases, viz. dihedral and quaternion groups, nilpotent groups, and groups of square free order. The work reported here is due to D. Webb (see [159–162]).

The following results generalize to higher G-theory of results in 2.2.5 of [79]. The notations are those of 2.2.4, 2.2.5 of [28].

7.1.18. THEOREM [159]. *Let π be an Abelian group and R a Noetherian ring (not necessarily commutative). Then*

$$G_n(R\pi) \simeq \bigoplus_{\rho \in X(\pi)} G_n(R\langle\rho\rangle) \quad \text{for all } n \geqslant 0,$$

where $X(\pi)$ is the set of cyclic quotients of π.

Next we have the following ramifications of 2.5.20 of [28].

7.1.19. THEOREM [159]. *Let $G = \pi \rtimes \Gamma$ be the semidirect product of an Abelian group π and a finite group Γ, R a Noetherian ring. Assume that the Γ-action stabilizes every* *cocyclic subgroup of π so that Γ acts on each cyclic quotient ρ of π. Let $R\langle\rho\rangle\#\Gamma$ be the twisted group-ring. Then for all $n > 0$*

$$G_n(RG) \simeq \bigoplus_{\rho\in X(\pi)} G_n\big(R\langle\rho\rangle\#\Gamma\big).$$

7.1.20. THEOREM [159]. *Let G be a non-Abelian group of order pq, $p|(q-1)$. Let Γ be the unique subgroup of* $\mathrm{Gal}(Q(\xi_q)/Q)$ *of order p. Then $\forall n \geqslant 0$*

$$G_n(\mathbb{Z}G) \simeq G_n(\mathbb{Z}) \oplus G_n\left(\mathbb{Z}\left[\varsigma_p, \frac{1}{p}\right]\right) \oplus G_n\left(\mathbb{Z}\left[\varsigma_q, \frac{1}{q}\right]\right)^{\Gamma}.$$

7.1.21. REMARKS. Note that the dihedral group D_{2s} of order $2s$ has the form $\pi \rtimes \Gamma$ where π is a cyclic group of order s and Γ has order 2. So, by 7.1.19

$$G_n(\mathbb{Z}D_{2s}) \simeq \bigoplus_{\rho\in X(\pi)} G_n\big(\mathbb{Z}\langle\rho\rangle\#\Gamma\big).$$

We now have a more explicit form of $G_n(\mathbb{Z}D_{2s})$.

7.1.22. THEOREM. *For all $n \geqslant 0$,*

$$G_n(\mathbb{Z}D_{2s}) \simeq \bigoplus_{\binom{d/s}{d>2}} G_n\left(\mathbb{Z}\left[\zeta_d, \frac{1}{d}\right]\right)_+ \oplus G_n\left(\mathbb{Z}\left[\frac{1}{2}\right]\right)^{\varepsilon} \oplus G_n(\mathbb{Z}),$$

$$\text{where } \varepsilon = \begin{cases} 1 & \text{if } s \text{ is odd,} \\ 2 & \text{if } s \text{ is even} \end{cases}$$

and $\mathbb{Z}[\zeta_d, \frac{1}{d}]_+$ is the complex conjugation invariant subring of $\mathbb{Z}[\zeta_d, \frac{1}{d}]$.

7.1.23. Now, let G be the generalized quaternion group of order 4.2^s, i.e.

$$G = \big\langle x, y \mid x^2 = y^2, y^4 = 1, yxy^{-1} = x^{-1}\big\rangle.$$

For $s \geqslant 0$, let $\Gamma = \{1, \gamma\}$ be a two-element group acting on $Q[\varsigma_{2^s}]$ by complex conjugation with fixed field $Q[\varsigma_{2^s}]_+$, the maximal real subfield.

Let $c: \Gamma \times \Gamma \to Q[\varsigma_{s^{s+1}}]$ be the normalized 2-cocycle given by $c(\gamma, \gamma) = -1$ and let $H = Q[\varsigma_{2^{s+1}}]\#_c\Gamma$ be the cross-product algebra, L a maximal Z-order in H (see [58]).

7.1.24. THEOREM [161]. *In the notation of* 7.1.23

$$G_n(\mathbb{Z}G) \simeq \bigoplus_{j=0}^{s} G_n\left(\mathbb{Z}\left[\xi_{2^j}, \frac{1}{2^j}\right]\right) \oplus G_n\left(l\left[\frac{1}{2^{s+1}}\right]\right) \oplus G_n\left(\mathbb{Z}\left(\frac{1}{2}\right)\right]^2.$$

7.1.25. Let π be a finite nilpotent group whose 2-Sylow subgroup has no subquotients isomorphic to the quaternion group of order 8, R a left Noetherian ring. The rational group algebra $Q\pi \sim \prod_{\rho\in X(\pi)} Q(\rho)$ where each $Q(\rho)$ is a simple algebra and $\rho:\pi\to \mathrm{GL}(V)$ is an irreducible rational representation of π and $X(\pi)$ is a complete set of inequivalent irreducible rational representation of π and $Q(\rho)\simeq M_{n(\rho)}(K_\rho)$ is a full matrix algebra over a field K_ρ (see [161]).

Let $Z(\rho)=M_{n(\rho)}(O_\rho)$ be a maximal A-order in $Q(\rho)$ where O_ρ is the ring of integers in K_ρ. Put $Z\langle\rho\rangle = Z(\rho)[\frac{1}{|\rho|}]$ where $|\rho| = \text{index}[\pi : \ker\rho]$. Then we have the following.

7.1.26. THEOREM [161]. *In the notation of* 7.1.25

$$G_n(R\pi)\simeq \bigoplus_{\rho\in X(\pi)} G_n\big(R\langle\rho\rangle\big), \quad \textit{where } R\langle\rho\rangle = R\otimes Z\langle\rho\rangle.$$

Hence

$$G_n(Z\pi)\simeq \bigoplus_{\rho\in X(\pi)} K_n\big(O_\rho[1/|\rho|]\big).$$

7.1.27. Now let G be a finite group of square free order. Then G is metacyclic and hence can be written as $G=\pi\rtimes\Gamma$ where π,Γ are cyclic of square free order. Then $QG\simeq Q\pi\#\Gamma\simeq\prod_{\rho\in X(\pi)} Q(\rho)\#\Gamma$ where $X(\pi)$ is the set of cyclic quotients of π.

7.1.28. THEOREM [162]. *In the notation of* 7.1.27 *we have*

$$G_n(RG)\simeq \bigoplus_{\rho\in X(\pi)} G_n\big(R\langle\rho\rangle\#\Gamma\big).$$

7.1.29. REMARKS. For further simplification of each $G_n(R\langle\rho\rangle\#\Gamma)$ in the context of a conjecture of Hambleton, Taylor and Williams, see [162].

7.2. *Higher class groups of orders and group-rings*

7.2.1. Let F be a number field with ring of integers O_F, Λ any R-order in a semi-simple F-algebra Σ. In this subsection, we review briefly some results on the higher class groups $Cl_n(\Lambda)$, $n\geqslant 0$, of Λ which constitute natural generalization to higher dimensions of the classical notion of class groups $Cl(\Lambda)$ of an order (see [19] or [79]).

The class groups $Cl(\Lambda)$, apart from generalizing the classical notion of class groups of integers in number fields, is also intimately connected with representation theory while $Cl(\mathbb{Z}G)$ is also known to house several topological/geometric invariants, see [156]. For various computations of $Cl(\Lambda)$, see [19].

7.2.2. DEFINITION. Let F be a number field with ring of integers O_F, Λ an O_F-order in a semi-simple algebra. The higher class groups are defined for all $n \geqslant 0$ by

$$Cl_n(\Lambda) = \operatorname{Ker}\left(SK_n(\Lambda) \to \bigoplus_{\underline{p}} SK_n(\hat{\Lambda}_{\underline{p}})\right),$$

where $\underline{p}$ runs through all prime ideals of O_F.

NOTE. One can show that $Cl_0(\Lambda) = Cl(\Lambda)$. Moreover, $Cl_1(O_F G)$ is intimately connected with the Whitehead groups of G.

The result below, due to Kuku, [73], says that $\forall n \geqslant 1$, $Cl_n(\Lambda)$ are finite groups for any O_F-orders Λ.

7.2.3. THEOREM [74]. *Let F be a number field with ring of integers O_F, Λ any O_F-order in a semi-simple F-algebra Σ. Then the groups $Cl_n(\Lambda)$ are finite.*

Note that $Cl_0(\Lambda) = Cl(\Lambda)$ is finite is a classical result. See [19].
The next result due to M.E. Keating says that $Cl_n(\Gamma)$ vanishes for maximal orders Γ.

7.2.4. THEOREM [60]. *Let F, O_F be as in 7.2.3, Γ a maximal O_F-order in a semi-simple algebra. Then for all $n \geqslant 1$, $Cl_n(\Gamma) = 0$.*

The next result 7.2.6 due to Kolster and Laubenbacher, [65], gives information on possible torsion in odd-dimensional class groups. First we make some observations.

7.2.5. REMARKS. Let F be a number field with ring of integers O_F, Λ any O_F-order in a semi-simple F-algebra Σ, and Γ a maximal order containing Λ. It is well known that $\hat{\Lambda}_{\underline{p}}$ is a maximal order for all except a finite number of prime ideals $\underline{p}$ of O_F at which $\Lambda_{\underline{p}}$ is not maximal. We denote by $\mathcal{P}_\Lambda$ the set of rational primes q lying below the primes ideals $\underline{p}$ in S_Λ, where S_Λ denotes the finite set of prime ideals $\underline{p}$ of O_F for which $\Lambda_{\underline{p}}$ is not maximal.

7.2.6. THEOREM [65]. *Let F, O_F, Λ be as in 7.2.3. Then for all $n \geqslant 1$, $Cl_{2n-1}(\Lambda)(q) = 0$ for $q \notin \mathcal{P}_\Lambda$.*

7.2.7. COROLLARY. *Let G be a finite group. Then for all $n \geqslant 1$, the only p-torsion that can possibly occur in $Cl_{2n-1}(O_F G)$ are for those p dividing the order of G.*

7.2.8. REMARKS. Finding out what p-torsion could occur in even-dimensional class groups of arbitrary orders in semi-simple algebras is still open. However, we have the following result 7.2.9 due to Guo and Kuku, [48], analogous to 7.2.6 above for Eichler and hereditary orders.

7.2.9. THEOREM [48]. *Let F be a number field with ring of integers O_F, Λ an Eichler order in a quaternion F-algebra or a hereditary order in a semi-simple F-algebra. Then, in the notation of* 7.2.3, *$Cl_{2n}(\Lambda)(q) = 0$ for $q \notin P_\Lambda$.*

7.3. *Profinite higher K-theory of orders and group-rings*

7.3.1. Let R be a Dedekind domain with quotient field F, Λ any R-order in a semi-simple F-algebra. In the notation of 3.1.6(xi) and [78], we shall write $K_n^{\mathrm{pr}}(\Lambda, \widehat{\mathbb{Z}}_\ell)$ for $K_n^{\mathrm{pr}}(\mathcal{P}(\Lambda), \widehat{\mathbb{Z}}_\ell)$ (resp. $G_n^{\mathrm{pr}}(\Lambda, \widehat{\mathbb{Z}}_\ell)$ for $K_n^{\mathrm{pr}}(\mathcal{M}(\Lambda), \widehat{\mathbb{Z}}_\ell)$) respectively for the profinite K-theory (resp. G-theory of Λ). Also, following the notations in 3.1.6(ix) and as in [78], we shall write

$$K_n(\Lambda, \widehat{\mathbb{Z}}_\ell) = \varprojlim K_n\big(\mathcal{P}(\Lambda), \mathbb{Z}/\ell^s\big), \qquad G_n(\Lambda, \widehat{\mathbb{Z}}_\ell) = \varprojlim K_n\big(\mathcal{M}(\Lambda), \mathbb{Z}/\ell^s\big).$$

Note that this situation applies to R being ring of integers in number fields and p-adic fields. Note that in the statements above and throughout this subsection ℓ is a prime.

Although several of the computations in [78] do apply to general exact categories we shall restrict ourselves in this section to results on orders and group rings. The results in this subsection are all due to Kuku, [78].

7.3.2. THEOREM [78]. *Let R be the ring of integers in a number field F, Λ an R-order in a semi-simple F-algebra. Then we have*

(1) *For all $n \geqslant 1$*

$$K_n(\Lambda)[\ell^s] \cong K_n^{\mathrm{pr}}(\Lambda, \widehat{\mathbb{Z}}_\ell)[\ell^s]; \qquad G_n(\Lambda)[\ell^s] \cong G_n^{\mathrm{pr}}(\Lambda, \widehat{\mathbb{Z}}_\ell)[\ell^s].$$

(2) *For all $n \geqslant 2$*

$$K_n(\Lambda)/\ell^s \cong K_n^{\mathrm{pr}}(\Lambda, \widehat{\mathbb{Z}}_\ell)/\ell^s; \qquad G_n(\Lambda)/\ell^s \cong G_n^{\mathrm{pr}}(\Lambda, \widehat{\mathbb{Z}}_\ell)/\ell^s.$$

(3) (a) $K_n(\Lambda) \otimes \widehat{\mathbb{Z}}_\ell \cong K_n^{\mathrm{pr}}(\Lambda, \widehat{\mathbb{Z}}_\ell) \simeq K_n(\Lambda, \widehat{\mathbb{Z}}_\ell)$ *are ℓ-complete profinite Abelian groups for all $n \geqslant 2$.*
(b) $G_n(\Lambda \otimes \widehat{\mathbb{Z}}_\ell) \simeq G_n^{\mathrm{pr}}(\Lambda, \widehat{\mathbb{Z}}_\ell) \simeq G_n(\Lambda, \widehat{\mathbb{Z}}_\ell)$ *are ℓ-complete profinite Abelian groups for all $n \geqslant 2$.*
(c) $K_{2n+1}(\Sigma) \otimes \widehat{\mathbb{Z}}_\ell \simeq K_{2n-1}^{\mathrm{pr}}(\Sigma, \widehat{\mathbb{Z}}_\ell) \simeq K_{2n-1}(\Sigma, \widehat{\mathbb{Z}}_\ell)$ *are ℓ-complete profinite Abelian groups for all $n \geqslant 2$.*

NOTE. An Abelian group G is said to be ℓ-complete if $G = \varprojlim(G/\ell^s)$.

We also have the following results in the local situation.

7.3.3. THEOREM [78]. *Let p be a rational prime, F a p-adic field (i.e. any finite extension of $\widehat{Q}_p$), R the ring of integers of F, Γ a maximal R-order in a semi-simple F-algebra Σ. Then for all $n \geqslant 2$ we have*

(1) $K_n^{\mathrm{pr}}(\Sigma, \widehat{\mathbb{Z}}_\ell) \simeq K_n(\Sigma, \widehat{\mathbb{Z}}_\ell)$ *is an ℓ-complete profinite Abelian group.*
(2) $K_n^{\mathrm{pr}}(\Gamma, \widehat{\mathbb{Z}}_\ell) \simeq K_n(\Gamma, \widehat{\mathbb{Z}}_\ell)$ *is an ℓ-complete profinite Abelian group.*

7.3.4. THEOREM [78]. *Let R be the ring of integers in a p-adic field F, ℓ a rational prime such that $l \neq p$, Λ an R-order in a semi-simple F-algebra Σ. Then, for all $n \geqslant 2$, $G_n^{\mathrm{pr}}(\Lambda, \widehat{\mathbb{Z}}_\ell)$ is an ℓ-complete profinite Abelian group.*

7.3.5. THEOREM [78]. *Let R be the ring of integers in a p-adic field F, Λ an R-order in a semi-simple F-algebra Σ, ℓ a rational prime such the $\ell \neq p$. Then for all $n \geqslant 1$,*

(i) *$G_n(\Lambda)_\ell$ are finite groups.*
(ii) *$K_n(\Sigma)_\ell$ are finite groups.*
(iii) *The kernel and cokernel of $G_n(\Lambda) \to G_n^{\mathrm{pr}}(\Lambda, \widehat{\mathbb{Z}}_\ell)$ are uniquely ℓ-divisible.*
(iv) *The kernel and cokernel of $K_n(\Sigma) \to K_n^{\mathrm{pr}}(\Sigma, \widehat{\mathbb{Z}}_\ell)$ are uniquely ℓ-divisible.*

7.3.6. THEOREM [78]. *Let R be the ring of integers in a number field F, Λ an R-order in a semi-simple F-algebra Σ satisfying the hypothesis of* 7.1.8, *$\underline{p}$ any prime ideal of R and ℓ a rational prime such that $\ell \neq \mathrm{char}(R/\underline{p})$. Then,*

(i) *$K_n(\hat{\Lambda}_{\underline{p}})_\ell$ is a finite group.*
(ii) *The map $\varphi : K_n(\hat{\Lambda}_{\underline{p}})_\ell \to K_n^{\mathrm{pr}}(\hat{\Lambda}_{\underline{p}}, \widehat{\mathbb{Z}}_\ell)_\ell$ is an isomorphism.*

7.4. *Higher K-theory of modules over EI categories*

7.4.1. An EI category $\mathcal{C}$ is a small category in which every endomorphism is an isomorphism.

$\mathcal{C}$ is said to be finite if the set $Is(\mathcal{C})$ of isomorphism classes of $\mathcal{C}$-objects is finite and for any two $\mathcal{C}$-objects X, Y the set $\mathrm{mor}_{\mathcal{C}}(X, Y)$ of $\mathcal{C}$-morphisms from X to Y is also finite.

Let R be a commutative ring with identity. An $R\mathcal{C}$ module is a contravariant functor from $\mathcal{C}$ to the category $\underline{R}$-mod of R-modules. There is a notion of finitely generated projective $R\mathcal{C}$-modules as well as a notion of finitely generated $R\mathcal{C}$-modules (see 7.4.3 below). So, for all $n > 0$, let $K_n(R\mathcal{C})$ be the (Quillen) K_n of the category $\underline{P}(R\mathcal{C})$ and $G_n(R\mathcal{C})$ the K_n of the category $\underline{M}(R\mathcal{C})$ when R is Noetherian.

The significance of the study of K-theory of $R\mathcal{C}$ modules lies mainly in the fact that several geometric invariants take values in the K-groups associated with $R\mathcal{C}$ where $\mathcal{C}$ is an appropriately defined EI-category and R could be $\mathbb{C}$, $\mathbb{Q}$, $\mathbb{R}$, etc. For example, if G is a finite group, $\mathcal{C} = \mathrm{orb}(G)$ (a finite EI-category), X a G-CW complex with round structure (see [96]), then the equivariant Reidemeister torsion takes values in $\mathrm{Wh}(\mathbb{Q}\,\mathrm{orb}(G))$ where $\mathrm{Wh}(\mathbb{Q}\,\mathrm{orb}(G))$ is the quotient of $K_1(\mathbb{Q}\,\mathrm{orb}(G))$ by the subgroups of "trivial units" – see [96]. For more invariants in the lower K-theory of modules over suitable EI-categories, see [96].

The study of modules over EI-categories is a natural generalization of study of modules over group rings.

7.4.2. EXAMPLES.

(i) Let G be a finite group, $\mathrm{ob}(\mathcal{C}) = \{G/H \mid H \leqslant G\}$ and let the morphisms be G-maps. Then $\mathcal{C}$ is a finite EI-category called the orbit category of G and denoted by $\mathrm{orb}(G)$. Note that here, $\mathcal{C}(G/H, G/H)$ ($\simeq \mathrm{Aut}(G/H)) \simeq N_G(H)/H$ where $N_G(H)$ is the normalizer of H in G (see [75]). We shall denote $N_G(H)/H$ by $W_G(H)$.
(ii) Let G be a Lie group, $\mathrm{ob}(\mathcal{C}) = \{G/H \mid H$ a compact subgroup of $G\}$. Again the morphisms are G-maps and $\mathcal{C}$ is also called the orbit category of G and denoted by $\mathrm{orb}(G)$.

(iii) Let G be a Lie group, $\mathrm{ob}(\mathcal{C}) = \{G/H \mid H \text{ compact subgroup of } G\}$. For G/H, $G/H' \in \mathrm{ob}(\mathcal{C})$ let $\mathcal{C}(G/H, G/H')$ be the set of homotopy classes of G-maps. Then $\mathcal{C}$ is an EI-category called the discrete orbit category of G and denoted by $\mathrm{orb}\,/(G)$.

(iv) Let G be a discrete group, $\mathcal{F}$ the family of all finite subgroups of G. Consider the category $\mathrm{Or}_{\mathcal{F}}(G)$ such that $\mathrm{ob}(\mathrm{Or}_{\mathcal{F}}(G)) = \{G/H \mid H \in \mathcal{F}\}$. $\mathrm{Or}_{\mathcal{F}}(G)$ is an EI category.

7.4.3. Let R be a commutative ring with identity, $\mathcal{C}$ an EI-category. A $\mathcal{C}$-module is a contravariant functor $\mathcal{C} \to \underline{R\text{-mod}}$.

An $\mathrm{ob}\,\mathcal{C}$-set is a functor N from $\mathcal{C}$ to the category of sets. Alternatively, an $\mathrm{ob}\,\mathcal{C}$-set could be visualized as a pair (N, β) where N is a set and $\beta : N \to \mathrm{ob}\,\mathcal{C}$ is a set map. Then $N\{\beta^{-1}(X) \mid X \in \mathrm{ob}\,\mathcal{C}\}$.

An $\mathrm{ob}\,\mathcal{C}$-set (N, β) is said to be finite if N is a finite set. If S is an (N, β)-subset of an $R\mathcal{C}$-module M, define span S as the smallest $R\mathcal{C}$-submodule of M containing S. Say that M is finitely generated if $M = \mathrm{span}\, S$ for some finite $\mathrm{ob}\,\mathcal{C}$-subset S of M.

If R is a Noetherian ring and $\mathcal{C}$ an EI-category, let $M(R\mathcal{C})$ be the category of finitely generated $R\mathcal{C}$-modules. Then $M(R\mathcal{C})$ is an exact category in the sense of Quillen, see [114].

An $R\mathcal{C}$-module P is said to be projective if any exact sequence of $R\mathcal{C}$-modules $0 \to M' \to M \to P \to 0$ splits or equivalently if $\mathrm{Hom}_{R\mathcal{C}}(P, -)$ is exact.

Let $\underline{P}(R\mathcal{C})$ be the category of finitely generated projective $(R\mathcal{C})$-modules. The $\underline{P}(R\mathcal{C})$ is also exact. We write $K_n(R\mathcal{C})$ for $K_n(P(R\mathcal{C}))$.

Now, let R be a commutative ring with identity, $\mathcal{C}$ an EI-category, $\underline{P}_R(R\mathcal{C})$ the category of finitely generated $R\mathcal{C}$-modules M such that for each $X \in \mathrm{ob}\,\mathcal{C}$ $M(X)$ is projective as R-module. Then $\underline{P}_R(R\mathcal{C})$ is an exact category and we write $G_n(R, \mathcal{C})$ for $K_n(\underline{P}_R(R\mathcal{C}))$. Note that if R is regular, then $G_n(R, \mathcal{C}) \simeq G_n(R\mathcal{C})$.

Finally, if R is a Dedekind domain with quotient field F, the inclusion functor $\mathcal{P}(R\mathcal{C}) \mapsto \mathcal{M}(R\mathcal{C})$ ($\mathcal{C}$ an EI-category) induces the Cartan maps $K_n(R\mathcal{C}) \to G_n(R\mathcal{C})$. Define $SK_n(R\mathcal{C}) := \mathrm{Ker}(K_n(R\mathcal{C}) \to K_n(F\mathcal{C}))$, $SG_n(R\mathcal{C}) = \mathrm{Ker}(G_n(R\mathcal{C}) \to G_n(F\mathcal{C}))$.

First we have the following splitting result:

7.4.4. THEOREM.

(i) *Let R be a commutative ring with identity; $\mathcal{C}$ an EI-category. Then*

$$K_n(R\mathcal{C}) \simeq \bigoplus_{X \in I_s(\mathcal{C})} K_n\big(R\big(\mathrm{Aut}(X)\big)\big).$$

(ii) *If R is commutative Noetherian ring and $\mathcal{C}$ a finite EI-category then*

$$G_n(R\mathcal{C}) \simeq \bigoplus_{X \in I_s(\mathcal{C})} G_n\big(R\big(\mathrm{Aut}(X)\big)\big).$$

We now record the following.

7.4.5. THEOREM. *Let R be the ring of integers in a number field F, $\mathcal{C}$ any finite EI-category. Then for all $n \geqslant 1$, we have*

(i) *$K_n(R\mathcal{C})$ is finitely generated Abelian group.*
(ii) *$G_n(R\mathcal{C})$ is a finitely generated Abelian group.*
(iii) *$SK_n(R\mathcal{C})$, $SK_n(\widehat{R}_{\underline{p}}\mathcal{C})$ are finite groups.*
(iv) *$SG_n(R\mathcal{C})$, $SG_n(\widehat{R}_{\underline{p}}\mathcal{C})$ are finite groups.*

Finally, we present the following result on the Cartan map.

7.4.6. THEOREM. *Let k be a field of characteristic p, $\mathcal{C}$ a finite EI-category. Then for all $n \geqslant 0$, the Cartan homomorphism $K_n(k\mathcal{C}) \to G_n(k\mathcal{C})$ induce an isomorphism*

$$\mathbb{Z}\left(\frac{1}{p}\right) \otimes K_n(k\mathcal{C}) \simeq \mathbb{Z}\left(\frac{1}{p}\right) \otimes G_n(k\mathcal{C}).$$

8. Equivariant higher algebraic K-theory together with relative generalizations

In order to economize on time and space, we restrict our discussion of equivariant higher K-theory in this section to finite group actions, with the remark that there are analogous theories for profinite group actions (see [70,80]) and compact Lie group actions (see [77,80]).

8.1. *Equivariant higher algebraic K-theory*

8.1.1. Let G be a finite group, S a G-set and $\underline{S}$ the category associated with S (or translation category of S), see [25,72].

If $\mathcal{D}$ is an exact category in the sense of Quillen, [114], then the category of $[\underline{S}, \mathcal{D}]$ of covariant functors from S to $\mathcal{D}$ is also exact (see [25]).

8.1.2. DEFINITION. Let $K_n^G(S, \mathcal{D})$ be the n-th algebraic K-group associated to the category $[S, \mathcal{D}]$ with respect to fibre-wise exact sequences.

We now have the following:

8.1.3. THEOREM. *$K_n^C(-\mathcal{D}) : GSet \to Ab$ is a Mackey functor.*

PROOF. See [24] and [25]. □

We now want to turn $K_0^G(-\mathcal{D})$ into a Green functor. We first recall the definition of a pairing of exact categories (see [110]).

8.1.4. DEFINITION. Let $\mathcal{D}_1$, $\mathcal{D}_2$, $\mathcal{D}_3$ be exact categories. An exact pairing $\langle\,,\rangle$: $\mathcal{D}_1 \times \mathcal{D}_2 \to \mathcal{D}_3$ given by $(X_1, X_2) \to \langle X_1 \circ X_2\rangle$ is a covariant functor such that

$$\begin{aligned}&\mathrm{Hom}\big((X_1, X_2), (X'_1, X'_2)\big)\\&\quad = \mathrm{Hom}(X_1, X'_1) \times \mathrm{Hom}(X_2, X'_2) \to \mathrm{Hom}\big(\langle X_1 \circ X_2\rangle, \langle X'_1 \circ X'_2\rangle\big)\end{aligned}$$

is biadditive and biexact (see 2.6.7 of [80]).

8.1.5. THEOREM. *Let $\mathcal{D}_1, \mathcal{D}_2, \mathcal{D}_3$ be exact categories and $\mathcal{D}_1 \times \mathcal{D}_2 \to D_3$ an exact pairing of exact categories. Then the pairing induces fibre-wise a pairing $[\underline{S}, \mathcal{D}_1] \times [\underline{S}, \mathcal{D}_2] \to [S, \mathcal{D}_3]$ and hence a pairing $K_0^G(S, \mathcal{D}_1) \times K_n^G(S, D_2) \to K_n^G(S, \mathcal{D}_3)$.*

Suppose $\mathcal{D}$ is an exact category such that the pairing $\mathcal{D} \times \mathcal{D} \to \mathcal{D}$ is naturally associative and commutative and there exists $E \in \mathcal{D}$ such that $\langle E \circ M\rangle = \langle M \circ E\rangle = M$. Then $K_0^G(-, \mathcal{D})$ is a Green functor and $K_n^G(-, D)$ is a unitary $K_0^G(-, \mathcal{D})$ module.

PROOF. See [24] or [25]. □

8.2. *Relative equivariant higher algebraic K-theory*

In this section, we discuss the relative version of the theory in 8.1.

8.2.1. DEFINITION. Let S, T be G-sets. Then the projection map $S \times T \to S$ gives rise to a functor $\underline{S \times T} \to \underline{S}$. Suppose that $\mathcal{D}$ is an exact category in the sense of Quillen, [78]. Then, a sequence $\varsigma_1 \to \varsigma_2 \to \varsigma_3$ of functors in $[\underline{S}, \mathcal{D}]$ is said to be T-exact if the sequence $\zeta'_1 \to \zeta'_2 \to \zeta'_3$ of restricted functors $\underline{S \times T} \to \underline{S} \to \underline{\mathcal{D}}$ is split exact.

If $\psi : S_1 \to S_2$ is a G-map, and $\varsigma_1 \to \varsigma_2 \to \varsigma_3$ is a T-exact sequence in $[\underline{S}_2, Q]$, then $\zeta'_1 \to \zeta'_2 \to \zeta'_3$ is a T-exact sequence in $[\underline{S}_1, \mathcal{D}]$ where $\varsigma'_i : S_1 \xrightarrow{\psi} S_2 \xrightarrow{\varsigma_i} \mathcal{D}$.

Let $K_n^G(S, \mathcal{D}, T)$ be the n-th algebraic K-group associated to the exact category $[S, \mathcal{D}]$ with respect to T-exact sequences.

8.2.2. DEFINITION. Let S, T be G-sets. A functor $\varsigma \in [\underline{S}, \mathcal{D}]$ is said to be T-projective if any T-exact sequence $\varsigma_1 \to \varsigma_2 \to \varsigma$ is split exact. Let $[\underline{S}, \mathcal{D}]_T$ be the additive category of T-projective functors in $[\underline{S}, \mathcal{D}]$ considered as an exact category with respect to split exact sequences. Note that the restriction functor associated to $S_1 \xrightarrow{\psi} S_2$ carries T-projective functors $\varsigma \in [S_2, \mathcal{D}]$ into T-projective functors $\varsigma \circ \psi \in [S_1, \mathcal{D}]$. Define $P_n^G(S, \mathcal{D}, T)$ as the n-th algebraic K-group associated to the exact category $[S, \mathcal{D}]_T$, with respect to split exact sequences.

8.2.3. THEOREM [35]. *$K_n^G(-, \mathcal{D}, T)$ and $P_n^G(-, \mathcal{D}, T)$ are Mackey functors from G-Set to Ab for all $n \geqslant 0$. If the pairing $\mathcal{D} \times \mathcal{D} \to \mathcal{D}$ is naturally associative and commutative and $\mathcal{D}$ contains a natural unit, then $K_0^G(-, D, T) : GSet \to Ab$ is a Green functor and $K_n^G(-, \mathcal{D}, T)$ and $P_n^G(-, \mathcal{D}, T)$ are $K_0^G(-, \mathcal{D}, T)$ modules.*

PROOF. See [24] or [25]. □

8.3. *Interpretation in terms of group-rings*

In this section, we discuss how to interpret the theories in the two previous sections in terms of group-rings.

8.3.1. Recall that any G-set S can be written as a finite disjoint union of transitive G-sets each of which is isomorphic to a quotient set G/H for some subgroup H of G. Since Mackey functors, by definition, take finite disjoint unions into finite direct sums, it will be enough to consider exact categories $[\underline{G/H}, \mathcal{D}]$ where $\mathcal{D}$ is an exact category in the sense of Quillen, [78].

For any ring A, let $\underline{M}(A)$ be the category of finitely generated A-modules and $\underline{P}(A)$ the category of finitely generated projective A modules.

8.3.2. THEOREM [72]. *Let G be a finite group, H a subgroup of G, A a commutative ring with identity, then there exists an equivalence of exact categories $[\underline{G/H}, \underline{M}(A)] \to \underline{M}(AH)$. Under this equivalence, $[\underline{G/H}, P(A)]$ is identified with the category of finitely generated A-projective left AH-modules.*

We also observe that a sequence of functors $\varsigma_1 \to \varsigma_2 \to \varsigma_3$ in $[G/H, \underline{P(A)}]$ or $[G/\underline{H}, \underline{M(A)}]$ is exact if the corresponding sequence $\varsigma_1(H) \to \varsigma_2(H) \to \varsigma_3(H)$ of AH-modules is exact.

8.3.3. REMARKS.

(i) It follows that for every $n \geqslant 0$, $K_n^G(\underline{G/H}, \underline{P}(A))$ can be identified with the n-th algebraic K-group of the category of finitely generated A-projective AH-modules while $K_n^G(\underline{G/H}, \underline{M}(A)) = G_n(AH)$ if A is Noetherian. It is well known that $K_n^G(\underline{G/H}, \underline{P}(A)) = K_n^G(\underline{G/H}, \underline{M}(A))$ is an isomorphism when A is regular.

(ii) Let $\phi : G/H_1 \to G/H_2$ be a G-map for $H_1 \leqslant H_2 \leqslant G$. We may restrict ourselves to the case $H_2 = G$ and so we have $\phi_* : [\underline{G/G}, \underline{M}(A)] \to [\underline{G/H}, \underline{M(A)}]$ corresponding to the restriction functor $\underline{M}(AG) \to \underline{M}(AH)$, while $\phi^* : [\underline{G/H}, \underline{M}(A)] \to [\underline{G/G}, \underline{M}(A)]$ corresponds to the induction functor $\underline{M}(AH) \to \underline{M}(AG)$ given by $N \to AG \otimes_{AH} N$. Similar situations hold for functor categories involving $P(A)$. So, we have corresponding restriction and induction homomorphisms for the respective K-groups.

(iii) If $\mathcal{D} = \underline{P}(A)$ and A is commutative, then the tensor product defines a naturally associative and commutative pairing $\underline{P}(A) \times \underline{P}(A) \to \underline{P}(A)$ with a natural unit and so $K_n^G(-, \underline{P}(A))$ are $K_0^G(-, \underline{P}(A))$-modules.

8.3.4. We now interpret the relative situation. So let T be a G-set. Note that a sequence $\zeta_1 \to \zeta_2 \to \zeta_3$ of functors in $[\underline{G/H}, \underline{P}(A)]$ or $[\underline{G/H}, \underline{M}(A)]$ is said to be T-exact if $\varsigma_i(H) \to \varsigma_2(H) \to \varsigma_3(H)$ is AH'-split exact for all $H' \leqslant H$ such that $T^{H'} \neq \emptyset$ where $T^{H'} = \{t \in T \mid gt = t \ \forall g \in H'\}$. In particular, the sequence is G/H-exact (resp. G/G exact) iff the corresponding sequence of AH-modules (resp. AG-modules) is split exact. If ε is the trivial subgroup of G, it is G/ε-exact if it is split exact as a sequence of A-modules.

So, $K_n^G(G/H, \underline{P}(A), T)$ (resp. $K_n^G(G/H, \underline{M}(A), T)$) is the n-th algebraic K-group of the category of finitely generated A-projective AH-modules (resp. category of finitely generated AH-modules) with respect to exact sequences which split when restricted to the various subgroups H' of H such that $T^{H'} \neq \emptyset$.

Moreover, observe that $P_n^G(G/H, \underline{P}(A), T)$ (resp. $P_n^G(G/H, \underline{M}(A), T)$) is an algebraic K-group of the category of finitely generated A-projective AH-modules (resp. finitely generated AH-modules) which are relatively H'-projective for subgroups H' of H such that $T^{H'} \neq \emptyset$ with respect to split exact sequences. In particular, $P_n^G(G/H, P(A), G/\varepsilon) = K_n(AH)$. If A is commutative, the $K_0^G(-, P(A), T)$ is a Green functor and $K_n^G(-, P(A), T)$ and $P_n^G(-, P(A), T)$ are $K_0^G(-, P(A), T)$-modules.

Now, let us interpret the maps associated to G-maps $S_1 \to S_2$. We may specialize to maps $\varphi: G/H_1 \to G/H_2$ for $H_1 \leqslant H_2 \leqslant G$ and for convenience we may restrict ourselves to the case $H_2 = G$, in which case we write $H_1 = H$. In this case $\varphi: [\underline{G/G}, \underline{M}(A)] \to [\underline{G/H}, \underline{M}(A)]$ corresponds to the restriction of AG-modules to AH-modules and $\varphi^*: [\underline{G/H}, \underline{M}(A)] \to [\underline{G/G}, \underline{M}(A)]$ corresponds to the induction of AH-modules to AG-modules (see [21]).

We hope that this wealth of equivariant higher algebraic K-groups will satisfy a lot of future needs, and moreover, that the way they have been produced systematically, will help to keep them in some order, and to produce new variants of them, whenever desired.

Since any G-set S can be written as a disjoint union of transitive G-sets, isomorphic to some coset-set G/H, and since all the above K-functors satisfy the additivity condition, the above identifications extend to K-groups, defined on an arbitrary G-set S.

8.4. *Some applications*

We are now in a position to draw various conclusions just by quoting well established induction theorems, concerning $K_0^G(-, \underline{P}(A))$ and $K_0^G(-, \underline{P}(A); T)$ and, more generally $R \otimes_Z K_0^G(-, \underline{P}(A))$ and $R \otimes_Z K_0^G(-, \underline{P}(A); T)$ for R a subring of $\mathbb{Q}$ or just any commutative ring (see [23,20–22]). Since any exact sequence in $\underline{P}(A)$ is split exact, we have a canonical identification $K_0^G(-, \underline{P}(A)) = K_0^G(-, \underline{P}(A); G/\varepsilon)$ ($\varepsilon \leqslant G$ the trivial subgroup) and, thus, may direct our attention to the relative case, only.

So let T be a G-set. For p a prime and q a prime or 0, let $\mathcal{D}(p, T, q)$ denote the set of subgroups $H \leqslant G$ such that the smallest normal subgroup H_1 of H with a q-factor group has a normal Sylow-subgroup H_2 with $T^{H_2} \neq \emptyset$ and a cyclic factor group H_1/H_2. Let $\mathcal{H}_q$ denote the set of subgroups $H \leqslant G$ which are q-hyperelementary, i.e. have a cyclic normal subgroup with a q-factor group (or are cyclic for $q = 0$).

For A and R commutative rings, let $\mathcal{D}(A, T, R)$ denote the union of all $\mathcal{D}(p, T, q)$ with $pA \neq A$ and $qR \neq R$ and let $\mathcal{H}_R$ denote the union of all $\mathcal{H}_q$ with $qR \neq R$.

Then it has been proved (see [20,21,23]), that $R \otimes_A K_0^G(-, P(A); T)$ is S-projective for some G-set S if $S^H \neq \emptyset$ for all $H \in \mathcal{D}(A, T, R) \cup H_R$. Moreover (see [21]), if A is a field of characteristic $p \neq 0$, then $K_0^G(-, P(A); T)$ is S-projective, already if $S^H \neq \emptyset$ for all $H \in \mathcal{D}(A, T, R)$.

8.4.1. Among the may possible applications of these results, we discuss just one special case. Let $A = k$ be a field a characteristic $p \neq 0$, let $R = \mathbb{Z}(\frac{1}{p})$ and let $S = \bigcup_{H \in \mathcal{D}(k,T,R)} G/H$. Then $R \otimes K_0^G(-, P(k), T)$ and thus $R \otimes K_i^G(-, \underline{P}(k), T)$ and $\otimes P_i(-, P(k); T)$ are S-projective. Moreover the Cartan map $P_i^G(X, P(k); T) \to K_i(X_i P(k); T)$ is an isomorphism for any G-set X, for which the Sylow-p-subgroups H of the stabilizers of the elements in X having a non-empty fixed point set T^H in T, since in this case T-exact sequences over X are split exact (see [23]) and thus all functors $\varsigma : \underline{X} \to \underline{P}(k)$ are T-projective, i.e. $[X, \underline{P}(k)]_T \hookrightarrow [X, \underline{P}(k)]$ is an isomorphism if $[X, \underline{P}(k)]$ is taken to be exact with respect to T-exact and thus split exact sequences.

This implies in particular that for all G-sets X the Cartan map

$$P_i^G\big(X \times S, P(k); T\big) \to K_i^G\big(X \times S, P(k); T\big)$$

is an isomorphism, since any stabilizer group of an element in $X \times S$ is a subgroup of a stabilizer group of an element in S and thus, by the very definition of S and $D(k, T, \mathbb{Z}(\frac{1}{p}))$, has a Sylow-$p$-subgroup H with $T^H \neq \emptyset$.

This finally implies that $P_i^G(-, \underline{P}(k); T)_S \to K_i^G(-, P(k), T)_S$ is an isomorphism, so by the general theory of Mackey functors,

$$\mathbb{Z}\left(\frac{1}{p}\right) \otimes P_i^G(-, P_{ki} T) \to \mathbb{Z}\left(\frac{1}{p}\right) \otimes K_i^G(-, P_{ki} T)$$

is an isomorphism. In the special case $T = G/\varepsilon$ this is just the K-theory of finitely generated projective kG-modules and $K_i^G(-, \underline{P}(k); G/\varepsilon)$ the K-theory of finitely generated G-modules with respect to exact sequences.

Thus, we have proved:

8.4.2. THEOREM. *Let k be a field of characteristic p, G a finite group. Then for all $\geqslant 0$ the Cartan map $K_n(kG) \to G_n(kG)$ induces isomorphisms*

$$\mathbb{Z}\left(\frac{1}{p}\right) \otimes K_n(kG) \simeq \mathbb{Z}\left(\frac{1}{P}\right) \otimes G_n(kG).$$

Finally, with the identification of Mackey functors: $\underline{GSet} \to \underline{Ab}$ *with Green's G-functors* $\underline{\delta}G \to \underline{Ab}$ *as in* [72], *and the above interpretations of our equivariant theory in terms of group rings, we now have from the foregoing, the following result* 8.4.3 *which says that higher algebraic K-groups are hyper-elementary computable.*

For the proof of this result, see [69].

8.4.3. THEOREM [69]. *Let R be a Dedekind domain, G a finite group, M any of the functors*

$$K_n(R-), G_n(R-), SG_n(R-), SK_n(R-); \delta G \to \underline{\mathbb{Z}\text{-mod}}.$$

For any commutative ring A with identity, define $(A \otimes M)(H) = A \otimes M(H)$, $H \in \delta G$. *Let P be a set of rational primes,* $\mathbb{Z}_P = \mathbb{Z}\{\frac{1}{q} \mid q \notin P\}$, $C(G)$ *the collection of all cyclic subgroups of* G, $h_p C(G) = \mathcal{A}$, *the collection of all hyper-elementary subgroups of* G, *i.e.* $\mathcal{A} = \{H \leqslant G \mid \exists H' \in C(G),\ H/H'$ *a p-group for some* $p \in P\}$.

Then $\mathbb{Z}_P \otimes M(G) = \varprojlim_H \mathbb{Z}_P \otimes M(H)$ *where* $\varprojlim \mathbb{Z}_P \otimes M(H)$ *is the subgroup of all* $(x_H) \in \prod_{H \in \mathcal{A}} \mathbb{Z}_P \otimes M(H)$ *such that for any* $H, H' \in \mathcal{A}$ *satisfying* $gH'g^{-1} \subset H$, $\varphi : H' \to H$ *give by* $\varphi(h) = ghg^{-1}$, *then* $\mathbb{Z}_P \otimes M(\varphi)(x_H) = x_{H'}$.

Acknowledgement

This chapter was written while I was a member of the Institute for Advanced Study, Princeton during the 2003–2004 Academic year. I am grateful to the Institute for hospitality and financial support.

References

[1] J.F. Adams, *Vector fields on spheres*, Ann. of Math. **75** (1962), 603–632.

[2] G. Allday and V. Puppe, *Cohomological Methods in Transformation Groups*, Cambridge Univ. Press (1993).

[3] D.W. Anderson, *Relations between K-Theories*, Lecture Notes in Math. **341**, Springer (1973).

[4] D.W. Anderson, M. Karoubi and J. Wagoner, *Relations between higher algebraic K-theory*, Lecture Notes in Math. **341**, Springer (1973), 68–76.

[5] S. Araki and H. Toda, *Multiplicative structures on* mod-q *cohomology theories*, Osaka J. Math. (1965), 71–115.

[6] D. Arlettaz, *Algebraic K-theory of rings from a topological view point*, Publ. Mat. **44** (2000), 3–84.

[7] M.F. Atiyah, *K-Theory*, W.A. Benjamin (1967).

[8] G. Banaszak and P. Zelewski, *Continuous K-theory*, K-Theory **9** (1995), 379–393.

[9] H. Bass, *K-theory and stable algebra*, Publ. Math. IHES **22** (1964), 1–60.

[10] H. Bass, *Algebraic K-Theory*, Benjamin (1968).

[11] H. Bass, *Lentra's calculation of* $G_0(R\pi)$ *and applications to Morse–Smale diffeomorphisms*, Lecture Notes in Math. **88**, Springer (1981), 287–291.

[12] H. Bass, A.O. Kuku and C. Pedrini (eds), *Algebraic K-Theory and its Applications*, ICTP K-Theory, Proceedings, World Scientific (1999).

[13] A. Borel, *Stable real cohomology of arithmetic groups*, Ann. Sci. École Norm. Sup. (4) **7** (1974), 235–272.

[14] A. Borel and J.P. Serre, *Le théorème de Riemann–Roch*, Bull. Soc. Math. France **8** (1958), 97–136.

[15] G. Bredon, *Introduction to Compact Transformation Groups*, Academic Press (1972).

[16] W. Browder, *Algebraic K-theory with coefficients* $\mathbb{Z}/p$, Lecture Notes in Math. **657**, Springer, 40–84.

[17] R. Charney, *A note on excision in K-theory*, Lecture Notes in Math. **1046**, Springer (1984), 47–54.

[18] A. Connes, *Non Commutative Geometry*, Academic Press (1994).

[19] C.W. Curtis and I. Reiner, *Methods of Representation Theory II*, Wiley (1987).

[20] A.W.M. Dress, *Contributions to the theory of induced representations*, Algebraic K-Theory II, Lecture Notes in Math. **343**, Springer (1973), 183–240.

[21] A.W.M. Dress, *On relative Grothendieck rings*, Lecture Notes in Math. **448**, Springer (1975), 79–131.

[22] A.W.M. Dress, *Induction and structure theorems for orthogonal representations of finite groups*, Ann. Math. **102** (1975), 291–325.

[23] A.W.M. Dress, *Vertices of integral representations*, Math. Z. **114** (1979), 159–169.

[24] A.W.M. Dress and A.O. Kuku, *A convenient setting for equivariant higher algebraic K-theory*, Proc. of the Oberwolfach Conf. (1980), Lecture Notes in Math. **966**, Springer (1982), 59–68.

[25] A.W.M Dress and A.O. Kuku, *The Cartan map for equivariant higher algebraic K-groups*, Comm. Algebra **9** (1981), 727–746.
[26] W. Dwyer and E. Friedlander, *Algebraic and étale K-theory*, Trans. Amer. Math. Soc. **292** (1985), 247–280.
[27] P. Elbaz-Vincent, H. Gangl and C. Soulé, *Quelques calculs de la cohomologie de $\mathrm{GL}_N(\mathbb{Z})$ et la K-theory de $\mathbb{Z}$*, Preprint (2002).
[28] F.T. Farell and L.E Jones, *Isomorphic conjectures in algebraic K-theory*, J. Amer. Math. Soc. **6** (1993), 249–297.
[29] F.T. Farell and L.E. Jones, *The lower algebraic K-theory of virtually infinite cyclic groups*, K-Theory **9** (1995), 13–30.
[30] Z. Fiedorowicz, H. Hauschild and J.P. May, *Equivariant algebraic K-theory*, Lecture Notes in Math. **966**, Springer, 23–80.
[31] E. Friedlander, *Étale Homotopy of Simplicial Schemes*, Ann. of Math. Studies **104**, Princeton Univ. Press (1982).
[32] E. Friedlander and C. Weibel, *An Overview of Algebraic K-Theory and its Applications*, World Scientific (1999).
[33] G. Garkusha, *On the homotopy cofibre of the spectrum $K(R) \to G(R)$*, Preprint.
[34] T. Geisser and M. Levine, *The K-theory of fields of characteristic p*, Invent. Math. **139** (2000), 459–493.
[35] T. Geisser and M. Levine, *The Bloch–Kato conjecture and a theorem of Suslin–Voevodsky*, J. Reine Angew. Math. **530** (2001), 55–103.
[36] S.M. Gersten, *On the functor K_2*, J. Algebra **17** (1971), 212–237.
[37] S.M. Gersten, *Higher K-theory for regular schemes*, Bull. Amer. Math. Soc. **79** (1973), 193–195.
[38] S.M. Gersten, *Higher K-theory of rings*, Lecture Notes in Math. **341**, Springer, 43–56.
[39] D. Grayson, *Higher algebraic K-theory II*, Lecture Notes in Math. **551**, Springer (1976), 217–240.
[40] D. Grayson, *Finite generation of K-groups of a curve over a finite field (after D. Quillen)*, Lecture Notes in Math. **966**, Springer (1982), 69–90.
[41] D. Grayson, *A survey of the K-theory of fields*, Contemp. Math. **83** (1989), 31–55.
[42] J.A. Green, *On the indecomposable representations of a finite group*, Math. Z. **70** (1959), 430–445.
[43] J.P.C. Greenleess, *Rational Mackey functors for compact Lie groups*, Proc. London Math. Soc. **76** (1998).
[44] J.P.C. Greenleess and J.P. May, *Generalized Tate Cohomology*, Mem. Amer. Math. Soc. **543** (1995).
[45] J.P.C. Greenleess and J.P. May, *Equivariant stable homotopy theory*, Handbook of Algebraic Topology, I.M. James (ed.), Elsevier (1999), 277–323.
[46] J.P.C. Greenleess and J.P. May, *Some remarks on the structure of Mackey functors*, Proc. Amer. Math. Soc. **115**, 237–243.
[47] X. Guo and A.O. Kuku, *Wild kernels for higher K-theory of division and semi-simple algebras*, Beitrage zür Algebra und Geometrie, to appear.
[48] X. Guo and A.O. Kuku, *Higher class groups of Eichler and hereditary orders*, Comm. Algebra **33** (2005), 709–718.
[49] X. Guo, A.O. Kuku and H. Qin, *K_2 of division algebras*, Comm. Algebra **33** (2005), 1073–1081.
[50] B. Harris and G. Segal, *K_i-groups of rings of algebraic integers*, Ann. of Math. **101** (1975), 20–33.
[51] A. Hatcher and J. Wagoner, *Pseudo-Isotopes on Compact Manifolds*, Astérisque **6**, Soc. Math. de France (1973).
[52] L. Hesselhalt and I. Madsin, *On the K-theory of local fields*, Ann. of Math. **158** (2003), 1–113.
[53] H. Innasaridze, *Algebraic K-Theory*, Kluwer Academic Publishers (1995).
[54] U. Jarsen, *Continuous étale cohomology*, Math. Ann. **280**, 207–245.
[55] B. Kahn, *On the Lichtenbaum–Quillen conjecture in algebraic K-theory and algebraic topology*, NATO ASI Series C **407**, Kluwer Academic Publishers (1993), 147–166.
[56] B. Kahn, *K-theory of semi-local rings with finite coefficients and étale cohomology*.
[57] M. Karoubi, *K-Theory. An Introduction*, Springer (1978).
[58] M. Karoubi, A.O. Kuku and C. Pedrini (eds), *Contemporary Developments in Algebraic K-Theory*, ICTP Lecture Notes Series **15** (2003), 536 pp.
[59] M. Karoubi and O. Villamayor, *K-theorie algebrique et K-theorie topologique*, Math. Scand. **28** (1971), 265–307.
[60] M.E. Keating, *Values of tame symbols on division algebras*, J. London Math. Soc. 2 **14** (1976), 25–30.

[61] M.E. Keating, *G_1 of integral group rings*, J. London Math. Soc. **14** (1976), 148–152.
[62] M.E. Keating, *K-theory of triangular rings and orders*, Lecture Notes in Math., Springer (1984), 178–192.
[63] B. Keller, *Derived categories and their uses*, Handbook of Algebra, Vol. 1, Elsevier (1996), 671–701.
[64] F. Keune, *Derived functors and algebraic K-theory*, Algebraic K-Theory I, Lecture Notes in Math. **341**, Springer (1973), 166–178.
[65] M. Kolster and R.C. Laubenbacher, *Higher class groups of orders*, Math. Z. **228** (1998), 229–246.
[66] A.O. Kuku, *Some finiteness theorems in the K-theory of orders in p-adic algebras*, J. London Math. Soc. **13** (1976), 122–128.
[67] A.O. Kuku, *SK_n of orders and G_n of finite rings*, Lecture Notes in Math. **551**, Springer, 60–68.
[68] A.O. Kuku, *SG_n of orders and group rings*, Math. Z. **165** (1979), 291–295.
[69] A.O. Kuku, *Higher algebraic K-theory of group rings and orders in algebras over number fields*, Comm. Algebra **10** (1982), 805–816.
[70] A.O. Kuku, *Equivariant K-theory and the cohomology of profinite groups*, Lecture Notes in Math. **1046**, Springer (1984), 234–244.
[71] A.O. Kuku, *K-theory of group rings of finite groups over maximal orders in division algebras*, J. Algebra **91** (1984), 18–31.
[72] A.O. Kuku, *Axiomatic theory of induced representations of finite groups*, Group Representations and its Applications, A.O. Kuku (ed.), Les Cours du CIMPA **4**, Nice (1985).
[73] A.O. Kuku, *K_n, SK_n of integral group rings*, Contemp. Math. **55** (1986), 333–338.
[74] A.O. Kuku, *Some finiteness results in the higher K-theory of orders and group rings*, Topology Appl. **25** (1987), 185–191.
[75] A.O. Kuku, *Higher K-theory of modules over EI categories*, Afrika Math. **3** (1996), 15–27.
[76] A.O. Kuku, *Ranks of K_n and G_n of orders and group rings of finite groups over integers in number fields*, J. Pure Appl. Algebra **138** (1999), 39–44.
[77] A.O. Kuku, *Equivariant higher K-theory for compact Lie group actions*, Beitrage Algebra Geom. **41** (2000), 141–150.
[78] A.O. Kuku, *Profinite and continuous higher K-theory of exact categories, orders and group rings*, K-Theory **22** (2001), 367–392.
[79] A.O. Kuku, *Classical algebraic K-theory: the functors of K_0, K_1, K_2*, Handbook of Algebra, Vol. 3, Elsevier (2003), 157–196.
[80] A.O. Kuku, *K-theory and representation theory*, Contemporary Developments in Algebraic K-Theory I, ICTP Lecture Notes Series **15** (2003), 259–356.
[81] A.O. Kuku, *Higher-dimensional class groups of orders and group-rings*, Preprint.
[82] A.O. Kuku and G. Tang, *Higher K-theory of group rings of virtually infinite cyclic groups*, Math. Ann. **325** (2003), 711–725.
[83] A.O. Kuku and G. Tang, *An explicit computation of "bar" homology groups of a non-unital ring*, Beitrage Algebra Geom. **44** (2003), 375–382.
[84] M. Kurihara, *Some remarks on conjectures about cyclotomic fields and K-groups of $\mathbb{Z}$*, Compositio Math. **81** (1992), 223–236.
[85] T.Y. Lam, *Induction techniques for Grothendieck groups and Whitehead groups of finite groups*, Ann. Sci. École Norm. Sup. (Paris) **1** (1968), 91–148.
[86] T.Y. Lam, *Artin exponent for finite groups*, J. Algebra **9** (1968), 94–119.
[87] R.C. Laubenbacher and D. Webb, *On SG_n of orders*, J. Algebra **133** (1990), 125–131.
[88] R.C. Laubenbacher, *Generalised Mayer–Vietoris sequences in algebraic K-theory*, J. Pure Appl. Algebra **51** (1988), 175–192.
[89] R. Lee and R.H. Szcarba, *The group $K_3(\mathbb{Z})$ is cyclic of order* 48, Ann. of Math. **104** (1976), 31–60.
[90] H.W. Lenstra, *Grothendieck groups of Abelian group rings*, J. Pure Appl. Algebra **20** (1981), 173–193.
[91] S. Lichtenbaum, *On the values of zeta and L-functions I*, Ann. of Math. **96** (1972), 338–360.
[92] S. Lichtenbaum, *Values of zeta functions, étale cohomology and algebraic K-theory*, Lecture Notes in Math. **342**, Springer (1973), 489–501.
[93] H. Lindner, *A remark on Mackey functors*, Manuscripta Math. **18** (1975), 273–278.
[94] J.L. Loday, *Cyclic Homology*, Springer (1992).
[95] J.L. Loday, *K-theorie algébrique et representations de groups*, Ann. Sci. École Norm. Sup. 9 **4** (1976), 309–377.

[96] W. Lück, *Transformation Groups and Algebraic K-Theory*, Lecture Notes in Math. **408**, Springer (1989).
[97] S. MacLane, *Categories for the Working Mathematician*, Springer (1971).
[98] B.A. Margurn, *Reviews in K-theory*, Math. Rev., Amer. Math. Soc. (1985).
[99] J.P. May et al., *Equivariant Homotopy and Cohomology Theories*, NSF–CBMS Regional Conference Series in Math. **91**, Amer. Math. Soc. (1996).
[100] A. Merkurjev and A. Suslin, *K-cohomology of Severi–Brauer varieties*, Math USSR Izv. **21** (1983), 307–340.
[101] J.S. Milne, *Étale Cohomology*, Princeton Univ. Press (1980).
[102] J. Milnor, *On axiomatic homology theory*, Pacific J. Math. **12** (1962), 337–345.
[103] J. Milnor, *Whitehead torsion*, Bull. Amer. Math. Soc. **72** (1966), 358–426.
[104] J. Milnor, *Algebraic K-theory and quadratic forms*, Invent. Math. **9** (1970), 318–344.
[105] J. Milnor, *Introduction to Algebraic K-Theory*, Princeton Univ. Press (1971).
[106] F. Morel, *Voevodsky's proof of Milnor's conjecture*, Bull. Amer. Math. Soc. **35** (1998), 123–143.
[107] J. Neisendorfer, *Primary Homotopy Theory*, Mem. Amer. Math. Soc. **232**, Amer. Math. Soc. (1980).
[108] R. Oliver, *Whitehead Groups of Finite Groups*, Cambridge Univ. Press (1988).
[109] F. Pederson, *Generalized cohomology groups*, Amer. J. Math. **78** (1956), 259–282.
[110] O. Pushin, *Higher Chern classes and Steenrod operations in motivic cohomology*, Preprint.
[111] D. Quillen, *Cohomology of groups*, Proc. ICM Nice (1970), Vol. 2, Gauthier-Villars (1971), 47–52.
[112] D. Quillen, *On the cohomology and K-theory of the general linear groups over a finite field*, Ann. of Math. **96** (1972), 552–586.
[113] D. Quillen, *Higher K-theory of categories with exact sequences*, New Developments in Topology, London Math. Soc. Lecture Notes **11**, Cambridge Univ. Press (1974), 95–110.
[114] D. Quillen, *Higher algebraic K-theory I*, K-Theory I, Lecture Notes in Math. **341**, Springer (1973), 85–147.
[115] D. Quillen, *Finite generation of the K groups of rings of algebraic integers*, Algebraic K-Theory I, Lecture Notes in Math. **341**, Springer (1973), 195–214.
[116] D.L. Rector, *K-theory of a space with coefficients in a (discrete) ring*, Bull. Amer. Math. Soc. **77** (1971), 571–575.
[117] J. Rognes, *$K_4(\mathbb{Z})$ is the trivial group*, Topology **39** (2000), 267–281.
[118] J. Rognes and C. Weibel, *Two primary algebraic K-theory of rings of integers in number fields*, J. Amer. Math. Soc. **13** (2000), 1–54.
[119] J. Rosenberg, *Algebraic K-Theory and its Applications*, Springer (1994).
[120] M. Schlichtung, *Negative K-theory of derived categories*, Preprint (2003).
[121] G. Segal, *Equivariant K-theory*, Publ. Math. IHES **34** (1968).
[122] G. Segal, *Representation ring of compact Lie group*, Publ. Math. IHES **34** (1968), 113–128.
[123] G. Segal, *Categories and cohomology theories*, Topology **13** (1974), 293–312.
[124] J.P. Serre, *Linear Representation of Finite Groups*, Springer (1977).
[125] J.P. Serre, *Local Fields*, Springer (1979).
[126] K. Shimakawa, *Multiple categories and algebraic K-theory*, J. Pure Appl. Algebra **41** (1986), 285–300.
[127] K. Shimakawa, *Mackey structures on equivariant algebraic K-theory*, K-Theory **5** (1992), 355–371.
[128] V. Srinivasa, *Algebraic K-Theory*, Progress in Math. **90**, Birkhäuser (1991).
[129] C. Soulé, *K-théorie des anneaux d'entiers de corps de nombres et cohomologie étale*, Invent. Math. **55** (1979), 252–295.
[130] A.A. Suslin, *On the K-theory of algebraically closed fields*, Invent. Math. **73** (1983), 241–245.
[131] A.A. Suslin, *On the K-theory of local fields*, J. Pure Appl. Algebra **34** (1984), 301–318.
[132] A.A. Suslin, *Algebraic K-theory of fields*, Proceedings of the ICM, Berkeley (1986) **1**, Amer. Math. Soc. (1987), 222–244.
[133] A.A. Suslin, *Excision in integral algebraic K-theory*, Proc. Steklov Inst. Math. **208** (1995), 255–279.
[134] A.A. Suslin, *Homology of GL_n, characteristic classes and Milnor K-theory*, Lecture Notes in Math. **1046**, Springer, 357–375.
[135] A.A. Suslin and A.V. Yufryakov, *K-theory of local division algebras*, Soviet Math. Dokl. **33** (1986).
[136] R.G. Swan, *Vector bundles and projective modules*, Trans. Amer. Math. Soc. **105** (1962), 264–277.
[137] R.G. Swan, *Algebraic K-Theory*, Lecture Notes in Math. **76**, Springer (1968).

[138] R.G. Swan, *Non-Abelian homological algebras and K-theory*, Proc. Sympos. Pure Math **XVII**, Amer. Math. Soc. (1970), 88–123.
[139] R.G. Swan, *K-Theory of Finite Groups and Orders*, Lecture Notes in Math. **149**, Springer (1970).
[140] R.G. Swan, *Algebraic K-theory*, Proc. ICM Nice (1970), Vol. 1, Gauthier-Villars (1971), 191–199.
[141] R.G. Swan, *Excision in algebraic K-theory*, J. Pure Appl. Algebra **1** (1971), 221–252.
[142] R.G. Swan, *Some relations between higher K-functors*, J. Algebra **21** (1972), 113–136.
[143] R.G. Swan, *Topological examples of projective modules*, Trans. Amer. Math. Soc. **230** (1977), 201–234.
[144] R.W. Thomason and T. Trobaugh, *Higher algebraic K-theory of schemes and of derived categories*, Grothendieck Festschrift II, Progress in Math. **88**, Birkhäuser (1990), 247–435.
[145] T. tom-Dieck, *Equivariant homology and Mackey functors*, Math. Ann. (1973).
[146] T. tom-Dieck, *Transformation Groups and Representation Theory*, Lecture Notes in Math. **766**, Springer (1979).
[147] T. tom-Dieck, *Transformation Groups*, Walter de Gruyter (1987).
[148] J.L. Verdier, *Catégories derivées et cohomologie étale (SGA 41/2)*, Lecture Notes in Math. **569**, Springer (1977).
[149] M.F. Vignéras, *Arithmétique des algébras de quaternions*, Lecture Notes in Math. **800**, Springer (1980).
[150] V. Voevodsky, A. Suslin and E.M. Friedlander (eds), Ann. of Math. Studies **143**, Princeton Univ. Press (2000).
[151] I. Volodin, *Algebraic K-theory as an extraordinary homology theory on the category of associative algebras with unit*, Izv. Akad. Nauk SSSR **35** (1971), 844–873.
[152] J.B. Wagoner, *Continuous cohomology and p-adic K-theory*, Lecture Notes in Math. **551**, Springer, 241–248.
[153] J.B. Wagoner, *Buildings, stratifications and higher K-theory*, Algebraic K-Theory I, Lecture Notes in Math. **341**, Springer (1973), 148–165.
[154] F. Waldhausen, *Algebraic K-theory of generalized free products*, Ann. of Math. **108** (1978), 135–256.
[155] F. Waldhausen, *Algebraic K-Theory of Spaces*, Lecture Notes in Math. **1126**, Springer (1985).
[156] C.T.C. Wall, *Finiteness conditions for CW-complexes*, Ann. of Math. **81** (1965), 56–69.
[157] C.T.C. Wall, *Survey of non-simply-connected manifolds*, Ann. of Math. **84** (1966), 217–276.
[158] D. Webb, *Grothendieck groups of dihedral and quaternion group rings*, J. Pure Appl. Algebra **35** (1985), 197–223.
[159] D. Webb, *The Lenstra map on classifying spaces and G-theory of group rings*, Invent. Math. **84** (1986), 73–89.
[160] D. Webb, *Quillen G-theory of Abelian group rings*, J. Pure Appl. Algebra **39** (1986), 177–465.
[161] D. Webb, *Higher G-theory of nilpotent group rings*, J. Algebra (1988), 457–465.
[162] D. Webb, *G-theory of group rings for groups of square-free order*, K-Theory **1** (1897), 417–422.
[163] C. Weibel, *Mayer–Vietoris sequence and* mod-p *K-theory*, Lecture Notes in Math. **966**, Springer (1982), 390–407.
[164] C. Weibel, *Mayer–Vietoris sequences and module structures on* NK_n, Lecture Notes in Math., Springer, 466–493.
[165] C. Weibel, *Introduction to Homological Algebra*, Cambridge Univ. Press (1998).
[166] C. Weibel, *Algebraic K-theory of rings of integers in local and global fields*, K-Theory Handbook, to appear.
[167] C. Weibel, *Virtual K-book*, http://math.rutgers.edu/~weibel/kbook.html.
[168] J.H.C. Whitehead, *Simplicial spaces, nuclei and M-groups*, Proc. London Math. Soc. **45** (1939), 234–327.
[169] J.H.C. Whitehead, *Simple homotopy types*, Amer. J. Math. **72** (1950), 1–57.
[170] A. Wiles, *The Iwasawa conjecture for totally real fields*, Ann. of Math. **131** (1990), 493–540.

Section 3B
Associative Rings and Algebras

Filter Dimension

V. Bavula
Department of Pure Mathematics, University of Sheffield, Hicks Building, Sheffield S3 7RH, UK
E-mail: *v.bavula@sheffield.ac.uk*

Contents

HANDBOOK OF ALGEBRA, VOL. 4
Edited by M. Hazewinkel

1. Introduction

Throughout the chapter, K is a field, a module M over an algebra A means a *left* module denoted ${}_AM$, $\otimes = \otimes_K$.

Intuitively, the filter dimension of an algebra or a module measures how 'close' standard filtrations of the algebra or the module are. In particular, for a simple algebra it also measures the growth of how 'fast' one can prove that the algebra is simple.

The filter dimension appears naturally when one wants to generalize the Bernstein's inequality for the Weyl algebras to the class of simple finitely generated algebras.

The n-th *Weyl algebra* A_n over the field K has $2n$ generators $X_1, \dots, X_n, \partial_1, \dots, \partial_n$ that satisfy the defining relations

$$\partial_i X_j - X_j \partial_i = \delta_{ij}, \quad \text{the Kronecker delta}, \quad X_i X_j - X_j X_i = \partial_i \partial_j - \partial_j \partial_i = 0,$$

for all $i, j = 1, \dots, n$. When $\operatorname{char} K = 0$ the Weyl algebra A_n is a simple Noetherian finitely generated algebra canonically isomorphic to the ring of differential operators $K[X_1, \dots, X_n, \frac{\partial}{\partial X_1}, \dots, \frac{\partial}{\partial X_n}]$ with polynomial coefficients ($X_i \leftrightarrow X_i, \partial_i \leftrightarrow \frac{\partial_i}{\partial X_i}$, $i = 1, \dots, n$).

Let K.dim and GK be the *(left) Krull* (in the sense of Rentschler and Gabriel, [22]) and the *Gelfand–Kirillov* dimension respectively.

THEOREM 1.1 (The Bernstein's inequality, [8]). *Let A_n be the n-th Weyl algebra over a field of characteristic zero. Then* $\mathrm{GK}(M) \geqslant n$ *for all non-zero finitely generated A_n-modules M.*

Let A be a simple finitely generated infinite-dimensional K-algebra. Then $\dim_K(M) = \infty$ for all non-zero A-modules M (the algebra A is simple, so the K-linear map $A \to \mathrm{Hom}_K(M, M)$, $a \mapsto (m \mapsto am)$, is injective, and so $\infty = \dim_K(A) \leqslant \dim_K(\mathrm{Hom}_K(M, M))$ hence $\dim_K(M) = \infty$). So, the Gelfand–Kirillov dimension (over K) $\mathrm{GK}(M) \geqslant 1$ for all non-zero A-modules M.

DEFINITION. $h_A := \inf\{\mathrm{GK}(M) \mid M$ is a non-zero finitely generated A-module$\}$ is called the *holonomic number* for the algebra A.

PROBLEM. For a simple finitely generated algebra find its holonomic number.

To find an approximation of the holonomic number for simple finitely generated algebras and to generalize the Bernstein inequality for these algebras was a main motivation for introducing the filter dimension, [4]. In this chapter d stands for the filter dimension fd or the left filter dimension lfd. The following two inequalities are central for the proofs of almost all results in this chapter.

THE FIRST FILTER INEQUALITY [4]. *Let A be a simple finitely generated algebra. Then*

$$\mathrm{GK}(M) \geqslant \frac{\mathrm{GK}(A)}{\mathrm{d}(A) + \max\{\mathrm{d}(A), 1\}}$$

for all non-zero finitely generated A-modules M.

THE SECOND FILTER INEQUALITY [5]. *Under certain mild conditions (Theorem 4.2) the (left) Krull dimension of the algebra A satisfies the following inequality*

$$\mathrm{K.dim}(A) \leqslant \mathrm{GK}(A)\left(1 - \frac{1}{\mathrm{d}(A) + \max\{\mathrm{d}(A), 1\}}\right).$$

The chapter is organized as follows. Both filter dimensions are introduced in Section 2. In Sections 3 and 4 the first and the second filter inequalities are proved respectively. In Section 4 we use both filter inequalities for giving short proofs of some classical results about the rings $\mathcal{D}(X)$ of differential operators on smooth irreducible affine algebraic varieties. The (left) filter dimension of $\mathcal{D}(X)$ is 1 (Section 5). A concept of multiplicity for the filter dimension and a concept of holonomic module for (simple) finitely generated algebras appear in Section 6. Every holonomic module has finite length (Theorem 6.8). In Section 7 an upper bound is given (i) for the Gelfand–Kirillov dimension of commutative subalgebras of simple finitely generated infinite-dimensional algebras (Theorem 7.2), and (ii) for the transcendence degree of subfields of quotient rings of (certain) simple finitely generated infinite-dimensional algebras (Theorems 7.4 and 7.5). In Section 8 a similar upper bound is obtained for the Gelfand–Kirillov dimension of *isotropic* subalgebras of strongly simple Poisson algebras (Theorem 8.1).

2. Filter dimension of algebras and modules

In this section, the filter dimension of algebras and modules will be defined.

2.1. *The Gelfand–Kirillov dimension*

Let $\mathcal{F}$ be the set of all functions from the set of natural numbers $\mathbb{N} = \{0, 1, \dots\}$ to itself. For each function $f \in \mathcal{F}$, the non-negative real number or ∞ defined as

$$\gamma(f) := \inf\{r \in \mathbb{R} \mid f(i) \leqslant i^r \text{ for } i \gg 0\}$$

is called the *degree* of f. The function f has *polynomial growth* if $\gamma(f) < \infty$. Let $f, g, p \in \mathcal{F}$, and $p(i) = p^*(i)$ for $i \gg 0$ where $p^*(t) \in \mathbb{Q}[t]$ (the polynomial algebra with coefficients from the field of rational numbers). Then

$$\gamma(f+g) \leqslant \max\{\gamma(f), \gamma(g)\}, \qquad \gamma(fg) \leqslant \gamma(f) + \gamma(g),$$
$$\gamma(p) = \deg_t(p^*(t)), \qquad \gamma(pg) = \gamma(p) + \gamma(g).$$

Let $A = K\langle a_1, \dots, a_s\rangle$ be a finitely generated K-algebra. The finite-dimensional filtration $F = \{A_i\}$ associated with algebra generators $a_1, \dots, a_s$:

$$A_0 := K \subseteq A_1 := K + \sum_{i=1}^{s} Ka_i \subseteq \cdots \subseteq A_i := A_1^i \subseteq \cdots$$

is called the *standard filtration* for the algebra A. Let $M = AM_0$ be a finitely generated A-module where M_0 is a finite-dimensional generating subspace. The finite-dimensional filtration $\{M_i := A_i M_0\}$ is called the *standard filtration* for the A-module M.

DEFINITION. $\mathrm{GK}(A) := \gamma(i \mapsto \dim_K(A_i))$ and $\mathrm{GK}(M) := \gamma(i \mapsto \dim_K(M_i))$ are called the *Gelfand–Kirillov* dimensions of the algebra A and the A-module M respectively.

It is easy to prove that the Gelfand–Kirillov dimension of the algebra (resp. the module) does not depend on the choice of the standard filtration of the algebra (resp. and the choice of the generating subspace of the module).

2.2. *The return functions and the (left) filter dimension*

DEFINITION [4]. The function $\nu_{F,M_0} : \mathbb{N} \to \mathbb{N} \cup \{\infty\}$,

$$\nu_{F,M_0}(i) := \min\{ j \in \mathbb{N} \cup \{\infty\}\colon\ A_j M_{i,\mathrm{gen}} \supseteq M_0 \text{ for all } M_{i,\mathrm{gen}} \}$$

is called the *return function* of the A-module M associated with the filtration $F = \{A_i\}$ of the algebra A and the generating subspace M_0 of the A-module M where $M_{i,\mathrm{gen}}$ runs through all generating subspaces for the A-module M such that $M_{i,\mathrm{gen}} \subseteq M_i$.

Suppose, in addition, that the finitely generated algebra A is a *simple* algebra. The *return function* $\nu_F \in \mathcal{F}$ and the *left return function* $\lambda_F \in \mathcal{F}$ for the algebra A with respect to the standard filtration $F := \{A_i\}$ for the algebra A are defined by the rules:

$$\nu_F(i) := \min\{ j \in \mathbb{N} \cup \{\infty\} \mid 1 \in A_j a A_j \text{ for all } 0 \neq a \in A_i \},$$
$$\lambda_F(i) := \min\{ j \in \mathbb{N} \cup \{\infty\} \mid 1 \in A a A_j \text{ for all } 0 \neq a \in A_i \},$$

where $A_j a A_j$ is the vector subspace of the algebra A spanned over the field K by the elements xay for all $x, y \in A_j$; and AaA_j is the left ideal of the algebra A generated by the set aA_j. The next result shows that under a mild restriction the maps $\nu_F(i)$ and $\lambda_F(i)$ are finite.

Recall that the centre of a simple algebra is a field.

LEMMA 2.1. *Let A be a simple finitely generated algebra such that its centre $Z(A)$ is an algebraic field extension of K. Then $\lambda_F(i) \leqslant \nu_F(i) < \infty$ for all $i \geqslant 0$.*

PROOF. The first inequality is evident.

The centre $Z = Z(A)$ of the simple algebra A is a field that contains K. Let $\{\omega_j \mid j \in J\}$ be a K-basis for the K-vector space Z. Since $\dim_K(A_i) < \infty$, one can find finitely many Z-linearly independent elements, say $a_1, \dots, a_s$, of A_i such that $A_i \subseteq Za_1 + \cdots + Za_s$. Next, one can find a finite subset, say J', of J such that $A_i \subseteq Va_1 + \cdots + Va_s$ where $V = \sum_{j \in J'} K\omega_j$. The field K' generated over K by the elements ω_j, $j \in J'$, is a finite field extension of K (i.e. $\dim_K(K') < \infty$) since Z/K is algebraic, hence $K' \subseteq A_n$ for some $n \geqslant 0$. Clearly, $A_i \subseteq K'a_1 + \cdots + K'a_s$.

The A-bimodule ${}_A A_A$ is simple with ring of endomorphisms $\mathrm{End}({}_A A_A) \simeq Z$. By the *Density* theorem, [21, 12.2], for each integer $1 \leqslant j \leqslant s$, there exist elements of the algebra A, say $x_1^j, \dots, x_m^j, y_1^j, \dots, y_m^j$, $m = m(j)$, such that for all $1 \leqslant l \leqslant s$

$$\sum_{k=1}^{m} x_k^j a_l y_k^j = \delta_{j,l}, \quad \text{the Kronecker delta.}$$

Let us fix a natural number, say $d = d_i$, such that A_d contains all the elements x_k^j, y_k^j, and the field K'. We claim that $\nu_F(i) \leqslant 2d$. Let $0 \neq a \in A_i$. Then $a = \lambda_1 a_1 + \cdots + \lambda_s a_s$ for some $\lambda_i \in K'$. There exists $\lambda_j \neq 0$. Then $\sum_{k=1}^{m} \lambda_j^{-1} x_k^j a_j y_k^j = 1$, and $\lambda_j^{-1} x_k^j, y_k^j \in A_{2d}$. This proves the claim and the lemma. □

REMARK. If the field K is *uncountable* then automatically the centre $Z(A)$ of a simple finitely generated algebra A is algebraic over K (since A has a countable K-basis and the rational function field $K(x)$ has uncountable basis over K since the elements $\frac{1}{x+\lambda}$, $\lambda \in K$, are K-linearly independent).

It is easy to see that for a finitely generated algebra A any two standard finite-dimensional filtrations $F = \{A_i\}$ and $G = \{B_i\}$ are *equivalent*, ($F \sim G$), that is, there exist natural numbers a, b, c, d such that

$$A_i \subseteq B_{ai+b} \quad \text{and} \quad B_i \subseteq A_{ci+d} \quad \text{for } i \gg 0.$$

If one of the inclusions holds, say the first, we write $F \leqslant G$.

LEMMA 2.2. *Let A be a finitely generated algebra equipped with two standard finite-dimensional filtrations $F = \{A_i\}$ and $G = \{B_i\}$.*

(1) *Let M be a finitely generated A-module. Then $\gamma(\nu_{F,M_0}) = \gamma(\nu_{G,N_0})$ for any finite-dimensional generating subspaces M_0 and N_0 of the A-module M.*
(2) *If, in addition, A is a simple algebra then $\gamma(\nu_F) = \gamma(\nu_G)$ and $\gamma(\lambda_F) = \gamma(\lambda_G)$.*

PROOF. (1) The module M has two standard finite-dimensional filtrations $\{M_i = A_i M_0\}$ and $\{N_i = B_i N_0\}$. Let $\nu = \nu_{F,M_0}$ and $\mu = \nu_{G,N_0}$.

Suppose that $F = G$. Choose a natural number s such that $M_0 \subseteq N_s$ and $N_0 \subseteq M_s$, so $N_i \subseteq M_{i+s}$ and $M_i \subseteq N_{i+s}$ for all $i \geqslant 0$. Let $N_{i,\mathrm{gen}}$ be any generating subspace for the A-module M such that $N_{i,\mathrm{gen}} \subseteq N_i$. Since $M_0 \subseteq A_{\nu(i+s)} N_{i,\mathrm{gen}}$ for all $i \geqslant 0$ and $N_0 \subseteq A_s M_0$, we have $N_0 \subseteq A_{\nu(i+s)+s} N_{i,\mathrm{gen}}$, hence, $\mu(i) \leqslant \nu(i+s)+s$ and finally $\gamma(\mu) \leqslant \gamma(\nu)$. By symmetry, the opposite inequality is true and so $\gamma(\mu) = \gamma(\nu)$.

Suppose that $M_0 = N_0$. The algebra A is a finitely generated algebra, so all standard finite-dimensional filtrations of the algebra A are equivalent. In particular, $F \sim G$ and so one can choose natural numbers a, b, c, d such that

$$A_i \subseteq B_{ai+b} \quad \text{and} \quad B_i \subseteq A_{ci+d} \quad \text{for } i \gg 0.$$

Then $N_i = B_i N_0 \subseteq A_{ci+d} M_0 = M_{ci+d}$ for all $i \geqslant 0$, hence $N_0 = M_0 \subseteq A_{\nu(ci+d)} N_{i,\mathrm{gen}} \subseteq B_{a\nu(ci+d)+b} N_{i,\mathrm{gen}}$, therefore $\mu(i) \leqslant a\nu(ci+d)+b$ for all $i \geqslant 0$, hence $\gamma(\mu) \leqslant \gamma(\nu)$. By symmetry, we get the opposite inequality which implies $\gamma(\mu)=\gamma(\nu)$. Now, $\gamma(\nu_{F,M_0})=\gamma(\nu_{F,N_0})=\gamma(\nu_{G,N_0})$.

(2) The algebra A is simple, equivalently, it is a simple (left) $A\otimes A^0$-module where A^0 is the opposite algebra to A. The opposite algebra has the standard filtration $F^0=\{A_i^0\}$, opposite to the filtration F. The tensor product of algebras $A\otimes A^0$, so-called, the enveloping algebra of A, has the standard filtration $F\otimes F^0=\{C_n\}$ which is the tensor product of the standard filtrations F and F^0, that is, $C_n=\sum\{A_i\otimes A_j^0,\ i+j\leqslant n\}$. Let $\nu_{F\otimes F^0,K}$ be the return function of the $A\otimes A^0$-module A associated with the filtration $F\otimes F^0$ and the generating subspace K. Then

$$\nu_F(i)\leqslant \nu_{F\otimes F^0,K}(i)\leqslant 2\nu_F(i) \quad \text{for all } i\geqslant 0,$$

and so

$$\gamma(\nu_F)=\gamma(\nu_{F\otimes F^0,K}), \tag{1}$$

and, by the first statement, we have $\gamma(\nu_F)=\gamma(\nu_{F\otimes F^0,K})=\gamma(\nu_{G\otimes G^0,K})=\gamma(\nu_G)$, as required. Using a similar argument as in the proof of the first statement one can prove that $\gamma(\lambda_F)=\gamma(\lambda_G)$. We leave this as an exercise. □

DEFINITION [4]. $\mathrm{fd}(M)=\gamma(\nu_{F,M_0})$ is the *filter dimension* of the A-module M, and $\mathrm{fd}(A):=\mathrm{fd}({}_{A\otimes A^0}A)$ is the *filter dimension* of the algebra A. If, in addition, the algebra A is simple, then $\mathrm{fd}(A)=\gamma(\nu_F)$, and $\mathrm{lfd}(A):=\gamma(\lambda_F)$ is called the *left filter dimension* of the algebra A.

By the previous lemma the definitions make sense (both filter dimensions do not depend on the choice of the standard filtration F for the algebra A).

By Lemma 2.1, $\mathrm{lfd}(A)\leqslant \mathrm{fd}(A)$.

QUESTION. What is the filter dimension of a polynomial algebra?

3. The first filter inequality

In this chapter, $\mathrm{d}(A)$ means either the filter dimension $\mathrm{fd}(A)$ or the left filter dimension $\mathrm{lfd}(A)$ of a simple finitely generated algebra A (i.e. $\mathrm{d}=\mathrm{fd},\mathrm{lfd}$). Both filter dimensions appear naturally when one tries to find a *lower* bound for the holonomic number (Theorem 3.1) and an *upper* bound (Theorem 4.2) for the (left and right) *Krull* dimension (in the sense of Rentschler and Gabriel, [22]) of simple finitely generated algebras.

The next theorem is a generalization of the *Bernstein's inequality* (Theorem 1.1) to the class of simple finitely generated algebras.

THEOREM 3.1 (The first filter inequality, [4,6]). *Let A be a simple finitely generated algebra. Then*

$$\mathrm{GK}(M)\geqslant \frac{\mathrm{GK}(A)}{\mathrm{d}(A)+\max\{\mathrm{d}(A),1\}}$$

for all non-zero finitely generated A-modules M where $\mathrm{d}=\mathrm{fd},\mathrm{lfd}$.

PROOF. Let $\lambda = \lambda_F$ be the left return function associated with a standard filtration F of the algebra A and let $0 \neq a \in A_i$. It suffices to prove the inequality for λ (since $\mathrm{fd}(A) \geqslant \mathrm{lfd}(A)$). It follows from the inclusion

$$AaM_{\lambda(i)} = AaA_{\lambda(i)}M_0 \supseteq 1M_0 = M_0$$

that the linear map

$$A_i \to \mathrm{Hom}(M_{\lambda(i)}, M_{\lambda(i)+i}), \qquad a \mapsto (m \mapsto am),$$

is injective, so $\dim A_i \leqslant \dim M_{\lambda(i)} \dim M_{\lambda(i)+i}$. Using the above elementary properties of the degree (see also [19, 8.1.7]), we have

$$\begin{aligned} \mathrm{GK}(A) &= \gamma(\dim A_i) \leqslant \gamma(\dim M_{\lambda(i)}) + \gamma(\dim M_{\lambda(i)+i}) \\ &\leqslant \gamma(\dim M_i)\gamma(\lambda) + \gamma(\dim M_i)\max\{\gamma(\lambda), 1\} \\ &= \mathrm{GK}(M)\big(\mathrm{lfd}A + \max\{\mathrm{lfd}A, 1\}\big) \\ &\leqslant \mathrm{GK}(M)\big(\mathrm{lfd}A + \max\{\mathrm{lfd}A, 1\}\big). \end{aligned}$$ □

The result above gives a lower bound for the holonomic number of a simple finitely generated algebra

$$h_A \geqslant \frac{\mathrm{GK}(A)}{\mathrm{d}(A) + \max\{\mathrm{d}(A), 1\}}.$$

THEOREM 3.2. *Let A be a finitely generated algebra. Then*

$$\mathrm{GK}(M) \leqslant \mathrm{GK}(A)\,\mathrm{fd}(M)$$

for any simple A-module M.

PROOF. Let $\nu = \nu_{F,Km}$ be the return function of the module M associated with a standard finite-dimensional filtration $F = \{A_i\}$ of the algebra A and a fixed non-zero element $m \in M$. Let $\pi : M \to K$ be a non-zero linear map satisfying $\pi(m) = 1$. Then, for any $i \geqslant 0$ and any $0 \neq u \in M_i$: $1 = \pi(m) \in \pi(A_{\nu(i)}u)$, and so the linear map

$$M_i \to \mathrm{Hom}(A_{\nu(i)}, K), \qquad u \mapsto \big(a \mapsto \pi(au)\big),$$

is an injective map hence $\dim M_i \leqslant \dim A_{\nu(i)}$ and finally $\mathrm{GK}(M) \leqslant \mathrm{GK}(A)\,\mathrm{fd}(M)$. □

COROLLARY 3.3. *Let A be a simple finitely generated infinite-dimensional algebra. Then*

$$\mathrm{fd}(A) \geqslant \frac{1}{2}.$$

PROOF. The algebra A is a finitely generated infinite-dimensional algebra hence $\mathrm{GK}(A) > 0$. Clearly, $\mathrm{GK}(A \otimes A^0) \leqslant \mathrm{GK}(A) + \mathrm{GK}(A^0) = 2\mathrm{GK}(A)$. Applying Theorem 3.2 to the simple $A \otimes A^0$-module $M = A$ we finish the proof:

$$\mathrm{GK}(A) = \mathrm{GK}({}_{A\otimes A^0}A) \leqslant \mathrm{GK}\big(A \otimes A^0\big)\,\mathrm{fd}({}_{A\otimes A^0}A) \leqslant 2\mathrm{GK}(A)\,\mathrm{fd}(A)$$

hence $\mathrm{fd}(A) \geqslant \frac{1}{2}$. □

QUESTION. Is $\mathrm{fd}(A) \geqslant 1$ for all simple finitely generated infinite-dimensional algebras A?

QUESTION. For which numbers $d \geqslant \frac{1}{2}$ there exists a simple finitely generated infinite-dimensional algebra A with $\mathrm{fd}(A) = d$?

COROLLARY 3.4. *Let A be a simple finitely generated infinite-dimensional algebra. Then*

$$\mathrm{fd}(M) \geqslant \frac{1}{\mathrm{fd}(A) + \max\{\mathrm{fd}(A), 1\}}$$

for all simple A-modules M.

PROOF. Applying Theorem 3.1 and Theorem 3.2, we have the result

$$\mathrm{fd}(M) \geqslant \frac{\mathrm{GK}(M)}{\mathrm{GK}(A)} \geqslant \frac{\mathrm{GK}(A)}{\mathrm{GK}(A)(\mathrm{fd}(A) + \max\{\mathrm{fd}(A), 1\})} = \frac{1}{\mathrm{fd}(A) + \max\{\mathrm{fd}(A), 1\}}.$$

□

4. Krull, Gelfand–Kirillov and filter dimensions of simple finitely generated algebras

In this section, we prove the second filter inequality (Theorem 4.2) and apply both filter inequalities for giving short proofs of some classical results about the rings of differential operators on a smooth irreducible affine algebraic varieties (Theorems 1.1, 4.4, 4.5, 4.7).

We say that an algebra A is (*left*) *finitely partitive*, [19, 8.3.17], if, given any finitely generated A-module M, there is an integer $n = n(M) > 0$ such that for every strictly descending chain of A-submodules of M:

$$M = M_0 \supset M_1 \supset \cdots \supset M_m$$

with $\mathrm{GK}(M_i/M_{i+1}) = \mathrm{GK}(M)$, one has $m \leqslant n$. McConnell and Robson write in their book [19, 8.3.17], that "*yet no examples are known which fail to have this property.*"

Recall that K.dim denotes the (left) Krull dimension in the sense of Rentschler and Gabriel, [22].

LEMMA 4.1. *Let A be a finitely partitive algebra with* $\mathrm{GK}(A) < \infty$. *Let* $a \in \mathbb{N}$, $b \geqslant 0$ *and suppose that* $\mathrm{GK}(M) \geqslant a + b$ *for all finitely generated A-modules M with* $\mathrm{K.dim}(M) = a$, *and that* $\mathrm{GK}(N) \in \mathbb{N}$ *for all finitely generated A-modules N with* $\mathrm{K.dim}(N) \geqslant a$. *Then* $\mathrm{GK}(M) \geqslant \mathrm{K.dim}(M) + b$ *for all finitely generated A-modules M with* $\mathrm{K.dim}(M) \geqslant a$. *In particular,* $\mathrm{GK}(A) \geqslant \mathrm{K.dim}(A) + b$.

REMARK. It is assumed that a module M with $\mathrm{K.dim}(M) = a$ exists.

PROOF. We use induction on $n = \mathrm{K.dim}(M)$. The base of induction, $n = a$, is true. Let $n > a$. There exists a descending chain of submodules $M = M_1 \supset M_2 \supset \cdots$ with $\mathrm{K.dim}(M_i/M_{i+1}) = n - 1$ for $i \geqslant 1$. By induction, $\mathrm{GK}(M_i/M_{i+1}) \geqslant n - 1 + b$ for $i \geqslant 1$. The algebra A is finitely partitive, so there exists i such that $\mathrm{GK}(M) > \mathrm{GK}(M_i/M_{i+1})$, so $\mathrm{GK}(M) - 1 \geqslant \mathrm{GK}(M_i/M_{i+1}) \geqslant n - 1 + b$, since $\mathrm{GK}(M) \in \mathbb{N}$, hence $\mathrm{GK}(M) \geqslant \mathrm{K.dim}(M) + b$. Since $\mathrm{K.dim}(A) \geqslant \mathrm{K.dim}(M)$ for all finitely generated A-modules M we have $\mathrm{GK}(A) \geqslant \mathrm{K.dim}(A) + b$. □

THEOREM 4.2 [5]. *Let A be a simple finitely generated finitely partitive algebra with* $\mathrm{GK}(A) < \infty$. *Suppose that the Gelfand–Kirillov dimension of every finitely generated A-module is a natural number. Then*

$$\mathrm{K.dim}(M) \leqslant \mathrm{GK}(M) - \frac{\mathrm{GK}(A)}{\mathrm{d}(A) + \max\{\mathrm{d}(A), 1\}}$$

for any non-zero finitely generated A-module M. In particular,

$$\mathrm{K.dim}(A) \leqslant \mathrm{GK}(A)\left(1 - \frac{1}{\mathrm{d}(A) + \max\{\mathrm{d}(A), 1\}}\right).$$

PROOF. Applying the lemma above to the family of finitely generated A-modules of Krull dimension 0, by Theorem 3.1, we can put $a = 0$ and

$$b = \frac{\mathrm{GK}(A)}{\mathrm{d}(A) + \max\{\mathrm{d}(A), 1\}},$$

and the result follows. □

Let K be a field of characteristic zero and B be a commutative K-algebra. The ring of (K-linear) *differential operators* $\mathcal{D}(B)$ on B is defined as $\mathcal{D}(B) = \bigcup_{i=0}^{\infty} \mathcal{D}_i(B)$ where $\mathcal{D}_0(B) = \mathrm{End}_R(B) \simeq B$ $((x \mapsto bx) \leftrightarrow b)$,

$$\mathcal{D}_i(B) = \{u \in \mathrm{End}_K(B)\colon [u, r] \in \mathcal{D}_{i-1}(B) \text{ for each } r \in B\}.$$

Note that the $\{\mathcal{D}_i(B)\}$ is, so-called, the *order filtration* for the algebra $\mathcal{D}(B)$:

$$\mathcal{D}_0(B) \subseteq \mathcal{D}_1(B) \subseteq \cdots \subseteq \mathcal{D}_i(B) \subseteq \cdots \quad \text{and} \quad \mathcal{D}_i(B)\mathcal{D}_j(B) \subseteq \mathcal{D}_{i+j}(B),$$
$$i, j \geqslant 0.$$

The subalgebra $\Delta(B)$ of $\mathrm{End}_K(B)$ generated by $B \equiv \mathrm{End}_R(B)$ and by the set $\mathrm{Der}_K(B)$ of all K-derivations of B is called the *derivation ring* of B. The derivation ring $\Delta(B)$ is a subring of $\mathcal{D}(B)$.

Let the *finitely generated* algebra B be a *regular commutative domain of Krull dimension* $n < \infty$. In geometric terms, B is the coordinate ring $\mathcal{O}(X)$ of a smooth irreducible affine algebraic variety X of dimension n. Then

- $\mathrm{Der}_K(B)$ *is a finitely generated projective* B*-module of rank* n;
- $\mathcal{D}(B) = \Delta(B)$;
- $\mathcal{D}(B)$ *is a simple (left and right) Noetherian domain with* $\mathrm{GK}\,\mathcal{D}(B) = 2n$ $(n = \mathrm{GK}(B) = \mathrm{K.dim}(B))$;
- $\mathcal{D}(B) = \Delta(B)$ *is an almost centralizing extension of* B;
- *the associated graded ring* $\mathrm{gr}\,\mathcal{D}(B) = \bigoplus \mathcal{D}_i(B)/\mathcal{D}_{i-1}(B)$ *is a commutative domain*;
- *the Gelfand–Kirillov dimension of every finitely generated* $\mathcal{D}(B)$*-module is a natural number.*

For the proofs of the statements above the reader is referred to [19, chapter 15]. So, the domain $\mathcal{D}(B)$ is a simple finitely generated infinite-dimensional Noetherian algebra, [19, chapter 15].

EXAMPLE. Let $P_n = K[X_1, \ldots, X_n]$ be a polynomial algebra. $\mathrm{Der}_K(P_n) = \bigoplus_{i=1}^n P_n \frac{\partial}{\partial X_i}$,

$$\mathcal{D}(P_n) = \Delta(P_n) = K\left[X_1, \ldots, X_n, \frac{\partial}{\partial X_1}, \ldots, \frac{\partial}{\partial X_n}\right]$$

is the ring of differential operators with polynomial coefficients, i.e. the n-th Weyl algebra A_n.

In Section 5, we prove the following result.

THEOREM 4.3 [5]. *The filter dimension and the left filter dimension of the ring of differential operators* $\mathcal{D}(B)$ *are both equal to* 1.

As an application we compute the Krull dimension of $\mathcal{D}(B)$.

THEOREM 4.4 [19, Chapter 15].

$$\mathrm{K.dim}\,\mathcal{D}(B) = \frac{\mathrm{GK}(\mathcal{D}(B))}{2} = \mathrm{K.dim}(B).$$

PROOF. The second equality is clear: $(\mathrm{GK}(\mathcal{D}(B)) = 2\mathrm{GK}(B) = 2\mathrm{K.dim}(B))$. It follows from Theorems 4.2 and 4.3 that

$$\mathrm{K.dim}\,\mathcal{D}(B) \leqslant \frac{\mathrm{GK}(\mathcal{D}(B))}{2} = \mathrm{K.dim}(B).$$

The map $I \to \mathcal{D}(B)I$ from the set of left ideals of B to the set of left ideals of $\mathcal{D}(B)$ is injective, thus $\mathrm{K.dim}(B) \leqslant \mathrm{K.dim}\,\mathcal{D}(B)$. □

This result shows that for the ring of differential operators on a smooth irreducible affine algebraic variety the inequality in Theorem 4.2 is *an equality*.

THEOREM 4.5 [19, 15.4.3]. *Let M be a non-zero finitely generated $\mathcal{D}(B)$-module. Then*

$$\mathrm{GK}(M) \geqslant \frac{\mathrm{GK}(\mathcal{D}(B))}{2} = \mathrm{K.dim}(B).$$

PROOF. By Theorems 3.1 and 4.3,

$$\mathrm{GK}(M) \geqslant \frac{\mathrm{GK}(\mathcal{D}(B))}{1+1} = \frac{2\mathrm{GK}(B)}{2} = \mathrm{GK}(B) = \mathrm{K.dim}(B). \qquad \square$$

So, for the ring of differential operators on a smooth affine algebraic variety the inequality in Theorem 3.1 is in fact *an equality*.

In general, it is difficult to find the exact value for the filter dimension but for the Weyl algebra A_n it is easy and one can find it directly.

THEOREM 4.6. *Both the filter dimension and the left filter dimension of the Weyl algebra A_n over a field of characteristic zero are equal to* 1.

PROOF. Denote by $a_1, \dots, a_{2n}$ the canonical generators of the Weyl algebra A_n and denote by $F = \{A_{n,i}\}_{i \geqslant 0}$ the standard filtration associated with the canonical generators. The associated graded algebra $\mathrm{gr}\, A_n := \bigoplus_{i \geqslant 0} A_{n,i}/A_{n,i-1}$ $(A_{n,-1} = 0)$ is a polynomial algebra in $2n$ variables, so

$$\mathrm{GK}(A_n) = \mathrm{GK}(\mathrm{gr}\, A_n) = 2n.$$

For every $i \geqslant 0$:

$$\mathrm{ad}\, a_j : A_{n,i} \to A_{n,i-1}, \; x \mapsto \mathrm{ad}\, a_j(x) := a_j x - x a_j.$$

The algebra A_n is central $(Z(A_n) = K)$, so

$$\mathrm{ad}\, a_j(x) = 0 \text{ for all } j = 1, \dots, 2n \quad \Leftrightarrow \quad x \in Z(A_n) = K = A_{n,0}.$$

These two facts imply $\nu_F(i) \leqslant i$ for $i \geqslant 0$, and so $\mathrm{d}(A_n) \leqslant 1$.

The A_n-module $P_n := K[X_1, \dots, X_n] \simeq A_n/(A_n\partial_1 + \dots + A_n\partial_n)$ has Gelfand–Kirillov dimension n. By Theorem 3.1 applied to the A_n-module P_n, we have

$$2n = \mathrm{GK}(A_n) \leqslant n\big(\mathrm{d}(A) + \max\{\mathrm{d}(A), 1\}\big),$$

hence $\mathrm{d}(A_n) \geqslant 1$, and so $\mathrm{d}(A_n) = 1$. $\square$

PROOF OF THE BERNSTEIN'S INEQUALITY (THEOREM 1.1). Since $\mathrm{GK}(A_n) = 2n$ and $d(A_n) = 1$, Theorem 3.1 gives $\mathrm{GK}(M) \geqslant \frac{2n}{2} = n$. $\square$

One also gets a short proof of the following result of Rentschler and Gabriel.

THEOREM 4.7 [22]. *If* char $K = 0$ *then the Krull dimension of the Weyl algebra* A_n *is*

$$\text{K.dim}(A_n) = n.$$

PROOF. Putting $\text{GK}(A_n) = 2n$ and $\text{d}(A_n) = 1$ into the second formula of Theorem 4.2 we have $\text{K.dim}(A_n) \leqslant \frac{2n}{2} = n$. The polynomial algebra $P_n = K[X_1, \dots, X_n]$ is the subalgebra of A_n such that A_n is a free right P_n-module. The map $I \mapsto A_n I$ from the set of left ideals of the polynomial algebra P_n to the set of left ideals of the Weyl algebra A_n is injective, thus $n = \text{K.dim}(P_n) \leqslant \text{K.dim}(A_n)$, and so $\text{K.dim}(A_n) = n$. □

5. Filter dimension of the ring of differential operators on a smooth irreducible affine algebraic variety (proof of Theorem 4.3)

Let K be a field of characteristic 0 and let the algebra B be as in the previous section, i.e. B is a finitely generated regular commutative algebra which is a domain. We keep the notations of the previous section. Recall that the derivation ring $\Delta = \Delta(B)$ coincides with the ring of differential operators $\mathcal{D}(B)$, [19, 15.5.6], and is a simple finitely generated finitely partitive K-algebra, [19, 15.3.8, 15.1.21]. We refer the reader to [19, Chapter 15] for basic definitions. We aim to prove Theorem 4.3.

Let $\{B_i\}$ and $\{\Delta_i\}$ be standard finite-dimensional filtrations on B and Δ respectively such that $B_i \subseteq \Delta_i$ for all $i \geqslant 0$. Then the enveloping algebra $\Delta^e := \Delta \otimes \Delta^0$ can be equipped with the standard finite-dimensional filtration $\{\Delta_i^e\}$ which is the tensor product of the filtrations $\{\Delta_i\}$ and $\{\Delta_i^0\}$ of the algebras Δ and Δ^0 respectively.

Then $B \simeq \Delta / \Delta \, \text{Der}_K B$ is a simple left Δ-module [19, 15.3.8] with $\text{GK}(\Delta) = 2\text{GK}(B)$, [19, 15.3.2]. By Theorem 3.1,

$$\text{d}(\Delta) + \max\{\text{d}(\Delta), 1\} \geqslant \frac{\text{GK}(\Delta)}{\text{GK}(B)} = \frac{2\text{GK}(B)}{\text{GK}(B)} = 2,$$

hence $\text{d}(\Delta) \geqslant 1$. It remains to prove the opposite inequality. For this, we recall some properties of Δ (see [19, Chapter 15], for details).

Given $0 \neq c \in B$, denote by B_c the localization of the algebra B at the powers of the element c, then $\Delta(B_c) \simeq \Delta(B)_c$ and the map $\Delta(B) \to \Delta(B)_c$, $d \to d/1$, is injective, [19, 5.1.25]. There is a finite subset $\{c_1, \dots, c_t\}$ of B such that the algebra $\prod_{i=1}^t \Delta(B_{c_i})$ is left and right faithfully flat over its subalgebra Δ,

$$\sum_{i=1}^{t} Bc_i = B \quad \text{(see the proof of 15.2.13, [19]).} \tag{2}$$

For each $c = c_i$, $\mathrm{Der}_K(B_c)$ is a free B_c-module with a basis $\partial_j = \frac{\partial}{\partial x_j}$, $j = 1, \ldots, n$, for some $x_1, \ldots, x_n \in B$, [19, 15.2.13]. Note that the choice of the x_j-th depends on the choice of the c_i. Then

$$\Delta(B)_c \simeq \Delta(B_c) = B_c\langle \partial_1, \ldots, \partial_n\rangle \supseteq K\langle x_1, \ldots, x_n, \partial_1, \ldots, \partial_n\rangle.$$

Fix $c = c_i$. We aim to prove the following statement:

- *there exist natural numbers a, b, α, and β, such that for any $0 \neq d \in \Delta_k$ there exists*

$$w \in \Delta^e_{ak+b}\colon\ wd = c^{\alpha k+\beta}. \tag{$*$}$$

Suppose that we are done. Then one can choose the numbers a, b, α, and β such that $(*)$ holds for all $i = 1, \ldots, t$. It follows from (2) that

$$\sum_{i=1}^{t} f_i c_i = 1 \quad \text{for some } f_i \in A.$$

Choose $\nu \in \mathbb{N}$: all $f_i c_i \in \Delta_\nu$, and set $N(k) = \alpha k + \beta$, then

$$1 = \left(\sum_{i=1}^{t} f_i c_i\right)^{tN(k)} = \sum_{i=1}^{t} g_i c_i^{N(k)} = \sum_{i=1}^{t} g_i w_i d = wd,$$

where the w_i are from $(*)$, i.e. $w_i \in \Delta^e_{ak+b}$, $w_i d = c_i^{N(k)}$. So, $w = \sum_{i=1}^{t} g_i w_i \in \Delta^e_{\nu t N(k)+ak+b}$ and so $\mathrm{d}(\Delta) \leqslant 1$, as required.

Fix $c = c_i$. By [19, 15.1.24], $\mathrm{Der}_K(B_c) \simeq \mathrm{Der}_K(B)_c$ and $\mathrm{Der}_K B$ can be seen as a finitely generated B-submodule of $\mathrm{Der}_K(B_c)$, [19, 15.1.7].

The algebra B contains the polynomial subalgebra $P = K[x_1, \ldots, x_n]$. The polynomial algebra P has the natural filtration $P = \bigcup_{i\geqslant 0} P_i$ by the total degree of the variables. Fix a natural number l such that $P_1 \subseteq B_l$, then $P_i \subseteq B_{li}$ for all $i \geqslant 0$. We denote by $Q = K(x_1, \ldots, x_n)$ the field of fractions of P. The field of fractions, say L, of the algebra B has the same transcendence degree n as the field of rational functions Q. The algebra B is a finitely generated algebra, hence L is a finite field extension of Q of dimension, say m, over Q. Let $e_1, \ldots, e_m \in B$ be a Q-basis for the vector space L over Q. Note that $L = QB$. One can find a natural number $\beta \geqslant 1$ and a non-zero polynomial $p \in P_\beta$ such that

$$\{B_1, e_j e_k \mid j, k = 1, \ldots, m\} \subseteq \sum_{\alpha=1}^{m} p^{-1} P_\beta e_\alpha.$$

Then $B_k \subseteq \sum_{j=1}^{m} p^{-2k} P_{2\beta k} e_j$ and $B_k e_i \subseteq \sum_{j=1}^{m} p^{-3k} P_{3\beta k} e_j$ for all $k \geqslant 1$ and $i = 1, \ldots, m$. Let $0 \neq d \in B_k$. The $m \times m$ matrix of the bijective Q-linear map $L \to L$, $x \mapsto dx$, with respect to the basis $e_1, \ldots, e_m$ has entries from the set $p^{-3k} P_{3\beta k}$. So, its characteristic polynomial

$$\chi_d(t) = t^m + \alpha_{m-1} t^{m-1} + \cdots + \alpha_0$$

has coefficients in $p^{-3mk}P_{3m\beta k}$, and $\alpha_0 \neq 0$ as $x \mapsto dx$ is a bijection. Now,

$$P_{6m\beta k} \ni p^{3mk}\alpha_0 = p^{3mk}\left(-d^{m-1} - \alpha_{m-1}d^{m-2} - \cdots - \alpha_1\right)d$$
$$\in B_{4m\beta k}P_{3m\beta k}d \subseteq B_{m\beta k(4+3l)}d. \tag{3}$$

Let $\delta_1, \ldots, \delta_t$ be a set of generators for the left B-module $\mathrm{Der}_K(B)$. Then

$$\partial_i \in \sum_{j=1}^{t} c^{-l'_1} B_{l_1}\delta_j \quad \text{for } i = 1, \ldots, n,$$

for some natural numbers $l'_1 \leqslant l_1$. Fix a natural number l_2 such that $\delta_j(B_1) \subseteq B_{l_2}$ and $\delta_j(c) \in B_{l_2}$ for $j = 1, \ldots, t$. Then

$$\partial^\alpha(B_k) \subseteq c^{-2|\alpha|(l'_1+1)}B_{k+3|\alpha|(l_1+l_2)} \quad \text{for all } \alpha \in \mathbb{N}^n, k \geqslant 1,$$

where $\alpha = (\alpha_1, \ldots, \alpha_n)$, $|\alpha| = \alpha_1 + \cdots + \alpha_n$, $\partial^\alpha = \partial_1^{\alpha_1} \cdots \partial_n^{\alpha_n}$. It follows from (3) that one can find $\alpha \in \mathbb{N}^n$ such that $|\alpha| \leqslant 6m\beta k$ and

$$1 \in K^*\partial^\alpha\left(p^{3mk}\alpha_0\right) \subseteq \partial^\alpha\left(B_{m\beta k(4+3l)}d\right)$$
$$\subseteq c^{-pk}\Delta^e_{qk+r}d,$$

where $p, q, r \in \mathbb{N}$ and $K^* = K\backslash\{0\}$. Now $(*)$ follows.

In fact we have proved the following corollary.

COROLLARY 5.1. *There exist natural numbers a and b such that for any $0 \neq d \in \Delta_k$ there exists an element $w \in \Delta^e_{ak+b}$ satisfying $wd = 1$.*

6. Multiplicity for the filter dimension, holonomic modules over simple finitely generated algebras

In this section, we introduce a concept of multiplicity for the filter dimension and a concept of holonomic module for (some) finitely generated algebras. We will prove that a holonomic module has finite length (Theorem 6.8). The multiplicity for the filter dimension is a key ingredient in the proof.

First we recall the definition of multiplicity in the commutative situation and then for certain non-commutative algebras (somewhat commutative algebras).

6.1. *Multiplicity in the commutative situation*

Let B be a commutative finitely generated K-algebra with a standard finite-dimensional filtration $F = \{B_i\}$, and let M be a finitely generated B-module with a finite-dimensional generating subspace, say M_0, and with the standard filtration $\{M_i = B_i M_0\}$ attached to it. Then there exists a polynomial $p(t) = lt^d + \cdots \in \mathbb{Q}[t]$ with rational coefficients of degree $d = \mathrm{GK}(M)$ such that

$$\dim_K(M_i) = p(i) \quad \text{for all } i \gg 0.$$

The polynomial $p(t)$ is called the *Hilbert polynomial* of the B-module M. The Hilbert polynomial does depend on the filtration $\{M_i\}$ of the module M but its leading coefficient l does not. The number $e(M) = d!l$ is called the *multiplicity* of the B-module M. It is a natural number which does depend on the filtration F of the algebra B.

In the case when $M = B$ is the homogeneous coordinate ring of a projective algebraic variety $X \subseteq \mathbb{P}^m$ equipped with the natural filtration that comes from the grading of the graded algebra B, the multiplicity is the *degree* of X, the number of points in which X meets a generic plane of complementary degree in $\mathbb{P}^m$ (K is an algebraically closed field).

6.2. *Somewhat commutative algebras*

A K-algebra R is called a *somewhat commutative algebra* if it has a finite-dimensional filtration $R = \bigcup_{i \geqslant 0} R_i$ such that the associated graded algebra $\mathrm{gr}\, R := \bigoplus_{i \geqslant 0} R_i / R_{i-1}$ is a commutative finitely generated K-algebra where $R_{-1} = 0$ and $R_0 = K$. Then the algebra R is a Noetherian finitely generated algebra since $\mathrm{gr}\, R$ is so. A finitely generated module over a somewhat commutative algebra has a Gelfand–Kirillov dimension which is a natural number. We refer the reader to the books [16,19] for the properties of somewhat commutative algebras.

DEFINITION. For a somewhat commutative algebra R we define the *holonomic number*,

$$h_R := \min\{\mathrm{GK}(M) \mid M \neq 0 \text{ is a finitely generated } R\text{-module}\}.$$

DEFINITION. A finitely generated R-module M is called a *holonomic* module if $\mathrm{GK}(M) = h_R$. In other words, a non-zero finitely generated R-module is holonomic iff it has least Gelfand–Kirillov dimension. If $h_R = 0$ then every holonomic R-module is finite-dimensional and vice versa.

EXAMPLES. (1) The holonomic number of the Weyl algebra A_n is n. The polynomial algebra $K[X_1, \ldots, X_n] \simeq A_n / \sum_{i=1}^n A_n \partial_i$ with the natural action of the ring of differential operators $A_n = K[X_1, \ldots, X_n, \frac{\partial}{\partial X_1}, \ldots, \frac{\partial}{\partial X_n}]$ is a simple holonomic A_n-module.

(2) Let X be a smooth irreducible affine algebraic variety of dimension n. The ring of differential operators $\mathcal{D}(X)$ is a simple somewhat commutative algebra of Gelfand–Kirillov dimension $2n$ with holonomic number $h_{\mathcal{D}(X)} = n$. The algebra $\mathcal{O}(X)$ of regular functions of the variety X is a simple $\mathcal{D}(X)$-module with respect to the natural action of the algebra $\mathcal{D}(X)$. In more detail, $\mathcal{O}(X) \simeq \mathcal{D}(X)/\mathcal{D}(X)\,\mathrm{Der}_K(\mathcal{O}(X))$ where $\mathrm{Der}_K(\mathcal{O}(X))$ is the $\mathcal{O}(X)$-module of derivations of the algebra $\mathcal{O}(X)$.

Let $R = \bigcup_{i \geqslant 0} R_i$ be a somewhat commutative algebra. The associated graded algebra $\operatorname{gr} R$ is a commutative affine algebra. Let us choose homogeneous algebra generators of the algebra $\operatorname{gr} R$, say $y_1, \ldots, y_s$, of graded degrees $1 \leqslant k_1, \ldots, k_s$ respectively (that is $y_i \in R_{k_i}/R_{k_i-1}$). A filtration $\Gamma = \{\Gamma_i,\ i \geqslant 0\}$ of an R-module $M = \bigcup_{i=0}^{\infty} \Gamma_i$ is called *good* if the associated graded $\operatorname{gr} R$-module $\operatorname{gr}_\Gamma M := \bigoplus_{i \geqslant 0} \Gamma_i/\Gamma_{i-1}$ is finitely generated. An R-module M has a good filtration iff it is finitely generated, and if $\{\Gamma_i\}$ and $\{\Omega_i\}$ are two good filtrations of M, then there exists a natural number t such that $\Gamma_i \subseteq \Omega_{i+t}$ and $\Omega_i \subseteq \Gamma_{i+t}$ for all i. If an R-module M is finitely generated and M_0 is a finite-dimensional generating subspace of M, then the standard filtration $\{\Gamma_i = R_i M_0\}$ is good (see [9–11,16,19,20] for details). The following two lemmas are well-known by specialists (see their proofs, for example, in [3, Theorem 3.2 and Proposition 3.3] respectively).

LEMMA 6.1. *Let $R = \bigcup_{i \geqslant 0} R_i$ be a somewhat commutative algebra, $k = \operatorname{lcm}(k_1, \ldots, k_s)$, and let M be a finitely generated R-module with good filtration $\Gamma = \{\Gamma_i\}$.*

(1) *There exist k polynomials $\gamma_0, \ldots, \gamma_{k-1} \in \mathbb{Q}[t]$ with coefficients from $[k^{\mathrm{GK}(M)}\mathrm{GK}(M)!]^{-1}\mathbb{Z}$ such that*

$$\dim \Gamma_i = \gamma_j(i) \quad \textit{for all } i \gg 0 \textit{ and } j \equiv i \pmod{k}.$$

(2) *The polynomials γ_j have the same degree $\mathrm{GK}(M)$ and the same leading coefficient $e(M)/\mathrm{GK}(M)!$ where $e(M)$ is called the* multiplicity *of M. The multiplicity $e(M)$ does not depend on the choice of the good filtration Γ.*

REMARK. A finitely generated R-module M has $e(M) = 0$ iff $\dim_K(M) < \infty$.

LEMMA 6.2. *Let $0 \to N \to M \to L \to 0$ be an exact sequence of modules over a somewhat commutative algebra R. Then $\mathrm{GK}(M) = \max\{\mathrm{GK}(N), \mathrm{GK}(L)\}$, and if $\mathrm{GK}(N) = \mathrm{GK}(M) = \mathrm{GK}(L)$ then $e(M) = e(N) + e(L)$.*

COROLLARY 6.3. *Let the algebra R be as in lemma* 6.1 *with holonomic number $h > 0$.*

(1) *Let M be a holonomic R-module with multiplicity $e(M)$. The R-module M has finite length $\leqslant e(M)k^h$.*
(2) *Every non-zero submodule or factor module of a holonomic R-module is a holonomic module.*

PROOF. This follows directly from Lemma 6.2. □

6.3. *Multiplicity*

Let f be a function from $\mathbb{N}$ to $\mathbb{R}_+ = \{r \in \mathbb{R}\colon r \geqslant 0\}$, the *leading coefficient* of f is a non-zero limit (if it exists)

$$\operatorname{lc}(f) = \lim \frac{f(i)}{i^d} \neq 0, \quad i \to \infty,$$

where $d = \gamma(f)$. If $d \in \mathbb{N}$, we define the *multiplicity* $e(f)$ of f by

$$e(f) = d!\,\mathrm{lc}(f).$$

The factor $d!$ ensures that the multiplicity $e(f)$ is a positive integer in some important cases. If $f(t) = a_d t^d + a_{d-1} t^{d-1} + \cdots + a_0$ is a polynomial of degree d with real coefficients then $\mathrm{lc}(f) = a_d$ and $e(f) = d! a_d$.

LEMMA 6.4. *Let A be a finitely generated algebra equipped with a standard finite-dimensional filtration $F = \{A_i\}$ and M be a finitely generated A-module with generating finite-dimensional subspaces M_0 and N_0.*

(1) *If* $\mathrm{lc}(\nu_{F,M_0})$ *exists then so does* $\mathrm{lc}(\nu_{F,N_0})$, *and* $\mathrm{lc}(\nu_{F,M_0}) = \mathrm{lc}(\nu_{F,N_0})$.

(2) *If* $\mathrm{lc}(\dim A_i M_0)$ *exists then so does* $\mathrm{lc}(\dim A_i N_0)$, *and* $\mathrm{lc}(\dim A_i M_0) = \mathrm{lc}(\dim A_i N_0)$.

PROOF. (1) The module M has two filtrations $\{M_i = A_i M_0\}$ and $\{N_i = A_i N_0\}$. Let $\nu = \nu_{F,M_0}$ and $\mu = \nu_{F,N_0}$. Choose a natural number s such that $M_0 \subseteq N_s$ and $N_0 \subseteq M_s$, so $N_i \subseteq M_{i+s}$ and $M_i \subseteq N_{i+s}$ for $i \geqslant 0$. Since $M_0 \subseteq A_{\nu(i+s)} N_{i,\mathrm{gen}}$ for each i and $N_0 \subseteq A_s M_0$, we have $N_0 \subseteq A_{\nu(i+s)+s} N_{i,\mathrm{gen}}$, hence, $\mu(i) \leqslant \nu(i+s)+s$. By symmetry, $\nu(i) \leqslant \mu(i+s)+s$, so if $\mathrm{lc}(\mu)$ exists then so does $\mathrm{lc}(\nu)$ and $\mathrm{lc}(\mu) = \mathrm{lc}(\nu)$.

(2) Since $\dim N_i \leqslant \dim M_{i+s}$ and $\dim M_i \leqslant \dim N_{i+s}$ for $i \geqslant 0$, the statement is clear. □

Lemma 6.4 shows that the leading coefficients of the functions $\dim A_i M_0$ and ν_{F,M_0} (if they exist) do not depend on the choice of the generating subspace M_0. So, denote them by

$$l(M) = l_F(M) \quad \text{and} \quad L(M) = L_F(M)$$

respectively (if they exist). If $\mathrm{GK}(M)$ (resp. $\mathrm{d}(A)$) is a natural number, then we denote by $e(M) = e_F(M)$ (resp. $E(M) = E_F(M)$) the multiplicity of the function $\dim A_i M_0$ (resp. ν_{F,M_0}).

We denote by $L(A) = L_F(A)$ the leading coefficient $L_F({}_{A\otimes A^0}A)$ of the return function $\nu_{F\otimes F^0,K}$ of the $A \otimes A^0$-module A.

6.4. *Holonomic modules*

DEFINITION. Let A be a finitely generated K-algebra, and h_A be its holonomic number. A non-zero finitely generated A-module M is called a *holonomic* A-module if $\mathrm{GK}(M) = h_A$. We denote by $\mathrm{hol}(A)$ the set of all the holonomic A-modules.

Since the holonomic number is an infimum it is not clear at the outset that there will be modules which achieve this dimension. Clearly, $\mathrm{hol}(A) \neq \emptyset$ if the Gelfand–Kirillov dimension of every finitely generated A-module is a natural number.

A non-zero submodule or a factor module of a holonomic is a holonomic module (since the Gelfand–Kirillov dimension of a submodule or a factor module does not exceed the

Gelfand–Kirillov of the module). If, in addition, the finitely generated algebra A is left Noetherian and finitely partitive then each holonomic A-module M has finite length and each simple sub-factor of M is a holonomic module.

Let us consider algebras A having the following properties:

- (S) A is a simple finitely generated infinite-dimensional algebra.
- (N) There exists a standard finite-dimensional filtration $F = \{A_i\}$ of the algebra A such that the associated graded algebra $\operatorname{gr} A := \bigoplus_{i\geqslant 0} A_i/A_{i-1}$, $A_{-1} = 0$, is left Noetherian.
- (D) $\mathrm{GK}(A) < \infty$, $\mathrm{fd}(A) < \infty$, both $l(A) = l_F(A)$ and $L(A) = L_F(A)$ exist.
- (H) For every holonomic A-module M there exists $l(M) = l_F(M)$.

In many cases we use a weaker form of the condition (D).

- (D′) $\mathrm{GK}(A) < \infty$, $d = \mathrm{fd}(A) < \infty$, there exist $l(A) = l_F(A)$ and a positive number $c > 0$ such that $\nu(i) \leqslant ci^d$ for $i \gg 0$ where ν is the return function $\nu_{F\otimes F^0,K}$ of the left $A \otimes A^0$-module A.

It follows from (N) that A is a left Noetherian algebra.

LEMMA 6.5 [4].

(1) *The Weyl algebra A_n over a field of characteristic zero with the standard finite-dimensional filtration $F = \{A_{n,i}\}$ associated with the canonical generators satisfies the conditions* (S), (N), (D), (H). *The return function $\nu_F(i) = i$ for $i \geqslant 0$, and so the leading coefficient of ν_F is $L_F(A_n) = 1$.*

(2) *$\nu_{G,K}(i) = i$ for $i \geqslant 0$ and $L_G(P_n) = 1$ where $\nu_{G,K}$ is the return function of the A_n-module $P_n = K[X_1, \dots, X_n] = A_n/(A_n\partial_1 + \dots + A_n\partial_n)$ with the usual filtration $G = \{P_{n,i}\}$ of the polynomial algebra.*

PROOF. (1) The only fact that we need to prove is that $\nu_F(i) = i$ for $i \geqslant 0$. We keep the notation of Theorem 4.6. In the proof of Theorem 4.6 we have seen that $\nu_F(i) \leqslant i$ for $i \geqslant 0$. It remains to prove the reverse inequality.

Each element u in A_n can be written in a unique way as a finite sum $u = \sum \lambda_{\alpha\beta} X^\alpha \partial^\beta$ where $\lambda_{\alpha\beta} \in K$ and X^α denotes the monomial $X_1^{\alpha_1} \cdots X_n^{\alpha_n}$ and similarly ∂^β denotes the monomial $\partial_1^{\beta_1} \cdots \partial_n^{\beta_n}$. The element u belongs to $A_{n,m}$ iff $|\alpha| + |\beta| \leqslant m$, where $|\alpha| = \alpha_1 + \dots + \alpha_n$. If $\alpha \in K[X_1, \dots, X_n]$, then

$$\partial_i^m \alpha = \sum_{j=0}^{m} \binom{m}{j} \frac{\partial^j \alpha}{\partial X_i^j} \partial_i^{m-j}, \qquad m \in \mathbb{N}.$$

It follows that for any $v \in \sum A_{n,i} \otimes A_{n,j}^0$, $i + j < m$, the element $vX_1^m = \sum \lambda_{\alpha\beta} X^\alpha \partial^\beta$ has the coefficient $\lambda_{0,0} = 0$, hence it could not be a non-zero scalar, and so $\nu(i) \geqslant i$ for all $i \geqslant 0$. Hence $\nu(i) = i$ all $i \geqslant 0$ and then $L_F(A_n) = 1$.

(2) The standard filtration of the A_n-module P_n associated with the generating subspace K coincides with the usual filtration of the polynomial algebra P_n. Since $\partial_j(P_{n,i}) \subseteq P_{n,i-1}$ for all $i \geqslant 0$ and j, $\nu_{G,K}(i) \leqslant i$ for $i \geqslant 0$. Using the same arguments as above we see that for any $u \in \sum_{j=0}^{i-1} A_{n,j} \otimes A_{n,i-j-1}^0$ the element uX_1^i belongs to the ideal of P_n generated by X_1, hence, $\nu_{G,K}(i) \geqslant i$, and so $\nu_{G,K}(i) = i$ for all $i \geqslant 0$ and $L_G(P_n) = 1$. $\square$

THEOREM 6.6 [4]. *Assume that an algebra A satisfies the conditions* (S), (H), (D), *resp.* (D′), *for some standard finite-dimensional filtration $F = \{A_i\}$ of A. Then for every holonomic A-module M its leading coefficient is bounded from below by a non-zero constant*:

$$l(M) \geqslant \sqrt{\frac{l(A)}{(L(A)L'(A))^{h_A}}},$$

where

$$L'(A) = \begin{cases} L(A), & \text{if } \mathrm{d}(A) > 1, \\ L(A)+1, & \text{if } \mathrm{d}(A) = 1, \\ 1, & \text{if } \mathrm{d}(A) < 1, \end{cases}$$

resp.

$$l(M) \geqslant \sqrt{\frac{l(A)}{(c(c+1))^{h_A}}}.$$

PROOF. Let M_0 be a generating finite-dimensional subspace of M and $\{M_i = A_i M_0\}$ be the standard finite-dimensional filtration of M. In the proof of Theorem 3.1 we proved that $\dim A_i \leqslant \dim M_{\lambda(i)} \dim M_{\lambda(i)+i}$ for $i \geqslant 0$ where λ is the left return function of the algebra A associated with the filtration F. Since $\lambda(i) \leqslant \nu(i)$ for $i \geqslant 0$ we have $\dim A_i \leqslant \dim M_{\nu(i)} \dim M_{\nu(i)+i}$, hence, if (D) holds then

$$l(A)i^{\mathrm{GK}(A)} + \cdots \leqslant l^2(M)\bigl(L(A)L'(A)\bigr)^{\mathrm{GK}(M)} i^{\mathrm{GK}(M)(\mathrm{fd}(A)+\max\{\mathrm{fd}(A),1\})} + \cdots,$$

where three dots denote smaller terms.

If (D′) holds then

$$l(A)i^{\mathrm{GK}(A)} + \cdots \leqslant l^2(M)\bigl(c(c+1)\bigr)^{\mathrm{GK}(M)} i^{\mathrm{GK}(M)(\mathrm{fd}(A)+\max\{\mathrm{fd}(A),1\})} + \cdots.$$

The module M is holonomic, i.e. $\mathrm{GK}(A) = \mathrm{GK}(M)(\mathrm{fd}(A) + \max\{\mathrm{fd}(A), 1\})$. Now, comparing the "leading" coefficients in the inequalities above we finish the proof. □

Let A be as in Theorem 6.6. We attach to the algebra A two positive numbers c_A and c'_A in the cases (D) and (D′) respectively:

$$c_A = \sqrt{\frac{l(A)}{(L(A)L'(A))^{h_A}}} \quad \text{and} \quad \bigl(c'_A\bigr) = \sqrt{\frac{l(A)}{(c(c+1))^{h_A}}}.$$

COROLLARY 6.7. *Assume that an algebra A satisfies the conditions* (S), (N), (H), (D) *or* (D′). *Let $0 \to N \to M \to L \to 0$ be an exact sequence of non-zero finitely generated A-modules. Then M is holonomic if and only if N and L are holonomic, in that case $l(M) = l(N) + l(L)$.*

PROOF. The algebra A is left Noetherian, so the module M is finitely generated iff both N and L are so. The proof of Proposition 3.11, [19, p. 295], shows that we can choose finite-dimensional generating subspaces N_0, M_0, L_0 of the modules N, M, L respectively such that the sequences

$$0 \to N_i = A_i N_0 \to M_i = A_i M_0 \to L_i = A_i L_0 \to 0$$

are exact for all i, hence, $\dim M_i = \dim N_i + \dim L_i$ and the results follow. □

THEOREM 6.8 [4]. *Suppose that the conditions* (S), (N), (H), (D) (*resp.* (D′)) *hold. Then each holonomic A-module M has finite length which is less or equal to $l(M)/c_A$ (resp. $l(M)/c'_A$).*

PROOF. If $M = M_1 \supset M_2 \supset \cdots \supset M_m \supset M_{m+1} = 0$ is a chain of distinct submodules, then by corollary 6.7 and theorem 6.6

$$l(M) = \sum_{i=1}^{m} l(M_i/M_{i+1}) \geqslant mc_A \quad (\text{resp. } l(M) \geqslant mc'_A),$$

thus $m \leqslant l(M)/c_A$ (resp. $m \leqslant l(M)/c'_A$). □

7. Filter dimension and commutative subalgebras of simple finitely generated algebras and their division algebras

In this section, using the first and the second filter inequalities, we obtain (i) an *upper* bound for the Gelfand–Kirillov dimension of commutative subalgebras of simple finitely generated infinite-dimensional algebras (Theorem 7.2), and (ii) an *upper* bound for the transcendence degree of subfields of quotient division rings of (certain) simple finitely generated infinite-dimensional algebras (Theorems 7.4 and 7.5).

For certain classes of algebras and their division algebras the maximum Gelfand–Kirillov dimension/transcendence degree over the commutative subalgebras/subfields were found in [1,12,17,13–15,2], and [23].

Recall that

$$\text{the Gelfand–Kirillov dimension } \mathrm{GK}(C) = \text{the Krull dimension } \mathrm{K.dim}(C)$$
$$= \text{the transcendence degree } \mathrm{tr.deg}_K(C)$$

for every commutative finitely generated algebra C which is a domain.

7.1. *An upper bound for the Gelfand–Kirillov dimensions of commutative subalgebras of simple finitely generated algebras*

PROPOSITION 7.1. *Let A and C be finitely generated algebras such that C is a commutative domain with field of fractions Q, $B := C \otimes A$, and $\mathcal{B} := Q \otimes A$. Let M be a finitely generated B-module such that $\mathcal{M} := \mathcal{B} \otimes_B M \neq 0$. Then* $\mathrm{GK}({}_B M) \geqslant \mathrm{GK}_Q({}_{\mathcal{B}}\mathcal{M}) + \mathrm{GK}(C)$.

REMARK. GK_Q stands for the Gelfand–Kirillov dimension over the field Q.

PROOF. Let us fix standard filtrations $\{A_i\}$ and $\{C_i\}$ for the algebras A and C respectively. Let $h(t) \in \mathbb{Q}[t]$ be the *Hilbert polynomial* for the algebra C, i.e. $\dim_K(C_i) = h(i)$ for $i \gg 0$. Recall that $\mathrm{GK}(C) = \deg_t(h(t))$. The algebra B has a standard filtration $\{B_i\}$ which is the tensor product of the standard filtrations $\{C_i\}$ and $\{A_i\}$ of the algebras C and A, i.e. $B_i := \sum_{j=0}^{i} C_j \otimes A_{i-j}$. By the assumption, the B-module M is finitely generated, so $M = BM_0$ where M_0 is a finite-dimensional generating subspace for M. Then the B-module M has a standard filtration $\{M_i := B_i M_0\}$. The Q-algebra $\mathcal{B}$ has a standard (finite-dimensional over Q) filtration $\{\mathcal{B}_i := Q \otimes A_i\}$, and the $\mathcal{B}$-module $\mathcal{M}$ has a standard (finite-dimensional over Q) filtration $\{\mathcal{M}_i := \mathcal{B}_i M_0' = QA_i M_0'\}$ where M_0' is the image of the vector space M_0 under the B-module homomorphism $M \to \mathcal{M}$, $m \mapsto m' := 1 \otimes_B m$.

For each $i \geqslant 0$, one can fix a K-subspace, say L_i, of $A_i M_0'$ such that $\dim_Q(QA_i M_0') = \dim_K(L_i)$. Now, $B_{2i} \supseteq C_i \otimes A_i$ implies $\dim_K(B_{2i}M_0) \geqslant \dim_K((C_i \otimes A_i)M_0)$, and $((C_i \otimes A_i)M_0)' \supseteq C_i L_i$ implies $\dim_K(((C_i \otimes A_i)M_0)') \geqslant \dim_K(C_i L_i) = \dim_K(C_i)\dim_K(L_i) = \dim_K(C_i)\dim_Q(\mathcal{M}_i)$. It follows that

$$
\begin{aligned}
\mathrm{GK}({}_B M) &= \gamma\big(\dim_K(M_i)\big) \geqslant \gamma\big(\dim_K(M_{2i})\big) = \gamma\big(\dim_K(B_{2i}M_0)\big) \\
&\geqslant \gamma\big(\dim_K\big((C_i \otimes A_i)M_0\big)\big) \\
&\geqslant \gamma\big(\dim_K\big(((C_i \otimes A_i)M_0)'\big)\big) \geqslant \gamma\big(\dim_K(C_i)\dim_Q(\mathcal{M}_i)\big) \\
&= \gamma\big(\dim_K(C_i)\big) + \gamma\big(\dim_Q(\mathcal{M}_i)\big) \\
&\quad \text{(since } \gamma\big(\dim_K(C_i)\big) = h(i), \text{ for } i \gg 0) \\
&= \mathrm{GK}(C) + \mathrm{GK}_Q({}_{\mathcal{B}}\mathcal{M}).
\end{aligned}
$$
□

Recall that $\mathrm{d} = \mathrm{fd}, \mathrm{lfd}$. A K-algebra A is called *central* if its centre $Z(A) = K$.

THEOREM 7.2 [7]. *Let A be a central simple finitely generated K-algebra of Gelfand–Kirillov dimension $0 < n < \infty$ (over K). Let C be a commutative subalgebra of A. Then*

$$
\mathrm{GK}(C) \leqslant \mathrm{GK}(A)\left(1 - \frac{1}{f_A + \max\{f_A, 1\}}\right),
$$

where $f_A := \max\{\mathrm{d}_{Q_m}(Q_m \otimes A) \mid 0 \leqslant m \leqslant n\}$, $Q_0 := K$, and $Q_m := K(x_1, \ldots, x_m)$ is a rational function field in indeterminates $x_1, \ldots, x_m$.

PROOF. Let $P_m = K[x_1, \ldots, x_m]$ be a polynomial algebra over the field K. Then Q_m is its field of fractions and $\mathrm{GK}(P_m) = m$. Suppose that P_m is a subalgebra of A. Then $m = \mathrm{GK}(P_m) \leqslant \mathrm{GK}(A) = n$. For each $m \geqslant 0$, $Q_m \otimes A$ is a central simple Q_m-algebra, [19, 9.6.9], of Gelfand–Kirillov dimension (over Q_m) $\mathrm{GK}_{Q_m}(Q_m \otimes A) = \mathrm{GK}(A) > 0$, hence

$$
\mathrm{GK}(A) = \mathrm{GK}({}_A A_A) \geqslant \mathrm{GK}({}_A A_{P_m}) = \mathrm{GK}({}_{P_m \otimes A} A) \quad (P_m \text{ is commutative})
$$

$$\geqslant \mathrm{GK}_{Q_m}\big({}_{Q_m \otimes A}(Q_m \otimes_{P_m} A)\big) + \mathrm{GK}(P_m) \quad \text{(Lemma 7.1)}$$

$$\geqslant \frac{\mathrm{GK}(A)}{\mathrm{d}_{Q_m}(Q_m \otimes A) + \max\{\mathrm{d}_{Q_m}(Q_m \otimes A), 1\}} + m \quad \text{(Theorem 3.1)}.$$

Hence,

$$m \leqslant \mathrm{GK}(A)\left(1 - \frac{1}{\mathrm{d}_{Q_m}(Q_m \otimes A) + \max\{\mathrm{d}_{Q_m}(Q_m \otimes A), 1\}}\right) \leqslant \mathrm{GK}(A),$$

and so

$$\mathrm{GK}(C) \leqslant \mathrm{GK}(A)\left(1 - \frac{1}{f_A + \max\{f_A, 1\}}\right).$$

□

As a consequence we have a short proof of the following well-known result.

COROLLARY 7.3. *Let K be an algebraically closed field of characteristic zero, X be a smooth irreducible affine algebraic variety of dimension $n := \dim(X) > 0$, and C be a commutative subalgebra of the ring of differential operators $\mathcal{D}(X)$. Then $\mathrm{GK}(C) \leqslant n$.*

PROOF. The algebra $\mathcal{D}(X)$ is central since K is an algebraically closed field of characteristic zero [19, chapter 15]. By Theorem 4.3, $f_{\mathcal{D}(X)} = 1$, and then, by Theorem 7.2,

$$\mathrm{GK}(C) \leqslant 2n\left(1 - \frac{1}{1+1}\right) = n.$$

□

REMARK. For the ring of differential operators $\mathcal{D}(X)$ the upper bound in Theorem 7.2 for the Gelfand–Kirillov dimension of commutative subalgebras of $\mathcal{D}(X)$ is an *exact* upper bound since as we mentioned above the algebra $\mathcal{O}(X)$ of regular functions on X is a commutative subalgebra of $\mathcal{D}(X)$ of Gelfand–Kirillov dimension n.

7.2. *An upper bound for the transcendence degree of subfields of quotient division algebras of simple finitely generated algebras*

Recall that the transcendence degree $\mathrm{tr.deg}_K(L)$ of a field extension L of a field K coincides with the Gelfand–Kirillov dimension $\mathrm{GK}_K(L)$, and, by the *Goldie's theorem*, a left Noetherian semiprime algebra A has a quotient algebra $D = D_A$ (i.e. $D = S^{-1}A$ where S is the set of *regular* elements = the set of non-zerodivisors of A). As a rule, the quotient algebra D has *infinite* Gelfand–Kirillov dimension and is *not* a finitely generated algebra (e.g., the quotient division algebra $D(X)$ of the ring of differential operators $\mathcal{D}(X)$ on *each* smooth irreducible affine algebraic variety X of dimension $n > 0$ over a field K of characteristic zero contains a *non-commutative free* subalgebra since $D(X) \supseteq D(\mathbb{A}^1)$ and the *first Weyl division algebra* $D(\mathbb{A}^1)$ has this property, [18]). So, if we want to find an upper bound for the transcendence degree of subfields in the quotient algebra D we can not apply Theorem 7.2. Nevertheless, imposing some natural (mild) restrictions on the algebra A one can obtain exactly the same upper bound for the transcendence degree of subfields in the quotient algebra D_A as the upper bound for the Gelfand–Kirillov dimension of commutative subalgebras in A.

THEOREM 7.4 [7]. *Let A be a simple finitely generated K-algebra such that $0 < n := \mathrm{GK}(A) < \infty$, all the algebras $Q_m \otimes A$, $m \geqslant 0$, are simple finitely partitive algebras where $Q_0 := K$, $Q_m := K(x_1, \dots, x_m)$ is a rational function field and, for each $m \geqslant 0$, the Gelfand–Kirillov dimension (over Q_m) of every finitely generated $Q_m \otimes A$-module is a natural number. Let $B = S^{-1}A$ be the localization of the algebra A at a left Ore subset S of A. Let L be a (commutative) subfield of the algebra B that contains K. Then*

$$\mathrm{tr.deg}_K(L) \leqslant \mathrm{GK}(A)\left(1 - \frac{1}{f_A + \max\{f_A, 1\}}\right),$$

where $f_A := \max\{\mathrm{d}_{Q_m}(Q_m \otimes A) \mid 0 \leqslant m \leqslant n\}$.

PROOF. It follows immediately from a definition of the Gelfand–Kirillov dimension that $\mathrm{GK}_{K'}(K' \otimes C) = \mathrm{GK}(C)$ for any K-algebra C and any field extension K' of K. In particular, $\mathrm{GK}_{Q_m}(Q_m \otimes A) = \mathrm{GK}(A)$ for all $m \geqslant 0$. By theorem 4.2,

$$\mathrm{K.dim}(Q_m \otimes A) \leqslant \mathrm{GK}(A)\left(1 - \frac{1}{\mathrm{d}_{Q_m}(Q_m \otimes A) + \max\{\mathrm{d}_{Q_m}(Q_m \otimes A), 1\}}\right).$$

Let L be a subfield of the algebra B that contains K. Suppose that L contains a rational function field (isomorphic to) Q_m for some $m \geqslant 0$.

$$\begin{aligned} m &= \mathrm{tr.deg}_K(Q_m) \leqslant \mathrm{K.dim}(Q_m \otimes Q_m) \\ &\leqslant \mathrm{K.dim}(Q_m \otimes B) \\ &\quad \text{(by [19, 6.5.3] since } Q_m \otimes B \text{ is a free } Q_m \otimes Q_m\text{-module)} \\ &= \mathrm{K.dim}\big(Q_m \otimes S^{-1}A\big) = \mathrm{K.dim}\big(S^{-1}(Q_m \otimes A)\big) \\ &\leqslant \mathrm{K.dim}(Q_m \otimes A) \quad \text{(by [19, 6.5.3(ii)(b)])} \\ &\leqslant \mathrm{GK}(A)\left(1 - \frac{1}{\mathrm{d}_{Q_m}(Q_m \otimes A) + \max\{\mathrm{d}_{Q_m}(Q_m \otimes A), 1\}}\right) \leqslant \mathrm{GK}(A). \end{aligned}$$

Hence

$$\mathrm{tr.deg}_K(L) \leqslant \mathrm{GK}(A)\left(1 - \frac{1}{f_A + \max\{f_A, 1\}}\right). \qquad \square$$

Recall that every somewhat commutative algebra A is a Noetherian finitely generated finitely partitive algebra of finite Gelfand–Kirillov dimension, the Gelfand–Kirillov dimension of every finitely generated A-modules is an integer, and (*Quillen's lemma*): *the ring* $\mathrm{End}_A(M)$ *is algebraic over* K (see [19, Chapter 8] or [16] for details).

THEOREM 7.5 [7]. *Let A be a central simple somewhat commutative infinite-dimensional K-algebra and let $D = D_A$ be its quotient algebra. Let L be a subfield of D that contains K. Then the transcendence degree of the field L (over K)*

$$\mathrm{tr.deg}_K(L) \leqslant \mathrm{GK}(A)\left(1 - \frac{1}{f_A + \max\{f_A, 1\}}\right),$$

where $f_A := \max\{\mathrm{d}_{Q_m}(Q_m \otimes A) \mid 0 \leqslant m \leqslant \mathrm{GK}(A)\}$.

PROOF. The algebra A is a somewhat commutative algebra, so it has a finite-dimensional filtration $A = \bigcup_{i \geqslant 0} A_i$ such that the associated graded algebra is a commutative finitely generated algebra. For each integer $m \geqslant 0$, the Q_m-algebra $Q_m \otimes A = \bigcup_{i \geqslant 0} Q_m \otimes A_i$ has the finite-dimensional filtration (over Q_m) such that the associated graded algebra $\mathrm{gr}(Q_m \otimes A) = \bigoplus_{i \geqslant 0} Q_m \otimes A_i / Q_m \otimes A_{i-1} \simeq Q_m \otimes \mathrm{gr}(A)$ is a commutative finitely generated Q_m-algebra. So, $Q_m \otimes A$ is a somewhat commutative Q_m-algebra.

By the assumption $\dim_K(A) = \infty$, hence $\dim_K(\mathrm{gr}(A)) = \infty$ which implies $\mathrm{GK}(\mathrm{gr}(A)) > 0$, and so $\mathrm{GK}(A) > 0$ (since $\mathrm{GK}(A) = \mathrm{GK}(\mathrm{gr}(A))$). The algebra A is a central simple K-algebra, so $Q_m \otimes A$ is a central simple Q_m-algebra, [19, 9.6.9]. Now, Theorem 7.5 follows from Theorem 7.4 applied to $B = D$. □

THEOREM 7.6. *Let K be an algebraically closed field of characteristic zero, $\mathcal{D}(X)$ be the ring of differential operators on a smooth irreducible affine algebraic variety X of dimension $n > 0$, and $D(X)$ be the quotient division ring for $\mathcal{D}(X)$. Let L be a (commutative) subfield of $D(X)$ that contains K. Then* $\mathrm{tr.deg}_K(L) \leqslant n$.

REMARK. This inequality is, in fact, an *exact* upper bound for the transcendence degree of subfields in $D(X)$ since the field of fractions $Q(X)$ for the algebra $\mathcal{O}(X)$ is a commutative subfield of the division ring $D(X)$ with $\mathrm{tr.deg}_K(Q(X)) = n$.

PROOF. Since $Q_m \otimes \mathcal{D}_K(\mathcal{O}(X)) \simeq \mathcal{D}_{Q_m}(Q_m \otimes \mathcal{O}(X))$ and $\mathrm{d}(\mathcal{D}(Q_m \otimes \mathcal{O}(X))) = 1$ for all $m \geqslant 0$ we have $f_{\mathcal{D}(X)} = 1$. Now, Theorem 7.6 follows from Theorem 7.5,

$$\mathrm{tr.deg}_K(L) \leqslant 2n\left(1 - \frac{1}{1+1}\right) = n.$$

□

Following [15] for a K-algebra A define the *commutative dimension*

$$\mathrm{Cdim}(A) := \max\bigl\{\mathrm{GK}(C) \mid C \text{ is a commutative subalgebra of } A\bigr\}.$$

The commutative dimension $\mathrm{Cdim}(A)$ (if finite) is the largest non-negative integer m such that the algebra A contains a polynomial algebra in m variables ([15, 1.1], or [19, 8.2.14]). So, $\mathrm{Cdim}(A) = \mathbb{N} \cup \{\infty\}$. If A is a subalgebra of B then $\mathrm{Cdim}(A) \leqslant \mathrm{Cdim}(B)$.

COROLLARY 7.7. *Let X and Y be smooth irreducible affine algebraic varieties of dimensions n and m respectively, let $D(X)$ and $D(Y)$ be quotient division rings for the rings of differential operators $\mathcal{D}(X)$ and $\mathcal{D}(Y)$. Then there is no K-algebra embedding $D(X) \to D(Y)$ if $n > m$.*

PROOF. By Theorem 7.6, $\mathrm{Cdim}(D(X)) = n$ and $\mathrm{Cdim}(D(Y)) = m$. Suppose that there is a K-algebra embedding $D(X) \to D(Y)$. Then $n = \mathrm{Cdim}(D(X)) \leqslant \mathrm{Cdim}(D(Y)) = m$. □

For the Weyl algebras $A_n = \mathcal{D}(\mathbb{A}^n)$ and $A_m = \mathcal{D}(\mathbb{A}^m)$ the result above was proved by Gelfand and Kirillov in [12]. They introduced a new invariant of an algebra A, the so-called (*Gelfand–Kirillov*) *transcendence degree* $\mathrm{GKtr.deg}(A)$, and proved that $\mathrm{GKtr.deg}(D_n) = 2n$. Recall that

$$\mathrm{GKtr.deg}(A) := \sup_V \inf_b \mathrm{GK}\bigl(K[bV]\bigr),$$

where V ranges over the finite-dimensional subspaces of A and b ranges over the regular elements of A. Another proofs of the corollary based on different ideas were given by A. Joseph, [14], and R. Resco, [23], see also [19, 6.6.19]. Joseph's proof is based on the fact that the centralizer of any isomorphic copy of the Weyl algebra A_n in its division algebra $D_n := D(\mathbb{A}^n)$ reduces to scalars ([15, 4.2]), Resco proved that $\mathrm{Cdim}(D_n) = n$ ([23, 4.2]) using the result of Rentschler and Gabriel, [22], that $\mathrm{K.dim}(A_n) = n$ (over an arbitrary field of characteristic zero).

8. Filter dimension and isotropic subalgebras of Poisson algebras

In this section, we apply Theorem 7.2 to obtain an upper bound for the Gelfand–Kirillov dimension of *isotropic* subalgebras of certain Poisson algebras (Theorem 8.1).

Let $(P, \{\cdot,\cdot\})$ be a *Poisson algebra* over the field K. Recall that P is an associative commutative K-algebra which is a Lie algebra with respect to the bracket $\{\cdot,\cdot\}$ for which *Leibniz's rule* holds:

$$\{a, xy\} = \{a, x\}y + x\{a, y\} \quad \text{for all } a, x, y \in P,$$

which means that the *inner derivation* $\mathrm{ad}(a) : P \to P$, $x \mapsto \{a, x\}$, of the Lie algebra P is also a derivation of the associative algebra P. Therefore, to each Poisson algebra P one can attach an associative subalgebra $A(P)$ of the ring of differential operators $\mathcal{D}(P)$ with coefficients from the algebra P which is generated by P and $\mathrm{ad}(P) := \{\mathrm{ad}(a) \mid a \in P\}$. If P is a finitely generated algebra then so is the algebra $A(P)$ with $\mathrm{GK}(A(P)) \leqslant \mathrm{GK}(\mathcal{D}(P)) < \infty$.

EXAMPLE. Let $P_{2n} = K[x_1, \ldots, x_{2n}]$ be the *Poisson polynomial algebra* over a field K of characteristic zero equipped with the *Poisson bracket*

$$\{f, g\} = \sum_{i=1}^{n} \left(\frac{\partial f}{\partial x_i} \frac{\partial g}{\partial x_{n+i}} - \frac{\partial f}{\partial x_{n+i}} \frac{\partial g}{\partial x_i} \right).$$

The algebra $A(P_{2n})$ is generated by the elements

$$x_1, \dots, x_{2n}, \quad \operatorname{ad}(x_i) = \frac{\partial}{\partial x_{n+i}}, \quad \operatorname{ad}(x_{n+i}) = -\frac{\partial}{\partial x_i}, \quad i = 1, \dots, n.$$

So, the algebra $A(P_{2n})$ is canonically isomorphic to the Weyl algebra A_{2n}.

DEFINITION. We say that a Poisson algebra P is a *strongly simple Poisson algebra* if

(1) P is a finitely generated (associative) algebra which is a domain,
(2) the algebra $A(P)$ is central simple, and
(3) for each set of algebraically independent elements $a_1, \dots, a_m$ of the algebra P such that $\{a_i, a_j\} = 0$ for all $i, j = 1, \dots, m$, the (commuting) elements $a_1, \dots, a_m$, $\operatorname{ad}(a_1), \dots, \operatorname{ad}(a_m)$ of the algebra $A(P)$ are algebraically independent.

THEOREM 8.1 [7]. *Let P be a strongly simple Poisson algebra, and C be an isotropic subalgebra of P, i.e. $\{C, C\} = 0$. Then*

$$\operatorname{GK}(C) \leqslant \frac{\operatorname{GK}(A(P))}{2} \left(1 - \frac{1}{f_{A(P)} + \max\{f_{A(P)}, 1\}}\right),$$

where $f_{A(P)} := \max\{\mathrm{d}_{Q_m}(Q_m \otimes A(P)) \mid 0 \leqslant m \leqslant \operatorname{GK}(A(P))\}$.

PROOF. By assumption the finitely generated algebra P is a domain, hence the finitely generated algebra $A(P)$ is a domain (as a subalgebra of the domain $\mathcal{D}(Q(P))$, the ring of differential operators with coefficients from the field of fractions $Q(P)$ for the algebra P). It suffices to prove the inequality for isotropic subalgebras of the Poisson algebra P that are polynomial algebras. So, let C be an isotropic polynomial subalgebra of P in m variables, say $a_1, \dots, a_m$. By assumption, the commuting elements $a_1, \dots, a_m, \operatorname{ad}(a_1), \dots, \operatorname{ad}(a_m)$ of the algebra $A(P)$ are algebraically independent. So, the Gelfand–Kirillov dimension of the subalgebra C' of $A(P)$ generated by these elements is equal to $2m$. By Theorem 7.2,

$$2\operatorname{GK}(C) = 2m = \operatorname{GK}(C') \leqslant \operatorname{GK}\big(A(P)\big)\left(1 - \frac{1}{f_{A(P)} + \max\{f_{A(P)}, 1\}}\right),$$

and this proves the inequality. □

COROLLARY 8.2.

(1) *The Poisson polynomial algebra $P_{2n} = K[x_1, \dots, x_{2n}]$ (with the Poisson bracket) over a field K of characteristic zero is a strongly simple Poisson algebra, the algebra $A(P_{2n})$ is canonically isomorphic to the Weyl algebra A_{2n}.*
(2) *The Gelfand–Kirillov dimension of every isotropic subalgebra of the polynomial Poisson algebra P_{2n} is $\leqslant n$.*

PROOF. (1) The third condition in the definition of strongly simple Poisson algebra is the only statement we have to prove. So, let $a_1, \dots, a_m$ be algebraically independent elements of the algebra P_{2n} such that $\{a_i, a_j\} = 0$ for all $i, j = 1, \dots, m$. One can find polynomials, say $a_{m+1}, \dots, a_{2n}$, in P_{2n} such that the elements $a_1, \dots, a_{2n}$ are algebraically independent, hence the determinant d of the Jacobian matrix $J := (\frac{\partial a_i}{\partial x_j})$ is a non-zero polynomial. Let $X = (\{x_i, x_j\})$ and $Y = (\{a_i, a_j\})$ be, so-called, the *Poisson matrices* associated with the elements $\{x_i\}$ and $\{a_i\}$. It follows from $Y = J^{\mathrm{T}} X J$ that $\det(Y) = d^2 \det(X) \neq 0$ since $\det(X) \neq 0$. The derivations

$$\delta_i := d^{-1} \det \begin{pmatrix} \{a_1, a_1\} & \dots & \{a_1, a_{i-1}\} & \{a_1, \cdot\} & \{a_1, a_{i+1}\} & \dots & \{a_1, a_{2n}\} \\ \{a_2, a_1\} & \dots & \{a_2, a_{i-1}\} & \{a_2, \cdot\} & \{a_2, a_{i+1}\} & \dots & \{a_2, a_{2n}\} \\ & & & \dots & & & \\ \{a_{2n}, a_1\} & \dots & \{a_{2n}, a_{i-1}\} & \{a_{2n}, \cdot\} & \{a_1, a_{i+1}\} & \dots & \{a_{2n}, a_{2n}\} \end{pmatrix},$$

$i = 1, \dots, 2n$, of the rational function field $Q_{2n} = K(x_1, \dots, x_{2n})$ satisfy the following properties: $\delta_i(a_j) = \delta_{i,j}$, the Kronecker delta. For each i and j, the kernel of the derivation $\Delta_{ij} := \delta_i \delta_j - \delta_j \delta_i \in \mathrm{Der}_K(Q_{2n})$ contains $2n$ algebraically independent elements $a_1, \dots, a_{2n}$. Hence $\Delta_{ij} = 0$ since the field Q_{2n} is algebraic over its subfield $K(a_1, \dots, a_{2n})$ and $\mathrm{char}(K) = 0$. So, the subalgebra, say W, of the ring of differential operators $\mathcal{D}(Q_{2n})$ generated by the elements $a_1, \dots, a_{2n}, \delta_1, \dots, \delta_{2n}$ is isomorphic to the Weyl algebra A_{2n}, and so $\mathrm{GK}(W) = \mathrm{GK}(A_{2n}) = 4n$.

Let U be the K-subalgebra of $\mathcal{D}(Q_{2n})$ generated by the elements $x_1, \dots, x_{2n}, \delta_1, \dots, \delta_{2n}$, and d^{-1}. Let P' be the localization of the polynomial algebra P_{2n} at the powers of the element d. Then $\delta_1, \dots, \delta_{2n} \in \sum_{i=1}^{2n} P' \operatorname{ad}(a_i)$ and $\operatorname{ad}(a_1), \dots, \operatorname{ad}(a_{2n}) \in \sum_{i=1}^{2n} P' \delta_i$, hence the algebra U is generated (over K) by P' and $\operatorname{ad}(a_1), \dots, \operatorname{ad}(a_{2n})$. The algebra U can be viewed as a subalgebra of the ring of differential operators $\mathcal{D}(P')$. Now, the inclusions, $W \subseteq U \subseteq \mathcal{D}(P')$ imply $4n = \mathrm{GK}(W) \leqslant \mathrm{GK}(U) \leqslant \mathrm{GK}(\mathcal{D}(P')) = 2\mathrm{GK}(P') = 4n$, therefore $\mathrm{GK}(U) = 4n$. The algebra U is a factor algebra of an iterated Ore extension $V = P'[t_1; \operatorname{ad}(a_1)] \cdots [t_{2n}; \operatorname{ad}(a_{2n})]$. Since P' is a domain, so is the algebra V. The algebra P' is a finitely generated algebra of Gelfand–Kirillov dimension $2n$, hence $\mathrm{GK}(V) = \mathrm{GK}(P') + 2n = 4n$ (by [19, 8.2.11]). Since $\mathrm{GK}(V) = \mathrm{GK}(U)$ and any proper factor algebra of V has Gelfand–Kirillov dimension strictly less than $\mathrm{GK}(V)$ (by [19, 8.3.5], since V is a domain), the algebras V and U must be isomorphic. Therefore, the (commuting) elements $a_1, \dots, a_m, \operatorname{ad}(a_1), \dots, \operatorname{ad}(a_m)$ of the algebra U (and $A(P)$) must be algebraically independent.

(2) Let C be an isotropic subalgebra of the Poisson algebra P_{2n}. Note that $f_{A(P_{2n})} = f_{A_{2n}} = 1$ and $\mathrm{GK}(A_{2n}) = 4n$. By Theorem 8.1,

$$\mathrm{GK}(C) \leqslant \frac{4n}{2}\left(1 - \frac{1}{1+1}\right) = n. \qquad \square$$

REMARK. This result means that for the Poisson polynomial algebra P_{2n} the right-hand side of the inequality of Theorem 8.1 is the *exact* upper bound for the Gelfand–Kirillov dimension of isotropic subalgebras in P_{2n} since the polynomial subalgebra $K[x_1, \dots, x_n]$ of P_{2n} is isotropic.

References

[1] S.A. Amitsur, *Commutative linear differential operators*, Pacific J. Math. **8** (1958), 1–10.

[2] S.A. Amitsur and L.W. Small, *Polynomials over division rings*, Israel J. Math. **31** (3–4) (1978), 353–358.

[3] V.V. Bavula, *Identification of the Hilbert function and Poincaré series, and the dimension of modules over filtered rings*, Russian Acad. Sci. Izv. Math. **44** (2) (1995), 225–246.

[4] V. Bavula, *Filter dimension of algebras and modules, a simplicity criterion for generalized Weyl algebras*, Comm. Algebra **24** (6) (1996), 1971–1992.

[5] V. Bavula, *Krull, Gelfand–Kirillov, and filter dimensions of simple affine algebras*, J. Algebra **206** (1) (1998), 33–39.

[6] V. Bavula, *Krull, Gelfand–Kirillov, filter, faithful and Schur dimensions*, Infinite Length Modules, Bielefeld, 1998, Trends Math., Birkhäuser (2000), 149–166.

[7] V. Bavula, *Gelfand–Kirillov dimension of commutative subalgebras of simple infinite dimensional algebras and their quotient division algebras*, J. Reine Angew. Math. **582** (2005), 61–85, ArXiv:math.RA/0401263.

[8] I.N. Bernstein, *Modules over a ring of differential operators. An investigation of the fundamental solutions of equations with constant coefficients*, Funkcional. Anal. i Prilozhen. **5** (2) (1971), 1–16.

[9] J.-E. Björk, *Rings of Differential Operators*, North-Holland Math. Library **21**, Amsterdam (1979).

[10] A. Borel et al., *Algebraic $\mathcal{D}$-Modules*, Perspectives in Math. **2**, Academic Press (1987).

[11] S.C. Coutinho, *A Primer of Algebraic $\mathcal{D}$-Modules*, London Math. Soc. Student Texts **33**, Cambridge Univ. Press (1995).

[12] I.M. Gelfand and A.A. Kirillov, *Sur les corps liés aux algèbres enveloppantes des algèbres de Lie*, Publ. Math. IHES **31** (1966), 5–19.

[13] A. Joseph, *Gelfand–Kirillov dimension for algebras associated with the Weyl algebra*, Ann. Inst. H. Poincaré **17** (1972), 325–336.

[14] A. Joseph, *Sur les algèbres de Weyl*, Lecture Notes (1974), unpublished.

[15] A. Joseph, *A generalization of Quillen's lemma and its application to the Weyl algebras*, Israel J. Math. **28** (3) (1977), 177–192.

[16] G. Krause and T. Lenagan, *Growth of Algebras and Gelfand–Kirillov Dimension*, revised ed., Graduate Studies in Math. **22**, Amer. Math. Soc. (2000).

[17] L. Makar-Limanov, *Commutativity of certain centralizers in the rings $R_{n,k}$*, Funkcional. Anal. i Prilozhen. **4** (4) (1970), 78.

[18] L. Makar-Limanov, *On subalgebras of the first Weyl skewfield*, Comm. Algebra **19** (7) (1991), 1971–1982.

[19] J.C. McConnell and J.C. Robson, *Noncommutative Noetherian Rings*, Wiley (1987).

[20] C. Nastasescu and F. Van Oystaeyen, *Graded Ring Theory*, North-Holland (1982).

[21] R. Pierce, *Associative Algebras*, Springer (1982).

[22] R. Rentschler and P. Gabriel, *Sur la dimension des anneaux et ensembles ordonnés*, C. R. Acad. Sci. Paris Sér. A-B **265** (1967), A712–A715 (French).

[23] R. Resco, *Transcendental division algebras and simple Noetherian rings*, Israel J. Math. **32** (2–3) (1979), 236–256.

Section 4E
Lie Algebras

Gelfand–Tsetlin Bases for Classical Lie Algebras

A.I. Molev

School of Mathematics and Statistics, University of Sydney, NSW 2006, Australia
E-mail: *alexm@maths.usyd.edu.au*

Contents

HANDBOOK OF ALGEBRA, VOL. 4
Edited by M. Hazewinkel

1. Introduction

The theory of semisimple Lie algebras and their representations lies in the heart of modern mathematics. It has numerous connections with other areas of mathematics and physics. The simple Lie algebras over the field of complex numbers were classified in the work of Cartan and Killing in the 1930s. There are four infinite series A_n, B_n, C_n, D_n which are called the *classical Lie algebras*, and five exceptional Lie algebras E_6, E_7, E_8, F_4, G_2. The structure of these Lie algebras is uniformly described in terms of certain finite sets of vectors in a Euclidean space called *root systems*. Due to the Weyl complete reducibility theorem, the theory of finite-dimensional representations of the semisimple Lie algebras is largely reduced to the study of irreducible representations. The irreducibles are parametrized by their *highest weights*. The characters and dimensions are explicitly known by the *Weyl formula*. The reader is referred to, e.g., the books of Bourbaki, [11], Dixmier, [19], Humphreys, [61], or Goodman and Wallach, [45], for a detailed exposition of the theory.

However, the Weyl formula for the dimension does not use any explicit construction of the representations. Such constructions remained unknown until 1950 when Gelfand and Tsetlin[1] published two short papers, [41] and [42] (in Russian), where they solved the problem for the general linear Lie algebras (type A_n) and the orthogonal Lie algebras (types B_n and D_n), respectively. Later, Baird and Biedenharn, [4] (1963), commented on [41] as follows:

> This paper is extremely brief (three pages) and does not appear to have been translated in either the usual journal translations or the translations on group-theoretical subjects of the American Mathematical Society, or even referred to in the review articles on group theory by Gelfand himself. Moreover, the results are presented without the slightest hint as to the methods employed and contain not a single reference or citation of other work. In an effort to understand the meaning of this very impressive work, we were led to develop the proofs ....

Baird and Biedenharn employed the calculus of *Young patterns* to derive the Gelfand–Tsetlin formulas.[2] Their interest to the formulas was also motivated by the connection with the fundamental *Wigner coefficients*; see Section 2.4 below.

A year earlier (1962) Zhelobenko published an independent paper, [167], where he derived the *branching rules* for all classical Lie algebras. In his approach the representations are realized in a space of polynomials satisfying the "indicator system" of differential equations. He outlined a method to construct the *lowering operators* and to derive the matrix element formulas for the case of the general linear Lie algebra $\mathfrak{gl}_n$. An explicit "infinitesimal" form for the lowering operators as elements of the enveloping algebra was found by Nagel and Moshinsky, [106] (1964), and independently by Hou Pei-yu, [59] (1966). The latter work relies on Zhelobenko's results, [167], and also contains a derivation of the Gelfand–Tsetlin formulas alternative to that of Baird and Biedenharn. This approach was further developed in the book by Zhelobenko, [168], which contains its detailed account.

The work of Nagel and Moshinsky was extended to the orthogonal Lie algebras $\mathfrak{o}_N$ by Pang and Hecht, [133], and Wong, [164], who produced explicit infinitesimal expres-

---

[1] Some authors and translators write this name in English as *Zetlin*, *Tzetlin*, *Cetlin*, or *Tseitlin*.

[2] An indication of the proof of the formulas of [41] is contained in a footnote in the paper of Gelfand and Graev, [39] (1965). It says that for the proof one has "to verify the commutation relations ...; this is done by direct calculation".

sions for the lowering operators and gave a derivation of the formulas of Gelfand and Tsetlin, [42].

During the half a century passed since the work of Gelfand and Tsetlin, many different approaches were developed to construct bases of the representations of the classical Lie algebras. New interpretations of the lowering operators and new proofs of the Gelfand–Tsetlin formulas were discovered by several authors. In particular, Gould, [46–48,50], employed the *characteristic identities* of Bracken and Green, [12,54], to calculate the Wigner coefficients and matrix elements of generators of $\mathfrak{gl}_n$ and $\mathfrak{o}_N$. The *extremal projector* discovered by Asherova, Smirnov and Tolstoy, [1–3], turned out to be a powerful instrument in the representation theory of simple Lie algebras. It plays an essential role in the theory of *Mickelsson algebras* developed by Zhelobenko which has a wide spectrum of applications from the branching rules and reduction problems to the classification of Harish-Chandra modules; see Zhelobenko's expository paper, [173], and his book, [174]. Two different *quantum minor* interpretations of the lowering and raising operators were given by Nazarov and Tarasov, [109], and the author, [96]. These techniques are based on the theory of quantum algebras called the *Yangians* and allow an independent derivation of the matrix element formulas. We shall discuss the above approaches in more detail in Sections 2.3, 2.4 and 2.5 below.

A quite different method to construct modules over the classical Lie algebras is developed in the papers by King and El-Sharkaway, [69], Berele, [6], King and Welsh, [70], Koike and Terada, [73], Proctor, [138], Nazarov, [107]. In particular, bases in the representations of the orthogonal and symplectic Lie algebras parametrized by $\mathfrak{o}_N$-standard or $\mathfrak{sp}_{2n}$-standard Young tableaux are constructed. This method provides an algorithm for the calculation of the representation matrices. It is based on the Weyl realization of the representations of the classical groups in tensor spaces; see Weyl, [159]. A detailed exposition of the theory of the classical groups together with many recent developments are presented in the book by Goodman and Wallach, [45].

Bases with special properties in the universal enveloping algebra for a simple Lie algebra $\mathfrak{g}$ and in some modules over $\mathfrak{g}$ were constructed by Lakshmibai, Musili and Seshadri, [75], Littelmann, [81,82], Chari and Xi, [15] (*monomial* bases); De Concini and Kazhdan, [18], Xi, [166] (*special* bases and their q-analogs); Gelfand and Zelevinsky, [44], Retakh and Zelevinsky, [140], Mathieu, [85] (*good* bases); Lusztig, [83], Kashiwara, [66], Du, [35,36] (*canonical* or *crystal* bases); see also Mathieu, [86], for a review and more references. Algorithms for computing the global crystal bases of irreducible modules for the classical Lie algebras were recently given by Leclerc and Toffin, [76], and Lecouvey, [77,78]. In general, no explicit formulas are known, however, for the matrix elements of the generators in such bases other than those of Gelfand and Tsetlin type. It is known, although, that for the canonical bases the matrix elements of the standard generators are nonnegative integers. Some classes of representations of the symplectic, odd orthogonal and the Lie algebras of type G_2 were explicitly constructed by Donnelly, [23,24,26], and Donnelly, Lewis and Pervine, [27]. The constructions were applied to establish combinatorial properties of the supporting graphs of the representations and were inspired by the earlier results of Proctor, [134,135,137]. Another graph-theoretic approach is developed by Wildberger, [160–163], to construct simple Lie algebras and their minuscule representations; see also Stembridge, [146].

We now discuss the main idea which leads to the construction of the Gelfand–Tsetlin bases. The first point is to regard a given classical Lie algebra not as a single object but as a part of a chain of subalgebras with natural embeddings. We illustrate this idea using representations of the symmetric groups $\mathfrak{S}_n$ as an example. Consider the chain of subgroups

$$\mathfrak{S}_1 \subset \mathfrak{S}_2 \subset \cdots \subset \mathfrak{S}_n, \tag{1.1}$$

where the subgroup $\mathfrak{S}_k$ of $\mathfrak{S}_{k+1}$ consists of the permutations which fix the index $k+1$ of the set $\{1, 2, \ldots, k+1\}$. The irreducible representations of the group $\mathfrak{S}_n$ are indexed by partitions λ of n. A partition $\lambda = (\lambda_1, \ldots, \lambda_l)$ with $\lambda_1 \geqslant \lambda_2 \geqslant \cdots \geqslant \lambda_l$ is depicted graphically as a *Young diagram* which consists of l left-justified rows of boxes so that the top row contains λ_1 boxes, the second row λ_2 boxes, etc. Denote by $V(\lambda)$ the irreducible representation of $\mathfrak{S}_n$ corresponding to the partition λ. One of the central results of the representation theory of the symmetric groups is the following *branching rule* which describes the restriction of $V(\lambda)$ to the subgroup $\mathfrak{S}_{n-1}$:

$$V(\lambda)|_{\mathfrak{S}_{n-1}} = \bigoplus_{\mu} V'(\mu),$$

summed over all partitions μ whose Young diagram is obtained from that of λ by removing one box. Here $V'(\mu)$ denotes the irreducible representation of $\mathfrak{S}_{n-1}$ corresponding to a partition μ. Thus, the restriction of $V(\lambda)$ to $\mathfrak{S}_{n-1}$ is *multiplicity-free*, i.e., it contains each irreducible representation of $\mathfrak{S}_{n-1}$ at most once. This makes it possible to obtain a natural parameterization of the basis vectors in $V(\lambda)$ by taking its further restrictions to the subsequent subgroups of the chain (1.1). Namely, the basis vectors will be parametrized by sequences of partitions

$$\lambda^{(1)} \to \lambda^{(2)} \to \cdots \to \lambda^{(n)} = \lambda,$$

where $\lambda^{(k)}$ is obtained from $\lambda^{(k+1)}$ by removing one box. Equivalently, each sequence of this type can be regarded as a *standard tableau of shape* λ which is obtained by writing the numbers $1, \ldots, n$ into the boxes of λ in such a way that the numbers increase along the rows and down the columns. In particular, the dimension of $V(\lambda)$ equals the number of standard tableaux of shape λ. There is only one irreducible representation of the trivial group $\mathfrak{S}_1$ therefore the procedure defines basis vectors up to a scalar factor. The corresponding basis is called the *Young basis*. The symmetric group $\mathfrak{S}_n$ is generated by the adjacent transpositions $s_i = (i, i+1)$. The construction of the representation $V(\lambda)$ can be completed by deriving explicit formulas for the action of the elements s_i in the basis which are also due to A. Young. This realization of $V(\lambda)$ is usually called *Young's orthogonal* (or *seminormal*) *form*. The details can be found, e.g., in James and Kerber, [63], and Sagan, [141]; see also Okounkov and Vershik, [112], where an alternative construction of the Young basis is produced. Branching rules and the corresponding Bratteli diagrams were employed by Halverson and Ram, [56], Leduc and Ram, [79], Ram, [139], to compute irreducible representations of the Iwahori–Hecke algebras and some families of centralizer algebras.

Quite a similar method can be applied to representations of the classical Lie algebras. Consider the general linear Lie algebra $\mathfrak{gl}_n$ which consists of complex $n \times n$-matrices with the usual matrix commutator. The chain (1.1) is now replaced by

$$\mathfrak{gl}_1 \subset \mathfrak{gl}_2 \subset \cdots \subset \mathfrak{gl}_n,$$

with natural embeddings $\mathfrak{gl}_k \subset \mathfrak{gl}_{k+1}$. The orthogonal Lie algebra $\mathfrak{o}_N$ can be regarded as a subalgebra of $\mathfrak{gl}_N$ which consists of skew-symmetric matrices. Again, we have a natural chain

$$\mathfrak{o}_2 \subset \mathfrak{o}_3 \subset \cdots \subset \mathfrak{o}_N. \tag{1.2}$$

Both restrictions $\mathfrak{gl}_n \downarrow \mathfrak{gl}_{n-1}$ and $\mathfrak{o}_N \downarrow \mathfrak{o}_{N-1}$ are multiplicity-free so that the application of the argument which we used for the chain (1.1) produces basis vectors in an irreducible representation of $\mathfrak{gl}_n$ or $\mathfrak{o}_N$. With an appropriate normalization, these bases are precisely those of Gelfand and Tsetlin given in [41] and [42]. Instead of the standard tableaux, the basis vectors here are parametrized by combinatorial objects called the *Gelfand–Tsetlin patterns*.

However, this approach does not work for the symplectic Lie algebras $\mathfrak{sp}_{2n}$ since the restriction $\mathfrak{sp}_{2n} \downarrow \mathfrak{sp}_{2n-2}$ is not multiplicity-free. The multiplicities are given by Zhelobenko's branching rule, [167], which was re-discovered later by Hegerfeldt, [58].[3] Various attempts to fix this problem were made by several authors. A natural idea is to introduce an intermediate Lie algebra "$\mathfrak{sp}_{2n-1}$" and try to restrict an irreducible representation of $\mathfrak{sp}_{2n}$ first to this subalgebra and then to $\mathfrak{sp}_{2n-2}$ in the hope to get simple spectra in the two restrictions. Such intermediate subalgebras and their representations were studied by Gelfand and Zelevinsky, [43], Proctor, [136], Shtepin, [142]. The drawback of this approach is the fact that the Lie algebra $\mathfrak{sp}_{2n-1}$ is not reductive so that the restriction of an irreducible representation of $\mathfrak{sp}_{2n}$ to $\mathfrak{sp}_{2n-1}$ is not completely reducible. In some sense, the separation of multiplicities can be achieved by constructing a filtration of $\mathfrak{sp}_{2n-1}$-modules; cf. Shtepin, [142].

Another idea is to use the restriction $\mathfrak{gl}_{2n} \downarrow \mathfrak{sp}_{2n}$. Gould and Kalnins, [51,53], constructed a basis for the representations of the symplectic Lie algebras parametrized by a subset of the Gelfand–Tsetlin $\mathfrak{gl}_{2n}$-patterns. Some matrix element formulas are also derived by using the $\mathfrak{gl}_{2n}$-action. A similar observation is made independently by Kirillov, [71], and Proctor, [136]. A description of the Gelfand–Tsetlin patterns for $\mathfrak{sp}_{2n}$ and $\mathfrak{o}_N$ can be obtained by regarding them as fixed points of involutions of the Gelfand–Tsetlin patterns for the corresponding Lie algebra $\mathfrak{gl}_N$.

The lowering operators in the symplectic case were given by Mickelsson, [91]; see also Bincer, [9,10]. The application of ordered monomials in the lowering operators to the highest vector yields a basis of the representation. However, the action of the Lie algebra generators in such a basis does not seem to be computable. The reason is the fact that, unlike the cases of $\mathfrak{gl}_n$ and $\mathfrak{o}_N$, the lowering operators do not commute so that the basis depends on the chosen ordering. A "hidden symmetry" has been needed (cf. Cherednik, [17]) to make

[3] Some western authors referred to Hegerfeldt's result as the original derivation of the rule.

a natural choice of an appropriate combination of the lowering operators. New ideas which led to a construction of a Gelfand–Tsetlin type basis for any irreducible finite-dimensional representation of $\mathfrak{sp}_{2n}$ came from the theory of *quantized enveloping algebras*. This is a part of the theory of *quantum groups* originating from the work of Drinfeld, [28,30], and Jimbo, [64]. A particular class of quantized enveloping algebras called *twisted Yangians* introduced by Olshanski, [118], plays the role of the hidden symmetries for the construction of the basis. We refer the reader to the book by Chari and Pressley, [14], and the review papers [103] and [104] for detailed expositions of the properties of these algebras and their origins. For each classical Lie algebra we attach the *Yangian* $\mathrm{Y}(N) = \mathrm{Y}(\mathfrak{gl}_N)$, or the *twisted Yangian* $\mathrm{Y}^{\pm}(N)$ as follows

type A_n	type B_n	type C_n	type D_n
$\mathrm{Y}(n+1)$	$\mathrm{Y}^+(2n+1)$	$\mathrm{Y}^-(2n)$	$\mathrm{Y}^+(2n)$.

The algebra $\mathrm{Y}(N)$ was first introduced in the work of Faddeev and the St.-Petersburg school in relation with the *inverse scattering method*; see for instance Takhtajan and Faddeev, [147], Kulish and Sklyanin, [74]. Olshanski, [118], introduced the twisted Yangians in relation with his *centralizer construction*; see also [105]. In particular, he established the following key fact which plays an important role in the basis construction. Given irreducible representations $V(\lambda)$ and $V'(\mu)$ of $\mathfrak{sp}_{2n}$ and $\mathfrak{sp}_{2n-2}$, respectively, there exists a natural irreducible action of the algebra $\mathrm{Y}^-(2)$ on the space $\mathrm{Hom}_{\mathfrak{sp}_{2n-2}}(V'(\mu), V(\lambda))$. The homomorphism space is isomorphic to the subspace $V(\lambda)^+_\mu$ of $V(\lambda)$ which is spanned by the highest vectors of weight μ for the subalgebra $\mathfrak{sp}_{2n-2}$. Finite-dimensional irreducible representations of the twisted Yangians were classified later in [97]. In particular, it turned out that the representation $V(\lambda)^+_\mu$ of $\mathrm{Y}^-(2)$ can be extended to the Yangian $\mathrm{Y}(2)$. Another proof of this fact was given recently by Nazarov, [107]. The algebra $\mathrm{Y}(2)$ and its representations are well-studied; see Tarasov, [149], Chari and Pressley, [13]. A large class of representations of $\mathrm{Y}(2)$ admits Gelfand–Tsetlin-type bases associated with the inclusion $\mathrm{Y}(1) \subset \mathrm{Y}(2)$; see [96]. This allows one to get a natural basis in the space $V(\lambda)^+_\mu$ and then by induction to get a basis in the entire space $V(\lambda)$. Moreover, it turns out to be possible to write down explicit formulas for the action of the generators of the symplectic Lie algebra in this basis; see the author's paper [98] for more details. This construction together with the work of Gelfand and Tsetlin thus provides explicit realizations of all finite-dimensional irreducible representations of the classical Lie algebras.

The same method can be applied to the pairs of the orthogonal Lie algebras $\mathfrak{o}_{N-2} \subset \mathfrak{o}_N$. Here the corresponding space $V(\lambda)^+_\mu$ is a natural $\mathrm{Y}^+(2)$-module which can also be extended to a $\mathrm{Y}(2)$-module. This leads to a construction of a natural basis in the representation $V(\lambda)$ and allows one to explicitly calculate the representation matrices; see [99,100]. This realization of $V(\lambda)$ is alternative to that of Gelfand and Tsetlin, [42]. To compare the two constructions, note that the basis of [42] in the orthogonal case lacks the *weight* property, i.e., the basis vectors are not eigenvectors for the Cartan subalgebra. The reason for that is the fact that the chain (1.2) involves Lie algebras of different types (B and D) and the embeddings are not compatible with the root systems. In the new approach we use instead the chains

$$\mathfrak{o}_2 \subset \mathfrak{o}_4 \subset \cdots \subset \mathfrak{o}_{2n} \quad \text{and} \quad \mathfrak{o}_3 \subset \mathfrak{o}_5 \subset \cdots \subset \mathfrak{o}_{2n+1}.$$

The embeddings here "respect" the root systems so that the basis of $V(\lambda)$ possesses the weight property in both the symplectic and orthogonal cases. However, the new weight bases, in turn, lack the *orthogonality* property of the Gelfand–Tsetlin bases: the latter are orthogonal with respect to the standard inner product in the representation space $V(\lambda)$. It is an open problem to construct a natural basis of $V(\lambda)$ in the B, C and D cases which would simultaneously accommodate the two properties.

This chapter is structured as follows. In Section 2 we review the construction of the Gelfand–Tsetlin basis for the general linear Lie algebra and discuss its various versions. We start by applying the most elementary approach which consists of using explicit formulas for the lowering operators in a way similar to the pioneering works of the sixties. Remarkably, these operators admit several other presentations which reflect different approaches to the problem developed in the literature. First, we outline the general theory of extremal projectors and Mickelsson algebras as a natural way to work with lowering operators. Next, we describe the $\mathfrak{gl}_n$-type *Mickelsson–Zhelobenko algebra* which is then used to prove the branching rule and derive the matrix element formulas. Further, we outline the Gould construction based upon the characteristic identities. Finally, we produce quantum minor formulas for the lowering operators inspired by the Yangian approach and describe the action of the Drinfeld generators in the Gelfand–Tsetlin basis.

In Section 3 we produce weight bases for representations of the orthogonal and symplectic Lie algebras. Here we describe the relevant Mickelsson–Zhelobenko algebra, formulate the branching rules and construct the basis vectors. Then we outline the properties of the (twisted) Yangians and their representations and explain their relationship with the lowering and raising operators. Finally, we sketch the main ideas in the calculation of the matrix element formulas.

Section 4 is devoted to the Gelfand–Tsetlin bases for the orthogonal Lie algebras. We outline the basis construction along the lines of the general method of Mickelsson algebras.

At the end of each section we give brief bibliographical comments pointing towards the original articles and to the references where the proofs or further details can be found.

2. Gelfand–Tsetlin bases for representations of $\mathfrak{gl}_n$

Let E_{ij}, $i, j = 1, \ldots, n$, denote the standard basis of the general linear Lie algebra $\mathfrak{gl}_n$ over the field of complex numbers. The subalgebra $\mathfrak{gl}_{n-1}$ is spanned by the basis elements E_{ij} with $i, j = 1, \ldots, n-1$. Denote by $\mathfrak{h} = \mathfrak{h}_n$ the diagonal Cartan subalgebra in $\mathfrak{gl}_n$. The elements $E_{11}, \ldots, E_{nn}$ form a basis of $\mathfrak{h}$.

Finite-dimensional irreducible representations of $\mathfrak{gl}_n$ are in a one-to-one correspondence with n-tuples of complex numbers $\lambda = (\lambda_1, \ldots, \lambda_n)$ such that

$$\lambda_i - \lambda_{i+1} \in \mathbb{Z}_+ \quad \text{for } i = 1, \ldots, n-1. \tag{2.1}$$

Here $\mathbb{Z}_+ = \{i \in \mathbb{Z}\colon\ i \geqslant 0\}$.

Such an n-tuple λ is called the *highest weight* of the corresponding representation which we shall denote by $L(\lambda)$. It contains a unique, up to a multiple, nonzero vector ξ (the *highest weight vector* (highest vector)) such that $E_{ii}\xi = \lambda_i \xi$ for $i = 1, \dots, n$ and $E_{ij}\xi = 0$ for $1 \leqslant i < j \leqslant n$.

The following theorem is the *branching rule* for the reduction $\mathfrak{gl}_n \downarrow \mathfrak{gl}_{n-1}$.

THEOREM 2.1. *The restriction of $L(\lambda)$ to the subalgebra $\mathfrak{gl}_{n-1}$ is isomorphic to the direct sum of pairwise inequivalent irreducible representations*

$$L(\lambda)|_{\mathfrak{gl}_{n-1}} \simeq \bigoplus_{\mu} L'(\mu),$$

summed over the highest weights μ satisfying the betweenness conditions

$$\lambda_i - \mu_i \in \mathbb{Z}_+ \quad \textit{and} \quad \mu_i - \lambda_{i+1} \in \mathbb{Z}_+ \quad \textit{for } i = 1, \dots, n-1. \tag{2.2}$$

The rule can presumably be attributed to I. Schur who was the first to discover the representation-theoretic significance of a particular class of symmetric polynomials which now bear his name. Without loss of generality we may regard λ as a partition: we can take the composition of $L(\lambda)$ with an appropriate automorphism of $\mathrm{U}(\mathfrak{gl}_n)$ which sends E_{ij} to $E_{ij} + \delta_{ij} a$ for some $a \in \mathbb{C}$. The *character* of $L(\lambda)$ regarded as a GL_n-module is the *Schur polynomial* $s_\lambda(x)$, $x = (x_1, \dots, x_n)$, defined by

$$s_\lambda(x) = \mathrm{tr}\big(g, L(\lambda)\big),$$

where $x_1, \dots, x_n$ are the eigenvalues of $g \in GL_n$. The Schur polynomial is symmetric in the x_i and can be given by the explicit combinatorial formula

$$s_\lambda(x) = \sum_T x^T, \tag{2.3}$$

summed over the *semistandard tableaux* T of shape λ (cf. Remark 2.2 below), where x^T is the monomial containing x_i with the power equal to the number of occurrences of i in T; see, e.g., Macdonald, [84, Chapter 1], or Sagan, [141, Chapter 4], for more details. To find out what happens when $L(\lambda)$ is restricted to GL_{n-1} we just need to put $x_n = 1$ into formula (2.3). The right-hand side will then be written as the sum of the Schur polynomials $s_\mu(x_1, \dots, x_{n-1})$ with μ satisfying (2.2).

On the other hand, the multiplicity-freeness of the reduction $\mathfrak{gl}_n \downarrow \mathfrak{gl}_{n-1}$ can be explained by the fact that the vector space $\mathrm{Hom}_{\mathfrak{gl}_{n-1}}(L'(\mu), L(\lambda))$ bears a natural irreducible representation of the centralizer $\mathrm{U}(\mathfrak{gl}_n)^{\mathfrak{gl}_{n-1}}$; see, e.g., Dixmier, [19, Section 9.1]. However, the centralizer is a commutative algebra and therefore if the homomorphism space is nonzero then it must be one-dimensional.

The branching rule is implicit in the formulas of Gelfand and Tsetlin, [41]. A proof based upon an explicit realization of the representations of GL_n was given by Zhelobenko, [167]. We outline a proof of Theorem 2.1 below in Section 2.3 which employs the modern theory of Mickelsson algebras following Zhelobenko, [174]. Two other proofs can be found in Goodman and Wallach, [45, Chapters 8 and 12].

Successive applications of the branching rule to the subalgebras of the chain

$$\mathfrak{gl}_1 \subset \mathfrak{gl}_2 \subset \cdots \subset \mathfrak{gl}_{n-1} \subset \mathfrak{gl}_n$$

yield a parameterization of basis vectors in $L(\lambda)$ by the combinatorial objects called the *Gelfand–Tsetlin patterns*. Such a pattern Λ (associated with λ) is an array of row vectors

$$\begin{array}{ccccccc} \lambda_{n1} & & \lambda_{n2} & & \cdots & & \lambda_{nn} \\ & \lambda_{n-1,1} & & \cdots & & \lambda_{n-1,n-1} & \\ & \cdots & & \cdots & & \cdots & \\ & & \lambda_{21} & & \lambda_{22} & & \\ & & & \lambda_{11} & & & \end{array}$$

where the upper row coincides with λ and the following conditions hold

$$\lambda_{ki} - \lambda_{k-1,i} \in \mathbb{Z}_+, \qquad \lambda_{k-1,i} - \lambda_{k,i+1} \in \mathbb{Z}_+, \quad i = 1, \dots, k-1, \tag{2.4}$$

for each $k = 2, \dots, n$.

REMARK 2.2. If the highest weight λ is a partition then there is a natural bijection between the patterns associated with λ and the *semistandard* λ-tableaux with entries in $\{1, \dots, n\}$. Namely, the pattern Λ can be viewed as the sequence of partitions

$$\lambda^{(1)} \subseteq \lambda^{(2)} \subseteq \cdots \subseteq \lambda^{(n)} = \lambda,$$

with $\lambda^{(k)} = (\lambda_{k1}, \dots, \lambda_{kk})$. Conditions (2.4) mean that the skew diagram $\lambda^{(k)}/\lambda^{(k-1)}$ is a *horizontal strip*; see, e.g., Macdonald, [84, Chapter 1]. The corresponding semistandard tableau is obtained by placing the entry k into each box of $\lambda^{(k)}/\lambda^{(k-1)}$.

The *Gelfand–Tsetlin basis* of $L(\lambda)$ is provided by the following theorem. Let us set $l_{ki} = \lambda_{ki} - i + 1$.

THEOREM 2.3. *There exists a basis $\{\xi_\Lambda\}$ in $L(\lambda)$ parametrized by all patterns Λ such that the action of generators of $\mathfrak{gl}_n$ is given by the formulas*

$$E_{kk}\xi_\Lambda = \left(\sum_{i=1}^{k} \lambda_{ki} - \sum_{i=1}^{k-1} \lambda_{k-1,i}\right)\xi_\Lambda, \tag{2.5}$$

$$E_{k,k+1}\xi_\Lambda = -\sum_{i=1}^{k} \frac{(l_{ki} - l_{k+1,1}) \cdots (l_{ki} - l_{k+1,k+1})}{(l_{ki} - l_{k1}) \cdots \wedge \cdots (l_{ki} - l_{kk})} \xi_{\Lambda+\delta_{ki}}, \tag{2.6}$$

$$E_{k+1,k}\xi_\Lambda = \sum_{i=1}^{k} \frac{(l_{ki} - l_{k-1,1}) \cdots (l_{ki} - l_{k-1,k-1})}{(l_{ki} - l_{k1}) \cdots \wedge \cdots (l_{ki} - l_{kk})} \xi_{\Lambda-\delta_{ki}}. \tag{2.7}$$

The arrays $\Lambda \pm \delta_{ki}$ *are obtained from* Λ *by replacing* λ_{ki} *by* $\lambda_{ki} \pm 1$. *It is supposed that* $\xi_\Lambda = 0$ *if the array* Λ *is not a pattern; the symbol* $\wedge$ *indicates that the zero factor in the denominator is skipped.*

A construction of the basis vectors is given in Theorem 2.7 below. A derivation of the matrix element formulas (2.5)–(2.7) is outlined in Section 2.3.

The vector space $L(\lambda)$ is equipped with a contravariant inner product $\langle\,,\rangle$. It is uniquely determined by the conditions

$$\langle \xi, \xi \rangle = 1 \quad \text{and} \quad \langle E_{ij}\eta, \zeta \rangle = \langle \eta, E_{ji}\zeta \rangle$$

for any vectors $\eta, \zeta \in L(\lambda)$ and any indices i, j. In other words, for the adjoint operator for E_{ij} with respect to the inner product we have $(E_{ij})^* = E_{ji}$.

PROPOSITION 2.4. *The basis* $\{\xi_\Lambda\}$ *is orthogonal with respect to the inner product* $\langle\,,\rangle$. *Moreover, we have*

$$\langle \xi_\Lambda, \xi_\Lambda \rangle = \prod_{k=2}^{n} \prod_{1 \leqslant i \leqslant j < k} \frac{(l_{ki} - l_{k-1,j})!}{(l_{k-1,i} - l_{k-1,j})!} \prod_{1 \leqslant i < j \leqslant k} \frac{(l_{ki} - l_{kj} - 1)!}{(l_{k-1,i} - l_{kj} - 1)!}.$$

The formulas of Theorem 2.3 can therefore be rewritten in the orthonormal basis

$$\zeta_\Lambda = \xi_\Lambda / \|\xi_\Lambda\|, \qquad \|\xi_\Lambda\|^2 = \langle \xi_\Lambda, \xi_\Lambda \rangle. \tag{2.8}$$

They were presented in this form in the original work by Gelfand and Tsetlin, [41]. A proof of Proposition 2.4 will be outlined in Section 2.3.

2.1. *Construction of the basis: lowering and raising operators*

For each $i = 1, \dots, n-1$ introduce the following elements of the universal enveloping algebra $\mathrm{U}(\mathfrak{gl}_n)$

$$z_{in} = \sum_{i > i_1 > \cdots > i_s \geqslant 1} E_{ii_1} E_{i_1 i_2} \cdots E_{i_{s-1} i_s} E_{i_s n} (h_i - h_{j_1}) \cdots (h_i - h_{j_r}), \tag{2.9}$$

$$z_{ni} = \sum_{i < i_1 < \cdots < i_s < n} E_{i_1 i} E_{i_2 i_1} \cdots E_{i_s i_{s-1}} E_{n i_s} (h_i - h_{j_1}) \cdots (h_i - h_{j_r}), \tag{2.10}$$

where s runs over nonnegative integers, $h_i = E_{ii} - i + 1$ and $\{j_1, \dots, j_r\}$ is the complementary subset to $\{i_1, \dots, i_s\}$ in the set $\{1, \dots, i-1\}$ or $\{i+1, \dots, n-1\}$, respectively. For instance,

$$z_{13} = E_{13}, \qquad z_{23} = E_{23}(h_2 - h_1) + E_{21}E_{13},$$
$$z_{32} = E_{32}, \qquad z_{31} = E_{31}(h_1 - h_2) + E_{21}E_{32}.$$

Consider now the irreducible finite-dimensional representation $L(\lambda)$ of $\mathfrak{gl}_n$ with the highest weight $\lambda = (\lambda_1, \dots, \lambda_n)$ and the highest vector ξ. Denote by $L(\lambda)^+$ the subspace of $\mathfrak{gl}_{n-1}$-highest vectors in $L(\lambda)$:

$$L(\lambda)^+ = \{\eta \in L(\lambda) \mid E_{ij}\eta = 0,\ 1 \leqslant i < j < n\}.$$

Given a $\mathfrak{gl}_{n-1}$-weight $\mu = (\mu_1, \dots, \mu_{n-1})$ we denote by $L(\lambda)^+_\mu$ the corresponding weight subspace in $L(\lambda)^+$:

$$L(\lambda)^+_\mu = \{\eta \in L(\lambda)^+ \mid E_{ii}\eta = \mu_i\eta,\ i = 1, \dots, n-1\}.$$

The main property of the elements z_{ni} and z_{in} is described by the following lemma.

LEMMA 2.5. *Let $\eta \in L(\lambda)^+_\mu$. Then for any $i = 1, \dots, n-1$ we have*

$$z_{in}\eta \in L(\lambda)^+_{\mu+\delta_i} \quad \text{and} \quad z_{ni}\eta \in L(\lambda)^+_{\mu-\delta_i},$$

where the weight $\mu \pm \delta_i$ is obtained from μ by replacing μ_i with $\mu_i \pm 1$.

This result allows us to regard the elements z_{in} and z_{ni} as operators in the space $L(\lambda)^+$. They are called the *raising* and *lowering operators*, respectively. By the branching rule (Theorem 2.1) the space $L(\lambda)^+_\mu$ is one-dimensional if the conditions (2.2) hold and it is zero otherwise. The following lemma will be proved in Section 2.3.

LEMMA 2.6. *Suppose that μ satisfies the betweenness conditions* (2.2). *Then the vector*

$$\xi_\mu = z_{n1}^{\lambda_1-\mu_1} \cdots z_{n,n-1}^{\lambda_{n-1}-\mu_{n-1}}\xi$$

is nonzero. Moreover, the space $L(\lambda)^+_\mu$ is spanned by ξ_μ.

The $\mathrm{U}(\mathfrak{gl}_{n-1})$-span of each nonzero vector ξ_μ is a $\mathfrak{gl}_{n-1}$-module isomorphic to $L'(\mu)$. Iterating the construction of the vectors ξ_μ for each pair of Lie algebras $\mathfrak{gl}_{k-1} \subset \mathfrak{gl}_k$ we shall be able to get a basis in the entire space $L(\lambda)$.

THEOREM 2.7. *The basis vectors ξ_Λ of Theorem* 2.3 *can be given by the formula*

$$\xi_\Lambda = \prod_{k=2,\dots,n}^{\rightarrow} \left(z_{k1}^{\lambda_{k1}-\lambda_{k-1,1}} \cdots z_{k,k-1}^{\lambda_{k,k-1}-\lambda_{k-1,k-1}}\right)\xi, \tag{2.11}$$

where the factors in the product are ordered according to increasing indices.

2.2. *Mickelsson algebra theory*

The lowering and raising operators z_{ni} and z_{in} in the space $L(\lambda)^+$ (see Lemma 2.5) satisfy some quadratic relations with rational coefficients in the parameters of the highest weights. These relations can be regarded in a representation independent form with a suitable interpretation of the coefficients as rational functions in the elements of the Cartan subalgebra $\mathfrak{h}$. In this abstract form the algebras of lowering and raising operators were introduced by Mickelsson, [92], who, however, did not use any rational extensions of the algebra $\mathrm{U}(\mathfrak{h})$. The importance of this extension was realized by Zhelobenko, [169,170], who developed a general structure theory of these algebras which he called *Mickelsson algebras*. Another important ingredient is the theory of *extremal projectors* which originated in the work of Asherova, Smirnov and Tolstoy, [1–3], and was further developed by Zhelobenko, [173,174].

Let $\mathfrak{g}$ be a Lie algebra over $\mathbb{C}$ and let $\mathfrak{k}$ be a subalgebra reductive in $\mathfrak{g}$. This means that the adjoint $\mathfrak{k}$-module $\mathfrak{g}$ is completely reducible. In particular, $\mathfrak{k}$ is a reductive Lie algebra. Fix a Cartan subalgebra $\mathfrak{h}$ of $\mathfrak{k}$ and a triangular decomposition

$$\mathfrak{k} = \mathfrak{k}^- \oplus \mathfrak{h} \oplus \mathfrak{k}^+.$$

The subalgebras $\mathfrak{k}^-$ and $\mathfrak{k}^+$ are respectively spanned by the negative and positive root vectors $e_{-\alpha}$ and e_α with α running over the set of positive roots Δ^+ of $\mathfrak{k}$ with respect to $\mathfrak{h}$. The root vectors will be assumed to be normalized in such a way that

$$[e_\alpha, e_{-\alpha}] = h_\alpha, \qquad \alpha(h_\alpha) = 2 \tag{2.12}$$

for all $\alpha \in \Delta^+$.

Let $\mathrm{J} = \mathrm{U}(\mathfrak{g})\mathfrak{k}^+$ be the left ideal of $\mathrm{U}(\mathfrak{g})$ generated by $\mathfrak{k}^+$. Its normalizer $\operatorname{Norm}\mathrm{J}$ is the subalgebra of $\mathrm{U}(\mathfrak{g})$ defined by

$$\operatorname{Norm}\mathrm{J} = \{u \in \mathrm{U}(\mathfrak{g}) \mid \mathrm{J}u \subseteq \mathrm{J}\}.$$

Then J is a two-sided ideal of $\operatorname{Norm}\mathrm{J}$ and the *Mickelsson algebra* $\mathrm{S}(\mathfrak{g}, \mathfrak{k})$ is defined as the quotient

$$\mathrm{S}(\mathfrak{g}, \mathfrak{k}) = \operatorname{Norm}\mathrm{J}/\mathrm{J}.$$

Let $\mathrm{R}(\mathfrak{h})$ denote the field of fractions of the commutative algebra $\mathrm{U}(\mathfrak{h})$. In what follows it is convenient to consider the extension $\mathrm{U}'(\mathfrak{g})$ of the universal enveloping algebra $\mathrm{U}(\mathfrak{g})$ defined by

$$\mathrm{U}'(\mathfrak{g}) = \mathrm{U}(\mathfrak{g}) \otimes_{\mathrm{U}(\mathfrak{h})} \mathrm{R}(\mathfrak{h}).$$

Let $\mathrm{J}' = \mathrm{U}'(\mathfrak{g})\mathfrak{k}^+$ be the left ideal of $\mathrm{U}'(\mathfrak{g})$ generated by $\mathfrak{k}^+$. Exactly as with the ideal J above, J' is a two-sided ideal of the normalizer $\mathrm{Norm}\,\mathrm{J}'$ and the *Mickelsson–Zhelobenko algebra*[4] $\mathrm{Z}(\mathfrak{g},\mathfrak{k})$ is defined as the quotient

$$\mathrm{Z}(\mathfrak{g},\mathfrak{k}) = \mathrm{Norm}\,\mathrm{J}'/\mathrm{J}'.$$

Clearly, $\mathrm{Z}(\mathfrak{g},\mathfrak{k})$ is an extension of the Mickelsson algebra $\mathrm{S}(\mathfrak{g},\mathfrak{k})$,

$$\mathrm{Z}(\mathfrak{g},\mathfrak{k}) = \mathrm{S}(\mathfrak{g},\mathfrak{k}) \otimes_{\mathrm{U}(\mathfrak{h})} \mathrm{R}(\mathfrak{h}).$$

An equivalent definition of the algebra $\mathrm{Z}(\mathfrak{g},\mathfrak{k})$ can be given by using the quotient space

$$\mathrm{M}(\mathfrak{g},\mathfrak{k}) = \mathrm{U}'(\mathfrak{g})/\mathrm{J}'.$$

The Mickelsson–Zhelobenko algebra $\mathrm{Z}(\mathfrak{g},\mathfrak{k})$ coincides with the subspace of $\mathfrak{k}$-highest vectors in $\mathrm{M}(\mathfrak{g},\mathfrak{k})$

$$\mathrm{Z}(\mathfrak{g},\mathfrak{k}) = \mathrm{M}(\mathfrak{g},\mathfrak{k})^+,$$

where

$$\mathrm{M}(\mathfrak{g},\mathfrak{k})^+ = \big\{v \in \mathrm{M}(\mathfrak{g},\mathfrak{k}) \mid \mathfrak{k}^+ v = 0\big\}.$$

The algebraic structure of the algebra $\mathrm{Z}(\mathfrak{g},\mathfrak{k})$ can be described with the use of the *extremal projector* for the Lie algebra $\mathfrak{k}$. In order to define it, suppose that the positive roots are $\Delta^+ = \{\alpha_1,\dots,\alpha_m\}$. Consider the vector space $\mathrm{F}_\mu(\mathfrak{k})$ of formal series of weight μ monomials

$$e_{-\alpha_1}^{k_1}\cdots e_{-\alpha_m}^{k_m} e_{\alpha_m}^{r_m}\cdots e_{\alpha_1}^{r_1}$$

with coefficients in $\mathrm{R}(\mathfrak{h})$, where

$$(r_1 - k_1)\alpha_1 + \cdots + (r_m - k_m)\alpha_m = \mu.$$

Introduce the space $\mathrm{F}(\mathfrak{k})$ as the direct sum

$$\mathrm{F}(\mathfrak{k}) = \bigoplus_\mu \mathrm{F}_\mu(\mathfrak{k}).$$

That is, the elements of $\mathrm{F}(\mathfrak{k})$ are finite sums $\sum x_\mu$ with $x_\mu \in \mathrm{F}_\mu(\mathfrak{k})$. It can be shown that $\mathrm{F}(\mathfrak{k})$ is an algebra with respect to the natural multiplication of formal series. The algebra

[4]Zhelobenko sometimes used the names *Z-algebra* or *extended Mickelsson algebra*. The author believes the new name is more appropriate and justified from the scientific point of view.

$\mathrm{F}(\mathfrak{k})$ is equipped with a Hermitian anti-involution (antilinear involutive anti-automorphism) defined by

$$e_\alpha^* = e_{-\alpha}, \quad \alpha \in \Delta^+.$$

Further, call an ordering of the positive roots *normal* if any composite root lies between its components. For instance, there are precisely two normal orderings for the root system of type B_2,

$$\Delta^+ = \{\alpha, \alpha+\beta, \alpha+2\beta, \beta\} \quad \text{and} \quad \Delta^+ = \{\beta, \alpha+2\beta, \alpha+\beta, \alpha\},$$

where α and β are the simple roots. In general, the number of normal orderings coincides with the number of reduced decompositions of the longest element of the corresponding Weyl group.

For any $\alpha \in \Delta^+$ introduce an element of $\mathrm{F}(\mathfrak{k})$ by

$$p_\alpha = 1 + \sum_{k=1}^{\infty} e_{-\alpha}^k e_\alpha^k \frac{(-1)^k}{k!(h_\alpha + \rho(h_\alpha) + 1) \cdots (h_\alpha + \rho(h_\alpha) + k)}, \tag{2.13}$$

where h_α is defined in (2.12) and ρ is the half sum of the positive roots. Finally, define the *extremal projector* $p = p_\mathfrak{k}$ by

$$p = p_{\alpha_1} \cdots p_{\alpha_m}$$

with the product taken in a normal ordering of the positive roots α_i.

THEOREM 2.8. *The element $p \in \mathrm{F}(\mathfrak{k})$ does not depend on the normal ordering on Δ^+ and satisfies the conditions*

$$e_\alpha p = p e_{-\alpha} = 0 \quad \textit{for all } \alpha \in \Delta^+. \tag{2.14}$$

Moreover, $p^* = p$ and $p^2 = p$.

In fact, the relations (2.14) uniquely determine the element p, up to a factor from $\mathrm{R}(\mathfrak{h})$. The extremal projector naturally acts on the vector space $\mathrm{M}(\mathfrak{g}, \mathfrak{k})$. The following corollary states that the Mickelsson–Zhelobenko algebra coincides with its image.

COROLLARY 2.9. *We have*

$$\mathrm{Z}(\mathfrak{g}, \mathfrak{k}) = p\mathrm{M}(\mathfrak{g}, \mathfrak{k}).$$

To get a more precise description of the algebra $\mathrm{Z}(\mathfrak{g}, \mathfrak{k})$ consider a $\mathfrak{k}$-module decomposition

$$\mathfrak{g} = \mathfrak{k} \oplus \mathfrak{p}.$$

Choose a weight basis $e_1, \dots, e_n$ (with respect to the adjoint action of $\mathfrak{h}$) of the complementary module $\mathfrak{p}$.

THEOREM 2.10. *The elements*

$$a_i = p e_i, \quad i = 1, \dots, n,$$

are generators of the Mickelsson–Zhelobenko algebra $\mathrm{Z}(\mathfrak{g}, \mathfrak{k})$. *Moreover, the monomials*

$$a_1^{k_1} \cdots a_n^{k_n}, \quad k_i \in \mathbb{Z}_+,$$

form a basis of $\mathrm{Z}(\mathfrak{g}, \mathfrak{k})$.

It can be proved that the generators a_i of $\mathrm{Z}(\mathfrak{g}, \mathfrak{k})$ satisfy quadratic defining relations; see [173]. For the pairs $(\mathfrak{g}, \mathfrak{k})$ relevant to the constructions of bases of Gelfand–Tsetlin type, the relations can be explicitly written down; cf. Sections 2.3 and 3.1 below.

Regarding $\mathrm{Z}(\mathfrak{g}, \mathfrak{k})$ as a right $\mathrm{R}(\mathfrak{h})$-module, it is possible to introduce the normalized elements

$$z_i = a_i \pi_i, \quad \pi_i \in \mathrm{U}(\mathfrak{h}),$$

by multiplying a_i by its *right denominator* π_i. Therefore the z_i can be viewed as elements of the Mickelsson algebra $\mathrm{S}(\mathfrak{g}, \mathfrak{k})$.

To formulate the final theorem of this section, for any $\mathfrak{g}$-module V set

$$V^+ = \{ v \in V \mid \mathfrak{k}^+ v = 0 \}.$$

THEOREM 2.11. *Let* $V = \mathrm{U}(\mathfrak{g}) v$ *be the cyclic* $\mathrm{U}(\mathfrak{g})$*-module generated by an element* $v \in V^+$. *Then the subspace* V^+ *is linearly spanned by the elements*

$$z_1^{k_1} \cdots z_n^{k_n} v, \quad k_i \in \mathbb{Z}_+.$$

2.3. *Mickelsson–Zhelobenko algebra* $\mathrm{Z}(\mathfrak{gl}_n, \mathfrak{gl}_{n-1})$

For any positive integer m consider the general linear Lie algebra $\mathfrak{gl}_m$. The positive roots of $\mathfrak{gl}_m$ with respect to the diagonal Cartan subalgebra $\mathfrak{h}$ (with the standard choice of the positive root system) are naturally enumerated by the pairs (i, j) with $1 \leqslant i < j \leqslant m$. In accordance with the general theory outlined in the previous section, for each pair introduce a formal series $p_{ij} \in \mathrm{F}(\mathfrak{gl}_m)$ by

$$p_{ij} = 1 + \sum_{k=1}^{\infty} (E_{ji})^k (E_{ij})^k \frac{(-1)^k}{k!(h_i - h_j + 1) \cdots (h_i - h_j + k)},$$

where, as before, $h_i = E_{ii} - i + 1$. Then define the element $p = p_m$ by

$$p = \prod_{i<j} p_{ij},$$

where the product is taken in a normal ordering on the pairs (i, j). By Theorem 2.8,

$$E_{ij}\, p = p E_{ji} = 0 \quad \text{for } 1 \leqslant i < j \leqslant m. \tag{2.15}$$

Now set $m = n - 1$. By Theorem 2.10, ordered monomials in the elements E_{nn}, pE_{in} and pE_{ni} with $i = 1, \dots, n-1$ form a basis of $\mathrm{Z}(\mathfrak{gl}_n, \mathfrak{gl}_{n-1})$ as a left or right $\mathrm{R}(\mathfrak{h})$-module. These elements can explicitly be given by

$$\begin{aligned} pE_{in} &= \sum_{i>i_1>\cdots>i_s\geqslant 1} E_{ii_1} E_{i_1 i_2} \cdots E_{i_{s-1} i_s} E_{i_s n} \frac{1}{(h_i - h_{i_1}) \cdots (h_i - h_{i_s})}, \\ pE_{ni} &= \sum_{i<i_1<\cdots<i_s<n} E_{i_1 i} E_{i_2 i_1} \cdots E_{i_s i_{s-1}} E_{n i_s} \frac{1}{(h_i - h_{i_1}) \cdots (h_i - h_{i_s})}, \end{aligned} \tag{2.16}$$

where $s = 0, 1, \dots$. Indeed, by choosing appropriate normal orderings on the positive roots, we can write

$$pE_{in} = p_{1i} \cdots p_{i-1,i} E_{in} \quad \text{and} \quad pE_{ni} = p_{i,i+1} \cdots p_{i,n-1} E_{ni}.$$

The lowering and raising operators introduced in Section 2.1 coincide with the normalized generators:

$$\begin{aligned} z_{in} &= pE_{in}(h_i - h_{i-1}) \cdots (h_i - h_1), \\ z_{ni} &= pE_{ni}(h_i - h_{i+1}) \cdots (h_i - h_{n-1}), \end{aligned} \tag{2.17}$$

which belong to the Mickelsson algebra $\mathrm{S}(\mathfrak{gl}_n, \mathfrak{gl}_{n-1})$. Thus, Lemma 2.5 is an immediate corollary of (2.15).

PROPOSITION 2.12. *The lowering and raising operators satisfy the following relations*

$$z_{ni} z_{nj} = z_{nj} z_{ni} \quad \textit{for all } i, j, \tag{2.18}$$

$$z_{in} z_{nj} = z_{nj} z_{in} \quad \textit{for } i \neq j, \tag{2.19}$$

and

$$z_{in} z_{ni} = \prod_{j=1,\ j\neq i}^{n} (h_i - h_j - 1) + \sum_{j=1}^{n-1} z_{nj} z_{jn} \prod_{k=1,\ k\neq j}^{n-1} \frac{h_i - h_k - 1}{h_j - h_k}. \tag{2.20}$$

PROOF. We use the properties of p. Assume that $i < j$. Then (2.15) and (2.16) imply that in $\mathrm{Z}(\mathfrak{gl}_n, \mathfrak{gl}_{n-1})$

$$pE_{ni}\,pE_{nj} = pE_{ni}\,E_{nj}, \qquad pE_{nj}\,pE_{ni} = pE_{ni}\,E_{nj}\frac{h_i - h_j + 1}{h_i - h_j}.$$

Now (2.18) follows from (2.17). The proof of (2.19) is similar. The "long" relation (2.20) can be verified by analogous but more complicated direct calculations. We give a different proof based upon the properties of the *Capelli determinant* $\mathcal{C}(u)$. Consider the $n \times n$-matrix E whose ij-th entry is E_{ij} and let u be a formal variable. Then $\mathcal{C}(u)$ is the polynomial with coefficients in the universal enveloping algebra $\mathrm{U}(\mathfrak{gl}_n)$ defined by

$$\mathcal{C}(u) = \sum_{\sigma \in \mathfrak{S}_n} \operatorname{sgn}\sigma \cdot (u + E)_{\sigma(1),1} \cdots (u + E - n + 1)_{\sigma(n),n}. \tag{2.21}$$

It is well known that all its coefficients belong to the center of $\mathrm{U}(\mathfrak{gl}_n)$ and generate the center; see, e.g., Howe and Umeda, [60]. This also easily follows from the properties of the *quantum determinant* of the Yangian for the Lie algebra $\mathfrak{gl}_n$; see, e.g., [104]. Therefore, these coefficients act in $L(\lambda)$ as scalars which can be easily found by applying $\mathcal{C}(u)$ to the highest vector ξ:

$$\mathcal{C}(u)|_{L(\lambda)} = (u + l_1) \cdots (u + l_n), \quad l_i = \lambda_i - i + 1. \tag{2.22}$$

On the other hand, the center of $\mathrm{U}(\mathfrak{gl}_n)$ is a subalgebra in the normalizer $\mathrm{Norm}\,\mathrm{J}$. We shall keep the same notation for the image of $\mathcal{C}(u)$ in the Mickelsson–Zhelobenko algebra $\mathrm{Z}(\mathfrak{gl}_n, \mathfrak{gl}_{n-1})$. To get explicit expressions of the coefficients of $\mathcal{C}(u)$ in terms of the lowering and raising operators we consider $\mathcal{C}(u)$ modulo the ideal J' and apply the projection p. A straightforward calculation yields two alternative formulas

$$\mathcal{C}(u) = (u + E_{nn}) \prod_{i=1}^{n-1} (u + h_i - 1) - \sum_{i=1}^{n-1} z_{in} z_{ni} \prod_{j=1,\ j \neq i}^{n-1} \frac{u + h_j - 1}{h_i - h_j} \tag{2.23}$$

and

$$\mathcal{C}(u) = \prod_{i=1}^{n} (u + h_i) - \sum_{i=1}^{n-1} z_{ni} z_{in} \prod_{j=1,\ j \neq i}^{n-1} \frac{u + h_j}{h_i - h_j}. \tag{2.24}$$

The formulas show that $\mathcal{C}(u)$ can be regarded as an interpolation polynomial for the products $z_{in} z_{ni}$ and $z_{ni} z_{in}$. Namely, for $i = 1, \ldots, n - 1$, we have

$$\mathcal{C}(-h_i + 1) = (-1)^{n-1} z_{in} z_{ni} \quad \text{and} \quad \mathcal{C}(-h_i) = (-1)^{n-1} z_{ni} z_{in} \tag{2.25}$$

with the agreement that when we evaluate u in $\mathrm{U}(\mathfrak{h})$ we write the coefficients of the polynomial to the left from powers of u. Comparing the values of (2.23) and (2.24) at $u = -h_i + 1$ we get (2.20). □

Note that the relation inverse to (2.20) can be obtained by comparing the values of (2.23) and (2.24) at $u = -h_i$.

Next we outline the proofs of the branching rule (Theorem 2.1) and the formulas for the basis elements of $L(\lambda)^+$ (Lemma 2.6). The module $L(\lambda)$ is generated by the highest vector ξ and we have

$$z_{in}\xi = 0, \quad i = 1, \dots, n-1.$$

So, by Theorem 2.11, the vector space $L(\lambda)^+$ is spanned by the elements

$$z_{n1}^{k_1} \cdots z_{n,n-1}^{k_{n-1}}\xi, \quad k_i \in \mathbb{Z}_+. \tag{2.26}$$

Let us set $\mu_i = \lambda_i - k_i$ for $1 \leqslant i \leqslant n-1$ and denote the vector (2.26) by ξ_μ. That is,

$$\xi_\mu = z_{n1}^{\lambda_1-\mu_1} \cdots z_{n,n-1}^{\lambda_{n-1}-\mu_{n-1}}\xi. \tag{2.27}$$

It is now sufficient to show that the vector ξ_μ is nonzero if and only if the betweenness conditions (2.2) hold. The linear independence of the vectors ξ_μ will follow from the fact that their weights are distinct. If $\xi_\mu \neq 0$ then using the relations (2.18) we conclude that each vector $z_{ni}^{\lambda_i-\mu_i}\xi$ is nonzero. On the other hand, $z_{ni}^{k_i}\xi$ is a $\mathfrak{gl}_{n-1}$-highest vector of the weight obtained from $(\lambda_1, \dots, \lambda_{n-1})$ by replacing λ_i with $\lambda_i - k_i$. Therefore, if $k_i \geqslant \lambda_i - \lambda_{i+1} + 1$ then the conditions (2.1) are violated for this weight which implies $z_{ni}^{k_i}\xi = 0$. Hence, $\lambda_i - \mu_i \leqslant \lambda_i - \lambda_{i+1}$ for each i, and μ satisfies (2.2).

For the proof of the converse statement we shall employ the following key lemma which will also be used for the proof of Theorem 2.3.

LEMMA 2.13. *We have for each* $i = 1, \dots, n-1$

$$z_{in}\xi_\mu = -(m_i - l_1) \cdots (m_i - l_n)\xi_{\mu+\delta_i}, \tag{2.28}$$

where

$$m_i = \mu_i - i + 1, \qquad l_i = \lambda_i - i + 1.$$

Here $\xi_{\mu+\delta_i} = 0$ *if* $\lambda_i = \mu_i$.

PROOF. The relation (2.19) implies that if $\lambda_i = \mu_i$ then $z_{in}\xi_\mu = 0$ which agrees with (2.28). Now let $\lambda_i - \mu_i \geqslant 1$. Using (2.18) and (2.25) we obtain

$$z_{in}\xi_\mu = z_{in}z_{ni}\xi_{\mu+\delta_i} = (-1)^{n-1}\mathcal{C}(-h_i + 1)\xi_{\mu+\delta_i} = (-1)^{n-1}\mathcal{C}(-m_i)\xi_{\mu+\delta_i}.$$

The relation (2.28) now follows from (2.22) and the centrality of $\mathcal{C}(u)$. □

If the betweenness conditions (2.2) hold then by Lemma 2.13, applying appropriate operators z_{in} repeatedly to the vector ξ_μ we can obtain the highest vector ξ with a nonzero coefficient. This gives $\xi_\mu \neq 0$.

Thus, we have proved that the vectors ξ_Λ defined in (2.11) form a basis of the representation $L(\lambda)$. The orthogonality of the basis vectors (Proposition 2.4) is implied by the fact that the operators pE_{ni} and pE_{in} are adjoint to each other with respect to the restriction of the inner product $\langle\,,\rangle$ to the subspace $L(\lambda)^+$. Therefore, for the adjoint operator to z_{ni} we have

$$z_{ni}^* = z_{in}\frac{(h_i - h_{i+1} - 1)\cdots(h_i - h_{n-1} - 1)}{(h_i - h_1)\cdots(h_i - h_{i-1})}$$

and Proposition 2.4 follows from Lemma 2.13 by induction.

Now we outline a derivation of formulas (2.5)–(2.7). First, since $E_{nn}z_{ni} = z_{ni}(E_{nn}+1)$ for any i, we have

$$E_{nn}\xi_\mu = \left(\sum_{i=1}^{n}\lambda_i - \sum_{i=1}^{n-1}\mu_i\right)\xi_\mu$$

which implies (2.5). To prove (2.6) is suffices to calculate $E_{n-1,n}\xi_{\mu\nu}$, where

$$\xi_{\mu\nu} = z_{n-1,1}^{\mu_1-\nu_1}\cdots z_{n-1,n-2}^{\mu_{n-2}-\nu_{n-2}}\xi_\mu$$

and the ν_i satisfy the betweenness conditions

$$\mu_i - \nu_i \in \mathbb{Z}_+ \quad \text{and} \quad \nu_i - \mu_{i+1} \in \mathbb{Z}_+ \quad \text{for } i = 1,\dots,n-2.$$

Since $E_{n-1,n}$ commutes with the $z_{n-1,i}$,

$$E_{n-1,n}\xi_{\mu\nu} = z_{n-1,1}^{\mu_1-\nu_1}\cdots z_{n-1,n-2}^{\mu_{n-2}-\nu_{n-2}}E_{n-1,n}\xi_\mu.$$

The following lemma is implied by the explicit formulas for the lowering and raising operators (2.9) and (2.10).

LEMMA 2.14. *We have the relation in* $\mathrm{U}'(\mathfrak{gl}_n)$ *modulo the ideal* J',

$$E_{n-1,n} = \sum_{i=1}^{n-1} z_{n-1,i}z_{in}\frac{1}{(h_i - h_1)\cdots\wedge_i\cdots(h_i - h_{n-1})},$$

where $z_{n-1,n-1} = 1$.

By Lemmas 2.13 and 2.14,

$$E_{n-1,n}\xi_{\mu\nu} = -\sum_{i=1}^{n-1}\frac{(m_i - l_1)\cdots(m_i - l_n)}{(m_i - m_1)\cdots\wedge_i\cdots(m_i - m_{n-1})}\xi_{\mu+\delta_i,\nu} \tag{2.29}$$

which proves (2.6). To prove (2.7) we use Proposition 2.4. Relation (2.29) implies that

$$E_{n,n-1}\xi_{\mu\nu} = \sum_{i=1}^{n-1} c_i(\mu,\nu)\xi_{\mu-\delta_i,\nu}$$

for some coefficients $c_i(\mu,\nu)$. Apply the operator $z_{j,n-1}$ to both sides of this relation. Since $z_{j,n-1}$ commutes with $E_{n,n-1}$ we obtain from Lemma 2.13 a recurrence relation for the $c_i(\mu,\nu)$: if $\mu_j - \nu_j \geqslant 1$ then

$$c_i(\mu,\nu+\delta_j) = c_i(\mu,\nu)\frac{m_i - \gamma_j - 1}{m_i - \gamma_j},$$

where $\gamma_j = \nu_j - j + 1$. The proof is completed by induction. The initial values of $c_i(\mu,\nu)$ are found by applying the relation

$$E_{n,n-1}z_{n-1,i} = z_{ni}\frac{1}{h_i - h_{n-1}} + z_{n-1,i}E_{n,n-1}\frac{h_i - h_{n-1} - 1}{h_i - h_{n-1}}$$

to the vector ξ_μ and taking into account that $E_{n,n-1} = z_{n,n-1}$. Performing the calculation we get

$$E_{n,n-1}\xi_{\mu\nu} = \sum_{i=1}^{n-1} \frac{(m_i - \gamma_1)\cdots(m_i - \gamma_{n-2})}{(m_i - m_1)\cdots\wedge_i\cdots(m_i - m_{n-1})}\xi_{\mu-\delta_i,\nu}$$

thus proving (2.7).

2.4. *Characteristic identities*

Denote by L the vector representation of $\mathfrak{gl}_n$ and consider its contragredient L^*. Note that L^* is isomorphic to $L(0,\dots,0,-1)$. Let $\{\varepsilon_1,\dots,\varepsilon_n\}$ denote the basis of L^* dual to the canonical basis $\{e_1,\dots,e_n\}$ of L. Introduce the $n\times n$-matrix E whose ij-th entry is the generator E_{ij}. We shall interpret E as the element

$$E = \sum_{i,j=1}^{n} e_{ij}\otimes E_{ij} \in \operatorname{End} L^*\otimes \mathrm{U}(\mathfrak{gl}_n),$$

where the e_{ij} are the standard matrix units acting on L^* by $e_{ij}\varepsilon_k = \delta_{jk}\varepsilon_i$. The basis element E_{ij} of $\mathfrak{gl}_n$ acts on L^* as $-e_{ji}$ and hence E may also be thought of as the image of the element

$$e = -\sum_{i,j=1}^{n} E_{ji}\otimes E_{ij} \in \mathrm{U}(\mathfrak{gl}_n)\otimes \mathrm{U}(\mathfrak{gl}_n).$$

On the other hand, using the standard coproduct Δ on $\mathrm{U}(\mathfrak{gl}_n)$ defined by

$$\Delta(E_{ij}) = E_{ij} \otimes 1 + 1 \otimes E_{ij},$$

we can write e in the form

$$e = \frac{1}{2}\big(z \otimes 1 + 1 \otimes z - \Delta(z)\big), \tag{2.30}$$

where z is the second-order Casimir element

$$z = \sum_{i,j=1}^{n} E_{ij} E_{ji} \in \mathrm{U}(\mathfrak{gl}_n).$$

We have the tensor product decomposition

$$L^* \otimes L(\lambda) \simeq L(\lambda - \delta_1) \oplus \cdots \oplus L(\lambda - \delta_n), \tag{2.31}$$

where $L(\lambda - \delta_i)$ is considered to be zero if $\lambda_i = \lambda_{i+1}$. On the level of characters this is a particular case of the *Pieri rule* for the expansion of the product of a Schur polynomial by an elementary symmetric polynomial; see, e.g., Macdonald, [84, Chapter 1]. The Casimir element z acts as a scalar operator in any highest weight representation $L(\lambda)$. The corresponding eigenvalue is given by

$$z|_{L(\lambda)} = \sum_{i=1}^{n} \lambda_i(\lambda_i + n - 2i + 1).$$

Regarding now E as an operator on $L^* \otimes L(\lambda)$ and using (2.30) we derive that the restriction of E to the summand $L(\lambda - \delta_r)$ in (2.31) is the scalar operator with the eigenvalue $\lambda_r + n - r$ which we shall denote by α_r. This implies the *characteristic identity* for the matrix E,

$$\prod_{r=1}^{n} (E - \alpha_r) = 0, \tag{2.32}$$

as an operator in $L^* \otimes L(\lambda)$. Moreover, the projection $P[r]$ of $L^* \otimes L(\lambda)$ to the summand $L(\lambda - \delta_r)$ can be written explicitly as

$$P[r] = \frac{(E - \alpha_1) \cdots \wedge_r \cdots (E - \alpha_n)}{(\alpha_r - \alpha_1) \cdots \wedge_r \cdots (\alpha_r - \alpha_n)}$$

with $\wedge_r$ indicating that the r-th factor is omitted. Together with (2.32) this yields the *spectral decomposition* of E,

$$E = \sum_{r=1}^{n} \alpha_r P[r]. \tag{2.33}$$

Consider the orthonormal Gelfand–Tsetlin bases $\{\zeta_\Lambda\}$ of $L(\lambda)$ and $\{\zeta_{\Lambda^{(r)}}\}$ of $L(\lambda-\delta_r)$ for $r=1,\dots,n$; see (2.8). Regarding the matrix element $P[r]_{ij}$ as an operator in $L(\lambda)$ we obtain

$$\big\langle \zeta_{\Lambda'}, P[r]_{ij}\zeta_\Lambda\big\rangle = \big\langle \varepsilon_i\otimes\zeta_{\Lambda'}, P[r](\varepsilon_j\otimes\zeta_\Lambda)\big\rangle, \tag{2.34}$$

where we have extended the inner products on L^* and $L(\lambda)$ to $L^*\otimes L(\lambda)$ by setting

$$\langle \eta\otimes\zeta, \eta'\otimes\zeta'\rangle = \langle\eta,\eta'\rangle\langle\zeta,\zeta'\rangle$$

with $\eta,\eta'\in L^*$ and $\zeta,\zeta'\in L(\lambda)$. Furthermore, using the expansions

$$\varepsilon_j\otimes\zeta_\Lambda = \sum_{s=1}^{n}\sum_{\Lambda^{(s)}}\langle \varepsilon_j\otimes\zeta_\Lambda, \zeta_{\Lambda^{(s)}}\rangle\zeta_{\Lambda^{(s)}},$$

brings (2.34) to the form

$$\sum_{\Lambda^{(r)}}\langle\varepsilon_i\otimes\zeta_{\Lambda'},\zeta_{\Lambda^{(r)}}\rangle\langle\varepsilon_j\otimes\zeta_\Lambda,\zeta_{\Lambda^{(r)}}\rangle,$$

where we have used the fact that $P[r]$ is the identity map on $L(\lambda-\delta_r)$, and zero on $L(\lambda-\delta_s)$ with $s\neq r$. The numbers $\langle\varepsilon_i\otimes\zeta_{\Lambda'},\zeta_{\Lambda^{(r)}}\rangle$ are the *Wigner coefficients* (a particular case of the *Clebsch–Gordan coefficients*). They can be used to express the matrix elements of the generators E_{ij} in the Gelfand–Tsetlin basis as follows. Using the spectral decomposition (2.33) we get

$$E_{ij} = \sum_{r=1}^{n}\alpha_r P[r]_{ij}.$$

Therefore, we derive the following result from (2.34).

THEOREM 2.15. *We have*

$$\langle\zeta_{\Lambda'}, E_{ij}\zeta_\Lambda\rangle = \sum_{r=1}^{n}\alpha_r\sum_{\Lambda^{(r)}}\langle\varepsilon_i\otimes\zeta_{\Lambda'},\zeta_{\Lambda^{(r)}}\rangle\langle\varepsilon_j\otimes\zeta_\Lambda,\zeta_{\Lambda^{(r)}}\rangle.$$

Employing the characteristic identities for both the Lie algebras $\mathfrak{gl}_{n+1}$ and $\mathfrak{gl}_n$ it is possible to determine the values of the Wigner coefficients and thus to get an independent derivation of the formulas of Theorem 2.3. In fact, explicit formulas for the matrix elements of E_{ij} with $|i-j|>1$ can also be given; see Gould, [48], for details.

The approach based upon the characteristic identities also leads to an alternative presentation of the lowering and raising operators. Taking ζ_Λ to be the highest vector ξ in (2.34) we conclude that $P[r]_{ij}\xi = 0$ for $j > r$. Consider now $\mathfrak{gl}_n$ as a subalgebra of $\mathfrak{gl}_{n+1}$. Suppose that ξ is a highest vector of weight λ in a representation $L(\lambda')$ of $\mathfrak{gl}_{n+1}$. The previous observation implies that the vector

$$\sum_{i=r}^{n} E_{n+1,i} P[r]_{ir}\xi$$

is again a $\mathfrak{gl}_n$-highest vector of weight $\lambda - \delta_r$.

PROPOSITION 2.16. *We have the following identity of operators on the space* $L(\lambda')^+_\lambda$:

$$pE_{n+1,r} = \sum_{i=r}^{n} E_{n+1,i} P[r]_{ir},$$

where p is the extremal projector for $\mathfrak{gl}_n$.

OUTLINE OF THE PROOF. Since both sides represent lowering operators they must be proportional. It is therefore sufficient to apply both sides to a vector $\xi \in L(\lambda')^+_\lambda$ and compare the coefficients at $E_{n+1,r}\xi$. For the calculation we use the explicit formula (2.16) for $pE_{n+1,r}$ and the relation

$$P[r]_{rr}\xi = \prod_{s=r+1}^{n} \frac{h_r - h_s - 1}{h_r - h_s}\xi$$

which can be derived from the characteristic identities. □

An analogous argument leads to a similar formula for the raising operators. Here one starts with the dual characteristic identity

$$\prod_{r=1}^{n} (\overline{E} - \bar{\alpha}_r) = 0,$$

where the ij-th matrix element of $\overline{E}$ is $-E_{ij}$, $\bar{\alpha}_r = -\lambda_r + r - 1$ and the powers of $\overline{E}$ are defined recursively by

$$(\overline{E}^p)_{ij} = \sum_{k=1}^{n} (\overline{E}^{p-1})_{kj} \overline{E}_{ik}.$$

For any $r = 1, \ldots, n$ the dual projection operator is given by

$$\overline{P}[r] = \frac{(\overline{E} - \bar{\alpha}_1) \cdots \wedge_r \cdots (\overline{E} - \bar{\alpha}_n)}{(\bar{\alpha}_r - \bar{\alpha}_1) \cdots \wedge_r \cdots (\bar{\alpha}_r - \bar{\alpha}_n)}.$$

PROPOSITION 2.17. *We have the following identity of operators on the space* $L(\lambda')^+_\lambda$:

$$pE_{r,n+1} = \sum_{i=1}^{r} E_{i,n+1}\overline{P}[r]_{ri}.$$

2.5. *Quantum minors*

For a complex parameter u introduce the $n \times n$-matrix $E(u) = u1 + E$. Given sequences $a_1, \ldots, a_s$ and $b_1, \ldots, b_s$ of elements of $\{1, \ldots, n\}$ the corresponding *quantum minor* of the matrix $E(u)$ is defined by the following equivalent formulas:

$$E(u)^{a_1\cdots a_s}_{b_1\cdots b_s} = \sum_{\sigma\in\mathfrak{S}_s} \operatorname{sgn}\sigma \cdot E(u)_{a_{\sigma(1)}b_1} \cdots E(u-s+1)_{a_{\sigma(s)}b_s} \tag{2.35}$$

$$= \sum_{\sigma\in\mathfrak{S}_s} \operatorname{sgn}\sigma \cdot E(u-s+1)_{a_1 b_{\sigma(1)}} \cdots E(u)_{a_s b_{\sigma(s)}}. \tag{2.36}$$

This is a polynomial in u with coefficients in $\mathrm{U}(\mathfrak{gl}_n)$. It is skew symmetric under permutations of the indices a_i, or b_i.

For any index $1 \leqslant i < n$ introduce the polynomials

$$\tau_{ni}(u) = E(u)^{i+1\cdots n}_{i\cdots n-1} \quad \text{and} \quad \tau_{in}(u) = (-1)^{i-1} E(u)^{1\cdots i}_{1\cdots i-1,n}.$$

For instance,

$$\tau_{13}(u) = E_{13}, \qquad \tau_{23}(u) = -E_{23}(u+E_{11}) + E_{21}E_{13},$$
$$\tau_{32}(u) = E_{32}, \qquad \tau_{31}(u) = E_{21}E_{32} - E_{31}(u+E_{22}-1).$$

PROPOSITION 2.18. *If* $\eta \in L(\lambda)^+_\mu$ *then*

$$\tau_{ni}(-\mu_i)\eta \in L(\lambda)^+_{\mu-\delta_i} \quad \textit{and} \quad \tau_{in}(-\mu_i+i-1)\eta \in L(\lambda)^+_{\mu+\delta_i}.$$

OUTLINE OF THE PROOF. The proof is based upon the following relations

$$\left[E_{ij}, E(u)^{a_1\cdots a_s}_{b_1\cdots b_s}\right] = \sum_{r=1}^{s}\left(\delta_{ja_r} E(u)^{a_1\cdots i\cdots a_s}_{b_1\cdots b_s} - \delta_{ib_r} E(u)^{a_1\cdots a_s}_{b_1\cdots j\cdots b_s}\right), \tag{2.37}$$

where i and j on the right-hand side are in the r-th slot. □

The relations (2.37) imply an important property of the quantum minors: for any indices i, j we have

$$\left[E_{a_i b_j}, E(u)^{a_1\cdots a_s}_{b_1\cdots b_s}\right] = 0.$$

In particular, this implies the centrality of the Capelli determinant $\mathcal{C}(u) = E(u)^{1\cdots n}_{1\cdots n}$; see (2.21).

The lowering and raising operators of Proposition 2.18 can be shown to essentially coincide with those defined in Section 2.1. To write down the formulas we shall need to evaluate the variable u in $\mathrm{U}(\mathfrak{h})$. To make this operation well-defined we use the agreement used in the evaluation of the Capelli determinant. See just below (2.25).

PROPOSITION 2.19. *We have the following identities for any $i = 1, \dots, n-1$*

$$\tau_{ni}(-h_i - i + 1) = z_{ni} \quad \textit{and} \quad \tau_{in}(-h_i) = z_{in}. \tag{2.38}$$

Using this interpretation of the lowering operators one can express the Gelfand–Tsetlin basis vector (2.11) in terms of the quantum minors $\tau_{ki}(u)$. The action of certain other quantum minors on these vectors can be explicitly found. This will provide one more independent proof of Theorem 2.3. For $m \geqslant 1$ introduce the polynomials $A_m(u)$, $B_m(u)$ and $C_m(u)$ by

$$A_m(u) = E(u)^{1\cdots m}_{1\cdots m}, \qquad B_m(u) = E(u)^{1\cdots m}_{1\cdots m-1,m+1},$$
$$C_m(u) = E(u)^{1\cdots m-1,m+1}_{1\cdots m}.$$

We use the notation $l_{mi} = \lambda_{mi} - i + 1$ and $l_i = \lambda_i - i + 1$.

THEOREM 2.20. *Let $\{\xi_\Lambda\}$ be the Gelfand–Tsetlin basis of $L(\lambda)$. Then*

$$A_m(u)\xi_\Lambda = (u + l_{m1}) \cdots (u + l_{mm})\xi_\Lambda, \tag{2.39}$$

$$B_m(-l_{mj})\xi_\Lambda = -\prod_{i=1}^{m+1}(l_{m+1,i} - l_{mj})\xi_{\Lambda+\delta_{mj}} \quad \textit{for } j = 1, \dots, m,$$

$$C_m(-l_{mj})\xi_\Lambda = \prod_{i=1}^{m-1}(l_{m-1,i} - l_{mj})\xi_{\Lambda-\delta_{mj}} \quad \textit{for } j = 1, \dots, m, \tag{2.40}$$

where $\Lambda \pm \delta_{mj}$ is obtained from Λ by replacing the entry λ_{mj} with $\lambda_{mj} \pm 1$.

Applying the Lagrange interpolation formula we can find the action of $B_m(u)$ and $C_m(u)$ for any u. Note that these polynomials have degree $m-1$ with leading coefficients $E_{m,m+1}$ and $E_{m+1,m}$, respectively. Theorem 2.3 is therefore an immediate corollary of Theorem 2.20.

Formula (2.40) prompts a quite different construction of the basis vectors of $L(\lambda)$ which uses the polynomials $C_m(u)$ instead of the traditional lowering operators z_{ni}. Indeed, for a particular value of u, $C_m(u)$ takes a basis vector into another one, up to a factor. Given a pattern Λ associated with λ, define the vector $\kappa_\Lambda \in L(\lambda)$ by

$$\kappa_\Lambda = \prod^{\rightarrow}_{k=1,\dots,n-1} \{C_{n-1}(-l_{n-1,k} - 1) \cdots C_{n-1}(-l_k + 1)C_{n-1}(-l_k)$$

$$\times C_{n-2}(-l_{n-2,k}-1)\cdots C_{n-2}(-l_k+1)C_{n-2}(-l_k)$$
$$\times \cdots \times C_k(-l_{kk}-1)\cdots C_k(-l_k+1)C_k(-l_k)\big\}\xi.$$

THEOREM 2.21. *The vectors κ_Λ with Λ running over all patterns associated with λ form a basis of $L(\lambda)$ and one has $\kappa_\Lambda = N_\Lambda \xi_\Lambda$ for a nonzero constant N_Λ.*

The value of the constant N_Λ can be found from (2.40). Using the relations between the elements $A_m(u)$, $B_m(u)$ and $C_m(u)$ one can derive Theorem 2.20 from Theorem 2.21 with the use of Proposition 2.22 below; see Nazarov and Tarasov, [109], for details.

Observe that $A_m(u)$ is the Capelli determinant (2.21) for the Lie algebra $\mathfrak{gl}_m$. Therefore, its coefficients a_{mi} defined by

$$A_m(u) = u^m + a_{m1}u^{m-1} + \cdots + a_{mm}$$

are generators of the center of the enveloping algebra $\mathrm{U}(\mathfrak{gl}_m)$. All together the elements a_{mi} with $1 \leqslant i \leqslant m \leqslant n$ generate a commutative subalgebra $\mathcal{A}_n$ of $\mathrm{U}(\mathfrak{gl}_n)$ which is called the *Gelfand–Tsetlin subalgebra*. By (2.39), the basis vectors ξ_Λ are simultaneous eigenvectors for the elements of the subalgebra $\mathcal{A}_n$. Introduce the corresponding eigenvalues of the generators a_{mi} by

$$a_{mi}\xi_\Lambda = \alpha_{mi}(\Lambda)\xi_\Lambda. \tag{2.41}$$

Thus, $\alpha_{mi}(\Lambda)$ is the i-th elementary symmetric polynomial in $l_{m1}, \ldots, l_{mm}$.

PROPOSITION 2.22. *For any pattern Λ associated with λ, the one-dimensional subspace of $L(\lambda)$ spanned by the basis vector ξ_Λ is uniquely determined by the set of eigenvalues $\{\alpha_{mi}(\Lambda)\}$.*

Bibliographical notes

The explicit formulas for the lowering and raising operators (2.9) and (2.10) first appeared in Nagel and Moshinsky, [106]; see also Hou Pei-yu, [59], and Zhelobenko, [168]. The derivation of the Gelfand–Tsetlin formulas outlined in Section 2.1 follows Zhelobenko, [168], and Asherova, Smirnov and Tolstoy, [2]. The extremal projectors were originally introduced by Asherova, Smirnov and Tolstoy, [1] (see also [3]). In a subsequent paper [2] the projectors were used to construct the lowering operators and derive the relations between them. A systematic study of the extremal projectors and the corresponding Mickelsson algebras was undertaken by Zhelobenko: a detailed exposition is given in his paper [173] and book [174]. The application to the Gelfand–Tsetlin formulas is contained in his paper [171]. Section 2.2 is a brief outline of the general results which are used in the basis constructions.

The first proof of Theorem 2.11 was given by van den Hombergh, [158], as an answer to the question posed by Mickelsson, [92]. A derivation of the relations in the Mickelsson–Zhelobenko algebra $\mathrm{Z}(\mathfrak{gl}_n, \mathfrak{gl}_m)$ with the use of the Capelli-type determinants is contained in the author's paper [101]. A proof of the formulas (2.23) and (2.24) is also given there. The results of Section 2.4 are due to Gould, [46–49]. The characteristic identity (2.32) was proved by Green, [54]. The significance of the Wigner coefficients in mathematical physics is discussed in the book by Biedenharn and Louck, [8]. The definition (2.35) of the quantum minors is inspired by the theory of "quantum" algebras called the *Yangians*; see [103, 104] for a review of the theory. The polynomials $A_m(u)$, $B_m(u)$ and $C_m(u)$ are essentially the images of the *Drinfeld generators* of the Yangian $\mathrm{Y}(n)$ under the evaluation homomorphism to the universal enveloping algebra $\mathrm{U}(\mathfrak{gl}_n)$. The quantum minor presentation of the lowering operators (2.38) is due to the author, [96]; see also [101]. The construction of the Gelfand–Tsetlin basis vectors κ_Λ with the use of the Drinfeld generators (Theorem 2.21) was devised by Nazarov and Tarasov, [109].

Analogs of the extremal projector were given by Tolstoy, [150–154], for a wide class of Lie (super)algebras and their quantized enveloping algebras. The corresponding super and quantum versions of the Mickelsson–Zhelobenko algebras are studied in [152–154]. An alternative "tensor formula" for the extremal projector was provided by Tolstoy and Draayer, [155]. The techniques of extremal projectors were applied by Khoroshkin and Tolstoy, [67], for calculation of the universal R-matrices for quantized enveloping algebras. A basis of Gelfand–Tsetlin type for representations of the exceptional Lie algebra G_2 was constructed by Sviridov, Smirnov and Tolstoy, [144,145].

Bases of Gelfand–Tsetlin type have been constructed for representations of various types of algebras. For the quantized enveloping algebra $\mathrm{U}_q(\mathfrak{gl}_n)$ such bases were constructed by Jimbo, [65], Ueno, Takebayashi and Shibukawa, [157], Nazarov and Tarasov, [109], Tolstoy, [153]. The results of [109] include q-analogs of Theorems 2.20 and 2.21, while [153] contains matrix element formulas for the generators corresponding to arbitrary roots. Gould and Biedenharn, [52], developed pattern calculus for representations of the quantum group $\mathrm{U}_q(u(n))$. Polynomial realizations of the Gelfand–Tsetlin basis for representations of $\mathrm{U}_q(\mathfrak{sl}_3)$ were given by Dobrev and Truini, [20,21], and Dobrev, Mitov and Truini, [22].

Gelfand–Tsetlin bases for 'generic' representations of the Yangian $\mathrm{Y}(n)$ were constructed in [96]. Theorem 2.20 was proved there in the more general context of representations of the *Yangian of level p* for $\mathfrak{gl}_n$ which was previously introduced by Cherednik, [17]. In particular, the enveloping algebra $\mathrm{U}(\mathfrak{gl}_n)$ coincides with the Yangian of level 1. A more general class of the *tame* Yangian modules was introduced and explicitly constructed by Nazarov and Tarasov, [110], via the *trapezium* or *skew* analogs of the Gelfand–Tsetlin patterns. Their approach was motivated by the so-called *centralizer construction* devised by Olshanski, [114,116,117], and also employed by Cherednik, [16,17]. Basis vectors in the tame Yangian modules are characterized in a way similar to Proposition 2.22. The skew Yangian modules were also studied in [101] with the use of the quantum Sylvester theorem and the Mickelsson algebras.

The center of $\mathrm{U}(\mathfrak{gl}_n)$ possesses several natural families of generators and so does the Gelfand–Tsetlin subalgebra $\mathcal{A}_n$. The corresponding eigenvalues in $L(\lambda)$ are known explicitly; see, e.g., [103] for a review. An alternative description of $\mathcal{A}_n$ was given by Gelfand, Krob, Lascoux, Leclerc, Retakh and Thibon, [40, Section 7.3], as an application of their theory of noncommutative symmetric functions and quasi-determinants.

The combinatorics of the skew Gelfand–Tsetlin patterns is employed by Berenstein and Zelevinsky, [7], to obtain multiplicity formulas for the skew representations of $\mathfrak{gl}_n$. Applications to continuous piecewise linear actions of the symmetric group were found by Kirillov and Berenstein, [72].

The explicit realization of irreducible finite-dimensional representations of $\mathfrak{gl}_n$ via the Gelfand–Tsetlin bases has important applications in the representation theory of quantum affine algebras and Yangians. In particular, Theorem 2.20 and its Yangian analog, [96], are crucial in the proof of the irreducibility criterion of the tensor products of the Yangian evaluation modules (a generalization to $\mathfrak{gl}_n$ of Theorem 3.8 below); see [102].

Analogs of the Gelfand–Tsetlin bases for representations of some Lie superalgebras and their quantum analogs were given by Ottoson, [119,120], Palev, [122–125], Palev, Stoilova and van der Jeugt, [131], Palev and Tolstoy, [132], Tolstoy, Istomina and Smirnov, [156]. Highest weight irreducible representations for the Lie (super)algebras of infinite matrices and their quantum analogs were constructed by Palev, [126,127], and Palev and Stoilova, [128–130], via bases of Gelfand–Tsetlin-type.

The explicit formulas of Theorem 2.3 make it possible to define a class of infinite-dimensional representations of $\mathfrak{gl}_n$ by altering the inequalities (2.4). Families of such representations were introduced by Gelfand and Graev, [39]. However, as was later observed by Lemire and Patera, [80], some necessary conditions were missing in [39], so that only a part of those families actually provides representations. A more general theory of the so-called *Gelfand–Tsetlin modules* is developed by Drozd, Futorny and Ovsienko, [31–34], Ovsienko, [121], and Mazorchuk, [87,88]. The starting point of the theory is to axiomatize the property of the basis vectors (2.41) and to consider the module generated by an eigenvector for the Gelfand–Tsetlin subalgebra with a given arbitrary set of eigenvalues $\{\alpha_{mi}\}$. Some q-analogs of such modules were constructed by Mazorchuk and Turowska, [90].

The formulas of Theorem 2.3 were applied by Olshanski, [113,115], to study unitary representations of the pseudo-unitary groups $U(p,q)$. In particular, he classified all irreducible unitarizable highest weight representations of the Lie algebra $\mathfrak{u}(p,q)$, [113]. This work was extended by the author to a family of the Enright–Varadarajan modules over $\mathfrak{u}(p,q)$, [95]. Analogs of the Gelfand–Tsetlin bases for the unitary highest weight modules were constructed in [94].

Applications of the Gelfand–Tsetlin bases in mathematical physics are reviewed in the books by Barut and Rączka, [5], and Biedenharn and Louck, [8].

3. Weight bases for representations of $\mathfrak{o}_N$ and $\mathfrak{sp}_{2n}$

Let $\mathfrak{g}_n$ denote the rank n simple complex Lie algebra of type B, C, or D. That is,

$$\mathfrak{g}_n = \mathfrak{o}_{2n+1},\ \mathfrak{sp}_{2n},\ \text{or}\ \mathfrak{o}_{2n}, \tag{3.1}$$

respectively. Let $V(\lambda)$ denote the finite-dimensional irreducible representation of $\mathfrak{g}_n$ with the highest weight λ. The restriction of $V(\lambda)$ to the subalgebra $\mathfrak{g}_{n-1}$ is not multiplicity-free in general. This means that if $V'(\mu)$ is the finite-dimensional irreducible representation of $\mathfrak{g}_{n-1}$ with the highest weight μ, then the space

$$\mathrm{Hom}_{\mathfrak{g}_{n-1}}\big(V'(\mu), V(\lambda)\big) \tag{3.2}$$

need not be one-dimensional. In order to construct a basis of $V(\lambda)$ associated with the chain of subalgebras

$$\mathfrak{g}_1 \subset \mathfrak{g}_2 \subset \cdots \subset \mathfrak{g}_n$$

we need to construct a basis of the space (3.2) which is isomorphic to the subspace $V(\lambda)^+_\mu$ of $\mathfrak{gl}_{n-1}$-highest vectors of weight μ in $V(\lambda)$. The subspace $V(\lambda)^+_\mu$ possesses a natural structure of a representation of the centralizer $\mathrm{C}_n = \mathrm{U}(\mathfrak{g}_n)^{\mathfrak{g}_{n-1}}$ of $\mathfrak{g}_{n-1}$ in the universal enveloping algebra $\mathrm{U}(\mathfrak{g}_n)$. It was shown by Olshanski, [118], that there exist natural homomorphisms

$$\mathrm{C}_1 \leftarrow \mathrm{C}_2 \leftarrow \cdots \leftarrow \mathrm{C}_n \leftarrow \mathrm{C}_{n+1} \leftarrow \cdots.$$

The projective limit of this chain turns out to be an extension of the *twisted Yangian* $\mathrm{Y}^+(2)$ or $\mathrm{Y}^-(2)$, in the orthogonal and symplectic case, respectively; see [118,104] and [105] for the definition and properties of the twisted Yangians. In particular, there is an algebra homomorphism $\mathrm{Y}^\pm(2) \to \mathrm{C}_n$ which allows one to equip the space $V(\lambda)^+_\mu$ with a $\mathrm{Y}^\pm(2)$-module structure. By the results of [97], the representation $V(\lambda)^+_\mu$ can be extended to a larger algebra, the *Yangian* $\mathrm{Y}(2)$. This is a key fact which allows us to construct a natural basis in each space $V(\lambda)^+_\mu$. In the C and D cases the $\mathrm{Y}(2)$-module $V(\lambda)^+_\mu$ is irreducible while in the B case it is a direct sum of two irreducible submodules. This does not lead, however, to major differences in the constructions, and the final formulas are similar in all the three cases.

The calculations of the matrix elements of the generators of $\mathfrak{g}_n$ are based on the relationship between the twisted Yangian $\mathrm{Y}^\pm(2)$ and the Mickelsson–Zhelobenko algebra $\mathrm{Z}(\mathfrak{g}_n, \mathfrak{g}_{n-1})$; see Section 2.2. We construct an algebra homomorphism $\mathrm{Y}^\pm(2) \to \mathrm{Z}(\mathfrak{g}_n, \mathfrak{g}_{n-1})$ which allows us to express the generators of the twisted Yangian, as operators in $V(\lambda)^+_\mu$, in terms of the lowering and raising operators.

3.1. *Raising and lowering operators*

Whenever possible we consider the three cases (3.1) simultaneously, unless otherwise stated. The rows and columns of $2n \times 2n$-matrices will be enumerated by the indices $-n, \dots, -1, 1, \dots, n$, while the rows and columns of $(2n+1) \times (2n+1)$-matrices will be enumerated by the indices $-n, \dots, -1, 0, 1, \dots, n$. Accordingly, the index 0 will usually be skipped in the former case. For $-n \leqslant i, j \leqslant n$ set

$$F_{ij} = E_{ij} - \theta_{ij} E_{-j,-i}, \tag{3.3}$$

where the E_{ij} are the standard matrix units, and

$$\theta_{ij} = \begin{cases} 1 & \text{in the orthogonal case,} \\ \operatorname{sgn} i \cdot \operatorname{sgn} j & \text{in the symplectic case.} \end{cases} \tag{3.4}$$

The matrices F_{ij} span the Lie algebra $\mathfrak{g}_n$. The subalgebra $\mathfrak{g}_{n-1}$ is spanned by the elements (3.3) with the indices i, j running over the set $\{-n+1, \dots, n-1\}$. Denote by $\mathfrak{h} = \mathfrak{h}_n$ the diagonal Cartan subalgebra in $\mathfrak{g}_n$. The elements $F_{11}, \dots, F_{nn}$ form a basis of $\mathfrak{h}$.

The finite-dimensional irreducible representations of $\mathfrak{g}_n$ are in a one-to-one correspondence with n-tuples $\lambda = (\lambda_1, \dots, \lambda_n)$ where the numbers λ_i satisfy the conditions

$$\lambda_i - \lambda_{i+1} \in \mathbb{Z}_+ \quad \text{for } i = 1, \dots, n-1, \tag{3.5}$$

and

$$\begin{aligned} -2\lambda_1 &\in \mathbb{Z}_+ \quad \text{for } \mathfrak{g}_n = \mathfrak{o}_{2n+1}, \\ -\lambda_1 &\in \mathbb{Z}_+ \quad \text{for } \mathfrak{g}_n = \mathfrak{sp}_{2n}, \\ -\lambda_1 - \lambda_2 &\in \mathbb{Z}_+ \quad \text{for } \mathfrak{g}_n = \mathfrak{o}_{2n}. \end{aligned} \tag{3.6}$$

Such an n-tuple λ is called the *highest weight* of the corresponding representation which we shall denote by $V(\lambda)$. It contains a unique, up to a constant factor, nonzero vector ξ (the *highest vector*) such that

$$\begin{aligned} F_{ii}\xi &= \lambda_i \xi \quad \text{for } i = 1, \dots, n, \\ F_{ij}\xi &= 0 \quad \text{for } -n \leqslant i < j \leqslant n. \end{aligned}$$

Denote by $V(\lambda)^+$ the subspace of $\mathfrak{g}_{n-1}$-highest vectors in $V(\lambda)$:

$$V(\lambda)^+ = \{\eta \in V(\lambda) \mid F_{ij}\eta = 0, \ -n < i < j < n\}.$$

Given a $\mathfrak{g}_{n-1}$-weight $\mu = (\mu_1, \dots, \mu_{n-1})$ we denote by $V(\lambda)^+_\mu$ the corresponding weight subspace in $V(\lambda)^+$:

$$V(\lambda)^+_\mu = \{\eta \in V(\lambda)^+ \mid F_{ii}\eta = \mu_i \eta, \ i = 1, \dots, n-1\}.$$

Consider the Mickelsson–Zhelobenko algebra $\mathrm{Z}(\mathfrak{g}_n, \mathfrak{g}_{n-1})$ introduced in Section 2.2. Let $p = p_{n-1}$ be the extremal projector for the Lie algebra $\mathfrak{g}_{n-1}$. It satisfies the conditions

$$F_{ij}p = pF_{ji} = 0 \quad \text{for } -n < i < j < n.$$

By Theorem 2.10, the elements

$$F_{nn}, \qquad pF_{ia}, \quad a = -n, n, \ i = -n+1, \dots, n-1, \tag{3.7}$$

are generators of $\mathrm{Z}(\mathfrak{g}_n, \mathfrak{g}_{n-1})$ in the orthogonal case. In the symplectic case, the algebra $\mathrm{Z}(\mathfrak{g}_n, \mathfrak{g}_{n-1})$ is generated by the elements (3.7) together with $F_{n,-n}$ and $F_{-n,n}$. To write down explicit formulas for the generators, introduce numbers ρ_i, where $i = 1, \dots, n$, by

$$\rho_i = \begin{cases} -i + 1/2 & \text{for } \mathfrak{g}_n = \mathfrak{o}_{2n+1}, \\ -i & \text{for } \mathfrak{g}_n = \mathfrak{sp}_{2n}, \\ -i + 1 & \text{for } \mathfrak{g}_n = \mathfrak{o}_{2n}. \end{cases}$$

The numbers $-\rho_i$ are coordinates of the half-sum of positive roots with respect to the upper triangular Borel subalgebra. Now set

$$f_i = F_{ii} + \rho_i, \quad f_{-i} = -f_i$$

for $i = 1, \ldots, n$. Moreover, in the case of $\mathfrak{o}_{2n+1}$ also set $f_0 = -1/2$. The generators pF_{ia} can be given by a uniform expression in all the three cases. Let $a \in \{-n, n\}$ and $i \in \{-n+1, \ldots, n-1\}$. Then we have modulo the ideal J',

$$pF_{ia} = F_{ia} + \sum_{i>i_1>\cdots>i_s>-n} F_{ii_1}F_{i_1i_2}\cdots F_{i_{s-1}i_s}F_{i_sa}\frac{1}{(f_i - f_{i_1})\cdots(f_i - f_{i_s})}, \tag{3.8}$$

summed over $s \geqslant 1$. It will be convenient to work with normalized generators of $\mathrm{Z}(\mathfrak{g}_n, \mathfrak{g}_{n-1})$. Set

$$z_{ia} = pF_{ia}(f_i - f_{i-1})\cdots(f_i - f_{-n+1})$$

in the B, C cases, and

$$z_{ia} = pF_{ia}(f_i - f_{i-1})\cdots\widehat{(f_i - f_{-i})}\cdots(f_i - f_{-n+1})$$

in the D case, where the hat indicates the factor to be omitted if it occurs. We shall also use the elements z_{ai} defined by

$$z_{ai} = (-1)^{n-i} z_{-i,-a} \quad \text{and} \quad z_{ai} = (-1)^{n-i}\,\mathrm{sgn}\,a \cdot z_{-i,-a},$$

in the orthogonal and symplectic case, respectively. The elements z_{ia} satisfy some quadratic relations which can be shown to be the defining relations of the algebra $\mathrm{Z}(\mathfrak{g}_n, \mathfrak{g}_{n-1})$. In particular, we have for all $a, b \in \{-n, n\}$ and $i + j \neq 0$,

$$z_{ia}z_{jb} + z_{ja}z_{ib}(f_i - f_j - 1) = z_{ib}z_{ja}(f_i - f_j). \tag{3.9}$$

Thus, z_{ia} and z_{ja} commute for $i + j \neq 0$. Also, z_{ia} and z_{ib} commute for $i \neq 0$ and all values of a and b. Analogs of the relation (2.20) in the algebra $\mathrm{Z}(\mathfrak{g}_n, \mathfrak{g}_{n-1})$ can be explicitly written down as well. However, we shall avoid using them in a way similar to the proof of Lemma 2.13.

The elements z_{ia} naturally act in the space $V(\lambda)^+$ by *raising* or *lowering* the weights. We have for $i = 1, \ldots, n-1$:

$$z_{ia} : V(\lambda)^+_\mu \to V(\lambda)^+_{\mu+\delta_i}, \qquad z_{ai} : V(\lambda)^+_\mu \to V(\lambda)^+_{\mu-\delta_i},$$

where $\mu \pm \delta_i$ is obtained from μ by replacing μ_i with $\mu_i \pm 1$. In the B case the operators z_{0a} preserve each subspace $V(\lambda)^+_\mu$.

We shall need the following element which can be checked to belong to the normalizer Norm J′, and so it can be regarded as an element of the algebra $\mathrm{Z}(\mathfrak{g}_n, \mathfrak{g}_{n-1})$:

$$z_{n,-n} = \sum_{n>i_1>\cdots>i_s>-n} F_{ni_1} F_{i_1 i_2} \cdots F_{i_s,-n} (f_n - f_{j_1}) \cdots (f_n - f_{j_k})$$

in the B, C cases, and

$$z_{n,-n} = \sum_{n>i_1>\cdots>i_s>-n} F_{ni_1} F_{i_1 i_2} \cdots F_{i_s,-n} \frac{(f_n - f_{j_1}) \cdots (f_n - f_{j_k})}{2 f_n}$$

in the D case, where $s = 0, 1, \ldots$ and $\{j_1, \ldots, j_k\}$ is the complement to the subset $\{i_1, \ldots, i_s\}$ in $\{-n+1, \ldots, n-1\}$. The following is a counterpart of Lemma 2.14 and is crucial in the calculation of the matrix elements of the generators in the bases.

LEMMA 3.1. *For $a \in \{-n, n\}$ we have*

$$F_{n-1,a} = \sum_{i=-n+1}^{n-1} z_{n-1,i} z_{ia} \frac{1}{(f_i - f_{-n+1}) \cdots \wedge_i \cdots (f_i - f_{n-1})}$$

in the B, C cases, and

$$F_{n-1,a} = \sum_{i=-n+1}^{n-1} z_{n-1,i} z_{ia} \frac{1}{(f_i - f_{-n+1}) \cdots \wedge_{-i,i} \cdots (f_i - f_{n-1})}$$

in the D case, where $z_{n-1,n-1} = 1$ and the equalities are considered in $\mathrm{U}'(\mathfrak{g}_n)$ modulo the ideal J′. The wedge indicates the indices to be skipped.

In order to write down the basis vectors, introduce the interpolation polynomials $Z_{n,-n}(u)$ with coefficients in the Mickelsson–Zhelobenko algebra $\mathrm{Z}(\mathfrak{g}_n, \mathfrak{g}_{n-1})$ by

$$Z_{n,-n}(u) = \sum_{i=1}^{n} z_{ni} z_{i,-n} \prod_{j=1,\ j\neq i}^{n} \frac{u^2 - g_j^2}{g_i^2 - g_j^2} \tag{3.10}$$

in the B, C cases, and

$$Z_{n,-n}(u) = \sum_{i=1}^{n-1} z_{ni} z_{i,-n} \prod_{j=1,\ j\neq i}^{n-1} \frac{u^2 - g_j^2}{g_i^2 - g_j^2} \tag{3.11}$$

in the D case, where $g_i = f_i + 1/2$. Accordingly, we have

$$Z_{n,-n}(g_i) = z_{ni} z_{i,-n} \tag{3.12}$$

with the agreement that when u is evaluated in $\mathrm{U}(\mathfrak{h})$, the coefficients of the polynomial $Z_{n,-n}(u)$ are written to the left of the powers of u, as is the case in the formulas (3.10) and (3.11).

3.2. *Branching rules, patterns and basis vectors*

The restriction of $V(\lambda)$ to the subalgebra $\mathfrak{g}_{n-1}$ is given by

$$V(\lambda)|_{\mathfrak{g}_{n-1}} \simeq \bigoplus_{\mu} c(\mu) V'(\mu),$$

where $V'(\mu)$ is the irreducible finite-dimensional representation of $\mathfrak{g}_{n-1}$ with highest weight μ. The multiplicity $c(\mu)$ coincides with the dimension of the space $V(\lambda)^+_\mu$, and its exact value is found from the Zhelobenko branching rules, [167]. We formulate them separately for each case recalling the conditions (3.5) and (3.6) on the highest weight λ. In the formulas below we use the notation

$$l_i = \lambda_i + \rho_i + 1/2, \qquad \gamma_i = \nu_i + \rho_i + 1/2,$$

where the ν_i are the parameters defined in the branching rules.

A parameterization of basis vectors in $V(\lambda)$ is obtained by applying the branching rules to its successive restrictions to the subalgebras of the chain

$$\mathfrak{g}_1 \subset \mathfrak{g}_2 \subset \cdots \subset \mathfrak{g}_{n-1} \subset \mathfrak{g}_n.$$

This leads to the definition of the Gelfand–Tsetlin patterns for the B, C and D types. Then we give formulas for the basis vectors of the representation $V(\lambda)$. We use the notation

$$l_{ki} = \lambda_{ki} + \rho_i + 1/2, \qquad l'_{ki} = \lambda'_{ki} + \rho_i + 1/2,$$

where the λ_{ki} and λ'_{ki} are the entries of the patterns defined below.

B type case. The multiplicity $c(\mu)$ equals the number of n-tuples $(\nu'_1, \nu_2, \ldots, \nu_n)$ satisfying the inequalities

$$-\lambda_1 \geqslant \nu'_1 \geqslant \lambda_1 \geqslant \nu_2 \geqslant \lambda_2 \geqslant \cdots \geqslant \nu_{n-1} \geqslant \lambda_{n-1} \geqslant \nu_n \geqslant \lambda_n,$$
$$-\mu_1 \geqslant \nu'_1 \geqslant \mu_1 \geqslant \nu_2 \geqslant \mu_2 \geqslant \cdots \geqslant \nu_{n-1} \geqslant \mu_{n-1} \geqslant \nu_n$$

with ν'_1 and all the ν_i being simultaneously integers or half-integers together with the λ_i. Equivalently, $c(\mu)$ equals the number of $(n+1)$-tuples $\nu = (\sigma, \nu_1, \ldots, \nu_n)$, with the entries given by

$$(\sigma, \nu_1) = \begin{cases} (0, \nu'_1) & \text{if } \nu'_1 \leqslant 0, \\ (1, -\nu'_1) & \text{if } \nu'_1 > 0. \end{cases}$$

LEMMA 3.2. *The vectors*

$$\xi_\nu = z_{n0}^{\sigma} \prod_{i=1}^{n-1} z_{ni}^{\nu_i-\mu_i} z_{i,-n}^{\nu_i-\lambda_i} \cdot \prod_{k=l_n}^{\gamma_n-1} Z_{n,-n}(k)\xi$$

form a basis of the space $V(\lambda)^+_\mu$.

Define the B *type pattern* Λ associated with λ as an array of the form

$$\begin{array}{cccccccc}
\sigma_n & & \lambda_{n1} & & \lambda_{n2} & & \cdots & & \lambda_{nn} \\
& \lambda'_{n1} & & \lambda'_{n2} & & \cdots & & \lambda'_{nn} & \\
\sigma_{n-1} & & \lambda_{n-1,1} & & \cdots & & \lambda_{n-1,n-1} & & \\
& \lambda'_{n-1,1} & & \cdots & & \lambda'_{n-1,n-1} & & & \\
\cdots & & \cdots & & \cdots & & & & \\
\sigma_1 & & \lambda_{11} & & & & & & \\
& \lambda'_{11} & & & & & & &
\end{array}$$

such that $\lambda = (\lambda_{n1}, \ldots, \lambda_{nn})$, each σ_k is 0 or 1, the remaining entries are all nonpositive integers or nonpositive half-integers together with the λ_i, and the following inequalities hold

$$\lambda'_{k1} \geqslant \lambda_{k1} \geqslant \lambda'_{k2} \geqslant \lambda_{k2} \geqslant \cdots \geqslant \lambda'_{k,k-1} \geqslant \lambda_{k,k-1} \geqslant \lambda'_{kk} \geqslant \lambda_{kk}$$

for $k = 1, \ldots, n$, and

$$\lambda'_{k1} \geqslant \lambda_{k-1,1} \geqslant \lambda'_{k2} \geqslant \lambda_{k-1,2} \geqslant \cdots \geqslant \lambda'_{k,k-1} \geqslant \lambda_{k-1,k-1} \geqslant \lambda'_{kk}$$

for $k = 2, \ldots, n$. In addition, in the case of the integer λ_i the condition

$$\lambda'_{k1} \leqslant -1 \quad \text{if } \sigma_k = 1$$

should hold for all $k = 1, \ldots, n$.

THEOREM 3.3. *The vectors*

$$\xi_\Lambda = \prod_{k=1,\ldots,n}^{\rightarrow} \left(z_{k0}^{\sigma_k} \cdot \prod_{i=1}^{k-1} z_{ki}^{\lambda'_{ki}-\lambda_{k-1,i}} z_{i,-k}^{\lambda'_{ki}-\lambda_{ki}} \cdot \prod_{j=l_{kk}}^{l'_{kk}-1} Z_{k,-k}(j) \right) \xi$$

parametrized by the patterns Λ *form a basis of the representation* $V(\lambda)$.

C type case. The multiplicity $c(\mu)$ equals the number of n-tuples of integers $\nu = (\nu_1, \dots, \nu_n)$ satisfying the inequalities

$$
\begin{aligned}
&0 \geqslant \nu_1 \geqslant \lambda_1 \geqslant \nu_2 \geqslant \lambda_2 \geqslant \cdots \geqslant \nu_{n-1} \geqslant \lambda_{n-1} \geqslant \nu_n \geqslant \lambda_n, \\
&0 \geqslant \nu_1 \geqslant \mu_1 \geqslant \nu_2 \geqslant \mu_2 \geqslant \cdots \geqslant \nu_{n-1} \geqslant \mu_{n-1} \geqslant \nu_n.
\end{aligned}
\tag{3.13}
$$

LEMMA 3.4. *The vectors*

$$
\xi_\nu = \prod_{i=1}^{n-1} z_{ni}^{\nu_i - \mu_i} z_{i,-n}^{\nu_i - \lambda_i} \cdot \prod_{k=l_n}^{\gamma_n - 1} Z_{n,-n}(k)\xi
$$

form a basis of the space $V(\lambda)_\mu^+$.

Define the *C type pattern* Λ associated with λ as an array of the form

$$
\begin{array}{cccccccc}
 & \lambda_{n1} & & \lambda_{n2} & & \cdots & & \lambda_{nn} \\
\lambda'_{n1} & & \lambda'_{n2} & & \cdots & & \lambda'_{nn} & \\
 & \lambda_{n-1,1} & & \cdots & & \lambda_{n-1,n-1} & & \\
\lambda'_{n-1,1} & & \cdots & & \lambda'_{n-1,n-1} & & & \\
 & \cdots & & \cdots & & & & \\
 & \lambda_{11} & & & & & & \\
\lambda'_{11} & & & & & & &
\end{array}
$$

such that $\lambda = (\lambda_{n1}, \dots, \lambda_{nn})$, the remaining entries are all non-positive integers and the following inequalities hold

$$
0 \geqslant \lambda'_{k1} \geqslant \lambda_{k1} \geqslant \lambda'_{k2} \geqslant \lambda_{k2} \geqslant \cdots \geqslant \lambda'_{k,k-1} \geqslant \lambda_{k,k-1} \geqslant \lambda'_{kk} \geqslant \lambda_{kk}
$$

for $k = 1, \dots, n$, and

$$
0 \geqslant \lambda'_{k1} \geqslant \lambda_{k-1,1} \geqslant \lambda'_{k2} \geqslant \lambda_{k-1,2} \geqslant \cdots \geqslant \lambda'_{k,k-1} \geqslant \lambda_{k-1,k-1} \geqslant \lambda'_{kk}
$$

for $k = 2, \dots, n$.

THEOREM 3.5. *The vectors*

$$
\xi_\Lambda = \prod_{k=1,\dots,n}^{\rightarrow} \left(\prod_{i=1}^{k-1} z_{ki}^{\lambda'_{ki} - \lambda_{k-1,i}} z_{i,-k}^{\lambda'_{ki} - \lambda_{ki}} \cdot \prod_{j=l_{kk}}^{l'_{kk}-1} Z_{k,-k}(j) \right) \xi
$$

parametrized by the patterns Λ *form a basis of the representation* $V(\lambda)$.

D type case. The multiplicity $c(\mu)$ equals the number of $(n-1)$-tuples $\nu = (\nu_1, \dots, \nu_{n-1})$ satisfying the inequalities

$$-|\lambda_1| \geqslant \nu_1 \geqslant \lambda_2 \geqslant \nu_2 \geqslant \lambda_3 \geqslant \cdots \geqslant \lambda_{n-1} \geqslant \nu_{n-1} \geqslant \lambda_n,$$
$$-|\mu_1| \geqslant \nu_1 \geqslant \mu_2 \geqslant \nu_2 \geqslant \mu_3 \geqslant \cdots \geqslant \mu_{n-1} \geqslant \nu_{n-1}$$

with all the ν_i being simultaneously integers or half-integers together with the λ_i. Set $\nu_0 = \max\{\lambda_1, \mu_1\}$.

LEMMA 3.6. *The vectors*

$$\xi_\nu = \prod_{i=1}^{n-1} z_{ni}^{\nu_{i-1}-\mu_i} z_{i,-n}^{\nu_{i-1}-\lambda_i} \cdot \prod_{k=l_n}^{\gamma_{n-1}-2} Z_{n,-n}(k)\xi$$

form a basis of the space $V(\lambda)^+_\mu$.

Define the *D type pattern* Λ associated with λ as an array of the form

$$\begin{array}{ccccccc}
\lambda_{n1} & & \lambda_{n2} & & \cdots & & \lambda_{nn} \\
& \lambda'_{n-1,1} & & \cdots & & \lambda'_{n-1,n-1} & \\
\lambda_{n-1,1} & & \cdots & & \lambda_{n-1,n-1} & & \\
& \cdots & & \cdots & & & \\
\lambda_{21} & & \lambda_{22} & & & & \\
& \lambda'_{11} & & & & & \\
\lambda_{11} & & & & & &
\end{array}$$

such that $\lambda = (\lambda_{n1}, \dots, \lambda_{nn})$, the remaining entries are all nonpositive integers or nonpositive half-integers together with the λ_i, and the following inequalities hold

$$-|\lambda_{k1}| \geqslant \lambda'_{k-1,1} \geqslant \lambda_{k2} \geqslant \lambda'_{k-1,2} \geqslant \cdots \geqslant \lambda_{k,k-1} \geqslant \lambda'_{k-1,k-1} \geqslant \lambda_{kk},$$
$$-|\lambda_{k-1,1}| \geqslant \lambda'_{k-1,1} \geqslant \lambda_{k-1,2} \geqslant \lambda'_{k-1,2} \geqslant \cdots \geqslant \lambda_{k-1,k-1} \geqslant \lambda'_{k-1,k-1}$$

for $k = 2, \dots, n$. Set $\lambda'_{k-1,0} = \max\{\lambda_{k1}, \lambda_{k-1,1}\}$.

THEOREM 3.7. *The vectors*

$$\xi_\Lambda = \prod_{k=2,\dots,n}^{\rightarrow} \left(\prod_{i=1}^{k-1} z_{ki}^{\lambda'_{k-1,i-1}-\lambda_{k-1,i}} z_{i,-k}^{\lambda'_{k-1,i-1}-\lambda_{ki}} \cdot \prod_{j=l_{kk}}^{l'_{k-1,k-1}-2} Z_{k,-k}(j) \right) \xi$$

parametrized by the patterns Λ *form a basis of the representation* $V(\lambda)$.

Proofs of Theorems 3.3, 3.5 and 3.7 will be outlined in the next two sections. These are based on the application of the representation theory of the twisted Yangians. Clearly, due to the branching rules, it is sufficient to construct a basis in the multiplicity space $V(\lambda)^+_\mu$.

3.3. *Yangians and their representations*

We start by introducing the *Yangian* $\mathrm{Y}(2)$ for the Lie algebra $\mathfrak{gl}_2$. In what follows it will be convenient to use the indices $-n, n$ to enumerate the rows and columns of 2×2-matrices. The Yangian $\mathrm{Y}(2)$ is the complex associative algebra with the generators $t_{ab}^{(1)}, t_{ab}^{(2)}, \dots$ where $a, b \in \{-n, n\}$, and the defining relations

$$(u - v)\big[t_{ab}(u), t_{cd}(v)\big] = t_{cb}(u)t_{ad}(v) - t_{cb}(v)t_{ad}(u), \tag{3.14}$$

where

$$t_{ab}(u) = \delta_{ab} + t_{ab}^{(1)}u^{-1} + t_{ab}^{(2)}u^{-2} + \dots \in \mathrm{Y}(2)\big[\big[u^{-1}\big]\big].$$

Introduce the series $s_{ab}(u)$, $a, b \in \{-n, n\}$ by

$$s_{ab}(u) = \theta_{nb}t_{an}(u)t_{-b,-n}(-u) + \theta_{-n,b}t_{a,-n}(u)t_{-b,n}(-u) \tag{3.15}$$

with θ_{ij} defined in (3.4). Write

$$s_{ab}(u) = \delta_{ab} + s_{ab}^{(1)}u^{-1} + s_{ab}^{(2)}u^{-2} + \cdots.$$

The *twisted Yangian* $\mathrm{Y}^{\pm}(2)$ is defined as the subalgebra of $\mathrm{Y}(2)$ generated by the elements $s_{ab}^{(1)}, s_{ab}^{(2)}, \dots$ where $a, b \in \{-n, n\}$. Also, $\mathrm{Y}^{\pm}(2)$ can be viewed as an abstract algebra with generators $s_{ab}^{(r)}$ and quadratic and linear defining relations which have the following form

$$\begin{aligned}
&\big(u^2 - v^2\big)\big[s_{ab}(u), s_{cd}(v)\big] \\
&\quad = (u + v)\big(s_{cb}(u)s_{ad}(v) - s_{cb}(v)s_{ad}(u)\big) \\
&\qquad - (u - v)\big(\theta_{c,-b}s_{a,-c}(u)s_{-b,d}(v) - \theta_{a,-d}s_{c,-a}(v)s_{-d,b}(u)\big) \\
&\qquad + \theta_{a,-b}\big(s_{c,-a}(u)s_{-b,d}(v) - s_{c,-a}(v)s_{-b,d}(u)\big)
\end{aligned}$$

and

$$\theta_{ab}s_{-b,-a}(-u) = s_{ab}(u) \pm \frac{s_{ab}(u) - s_{ab}(-u)}{2u}.$$

Whenever the double sign $\pm$ or $\mp$ occurs, the upper sign corresponds to the orthogonal case and the lower sign to the symplectic case. In particular, we have the relation

$$\big[s_{n,-n}(u), s_{n,-n}(v)\big] = 0.$$

The Yangian $\mathrm{Y}(2)$ is a Hopf algebra with the coproduct

$$\Delta\big(t_{ab}(u)\big) = t_{an}(u) \otimes t_{nb}(u) + t_{a,-n}(u) \otimes t_{-n,b}(u). \tag{3.16}$$

The twisted Yangian $\mathrm{Y}^{\pm}(2)$ is a left coideal in $\mathrm{Y}(2)$ with

$$\Delta\big(s_{ab}(u)\big) = \sum_{c,d\in\{-n,n\}} \theta_{bd} t_{ac}(u) t_{-b,-d}(-u) \otimes s_{cd}(u). \tag{3.17}$$

Given a pair of complex numbers (α, β) such that $\alpha - \beta \in \mathbb{Z}_+$ we denote by $L(\alpha, \beta)$ the irreducible representation of the Lie algebra $\mathfrak{gl}_2$ with highest weight (α, β) with respect to the upper triangular Borel subalgebra. Then $\dim L(\alpha, \beta) = \alpha - \beta + 1$. We equip $L(\alpha, \beta)$ with a $\mathrm{Y}(2)$-module structure by using the algebra homomorphism $\mathrm{Y}(2) \to \mathrm{U}(\mathfrak{gl}_2)$ given by

$$t_{ab}(u) \mapsto \delta_{ab} + E_{ab} u^{-1}, \quad a, b \in \{-n, n\}.$$

The coproduct (3.16) allows us to construct representations of $\mathrm{Y}(2)$ of the form

$$L = L(\alpha_1, \beta_1) \otimes \cdots \otimes L(\alpha_k, \beta_k). \tag{3.18}$$

Any finite-dimensional irreducible $\mathrm{Y}(2)$-module is isomorphic to a representation of this type twisted by an automorphism of $\mathrm{Y}(2)$ of the form

$$t_{ab}(u) \mapsto \big(1 + \varphi_1 u^{-1} + \varphi_2 u^{-2} + \cdots\big) t_{ab}(u), \quad \varphi_i \in \mathbb{C}.$$

There is an explicit irreducibility criterion for the $\mathrm{Y}(2)$-module L. To formulate the result, with each $L(\alpha, \beta)$ associate the *string*

$$S(\alpha, \beta) = \{\beta, \beta + 1, \ldots, \alpha - 1\} \subset \mathbb{C}.$$

We say that two strings S_1 and S_2 are *in general position* if

$$\text{either} \quad S_1 \cup S_2 \quad \text{is not a string, or} \quad S_1 \subseteq S_2, \quad \text{or} \quad S_2 \subseteq S_1.$$

THEOREM 3.8. *The representation* (3.18) *of* $\mathrm{Y}(2)$ *is irreducible if and only if the strings* $S(\alpha_i, \beta_i)$, $i = 1, \ldots, k$, *are pairwise in general position.*

Note that the generators $t_{ab}^{(r)}$ with $r > k$ act as zero operators in L. Therefore, the operators $T_{ab}(u) = u^k t_{ab}(u)$ are polynomials in u:

$$T_{ab}(u) = \delta_{ab} u^k + t_{ab}^{(1)} u^{k-1} + \cdots + t_{ab}^{(k)}. \tag{3.19}$$

Let ξ_i denote the highest vector of the $\mathfrak{gl}_2$-module $L(\alpha_i, \beta_i)$. Suppose that the $\mathrm{Y}(2)$-module L given by (3.18) is irreducible and the strings $S(\alpha_i, \beta_i)$ are pairwise disjoint. Set

$$\eta = \xi_1 \otimes \cdots \otimes \xi_k. \tag{3.20}$$

Then using (3.16) we easily check that η is the highest vector of the $\mathrm{Y}(2)$-module L. That is, η is annihilated by $T_{-n,n}(u)$, and it is an eigenvector for the operators $T_{nn}(u)$ and $T_{-n,-n}(u)$. Explicitly,

$$\begin{aligned} T_{-n,-n}(u)\eta &= (u+\alpha_1)\cdots(u+\alpha_k)\eta, \\ T_{nn}(u)\eta &= (u+\beta_1)\cdots(u+\beta_k)\eta. \end{aligned} \tag{3.21}$$

Let a k-tuple $\gamma = (\gamma_1, \dots, \gamma_k)$ satisfy the following conditions: for each i

$$\alpha_i - \gamma_i \in \mathbb{Z}_+, \qquad \gamma_i - \beta_i \in \mathbb{Z}_+. \tag{3.22}$$

Set

$$\eta_\gamma = \prod_{i=1}^{k} T_{n,-n}(-\gamma_i+1)\cdots T_{n,-n}(-\beta_i-1)T_{n,-n}(-\beta_i)\eta.$$

The following theorem provides a Gelfand–Tsetlin type basis for representations of the Yangian $\mathrm{Y}(2)$ associated with the embedding $\mathrm{Y}(1) \subset \mathrm{Y}(2)$. Here $\mathrm{Y}(1)$ is the (commutative) subalgebra of $\mathrm{Y}(2)$ generated by the elements $t_{nn}^{(r)}$, $r \geqslant 1$.

THEOREM 3.9. *Let the* $\mathrm{Y}(2)$*-module* L *given by* (3.18) *be irreducible and let the strings* $S(\alpha_i, \beta_i)$ *be pairwise disjoint. Then the vectors* η_γ *with* γ *satisfying* (3.22) *form a basis of* L*. Moreover, the generators of* $\mathrm{Y}(2)$ *act in this basis by the rules*

$$\begin{aligned} T_{nn}(u)\eta_\gamma &= (u+\gamma_1)\cdots(u+\gamma_k)\eta_\gamma, \\ T_{n,-n}(-\gamma_i)\eta_\gamma &= \eta_{\gamma+\delta_i}, \\ T_{-n,n}(-\gamma_i)\eta_\gamma &= -\prod_{m=1}^{k}(\alpha_m-\gamma_i+1)(\beta_m-\gamma_i)\eta_{\gamma-\delta_i}, \\ T_{-n,-n}(u)\eta_\gamma &= \prod_{i=1}^{k}\frac{(u+\alpha_i+1)(u+\beta_i)}{u+\gamma_i+1}\eta_\gamma \\ &\quad + \prod_{i=1}^{k}\frac{1}{u+\gamma_i+1}T_{-n,n}(u)T_{n,-n}(u+1)\eta_\gamma. \end{aligned} \tag{3.23}$$

These formulas are derived from the defining relations for the Yangian (3.14) with the use of the *quantum determinant*

$$d(u) = T_{-n,-n}(u+1)T_{nn}(u) - T_{n,-n}(u+1)T_{-n,n}(u) \tag{3.24}$$

$$= T_{-n,-n}(u)T_{nn}(u+1) - T_{-n,n}(u)T_{n,-n}(u+1). \tag{3.25}$$

The coefficients of the quantum determinant belong to the center of $\mathrm{Y}(2)$ and so $d(u)$ acts in L as a scalar which can be found by the application of (3.24) to the highest vector η. Indeed, by (3.21)

$$d(u)\eta = (u+\alpha_1+1)\cdots(u+\alpha_k+1)(u+\beta_1)\cdots(u+\beta_k)\eta.$$

This allows us to derive the last formula in (3.23) from (3.25). The operators $T_{-n,n}(u)$ and $T_{n,-n}(u)$ are polynomials in u of degree $\leqslant k-1$; see (3.19). Therefore, their action can be found from (3.23) by using the Lagrange interpolation formula.

We can regard (3.18) as a module over the twisted Yangian $\mathrm{Y}^-(2)$ obtained by restriction. An irreducibility criterion for such a module is provided by the following theorem.

THEOREM 3.10. *The representation* (3.18) *of* $\mathrm{Y}^-(2)$ *is irreducible if and only if each pair of strings*

$$S(\alpha_i,\beta_i),\ S(\alpha_j,\beta_j) \quad \text{and} \quad S(\alpha_i,\beta_i),\ S(-\beta_j,-\alpha_j)$$

is in general position for all $i<j$.

The defining relations (3.14) allow us to rewrite formula (3.15) for $s_{n,-n}(u)$ in the form

$$s_{n,-n}(u) = \frac{u+1/2}{u}\big(t_{n,-n}(u)t_{nn}(-u) - t_{n,-n}(-u)t_{nn}(u)\big).$$

Therefore the operator in L defined by

$$\begin{aligned} S_{n,-n}(u) &= \frac{u^{2k}}{u+1/2} s_{n,-n}(u) \\ &= \frac{(-1)^k}{u}\big(T_{n,-n}(u)T_{nn}(-u) - T_{n,-n}(-u)T_{nn}(u)\big) \end{aligned} \tag{3.26}$$

is an even polynomial in u of degree $\leqslant 2k-2$. Its action in the basis of L provided in Theorem 3.9 is given by

$$S_{n,-n}(\gamma_i)\eta_\gamma = 2\prod_{a=1,\ a\neq i}^{k}(-\gamma_i-\gamma_a)\eta_{\gamma+\delta_i}, \qquad i=1,\dots,k.$$

We have thus proved the following corollary.

COROLLARY 3.11. *Suppose that the* $\mathrm{Y}^-(2)$*-module* L *is irreducible and we have*

$$S(\alpha_i,\beta_i)\cap S(\alpha_j,\beta_j)=\emptyset \quad \text{and} \quad S(\alpha_i,\beta_i)\cap S(-\beta_j,-\alpha_j)=\emptyset$$

for all $i < j$.[5] *Then the vectors*

$$\xi_\gamma = \prod_{i=1}^{k} S_{n,-n}(\gamma_i - 1) \cdots S_{n,-n}(\beta_i + 1) S_{n,-n}(\beta_i) \eta$$

with γ satisfying (3.22) *form a basis of L.*

Let us now turn to the orthogonal twisted Yangian $\mathrm{Y}^+(2)$. For any $\delta \in \mathbb{C}$ denote by $W(\delta)$ the one-dimensional representation of $\mathrm{Y}^+(2)$ spanned by a vector w such that

$$s_{nn}(u)w = \frac{u+\delta}{u+1/2} w, \qquad s_{-n,-n}(u)w = \frac{u-\delta+1}{u+1/2} w,$$

and $s_{a,-a}(u)w = 0$ for $a = -n, n$. By (3.17) we can regard the tensor product $L \otimes W(\delta)$ as a representation of $\mathrm{Y}^+(2)$. The representations of $\mathrm{Y}^+(2)$ of this type, and the representations of $\mathrm{Y}^-(2)$ of type (3.18) essentially exhaust all finite-dimensional irreducible representations of $\mathrm{Y}^\pm(2)$, [97].

The following is an analog of Theorem 3.10.

THEOREM 3.12. *The representation $L \otimes W(\delta)$ of $\mathrm{Y}^+(2)$ is irreducible if and only if each pair of strings*

$$S(\alpha_i, \beta_i),\ S(\alpha_j, \beta_j) \quad \textit{and} \quad S(\alpha_i, \beta_i),\ S(-\beta_j, -\alpha_j)$$

is in general position for all $i < j$, and none of the strings $S(\alpha_i, \beta_i)$ or $S(-\beta_i, -\alpha_i)$ contains $-\delta$.

Using the vector space isomorphism

$$L \otimes W(\delta) \to L, \qquad v \otimes w \mapsto v, \qquad v \in L, \tag{3.27}$$

we can regard L as a $\mathrm{Y}^+(2)$-module. Accordingly, using the defining relations (3.14) and the coproduct formula (3.17) we can write $s_{n,-n}(u)$, as an operator in L, in the form

$$s_{n,-n}(u) = \frac{u-\delta}{u} t_{n,-n}(u) t_{nn}(-u) + \frac{u+\delta}{u} t_{n,-n}(-u) t_{nn}(u).$$

Therefore the operator in L defined by

$$\begin{aligned} S_{n,-n}(u) &= u^{2k} s_{n,-n}(u) \\ &= \frac{(-1)^k}{u} \big((u-\delta) T_{n,-n}(u) T_{nn}(-u) + (u+\delta) T_{n,-n}(-u) T_{nn}(u) \big) \end{aligned} \tag{3.28}$$

[5] The second condition was erroneously omitted in the formulation of [98, Proposition 4.2] although it is implicit in the proof.

is an even polynomial in u of degree $\leqslant 2k-2$. Its action in the basis of L provided in Theorem 3.9 is given by

$$S_{n,-n}(\gamma_i)\eta_\gamma = 2(-\delta-\gamma_i)\prod_{a=1,\ a\neq i}^{k}(-\gamma_i-\gamma_a)\eta_{\gamma+\delta_i}, \qquad i=1,\dots,k.$$

We have thus proved the following corollary.

COROLLARY 3.13. *Suppose that the* $\mathrm{Y}^+(2)$*-module* $L\otimes W(\delta)$ *is irreducible and we have*

$$S(\alpha_i,\beta_i)\cap S(\alpha_j,\beta_j)=\emptyset \quad \textit{and} \quad S(\alpha_i,\beta_i)\cap S(-\beta_j,-\alpha_j)=\emptyset$$

for all $i<j$. *Then the vectors*

$$\xi_\gamma = \prod_{i=1}^{k} S_{n,-n}(\gamma_i-1)\cdots S_{n,-n}(\beta_i+1)S_{n,-n}(\beta_i)\eta$$

with γ *satisfying* (3.22) *form a basis of* L.

3.4. *Yangian action on the multiplicity space*

Now we construct an algebra homomorphism $\mathrm{Y}^\pm(2)\to \mathrm{Z}(\mathfrak{g}_n,\mathfrak{g}_{n-1})$ and then use it to define an action of $\mathrm{Y}^\pm(2)$ on the multiplicity space $V(\lambda)^+_\mu$.

For $a,b\in\{-n,n\}$ and a complex parameter u introduce the elements $Z_{ab}(u)$ of the Mickelsson–Zhelobenko algebra $\mathrm{Z}(\mathfrak{g}_n,\mathfrak{g}_{n-1})$ by

$$Z_{ab}(u) = -\left(\delta_{ab}\left(u+\rho_n+\frac{1}{2}\right)+F_{ab}\right)\prod_{i=-n+}^{n-1}(u+g_i) + \sum_{i=-n+1}^{n-1} z_{ai}z_{ib}\prod_{j=-n+1,\ j\neq i}^{n-1}\frac{u+g_j}{g_i-g_j} \tag{3.29}$$

in the B case,

$$Z_{ab}(u) = \left(\delta_{ab}\left(u+\rho_n+\frac{1}{2}\right)+F_{ab}\right)\prod_{i=-n+1}^{n-1}(u+g_i) - \sum_{i=-n+1}^{n-1} z_{ai}z_{ib}\prod_{j=-n+1,\ j\neq i}^{n-1}\frac{u+g_j}{g_i-g_j} \tag{3.30}$$

in the C case, and

$$Z_{ab}(u) = -\Bigg(\bigg(\delta_{ab}\Big(u + \rho_n + \frac{1}{2}\Big) + F_{ab}\bigg) \prod_{i=-n+1}^{n-1} (u + g_i)$$

$$- \sum_{i=-n+1}^{n-1} z_{ai} z_{ib}(u + g_{-i}) \prod_{j=-n+1,\ j\neq\pm i}^{n-1} \frac{u + g_j}{g_i - g_j}\Bigg) \frac{1}{2u+1} \tag{3.31}$$

in the D case, where $g_i = f_i + 1/2$ for all i. In particular, it can be verified that each $Z_{n,-n}(u)$ coincides with the corresponding interpolation polynomial given in (3.10) or (3.11).

Consider now the three cases separately. We shall assume $\mu_n = -\infty$ in the notation below.

B type case.

THEOREM 3.14.

(i) *The mapping*

$$s_{ab}(u) \mapsto -u^{-2n} Z_{ab}(u), \quad a, b \in \{-n, n\}, \tag{3.32}$$

defines an algebra homomorphism $\mathrm{Y}^+(2) \to \mathrm{Z}(\mathfrak{g}_n, \mathfrak{g}_{n-1})$.

(ii) *The* $\mathrm{Y}^+(2)$*-module* $V(\lambda)^+_\mu$ *defined via the homomorphism* (3.32) *is isomorphic to the direct sum of two irreducible submodules,* $V(\lambda)^+_\mu \simeq U \oplus U'$, *where*

$$U = L(0, \beta_1) \otimes L(\alpha_2, \beta_2) \otimes \cdots \otimes L(\alpha_n, \beta_n) \otimes W(1/2),$$
$$U' = L(-1, \beta_1) \otimes L(\alpha_2, \beta_2) \otimes \cdots \otimes L(\alpha_n, \beta_n) \otimes W(1/2),$$

if the λ_i *are integers (it is supposed that* $U' = \{0\}$ *if* $\beta_1 = 0$); *or*

$$U = L(-1/2, \beta_1) \otimes L(\alpha_2, \beta_2) \otimes \cdots \otimes L(\alpha_n, \beta_n) \otimes W(0),$$
$$U' = L(-1/2, \beta_1) \otimes L(\alpha_2, \beta_2) \otimes \cdots \otimes L(\alpha_n, \beta_n) \otimes W(1),$$

if the λ_i *are half-integers, and the following notation is used*

$$\alpha_i = \min\{\lambda_{i-1}, \mu_{i-1}\} - i + 1, \quad i = 2, \ldots, n,$$
$$\beta_i = \max\{\lambda_i, \mu_i\} - i + 1, \quad i = 1, \ldots, n.$$

C type case.

THEOREM 3.15.
(i) *The mapping*

$$s_{ab}(u) \mapsto (u + 1/2)u^{-2n} Z_{ab}(u), \quad a, b \in \{-n, n\}, \tag{3.33}$$

defines an algebra homomorphism $\mathrm{Y}^-(2) \to \mathrm{Z}(\mathfrak{g}_n, \mathfrak{g}_{n-1})$.
(ii) *The* $\mathrm{Y}^-(2)$*-module* $V(\lambda)^+_\mu$ *defined via the homomorphism* (3.33) *is irreducible and isomorphic to the tensor product*

$$L(\alpha_1, \beta_1) \otimes \cdots \otimes L(\alpha_n, \beta_n),$$

where $\alpha_1 = -1/2$ *and*

$$\alpha_i = \min\{\lambda_{i-1}, \mu_{i-1}\} - i + 1/2, \quad i = 2, \dots, n,$$

$$\beta_i = \max\{\lambda_i, \mu_i\} - i + 1/2, \quad i = 1, \dots, n.$$

D type case.

THEOREM 3.16.
(i) *The mapping*

$$s_{ab}(u) \mapsto -2u^{-2n+2} Z_{ab}(u), \quad a, b \in \{-n, n\}, \tag{3.34}$$

defines an algebra homomorphism $\mathrm{Y}^+(2) \to \mathrm{Z}(\mathfrak{g}_n, \mathfrak{g}_{n-1})$.
(ii) *The* $\mathrm{Y}^+(2)$*-module* $V(\lambda)^+_\mu$ *defined via the homomorphism* (3.34) *is irreducible and isomorphic to the tensor product*

$$L(\alpha_1, \beta_1) \otimes \cdots \otimes L(\alpha_{n-1}, \beta_{n-1}) \otimes W(-\alpha_0),$$

where $\alpha_1 = \min\{-|\lambda_1|, -|\mu_1|\} - 1/2$, $\alpha_0 = \alpha_1 + |\lambda_1 + \mu_1|$,

$$\alpha_i = \min\{\lambda_i, \mu_i\} - i + 1/2, \quad i = 2, \dots, n - 1,$$

$$\beta_i = \max\{\lambda_{i+1}, \mu_{i+1}\} - i + 1/2, \quad i = 1, \dots, n - 1.$$

OUTLINE OF THE PROOF. Part (i) of Theorems 3.14–3.16 is verified by using the composition of homomorphisms

$$\mathrm{Y}^{\pm}(2) \to \mathrm{C}_n \to \mathrm{Z}(\mathfrak{g}_n, \mathfrak{g}_{n-1}),$$

where C_n is the centralizer $\mathrm{U}(\mathfrak{g}_n)^{\mathfrak{g}_{n-1}}$. The first arrow is the homomorphism provided by the centralizer construction (see [105,118]) while the second is the natural projection.

By the results of [97], every irreducible finite-dimensional representation of the twisted Yangian is a highest weight representation. It contains a unique, up to a constant factor, vector which is annihilated by $s_{-n,n}(u)$ and which is an eigenvector of $s_{nn}(u)$. The corresponding eigenvalue (the highest weight) uniquely determines the representation. The vectors in $V(\lambda)^+_\mu$ annihilated by $s_{-n,n}(u)$ can be explicitly constructed by using the lowering operators. One of these vectors is given by

$$\xi_\mu = \prod_{i=1}^{n-1} \left(z_{ni}^{\max\{\lambda_i,\mu_i\}-\mu_i} z_{i,-n}^{\max\{\lambda_i,\mu_i\}-\lambda_i} \right) \xi,$$

where ξ is the highest vector of $V(\lambda)$. This is the only vector in the C, D cases, while in the B case there is another one defined by

$$\xi'_\mu = z_{n0}\xi_\mu.$$

Calculating the eigenvalues of these vectors we conclude that they respectively coincide with the eigenvalues of the tensor product of the highest vectors of the modules $L(\alpha_i, \beta_i)$; see (3.20). □

REMARK 3.17. Theorems 3.14–3.16 can be proved without using the branching rules for the reductions $\mathfrak{sp}_{2n} \downarrow \mathfrak{sp}_{2n-2}$ and $\mathfrak{o}_N \downarrow \mathfrak{o}_{N-2}$. Therefore, the reduction multiplicities can be found by calculating the dimension of the space $V(\lambda)^+_\mu$. For instance, in the symplectic case, Theorem 3.15 gives

$$c(\mu) = \prod_{i=1}^{n} (\alpha_i - \beta_i + 1)$$

which, of course, coincides with the value provided by the C type branching rule; see Section 3.2.

While keeping λ and μ fixed we let ν run over the values determined by the branching rules; see Section 3.2. Using the homomorphisms of Theorems 3.14–3.16 we conclude from (3.26) and (3.28) that the element $S_{n,-n}(u)$ acts in the representation $V(\lambda)^+_\mu$ precisely as the operator $-Z_{n,-n}(u)$, $Z_{n,-n}(u)$, or $-2Z_{n,-n}(u)$, in the B, C or D cases, respectively. Thus, by Corollaries 3.11 and 3.13, the following vectors ξ_ν form a basis of the space $V(\lambda)^+_\mu$, where

$$\xi_\nu = z_{n0}^{\sigma} \prod_{i=1}^{n} Z_{n,-n}(\gamma_i - 1) \cdots Z_{n,-n}(\beta_i + 1) Z_{n,-n}(\beta_i) \xi_\mu$$

in the B case,

$$\xi_\nu = \prod_{i=1}^{n} Z_{n,-n}(\gamma_i - 1) \cdots Z_{n,-n}(\beta_i + 1) Z_{n,-n}(\beta_i) \xi_\mu \tag{3.35}$$

in the C case, and

$$\xi_\nu = \prod_{i=1}^{n-1} Z_{n,-n}(\gamma_i - 1) \cdots Z_{n,-n}(\beta_i + 1) Z_{n,-n}(\beta_i) \xi_\mu$$

in the D case. Applying the interpolation properties of the polynomials $Z_{n,-n}(u)$ we bring the above formulas to the form given in Lemmas 3.2, 3.4 and 3.6, respectively. Clearly, Theorems 3.3, 3.5 and 3.7 follow.

3.5. *Calculation of the matrix elements*

Without writing down all explicit formulas we shall demonstrate how the matrix elements of the generators of $\mathfrak{g}_n$ in the basis ξ_Λ provided by Theorems 3.3, 3.5 and 3.7 can be calculated. The interested reader is referred to the papers [98–100] for details. We choose the following generators

$$F_{k-1,-k}, \qquad F_{k-1,k}, \quad k = 1, \ldots, n,$$

in the B case,

$$F_{k-1,-k}, \quad k = 2, \ldots, n, \quad \text{and} \quad F_{-k,k}, \qquad F_{k,-k}, \quad k = 1, \ldots, n,$$

in the C case, and

$$F_{k-1,-k}, \qquad F_{k-1,k}, \quad k = 2, \ldots, n, \quad \text{and} \quad F_{21}, \qquad F_{-2,1}$$

in the D case.

In the symplectic case the elements F_{kk}, $F_{k,-k}$, $F_{-k,k}$ commute with the subalgebra $\mathfrak{g}_{k-1}$ in $\mathrm{U}(\mathfrak{g}_k)$. Therefore, these operators preserve the subspace of $\mathfrak{g}_{k-1}$-highest vectors in $V(\lambda)$. So, it suffices to compute the action of these operators with $k = n$ in the basis $\{\xi_\nu\}$ of the space $V(\lambda)^+_\mu$; see Lemma 3.4. For F_{nn} we immediately get

$$F_{nn}\xi_\nu = \left(2\sum_{i=1}^{n} \nu_i - \sum_{i=1}^{n} \lambda_i - \sum_{i=1}^{n-1} \mu_i\right)\xi_\nu.$$

Further, by (3.35)

$$Z_{n,-n}(\gamma_i)\xi_\nu = \xi_{\nu+\delta_i}, \quad i = 1, \ldots, n.$$

However, $Z_{n,-n}(u)$ is a polynomial in u^2 of degree $n-1$ with the highest coefficient $F_{n,-n}$. Applying the Lagrange interpolation formula with the interpolation points γ_i, $i = 1, \ldots, n$, we obtain

$$Z_{n,-n}(u)\xi_\nu = \sum_{i=1}^{n} \prod_{a=1,\, a\neq i}^{n} \frac{u^2 - \gamma_a^2}{\gamma_i^2 - \gamma_a^2} \xi_{\nu+\delta_i}.$$

Taking here the coefficient at u^{2n-2} we get

$$F_{n,-n}\xi_\nu = \sum_{i=1}^{n} \prod_{a=1,\ a\neq i}^{n} \frac{1}{\gamma_i^2 - \gamma_a^2} \xi_{\nu+\delta_i}. \tag{3.36}$$

The action of $F_{-n,n}$ is found in a similar way with the use of Theorem 3.9.

In the orthogonal case the action of F_{nn} is found in the same way. However, the elements $F_{n,-n}$ and $F_{-n,n}$ are zero. We shall use second-order elements of the enveloping algebra instead. These are given by

$$\Phi_{-a,a} = \frac{1}{2} \sum_{i=-n+1}^{n-1} F_{-a,i} F_{ia}$$

with $a \in \{-n, n\}$. The elements $\Phi_{-a,a}$ commute with the subalgebra $\mathfrak{g}_{n-1}$ so that, like in the symplectic case, they preserve the space $V(\lambda)^+_\mu$ and their action in the basis $\{\xi_\nu\}$ is given by formulas similar to those for $F_{-a,a}$.

The calculation of the matrix elements of the generators $F_{k-1,-k}$ is similar in all three cases. We may assume $k = n$. The operator $F_{n-1,-n}$ preserves the subspace of $\mathfrak{g}_{n-2}$ highest vectors in $V(\lambda)$. Consider the symplectic case as an example. Suppose that μ' is a fixed $\mathfrak{g}_{n-2}$ highest weight, ν' is an $(n-1)$-tuple of integers such that the inequalities (3.13) are satisfied with λ, ν, μ respectively replaced by μ, ν', μ', and set $\gamma_i' = \nu_i' + \rho_i + 1/2$. It suffices to calculate the action of $F_{n-1,-n}$ on the basis vectors of the form

$$\xi_{\nu\mu\nu'} = X_{\mu\nu'}\xi_{\nu\mu},$$

where $\xi_{\nu\mu} = \xi_\nu$ and $X_{\mu\nu'}$ denotes the operator

$$X_{\mu\nu'} = \prod_{i=1}^{n-2} z_{n-1,i}^{\nu_i'-\mu_i'} z_{n-1,-i}^{\nu_i'-\mu_i} \cdot \prod_{a=m_{n-1}}^{\gamma_{n-1}'-1} Z_{n-1,-n+1}(a),$$

where we have used the notation $m_i = \mu_i + \rho_i + 1/2$. The operator $F_{n-1,-n}$ is permutable with the elements $z_{n-1,i}$ and $Z_{n-1,-n+1}(u)$. Hence, we can write

$$F_{n-1,-n}\xi_{\nu\mu\nu'} = X_{\mu\nu'} F_{n-1,-n}\xi_{\nu\mu}.$$

Now we apply Lemma 3.1. It remains to calculate $z_{ni}\xi_{\nu\mu}$ and $X_{\mu\nu'} z_{n-1,-i}$. Using the relations between the elements of the Mickelsson–Zhelobenko algebra $\mathrm{Z}(\mathfrak{g}_n, \mathfrak{g}_{n-1})$ given in (3.9), we find that

$$z_{ni}\xi_{\nu\mu} = \xi_{\nu,\mu-\delta_i}$$

if $i > 0$. Otherwise, if $i = -j$ with positive j, write

$$z_{n,-j}\xi_{\nu\mu} = z_{n,-j} z_{nj} \xi_{\nu,\mu+\delta_j} = Z_{n,-n}(m_j)\xi_{\nu,\mu+\delta_j}, \tag{3.37}$$

where we have used the interpolation properties (3.12) of the polynomials $Z_{n,-n}(u)$. Finally, we use the expression (3.35) of the basis vectors and Theorem 3.9 to present (3.37) as a linear combination of basis vectors. The same argument applies to calculate $X_{\mu\nu'} z_{n-1,-i}$.

The final formulas for the matrix elements of the generators $F_{n-1,-n}$ in all the three cases are given by multiplicative expressions in the entries of the patterns which exhibit some similarity to the formulas of Theorem 2.3.

In the orthogonal case we also need to find the action of the generators $F_{n-1,n}$. Unlike the case of the generators $F_{n-1,-n}$, the corresponding matrix elements will be given by certain combinations of multiplicative expressions for which it does not seem to be possible to bring them into a product form. There are two alternative ways to calculate these combinations which we briefly outline below. First, as in the previous calculation, we can write

$$F_{n-1,n}\xi_{\nu\mu\nu'} = X_{\mu\nu'} F_{n-1,n}\xi_{\nu\mu}.$$

Applying again Lemma 3.1, we come to the calculation of $z_{in}\xi_{\nu\mu}$. This time the interpolation property of $Z_{-n,-n}(u)$ (see (3.29) and (3.31)) allows us to write, e.g., for $i > 0$

$$z_{in}\xi_{\nu\mu} = z_{in} z_{ni} \xi_{\nu,\mu+\delta_i} = z_{-n,-i} z_{-i,-n} \xi_{\nu,\mu+\delta_i} = Z_{-n,-n}(m_i)\xi_{\nu,\mu+\delta_i}.$$

Now, as $Z_{-n,-n}(u)$ is, up to a multiple, the image of $S_{-n,-n}(u)$ under the homomorphism $\mathrm{Y}^+(2) \to \mathrm{Z}(\mathfrak{g}_n, \mathfrak{g}_{n-1})$, we can express this operator in terms of the Yangian operators $T_{ab}(u)$ and then apply Theorem 3.9 to calculate its action.

Alternatively, the generator $F_{n-1,n}$ can be written modulo the left ideal J' of $\mathrm{U}'(\mathfrak{g}_n)$ as

$$F_{n-1,n} = \Phi_{n-1,-n}(2)\Phi_{-n,n} - \Phi_{-n,n}\Phi_{n-1,-n}(0), \tag{3.38}$$

where

$$\Phi_{n-1,-n}(u) = \sum_{i=-n+1}^{n-1} z_{n-1,i} z_{i,-n} \prod_{a=-n+1,\ a\neq i}^{n-1} \frac{1}{f_i - f_a} \cdot \frac{1}{u + f_i + F_{nn}} \tag{3.39}$$

in the B case, and

$$\Phi_{n-1,-n}(u) = \sum_{i=-n+1}^{n-1} z_{n-1,i} z_{i,-n} \prod_{a=-n+1,\ a\neq\pm i}^{n-1} \frac{1}{f_i - f_a} \cdot \frac{1}{u + f_i + F_{nn}} \tag{3.40}$$

in the D case. The action of $\Phi_{n-1,-n}(u)$ is found exactly as that of $F_{n-1,-n}$ and the matrix elements have a similar multiplicative form. Note, however, that formula (3.38), regarded as an equality of operators acting on $V(\lambda)^+$, is only valid provided the denominators in (3.39) or (3.40) do not vanish. Therefore, in order to use (3.38), we first consider $V(\lambda)$ with 'generic' entries of λ and calculate the matrix elements of $F_{n-1,n}$ as functions in the entries of the patterns Λ. The final explicit formulas can be written in a singularity-free form and they are valid in the general case.

Bibliographical notes

The exposition here is based upon the author's papers [98–100]. Slight changes in the notation were made in order to present the results in a uniform manner for all the three cases. The branching rules for all classical reductions $\mathfrak{o}_N \downarrow \mathfrak{o}_{N-1}$ and $\mathfrak{sp}_{2n} \downarrow \mathfrak{sp}_{2n-2}$ are due to Zhelobenko, [167]; see also Hegerfeldt, [58], King, [68], Proctor, [138], Okounkov, [111], Goodman and Wallach, [45]. The lowering operators for the symplectic Lie algebras were first constructed by Mickelsson, [91]; see also Bincer, [9]. The explicit relations in the algebra $\mathrm{Z}(\mathfrak{sp}_{2n}, \mathfrak{sp}_{2n-2})$ were calculated by Zhelobenko, [170].

The algebra $\mathrm{Y}(n)$ was first studied in the work of Faddeev and the St.-Petersburg school in relation with the inverse scattering method; see, for instance, Takhtajan and Faddeev, [147], Kulish and Sklyanin, [74]. The term "Yangian" was introduced by Drinfeld in [28]. In that paper he defined the Yangian $\mathrm{Y}(\mathfrak{a})$ for each simple finite-dimensional Lie algebra $\mathfrak{a}$. Finite-dimensional irreducible representations of $\mathrm{Y}(\mathfrak{a})$ were classified by Drinfeld, [29], with the use of a previous work by Tarasov, [148,149]. Theorem 3.9 goes back to this work of Tarasov; see also [96,110]. The criterion of Theorem 3.8 is due to Chari and Pressley, [13]. It can also be deduced from the results of [148,149]; see [97]. The twisted Yangians were introduced by Olshanski, [118]; see also [104]. Their finite-dimensional irreducible representations were classified in the author's paper [97] which, in particular, contains the criteria of Theorems 3.10 and 3.12. For more details on the (twisted) Yangians and their applications in classical representation theory see the expository papers [104,103] and the recent work of Nazarov, [107,108], where, in particular, the skew representations of twisted Yangians were studied.

In some particular cases, bases in $V(\lambda)$ were constructed, e.g., by Wong and Yeh, [165], Smirnov and Tolstoy, [143].

Weight bases for the fundamental representations of $\mathfrak{o}_{2n+1}$ and $\mathfrak{sp}_{2n}$ were independently constructed by Donnelly, [24–26], in a different way. He also demonstrated that these bases of his coincide with those of Theorems 3.3 and 3.5, up to a diagonal equivalence.

Harada, [57], employed the results of [98] to construct a new integrable (Gelfand–Tsetlin) system on the coadjoint orbits of the symplectic groups. This provides an analog of the Guillemin–Sternberg construction, [55], for the unitary groups.

4. Gelfand–Tsetlin bases for representations of $\mathfrak{o}_N$

In this section we sketch the construction of the bases proposed originally by Gelfand and Tsetlin in [42]. It is based upon the fact that the restriction $\mathfrak{o}_N \downarrow \mathfrak{o}_{N-1}$ is multiplicity-free. This makes the construction similar to the $\mathfrak{gl}_n$ case. We shall be applying the general method of Mickelsson algebras outlined in Section 2.2. In particular, the corresponding branching rules can be derived from Theorem 2.11; cf. Section 2.3.

It will be convenient to change the notation for the elements of the orthogonal Lie algebra $\mathfrak{o}_N$ used in Section 3. We shall now use the standard enumeration of the rows and columns of $N \times N$-matrices by the numbers $\{1, \dots, N\}$. Define an involution of this set of indices by setting $i' = N - i + 1$. The Lie algebra $\mathfrak{o}_N$ is spanned by the elements

$$F_{ij} = E_{ij} - E_{j'i'}, \quad i, j = 1, \dots, N. \tag{4.1}$$

We shall keep the notation $\mathfrak{g}_n$ for $\mathfrak{o}_N$ with $N = 2n + 1$ or $N = 2n$.

The finite-dimensional irreducible representations of $\mathfrak{g}_n$ are now parametrized by n-tuples $\lambda = (\lambda_1, \ldots, \lambda_n)$ where the numbers λ_i satisfy the conditions

$$\lambda_i - \lambda_{i+1} \in \mathbb{Z}_+ \quad \text{for } i = 1, \ldots, n-1, \tag{4.2}$$

and

$$\begin{aligned} 2\lambda_n &\in \mathbb{Z}_+ \quad \text{for } \mathfrak{g}_n = \mathfrak{o}_{2n+1}, \\ \lambda_{n-1} + \lambda_n &\in \mathbb{Z}_+ \quad \text{for } \mathfrak{g}_n = \mathfrak{o}_{2n}. \end{aligned} \tag{4.3}$$

Such an n-tuple λ is called the *highest weight* of the corresponding representation which we shall denote by $V(\lambda)$. It contains a unique, up to a constant factor, nonzero vector ξ (the *highest vector*) such that

$$\begin{aligned} F_{ii}\xi &= \lambda_i \xi \quad && \text{for } i = 1, \ldots, n, \\ F_{ij}\xi &= 0 \quad && \text{for } 1 \leqslant i < j \leqslant N. \end{aligned}$$

4.1. *Lowering operators for the reduction* $\mathfrak{o}_{2n+1} \downarrow \mathfrak{o}_{2n}$

Taking $N = 2n + 1$ in the definition (4.1), we shall consider $\mathfrak{o}_{2n}$ as the subalgebra of $\mathfrak{o}_{2n+1}$ spanned by the elements (4.1) with $i, j \neq n + 1$. In accordance with the branching rule, the restriction of $V(\lambda)$ to the subalgebra $\mathfrak{o}_{2n}$ is given by

$$V(\lambda)|_{\mathfrak{o}_{2n}} \simeq \bigoplus_{\mu} V'(\mu),$$

where $V'(\mu)$ is the irreducible finite-dimensional representation of $\mathfrak{o}_{2n}$ with highest weight μ and the sum is taken over the weights μ satisfying the inequalities

$$\lambda_1 \geqslant \mu_1 \geqslant \lambda_2 \geqslant \mu_2 \geqslant \cdots \geqslant \lambda_{n-1} \geqslant \mu_{n-1} \geqslant \lambda_n \geqslant |\mu_n|, \tag{4.4}$$

with all the μ_i being simultaneously integers or half-integers together with the λ_i.

The elements $F_{n+1,i}$ span the $\mathfrak{o}_{2n}$-invariant complement to $\mathfrak{o}_{2n}$ in $\mathfrak{o}_{2n+1}$. Therefore, by the general theory of Section 2.2, the Mickelsson–Zhelobenko algebra $\mathrm{Z}(\mathfrak{o}_{2n+1}, \mathfrak{o}_{2n})$ is generated by the elements

$$pF_{n+1,i}, \quad i = 1, \ldots, n, n', \ldots, 1', \tag{4.5}$$

where p is the extremal projector for the Lie algebra $\mathfrak{o}_{2n}$. Let $\{\varepsilon_1, \ldots, \varepsilon_n\}$ be the basis of $\mathfrak{h}^*$ dual to the basis $\{F_{11}, \ldots, F_{nn}\}$ of the Cartan subalgebra $\mathfrak{h}$ of $\mathfrak{o}_{2n}$. Set $\varepsilon_{i'} = -\varepsilon_i$ for $i = 1, \ldots, n$. Denote by p_{ij} the element p_α given by (2.13) for the positive root $\alpha = \varepsilon_i - \varepsilon_j$.

Choosing an appropriate normal ordering on the positive roots, for any $i = 1, \ldots, n$ we can write the elements (4.5) in the form

$$pF_{n+1,i} = p_{i,i+1} \cdots p_{in} p_{in'} \cdots p_{i1'} F_{n+1,i}, \tag{4.6}$$

where the factor $p_{ii'}$ is skipped in the product. Therefore the right denominator of this fraction is

$$\pi_i = f_{i,i+1} \cdots f_{in} f_{in'} \cdots f_{i1'},$$

where

$$f_{ij} = \begin{cases} F_{ii} - F_{jj} + j - i & \text{if } j = 1, \ldots, n, \\ F_{ii} - F_{jj} + j - i - 2 & \text{if } j = 1', \ldots, n'. \end{cases}$$

Hence, the elements $s'_{ni} = pF_{n+1,i}\pi_i$ with $i = 1, \ldots, n$ belong to the Mickelsson algebra $\mathrm{S}(\mathfrak{o}_{2n+1}, \mathfrak{o}_{2n})$. One can verify that they are pairwise commuting.

Denote by $V(\lambda)^+$ the subspace of $\mathfrak{o}_{2n}$-highest vectors in $V(\lambda)$. Given a $\mathfrak{o}_{2n}$-highest weight $\mu = (\mu_1, \ldots, \mu_n)$ we denote by $V(\lambda)^+_\mu$ the corresponding weight subspace in $V(\lambda)^+$:

$$V(\lambda)^+_\mu = \{\eta \in V(\lambda)^+ \mid F_{ii}\eta = \mu_i \eta,\ i = 1, \ldots, n\}.$$

By the branching rule, the space $V(\lambda)^+_\mu$ is one-dimensional if the condition (4.4) is satisfied. Otherwise, it is zero.

THEOREM 4.1. *Suppose that the inequalities* (4.4) *hold. Then the space* $V(\lambda)^+_\mu$ *is spanned by the vector*

$$s'^{\lambda_1 - \mu_1}_{n1} \cdots s'^{\lambda_n - \mu_n}_{nn} \xi.$$

4.2. *Lowering operators for the reduction* $\mathfrak{o}_{2n} \downarrow \mathfrak{o}_{2n-1}$

Taking $N = 2n$ in the definition (4.1), we shall consider $\mathfrak{o}_{2n-1}$ as the subalgebra of $\mathfrak{o}_{2n}$ spanned by the elements (4.1) with $i, j \neq n, n'$ together with

$$\frac{1}{\sqrt{2}}(F_{ni} - F_{n'i}), \quad i = 1, \ldots, n-1, (n-1)', \ldots, 1'.$$

In accordance with the branching rule, the restriction of $V(\lambda)$ to the subalgebra $\mathfrak{o}_{2n-1}$ is given by

$$V(\lambda)|_{\mathfrak{o}_{2n-1}} \simeq \bigoplus_\mu V'(\mu),$$

where $V'(\mu)$ is the irreducible finite-dimensional representation of $\mathfrak{o}_{2n-1}$ with the highest weight μ and the sum is taken over the weights μ satisfying the inequalities

$$\lambda_1 \geqslant \mu_1 \geqslant \lambda_2 \geqslant \mu_2 \geqslant \cdots \geqslant \lambda_{n-1} \geqslant \mu_{n-1} \geqslant |\lambda_n|, \tag{4.7}$$

with all the μ_i being simultaneously integers or half-integers together with the λ_i.

The elements

$$F_{nn}, \qquad F'_{ni} = \frac{1}{\sqrt{2}}(F_{ni} + F_{n'i}), \quad i = 1, \dots, n-1, (n-1)', \dots, 1', \tag{4.8}$$

span the $\mathfrak{o}_{2n-1}$-invariant complement to $\mathfrak{o}_{2n-1}$ in $\mathfrak{o}_{2n}$. Therefore, by the general theory of Section 2.2, the Mickelsson–Zhelobenko algebra $\mathrm{Z}(\mathfrak{o}_{2n}, \mathfrak{o}_{2n-1})$ is generated by the elements

$$pF_{nn}, \qquad pF'_{ni}, \quad i = 1, \dots, n-1, (n-1)', \dots, 1', \tag{4.9}$$

where p is the extremal projector for the Lie algebra $\mathfrak{o}_{2n-1}$. Let $\{\varepsilon_1, \dots, \varepsilon_{n-1}\}$ be the basis of $\mathfrak{h}^*$ dual to the basis $\{F_{11}, \dots, F_{n-1,n-1}\}$ of the Cartan subalgebra $\mathfrak{h}$ of $\mathfrak{o}_{2n-1}$. Set $\varepsilon_{i'} = -\varepsilon_i$ for $i = 1, \dots, n-1$. Denote by p_{ij} and p_i the elements p_α given by (2.13) for the positive roots $\alpha = \varepsilon_i - \varepsilon_j$ and $\alpha = \varepsilon_i$, respectively. Choosing an appropriate normal ordering on the positive roots, for any $i = 1, \dots, n-1$ we can write the elements (4.9) in the form

$$pF'_{ni} = p_{i,i+1} \cdots p_{i,n-1} p_i \, p_{i,(n-1)'} \cdots p_{i1'} F'_{ni}, \tag{4.10}$$

where the factor $p_{ii'}$ is skipped in the product. Therefore the right denominator of this fraction is

$$\pi_i = f_{i,i+1} \cdots f_{i,n-1} f_i f'_i f_{i,(n-1)'} \cdots f_{i1'},$$

where

$$f_{ij} = \begin{cases} F_{ii} - F_{jj} + j - i & \text{if } j = 1, \dots, n-1, \\ F_{ii} - F_{jj} + j - i - 2 & \text{if } j = 1', \dots, (n-1)' \end{cases}$$

and $f_i = f'_i - 1 = 2(F_{ii} + n - i)$. Hence, the elements $s_{ni} = pF'_{ni}\pi_i$ with $i = 1, \dots, n-1$ belong to the Mickelsson algebra $\mathrm{S}(\mathfrak{o}_{2n}, \mathfrak{o}_{2n-1})$. One can verify that they are pairwise commuting.

Denote by $V(\lambda)^+$ the subspace of $\mathfrak{o}_{2n-1}$-highest vectors in $V(\lambda)$. Given a $\mathfrak{o}_{2n-1}$-highest weight $\mu = (\mu_1, \dots, \mu_{n-1})$ we denote by $V(\lambda)^+_\mu$ the corresponding weight subspace in $V(\lambda)^+$:

$$V(\lambda)^+_\mu = \{\eta \in V(\lambda)^+ \mid F_{ii}\eta = \mu_i \eta, \ i = 1, \dots, n-1\}.$$

By the branching rule, the space $V(\lambda)^+_\mu$ is one-dimensional if the condition (4.7) is satisfied. Otherwise, it is zero.

THEOREM 4.2. *Suppose that the inequalities* (4.7) *hold. Then the space* $V(\lambda)^+_\mu$ *is spanned by the vector*

$$s_{n1}^{\lambda_1-\mu_1}\cdots s_{n,n-1}^{\lambda_{n-1}-\mu_{n-1}}\xi.$$

Note that the generator pF_{nn} of the algebra $\mathrm{Z}(\mathfrak{o}_{2n},\mathfrak{o}_{2n-1})$ does not occur in the formula for the basis vector as it has zero weight with respect to $\mathfrak{h}$.

4.3. *Basis vectors*

The representation $V(\lambda)$ of the Lie algebra $\mathfrak{g}_n=\mathfrak{o}_{2n+1}$ or $\mathfrak{o}_{2n}$ is equipped with a contravariant inner product which is uniquely determined by the conditions

$$\langle\xi,\xi\rangle=1\quad\text{and}\quad\langle F_{ij}u,v\rangle=\langle u,F_{ji}v\rangle$$

for all $u,v\in V(\lambda)$ and any indices i,j.

Combining Theorems 4.1 and 4.2 we can construct another basis for each representation $V(\lambda)$ of $\mathfrak{g}_n$; cf. Section 3.2.

B type case. We need to modify the definition of the B type pattern Λ introduced in section 3.2. Here Λ is an array of the form

$$\begin{array}{ccccccccccc}
\lambda_{n1} & & \lambda_{n2} & & & \cdots & & & \lambda_{nn} & & \\
 & \lambda'_{n1} & & \lambda'_{n2} & & & \cdots & & & \lambda'_{nn} & \\
 & & \lambda_{n-1,1} & & & \cdots & & \lambda_{n-1,n-1} & & & \\
 & & & \lambda'_{n-1,1} & & & \cdots & & \lambda'_{n-1,n-1} & & \\
 & & & & & \cdots & & \cdots & & & \\
 & & & & & & & & \lambda_{11} & & \\
 & & & & & & & & & \lambda'_{11} &
\end{array}$$

such that $\lambda=(\lambda_{n1},\dots,\lambda_{nn})$, the remaining entries are all integers or half-integers together with the λ_i, and the following inequalities hold

$$\lambda_{k1}\geqslant\lambda'_{k1}\geqslant\lambda_{k2}\geqslant\lambda'_{k2}\geqslant\cdots\geqslant\lambda'_{k,k-1}\geqslant\lambda_{kk}\geqslant|\lambda'_{kk}|$$

for $k=1,\dots,n$, and

$$\lambda'_{k1}\geqslant\lambda_{k-1,1}\geqslant\lambda'_{k2}\geqslant\lambda_{k-1,2}\geqslant\cdots\geqslant\lambda'_{k,k-1}\geqslant\lambda_{k-1,k-1}\geqslant|\lambda'_{kk}|$$

for $k=2,\dots,n$.

THEOREM 4.3. *The vectors*

$$\eta_\Lambda = s_{11}^{\prime\lambda_{11}-\lambda'_{11}} \prod_{k=2,\dots,n}^{\rightarrow} \left(s_{k1}^{\prime\lambda_{k1}-\lambda'_{k1}} \cdots s_{kk}^{\prime\lambda_{kk}-\lambda'_{kk}} s_{k1}^{\lambda'_{k1}-\lambda_{k-1,1}} \cdots s_{k,k-1}^{\lambda'_{k,k-1}-\lambda_{k-1,k-1}}\right)\xi$$

parametrized by the patterns Λ form an orthogonal basis of the representation $V(\lambda)$.

D type case. Here we define the D type patterns Λ as arrays of the form

$$\begin{array}{ccccccccc}
\lambda_{n1} & & \lambda_{n2} & & \cdots & & \lambda_{nn} & & \\
& \lambda'_{n-1,1} & & \cdots & & \lambda'_{n-1,n-1} & & & \\
& & \lambda_{n-1,1} & & \cdots & & \lambda_{n-1,n-1} & & \\
& & & \cdots & & \cdots & & & \\
& & & & & \lambda'_{11} & & & \\
& & & & & & \lambda_{11} & &
\end{array}$$

such that $\lambda = (\lambda_{n1}, \dots, \lambda_{nn})$, the remaining entries are all integers or half-integers together with the λ_i, and the following inequalities hold

$$\lambda_{k1} \geqslant \lambda'_{k-1,1} \geqslant \lambda_{k2} \geqslant \lambda'_{k-1,2} \geqslant \cdots \geqslant \lambda_{k,k-1} \geqslant \lambda'_{k-1,k-1} \geqslant |\lambda_{kk}|$$

for $k = 2, \dots, n$, and

$$\lambda'_{k1} \geqslant \lambda_{k1} \geqslant \lambda'_{k2} \geqslant \lambda_{k2} \geqslant \cdots \geqslant \lambda_{k,k-1} \geqslant \lambda'_{kk} \geqslant |\lambda_{kk}|$$

for $k = 1, \dots, n-1$.

THEOREM 4.4. *The vectors*

$$\eta_\Lambda = \prod_{k=1,\dots,n-1}^{\rightarrow} \left(s_{k+1,1}^{\lambda_{k+1,1}-\lambda'_{k1}} \cdots s_{k+1,k}^{\lambda_{k+1,k}-\lambda'_{kk}} s_{k1}^{\prime\lambda'_{k1}-\lambda_{k1}} \cdots s_{kk}^{\prime\lambda'_{kk}-\lambda_{kk}}\right)\xi$$

parametrized by the patterns Λ form an orthogonal basis of the representation $V(\lambda)$.

The norms of the basis vectors η_Λ can be found in an explicit form. The formulas for the matrix elements of the generators of the Lie algebra $\mathfrak{o}_N$ in the original paper by Gelfand and Tsetlin, [42], are given in the orthonormal basis

$$\zeta_\Lambda = \eta_\Lambda / \|\eta_\Lambda\|, \qquad \|\eta_\Lambda\|^2 = \langle \eta_\Lambda, \eta_\Lambda \rangle.$$

Bibliographical notes

The exposition of this section follows Zhelobenko, [171]. The branching rules were previously derived by him in [167]. The lowering operators for the reduction $\mathfrak{o}_N \downarrow \mathfrak{o}_{N-1}$ were constructed by Pang and Hecht, [133], and Wong, [164]; see also Mickelsson, [93]. They are presented in a form similar to (2.9) and (2.10) although more complicated. A derivation of the matrix element formulas of Gelfand and Tsetlin, [42], was also given in [133] and [164] which basically follows the approach outlined in Section 2.1. The defining relations for the algebra $\mathrm{Z}(\mathfrak{o}_N, \mathfrak{o}_{N-1})$ were given in an explicit form by Zhelobenko, [170]. Gould's approach based upon the characteristic identities of Bracken and Green, [12,54], for the orthogonal Lie algebras is also applicable; see Gould, [46,47,50]. It produces an independent derivation of the matrix element formulas. Although the quantum minor approach has not been developed so far for the Gelfand–Tsetlin basis for the orthogonal Lie algebras, it seems to be plausible that the corresponding analogs of the results outlined in Section 2.5 can be obtained.

Analogs of the Gelfand–Tsetlin bases, [42], for representations of a nonstandard deformation $\mathrm{U}'_q(\mathfrak{o}_N)$ of $\mathrm{U}(\mathfrak{o}_N)$ were given by Gavrilik and Klimyk, [38], Gavrilik and Iorgov, [37], and Iorgov and Klimyk, [62].

The Gelfand–Tsetlin modules over the orthogonal Lie algebras were studied by Mazorchuk, [89], with the use of the matrix element formulas from [42].

Acknowledgements

It gives me pleasure to thank I.M. Gelfand for his comment on the preliminary version of the chapter. My thanks also extend to V.K. Dobrev, V.M. Futorny, M.D. Gould, M. Harada, M.L. Nazarov, G.I. Olshanski, S.A. Ovsienko, T.D. Palev, V.S. Retakh, and V.N. Tolstoy who sent me remarks and references.

References

[1] R.M. Asherova, Yu.F. Smirnov and V.N. Tolstoy, *Projection operators for simple Lie groups*, Theor. Math. Phys. **8** (1971), 813–825.

[2] R.M. Asherova, Yu.F. Smirnov and V.N. Tolstoy, *Projection operators for simple Lie groups. II. General scheme for constructing lowering operators. The groups* SU(n), Theor. Math. Phys. **15** (1973), 392–401.

[3] R.M. Asherova, Yu.F. Smirnov and V.N. Tolstoy, *Description of a certain class of projection operators for complex semisimple Lie algebras*, Math. Notes **26** (1–2) (1979), 499–504.

[4] G.E. Baird and L.C. Biedenharn, *On the representations of the semisimple Lie groups. II*, J. Math. Phys. **4** (1963), 1449–1466.

[5] A.O. Barut and R. Rączka, *Theory of Group Representations and Applications*, 2nd ed., World Scientific (1986).

[6] A. Berele, *Construction of Sp-modules by tableaux*, Linear and Multilinear Algebra **19** (1986), 299–307.

[7] A.D. Berenstein and A.V. Zelevinsky, *Involutions on Gel'fand–Tsetlin schemes and multiplicities in skew* GL_n*-modules*, Soviet Math. Dokl. **37** (1988), 799–802.

[8] L.C. Biedenharn and J.D. Louck, *Angular Momentum in Quantum Physics: Theory and Application*, Addison-Wesley (1981).

[9] A. Bincer, *Missing label operators in the reduction $Sp(2n) \downarrow Sp(2n-2)$*, J. Math. Phys. **21** (1980), 671–674.

[10] A. Bincer, *Mickelsson lowering operators for the symplectic group*, J. Math. Phys. **23** (1982), 347–349.

[11] N. Bourbaki, *Groupes et algèbres de Lie, Chapitres 4, 5 et 6*, Hermann (1968).

[12] A.J. Bracken and H.S. Green, *Vector operators and a polynomial identity for* $\mathrm{SO}(n)$, J. Math. Phys. **12** (1971), 2099–2106.

[13] V. Chari and A. Pressley, *Yangians and R-matrices*, Enseign. Math. **36** (1990), 267–302.

[14] V. Chari and A. Pressley, *A Guide to Quantum Groups*, Cambridge Univ. Press (1994).

[15] V. Chari and N. Xi, *Monomial bases of quantized enveloping algebras*, Recent Developments in Quantum Affine Algebras and Related Topics, Raleigh, NC, 1998, Contemp. Math. **248**, Amer. Math. Soc. (1999), 69–81.

[16] I.V. Cherednik, *A new interpretation of Gelfand–Tzetlin bases*, Duke Math. J. **54** (1987), 563–577.

[17] I.V. Cherednik, *Quantum groups as hidden symmetries of classic representation theory*, Differential Geometric Methods in Physics, A.I. Solomon (ed.), World Scientific (1989), 47–54.

[18] C. De Concini and D. Kazhdan, *Special bases for S_N and* $\mathrm{GL}(n)$, Israel J. Math. **40** (1981), 275–290.

[19] J. Dixmier, *Algèbres enveloppantes*, Gauthier-Villars (1974).

[20] V.K. Dobrev and P. Truini, *Irregular $U_q(\mathrm{sl}(3))$ representations at roots of unity via Gelfand–(Weyl)–Zetlin basis*, J. Math. Phys. **38** (1997), 2631–2651.

[21] V.K. Dobrev and P. Truini, *Polynomial realization of the $U_q(\mathrm{sl}(3))$ Gelfand–(Weyl)–Zetlin basis*, J. Math. Phys. **38** (1997), 3750–3767.

[22] V.K. Dobrev, A.D. Mitov and P. Truini, *Normalized $U_q(\mathrm{sl}(3))$ Gelfand–(Weyl)–Zetlin basis and new summation formulas for q-hypergeometric functions*, J. Math. Phys. **41** (2000), 7752–7768.

[23] R.G. Donnelly, *Symplectic analogs of the distributive lattices $L(m,n)$*, J. Combin. Theory Ser. A **88** (1999), 217–234.

[24] R.G. Donnelly, *Explicit constructions of the fundamental representations of the symplectic Lie algebras*, J. Algebra **233** (2000), 37–64.

[25] R.G. Donnelly, *Explicit constructions of the fundamental representations of the odd orthogonal Lie algebras*, to appear.

[26] R.G. Donnelly, *Extremal properties of bases for representations of semisimple Lie algebras*, J. Algebraic Combin. **17** (2003), 255–282.

[27] R.G. Donnelly, S.J. Lewis and R. Pervine, *Constructions of representations of $o(2n+1, C)$ that imply Molev and Reiner–Stanton lattices are strongly Sperner*, Discrete Math. **263** (2003), 61–79.

[28] V.G. Drinfeld, *Hopf algebras and the quantum Yang–Baxter equation*, Soviet Math. Dokl. **32** (1985), 254–258.

[29] V.G. Drinfeld, *A new realization of Yangians and quantized affine algebras*, Soviet Math. Dokl. **36** (1988), 212–216.

[30] V.G. Drinfeld, *Quantum groups*, Proc. Int. Congress Math., Berkeley, 1986, Amer. Math. Soc. (1987), 798–820.

[31] Yu.A. Drozd, V.M. Futorny and S.A. Ovsienko, *Irreducible weighted* $\mathrm{sl}(3)$*-modules*, Funct. Anal. Appl. **23** (1989), 217–218.

[32] Yu.A. Drozd, V.M. Futorny and S.A. Ovsienko, *On Gel'fand–Zetlin modules*, Proceedings of the Winter School on Geometry and Physics, Srní, 1990, Rend. Circ. Mat. Palermo (2) **26** (Suppl.) (1991), 143–147.

[33] Yu.A. Drozd, V.M. Futorny and S.A. Ovsienko, *Gelfand–Zetlin modules over Lie algebra* $\mathrm{sl}(3)$, Contemp. Math. **131** (1992), 23–29.

[34] Yu.A. Drozd, V.M. Futorny and S.A. Ovsienko, *Harish-Chandra subalgebras and Gel'fand–Zetlin modules*, Finite-Dimensional Algebras and Related Topics, Ottawa, ON, 1992, NATO Adv. Sci. Inst. Ser. C Math. Phys. Sci. **424**, Kluwer Acad. Publ. (1994), 79–93.

[35] J. Du, *Canonical bases for irreducible representations of quantum* GL_n, Bull. London Math. Soc. **24** (1992), 325–334.

[36] J. Du, *Canonical bases for irreducible representations of quantum* GL_n *II*, J. London Math. Soc. **51** (1995), 461–470.

[37] A.M. Gavrilik and N.Z. Iorgov, *q-deformed algebras $U_q(\mathrm{so}_n)$ and their representations*, Methods Funct. Anal. Topology **3** (1997), 51–63.

[38] A.M. Gavrilik and A.U. Klimyk, *q-deformed orthogonal and pseudo-orthogonal algebras and their representations*, Lett. Math. Phys. **21** (1991), 215–220.

[39] I.M. Gelfand and M.I. Graev, *Finite-dimensional irreducible representations of the unitary and the full linear groups, and related special functions*, Izv. Akad. Nauk SSSR, Ser. Mat. **29** (1965), 1329–1356 (Russian). English transl. I.M. Gelfand, *Collected Papers*, Vol. II, Springer-Verlag (1988), 662–692.

[40] I.M. Gelfand, D. Krob, A. Lascoux, B. Leclerc, V.S. Retakh and J.-Y. Thibon, *Noncommutative symmetric functions*, Adv. Math. **112** (1995), 218–348.

[41] I.M. Gelfand and M.L. Tsetlin, *Finite-dimensional representations of the group of unimodular matrices*, Dokl. Akad. Nauk SSSR **71** (1950), 825–828 (Russian). English transl. I.M. Gelfand, *Collected Papers*, Vol. II, Springer-Verlag (1988), 653–656.

[42] I.M. Gelfand and M.L. Tsetlin, *Finite-dimensional representations of groups of orthogonal matrices*, Dokl. Akad. Nauk SSSR **71** (1950), 1017–1020 (Russian). English transl. I.M. Gelfand, *Collected Papers*, Vol. II, Springer-Verlag (1988), 657–661.

[43] I.M. Gelfand and A. Zelevinsky, *Models of representations of classical groups and their hidden symmetries*, Funct. Anal. Appl. **18** (1984), 183–198.

[44] I.M. Gelfand and A. Zelevinsky, *Multiplicities and proper bases for gl_n*, Group Theoretical Methods in Physics, Vol. II, Yurmala, 1985, VNU Sci. Press (1986), 147–159.

[45] R. Goodman and N.R. Wallach, *Representations and Invariants of the Classical Groups*, Cambridge Univ. Press (1998).

[46] M.D. Gould, *The characteristic identities and reduced matrix elements of the unitary and orthogonal groups*, J. Austral. Math. Soc. B **20** (1978), 401–433.

[47] M.D. Gould, *On an infinitesimal approach to semisimple Lie groups and raising and lowering operators of $O(n)$ and $U(n)$*, J. Math. Phys. **21** (1980), 444–453.

[48] M.D. Gould, *On the matrix elements of the $U(n)$ generators*, J. Math. Phys. **22** (1981), 15–22.

[49] M.D. Gould, *General $U(N)$ raising and lowering operators*, J. Math. Phys. **22** (1981), 267–270.

[50] M.D. Gould, *Wigner coefficients for a semisimple Lie group and the matrix elements of the $O(n)$ generators*, J. Math. Phys. **22** (1981), 2376–2388.

[51] M.D. Gould, *Representation theory of the symplectic groups I*, J. Math. Phys. **30** (1989), 1205–1218.

[52] M.D. Gould and L.C. Biedenharn, *The pattern calculus for tensor operators in quantum groups*, J. Math. Phys. **33** (1992), 3613–3635.

[53] M.D. Gould and E.G. Kalnins, *A projection-based solution to the $Sp(2n)$ state labeling problem*, J. Math. Phys. **26** (1985), 1446–1457.

[54] H.S. Green, *Characteristic identities for generators of* $\mathrm{GL}(n)$, *$O(n)$ and $Sp(n)$*, J. Math. Phys. **12** (1971), 2106–2113.

[55] V. Guillemin and S. Sternberg, *The Gelfand–Cetlin system and quantization of the complex flag manifolds*, J. Funct. Anal. **52** (1983), 106–128.

[56] T. Halverson and A. Ram, *Characters of algebras containing a Jones basic construction: the Temperley–Lieb, Okada, Brauer, and Birman–Wenzl algebras*, Adv. Math. **116** (1995), 263–321.

[57] M. Harada, *The symplectic geometry of the Gel'fand–Cetlin–Molev basis for representations of $Sp(2n, \mathbb{C})$*, Preprint math.SG/0404485.

[58] G.C. Hegerfeldt, *Branching theorem for the symplectic groups*, J. Math. Phys. **8** (1967), 1195–1196.

[59] Hou Pei-yu, *Orthonormal bases and infinitesimal operators of the irreducible representations of group U_n*, Sci. Sinica **15** (1966), 763–772.

[60] R. Howe and T. Umeda, *The Capelli identity, the double commutant theorem, and multiplicity-free actions*, Math. Ann. **290** (1991), 569–619.

[61] J.E. Humphreys, *Introduction to Lie Algebras and Representation Theory*, Graduate Texts in Math. **9**, Springer (1972).

[62] N.Z. Iorgov and A.U. Klimyk, *The nonstandard deformation $U_q'(\mathrm{so}_n)$ for q a root of unity*, Methods Funct. Anal. Topology **6** (2000), 56–71.

[63] G. James and A. Kerber, *The Representation Theory of the Symmetric Group*, Addison-Wesley (1981).

[64] M. Jimbo, *A q-difference analogue of* $\mathrm{U}(\mathfrak{g})$ *and the Yang–Baxter equation*, Lett. Math. Phys. **10** (1985), 63–69.

[65] M. Jimbo, *Quantum R-matrix for the generalized Toda system*, Comm. Math. Phys. **102** (1986), 537–547.

[66] M. Kashiwara, *Crystallizing the q-analogue of universal enveloping algebras*, Comm. Math. Phys. **133** (1990), 249–260.

[67] S.M. Khoroshkin and V.N. Tolstoy, *Extremal projector and universal R-matrix for quantum contragredient Lie (super)algebras*, Quantum Group and Related Topics, Wrocław, 1991, Math. Phys. Stud. **13**, Kluwer Acad. Publ. (1992), 23–32.

[68] R.C. King, *Weight multiplicities for the classical groups*, Group Theoretical Methods in Physics, Fourth Internat. Colloq., Nijmegen, 1975, Lecture Notes in Phys. **50**, Springer (1976), 490–499.

[69] R.C. King and N.G.I. El-Sharkaway, *Standard Young tableaux and weight multiplicities of the classical Lie groups*, J. Phys. A **16** (1983), 3153–3177.

[70] R.C. King and T.A. Welsh, *Construction of orthogonal group modules using tableaux*, Linear and Multilinear Algebra **33** (1993), 251–283.

[71] A.A. Kirillov, *A remark on the Gelfand–Tsetlin patterns for symplectic groups*, J. Geom. Phys. **5** (1988), 473–482.

[72] A.N. Kirillov and A.D. Berenstein, *Groups generated by involutions, Gel'fand–Tsetlin patterns, and combinatorics of Young tableaux*, St. Petersburg Math. J. **7** (1996), 77–127.

[73] K. Koike and I. Terada, *Young-diagrammatic methods for the representation theory of the classical groups of type B_n, C_n, D_n*, J. Algebra **107** (1987), 466–511.

[74] P.P. Kulish and E.K. Sklyanin, *Quantum spectral transform method: recent developments*, Integrable Quantum Field Theories, Lecture Notes in Phys. **151**, Springer (1982), 61–119.

[75] V. Lakshmibai, C. Musili and C.S. Seshadri, *Geometry of G/P. IV. Standard monomial theory for classical types*, Proc. Indian Acad. Sci. Sect. A Math. Sci. **88** (1979), 279–362.

[76] B. Leclerc and P. Toffin, *A simple algorithm for computing the global crystal basis of an irreducible $U_q(sl_n)$-module*, Internat. J. Alg. Comput. **10** (2000), 191–208.

[77] C. Lecouvey, *An algorithm for computing the global basis of a finite dimensional irreducible $U_q(so_{2n+1})$ or $U_q(so_{2n})$-module*, Comm. Algebra **32** (2004), 1969–1996.

[78] C. Lecouvey, *An algorithm for computing the global basis of an irreducible $U_q(sp_{2n})$-module*, Adv. in Appl. Math. **29** (2002), 46–78.

[79] R. Leduc and A. Ram, *A ribbon Hopf algebra approach to the irreducible representations of centralizer algebras: the Brauer, Birman–Wenzl, and type A Iwahori–Hecke algebras*, Adv. Math. **125** (1997), 1–94.

[80] F. Lemire and J. Patera, *Formal analytic continuation of Gelfand's finite-dimensional representations of $gl(n, \mathrm{C})$*, J. Math. Phys. **20** (1979), 820–829.

[81] P. Littelmann, *An algorithm to compute bases and representation matrices for SL_{n+1}-representations*, J. Pure Appl. Algebra **117/118** (1997), 447–468.

[82] P. Littelmann, *Cones, crystals, and patterns*, Transformation Groups **3** (1998), 145–179.

[83] G. Lusztig, *Canonical bases arising from quantized enveloping algebras*, J. Amer. Math. Soc. **3** (1990), 447–498.

[84] I.G. Macdonald, *Symmetric Functions and Hall Polynomials*, 2nd ed., Oxford Univ. Press (1995).

[85] O. Mathieu, *Good bases for G-modules*, Geom. Dedicata **36** (1990), 51–66.

[86] O. Mathieu, *Bases des représentations des groupes simples complexes (d'après Kashiwara, Lusztig, Ringel et al.)*, Sémin. Bourbaki, Vol. 1990/91, Astérisque **201–203**, Exp. no. 743 (1992), 421–442.

[87] V. Mazorchuk, *Generalized Verma Modules*, Mathematical Studies Monograph Series **8**, VNTL Publishers, L'viv (2000).

[88] V. Mazorchuk, *On categories of Gelfand–Zetlin modules*, Noncommutative Structures in Mathematics and Physics, Kiev, 2000, NATO Sci. Ser. II Math. Phys. Chem. **22**, Kluwer Acad. Publ. (2001), 299–307.

[89] V. Mazorchuk, *On Gelfand–Zetlin modules over orthogonal Lie algebras*, Algebra Colloq. **8** (2001), 345–360.

[90] V. Mazorchuk and L. Turowska, *On Gelfand–Zetlin modules over $U_q(\mathrm{gl}_n)$*, Quantum Groups and Integrable Systems, Prague, 1999, Czechoslovak J. Phys. **50** (2000), 139–144.

[91] J. Mickelsson, *Lowering operators and the symplectic group*, Rep. Math. Phys. **3** (1972), 193–199.

[92] J. Mickelsson, *Step algebras of semi-simple subalgebras of Lie algebras*, Rep. Math. Phys. **4** (1973), 307–318.

[93] J. Mickelsson, *Lowering operators for the reduction $U(n) \downarrow SO(n)$*, Rep. Math. Phys. **4** (1973), 319–332.

[94] A.I. Molev, *Gelfand–Tsetlin basis for irreducible unitarizable highest weight representations of $u(p, q)$*, Funct. Anal. Appl. **23** (1990), 236–238.

[95] A.I. Molev, *Unitarizability of some Enright–Varadarajan $u(p,q)$-modules*, Topics in Representation Theory, A.A. Kirillov (ed.), Advances in Soviet Math. **2**, Amer. Math. Soc. (1991), 199–219.

[96] A.I. Molev, *Gelfand–Tsetlin basis for representations of Yangians*, Lett. Math. Phys. **30** (1994), 53–60.

[97] A.I. Molev, *Finite-dimensional irreducible representations of twisted Yangians*, J. Math. Phys. **39** (1998), 5559–5600.

[98] A.I. Molev, *A basis for representations of symplectic Lie algebras*, Comm. Math. Phys. **201** (1999), 591–618.

[99] A.I. Molev, *A weight basis for representations of even orthogonal Lie algebras*, Combinatorial Methods in Representation Theory, Adv. Studies in Pure Math. **28** (2000), 223–242.

[100] A.I. Molev, *Weight bases of Gelfand–Tsetlin type for representations of classical Lie algebras*, J. Phys. A: Math. Gen. **33** (2000), 4143–4168.

[101] A.I. Molev, *Yangians and transvector algebras*, Discrete Math. **246** (2002), 231–253.

[102] A.I. Molev, *Irreducibility criterion for tensor products of Yangian evaluation modules*, Duke Math. J. **112** (2002), 307–341.

[103] A.I. Molev, *Yangians and their applications*, Handbook of Algebra, Vol. 3, M. Hazewinkel (ed.), Elsevier (2003), 907–960.

[104] A. Molev, M. Nazarov and G. Olshanski, *Yangians and classical Lie algebras*, Russian Math. Surveys **51** (2) (1996), 205–282.

[105] A. Molev and G. Olshanski, *Centralizer construction for twisted Yangians*, Selecta Math. (N.S.) **6** (2000), 269–317.

[106] J.G. Nagel and M. Moshinsky, *Operators that lower or raise the irreducible vector spaces of U_{n-1} contained in an irreducible vector space of U_n*, J. Math. Phys. **6** (1965), 682–694.

[107] M. Nazarov, *Representations of twisted Yangians associated with skew Young diagrams*, Selecta Math. (N.S.) **10** (2004), 71–129.

[108] M. Nazarov, *Representations of Yangians associated with skew Young diagrams*, Proceeding of the ICM-2002, Vol. II, 643–654.

[109] M. Nazarov and V. Tarasov, *Yangians and Gelfand–Zetlin bases*, Publ. RIMS, Kyoto Univ. **30** (1994), 459–478.

[110] M. Nazarov and V. Tarasov, *Representations of Yangians with Gelfand–Zetlin bases*, J. Reine Angew. Math. **496** (1998), 181–212.

[111] A. Okounkov, *Multiplicities and Newton polytopes*, Kirillov's Seminar on Representation Theory, G. Olshanski (ed.), Amer. Math. Soc. Transl. **181**, Amer. Math. Soc. (1998), 231–244.

[112] A. Okounkov and A. Vershik, *A new approach to representation theory of symmetric groups*, Selecta Math. (N.S.) **2** (1996), 581–605.

[113] G.I. Olshanski, *Description of unitary representations with highest weight for the groups $U(p,q)^{\sim}$*, Funct. Anal. Appl. **14** (1981), 190–200.

[114] G.I. Olshanski, *Extension of the algebra $U(g)$ for infinite-dimensional classical Lie algebras g, and the Yangians $Y(gl(m))$*, Soviet Math. Dokl. **36** (1988), 569–573.

[115] G.I. Olshanski, *Irreducible unitary representations of the groups $U(p,q)$ sustaining passage to the limit as $q \to \infty$*, Zapiski Nauchn. Semin. LOMI **172** (1989), 114–120 (Russian). English transl. J. Soviet Math. **59** (1992), 1102–1107.

[116] G.I. Olshanski, *Yangians and universal enveloping algebras*, J. Soviet Math. **47** (1989), 2466–2473.

[117] G.I. Olshanski, *Representations of infinite-dimensional classical groups, limits of enveloping algebras, and Yangians*, Topics in Representation Theory, A.A. Kirillov (ed.), Advances in Soviet Math. **2**, Amer. Math. Soc. (1991), 1–66.

[118] G. Olshanski, *Twisted Yangians and infinite-dimensional classical Lie algebras*, Quantum Groups, P.P. Kulish (ed.), Lecture Notes in Math. **1510**, Springer (1992), 103–120.

[119] U. Ottoson, *A classification of the unitary irreducible representations of* $\mathrm{SO}_0(N,1)$, Comm. Math. Phys. **8** (1968), 228–244.

[120] U. Ottoson, *A classification of the unitary irreducible representations of* $\mathrm{SU}(N,1)$, Comm. Math. Phys. **10** (1968), 114–131.

[121] S. Ovsienko, *Finiteness statements for Gelfand–Tsetlin modules*, Algebraic Structures and their Applications, Math. Inst., Kiev (2002).

[122] T.D. Palev, *Finite-dimensional representations of the special linear Lie superalgebra* $sl(1|n)$ *I. Typical representations*, J. Math. Phys. **28** (1987), 2280–2303.

[123] T.D. Palev, *Finite-dimensional representations of the special linear Lie superalgebra* $sl(1|n)$ *II. Nontypical representations*, J. Math. Phys. **29** (1988), 2589–2598.

[124] T.D. Palev, *Irreducible finite-dimensional representations of the Lie superalgebra* $gl(n|1)$ *in a Gel'fand–Zetlin basis*, J. Math. Phys. **30** (1989), 1433–1442.

[125] T.D. Palev, *Essentially typical representations of the Lie superalgebras* $gl(n/m)$ *in a Gel'fand–Zetlin basis*, Funct. Anal. Appl. **23** (1989), 141–142.

[126] T.D. Palev, *Highest weight irreducible unitary representations of the Lie algebras of infinite matrices I. The algebra* $gl(\infty)$, J. Math. Phys. **31** (1990), 579–586.

[127] T.D. Palev, *Highest weight irreducible unitarizable representations of the Lie algebras of infinite matrices. The algebra* A_∞, J. Math. Phys. **31** (1990), 1078–1084.

[128] T.D. Palev and N.I. Stoilova, *Highest weight representations of the quantum algebra* $U_h(gl_\infty)$, J. Phys. A **30** (1997), L699–L705.

[129] T.D. Palev and N.I. Stoilova, *Highest weight irreducible representations of the quantum algebra* $U_h(A_\infty)$, J. Math. Phys. **39** (1998), 5832–5849.

[130] T.D. Palev and N.I. Stoilova, *Highest weight irreducible representations of the Lie superalgebra* $gl(1|\infty)$, J. Math. Phys. **40** (1999), 1574–1594.

[131] T.D. Palev, N.I. Stoilova and J. van der Jeugt, *Finite-dimensional representations of the quantum superalgebra* $U_q[\mathrm{gl}(n/m)]$ *and related* q*-identities*, Comm. Math. Phys. **166** (1994), 367–378.

[132] T.D. Palev and V.N. Tolstoy, *Finite-dimensional irreducible representations of the quantum superalgebra* $U_q[\mathrm{gl}(n/1)]$, Comm. Math. Phys. **141** (1991), 549–558.

[133] S.C. Pang and K.T. Hecht, *Lowering and raising operators for the orthogonal group in the chain* $O(n) \supset O(n-1) \supset \cdots$, *and their graphs*, J. Math. Phys. **8** (1967), 1233–1251.

[134] R. Proctor, *Representations of* $\mathfrak{sl}(2, C)$ *on posets and the Sperner property*, SIAM J. Algebraic Discrete Methods **3** (1982), 275–280.

[135] R. Proctor, *Bruhat lattices, plane partition generating functions, and minuscule representations*, European J. Combin. **5** (1984), 331–350.

[136] R. Proctor, *Odd symplectic groups*, Invent. Math. **92** (1988), 307–332.

[137] R. Proctor, *Solution of a Sperner conjecture of Stanley with a construction of Gel'fand*, J. Combin. Theory Ser. A **54** (1990), 225–234.

[138] R. Proctor, *Young tableaux, Gelfand patterns, and branching rules for classical groups*, J. Algebra **164** (1994), 299–360.

[139] A. Ram, *Seminormal representations of Weyl groups and Iwahori–Hecke algebras*, Proc. London Math. Soc. **75** (1997), 99–133.

[140] V. Retakh and A. Zelevinsky, *Base affine space and canonical basis in irreducible representations of* $Sp(4)$, Dokl. Akad. Nauk USSR **300** (1988), 31–35.

[141] B.E. Sagan, *The Symmetric Group. Representations, Combinatorial Algorithms, and Symmetric Functions*, 2nd ed., Graduate Texts in Math. **203**, Springer-Verlag (2001).

[142] V.V. Shtepin, *Intermediate Lie algebras and their finite-dimensional representations*, Russian Akad. Sci. Izv. Math. **43** (1994), 559–579.

[143] Yu.F. Smirnov and V.N. Tolstoy, *A new projected basis in the theory of five-dimensional quasi-spin*, Rep. Math. Phys. **4** (1973), 97–111.

[144] D.T. Sviridov, Yu.F. Smirnov and V.N. Tolstoy, *The construction of the wave functions for quantum systems with the* G_2 *symmetry*, Dokl. Akad. Nauk SSSR **206** (1972), 1317–1320 (Russian).

[145] D.T. Sviridov, Yu.F. Smirnov and V.N. Tolstoy, *On the structure of the irreducible representation basis for the exceptional group* G_2, Rep. Math. Phys. **7** (1975), 349–361.

[146] J.R. Stembridge, *On minuscule representations, plane partitions and involutions in complex Lie groups*, Duke Math. J. **73** (1994), 469–490.

[147] L.A. Takhtajan and L.D. Faddeev, *Quantum inverse scattering method and the Heisenberg* XYZ*-model*, Russian Math. Surveys **34** (1979), 11–68.

[148] V.O. Tarasov, *Structure of quantum* L*-operators for the* R*-matrix of the* XXZ*-model*, Theor. Math. Phys. **61** (1984), 1065–1071.

[149] V.O. Tarasov, *Irreducible monodromy matrices for the R-matrix of the XXZ-model and lattice local quantum Hamiltonians*, Theor. Math. Phys. **63** (1985), 440–454.

[150] V.N. Tolstoy, *Extremal projectors for reductive classical Lie superalgebras with non-degenerate generalized Killing form*, Uspekhi Mat. Nauk **40** (1985), 225–226 (Russian).

[151] V.N. Tolstoy, *Extremal projectors for contragredient Lie algebras and superalgebras of finite growth*, Russian Math. Surveys **44** (1989), 257–258.

[152] V.N. Tolstoy, *Extremal projectors and reduction superalgebras over Lie superalgebras*, Group Theoretical Methods in Physics, Vol. 2, M.A. Markov (ed.), Nauka (1986), 46–55 (Russian).

[153] V.N. Tolstoy, *Extremal projectors for quantized Kac–Moody superalgebras and some of their applications*, Quantum Groups, Clausthal, 1989, Lecture Notes in Phys. **370**, Springer (1990), 118–125.

[154] V.N. Tolstoy, *Projection operator method for quantum groups*, Special Functions 2000: Current Perspective and Future Directions, Proceedings of the NATO Advance Study Institute, J. Bustoz, M.E.H. Ismail and S.K. Suslov (eds), NATO Sci. Series II **30**, Kluwer Acad. Publ. (2001), 457–488.

[155] V.N. Tolstoy and J.P. Draayer, *New approach in theory of Clebsch–Gordan coefficients for $u(n)$ and $U_q(u(n))$*, Czech J. Phys. **50** (2000), 1359–1370.

[156] V.N. Tolstoy, I.F. Istomina and Yu.F. Smirnov, *The Gel'fand–Tseĭtlin basis for the Lie superalgebra* $\mathrm{gl}(n/m)$, Group Theoretical Methods in Physics, Vol. I, Yurmala, 1985, VNU Sci. Press (1986), 337–348.

[157] K. Ueno, T. Takebayashi and Y. Shibukawa, *Gelfand–Zetlin basis for $U_q(gl(N+1))$-modules*, Lett. Math. Phys. **18** (1989), 215–221.

[158] A. van den Hombergh, *A note on Mickelsson's step algebra*, Indag. Math. **37** (1975), 42–47.

[159] H. Weyl, *Classical Groups, their Invariants and Representations*, Princeton Univ. Press (1946).

[160] N.J. Wildberger, *A combinatorial construction for simply-laced Lie algebras*, Adv. in Appl. Math. **30** (2003), 385–396.

[161] N.J. Wildberger, *A combinatorial construction of G_2*, J. Lie Theory **13** (2003), 155–165.

[162] N.J. Wildberger, *Minuscule posets from neighbourly graph sequences*, European J. Combin. **24** (6) (2003), 741–757.

[163] N.J. Wildberger, *Quarks, diamonds and representations of $sl(3)$*, to appear.

[164] M.K.F. Wong, *Representations of the orthogonal group I. Lowering and raising operators of the orthogonal group and matrix elements of the generators*, J. Math. Phys. **8** (1967), 1899–1911.

[165] M.K.F. Wong and H.-Y. Yeh, *The most degenerate irreducible representations of the symplectic group*, J. Math. Phys. **21** (1980), 630–635.

[166] N. Xi, *Special bases of irreducible modules of the quantized universal enveloping algebra* $U_v(\mathrm{gl}(n))$, J. Algebra **154** (1993), 377–386.

[167] D.P. Zhelobenko, *The classical groups. Spectral analysis of their finite-dimensional representations*, Russian Math. Surveys **17** (1962), 1–94.

[168] D.P. Želobenko, *Compact Lie Groups and their Representations*, Transl. Math. Monographs **40**, Amer. Math. Soc. (1973).

[169] D.P. Zhelobenko, *S-algebras and Verma modules over reductive Lie algebras*, Soviet Math. Dokl. **28** (1983), 696–700.

[170] D.P. Zhelobenko, *Z-algebras over reductive Lie algebras*, Soviet Math. Dokl. **28** (1983), 777–781.

[171] D.P. Zhelobenko, *On Gelfand–Zetlin bases for classical Lie algebras*, Representations of Lie Groups and Lie Algebras, A.A. Kirillov (ed.), Akademiai Kiado (1985), 79–106.

[172] D.P. Zhelobenko, *Extremal projectors and generalized Mickelsson algebras on reductive Lie algebras*, Math. USSR-Izv. **33** (1989), 85–100.

[173] D.P. Zhelobenko, *An introduction to the theory of S-algebras over reductive Lie algebras*, Representations of Lie Groups and Related Topics, A.M. Vershik and D.P. Zhelobenko (eds), Adv. Studies in Contemp. Math. **7**, Gordon and Breach Science Publishers (1990), 155–221.

[174] D.P. Zhelobenko, *Representations of Reductive Lie Algebras*, Nauka (1994) (Russian).

Section 4H
Rings and Algebras with Additional Structure

Hopf Algebras

Miriam Cohen[1]

Department of Mathematics, Ben Gurion University of the Negev, Beer Sheva, Israel
E-mail: mia@cs.bgu.ac.il

Shlomo Gelaki[2]

Faculty of Mathematics, Technion, Haifa, Israel
E-mail: gelaki@tx.technion.ac.il

Sara Westreich

Interdisciplinary Department of the Social Sciences, Bar-Ilan University, Ramat-Gan, Israel
E-mail: swestric@mail.biu.ac.il

Contents

[1]This research was supported by a grant from the United States–Israel Binational Science Foundation (BSF), Jerusalem, Israel.

[2]This research was supported by The Israel Science Foundation founded by the Israel Academy of Sciences and Humanities.

HANDBOOK OF ALGEBRA, VOL. 4
Edited by M. Hazewinkel

Introduction

Hopf algebras became an object of study from an algebraic standpoint only in the late 1960s. It soon became evident that applications of this theory are abundant in a wide variety of fields. These applications range from topology, knot theory, algebraic geometry, C^*-algebras and combinatorics to statistical mechanics, quantum field theory, language theory in computer science, robotics, telecommunications and even chemistry.

The basic idea was developed in the work of H. Hopf, [108], on topological groups. The (co)homology of such groups form what is now termed: graded Hopf algebras. Algebraic properties of such Hopf algebras were first studied in Milnor and Moore's [163] fundamental work in the mid 60s.

Let us start by introducing an operation termed "comultiplication" which Hopf algebras are endowed with and which are their main novelty. Comultiplication is in a sense going into the "opposite" direction of multiplication. When multiplying, one takes a pair of elements and gets a single element, while when comultiplying one starts with a single element which "opens up" to a sum of pairs.

Explicitly, if $(A, \mu, 1)$ is an algebra over the base field k then 1 can be considered as a map $1: k \to A$ satisfying $\mu \circ (1 \otimes \mathrm{id}) = \mathrm{id}$ and the multiplication $\mu: A \otimes A \to A$ satisfies associativity. Dualizing this: A coalgebra (C, Δ, ε) has a counit ε satisfying $(\varepsilon \otimes \mathrm{id}) \circ \Delta = \mathrm{id}$ and a comultiplication $\Delta: C \to C \otimes C$ which is coassociative, namely, $(\Delta \otimes \mathrm{id}) \circ \Delta = (\mathrm{id} \otimes \Delta) \circ \Delta$. What coassociativity means is that after "opening up" C for the first time, one can either "open up" the left or the right tensorands and get the same result.

A bialgebra is an algebra over k which is also a coalgebra such that Δ and ε are multiplicative. A Hopf algebra is a bialgebra with an additional special map called the antipode.

The special way in which coalgebras or better yet Hopf algebras arise in the study of the variety of fields is best displayed in combinatorics, where these notions serve as a valuable formal framework. The coproduct displays all ways of decomposing a structure into appropriate parts, while the antipode replaces the role usually played by Möbius inversion.

There are two basic examples of Hopf algebras. The group algebra kG of a group G, where Δ is the diagonal map $g \mapsto g \otimes g$, $\varepsilon(g) = 1$ and $S(g) = g^{-1}$ for all $g \in G$. The second example is $U(\mathfrak{g})$, the enveloping algebra of a Lie algebra $\mathfrak{g}$ where $\Delta(x) = x \otimes 1 + 1 \otimes x$, $\varepsilon(x) = 0$ and $S(x) = -x$ for each $x \in \mathfrak{g}$ and extend these maps to $U(\mathfrak{g})$.

A geometrical motivation for the definition of a Hopf algebra is the following. Let G be a finite group and let k be any field. Consider the k-vector space $H := \mathrm{F}un(G)$ of all k-valued functions on G. Then H, equipped with the pointwise multiplication, is a *commutative* and *associative* algebra with unit 1 (the function which assigns the value $1 \in k$ to every $g \in G$). Note that the multiplication and unit in H can be regarded as k-linear maps $\mu: H \otimes H \to H$ and $\eta: k \to H$

$$\mu(\alpha \otimes \beta)(g) := \alpha(g)\beta(g) \quad \text{and} \quad \eta(a) := a1.$$

The multiplication in G gives rise to a comultiplication map on H:

$$\Delta: H \to H \otimes H, \quad \Delta(\alpha)(g, h) = \alpha(gh)$$

(here we identify $Fun(G \times G)$ with $H \otimes H$, which is allowed since G is finite). Since the multiplication in G is associative, one sees easily that Δ is coassociative. Note also that Δ is an algebra homomorphism.

Second, the unit element $e \in G$ gives rise to a *counit* map on H:

$$\varepsilon : H \to k, \quad \varepsilon(\alpha) = \alpha(e).$$

Note that ε is an algebra homomorphism.

Finally, the inverse operation in G gives rise to an *antipode* map on H:

$$S : H \to H, \quad S(\alpha)(g) = \alpha\left(g^{-1}\right).$$

The axioms of the inverse operation in G translate to

$$(S \otimes \mathrm{id}) \circ \Delta = \eta \circ \varepsilon = (\mathrm{id} \otimes S) \circ \Delta.$$

Summarizing, we see that the concept 'G is a group' can be expressed in terms of its algebra of functions by saying that this algebra admits three additional structure maps Δ, ε, S making it a commutative Hopf algebra.

More generally, if G is an affine algebraic group, one can similarly define a structure of a (commutative) Hopf algebra on the coordinate ring $H := Fun(G)$ of all regular functions on G. The Hopf algebra H is affine (i.e. finitely generated and commutative with 0 radical). Moreover, if G is an affine variety, then its coordinate ring $H := Fun(G)$ of all regular functions on G is an affine Hopf algebra if and only if G is an affine algebraic group. This observation was used to develop the structure theory of algebraic groups (e.g., reductive, solvable) using the theory of Hopf algebras.

In the language of schemes introduced by Grothendieck, one may think of affine Hopf algebras as *affine group schemes*.

It is thus not surprising that much of the work on Hopf algebras in the 70s was inspired by group theory. Kaplansky's conjectures are good examples of this approach. It was soon found that although quite a few properties can be generalized from group theory to the theory of Hopf algebras, some others are either false in general or hard to translate.

During the beginning of the 80s Hopf algebras have entered in a fundamental way as a unifying tool to analyze the theory of actions of various algebraic structures on algebras. Group actions, group-gradings and actions of Lie algebras are all examples of the theory. However the most striking boost to the theory was given in the beginning of the 80s with the introduction of quantum groups. These are Hopf algebras arising from solutions to the quantum Yang–Baxter equation from statistical mechanics. Another important connection to physics is the realization that standard notions of renormalization theory are derived from the Hopf algebra of rooted trees. This Hopf algebra acts on the Feinman diagrams.

During the 90s there has been a surge of interest in the structure of general Hopf algebras (mainly finite-dimensional) and fundamental examples of quantum groups. Techniques from other areas such as representation theory, algebraic geometry, category theory, Lie theory and ring theory were employed to answer some basic questions. The best results are attained for semisimple or pointed Hopf algebras.

Categorical considerations enter in a fundamental way in the general study of Hopf algebras since one of their striking properties is that they can be characterized by their categories of modules or comodules. Most of the basic Hopf algebraic concepts can be expressed in categorical terms. Furthermore, the theory of finite-dimensional (semisimple) Hopf algebras is one of the main motivations for the study of finite (fusion) categories, which is also motivated by physics (conformal field theory in the semisimple case and logarithmic conformal field theories in the non-semisimple case).

In fact the theory of Hopf algebras also motivated the definition and study of several other central algebraic objects; e.g., quasi-Hopf algebras and weak Hopf algebras.

Part 1. Basic concepts

Loosely speaking, a Hopf algebra is an algebra over a field k also equipped with a "dual" structure, such that the two structures are compatible. In what follows we give precise definitions of the concepts involved.

Throughout, we let k be a field. Vector spaces, algebras and tensor products are assumed to be over k unless stated otherwise. Algebras A are assumed to have a unit $1 = 1_A$ and $u : k \to A$ is the unit map $\alpha \mapsto \alpha 1_A$ for all $\alpha \in k$.

Our basic references are [1,166,221]. Another reference with more emphasis on the theory of coalgebras is [54].

1.1. Coalgebras and comodules

DEFINITION 1.1.1. A *coalgebra* over k (or simply a coalgebra) is a vector space C together with two linear maps, comultiplication $\Delta : C \to C \otimes C$ and counit $\varepsilon : C \to k$, such that:

(1) Δ is coassociative: $(\Delta \otimes \mathrm{id}) \circ \Delta = (\mathrm{id} \otimes \Delta) \circ \Delta$.

(2) ε satisfies the counit property: $(\varepsilon \otimes \mathrm{id}) \circ \Delta = \mathrm{id} = (\mathrm{id} \otimes \varepsilon) \circ \Delta$.

If (C, Δ, ε) is a coalgebra then C^* is an algebra with multiplication given by Δ^*, the dual map of Δ. However, if we begin with an algebra A and try to dualize then difficulties arise for infinite-dimensional A. But if (A, μ, u) is a finite-dimensional algebra then (A^*, μ^*, u^*) is a coalgebra with $\Delta = \mu^*$ and $\varepsilon = u^*$. Explicitly: $\Delta f(a \otimes b) = f(ab)$ and $\varepsilon(f) = f(1)$ for all $f \in A^*$, $a, b \in A$.

Inductively, one can apply Δ n times to any one of the tensorands and get the same result. Trying to keep track of the multitude of indices involved in an n-fold application of Δ would be prohibitive. It is in order to simplify this that Heyneman and Sweedler, [106], introduced the extremely successful, so-called, sigma-notation.

SIGMA-NOTATION. For any $c \in C$ write: $\Delta(c) = \sum c_1 \otimes c_2$. The subscripts 1 and 2 are symbols and do not indicate particular elements. When Δ is applied again to the left tensorand this would symbolically be written as $\sum c_{1_1} \otimes c_{1_2} \otimes c_2$, while applying Δ to the right as: $\sum c_1 \otimes c_{2_1} \otimes c_{2_2}$. Coassociativity means that these two expressions are equal and hence it makes sense to write this element as $\sum c_1 \otimes c_2 \otimes c_3$. Iterating this procedure gives $\Delta_{n-1}(c) = \sum c_1 \otimes \cdots \otimes c_n$ where $\Delta_{n-1}(c)$ is the unique element obtained by applying coassociativity $(n-1)$ times. In this notation the counit property says that $c = \sum \varepsilon(c_1)c_2 = \sum c_1\varepsilon(c_2)$ for all $c \in C$.

EXAMPLE 1.1.2. Let G be a group and $C = kG$, the group algebra. Define Δ, ε by $\Delta(g) = g \otimes g$ and $\varepsilon(g) = 1$, for all $g \in G$ and extend linearly. It is obvious that Δ is coassociative (in fact $\Delta_{n-1}(g) = g \otimes g \otimes \cdots \otimes g$) and that ε is a counit. When referring to kG in the sequel we regard it as a coalgebra equipped with these Δ and ε.

This inspired a general terminology for coalgebras:

DEFINITION 1.1.3. Let (C, Δ, ε) be a coalgebra. Then $0 \neq c \in C$ is called a *grouplike element* if $\Delta(c) = c \otimes c$. Denote by $G(C)$ the set of all grouplike elements of C.

REMARK 1.1.4 [221, p. 55]. $G(C)$ is a linearly independent set over k.

Let V, W be vector spaces. Then the *flip* (twist) map $\tau : V \otimes W \to W \otimes V$ is defined by $v \otimes w \mapsto w \otimes v$.

DEFINITION 1.1.5. Let (C, Δ, ε) be a coalgebra. An element $c \in C$ is *cocommutative* if $\Delta(c) = \tau \circ \Delta(c)$. The coalgebra C is cocommutative if all its elements are cocommutative. (Compare: an algebra A with multiplication μ is commutative if $\mu = \mu \circ \tau$.)

Observe that the group algebra kG is cocommutative, while the following are examples of a non-cocommutative coalgebras.

EXAMPLE 1.1.6. Let $C = sp_k\{1, g, x\}$ with coalgebra structure given by $\Delta(1) = 1 \otimes 1$, $\Delta(g) = g \otimes g$, $\Delta(x) = x \otimes 1 + g \otimes x$ and $\varepsilon(1) = 1$, $\varepsilon(g) = 1$, $\varepsilon(x) = 0$.

EXAMPLE 1.1.7. Let V be a finite-dimensional vector space and let $\mathrm{End}(V) = M_n(k)$ with $\{e_{ij}\}$ its standard basis. Then the coalgebra structure on $\mathrm{End}(V)^*$ which is dual to the algebra structure of $\mathrm{End}(V)$ is given explicitly by

$$\Delta\left(T_i^j\right) = \sum_{k=1}^{n} T_i^k \otimes T_k^j \quad \text{and} \quad \varepsilon\left(T_i^j\right) = \delta_{ij},$$

where T_i^j are the coordinate functions given by $T_i^j(e_{kl}) = \delta_{ik}\delta_{jl}$.

DEFINITION 1.1.8. Let $(C, \Delta_C, \varepsilon_C)$ and $(D, \Delta_D, \varepsilon_D)$ be coalgebras. A (linear) map $f : C \to D$ is a *coalgebra map* if

$$\Delta_D \circ f = (f \otimes f) \circ \Delta_C \quad \text{and} \quad \varepsilon_C = \varepsilon_D \circ f.$$

DEFINITION 1.1.9. Let $C \neq 0$ be a coalgebra.

C is *simple* if it has no proper subcoalgebras.

C is *irreducible* if any two non-zero subcoalgebras of C have a non-zero intersection.

It will be evident from local finiteness, described next, that C is irreducible if and only if it contains a unique simple subcoalgebra. This is the reason for alternatively calling such a coalgebra "colocal".

C is *indecomposable* if C can not be expressed as a non-trivial coalgebra direct sum.

Coalgebras C have the following striking properties.

(1) Local finiteness: Every element of C is contained in a finite-dimensional subcoalgebra. Thus, in particular, C contains a simple subcoalgebra and every simple subcoalgebra of C is finite-dimensional.

(2) Distributivity: If $C = \bigoplus C_i$, where each C_i is a subcoalgebra of C, and D is a subcoalgebra of C, then $D = \bigoplus (D \cap C_i)$.

Consequences of the above properties are the following:

THEOREM 1.1.10 [116,101]. *Every coalgebra is uniquely a direct sum of indecomposable subcoalgebras.*

Moreover, [116], if C is a cocommutative coalgebra then every indecomposable coalgebra is irreducible. Hence C is a unique direct sum of irreducible subcoalgebras. These irreducible subcoalgebras are maximal with respect to the irreducibility property, they are the so-called irreducible components of C.

The second part of the theorem above is not necessarily true if C in not cocommutative. For example, the coalgebra defined in Example 1.1.6, is not a sum of irreducible subcoalgebras since its only irreducible components are $k1$ and kg.

REMARK 1.1.11. If $g \in G(C)$, then kg is a simple subcoalgebra of C of dimension 1. Conversely, any 1-dimensional subcoalgebra of C is of the form kg where $g \in G(C)$.

If k is algebraically closed and C is cocommutative then every simple subcoalgebra is 1-dimensional (this follows easily from considering the dual of C).

DEFINITION 1.1.12. Let C be a coalgebra.

The *coradical* C_0 of C is the sum of all simple subcoalgebras of C.

C is *cosemisimple* if $C_0 = C$.

C is *pointed* if every simple subcoalgebra of C is 1-dimensional (by Remark 1.1.11 this means that $C_0 = kG(C)$).

C is *connected* if C_0 is one-dimensional.

The coradical C_0 induces a filtration on C by the so-called *wedge product* as follows: for each $n \geqslant 1$ define inductively

$$C_n = \Delta^{-1}(C \otimes C_{n-1} + C_0 \otimes C) := C_0 \wedge C_{n-1}.$$

In fact $C_n = (J^n)^{\perp}$ where J is the Jacobson radical of the algebra C^*. The following hold:

THEOREM 1.1.13. *$\{C_n \mid n \geqslant 0\}$ is a family of subcoalgebras of C satisfying*

(1) *$C_n \subseteq C_{n+1}$ and $C = \bigcup_{n \geqslant 0} C_n$,*

(2) *$\Delta(C_n) \subseteq \sum_{i=0}^{n} C_i \otimes C_{n-i}$.*

The filtration $\{C_n\}$ is called the *coradical filtration* of C.

The following theorem has important implications:

THEOREM 1.1.14 [105]. *Let C and D be coalgebras and $f : C \to D$ a coalgebra morphism so that the restriction of f to C_1 is injective. Then f is injective.*

DEFINITION 1.1.15. Let C be a coalgebra with a distinguished group like element 1 (which will be the unit element in the case C is the underlying coalgebra of a bialgebra or Hopf algebra). Then $x \in C$ is called a *primitive element* of C if $\Delta(x) = x \otimes 1 + 1 \otimes x$. The set of primitive elements of C is denoted by $P(C)$. More generally, an element $x \in C$ is called *(σ, τ)-primitive* if $\Delta(x) = x \otimes \sigma + \tau \otimes x$ for some $\sigma, \tau \in G(C)$.

The set of (σ, τ)-primitive elements is denoted by $P_{\sigma,\tau}(C)$. Such elements are also called *skew-primitive* elements. By definition, skew-primitive elements of C are in C_1.

The dual of the notion of an ideal is that of a coideal.

DEFINITION 1.1.16. A subspace $I \subseteq C$ is a *coideal* if

$$\Delta(I) \subseteq I \otimes C + C \otimes I \quad \text{and} \quad \varepsilon(I) = 0.$$

A subspace $I \subseteq C$ is a left coideal if

$$\Delta(I) \subseteq I \otimes C.$$

Right coideals are defined analogously.

When I is a coideal of C, C/I is a coalgebra in a natural way.

Just as coalgebras and algebras are dual concepts so are comodules and modules.

DEFINITION 1.1.17. For a coalgebra (C, Δ, ε), a (right) *C-comodule* is a vector space M with a k-linear map $\rho : M \to M \otimes C$ such that

$$(\rho \otimes \mathrm{id}) \circ \rho = (\mathrm{id} \otimes \Delta) \circ \rho \quad \text{and} \quad \mathrm{id} = (\mathrm{id} \otimes \varepsilon) \circ \rho.$$

Left C-comodules are defined analogously.

We sometimes write (M, ρ) or (M, ρ_M) for a C-comodule. There is also a sigma-notation for right C-comodules. One writes

$$\rho(m) = \sum m_0 \otimes m_1 \in M \otimes C.$$

EXAMPLE 1.1.18.

(1) Every coalgebra (C, Δ, ε) is a right C-comodule by choosing $\rho = \Delta$. The right (left) C-subcomodules of C are precisely the *right* (*left*) *coideals*.

(2) A somewhat less obvious example is the following: Let G be a group and $M = \sum_{g\in G}^{\oplus} M_g$ be a G-graded vector space. Then M is a right kG-comodule by setting $\rho(m) = m \otimes g$ for each $m \in M_g$.

(3) Recall Example 1.1.7 that if $V = Sp_k\{v_1, \dots, v_n\}$ is a finite-dimensional vector space then $(\mathrm{End}(V))^*$ is a coalgebra. Now, V is a right $\mathrm{End}(V)^*$-comodule via

$$\rho(v_i) = \sum_j v_j \otimes T_i^j.$$

Just as for coalgebras we have local-finiteness for comodules M. That is, a comodule M contains a finite-dimensional subcomodule; in particular, M contains a simple subcomodule and every simple subcomodule of M is finite-dimensional. Moreover, if (M, ρ) is a simple C-comodule then $\rho^{-1}(M \otimes C)$ is a simple subcoalgebra of C.

DEFINITION 1.1.19. Let (M, ρ_M) and (N, ρ_N) be right C-comodules. Then a map $f : M \to N$ is a *comodule-map* if

$$\rho_N \circ f = (f \otimes \mathrm{id}) \circ \rho_M.$$

DEFINITION 1.1.20 [163]. Let C be a coalgebra and let (M, ρ_M) and (N, ρ_N) be right and left C-comodules respectively. Then the *cotensor product* $M\Box_C N$ of M and N over C is the equalizer

$$M \otimes N \underset{\rho_M \otimes \mathrm{id}_N}{\overset{\mathrm{id}_M \otimes \rho_N}{\rightrightarrows}} M \otimes C \otimes N.$$

That is

$$M\Box_C N = \{x \in M \otimes N \mid (\mathrm{id}_M \otimes \rho_N)(x) = (\rho_M \otimes \mathrm{id}_N)(x)\}.$$

EXAMPLE 1.1.21. Let C be a coalgebra and M a right C-comodule then $M\Box_C C = \rho(M) \cong M$.

Let V be a vector space M, N and C as above. Then there exists a canonical isomorphism

$$V \otimes (M\Box_C N) = (V \otimes M)\Box_C N$$

(where $V \otimes M$ is a right C-comodule via $\mathrm{id} \otimes \rho_M$).

The dual notion of a bi-module over two algebras is a bi-comodule over two coalgebras.

DEFINITION 1.1.22. Let H, L be coalgebras and let M be a left H-comodule via ρ and a right L-comodule via η. Then M is an (H, L)-*bicomodule* if

$$(\rho \otimes \mathrm{id}) \circ \eta = (\mathrm{id} \otimes \eta) \circ \rho.$$

DEFINITION 1.1.23. Let (C, Δ, ε) be a coalgebra and A an algebra. Then $\mathrm{Hom}_k(C, A)$ becomes an algebra under the *convolution product* $(f * g)(c) = \sum f(c_1)g(c_2)$ for all $f, g \in \mathrm{Hom}_k(C, A)$ and $c \in C$. The unit element is $u\varepsilon$.

In particular, as was already mentioned, $C^* = \mathrm{Hom}(C, k)$ is an algebra. In fact the convolution product in C^* is Δ^*.

We denote the evaluation $p(c)$ by $\langle p, c\rangle$ for any $p \in C^*$ and $c \in C$.

Observe that if I is a coideal of C then $I^\perp = \{f \in C^* \mid \langle f, I\rangle = 0\}$ is a subalgebra of C^*. While if J is a right coideal of C then $J^\perp$ is a right ideal of C^*. If D is a subcoalgebra of C then $D^\perp$ is a two-sided ideal of C^*.

1.2. Hopf algebras

Definitions and basic examples

DEFINITION 1.2.1. Let H be an algebra with multiplication μ and unit map u, and a coalgebra with comultiplication Δ and counit ε. Then H is a *bialgebra* if Δ and ε are algebra maps or equivalently μ, u are coalgebra maps.

DEFINITION 1.2.2. Let H be a bialgebra. Then H is a *Hopf algebra* if there exists an element $S \in \mathrm{End}_k(H)$ so that

$$\sum S(h_1)h_2 = \varepsilon(h)1_H = \sum h_1 S(h_2)$$

for all $h \in H$. The map S is called an *antipode* for H. It is the inverse of the identity map under convolution.

A map $f: H \to H'$ of bialgebras (Hopf algebras) H, H' is called a *bialgebra (Hopf algebra) homomorphism* if it is both an algebra and a coalgebra homomorphism (and $f \circ S_H = S_{H'} \circ f$).

The kernel of a bialgebra (Hopf algebra) map is a *biideal* (*Hopf ideal*). That is, it is both an ideal and a coideal (and stable under S).

H is a *cocommutative* Hopf algebra if it is cocommutative as a coalgebra.

H is a *pointed* Hopf algebra if it is pointed as a coalgebra.

H is a *semisimple* Hopf algebra if it is semisimple as an algebra.

H is a *cosemisimple* Hopf algebra if it is cosemisimple as a coalgebra.

It is easy to see that if H is either a commutative or a cocommutative Hopf algebra then $S^2 = \mathrm{id}$.

EXAMPLE 1.2.3.

(1) The group algebra kG with antipode S defined by $S(g) = g^{-1}$ for all $g \in G$. It is a pointed cosemisimple cocommutative Hopf algebra. If G is finite then $(kG)^*$ is a commutative semisimple Hopf algebra.

(2) Let $\mathfrak{g}$ be a Lie algebra and $U(\mathfrak{g})$ its enveloping algebra. For each $x \in \mathfrak{g}$ define $\Delta(x) = 1 \otimes x + x \otimes 1$ and $S(x) = -x$. Then $U(\mathfrak{g})$ is a connected, thus pointed, cocommutative Hopf algebra.

(3) Let G be a group. Let $H = R(G)$ be the Hopf algebra of all real-valued representative functions on G with pointwise multiplication, coproduct given by $\Delta(f)(x, y) = f(xy)$, counit given by $\varepsilon(f) = f(e)$ and the antipode is given by $S(f)(x) = f(x^{-1})$ where $f \in R(G)$, $x, y \in G$ and e is the identity of G (see [1, 2.2] for details).

REMARK 1.2.4.

(1) If H is a Hopf algebra then $G(H)$ is in fact a group, where $g^{-1} = S(g)$.

(2) A combinatorial calculation implies that if x is a primitive element of H and k is of characteristic zero then the set $\{x^i \mid i > 0\}$ is linearly independent. Hence if H is finite-dimensional over k then $P(H) = 0$.

The following Hopf algebra is the smallest non-commutative, non-cocommutative Hopf algebra. It is again pointed. This is Sweedler's 4-dimensional Hopf algebra:

EXAMPLE 1.2.5. Let $\operatorname{ch} k \neq 2$ and

$$H_4 := k\langle 1, g, x, gx \mid g^2 = 1, x^2 = 0, xg = -gx\rangle$$

with coalgebra structure as in Example 1.1.6, $S(g) = g$ and $S(x) = -gx$. Then H_4 is a Hopf algebra. Note that S has order 4.

Here are basic properties of the antipode:

THEOREM 1.2.6. *Let H be a Hopf algebra with antipode S. Then*

(1) *S is an anti-algebra morphism; that is, $S(hh') = S(h')S(h)$, for all $h, h' \in H$ and $S(1_H) = 1_H$.*

(2) *S is an anti-coalgebra morphism; that is, $\Delta(S(h)) = \sum S(h_2) \otimes S(h_1)$ and $\varepsilon(S(h)) = \varepsilon(h)$, for all $h \in H$.*

(3) *If H is finite-dimensional then S is bijective.*

REMARK 1.2.7. Let H be a Hopf algebra with bijective antipode S and let $H^{\mathrm{cop}} = H$ as an algebra with comultiplication $\Delta^{\mathrm{cop}} := \tau \circ \Delta$ (where τ is the standard flip map). Then H^{cop} is a Hopf algebra with antipode S^{-1}.

Integrals. Integrals for Hopf algebras H are classically defined as certain elements in H^*. The name being motivated by the following example:

EXAMPLE 1.2.8. Let G be a compact topological group and $H = R(G)$ as in Example 1.2.3(3). Suppose η is a Haar measure on G and set $T(f) = \int_G f \, d\eta$, $f \in H$. Then $T \in H^*$ with an invariance property induced from the left invariance of the Haar measure. Specifically: $xT = \langle x, 1\rangle T$ for all $x \in H^*$.

The classical definition of a left integral *for* H is an element $T \in H^*$ so that for each $x \in H^*$, $xT = \langle x, 1\rangle T$. If H is finite-dimensional one can thus define an integral for H^* (which is an element of $(H^*)^* = H$). By an abuse of notation it is called a left integral *in* H.

DEFINITION 1.2.9. Let H be a finite-dimensional Hopf algebra. A *left integral* in H is an element $t \in H$ such that $ht = \varepsilon(h)t$ for all $h \in H$. A *right integral* is an element $t' \in H$ such that $t'h = \varepsilon(h)t'$ for all $h \in H$. The space of left (right) integrals is denoted by $\int_H^l$ ($\int_H^r$) respectively. H is called *unimodular* if $\int_H^l = \int_H^r$.

Note that kt is a left ideal of H (it is in fact an ideal of H as seen in Theorem 2.2.1).

EXAMPLE 1.2.10. A prime example of an integral is the "averaging element". Specifically, if G is a finite group and $H = kG$, then $t = \sum_{g \in G} g$ generates the space of left and right integrals in H.

EXAMPLE 1.2.11. Let $H = H_4$ of Example 1.2.5. Then $x + gx \in \int_H^l$ and $x - gx \in \int_H^r$. Thus H is not unimodular.

The finite dual

One of the important features of Hopf algebras is that its definition is in a sense self-dual. Namely, if H is a finite-dimensional Hopf algebra then its linear dual H^* has a canonical structure of a Hopf algebra with structure maps the transposes of the structure maps of H. It is called the *dual Hopf algebra* of H.

When H is an infinite-dimensional Hopf algebra, H^* is an algebra but no longer a Hopf algebra. However H^* contains a Hopf algebra (which is maximal with respect to this property) which is called the finite-dual of H.

DEFINITION 1.2.12. Let H be a Hopf algebra. Then the *finite-dual* of H is defined to be

$$H^0 = \{p \in H^* \mid p \text{ vanishes on an ideal of } H \text{ of finite codimension}\}.$$

There are equivalent conditions for $p \in H^*$ to belong to H^0. Here is a typical one.

PROPOSITION 1.2.13 [221, p. 115]. *Let H be a Hopf algebra, then*

$$H^0 = \{p \in H^* \mid \dim(H \rightharpoonup p) < \infty\},$$

where $h \rightharpoonup p$ *is defined by* $\langle h \rightharpoonup p, h'\rangle = \langle p, h'h\rangle$ *for all* $h, h' \in H$ *and* $p \in H^*$.

THEOREM 1.2.14 [221, p. 122]. *Let H be a Hopf algebra, then* H^0 *is a Hopf algebra with structure maps dual to those in* H.

H^* is a topological space with the finite discrete topology on $\mathrm{Hom}(H,k)$, where k has the discrete topology. A subspace V of H^* is dense in H^* if it separates the points of H.

The finite-dual can be dense in H^*, on the other hand it may be trivial.

EXAMPLE 1.2.15.

(1) If H is an affine commutative Hopf algebra then H^0 is dense in H, [221, p. 121].

(2) If $\mathfrak{g}$ is a finite-dimensional semisimple Lie algebra and q is not a root of unity then $(U_q(\mathfrak{g}))^0$ is dense in $(U_q(\mathfrak{g}))^*$, [111].

(3) If K is an infinite field of cardinality greater than that of k and $G = PSL_2(K)$, then $(kG)^0 = k\varepsilon$, [24].

An equivalent criterion for density of H^0 in H^* is that H is *residually finite-dimensional*. That is, there exists a family $\{\pi_\alpha\}$ of finite-dimensional k-representations of A such that $\bigcap_\alpha \mathrm{Ker}\,\pi_\alpha = \{0\}$, [166].

1.3. Modules and comodules for Hopf algebras

The representation and co-representation theories for Hopf algebras are particularly rich, as a result of the various structures involved. For any Hopf algebra H, there are several ways in which H is acted upon by H or by H^*. Here are some:

DEFINITION 1.3.1. If H is a Hopf algebra then:

(1) H is a left H^*-module by

$$p \rightharpoonup h = \sum \langle p, h_2\rangle h_1$$

and a right H^*-module by

$$h \leftharpoonup p = \sum \langle p, h_1\rangle h_2$$

for all $h \in H$ and $p \in H^*$.

(2) H is a left H-module via the left adjoint action ad_l:

$$h \underset{\mathrm{ad}_l}{\cdot} x = \sum h_1 x S(h_2)$$

and a right H-module via the right adjoint action ad_r:

$$x \underset{\mathrm{ad}_r}{\cdot} h = \sum S(h_1) x h_2$$

for all $h, x \in H$.

Notice that the adjoint actions boil down to the usual adjoint actions of groups and Lie algebras for the Hopf algebras kG and $U(\mathfrak{g})$, respectively.

There is more to say about the actions just described. The adjoint actions make H into an H-module algebra while when H is finite-dimensional then $\rightharpoonup$ ($\leftharpoonup$) makes H into a left (right) H^*-module algebra, respectively.

DEFINITION 1.3.2. Let A be an algebra and H a Hopf algebra. If A is left H-module then A is an *H-module algebra* if

$$h \cdot (ab) = \sum (h_1 \cdot a)(h_2 \cdot b)$$

for all $h \in H$, $a, b \in A$.

Analogously one can define module coalgebras and comodule (co)algebras. In the language of categories these mean that the structure maps of the algebras or the coalgebras are maps in the category of H-modules or in the category of H-comodules. (For more details see Part 4.)

EXAMPLE 1.3.3. H is a right H-comodule coalgebra via the right adjoint coaction: $\rho: H \to H \otimes H$ given by $h \mapsto \sum h_2 \otimes S(h_1)h_3$.

REMARK 1.3.4. If H is a finite-dimensional Hopf algebra then (A, ρ) is a right H-comodule (algebra) if and only if $(A, \cdot)$ is a left H^*-module (algebra), where $\cdot$ and ρ are transposes of each other. That is, if (A, ρ) is a right H-comodule (algebra) then for $a \in A$ with $\rho(a) = \sum a_0 \otimes a_1 \in A \otimes H$ define

$$p \cdot a = \sum \langle p, a_1 \rangle a_0$$

for all $p \in H^*$. Conversely, given an action $\cdot$ of H^* on A define for $a \in A$,

$$\rho(a) = \sum (h_i^* \cdot a) \otimes h_i,$$

where $\{h_i\}$, $\{h_i^*\}$ are dual bases of H and H^*.

EXAMPLE 1.3.5. If G is a finite group and $A = \sum_{g \in G}^{\oplus} A_g$ is a G-graded algebra (that is, $A_g A_h \subset A_{gh}$) then as in Example 1.1.18(2), A is a right kG-comodule algebra. By the remark above, A becomes a left $(kG)^*$-module algebra by defining for all $a = \sum a_g \in A$, $p_g \cdot a = a_g$, where $\{p_g\}_{g \in G}$ is the basis of $(kG)^*$ dual to the basis $\{g\}_{g \in G}$ of kG.

DEFINITION 1.3.6. Let $(M, \cdot)$ be a left H-module. Then the *H-invariants* are

$$M^H = \{m \in M \mid h \cdot m = \varepsilon(h)m, \text{ for all } h \in H\}.$$

Let (M, ρ) be a right H-comodule. Then the *H-coinvariants* are

$$M^{\operatorname{co} H} = \{m \in M \mid \rho(m) = m \otimes 1\}.$$

${}^{\operatorname{co} H}M$ is defined similarly for left H-comodules.

EXAMPLE 1.3.7.
(1) [24]. Let H be a left H-module by the left adjoint action ad, then H^H is the center of H.
(2) Let H be a right H-comodule via Δ. Then $H^{\mathrm{co}\,H} = k$.

REMARK 1.3.8. Just as the averaging map for the theory of group actions so do integrals play a central role in the theory of actions of finite-dimensional Hopf algebras. For if M is a left H-module then it is immediate that $t \cdot M \subset M^H$ for $t \in \int_l^H$. If $\varepsilon(t) \neq 0$ then this is actually an equality (for then, $t \cdot m = \varepsilon(t)m$ for $m \in M^H$ implies that $m = t \cdot (\frac{1}{\varepsilon(t)} m)$).

1.4. Normal Hopf subalgebras, quotients and extensions

A basic concept in group theory is that of a normal subgroup, which is a subgroup stable under the adjoint action of the group on itself. It is characterized by the property that every right coset is a left coset as well. Equivalently, a normal subgroup is a kernel of a group homomorphism, and thus the quotient group is defined.

The Hopf algebra analogue of normality is given by:

DEFINITION 1.4.1. Let K be a Hopf subalgebra of H. Then K is a normal in H if

$$(\mathrm{ad}_l\, H)(K) \subseteq K \quad \text{and} \quad (\mathrm{ad}_r\, H)(K) \subseteq K,$$

where ad_l, ad_r are the left and right adjoint actions of H on itself (Definition 1.3.1(2)).

It is straightforward to verify that if K is a normal Hopf subalgebra of H then $HK^+ = K^+H$, where $K^+ = \{h \in K \mid \varepsilon(h) = 0\}$. Since HK^+ is a Hopf ideal of H it follows that H/HK^+ is a Hopf algebra (while H/K is usually meaningless). Let $\overline{H} := H/K^+H$ then we have the following exact sequence of Hopf algebras:

$$K \hookrightarrow H \twoheadrightarrow \overline{H}$$

and H is called a Hopf extension (of $\overline{H}$ by K).

However, in the converse direction, if $\pi : H \to \overline{H}$ is a Hopf algebra epimorphism then $\operatorname{Ker}\pi$ is not necessarily of the form K^+H for some normal Hopf subalgebra K. Furthermore, even if K is a Hopf subalgebra of H such that $HK^+ = K^+H$ (and thus H/K^+H is a Hopf algebra) it is not always true that K is a normal Hopf subalgebra of H. To discuss this (following [206,207]) we introduce the following notion:

Given any Hopf algebra epimorphism $\pi : H \to \overline{H}$, H becomes both a left and a right $\overline{H}$-comodule via

$$\rho_l = (\pi \otimes \mathrm{id}) \circ \Delta, \qquad \rho_r = (\mathrm{id} \otimes \pi) \circ \Delta.$$

Denote by $H^{\mathrm{co}\,\overline{H}}$ and ${}^{\mathrm{co}\,\overline{H}}H$ the algebras of coinvariants for those coactions (as defined in Definition 1.3.6).

If K satisfies $HK^+ = K^+H$ and we let $\overline{H} = H/K^+H$, then clearly $K \subset H^{\mathrm{co}\,\overline{H}}$ and $K \subset {}^{\mathrm{co}\,\overline{H}}H$.

PROPOSITION 1.4.2 [207]. *If H is faithfully flat over K then*

$$K = H^{\mathrm{co}\,\overline{H}} = {}^{\mathrm{co}\,\overline{H}}H.$$

In this case K is indeed a normal Hopf subalgebra of H.

Examples for which H is faithfully flat over any Hopf subalgebra K are:

(1) The Hopf algebra H is finite-dimensional, [178] (see also Remark 3.2.8).

(2) If H is commutative, [59].

(3) If H_0, the coradical of H, is cocommutative, [224]. In particular if H is cocommutative or pointed.

Here is an example of a non-commutative non-cocommutative Hopf algebra which is an extension of a commutative and cocommutative Hopf algebra (in fact the unique non-commutative non-cocommutative semisimple Hopf algebra of dimension 8).

EXAMPLE 1.4.3 [115, Example 4.1]. Let k be of characteristic 0 and let

$$H_8 := k\left\langle x, y, z \mid x^2 = y^2 = 1, xy = yx, zx = yz, xz = zy, \right.$$
$$\left. z^2 = \frac{1}{2}(1 + x + y - xy)\right\rangle$$

with a coalgebra structure given by

$$\Delta(x) = x \otimes x, \qquad \Delta(y) = y \otimes y,$$
$$\Delta(z) = \frac{1}{2}\big[(1 + y) \otimes 1 + (1 - y) \otimes x\big](z \otimes z).$$

Then $K := k\langle x, y\rangle \cong k(\mathbb{Z}_2 \times \mathbb{Z}_2)$ is a normal Hopf subalgebra of H and $\overline{H} = H/HK^+$ is isomorphic to $k\mathbb{Z}_2$. So H is the Hopf extension

$$K \cong k(\mathbb{Z}_2 \times \mathbb{Z}_2) \hookrightarrow H \xrightarrow{\pi} k(\bar{z}) \cong k(\mathbb{Z}_2),$$

where $\pi(x) = \pi(y) = 1$, $\pi(z) = \bar{z}$ and $\Delta(\bar{z}) = \bar{z} \otimes \bar{z}$.

1.5. Special Hopf algebras

A generalization of cocommutative Hopf algebras are quasitriangular Hopf algebras, introduced by Drinfeld. The dual notion is that of coquasitriangular Hopf algebras (sometimes called braided Hopf algebras). Though quasitriangular Hopf algebras were introduced in the context of solutions of the quantum Yang–Baxter equation, they play an important role in the general theory of Hopf algebras.

DEFINITION 1.5.1 [72]. A *quasitriangular* Hopf algebra is a pair (H, R), where H is a Hopf algebra and $R = \sum R^1 \otimes R^2 \in H \otimes H$ is invertible, such that the following hold:

(QT1) $(\Delta \otimes \mathrm{id})(R) = R^{13}R^{23} = \sum R^1 \otimes r^1 \otimes R^2 r^2$,

(QT2) $(\mathrm{id} \otimes \Delta)(R) = R^{13}R^{12} = \sum R^1 r^1 \otimes r^2 \otimes R^2$,

(QT3) $(\tau \circ \Delta)(h) = R\Delta(h)R^{-1}$ for all $h \in H$,

where $r = R$ and $R^{13} = \sum R^1 \otimes 1 \otimes R^2 \in H^{\otimes 3}$ etc.

R is sometimes called a universal R-matrix.

A consequence of the above is that $R^{-1} = \sum S(R^1) \otimes R^2$, $\sum \varepsilon(R^1)R^2 = \sum R^1 \varepsilon(R^2) = 1$ and $(S \otimes S)(R) = R$.

If $R^{-1} = R^{\tau} (:= \sum R^2 \otimes R^1)$ then (H, R) is called a *triangular* Hopf algebra and R is called unitary.

It is property (QT3) which generalizes cocommutativity. In fact, Drinfeld termed Hopf algebras satisfying this property *almost cocommutative*.

EXAMPLE 1.5.2. The following are basic examples of quasitriangular Hopf algebras:

(1) Every cocommutative Hopf algebra is triangular with $R = 1 \otimes 1$.

(2) Let k be of characteristic $\neq 2$, then the group algebra $k\mathbb{Z}_2 = k\langle 1, g\rangle$ is triangular with

$$R = \frac{1}{2}(1 \otimes 1 + 1 \otimes g + g \otimes 1 - g \otimes g).$$

(3) [192]. Let $H = H_4$ be as in Example 1.2.5. Then H is quasitriangular with a family of quasitriangular structures R_α, $\alpha \in k$, given by

$$\begin{aligned} R_\alpha = {} & \frac{1}{2}(1 \otimes 1 + 1 \otimes g + g \otimes 1 - g \otimes g) \\ & + \frac{\alpha}{2}(x \otimes x + x \otimes gx - gx \otimes x + xg \otimes xg). \end{aligned}$$

An important element in (H, R) is the so-called *Drinfeld element*

$$u = \sum S(R^2)R^1.$$

It is shown in [72] that u is an invertible element in H, $\Delta(u) = (u \otimes u)(R^{\tau}R)^{-1}$ and S^2 is an inner automorphism of H induced by u; that is, $S^2(h) = uhu^{-1}$, all $h \in H$. Moreover, S^4 is induced by the grouplike element $uS(u)^{-1}$.

Another important ingredient of a quasitriangular Hopf algebra (H, R) is the *Drinfeld map* $f : H^* \to H$ given by

$$f(p) = (\mathrm{id} \otimes p)R^{\tau}R.$$

Usually, it is not an algebra map, however its restriction to

$$O(H^*) := \{p \in H^* \mid \langle p, hh'\rangle = \langle p, S^2(h')h\rangle \ \forall h, h' \in H\}$$

is an algebra map to the center of H. If H is semisimple and k is algebraically closed of characteristic 0 then $O(H^*)$ coincides with the character ring of H (see Section 5.1 for definition and properties) hence $O(H^*)$ is called the algebra of generalized characters.

DEFINITION 1.5.3. A finite-dimensional quasitriangular Hopf algebra for which the Drinfeld map f is injective (and thus bijective) is called *factorizable*.

It was proved in [194] that every factorizable Hopf algebra is unimodular.

We have:

THEOREM 1.5.4. *Let (H, R) be a finite-dimensional quasitriangular Hopf algebra. Then the following are equivalent*:

(1) *H is factorizable.*

(2) [97]. *$f(T') \neq 0$ where $T' \neq 0$ is a right integral for H^*.*

(3) [52]. *f restricted to $O(H^*)$ is injective.*

EXAMPLE 1.5.5. The *Drinfeld double*, $D(H)$, is defined for any finite-dimensional Hopf algebra H. It is a factorizable Hopf algebra which contains H (see Section 4.3 for details).

When (H, R) is finite-dimensional, $R \in H \otimes H \cong (H^* \otimes H^*)^*$, hence R defines a bilinear form (or an R-form) $\langle\ |\ \rangle_R$ on H^*. The properties of this form define the axioms for coquasitriangular Hopf algebras. Thus if (H, R) is a finite-dimensional Hopf algebra then $(H^*, \langle\ |\ \rangle_R)$ is a coquasitriangular Hopf algebra. Coquasitriangular Hopf algebras have been studied by a number of people (cf. [142,136,155,202]).

DEFINITION 1.5.6. A *coquasitriangular* Hopf algebra is a pair $(H, \langle\ |\ \rangle)$ where H is a Hopf algebra over k and $\langle\ |\ \rangle : H \otimes H \to k$ is a k-linear form (braiding) which is convolution invertible in $\mathrm{Hom}_k(H \otimes H, k)$ such that the following hold for all $h, g, l \in H$:

(CQT1) $\langle h|gl\rangle = \sum \langle h_1|g\rangle\langle h_2|l\rangle$.

(CQT2) $\langle hg|l\rangle = \sum \langle g|l_1\rangle\langle h|l_2\rangle$.

(CQT3) $\sum \langle h_1|g_1\rangle g_2 h_2 = \sum h_1 g_1 \langle h_2|g_2\rangle$.

If $\sum \langle h_1|g_1\rangle\langle g_2|h_2\rangle = \varepsilon(g)\varepsilon(h)$ then $(H, \langle\ |\ \rangle)$ is called a *cotriangular* Hopf algebra.

A non-trivial example of a cotriangular Hopf algebra is kG where G is an Abelian group with a symmetric bicharacter $\langle\ |\ \rangle$. This Hopf algebra is commutative and cocommutative and it arises in the context of Lie color algebras (cf. [14]).

A special class of quasitriangular Hopf algebras is that of ribbon Hopf algebras.

DEFINITION 1.5.7. A finite-dimensional *ribbon* Hopf algebra over k is a triple (H, R, v), where (H, R) is a finite-dimensional quasitriangular Hopf algebra over k and $v \in H$ satisfies the following axioms:

(R1) v is in the center of H,

(R2) $S(v) = v$, and

(R3) $\Delta(v) = (v \otimes v)(R^\tau R)^{-1} = (R^\tau R)^{-1}(v \otimes v)$.

It follows from these axioms that $\varepsilon(v) = 1$ and $v^2 = uS(u)$. Also, observe that $G := u^{-1}v$ is a grouplike element of H. It is called *the special grouplike element* of H. Ribbon Hopf algebras were introduced and studied by Reshetikhin and Turaev in connection with invariants of links and 3-manifolds, [198]. Here u is the Drinfeld element.

REMARK 1.5.8. Any triangular Hopf algebra is ribbon with 1 as the ribbon element and u^{-1} as the special grouplike element. Also, any semisimple quasitriangular Hopf algebra such that $S^2 = \mathrm{id}$ (e.g., if the characteristic of k is zero) is ribbon with u as the ribbon element and 1 as the special grouplike element.

1.6. Twisting in Hopf algebras

Let H be a Hopf algebra. It is possible to twist either the comultiplication or the multiplication of H and thereby construct a new Hopf algebra. The following fundamental definition is due to Drinfeld, [71].

DEFINITION 1.6.1. A dual *Hopf 2-cocycle* (= twist) for H is an invertible element $J \in H \otimes H$ which satisfies:

$$(\Delta \otimes \mathrm{id})(J)(J \otimes 1) = (\mathrm{id} \otimes \Delta)(J)(1 \otimes J),$$

$$(\varepsilon \otimes \mathrm{id})(J) = (\mathrm{id} \otimes \varepsilon)(J) = 1.$$

Given a twist J for H, one can twist the comultiplication and define a new Hopf algebra structure $(H^J, m, 1, \Delta^J, \varepsilon, S^J)$ on the algebra $(H, m, 1)$. The coproduct and antipode are determined by

$$\Delta^J(a) = J^{-1}\Delta(a)J, \qquad S^J(a) = Q^{-1}S(a)Q$$

for every $a \in H$, where $Q := m \circ (S \otimes \mathrm{id})(J)$.

Suppose (H, R) is also (quasi)triangular. Then it is straightforward to verify that (H^J, R^J) is quasi(triangular) with universal R-matrix $R^J := (J^\tau)^{-1}RJ$ (where $J^\tau = \tau(J)$).

The tensor category of left (right) H-modules is equivalent to that of left (right) H^J-modules.

Dually, one can twist the multiplication on H.

DEFINITION 1.6.2. A linear form $\sigma : H \otimes H \to k$ is called a *Hopf 2-cocycle* for H (see [63]) if it has an inverse σ^{-1} under the convolution product $*$ in $\mathrm{Hom}_k(H \otimes H, k)$, and satisfies the cocycle condition:

$$\sum \sigma(a_1b_1, c)\sigma(a_2, b_2) = \sum \sigma(a, b_1c_1)\sigma(b_2, c_2),$$

$$\sigma(a, 1) = \varepsilon(a) = \sigma(1, a)$$

for all $a, b, c \in H$.

Observe that any 2-cocycle σ on a group G can be extended to a Hopf 2-cocycle for kG.

Given a Hopf 2-cocycle σ for H, one can construct a new Hopf algebra structure $(H^\sigma, m^\sigma, 1, \Delta, \varepsilon, S^\sigma)$ on the coalgebra (H, Δ, ε). The new multiplication is given by

$$m^\sigma(a \otimes b) = \sum \sigma^{-1}(a_1, b_1) a_2 b_2 \sigma(a_3, b_3)$$

for all $a, b \in H$. The new antipode is given by

$$S^\sigma(a) = \sum \sigma^{-1}\big(a_1, S(a_2)\big) S(a_3) \sigma\big(S(a_4), a_5\big)$$

for all $a \in H$.

Suppose H is also co(quasi)triangular with universal R-form $\langle\ |\ \rangle : H \otimes H \to k$. Then H^σ is co(quasi)triangular with universal R-form $(\sigma \circ \tau)^{-1} * \langle\ |\ \rangle * \sigma$.

Observe that a twist J on H yields a Hopf 2-cocycle σ_J on H^* by identifying $H \otimes H$ with a subalgebra of $(H^* \otimes H^*)^*$. If H is finite-dimensional then the existence of a twist J for H is equivalent to the existence of a Hopf 2-cocycle σ_J for H^*.

EXAMPLE 1.6.3.

(1) If (H, R) is quasitriangular then R is a twist for $(H^{\mathrm{cop}}, R^\tau)$ and $((H^{\mathrm{cop}})^R, (R^\tau)^R) = (H, R)$. Dually, if $(H, \langle\ |\ \rangle)$ is coquasitriangular then $\langle\ |\ \rangle \circ \tau$ is a Hopf 2-cocycle for H.

(2) Let G be a non-Abelian finite group, $H = kG$ and J a twist for H. Since every cocommutative Hopf algebra is triangular with $R = 1 \otimes 1$ we have that $(kG)^J$ is a triangular Hopf algebra with non-trivial triangular structure $R^J := (J^\tau)^{-1} J$.

A dual Hopf 2-cocycle is a special case of what is known as a (dual) pseudo-cocycle. A (dual) pseudo-cocycle for H is an invertible element $J \in H \otimes H$ satisfying necessary and sufficient conditions that make $\Delta^J = J^{-1} \Delta J$ coassociative.

EXAMPLE 1.6.4 [180]. The Hopf algebra H_8 in Example 1.4.3 can be obtained from the group algebra of either D_4 or Q by pseudo-twists (where D_4 is the dihedral group and Q is the quaternion group). However H_8 has no non-trivial Hopf 2-cocycle or dual Hopf 2-cocycle, [229].

1.7. Constructing new Hopf algebras from known ones

Techniques in developing the structure theory for Hopf algebras entail constructing new Hopf algebras from known ones by using several methods. Some of these methods have been described already. We list them briefly.

(1) For two Hopf algebras H_1 and H_2 one can always define $H_1 \otimes H_2$, the tensor Hopf algebra with tensor structure maps.

(2) Given a Hopf algebra H one can twist either the multiplication via a 2-cocycle or the comultiplication via a dual 2-cocycle and obtain another Hopf algebra H_σ or H^J (Section 1.6).

(3) Given Hopf algebras H and K one can sometimes define an extension $K\#_\sigma^\tau H$ of H by K by using a 2-cocycle σ and a dual 2-cocycle τ (Remark 3.2.8).

(4) Another possible product of two Hopf algebras H_1 and H_2 satisfying certain conditions is the double crossed product $H_1 \bowtie H_2$. In particular $D(H) = H^{*\mathrm{cop}} \bowtie H$, the Drinfeld double of H, is defined for any finite-dimensional Hopf algebra H (Definition 4.3.8).

(5) In certain cases the smash product $A\#H$ can be turned into a Hopf algebra in a process called biproduct or bosonization (Theorem 4.4.6).

Part 2. Fundamental theorems

2.1. The fundamental theorem for Hopf modules

DEFINITION 2.1.1. Let H be a Hopf algebra. Then $(M, \cdot, \rho)$ is a *right H-Hopf module* if

(HM1) $(M, \cdot)$ is a right H-module.

(HM2) (M, ρ) is a right H-comodule.

(HM3) The compatibility condition

$$\rho(m \cdot h) = \sum m_0 \cdot h_1 \otimes m_1 h_2$$

for all $m \in M$, $h \in H$, is satisfied.

EXAMPLE 2.1.2. Let V be a vector space and $M = V \otimes H$. $(M, \cdot)$ is a right H-module by $(m \otimes x) \cdot h = m \otimes xh$. (M, ρ) is a right H-comodule by $\rho(m \otimes x) = \sum m \otimes x_1 \otimes x_2$ for all $m \in M$, $x \in H$. These make M into a right H-Hopf module. It is called a *trivial Hopf module*.

The fundamental theorem says that all Hopf modules are trivial with $V = M^{\mathrm{co}\,H}$. Explicitly:

THEOREM 2.1.3 [135]. *Let M be a right H-Hopf module. Then $M \cong M^{\mathrm{co}\,H} \otimes H$ as right H-Hopf modules, where $M^{\mathrm{co}\,H} \otimes H$ is a trivial Hopf module.*

The isomorphism $M \to M^{\mathrm{co}\,H} \otimes H$ is given by:

$$m \mapsto \sum m_0 \cdot \bigl(S(m_1)\bigr) \otimes m_2.$$

The crucial steps are to prove that $\sum m_0 \cdot (S(m_1)) \in M^{\mathrm{co}\,H}$ for all $m \in M$ and that this map is an H-module isomorphism.

This theorem is essential in the proof about the uniqueness of the integral and in the Nichols–Zoeller theorem (see Sections 2.2 and 2.5).

2.2. Integrals

The following important theorem about integrals is due to Larson and Sweedler:

THEOREM 2.2.1 [135]. *Let H be a finite-dimensional Hopf algebra. Then*

(1) $\int_H^l$ *and* $\int_H^r$ *are one-dimensional ideals (with generators $0 \neq t$ and $0 \neq t'$ respectively).*

(2) $H^* \rightharpoonup t = H = t' \leftharpoonup H^*$. *That is, H is a free left (right) H^*-module of rank* 1.

The major step in the proof of the theorem is to show that H^* is a right H-Hopf module. Now, H^* is a left H^*-module via left multiplication. The transpose of this action makes H^* into a right H-comodule as in Remark 1.3.4. Moreover, H^* is also a right H-module via $p \leftharpoonup h = \sum \langle S(h), p_2 \rangle p_1$ for all $h \in H$, $p \in H^*$. Once it is proved that this action and coaction satisfy the compatibility condition for Hopf modules (HM3), the fundamental Theorem 2.1.3 implies that $H^* \cong (H^*)^{\mathrm{co}\,H} \otimes H$. However, $(H^*)^{\mathrm{co}\,H} = \int_{H^*}^l$. Thus since $\dim H^* = \dim H$, this implies that $\int_{H^*}^l$ is one-dimensional.

It is quite surprising that the existence of $0 \neq t \in H$ so that $xt = \varepsilon(x)t$ for all $x \in H$ implies that H is finite-dimensional. This follows from the following theorem and its corollary

THEOREM 2.2.2 [220]. *Let H be any Hopf algebra and $0 \neq I$ a right ideal in H. Then $H^* \rightharpoonup I = H$.*

COROLLARY 2.2.3. *If a Hopf algebra H contains a non-zero finite-dimensional left (right) ideal then H is finite-dimensional.*

As we have seen in Example 1.2.8, when the Hopf algebra H is infinite-dimensional there may still exist $0 \neq T \in H^*$ so that $xT = \langle x, 1 \rangle T$ for all $x \in H^*$. (Recall that H^* is no longer a Hopf algebra.) It is straightforward to see that

$$xT = \langle x, 1 \rangle T \Leftrightarrow \langle T, h \rangle = \sum \langle T, h_2 \rangle h_1$$

for all $h \in H$, which is equivalent to T being a left H-comodule map.

Here again the dimension of the space of such T is 1, [219], which means uniqueness of the integral. Moreover,

THEOREM 2.2.4 [220]. *Let H be a Hopf algebra, then the following are equivalent:*

(1) *H is cosemisimple.*

(2) *There exists $T \in H^*$ so that $\langle T, 1 \rangle \neq 0$ and $xT = \langle x, 1 \rangle T$ for all $x \in H^*$.*

Some quantized coordinate algebras of algebraic groups G are known to be cosemisimple (e.g., $O_q(SL_n)$ where q is not a root of unity, [104]).

Back to finite-dimensional Hopf algebras H. The space $\int_H^l$ is a 2-sided 1-dimensional ideal of H. Hence for any $h \in H$, $th = \langle \alpha, h \rangle t$, where $\langle \alpha, h \rangle \in k$. Since the map α is an algebra map it follows that $\alpha \in G(H^*)$. This element α is called the left *distinguished grouplike* or left *modular element* of H (of course H is unimodular if and only if $\alpha = \varepsilon$). The right modular element of H is defined analogously and equals α^{-1}.

Another consequence of Theorem 2.2.1 is that any finite-dimensional Hopf algebra is a Frobenius algebra.

THEOREM 2.2.5 [183]. *Let H be a finite-dimensional Hopf algebra, $T \in \int_l^{H^*}$ and $t' \in \int_r^H$ such that $\langle T, t'\rangle = 1$. Then T is a Frobenius homomorphism with dual bases $(S(t'_1), t'_2)$ (that is, for all $h \in H$, $h = \sum S(t'_1)\langle T, t'_2 h\rangle$).*

An important application is the use of the integral in computations of traces (Tr) of linear endomorphisms.

THEOREM 2.2.6 [191]. *Let H be a finite-dimensional Hopf algebra, $t \in \int_H^l$ and $T' \in \int_{H^*}^r$ so that $\langle T', t\rangle = 1$. Then $\mathrm{Tr}(f) = \sum \langle T', S(t_2)\rangle f(t_1)$ for all $f \in \mathrm{End}_k(H)$.*

As a corollary we have

THEOREM 2.2.7. *Let H, T' and t be as in Theorem 2.2.6, then*

$$\mathrm{Tr}(S^2) = \langle \varepsilon, t\rangle \langle T', 1\rangle.$$

2.3. Maschke's theorem

A classical result about finite groups G is Maschke's theorem: kG is a semisimple algebra if and only if $|G|^{-1} \in k$. In Hopf algebraic terms: Let $t := \sum_{g \in G} g$, then $t \in \int_{kG}$ and $\varepsilon(t) = |G|$. Thus $|G|^{-1} \in k$ if and only if $\varepsilon(t) \neq 0$ in k. Hence kG is a semisimple algebra if and only if $\varepsilon(t) \neq 0$. Inspired by this, Larson and Sweedler showed the last statement to be true for any finite-dimensional Hopf algebra.

THEOREM 2.3.1 [135]. *Let H be any finite-dimensional Hopf algebra. Then H is semisimple if and only if $\varepsilon(\int_H^l) \neq 0$ (if and only if $\varepsilon(\int_H^r) \neq 0$).*

One direction is easily proved. Assume H is a semisimple algebra. Since $\operatorname{Ker} \varepsilon$ is an ideal of H, there exists a left ideal $0 \neq I$ of H so that $H = I \oplus \operatorname{Ker} \varepsilon$. Then it is shown directly that $I \subset \int_H^l$ and hence $\varepsilon(\int_H^l) \neq 0$.

For the converse choose $t \in \int_H^l$ so that $\varepsilon(t) = 1$. Let M be any left H-module and N be an H-submodule of M. The essence of the proof is to use the integral to produce an H-complement of N from a mere k-complement of N. Let $\pi : M \to N$ be any k-linear projection and define $\tilde{\pi} : M \to N$ by $\tilde{\pi}(m) = \sum t_1 \cdot \pi(S(t_2) \cdot m)$ for all $m \in M$. Then $\tilde{\pi}$ is an H-projection of M onto N.

REMARK 2.3.2. If H is semisimple then H is unimodular (since if $t \in \int_l^H$ with $\varepsilon(t) = 1$ and $t' \in \int_r^H$ with $\varepsilon(t') = 1$, then $t = t't = t'$).

2.4. The antipode

Let H be a finite-dimensional Hopf algebra with antipode S. The antipode, intertwined with the integral and the modular elements play a central role in the theory, [131,193].

THEOREM 2.4.1 [189]. *Let $\alpha \in G(H^*)$ be the left modular element of H and $g \in G(H)$ be the right modular element of H^*. Then for all $h \in H$*

$$S^4(h) = g(\alpha \rightharpoonup h \leftharpoonup \alpha^{-1})g^{-1} = \alpha \rightharpoonup (ghg^{-1}) \leftharpoonup \alpha^{-1}.$$

A simplified proof of this theorem appears in [209]. It depends on treating H as a Frobenius algebra as in Theorem 2.2.5. It can also be derived from the trace formula (Theorem 2.2.6). Another way of proving this formula appears in [72] via the Drinfeld double.

Since H is finite-dimensional and since grouplike elements are linearly independent, there exist only a finite number of powers of the grouplike elements a and α. Thus, a corollary of the previous theorem is:

THEOREM 2.4.2 [189]. *Let H be a finite-dimensional Hopf algebra, then S has finite order (necessarily even).*

It is worth mentioning that for any n there exists a Hopf algebra with an antipode of order $2n$; they are called the Taft algebras H_n.

EXAMPLE 2.4.3 [222]. Let $n \geqslant 1$ and $\omega \in k$ be a primitive n-th root of unity. Then $H_{n,\omega}$ is a Hopf algebra defined as follows. As an algebra $H_{n,\omega}$ is generated by g, x subject to the relations

$$g^n = 1, \quad x^n = 0, \quad \text{and} \quad xg = \omega g x.$$

The coalgebra structure of $H_{n,\omega}$ is determined by

$$\Delta(g) = g \otimes g \quad \text{and} \quad \Delta(x) = x \otimes g + 1 \otimes x.$$

Observe that $\dim H_{n,\omega} = n^2$; indeed $\{g^i x^j\}_{0 \leqslant i,j < n}$ is a linear basis for $H_{n,\omega}$. The antipode of $H_{n,\omega}$ is determined by $S(g) = g^{-1}$ and $S(x) = -xg^{-1}$. Hence S^2 is the algebra automorphism of $H_{n,\omega}$ determined by $S^2(g) = g$ and $S^2(x) = gxg^{-1} = \omega^{-1}x$ and so $S^{2n} = \mathrm{id}$.

2.5. The Nichols–Zoeller theorem

One of the fundamental results in the theory of finite-dimensional Hopf algebras is the Nichols–Zoeller theorem which is a positive answer to one of Kaplansky's conjectures, [116]. This conjecture was inspired by the famous "Lagrange theorem" in group theory, and boils down to it for $H = kG$.

THEOREM 2.5.1 [178]. *Let H be a finite-dimensional Hopf algebra and K a Hopf subalgebra of H. Then H is free as a right K-module. In particular,* $\dim K$ *divides* $\dim H$.

The theorem is actually more general. It says that every $M \in \mathcal{M}_K^H$ is free as a right K-module. Here $M \in \mathcal{M}_K^H$ if M satisfies the conditions in Definition 2.1.1 with a modified (HM1) in which K replaces H.

The essential part of the proof is based on the fact that any finite-dimensional Hopf algebra is a Frobenius algebra, on the fundamental theorem for Hopf modules and on the Krull–Schmidt theorem. By using these it is shown that if W is a finitely generated right K-module and V is a finitely generated faithful right K-module so that $W \otimes V \cong W^{\dim V}$ as right K-modules, then W is free over K. Next it is proved that $M \otimes H \cong M^{\dim H}$ for any $M \in \mathcal{M}_K^H$. Both imply that M is free over K.

A simple corollary of this theorem is

COROLLARY 2.5.2. *Let H be a finite-dimensional Hopf algebra over any field. Then the order of $G(H)$ divides* $\dim(H)$.

Another corollary is

COROLLARY 2.5.3. *Let H and K be as in Theorem* 2.5.1 *and assume H is semisimple, then so is K.*

2.6. Kac–Zhu theorem

The following theorem was conjectured in [116] and proved by Zhu using the Nichols–Zoeller theorem and extending work of G.I. Kac, [114], who worked in the framework of C^*-algebras.

THEOREM 2.6.1 [114,250]. *Let H be a Hopf algebra over an algebraically closed field k of characteristic* 0. *If* $\dim(H) = p$ *is prime then H is isomorphic to the group algebra $k\mathbb{Z}_p$.*

If $G(H)$ or $G(H^*)$ is non-trivial then by the Nichols–Zoeller theorem $H = k\mathbb{Z}_p$ and we are done. Otherwise, $S^4 = \mathrm{id}$ by the formula for S^4 (Theorem 2.4.1). If p is odd then it is easy to see that $\mathrm{Tr}(S^2) \neq 0$ which implies by Theorem 2.2.7 that H is semisimple. If $p = 2$ then semisimplicity can be proved directly. Now, the character theory of semisimple Hopf algebras and the appropriate "class equation" described in Thorem 5.1.6 imply that $H = k\mathbb{Z}_p$.

Part 3. Actions and coactions

3.1. Smash products, crossed products and invariants

One of the important topics in ring theory during the 70s was, so-called, non-commutative Galois theory. Specifically, if A is an algebra, G is a group of automorphisms of A and A^G

is the subalgebra of G-invariants, then the study concerned connections between the ideal structure of A and A^G. Much of the information is encoded in a generalized semidirect product, the skew group algebra $A * G$, [165], and the connection $A^G \subset A \subset A * G$.

Another setup with a similar flavor is: A is an algebra, L is a Lie algebra of derivations of A and $A^L = \{a \in A \mid l(a) = 0 \text{ for all } l \in L\}$. The analogue of $A * G$ is not so obvious. A third setup of the same nature is that of group-graded algebras, [48,176], where G is a finite group. The analogue of A^G and A^L is A_1 (where 1 is the identity element of G), but the analogue of $A * G$ is even less obvious.

It turns out that all three setups are unified by the fact that all the algebras A are H-module algebras for appropriate H, [47,21], and once this is understood there exists a generalization of the semidirect product which is $A * G$ for $H = kG$. This is the smash product $A\#H$. It plays a central role in the theory as did $A * G$ for non-commutative Galois theory.

DEFINITION 3.1.1. Let H be a Hopf algebra and let A be a left H-module algebra. Then the *smash product algebra* $A\#H$ is defined as follows:

(1) As vector spaces, $A\#H = A \otimes H$. Write $a\#h$ for $a \otimes h$.

(2) Multiplication is given by:

$$(a\#h)(b\#g) = \sum a(h_1 \cdot b)\#h_2k$$

for all $a, b \in A$, $h, g \in H$.

The smash product $A\#H$ is an algebra which contains A via $a \mapsto a\#1$, for $a \in A$, and H via $h \mapsto 1\#h$, for $h \in H$.

A generalization of $A\#H$ is, as for group actions, the notion of a crossed product. It is not necessary for the algebra A to be an H-module algebra, but only an *H-measured algebra*. That is, there exists a linear map $H \otimes A \to A$ given by $h \otimes a \mapsto h \cdot a$, such that $h \cdot 1 = \varepsilon(h)1$ and $h \cdot (ab) = \sum (h_1 \cdot a)(h_2 \cdot b)$ for all $h \in H$, $a, b \in A$.

DEFINITION 3.1.2. Let H be a Hopf algebra and A an H-measured algebra. Assume that σ is an invertible map (under convolution) in $\mathrm{Hom}_k(H \otimes H, A)$. The *crossed product* $A\#_\sigma H$ is $A \otimes H$ as a vector space. We write $a\#h$ for element in $A \otimes H$. Multiplication is given by:

$$(a\#h)(b\#g) = \sum a(h_1 \cdot b)\sigma(h_2, g_1)\#h_3 g_2$$

for all $a, b \in A$, $h, g \in H$.

THEOREM 3.1.3 [67,24]. *$A\#_\sigma H$ is an associative algebra with identity element $1\#1$ if and only if:*

(1) *A is a twisted H-module; that is, $1 \cdot a = a$ and*

$$h \cdot (g \cdot a) = \sum \sigma(h_1, g_1)(h_2 g_2 \cdot a)\sigma^{-1}(h_3, g_3)$$

for all $h, g \in H, a \in A$.

(2) σ *is a 2-cocycle; that is,* $\sigma(h,1)=\sigma(1,h)=\varepsilon(h)1$ *and*

$$\sum h_1\cdot\sigma(g_1,m_1)\sigma(h_2,g_2m_2)=\sum\sigma(h_1,g_1)\sigma(h_2g_2,m_2)$$

for all $h,g,m\in H$.

REMARK 3.1.4. Observe that for the special case of $A=k$ (which is an H-module algebra via $h\cdot\alpha=\varepsilon(h)\alpha$ for all $h\in H$, $\alpha\in k$) the definition of a 2-cocycle coincides with Hopf 2-cocycle (see Definition 1.6.2). In this case we can construct the crossed product $k\#_\sigma H$ which is denoted also by ${}_\sigma H$ and is called a *twisted Hopf algebra* (reminiscent of twisted group rings; it is not a Hopf algebra though).

One can similarly define crossed product for right actions and can obtain a right twist of H. The twist for H defined in Section 1.6 has then the form ${}_\sigma H_{\sigma^{-1}}$ (which is a Hopf algebra).

Let A be an H-module algebra. By Remark 1.3.8 if $t\in\int_l^H$ then $t\cdot A\subset A^H$ and equality holds when H is semisimple. A^H is connected to $A\#H$ in several ways:

[20]. Let H be finite-dimensional. Then $A^H\cong(\mathrm{End}_{A\#H}(A))^{\mathrm{op}}$ as algebra (op = opposite algebra).

[44]. Let H be finite-dimensional and assume $t\cdot A=A^H$. Then there exists an idempotent $e\in A\#H$ such that $e(A\#H)e=A^He\cong A^H$ as algebras.

The close relationship between A^H and $A\#H$ can be expressed in terms of a Morita context.

THEOREM 3.1.5 [43,44]. *Let* H *be a finite-dimensional Hopf algebra,* $0\neq t\in\int_l^H$ *and* α *the distinguished grouplike element associated with it. Consider* A *as a left (right)* A^H*-module via left (right) multiplication, as a left* $A\#H$*-module via* $(a\#h)\cdot b=a(h\cdot b)$ *and as a right* $A\#H$*-module via* $b\cdot(a\#h)=[S^{-1}(\alpha\rightharpoonup h)]\cdot(ba)$*, for all* $a,b\in A$, $h\in H$. *Then* $M:={}_{A^H}A_{A\#H}$ *and* $N:={}_{A\#H}A_{A^H}$ *together with the maps:*

$$[\ ,\ ]: A_{A^H}\otimes A\to A\#H\quad\text{given by } [a,b]=(a\#t)(b\#1),$$

$$(\ ,\ ): A_{A\#H}\otimes A\to A^H\quad\text{given by } (a,b)=t\cdot(ab)$$

give a Morita context for A^H *and* $A\#H$.

This extends earlier work on group actions by [37]. Note that $(A,A)=t\cdot A$.

COROLLARY 3.1.6. *If both* $t\cdot A=A^H$ *and* $(A\#t)(A\#1)=A\#H$ *then* A^H *is Morita-equivalent to* $A\#H$.

The surjectivity of the Morita map [,] has strong implications as will be seen in the next section.

Using ring theoretic properties deduced from the theory of Morita contexts we have:

COROLLARY 3.1.7 [44]. *The following are equivalent*:

(1) *$A\#H$ is a prime ring.*
(2) *A is a left and right faithful $A\#H$-module and A is a prime ring.*
(3) *If A is left and right faithful $A\#H$-module then A^H is a primitive ring if and only if $A\#H$ is a primitive ring.*

As for group actions, semiprimeness of $A\#H$ insures that the ideal structures of A and A^H are closely related. If A is semiprime it is known that $A\#kG$ is semiprime if $|G|^{-1} \in k$, [94], in this case kG is semisimple. This suggests the still open generalized Maschke-type theorems:

QUESTION 3.1.8.

(1) [43]. If H is semisimple and A is a semiprime H-module algebra, is $A\#H$ semiprime?
(2) [25]. More generally: under the conditions of (1), is $A\#_\sigma H$ semiprime for any 2-cocycle σ?
(3) [42]. In a similar spirit, let $B \bowtie H$ be a double crossproduct of B and H ([153]). If B and H are semisimple Hopf algebras is $B \bowtie H$ semisimple?

The answer to the first and second questions is positive in the following cases: (a) If A is also Artinian, [43,25]. (b) If A is any k-affine PI algebra and the characteristic of $k = 0$ or with a suitable assumption on the PI degree if the characteristic of $k > 0$, [138]. (c) If the action of H on A is inner (see Definition 3.4.5), [24,26]. (d) If H is commutative (for then H is essentially $(kG)^*$ and the answer is positive in this situation by [47]). (e) If H is pointed cocommutative, [39,172].

Question (3) has a positive answer when the field k is algebraically closed of characteristic zero. It also has a positive answer over arbitrary fields for the Drinfeld double, namely, for $B = H^{*\mathrm{cop}}$, [192].

A dual notion of the surjectivity of the form $(\, , \,)$ in the Morita context is better suited for infinite-dimensional Hopf algebras H and right H-comodule algebras A. This is the notion of a total integral, [61].

DEFINITION 3.1.9 [61]. Let A be a right H-comodule algebra. Then a (right) *total integral* for A is a right H-comodule map $\Phi : H \to A$ such that $\Phi(1) = 1$.

REMARK 3.1.10 [43,61]. When H is finite-dimensional then surjectivity of the Morita map $(\, , \,)$ on an H-module algebra A is equivalent to the existence of a total integral for A considered as a right H^*-comodule algebra.

REMARK 3.1.11. Note that the existence of a total integral for k is equivalent to H being cosemisimple by Theorem 2.2.4.

The existence of a total integral is also related to Galois extensions (see next section) by the following:

REMARK 3.1.12 [61]. If $A^{\operatorname{co} H} \subset A$ is a faithfully flat H-Galois extension then there exists a total integral for A.

3.2. Galois extensions

Crossed products are examples of Hopf Galois extensions. The definition of these extensions has its roots in the Chase, Harrison and Rosenberg, [37], approach to Galois theory for groups acting on commutative rings. The definition is given in terms of coactions.

DEFINITION 3.2.1. Let (A, ρ) be a right H-comodule algebra. Then the extension $A^{\operatorname{co} H} \subset A$ is *right H-Galois* if the map

$$\beta : A \otimes_{A^{\operatorname{co} H}} A \to A \otimes H \quad \text{given by} \quad a \otimes b \mapsto (a \otimes 1)\rho(b)$$

is bijective.

EXAMPLE 3.2.2.
1. Classical Galois field extensions are examples of Hopf–Galois extensions, [165].
2. Let a Hopf algebra H be a right H-comodule algebra via Δ. Then $H^{\operatorname{co} H} = k$ and $k \subset H$ is right H-Galois with $\beta^{-1} : H \otimes H \to H \otimes H$ given by $x \otimes y \mapsto \sum x(S(y_1)) \otimes y_2$.
3. Let H be a finite-dimensional Hopf algebra, K a normal Hopf subalgebra and $\overline{H} = H/K^+H$. Then $K \subset H$ is $\overline{H}$-Galois. This is true since $H^{\operatorname{co} \overline{H}} = K$ by Proposition 1.4.2, and the Galois map $\beta : H \otimes_K H \to H \otimes \overline{H}$ has as an inverse defined similarly to β^{-1} of part (2).
4. Let $B = A \#_\sigma H$ be any crossed product. Then B is an H-comodule algebra via $\rho = \mathrm{id}_A \otimes \Delta$ and $B^{\operatorname{co} H} = A \#_\sigma k1 \cong A$. The fact that $A \subset B$ is Galois follows since

$$(A \#_\sigma H) \otimes_A (A \#_\sigma H) \cong (A \#_\sigma H) \otimes H$$

and the Galois map has the form $\mathrm{id}_A \otimes \beta$ with an inverse $\mathrm{id}_A \otimes \beta^{-1} : (A \#_\sigma H) \otimes H \to (A \#_\sigma H) \otimes H$ where β, β^{-1} are defined as in (2) above.

Recall Example 1.1.18(2). If $A = \sum_{g \in G}^{\oplus} A_g$ then A is a kG-comodule algebra by $x \mapsto x \otimes g$ for all $x \in A_g$. In this case:

THEOREM 3.2.3 [234]. *$A_1 \subset A$ is kG-Galois if and only if $A_g A_h = A_{gh}$ for all $g, h \in G$ (A is then called strongly graded).*

Of special interest is the case where $A^{\operatorname{co} H} = k$.

DEFINITION 3.2.4. If the extension $k \subset A$ is H-Galois then A is called an *H-Galois object* (sometimes the extension is called an H-Galois extension).

There also exists a two-sided analogue which plays an important role in the representation theory of H.

DEFINITION 3.2.5. Let H, L be Hopf algebras and let A be an (H, L)-bicomodule algebra. Then A is called an (H, L)-bi-Galois object if A is both a left H-Galois object and a right L-Galois object.

When $A^{\operatorname{co} H} \subset A$ is right H-Galois then $C_A(A^{\operatorname{co} H})$, the centralizer of $A^{\operatorname{co} H}$ in A, becomes a right H-module algebra via the so-called Miyashita–Ulbrich action.

DEFINITION 3.2.6 [235,68]. Let $A^{\operatorname{co} H} \subset A$ be a right H-Galois extension with Galois map β. Define the *right Miyashita–Ulbrich action* as follows:

For $h \in H$ write $\beta^{-1}(1 \otimes h) = \sum a_i \otimes b_i$ and then define

$$x \leftarrow h = \sum a_i x b_i, \quad \text{for all } x \in C_A\left(A^{\operatorname{co} H}\right).$$

THEOREM 3.2.7. *The above $\leftarrow$ defines a right action of H on $C_A(A^{\operatorname{co} H})$. The algebra of invariants of this action is $Z(A)$, the center of A.*

Back to crossed products. While all crossed products are Galois extensions, in order for a Galois extension $A^{\operatorname{co} H} \subset A$ to become a crossed product $A^{\operatorname{co} H} \#_\sigma H$, the extension must have the *normal basis property*, that is

$$A \cong A^{\operatorname{co} H} \otimes H$$

as a left $A^{\operatorname{co} H}$-module and a right H-comodule. This is the essence of the work done by [129,67,25].

REMARK 3.2.8. Normal Hopf subalgebras K of finite-dimensional H (Example 3.2.2(3)) are not only Galois but $K \subset H$ satisfies the normal basis property, [207,66], hence $H = K \#_\sigma \overline{H}$ as algebras.

In fact, $H = K \#_\sigma^\tau \overline{H}$ as Hopf algebras where τ is a dual cocycle which twists the coalgebra structure of $K \otimes H$.

Progress in the theory of such extensions has been made in the case that $K = (kG)^*$ and $\overline{H} = kG'$, where G and G' are finite groups. These extensions can be described in terms of actions, coactions, a cocycle and a dual cocycle relating the groups G and G' and satisfying certain compatibility conditions. Such groups were considered in [225] and were named *matched pairs*. The subject was further discussed in [146,148,5,3].

Other instances in which Galois extensions $A^{\operatorname{co} H} \subset A$ are crossed products $A \cong A^{\operatorname{co} H} \#_\sigma H$ are:

THEOREM 3.2.9.

(1) [128]. *If H is a finite-dimensional Hopf algebra and A is an H-Galois object then A is a crossed product over k.*

(2) [196]. *If H is a finite-dimensional Hopf algebra such that A and $A \otimes H^*$ satisfy Krull–Schmidt for projectives and $A_{A^{\operatorname{co} H}}$ is free then A is a crossed product over $A^{\operatorname{co} H}$.*

As a corollary, if H is a finite-dimensional Hopf algebra over an algebraically closed field k, any H-Galois extension $A^{\mathrm{co}\,H} \subset A$ with $A^{\mathrm{co}\,H}$ a local ring is a crossed product over $A^{\mathrm{co}\,H}$.

(3) [19]. *If H and A satisfy the equivalent conditions of the following Theorem* 3.2.12 *and H is connected (as a coalgebra) then A is a crossed product over $A^{\mathrm{co}\,H}$.*

When H is finite-dimensional then a left H-action on A gives rise to a right H^*-coaction on A so that $A^H = A^{\mathrm{co}\,H^*}$. Thus it makes sense to ask when is the extension $A^H \subset A$ right H^*-Galois. Here are some equivalences:

THEOREM 3.2.10 [44,129]. *Let H be a finite-dimensional Hopf algebra and A a left H-module algebra. Then the following are equivalent:*

(1) *$A^H \subset A$ is right H^* Galois.*
(2) *The Morita map* [,] *of Theorem* 3.1.5 *is surjective.*
(3) *For any $M \in {}_{A\#H}\mathcal{M}od$, $M \cong A \otimes_{A^H} M^H$ as an $A\#H$-module (via $a \otimes m \mapsto a \cdot m$).*
(4) *A is a generator for ${}_{A\#H}\mathcal{M}od$.*

When $A = D$ is a division algebra more can be said.

THEOREM 3.2.11 [44]. *If $A = D$ is a division algebra then the equivalent conditions of Theorem* 3.2.10 *are also equivalent to each of the following:*

(1) *The right (left) dimension of D over D^H equals* $\dim H$.
(2) *$D\#H$ is a simple ring.*
(3) *D is a faithful left or right $D\#H$-module.*
(4) *$D \cong D^H \#_\sigma H^*$.*

An extension of Theorem 3.2.10, which also generalizes a theorem for algebraic groups on induction of modules and affine quotients is:

THEOREM 3.2.12 [206]. *Let H be a Hopf algebra with bijective antipode and A a right H-comodule algebra. Then the following are equivalent:*

(1) *$A^{\mathrm{co}\,H} \subset A$ is a right H-Galois extension and A is a faithfully flat left (or right) $A^{\mathrm{co}\,H}$-module.*
(2) *The Galois map β is surjective and A is an injective H-comodule.*
(3) *The map $\mathcal{M}od_{A^{\mathrm{co}\,H}} \to \mathcal{M}_A^H$ given by $N \mapsto N \otimes_{A^{\mathrm{co}\,H}} A$ is an equivalence (where $\mathcal{M}_A^H$ is the subcategory of A-modules in the category of H-comodules,* [60]).

The ideal structures of $A^{\mathrm{co}\,H}$ and A are closely related for some Galois extensions.

THEOREM 3.2.13 [172]. *Let A and H be as in Theorem* 3.2.12 *and assume $A^{\mathrm{co}\,H} \subset A$ is a right H-Galois extension and A is a faithfully flat $A^{\mathrm{co}\,H}$-module. Then*

(1) *There exists a bijection between the following sets of ideals:*

$$\{I \subset A^{\mathrm{co}\,H} \text{ satisfying } IA = AI\} \underset{\Psi}{\overset{\Phi}{\rightleftarrows}} \{J \subset A \text{ which are } H\text{-subcomodules}\},$$

$$\Phi : I \to IA \quad \text{and} \quad \Psi : J \to J \cap A^{\mathrm{co}\,H}.$$

(2) *If H is also finite-dimensional then there is a bijective correspondence between H-equivalent primes of $A^{\mathrm{co}\,H}$ and H^*-equivalent primes of A (the equivalence relation is quite natural; see* [172] *for the exact definition).*

If $A^{\mathrm{co}\,H} \subset Z(A)$, more can be said.

THEOREM 3.2.14. *Let H be a finite-dimensional Hopf algebra and A an H-module algebra. If A/A^H is H^*-Galois and $A^H \subset Z(A)$ then:*

(1) *By* [13,129,61], *A is a faithfully flat right $A^{\mathrm{co}\,H^*}$-module and the Morita map $(\ ,\ )$ is surjective. In particular A^H and $A\#H$ are Morita equivalent,* [44].

(2) [49]. *For every ideal I of $A\#H$*

$$I = (I \cap A)\#H = \big((I \cap A^H)A\big)\#H.$$

Consequently, if A is an H-Galois object then $A\#H$ is a simple ring.

(3) [172]. *If H^* is pointed and $G = G(H^*)$, then there exists a bijection*

$$\mathrm{Spec}(A)/G \to \mathrm{Spec}\big(A^{\mathrm{co}\,H^*}\big)$$

given by $[P] \mapsto P \cap A^{\mathrm{co}\,H^*}$ (where $\mathrm{Spec}(A)/G$ is the set of G-orbits $[P]$ in $\mathrm{Spec}(A)$).

3.3. Duality theorems

Let A be a G-graded algebra, where G is a finite group; then A is a left $(kG)^*$-module algebra (Example 1.3.5) and $A\#(kG)^*$ is a left kG-module where kG acts trivially on A and by $\rightharpoonup$ (Definition 1.3.1) on kG. The first "duality" theorem was proved in this setup and considerably generalized independently by [25] and [239].

THEOREM 3.3.1 [47]. *Let G be a finite group and A be a G-graded algebra. Then*

$$\big(A\#(kG)^*\big)\#kG \cong M_n(A),$$

where $M_n(A)$ is the algebra of $n \times n$ matrices over A.

The following theorem deals with a general H. Again H^* acts on $A\#H$ as for the G-graded case.

THEOREM 3.3.2 [25,239]. *Let H be a finite-dimensional Hopf algebra and A an H-module algebra. Then*

$$(A\#H)\#H^* \cong M_n(A).$$

This theorem is in fact a corollary of the most general result:

THEOREM 3.3.3 [25]. *Let H be a Hopf algebra and U a Hopf subalgebra of H^0 such that both H and U have bijective antipodes, and assume that U satisfies the RL-condition with respect to H (see* [25] *for the definition). Let A be a U-comodule algebra and define an action of H on A by: $h \cdot a = \sum \langle a_1, h \rangle a_0$. Then*

$$(A\#H)\#U \cong A \otimes (H\#U).$$

In particular:

COROLLARY 3.3.4 [25]. *Specifying Theorem* 3.3.3 *to a residually finite-dimensional Hopf algebra H and U a dense Hopf subalgebra of H^0, then*

$$(A\#H)\#U \cong A \otimes L,$$

where L is a dense subring of $\mathrm{End}_k(H)$.

The duality theorem has been reproved using other methods. For example, by using the right smash product of a comodule algebra with a Hopf algebra, [16]. It has been generalized by using other constructions. For example, by using "opposite smash products" for right H-comodule algebras, [127].

There also exist duality theorems for crossed coproducts of Hopf algebras coacting (weakly) on coalgebras, [55].

The duality theorem has its origin in operator algebra theory for actions and coactions of locally compact groups on von Neumann algebras, [175], and Kac algebras, [75]. It has been generalized to weak Kac algebras, [181].

3.4. Analogues of two theorems of E. Noether; inner and outer actions

A classical theorem of E. Noether on invariants states that if A is a commutative k-affine algebra and G is a finite group of automorphisms on A then A^G is k-affine. This theorem has the following generalizations. Part (1) of the generalization is a consequence of a result of Grothendieck, [59, p. 309]. A more explicit proof which uses determinants is due to [92].

THEOREM 3.4.1 [92]. *Let H be a finite-dimensional cocommutative Hopf algebra and let A be a commutative H-module algebra. Then*

(1) *A is integral over A^H.*
(2) *If A is k-affine then so is A^H.*

This was generalized as follows:

THEOREM 3.4.2 [51]. *Let (H, R) be a triangular semisimple Hopf algebra in characteristic* 0. *Let A be an H-commutative H-module algebra (see Definition* 4.3.3), *then*:

(1) *A is integral over A^H.*

(2) *A is a PI algebra.*
(3) *If A is k-affine then so is* A^H.

Another proof of the above theorem follows from [77] who proved that H in Theorem 3.4.2 is a twisting of a group algebra kG, and from [170] who showed various algebra properties invariant under twisting.

A result similar in flavor is:

THEOREM 3.4.3 [166]. *Let H be a finite-dimensional Hopf algebra and A a left Noetherian H-module algebra so that the Morita map $(\ ,\ )$ is surjective. If A is k-affine then so is A^H.*

The following is an infinite-dimensional Noether-type theorem for coactions.

THEOREM 3.4.4 [74]. *Let $A = A_0 \oplus A_1 \oplus \cdots$ be a right Noetherian $\mathbb{N}$-graded algebra with $A_0 = k$. Let H be a cosemisimple Hopf algebra and suppose A is a right H-comodule so that each A_i is a subcomodule of A. Then the subalgebra $A^{\mathrm{co}\,H}$ is k-affine.*

The classical *Noether–Skolem theorem* asserts that if A is a simple Artinian ring with center Z and $B \supset Z$ is a simple subalgebra of A with Z-finite dimension then any isomorphism of B into A extends to an inner automorphism of A. This can be generalized to Hopf algebras as in the next theorem. First a definition.

DEFINITION 3.4.5. Let C be a coalgebra and $B \subset A$ be algebras. Consider a left action $C \otimes B \to A$, given by $c \otimes b \mapsto c \cdot b$ which measures B to A (that is $x \cdot (ab) = \sum (x_1 \cdot a)(x_2 \cdot b)$ and $x \cdot 1 = \varepsilon(x)1$ for all $x \in C$, $a, b \in B$). Then the *measuring is inner* if there exists a convolution invertible map $u \in \mathrm{Hom}(C, A)$ such that for all $x \in C$, $b \in B$,

$$x \cdot b = \sum u(x_1) b u^{-1}(x_2).$$

This definition boils down to the usual definition of inner automorphisms and inner derivations in the appropriate settings. For if σ is an automorphism of A and there exists $a \in A$ such that $\sigma \cdot x = axa^{-1}$ for all $x \in A$ then define $u(\sigma) = a$ and extend the definition to $C := k\langle\sigma\rangle$. Similarly, if a derivation δ satisfies $\delta(x) = ax - xa$ for some $a \in A$ all $x \in A$, then define $u(\delta) = a$, $u(1) = 1$ and extend it to $C := k\langle 1, \delta\rangle$.

EXAMPLE 3.4.6. The left adjoint action of H on A is inner (with $u = \mathrm{id}$ and $u^{-1} = S$).

THEOREM 3.4.7 [126]. *Let A be a simple Artinian algebra with center k and let B be a finite-dimensional simple subalgebra of A. Let C be a coalgebra which measures B to A. Assume also that $B^{\mathrm{op}} \otimes D^*$ is a simple algebra for each simple subcoalgebra D of C. Then the measuring is inner.*

Important applications of this theorem are:

COROLLARY 3.4.8. *Let A be a simple Artinian algebra with center k and B be a finite-dimensional simple subalgebra of A. Let C be a pointed coalgebra. Then any measuring of B to A by C is inner.*

The following was proved independently in [144].

COROLLARY 3.4.9. *Let A be a simple algebra which is finite-dimensional over its center. Let C be a coalgebra. Then any measuring of A by C is inner.*

If H is a Hopf algebra and A is an H-module algebra we say that H is inner on A if the measuring of A by H is inner.

If G is a group of automorphisms of A then the set N of all $g \in G$ which are inner automorphisms is a normal subgroup of G and kN is the maximal sub-Hopf algebra of kG which is inner on A. Moreover, G/N acts on A^N. This can be generalized as follows:

THEOREM 3.4.10. *Let H be a finite-dimensional pointed Hopf algebra and A an H-module algebra. Then*

(1) [145]. *There exists a unique sub-Hopf algebra H_{inn} of H which is inner on A and is maximal with respect to this property.*
(2) [208]. *If H is also cocommutative then H_{inn} is a normal sub-Hopf algebra and $\overline{H} = H/H(H_{\mathrm{inn}})^+$ acts on $A^{H_{\mathrm{inn}}}$.*

When a group G acts by automorphisms on a prime ring A then some $g \in G$ may fail to be inner on A, but extending the action of G to Q, the symmetric Martindale ring of fractions of A, it may be inner on Q. When this happens g is called X-inner (where X stands for Kharchenko, see [123]). Extending these ideas to Hopf algebra actions on algebras, [41], constructed H-quotient rings on which the action may be inner.

The best results are obtained for pointed Hopf algebras H; it was proved that the action of H can be extended to Q as for group actions, [167,171]. In this case one can define X-inner actions to be actions which become inner when extended to Q.

While for actions of groups by automorphisms (or of restricted Lie algebras by derivations) we define X-outer actions as those for which the only inner automorphism (derivation) is trivial, it is rather unsatisfactory to extend this definition for general Hopf algebras.

The following definition for outer actions of pointed Hopf algebras is due to [161]. Let H be a pointed Hopf algebra acting on a prime algebra A, let Q be the symmetric Martindale ring of quotients of A and let K be the center of Q. Set $E := C_{Q\#H}(A)$, the centralizer of A in $Q\#H$.

DEFINITION 3.4.11. Let H be a pointed Hopf algebra acting on a prime algebra A. Then the action of H on A is *X-outer* if $E = K$.

Previous results obtained in [124] for X-outer actions of groups and restricted Lie algebras can be generalized for X-outer actions of pointed Hopf algebras. We list some of them below.

THEOREM 3.4.12 [161]. *Let H be a finite-dimensional pointed Hopf algebra acting on a prime algebra A. If the action of H on A is X-outer, then*:

(1) *$Q\#H$ is a prime algebra, and if Q is H-simple then $Q\#H$ is a simple algebra.*
(2) *If $A\#H$ is a prime algebra and A is H-simple then $A\#H$ is a simple algebra.*
(3) *A^H is a prime algebra.*
(4) *$C_Q(A^H) = K$, where $C_Q(A^H)$ is the centralizer of A^H in Q.*
(5) *A and A^H satisfy the same multilinear identities.*

Based on these results, a Galois type correspondence theory for X-outer actions of finite-dimensional pointed Hopf algebras on prime algebras was proved in [247,244,245].

THEOREM 3.4.13. *Let k be a field of characteristic zero and let H be a finite-dimensional pointed Hopf algebra over k acting on a prime algebra A such that the action is X-outer. Consider A and $K\#H$ as subalgebras of $Q\#H$. For a rationally complete intermediate subalgebra U of A, let $\Phi(U)$ denote the centralizer of U in $K\#H$, and for a right coideal subalgebra Λ of $K\#H$ containing K, let R^Λ denote the centralizer of Λ in A. Then $U = R^{\Phi(U)}$ and $\Lambda = \Phi(R^\Lambda)$. Thus $U \to \Phi(U)$ determines a one to one correspondence between the set of rationally complete intermediate subalgebras U of A and the set of right subcomodule algebras of $K\#H$ containing K.*

Part 4. Categories of representations of Hopf algebras

The abundance of structures related to Hopf algebras give rise to a number of new constructions. Many constructions are related to quantum groups or to solutions of the quantum Yang–Baxter equation, some are related to Hopf algebras in categories and others are deformations of known objects.

4.1. Rigid tensor categories and Hopf algebras

An important point of view of Hopf algebras arises from categorical considerations.

Let $(A, m, 1)$ be a finite-dimensional unital associative algebra, and let $\mathcal{C} := \mathrm{Rep}(A)$ be the category of finite-dimensional left A-modules. Clearly, $\mathcal{C}$ is a k-linear Abelian category. Also, the algebra A can be reconstructed from $\mathcal{C}$ and the forgetful functor $\mathcal{C} \to \mathrm{Vec}$.

Suppose that $A = H$ is a Hopf algebra. Then $\mathcal{C}$ turns out to have a rich structure as seen in the following:

(1) Since ε is an algebra map, k becomes an object of $\mathcal{C}$ by
$$a \cdot x := \varepsilon(a)x$$
for any $a \in H$, $x \in k$.

(2) Since Δ is an algebra map it follows that for any $V, W \in \mathcal{C}$, $V \otimes W$ becomes an object of $\mathcal{C}$ by:
$$a \cdot (v \otimes w) := \sum a_1 \cdot v \otimes a_2 \cdot w$$
for any $a \in H$, $v \in V$ and $w \in W$.

(3) The coassociativity of Δ implies that the standard associativity isomorphism $(U \otimes V) \otimes W \to U \otimes (V \otimes W)$ is an H-module map.
(4) Since S is an anti-algebra isomorphism it follows that for any $V \in \mathcal{C}$, its linear dual V^* becomes an object of $\mathcal{C}$ by:

$$(a \cdot f)(v) := f\big(S(a) \cdot v\big)$$

for any $a \in H$, $v \in V$ and $f \in V^*$.
(5) Let $V \in \mathcal{C}$ then it is straightforward to verify that the maps

$$ev_V : V^* \otimes V \to k, \quad f \otimes v \mapsto f(v)$$

and $coev_V : k \to V \otimes V^*$ determined by

$$1 \mapsto \sum_i v_i \otimes f_i$$

are in fact H-module maps, where $\{v_i\}$ and $\{f_i\}$ are dual bases of V, V^* respectively.
The above indicate that $(\mathrm{Rep}(H), \otimes, k, a, l, r)$ is a rigid tensor (also called monoidal) category where k is the unit object and a, l, r are the standard associativity and unit constrains as in Vec. (See [150,151,200] for definitions and properties of such categories.)

In the following Tannaka–Krein type theorem it is shown that corresponding conditions on $\mathrm{Rep}(A)$ when A is an algebra, induce a Hopf algebra structure on A (see, e.g., [89]):

THEOREM 4.1.1. *Let* $(A, m, 1)$ *be an algebra and* $\mathcal{C} := \mathrm{Rep}(A)$. *Then there exists a bijection between*:
(1) *rigid tensor structures on* $\mathcal{C}$, *together with a compatible tensor structure on the forgetful functor* Forget : $\mathcal{C} \to$ Vec, *and*
(2) *Hopf algebra structures on* $(A, m, 1)$.

REMARK 4.1.2. If we omit the rigidity requirement from $\mathcal{C}$ then the corresponding structure on A is that of a bialgebra.

Similarly, the category ${}^H\mathcal{M}$ of left H-comodules has a structure of a rigid tensor category as well. It is thus useful to think of a Hopf algebra as an algebra (coalgebra) whose representation category (the category of H-comodules) has a structure of a rigid tensor category. The Hopf algebra structure is unique only up to twisting as discussed in the following Theorem 4.1.4.

DEFINITION 4.1.3. Let H and L be Hopf algebras. Then H and L are monoidally co-Morita equivalent (or Morita–Takeuchi equivalent) if the categories ${}^L\mathcal{M}$ and ${}^H\mathcal{M}$ are equivalent as tensor categories.

If H can be obtained from L by twisting the algebra structure then ${}^L\mathcal{M}$ and ${}^H\mathcal{M}$ are monoidally co-Morita equivalent. But more is true:

THEOREM 4.1.4 [204]. *Let H and L be Hopf algebras. Then:*

(1) *H and L are monoidally co-Morita equivalent $\Leftrightarrow$ there exists an (H, L)-bi-Galois object M.*
(2) *If a crossed product over k is an (H, L)-bi-Galois object then H can be obtained from L by twisting the algebra structure.*
(3) *As a corollary of (2) and Theorem 3.2.9(1) if H and L are finite-dimensional monoidally co-Morita equivalent then H can be obtained from L by twisting the algebra structure.*

The following is a very useful criteria for co-Morita equivalence.

THEOREM 4.1.5 [149]. *Suppose K is a sub-Hopf algebra of a Hopf algebra H. If I, J are Hopf ideals in K so that $I = g \rightharpoonup J \leftharpoonup g^{-1}$ for some $g \in G(K)^*$, then $H/(I)$ and $H/(J)$ are monoidally co-Morita equivalent* (*where (I) and (J) are the Hopf ideals in H generated by I and J respectively*).

The striking property of quasitriangular (coquasitriangular) Hopf algebras H is that the tensor category $\mathcal{C} = \operatorname{Rep}(H)$ (${}^H\mathcal{M}$) of their category of representations (comodules) is also *braided*. Namely, just like in group representations, there is a natural isomorphism between $X \otimes Y$ and $Y \otimes X$ for any two representations (comodules) X, Y (this is not necessarily true in general). Specifically, if (H, R) is quasitriangular where $R = \sum R^1 \otimes R^2$, we can define for any $X, Y \in \mathcal{C}$ an isomorphism

$$c_{X,Y} : X \otimes Y \to Y \otimes X$$

given by:

$$x \otimes y \mapsto \tau\big(R \cdot (x \otimes y)\big) = \sum R^2 \cdot y \otimes R^1 \cdot x.$$

Analogously we can define for comodules over a coquasitriangular Hopf algebra $(H, \langle | \rangle)$

$$x \otimes y \mapsto \sum \langle x_1 | y_1 \rangle y_0 \otimes x_0$$

for all $x \in X$, $y \in Y$.

The map $c_{X,Y}$ above is in fact an H-module map by property (QT2) of R (see Definition 1.5.1). The collection $c := \{c_{X,Y} \mid X, Y \in \mathcal{C}\}$ determines a braided structure on $\mathcal{C}$. The meaning of (QT1) in the definition is that to move Z to the left or to the right of $X \otimes Y$ is the same thing as to permute X, Y separately with Z. (The reader is referred to [112,113] for a detailed discussion of braided tensor categories.)

If (H, R) is triangular, then R determines a *symmetric* structure on $\mathcal{C}$. The meaning of this is that the composition $X \otimes Y \overset{c_{X,Y}}{\to} Y \otimes X \overset{c_{Y,X}}{\to} X \otimes Y$ is the identity for every X, Y. This is a generalization of the standard flip τ in the category of representations of cocommutative Hopf algebras (e.g., the universal enveloping algebra $U(\mathfrak{g})$ or the group algebra kG).

REMARK 4.1.6. If $\mathcal{C}$ is a braided tensor category then the braiding gives rise to a representation of the braid group $\mathbb{B}_n$ on $V^{\otimes n}$, $V \in \mathcal{C}$, in the following sense: Each generator σ_i of $\mathbb{B}_n$ acts on $V^{\otimes n}$ by applying the braiding $c_{V,V}$ to the $(i, i+1)$ component of $V^{\otimes n}$. Explicitly, the representation $\rho_n : \mathbb{B}_n \to \mathrm{Aut}(V^{\otimes n})$ is given by:

$$\rho_n(\sigma_i) = \mathrm{id}^{\otimes i-1} \otimes c_{V,V} \otimes \mathrm{id}^{\otimes n-i-1}.$$

If the category is symmetric then it gives rise to a non-trivial representation of the symmetric group S_n on the n-th tensor product of objects of $\mathcal{C}$.

REMARK 4.1.7. Theorem 4.1.1 extends to a bijective assignment between (quasi)triangular structures on H, and (braided) symmetric rigid tensor structures on $\mathcal{C}$.

Going a step further, just as algebras can be reconstructed from their category of representations one can reconstruct a ((co)quasitriangular) Hopf algebra from a rigid tensor category which admits a fiber (= exact, faithful and tensor) functor to the category of vector spaces. This problem has been studied extensively by many authors who have considered various possible set-ups and have accordingly reconstructed various structures (cf. [200,58,187,236,112,154,155,157,202]). Here is one example:

THEOREM 4.1.8 [236]. *Let $\mathcal{C}$ be a small Abelian rigid tensor category and F a k-linear fiber functor to* Vec. *Then there exists a Hopf algebra H such that $\mathcal{C}$ is equivalent to ${}^H\mathcal{M}$, the category of left H-comodules, and F is isomorphic to the forgetful functor.*

The basic idea of the proof is the following: For each $V \in \mathrm{Vec}$ define a functor $F_V : \mathcal{C} \to \mathrm{Vec}$ by $X \mapsto F(X) \otimes V$ for all $X \in \mathcal{C}$. The finiteness assumptions on the category imply that the functor $V \mapsto \mathrm{Mor}(F, F_V)$ is representable, that is, there exists an object $H \in \mathrm{Vec}$ such that

$$\mathrm{Hom}_k(H, V) = \mathrm{Mor}(F, F_V)$$

for all $V \in \mathrm{Vec}$. Then H is our desired Hopf algebra where the structure maps of H are reconstructed as well.

To the notion of a ribbon Hopf algebra there corresponds the notion of a ribbon category. A *ribbon category* $\mathcal{C}$ is a rigid braided tensor category with the following extra structure: There exists an automorphism of the identity functor id of $\mathcal{C}$, which is compatible with the tensor product, braiding and taking duals in a certain natural sense. In ribbon categories it is possible to define dimensions of objects (sometimes called quantum dimensions), and more generally to define traces of endomorphisms. This allows to associate link invariants to any ribbon category. In particular, all classical polynomial invariants (e.g., Jones polynomial) can be constructed in this fashion. (The reader is referred to [120–122,194] for an extensive study of ribbon Hopf algebras and their connections with Hennings' and Kauffman's invariants of knots, links and 3-manifolds.)

An important class of ribbon categories is the class of modular categories, [15,232]. A *modular category* $\mathcal{C}$ is a semisimple ribbon category with finitely many (up to isomorphism) irreducible objects $\{V_i \mid 0 \leqslant i \leqslant m\}$ with V_0 as the unit object, so that the matrix $s := (s_{ij})$, where $s_{ij} := \mathrm{tr}(c_{V_{J^*},V_i} c_{V_i,V_{J^*}})$, is invertible.

EXAMPLE 4.1.9 [77]. Let H be a semisimple Hopf algebra over an algebraically closed field k of characteristic 0. Then $\mathrm{Rep}(D(H))$ is a modular category. This is true essentially since $D(H)$ is factorizable.

Modular categories arise naturally in physics in the framework of quantum field theory, and in topology in the framework of invariants of 3-manifolds (see, e.g., [232]).

4.2. The FRT construction

Let G be an affine algebraic group over k; we then associate to G in the usual way two k-Hopf algebras: $A(G)$ (sometimes denoted by $O(G)$), whose elements are representative functions on G, and $U(\mathfrak{g})$, whose underlying k-algebra is the enveloping algebra of the Lie algebra $\mathfrak{g}$ of G. Those Hopf algebras certified by workers in the field as being "quantum groups", fall into two main classes: those deforming the type $U(\mathfrak{g})$, the quantum enveloping algebras, and those deforming the type $A(G)$.

Perhaps the earliest systematic construction of infinite families of these two types of Hopf algebras, was furnished by the following seminal work of Faddeev, Reshetikhin and Takhtajan, [90,91]. It was studied later extensively by many authors (e.g., [12,40,103,136, 230,202,218,226] and others in quantum group theory).

Let V be a finite-dimensional vector space and let $R : V \otimes V \to V \otimes V$ be a linear isomorphism that satisfies the braid relation, namely:

$$(R \otimes I_V) \circ (I_V \otimes R) \circ (R \otimes I_V) = (I_V \otimes R) \circ (R \otimes I_V) \circ (I_V \otimes R).$$

Let $\{v_1, \ldots, v_n\}$ be a basis of V and assume R as above is given by:

$$R(v_i \otimes v_j) = \sum_{k,l=1}^{n} R_{i,j}^{k,l} v_k \otimes v_l.$$

Recall, Example 1.1.18, that $\mathrm{End}(V)^*$ is a coalgebra and V is a right $\mathrm{End}(V)^*$-comodule. Let T be the tensor algebra of $\mathrm{End}(V)^*$. Then T is a bialgebra by extending the coproduct on $\mathrm{End}(V)^*$ to T multiplicatively. We wish to construct a bialgebra $A(R) = T/I$, for some biideal I, so that the map R will induce a braiding in the category of right $A(R)$-comodules. This is given by the following defining relations which are precisely those needed to make R into a right $A(R)$-comodule map.

$$\sum_{I,J=1}^{n} R_{ij}^{IJ} T_I^{i'} T_J^{j'} = \sum_{I,J=1}^{n} R_{IJ}^{i'j'} T_j^I T_i^J, \tag{1}$$

all $1 \leqslant i, i', j, j' \leqslant n$.

We have:

THEOREM 4.2.1 [90]. *Let T be as above and let I be the ideal of T generated by formula* (1). *Then I is a biideal and thus $A(R)$ is a bialgebra.*

A Hopf algebra version $H(R)$ of the FRT-construction is given in [202] with regard to a rigid tensor category.

It was proved in [136] that $A(R)$ is coquasitriangular with a braiding structure given on generators by

$$\langle T_i^j | T_k^l \rangle_R = R_{ik}^{lj}.$$

The paper [90] goes on to construct inside $(A(R))^\circ$ a bialgebra, yet not a Hopf algebra. This result was improved by Faddeev, Reshetikhin and Takhtajan in a later paper [91], where they construct inside $(A(R))^\circ$ a Hopf algebra $\widehat{U}(R)$, properly containing their earlier construction. The Hopf algebra $\widehat{U}(R)$ is defined as follows:

Let l^+, l^-, r^+, r^- be the following maps from $A(R)$ to $A(R)^0$.

$$l^+(a) = \langle a|-\rangle_R, \qquad r^+(a) = \langle -|a\rangle_R,$$
$$l^-(a) = \langle a|-\rangle_R^*, \qquad r^+(a) = \langle -|a\rangle_R^*,$$

where $\langle\ |\ \rangle_R^*$ is the convolution inverse of $\langle\ |\ \rangle_R$. Then the following Hopf algebra is constructed:

$$\widehat{U}(R) = Im\big(l^+\big) + Im(l^-) + Im\big(r^+\big) + Im(r^-) \subset \big(A(R)\big)^0.$$

The Hopf algebra $(\widehat{U}(R), \mathcal{R})$ is essentially quasitriangular in the following sense: The braiding $\langle\ |\ \rangle_R = \mathcal{R}$ is an element of $(A(R) \otimes A(R))^*$. But more is true, $\mathcal{R}$ is an element of $\widehat{U}(R) \hat{\otimes} \widehat{U}(R)$ which is the topological completion of $\widehat{U}(R) \otimes \widehat{U}(R)$ in the topological space $(A(R) \otimes A(R))^*$. (The topological aspects are discussed in details in [132].)

4.3. Yetter–Drinfeld categories and the Drinfeld double

One of the most significant categorical aspects of bialgebras H was introduced by Yetter in [248]. The coalgebra and algebra structure of H are taken into account simultaneously.

DEFINITION 4.3.1. Let H be a bialgebra over k. The "Yetter–Drinfeld" category ${}_H^H\mathcal{YD}$ (${}_H\mathcal{YD}^H$) is the category of objects which are left H-modules, left (right) H-comodules, and each $M \in {}_H^H \mathcal{YD}({}_H\mathcal{YD}^H)$ satisfies the compatibility condition, namely, for all $h \in H$, $m \in M$

$$\sum h_1 m_{-1} \otimes h_2 \cdot m_0 = \sum (h_1 \cdot m)_{-1} h_2 \otimes (h_1 \cdot m)_0,$$

where $\rho(m) = \sum m_{-1} \otimes m_0$ (or respectively

$$\sum h_1 \cdot m_0 \otimes h_2 m_1 = \sum (h_2 \cdot m)_0 \otimes (h_2 \cdot m)_1 h_1,$$

where $\rho(m) = \sum m_0 \otimes m_1$).

Similar compatibility conditions are given for right–right and right–left Yetter–Drinfeld categories.

EXAMPLE 4.3.2.

(1) A particular example of an object in ${}_H^H\mathcal{YD}$ ($\mathcal{YD}_H^H$) is H itself considered as a left (right) H-comodule via Δ and a left (right) H-module via the left (right) adjoint action.

(2) If (H, R) is a quasitriangular Hopf algebra then every left H-module M is in ${}_H^H\mathcal{YD}$ by defining $\rho : M \mapsto H \otimes M$ by

$$\rho(m) = \sum S(R^1) \otimes R^2 \cdot m.$$

Similarly, if $(H, \langle\ |\ \rangle)$ is a coquasitriangular Hopf algebra then every right H-comodule M is in ${}_H\mathcal{YD}^H$ by defining

$$h \cdot m = \sum \langle m_1 | h \rangle m_0$$

for all $h \in H$, $m \in M$.

The Yetter–Drinfeld category ${}_H^H\mathcal{YD}$ has the following natural pre-braiding structure: Given $M, N \in {}_H^H\mathcal{YD}$, define $c_{M,N} : M \otimes N \mapsto N \otimes M$ by:

$$c_{M,N}(m \otimes n) = \sum m_{-1} \cdot n \otimes m_0$$

for $m \in M$, $n \in N$. When H is a Hopf algebra with an invertible antipode then ${}_H^H\mathcal{YD}$ is a braided tensor category with a braiding structure defined by c.

The category ${}_H^H\mathcal{M}_H^H$ of two-sided two-cosided Hopf-modules satisfying six compatibility relations (also called tetramodules) was considered in [246]. He discussed the interrelation between ${}_H^H\mathcal{YD}$ and ${}_H^H\mathcal{M}_H^H$. An equivalence between the pre-braided categories of tetramodules and that of Yetter–Drinfeld modules over H was stated in [203].

Another related category is the category of Doi–Hopf modules, [62]. This category includes a variety of modules as special cases; for example Hopf modules and graded modules are Doi–Hopf modules. Furthermore, it was proved in [33] that ${}_H^H\mathcal{YD}$ can be considered as a special case of Doi–Hopf modules and in [18] that the same holds for two-sided two-cosided Hopf modules, illustrating the "unifying" property of Doi's concept. In fact, the above mentioned category equivalence between Yetter–Drinfeld modules and tetramodules can be described in terms of an adjoint pair of functors between categories of Doi–Hopf modules (in the sense of [34]).

In a Yetter–Drinfeld category we can consider *commutativity in the category*. This is defined in general as follows:

DEFINITION 4.3.3. A left H-module and left H-comodule algebra $(A, \cdot, \rho)$ is called *H-commutative* (or *quantum commutative*) if

$$ab = \sum (a_{-1} \cdot b)a_0 \quad \text{for all } a, b \in A.$$

H-commutativity is defined similarly for a right H-module and right H-comodule algebras, etc.

REMARK 4.3.4. If $(A, \cdot, \rho)$ is H-commutative then $A^{\mathrm{co}\,H}$ and A^H are contained in $Z(A)$.

EXAMPLE 4.3.5. The following are examples of H-commutative algebras.

(1) Let H be as in Example 4.3.2(1). Then $(H, \mathrm{ad}_l, \Delta)$ $((H, \mathrm{ad}_r, \Delta)$ resp.) is H-commutative.
(2) [49,160]. Let A be a commutative *superalgebra*, that is A is a $\mathbb{Z}_2$-graded algebra and $ab = (-1)^{\deg a \deg b} ba$ for homogeneous elements $a, b \in A$. Let $G = \{1, g\} \cong \mathbb{Z}_2$ and $H = kG$. Consider A as an H-comodule by the $\mathbb{Z}_2$-grading and an H-module by defining $g \cdot a = (-1)^{\deg a} a$ for homogeneous $a \in A$. Then A is H-commutative.
(3) [49]. Let $A := \mathbb{C}_q[x, y]$, the *quantum plane*, that is A equals the free algebra $\mathbb{C}\langle x, y\rangle$ modulo the relation $xy = qyx$, where q is an n-th root of 1. Let $G := \mathbb{Z}_n \times \mathbb{Z}_n$ and $H = kG$, then A is H-commutative for a certain action and coaction of H. If we localize and obtain $B := A[x^{-1}, y^{-1}]$ then B/B^H is H-commutative and H^*-Galois.

Recall the Miyashita–Ulbrich action defined in Definition 3.2.6. The following theorem describes how H-commutativity is related to H-Galois extensions and to objects in $\mathcal{YD}_H^H$ via the Miyashita–Ulbrich action $\leftarrow$.

THEOREM 4.3.6. *Let $A^{\mathrm{co}\,H} \subset A$ be a right H-Galois extension, then:*

(1) [235,68]. *$(C_A(A^{\mathrm{co}\,H}), \leftarrow, \rho)$ is H-commutative.*
(2) [35]. *If H has a bijective antipode then*

$$\left(C_A\left(A^{\mathrm{co}\,H}\right), \leftarrow, \rho\right) \in \mathcal{YD}_H^H.$$

In particular, if $A^{\mathrm{co}\,H} \subset Z(A)$ then $(A, \leftarrow, \rho) \in \mathcal{YD}_H^H$.
(3) [45]. *If A is a right H-commutative H-module algebra, then $C_A(A^{\mathrm{co}\,H}) = A$, the given action $\cdot$ coincides with $\leftarrow$ and $(A, \cdot, \rho) \in \mathcal{YD}_H^H$.*

EXAMPLE 4.3.7. Let $A = H$ be a right H-comodule with $\rho = \Delta$. Then by Example 3.2.2(2), $k \subset H$ is a right H-Galois with $\beta^{-1}(1 \otimes h) = \sum S(h_1) \otimes h_2$ and so the Miyashita–Ulbrich action $x \leftarrow h = \sum S(h_1)xh_2$ is the right adjoint action.

Theorem 4.3.6(3) generalizes now the right version of Example 4.3.2(1).

The Drinfeld double

If H is finite-dimensional then ${}_H\mathcal{YD}^H$ has a nice realization related to the so-called Drinfeld double of H constructed by Drinfeld, [71]. The Drinfeld double is a double crossproduct, a construction described in [153] and modified in [192]. The double crossproduct of the bialgebras H and B is defined when B is a left H-module coalgebra and H is a right B-module coalgebra satisfying some compatibility conditions. A special case is when H is a finite-dimensional Hopf algebra, $B = H^*$ and the actions are given by the left coadjoint action of H on H^*

$$h \rightharpoondown p = \sum h_1 \rightharpoonup p \leftharpoonup S^{-1}(h_2)$$

and the right coadjoint action of H^* on H

$$h \leftharpoondown p = \sum S^{*-1}(p_1) \rightharpoonup h \leftharpoonup p_2$$

for $h \in H$, $p \in H^*$.

DEFINITION 4.3.8. Let H be a finite-dimensional Hopf algebra. The Drinfeld double $D(H) = H^{*\mathrm{cop}} \bowtie H$ is defined as the vector space $H^{*\mathrm{cop}} \otimes H$ with multiplication defined by

$$(p \bowtie h)(p' \bowtie h') = p(h_1 \rightharpoondown p'_2) \bowtie (h_2 \leftharpoondown p'_1)h'$$

and comultiplication given by the tensor comultiplication in the tensor coalgebra $H^{*\mathrm{cop}} \otimes H$, that is:

$$\Delta_{D(H)}(p \bowtie h) = (p_2 \bowtie h_1) \otimes (p_1 \bowtie h_2)$$

for all $h \in H$, $p \in H^*$.

The antipode is given by

$$S_{D(H)}(p \bowtie h) = \bigl(1 \bowtie S(h)\bigr)\bigl(S^*(p) \bowtie 1\bigr).$$

More precise formulas for the multiplication in $D(H)$ are given by:

$$(p \bowtie h)(p' \bowtie h') = \sum p\bigl(h_1 \rightharpoonup p' \leftharpoonup S^{-1}(h_3)\bigr) \bowtie h_2 p',$$

$$(p \bowtie h)(p' \bowtie h') = \sum pp' \bowtie \bigl(S^{*-1}(p_1) \rightharpoonup h \leftharpoonup p'_3\bigr)h'.$$

Observe that the Hopf algebras H and $H^{*\mathrm{cop}}$ are contained in $D(H)$ hence a left $D(H)$-module M is in particular a left H-module and a left H^*-module. Thus, by Remark 1.3.4,

M is also a right H-comodule. Now, a straightforward (long) verification shows that the definition of the multiplication in $D(H)$ implies that $h \cdot (p \cdot m) = ((\varepsilon \bowtie h)(p \bowtie 1)) \cdot m$ for all $m \in M$, $h \in H$ is equivalent to M being an object in ${}_H\mathcal{YD}^H$. We summarize:

THEOREM 4.3.9 [156]. *Let H be a finite-dimensional Hopf algebra. Then the Yetter–Drinfeld category ${}_H\mathcal{YD}^H$ is equivalent to the category of left modules over the Drinfeld double $D(H)$.*

REMARK 4.3.10. The process of taking the double is mostly effective if it is done just once; the double of $D(H)$ can be obtained from the tensor product $D(H) \otimes D(H)$ by twisting the comultiplication, [197,209].

The Drinfeld double $D(H)$ is naturally quasitriangular by letting

$$R := \sum (\varepsilon \bowtie h_i) \otimes (h_i^* \bowtie 1),$$

where $\{h_i\}$ and $\{h_i^*\}$ are any dual bases of H and H^*. Consequently, $(D(H)^*, \langle\ |\ \rangle_R)$ is coquasitriangular. Moreover, $D(H)$ is factorizable (hence unimodular), [71,197,194].

REMARK 4.3.11.

(1) [140]. Recall, Example 1.6.3, that $\sigma = \langle\ |\ \rangle_R \circ \tau$ is a Hopf 2-cocycle on $D(H)^*$. Then

$${}_\sigma D(H)^* \cong \mathcal{H}(H) = H\#H^*$$

as algebras, where ${}_\sigma D(H)^*$ is a twisted Hopf algebra as defined in Remark 3.1.4 and $\mathcal{H}(H)$ is the so-called "Heisenberg double" of H; it is a simple algebra.

(2) It is straightforward to check that

$$J := \sum_i (h_i^* \otimes 1) \otimes (\varepsilon \otimes h_i)$$

is a twist for the Hopf algebra $H^{*\mathrm{op}} \otimes H$ where $\{h_i\}$ and $\{h_i^*\}$ are dual bases in H and H^* respectively. Then $(H^{*\mathrm{op}} \otimes H)^J$, the Hopf algebra obtained by twisting the comultiplication via J is isomorphic to $D(H)^{*\mathrm{op}}$, the (opposite) dual of the Drinfeld double of H.

For a braided monoidal category $\mathcal{C}$, a Brauer group $\mathrm{Br}(\mathcal{C})$ was defined by [240] so that many Brauer group constructions are particular cases. For example, the classical Brauer group of a commutative ring k is $\mathrm{Br}(\mathcal{C})$ for $\mathcal{C}$ the category of k-modules; the Brauer group of a scheme (X, O_X) is $\mathrm{Br}(\mathcal{C})$ where is $\mathcal{C}$ is the category of O_X-module sheaves; the Brauer–Long group is $\mathrm{Br}(\mathcal{C})$ for $\mathcal{C}$ the category of H-dimodules (where H is a commutative cocommutative Hopf algebra).

The Brauer–Long group was generalized and denoted by $\mathrm{BQ}(k, H)$ where H is any Hopf algebra with a bijective antipode and the dimodules are replaced by Yetter–Drinfeld modules. It is proved:

THEOREM 4.3.12 [241]. *Let H be a finite-dimensional Hopf algebra then there is an exact sequence*

$$1 \to G(D(H)^*) \to G(D(H)) \to \mathrm{Aut}_{\mathrm{Hopf}}(H) \to \mathrm{BQ}(k, H),$$

where $D(H)$ is the Drinfeld double of H.

As a consequence, it follows that the Brauer group of Sweedler's four-dimensional Hopf algebra H_4 contains $k^*/\{-1, 1\}$ as a subgroup and thus $\mathrm{BQ}(k, H)$ is in general highly non-torsion.

4.4. Hopf algebras in braided categories, biproducts and bosonizations

One of the first comprehensive steps to abstract the notion of a Hopf algebra from the work of Hopf in topology was carried out by Milnor and Moore, [163]. Though their notion of a Hopf algebra is not the one used in this chapter, it turns out to be a Hopf algebra in the category of $k\mathbb{Z}$-comodules. Specifically,

EXAMPLE 4.4.1 [163]. Let $A = A_0 \oplus A_1 \oplus \cdots$ be a $\mathbb{N}$-graded vector space over a field k. If A and B are $\mathbb{N}$-graded then the "twisting morphism" $T : A \otimes B \to B \otimes A$ is the morphism defined by:

$$T(a \otimes b) = (-1)^{pq} b \otimes a$$

for $a \in A_p$, $b \in B_q$.

A is called a graded Hopf algebra over k if

(1) $(A, \mu, 1)$ is a graded k-algebra (as usual).
(2) (A, Δ, ε) is a graded k-coalgebra, viz., $\Delta(A_n) \subset \sum_{i=0}^{n} A_i \otimes A_{n-i}$ and $\sum_{i>0} A_i \subset \operatorname{Ker} \varepsilon$.
(3) $\Delta \circ \mu = (\mu \otimes \mu) \circ (\mathrm{id} \otimes T \otimes \mathrm{id}) \circ (\Delta \otimes \Delta)$ and ε is an algebra map.
(4) The identity map id_A is invertible under convolution. In particular, its inverse S satisfies $S \circ \mu = \mu \circ T \circ (S \otimes S)$.

It is condition (3) in the above example that reflects the basic idea of a bialgebra in a braided category, where the braiding plays the role of the "twisting operator".

Additional examples in the same spirit are enveloping algebras of Lie superalgebras and more generally, of Lie color algebras, [205].

These examples were known without the formalism of category theory. The more general notion of Hopf algebras in braided tensor categories was introduced in [159] and have been since extensively studied by many authors. A comprehensive survey is given in [228].

Notions like algebras, coalgebras and bialgebras can be considered categorically, that is, all structure maps are required to be maps *in the category*. (H-module algebras and H-comodule algebras are examples that have already been mentioned.) To define a bialgebra in a category requires an appropriate braiding. Explicitly:

Given a braided tensor category with a braiding structure c, one can define an algebra structure on the tensor product of two algebras as follows: For any (A, μ_A) and (B, μ_B) in the category define $\mu_{A\otimes B} : (A \otimes B) \otimes (A \otimes B) \mapsto A \otimes B$ by:

$$\mu_{A\otimes B} = (\mu_A \otimes \mu_B)(\mathrm{id} \otimes c_{B,A} \otimes \mathrm{id}).$$

Once the tensor product of two algebras is an algebra we can define:

DEFINITION 4.4.2. A *bialgebra in a braided tensor category* is a 5-tuple $(A, \mu, 1, \Delta, \varepsilon)$ where $(A, \mu, 1)$ is an algebra in the category and (A, Δ, ε) is a coalgebra in the category so that

$$\Delta \circ \mu_A = \mu_{A\otimes A} \circ (\Delta \otimes \Delta).$$

A is a *Hopf algebra in the category* if moreover the identity id_A is invertible under convolution; its inverse is the antipode of A.

Hopf algebras in the category are also called *braided Hopf algebras*.

EXAMPLE 4.4.3 [11]. Let H be any Hopf algebra and let $V \in {}^H_H\mathcal{YD}$. Then the tensor algebra $T(V) = \bigoplus_{n\geqslant 0} T(V)(n)$, where $T(V)(n) = V^{\otimes n}$ is also an object in ${}^H_H\mathcal{YD}$. If we define $\Delta_V(v) = 1 \otimes v + v \otimes 1$ for all $v \in V$, then there is a unique extension of Δ_V to a map $\Delta : T(V) \to T(V) \otimes T(V)$ which is an algebra map in ${}^H_H\mathcal{YD}$. The counit ε is defined by $\varepsilon(v) = 0$ for all $v \in V$ and thus $T(V)$ is a (graded) bialgebra in the category. Moreover, it can be proved that $T(V)$ is actually a Hopf algebra in ${}^H_H\mathcal{YD}$.

Let I be the largest Hopf ideal generated by homogeneous elements of degree > 1. Then $B(V) = T(V)/I$ is a (graded) Hopf algebra in ${}^H_H\mathcal{YD}$. The graded braided Hopf algebra $B(V)$ is unique with respect to the following properties: $B(V)$ is connected as a coalgebra, generated as an algebra by elements of degree 1 and $V = B(V)(1) = P(B(V))$, the space of primitive elements of $B(V)$.

The braided Hopf algebra $B(V)$ was termed the *Nichols algebra* of V honoring Nichols who described $B(V)$ in a different setting, [177].

Nichols algebras were rediscovered and studied independently by several authors. They were treated as the invariant parts of "algebras of quantum differential forms" in [246] and as "quantum symmetric algebras" in [199].

It has been proved that most of the fundamental properties of ordinary finite-dimensional Hopf algebras can be generalized to braided Hopf algebra theory, even in a more generalized form, [228]. We summarize:

THEOREM 4.4.4. *The following is true for Hopf algebras in a category*:

(1) [143]. *The fundamental theorem for Hopf modules* (*Theorem* 2.1.3).
(2) [143,227]. *The bijectivity of the antipode* (*Theorem* 2.2.6).
(3) [143,227]. *The uniqueness of the integral* (*Theorem* 2.2.1).
(4) [95]. *Frobenius property* (*Theorem* 2.2.5).

(5) [65]. *Trace formulas* (*Theorem* 2.2.6).
(6) [23]. *A variation of the S^4 formula* (*Theorem* 2.4.1).
(7) [201]. *Nichols–Zoeller theorem* (*Theorem* 2.5.1).
(8) [22]. *Characterizations of the Yetter–Drinfeld category and generalizations of biproducts and bosonizations* (*see next section*).

Starting from a Hopf algebra B in the braided tensor category ${}_H^H\mathcal{YD}$ it is possible to "lift" B to an ordinary Hopf algebra. The process was given in [190] (without using the notion of a Hopf algebra in the category) and in [158].

To describe the process we need first to introduce the smash coproduct.

Let H be a bialgebra and A a left H-comodule coalgebra. To avoid confusion we write for all $a \in A$:

$$\Delta_A(a) = \sum a^1 \otimes a^2 \quad \text{and} \quad \rho_H(a) = \sum a_{-1} \otimes a_0.$$

Then the tensor product $A \otimes H$ can be equipped with a coalgebra structure via the *smash coproduct* as follows:

PROPOSITION 4.4.5 [164]. *Let H be a Hopf algebra and let A be a left H-comodule coalgebra. Let $A\#H$ be the vector space $A \otimes H$ with coproduct given by*

$$\Delta(a\#h) = \sum a^1\#(a^2)_{-1}h_1 \otimes (a^2)_0\#h_2$$

and counit given by $\varepsilon(a\#h) = \varepsilon_A(a)\varepsilon_H(h)$. Then the above structure maps make $A\#H$ into a coalgebra.

Let $A \otimes H$ be equipped with the smash product and the smash coproduct. We call it a *biproduct* and denote it by $A * H$. Necessary and sufficient conditions for the biproduct to be a Hopf algebra were given by Radford in [190]. We give a categorical version of the theorem.

THEOREM 4.4.6. *Let H be a bialgebra and let A be an algebra in ${}_H\mathcal{M}od$ and a coalgebra in ${}^H\mathcal{C}om$. Then $A * H$ is a bialgebra if and only if A is a bialgebra in ${}_H^H\mathcal{YD}$.*

If H is a Hopf algebra with a bijective antipode S_H and A is a Hopf algebra in ${}_H^H\mathcal{YD}$ with an antipode S_A then $A * H$ is a Hopf algebra with antipode given by:

$$S(a * h) = \bigl(1 * S_H^{-1}(a_{-1}h)\bigr)\bigl(S_A(a_0) * 1\bigr).$$

A similar process was named *bosonization*, [158]. He considers the braided category ${}_H\mathcal{M}od$ over a quasitriangular Hopf algebra H and proves moreover that if H is triangular and A is quasitriangular in the category, then $A * H$ is quasitriangular.

EXAMPLE 4.4.7. Let $H = H_4$ be Sweedler's 4-dimensional Hopf algebra (Example 1.2.5) then $H = A * kG$ where $G = \{1, g\} \cong \mathbb{Z}_2$ and $A = sp_k\{1, x\}$.

The following is a version of a structure theorem about biproducts. It is most useful in the classification theory of finite-dimensional Hopf algebras.

THEOREM 4.4.8 [190]. *If $H \overset{i}{\hookrightarrow} B \overset{\pi}{\to} H$ is a sequence of finite-dimensional Hopf algebra maps where i is injective, π is surjective and $\pi \circ i = \mathrm{id}_H$, then there exists a coideal subalgebra $A \subset B$ such that:*

(1) *A is a left H-module algebra via the adjoint action.*
(2) *A is a left H-comodule algebra via $\rho(b) = \sum a^{(1)} \otimes a^{(2)} = \sum \pi(a_{(1)}) \otimes a_{(2)}$.*
(3) *$A \cong B/BH^+$ as Hopf algebras (where $H^+ = \operatorname{Ker} \varepsilon$).*
(4) *A is a Hopf algebra in the category ${}^H_H\mathcal{YD}$.*
(5) *$B \cong A * H$ as a bialgebra.*

Bosonizations and biproducts were used in [46,93] to prove a generalized Schur double centralizer theorem for Lie algebras in certain symmetric tensor categories $\mathcal{C}$. Explicitly, for a finite-dimensional object $V \in \mathcal{C}$ one can define the $\mathcal{C}$-analogue, $U(gl_{\mathcal{C}}(V))$, of the enveloping Lie algebra $U(gl(V))$ by using the braiding structure. Then $U(gl_{\mathcal{C}}(V))$ is a Hopf algebra in the category $\mathcal{C}$ and thus by Theorem 4.4.6, $\widehat{H} = U(gl_{\mathcal{C}}(V)) * H$ is an ordinary Hopf algebra. The Hopf algebra $\widehat{H}$ acts on $V^{\otimes n}$ via Δ while the symmetric group S_n acts on $V^{\otimes n}$ via the usual flip map. It was proved:

THEOREM 4.4.9 [46]. *Let $\mathcal{C}$, V, $U(gl_{\mathcal{C}}(V))$ and $\widehat{H}$ be as above and assume that the characteristic of k is 0. Then the actions of kS_n and $\widehat{H}$ on $V^{\otimes n}$ centralize each other.*

[93] have proved in the same spirit a double centralizer theorem for Lie color algebras.

Part 5. Structure theory for special classes of Hopf algebras

5.1. Semisimple Hopf algebras

There are several surveys regarding semisimple Hopf algebras, the reader is referred to [168,169,4].

Observe first that since $\operatorname{Ker} \varepsilon$ is a non-zero ideal of H, it follows that there exist no Hopf algebras which are simple as algebras. Thus the simplest objects are the semisimple Hopf algebras.

A consequence of Corollary 2.2.3 is that all semisimple Hopf algebras are finite-dimensional. For if H is semisimple then $H = I \oplus \operatorname{Ker} \varepsilon$ where I is a 1-dimensional left ideal of H (since $\operatorname{Ker} \varepsilon$ has codimension 1). Moreover, if H is a semisimple Hopf algebra then it is a separable algebra (i.e. for any field extension $E \supseteq k$, $H \otimes E$ is semisimple). This is easily seen from Maschke's theorem (Theorem 2.3.1) and from the fact that the extensions of Δ, ε, S to $\overline{H} = H \otimes E$ make $\overline{H}$ a Hopf algebra over E with integral $\int_H^l \otimes E$.

Semisimple Hopf algebras in characteristic 0 are close in spirit to kG, G a finite group as will be seen in this section. Some of Kaplansky's conjectures, [116], are inspired by this resemblance (see [213] for a detailed exposition).

The square of the antipode

KAPLANSKY'S 5TH CONJECTURE. *Let H be a semisimple Hopf algebra. Then* $S^2 = \mathrm{id}$.

In [131] it is proved that if H is semisimple over an algebraically closed field then S^2 is an inner automorphism, and in [183] the same is proved over any base field k.

Observe that from Theorems 2.2.7 and 2.3.1, we deduce:

THEOREM 5.1.1. *Let H be a finite-dimensional Hopf algebra then H is semisimple and cosemisimple if and only if* $\mathrm{Tr}(S^2) \neq 0$.

By using Theorem 5.1.1, a positive answer to Kaplansky's 5-th conjecture in characteristic 0 was given by Larson and Radford:

THEOREM 5.1.2 [133,134]. *Let H be a finite-dimensional Hopf algebra and assume that the base field k has characteristic* 0. *Then the following are equivalent*:

(1) $S^2 = \mathrm{id}$.
(2) *H and H^* are semisimple.*
(3) *H is semisimple.*
(4) *H^* is semisimple.*

When the characteristic of k is positive then (1) $\Rightarrow$ (3) is trivially false, for example, if $H = kG$ and the characteristic of k divides the order of G.

In positive characteristic it is thus natural to consider Hopf algebras which are *both* semisimple and cosemisimple. This was manifested in [76] where it was proved that any such Hopf algebra in positive characteristic can be lifted to a semisimple (and hence cosemisimple) Hopf algebra in characteristic 0 of the same dimension. This implies that, essentially, it is enough to study semisimple Hopf algebras in characteristic 0, and lift the results to positive characteristic. The proof of the lifting theorem uses Witt vectors (see, e.g., [211]), and the so-called Gerstenhaber–Schack cohomology, [99]. In particular it allowed to prove:

THEOREM 5.1.3 [77]. *Let H be a semisimple and cosemisimple Hopf algebra over any field k. Then* $S^2 = \mathrm{id}$. *Moreover, if H is any finite-dimensional Hopf algebra over any field k, then H is semisimple and cosemisimple* $\Leftrightarrow$ $S^2 = \mathrm{id}$ *and* $\dim(H) \neq 0$ *in k.*

If H is a semisimple Hopf algebra and the characteristic of k is large enough in comparison to the dimension of H then H is also cosemisimple and hence $S^2 = \mathrm{id}$ (see [212,76]).

Character theory

Motivated by group representation theory, a basic tool in the theory of semisimple Hopf algebras H over an algebraically closed field of characteristic 0 is the character ring of H.

Let V be a finite-dimensional left H-module and let $\rho_V : H \to \mathrm{End}(V)$ be the corresponding representation. Then $\chi_V \in H^*$ is defined by

$$\chi_V(h) := \mathrm{Tr}\big(\rho_V(h)\big)$$

for any $h \in H$. Since $\mathrm{Tr}(\rho_V(hh')) = \mathrm{Tr}(\rho_V(h'h))$ for all $h, h' \in H$ it follows that χ_V is a cocommutative element of H^*.

If V is an irreducible module we say that χ_V is an irreducible character. It is easily seen that

(1) $\chi_{V\oplus W} = \chi_V + \chi_W$,
(2) $\chi_{V\otimes W} = \chi_V * \chi_W$,
(3) $\chi_{V^*} = S(\chi_V)$.

Define the character ring $R(H)$ of H to be the k-span in H^* of all the characters on H. Since H is semisimple it follows that $R(H)$ is generated over k by the finite set of its irreducible characters. Actually, $R(H)$ is the subalgebra of all cocommutative elements of H^*.

Let $t \in \int_l^H$ be such that $\varepsilon(t) = 1$. Define a form $\langle\ |\ \rangle$ on $R(H)$ by:

$$\langle\varphi|\psi\rangle := \big\langle\varphi * S(\psi), t\big\rangle = \sum\langle\varphi, t_1\rangle\big\langle S(\psi), t_2\big\rangle$$

for any characters $\varphi, \psi \in R(H)$.

As for characters of finite groups, there are orthogonality relations for irreducible characters via this form. Let $\{V_0, V_1, \ldots, V_m\}$ be a complete set of irreducible left H-modules, where V_0 is the trivial module. Let $n_i = \dim(V_i)$ and let χ_i denote the character χ_{V_i}. We have:

THEOREM 5.1.4 [131]. *Let H be a semisimple Hopf algebra over an algebraically closed field and let $\{\chi_0, \ldots, \chi_m\}$ be the set of irreducible characters of H. Then $\langle\chi_i|\chi_j\rangle = \delta_{ij}$.*

A consequence of the above theorem is that $R(H)$ is a semisimple algebra.

Let $R_{\mathbb{Z}}(H) = \sum_i \mathbb{Z}\chi_i \subset R(H)$ where the $\{\chi_i\}$ are the irreducible characters on H. Since the $\{\chi_i\}$ are $\mathbb{Z}$-independent by orthogonality, $R_{\mathbb{Z}}(H)$ is a finite free $\mathbb{Z}$-module. In fact $R_{\mathbb{Z}}(H) \cong K_0(H)$, the Grothendieck ring of H.

THEOREM 5.1.5 [180]. *Two semisimple Hopf algebras have isomorphic Grothendieck rings if and only if they are pseudo-twists of each other.*

An important generalization from group theory is the *class equation* for semisimple Hopf algebras.

THEOREM 5.1.6 [114,250]. *Let H be a semisimple Hopf algebra over an algebraically close field of characteristic* 0. *Let $\{e_0, e_1, \ldots, e_m\}$ be a complete set of primitive orthogonal idempotents in $R(H)$, where e_0 is an integral for H^*. Then*

$$\dim(H) = 1 + \sum_{i=1}^{m} \dim(e_i H^*)$$

and $\dim(e_i H^*)$ *divides* $\dim(H)$ *for all* $0 \leqslant i \leqslant m$.

Many results in the classification theory of semisimple Hopf algebras are due to the class equation. An immediate one is the Kac–Zhu theorem (Theorem 2.6.1).

When $H = kG$, the theorem boils down to the usual class equation for finite groups. When $H = (kG)^*$ then $R(H) = H^* = kG$ and the class equation says that the dimension of an irreducible G-module divides the order of G. This is the classical theorem of Frobenius for finite groups that motivated Kaplansky's 6th conjecture:

KAPLANSKY'S 6TH CONJECTURE [116]. *Let H be a semisimple Hopf algebra. Then the dimension of any irreducible H-module divides the dimension of H.*

We say that H is of *Frobenius type* if it satisfies this conjecture. Let the base field k be algebraically closed of characteristic 0, then H is of Frobenius type in the following cases:

(1) If (H, R) is quasitriangular, [78].
(2) If H is semisolvable, that is, H has a normal series of Hopf subalgebras such that each Hopf quotient is either commutative or cocommutative, [173].
(3) If $R(H)$ is central in H^*, [249].
(4) If H is cotriangular, [81]. This was proved using Theorem 5.1.7 below.

In fact, for the case of semisimple quasitriangular Hopf algebra it is proved that the dimension of any irreducible $D(H)$-module divides the dimension of H. The proof uses the theory of modular categories (the representation category of $D(H)$ is modular), and in particular the Verlinde formula, [242], applies. See [210,231] for later proofs in the quasitriangular case.

Other important results in this direction are that if H has an irreducible module of dimension 2 then H has even dimension, [179], and more generally that if H has an even-dimensional irreducible module then H has even dimension, [119].

Semisimple triangular Hopf algebras

The structure of triangular Hopf algebras is far from trivial, and yet is more tractable than that of general Hopf algebras, due to their proximity to groups and Lie algebras.

THEOREM 5.1.7 [77]. *Any semisimple triangular Hopf algebra over an algebraically closed field k of characteristic* 0 *is isomorphic to $(kG)^J$ for a unique* (*up to isomorphism*) *finite group G and a unique* (*up to gauge equivalence*) *twist J.*

The proof of this theorem is based on a deep theorem of Deligne on Tannakian categories, [56]. The idea is that if (H, R) is a semisimple triangular Hopf algebra then one can modify R to get an element $\widetilde{R}$ such that the category $\mathrm{Rep}(H, \widetilde{R})$ is not only semisimple and symmetric but also has the property that the categorical dimensions of its objects are non-negative integers. Thus by the Deligne theorem, $\mathrm{Rep}(H, \widetilde{R})$ is Tannakian; that is, it is equivalent to $\mathrm{Rep}(G)$ for a unique finite group G, [58].

This theorem is the key step in the complete classification of triangular semisimple Hopf algebras described in the following theorem.

THEOREM 5.1.8 [80]. *Triangular semisimple Hopf algebras of dimension N over k are in one to one correspondence with quadruples (G, H, V, u), where G is a finite group of order N, $H < G$, V is an irreducible projective representation of H of dimension $|H|^{1/2}$, and $u \in G$ a central element of order $\leqslant 2$.*

By the previous theorem one needs to classify twists for a given finite group G, up to gauge equivalence. It turns out that any twist J for G is gauge equivalent to a "minimal" twist coming from a subgroup H of G. Finally, using Movshev's theory, [174], one shows that equivalence classes of minimal twists for a finite group H are in bijection with isomorphism classes of irreducible projective representation of H of dimension $|H|^{1/2}$. It is interesting to note that H is a central type group, so in particular solvable, [109].

For *non-semisimple* finite-dimensional triangular Hopf algebras H over an algebraically closed field k of characteristic 0 it is no longer true that the categorical dimensions of objects in $\mathrm{Rep}(H)$ are non-negative integers; so the Deligne theorem, [56], cannot be applied. Nevertheless using a recent theorem of Deligne, [57], it was proved in [84] that any finite-dimensional triangular Hopf algebra has the Chevalley property; namely, the semisimple part of H is itself a Hopf algebra. This leads to the complete and explicit classification of finite-dimensional triangular Hopf algebras over k. As a consequence, for example, it is proved that in any finite-dimensional triangular Hopf algebra H, $u^2 = 1$ and hence $S^4 = \mathrm{id}$.

Results for special dimensions

Another conjecture of Kaplansky is the following:

KAPLANSKY'S 10TH CONJECTURE. *For each integer $n > 0$ there are only finitely many isomorphism classes of n-dimensional Hopf algebras.*

A positive answer to this conjecture was given in the semisimple case, any characteristic:

THEOREM 5.1.9 [216]. *Let k be an algebraically closed field. Then for each integer $n > 0$ there are only finitely many isomorphism classes of n-dimensional semisimple and cosemisimple Hopf algebras.*

Using three independent methods the conjecture was shown to be false in the non-semisimple case, [8,17,96] (though all these Hopf algebras are twists of each other, [149]).

In what follows we list results for special dimensions over an algebraically closed field of characteristic zero. Let $p \neq q$ be prime numbers, then:

(1) If $\dim H = pq$ then H is either kG or $(kG)^*$ for some group G, [77,98].
(2) If $\dim H = p^2$ then $H = kG$ for some finite group G, [147].
(3) Let $p \neq 2$ and $\dim H = p^3$. Then there exist exactly $p + 8$ isomorphism classes. Seven of these are of the form kG or $(kG)^*$ and the other $p + 1$ are all non-commutative, non-cocommutative and self dual, [146].
(4) If $\dim H = p^n$, then H is solvable, in particular H is of Frobenius type, [173].

It follows that the first possible dimension for a semisimple Hopf algebra to be neither commutative nor cocommutative is 8. An example of such a Hopf algebra was constructed already in 1966, [115].

Using the above results semisimple Hopf algebras of odd dimension less then 60 over an algebraically closed field of characteristic zero are all classified. Even-dimensional Hopf algebras are less known. For specific dimensions see the surveys mentioned in the beginning of this section.

It was proved in [85].

THEOREM 5.1.10. *Let H be a Hopf algebra over an algebraically closed field of characteristic 0 whose dimension is pq, where p, q are prime and $p < q < 2p + 2$. Then H is semisimple, hence is either kG or $(kG)^*$ for some group G.*

The exponent and the Schur indicator

Motivated by other aspects of group theory two other notions were generalized to Hopf algebras: the exponent and the Schur indicator.

The classical notion of the exponent of a group is generalized in [79], motivated by [117, 118], in which the exponent of Hopf algebras whose antipode is involutive is studied. In fact, for such Hopf algebras the notion of exponent has existed for over 30 years.

DEFINITION 5.1.11 [79]. The *exponent* $\exp(H)$ of H is the smallest positive integer n such that $m_n \circ (\mathrm{id} \otimes S^{-2} \otimes \cdots \otimes S^{-2n+2}) \circ \Delta_n = \varepsilon \cdot 1$, where m_n, Δ_n are the iterated product and coproduct.

If H is involutive (for example, H is semisimple and cosemisimple), then $\exp(H)$ equals the smallest positive integer n so that $m_n \circ \Delta_n = \varepsilon \cdot 1$.

In [79] it is shown that $\exp(H)$ equals the order of the Drinfeld element u of the quantum double $D(H)$, and the order of $R^\tau R$, where R is the universal R-matrix of $D(H)$. This was motivated by a theorem in conformal field theory, [237].

In [118] it was conjectured that if H is semisimple and cosemisimple then $\exp(H)$ is always finite and divides $\dim(H)$.

THEOREM 5.1.12 [79]. *For a semisimple and cosemisimple Hopf algebra H, $\exp(H)$ is finite and divides $\dim(H)^3$.*

In [119] it is proved that if 2 divides $\dim(H)$ then 2 divides $\exp(H)$. Whether this is true for any odd prime p is still an open question.

For non-semisimple finite-dimensional Hopf algebras the exponent is usually infinite. However, it was proved in [82] that the order of unipotency of u is always finite, since all the eigenvalues of u are roots of unity. This order of unipotency of u was termed the *quasi-exponent* of H, and is denoted by $\mathrm{qexp}(H)$; it reduces to $\exp(H)$ when H is semisimple. In [82] equivalent definitions of $\mathrm{qexp}(H)$, generalizing the ones in the exponent case, are given and it is proved that $\mathrm{qexp}(H)$ is an invariant of the tensor category $\mathrm{Rep}(H)$.

The theory of quasi-exponents was applied to study the group of grouplike elements of twisted quantum groups at roots of unity, [82].

Another generalization of group theory is the *Schur indicator* ν.

DEFINITION 5.1.13 [137]. Let H be a semisimple Hopf algebra over an algebraically closed field of characteristic 0, and $t \in \int_H$ such that $\langle \varepsilon, t\rangle = 1$. For any irreducible character χ define the Schur indicator by

$$\nu(\chi) = \sum \chi(t_1 t_2).$$

It is proved:

THEOREM 5.1.14 [137]. *The Schur indicator satisfies the following*:

(1) $\nu(\chi) \in \{0, 1, -1\}$ *for all* $\chi \in \mathrm{Irr}(H)$.

(2) $\nu(\chi) \neq 0$ *if and only if* $V_\chi \cong V_{\chi^*}$

Moreover, $\nu(\chi) = 1$ (*resp.* -1) *if and only if* V_χ *admits a symmetric* (*resp. skew symmetric*) *non-degenerate bilinear* H*-invariant form.*

Theorems 5.1.12 and 5.1.14 found an interesting applications in [119] where it is proved that a semisimple Hopf algebra over $\mathbb{C}$ with a non-trivial self-dual irreducible representation or with an even-dimensional irreducible representation, must have even dimension.

Hopf algebras with positive bases

A finite-dimensional Hopf algebra H over $\mathbb{C}$ is said to have a positive base if it has a linear basis with respect to which all the structure constants are positive. For example, a bicrossproduct Hopf algebra arising from a finite group G and an exact factorization $G = G_+ G_-$ of G is a Hopf algebra with a positive base (e.g., $\mathbb{C}[G]$ and $D(G)$). In fact, if H is a Hopf algebra with a positive base then H is of this form, [141]. Note that in particular H is semisimple.

5.2. Pointed Hopf algebras

The reader is referred to [4,9,11] for more results, explanations and details.

Pointed Hopf algebras are of special interest as many important examples of Hopf algebras are such. In particular group algebras, enveloping algebras of Lie algebras, quantized enveloping algebras and many quantum groups are all pointed.

Moreover, by Remark 1.1.11, every cocommutative coalgebra C over an algebraically closed field k is pointed. Group algebras and enveloping algebras of Lie algebras are clearly cocommutative, but more is true. They serve as the building blocks in one of the first fundamental theorems about cocommutative Hopf algebras. This theorem was proved independently by Cartier, [36] and Kostant, unpublished (see [221, Preface]).

THEOREM 5.2.1. *A cocommutative Hopf algebra over an algebraically closed field of characteristic* 0 *is a smash product of the group algebra* kG *and the enveloping algebra* $U(\mathfrak{g})$, *where* G *is the group of grouplike elements of* H *and* $\mathfrak{g}$ *is the Lie algebra of primitive elements of* H.

The converse is obviously true. Furthermore, the converse can be generalized to any Hopf algebra H as follows: If H is generated as an algebra by $G(H)$ and the skew-primitive elements of H (see Definition 1.1.15) then H is pointed. In view of Theorem 5.2.1 and the above generalized converse, it was conjectured, [9], that all finite-dimensional pointed Hopf algebras over an algebraically closed field of characteristic 0 are generated as algebras by grouplike and skew-primitive elements. The conjecture is false in the infinite-dimensional case, [11, Example 3.6].

Finite-dimensional pointed Hopf algebras were characterized in the following cases:

(1) For a prime number $p > 2$. The only pointed Hopf algebras of dimension p^2 are the Taft algebras. This was already shown in [177].

(2) For $p = 2$ there is exactly one isomorphism class of dimension 2^n, [177,31]. These pointed Hopf algebras are generalizations of Sweedler's four-dimensional Hopf algebra H_4 and were investigated by, e.g., [17,185,186].

(3) Hopf algebras of dimension p^3, p^4, p^5 and pq^2, q another prime, are fully characterized, [8,6,30,217,10,100].

(4) Partial results are known for pointed Hopf algebras with some special properties.

An essential tool in the study of a pointed coalgebra C is its coradical filtration C_n. This structure, given in the fundamental theorem of Taft and Wilson, allows induction starting from $C_0 = kG(C)$. The following version is somewhat stronger than the original one:

THEOREM 5.2.2 [223]. *Let* C *be a pointed coalgebra. Then for any* $c \in C_n$, $n \geqslant 1$, *we have*

$$c = \sum_{g,h \in G(C)} c_{g,h}, \quad \text{where } \Delta(c_{g,h}) = c_{g,h} \otimes g + h \otimes c_{g,h} + w$$

for some $w \in C_{n-1} \otimes C_{n-1}$.

If H is a pointed Hopf algebra with coradical filtration H_n then $H_0 = kG(H)$ and

$$H_1 = kG(H) \oplus \left(\bigoplus_{\sigma,\tau \in G(H)} P_{\sigma,\tau}(H) \right).$$

Moreover, the coradical filtration is a Hopf algebra filtration, that is $H_i H_j \subset H_{i+j}$ and $S(H_i) \subset H_i$ for all $i, j \geqslant 0$.

The coradical filtration is the starting point for the *lifting method* which is a powerful tool in the structure theory of pointed Hopf algebras. This method was used in a series of works of Andruskiewitsch and Schneider (for more details see the references at the beginning of this section). We give a brief overview of this method and the main results.

Let H be a pointed Hopf algebra, let $\{H_n \mid n \geqslant 0\}$ denote the coradical filtration of H and set $H_{-1} = k$. Let

$$\operatorname{gr} H = \bigoplus_{n \geqslant 0} \operatorname{gr} H(n),$$

where $\operatorname{gr} H(n) = H_n / H_{n-1}$ for all $n \geqslant 0$. Since the coradical of H is a Hopf subalgebra it follows by [166, 5.2.8] that $\operatorname{gr} H$ is a graded Hopf algebra. Now, there is a Hopf algebra projection $\pi : \operatorname{gr} H \to \operatorname{gr} H(0) = kG(H)$ and a Hopf algebra injection $i : \operatorname{gr} H(0) \to \operatorname{gr} H$. By Theorem 4.4.6 this implies that we have a biproduct

$$\operatorname{gr} H \cong R \# kG(H),$$

where $R = \{x \in \operatorname{gr} H \mid (\mathrm{id} \otimes \pi)\Delta(x) = x \otimes 1\}$ is a Hopf algebra in the category ${}^{kG(H)}_{kG(H)}\mathcal{YD}$.

The structure of R is the key to understanding the structure of the original Hopf algebra H. The vector space V of all primitive elements of R is also an object in ${}^{kG(H)}_{kG(H)}\mathcal{YD}$ and thus has a braiding

$$c : V \otimes V \to V \otimes V.$$

This braiding is called the *infinitesimal braiding* of H.

The subalgebra of R generated by V turns out to be $B(V)$, the so-called Nichols algebra of V (see Example 4.4.3).

Given a group G and a vector space $(V, c) \in {}^{kG}_{kG}\mathcal{YD}$ the first problem is to study the structure of the Nichols algebras $B(V)$ and to determine when $B(V)$ is finite-dimensional. The second problem is to determine all pointed Hopf algebras H such that $\operatorname{gr} H \cong B(V) \# kG$ (the lifting problem).

The best results for this method were achieved in the case when G is Abelian and the braiding is of *Cartan type*. That is, for a certain basis $\{v_1, \ldots, v_n\}$ of V we have

$$c(v_i \otimes v_j) = q_{ij}(v_j \otimes v_i),$$

where $q_{ij} q_{ji} = q^{d_i a_{ij}}$, $q \neq 0$ and (a_{ij}) is a generalized symmetrizable Cartan matrix with positive integers $\{d_1, \ldots, d_n\}$ so that $d_i a_{ij} = d_j a_{ji}$.

The Cartan type of the pointed Hopf algebra is *invariant under twisting*.

Pointed Hopf algebras H (finite- or infinite-dimensional) such that $G(H)$ is Abelian and the braiding on V is of Cartan type (plus some additional requirements) were fully characterized by Andruskiewitsch and Schneider. They are generalizations of quantized Lie algebras.

The order of S^2 for pointed Hopf algebras

By Theorem 2.4.1, the order of S^2 divides $2\dim(H)$, when H is a finite-dimensional Hopf algebra. Using the coradical filtration the following was proved:

THEOREM 5.2.3 [197]. *Let H be a finite-dimensional pointed Hopf algebra over $\mathbb{C}$. Then $|S^2|$ divides* $\dim(H)/|G(H)|$.

THEOREM 5.2.4 [82]. *Let H be a finite-dimensional pointed Hopf algebra over $\mathbb{C}$. Then $|S^2|$ divides* $\exp(G(H))$ (*and hence* $\dim(H)$).

This theorem was used to prove

THEOREM 5.2.5 [82]. *Let H be a finite-dimensional pointed Hopf algebra over $\mathbb{C}$. Then* $\operatorname{qexp}(H) = \exp(G(H))$.

Hopf algebras of rooted trees

The $\mathbb{Z}$-graded Hopf algebra A of rooted trees was introduced in [102], in connection with numerical algorithms for ordinary differential equations; this Hopf algebra is cocommutative but not commutative.

On the other hand Kreimer, [125], has discovered the interesting fact that the process of renormalization in quantum field theory may be described by means of Hopf algebras related to operads of rooted trees. The Hopf algebra L of decorated rooted trees described by A. Connes and D. Kreimer, [53], arises from the combinatorics of perturbative renormalization, and is related to cyclic cohomology and non-commutative geometry. It is $\mathbb{Z}$-graded and commutative but not cocommutative.

In [184] the author proves that the Hopf algebras A and L are dual to one another. Moreover, he identifies a certain linear operator on A as a dual operator to L.

Related structures

There are several algebraic structures generalizing the notion of a Hopf algebra, which are very important and interesting in their own right and have been studied extensively. For example, quasi-Hopf algebras, weak Hopf algebras, Hopf algebroids, infinitesimal Hopf algebras, [2], multiplier Hopf algebras, [238], and dendriform Hopf algebras, [139], are such algebraic structures. In the following we briefly discuss some of them.

(1) Quasi-Hopf algebras. The notion of a quasi-Hopf algebra H, due to V. Drinfeld, generalizes the notion of a Hopf algebra in that the associativity constraint $(U \otimes V) \otimes W \simeq U \otimes (V \otimes W)$ in the tensor category $\operatorname{Rep}(H)$ can be non-trivial, [28,29]. More precisely, a quasi-Hopf algebra is a unital associative algebra with comultiplication Δ, counit ε, and antipode S satisfying some axioms, where the main difference is that Δ is only coassociative up to conjugation by an invertible element in $H \otimes H \otimes H$, [73]. See also [215].

The importance of these algebras lies in the fact that their representation category is tensor (usually with a non-standard associativity). For example, they produce solutions to the Knizhnik–Zamolodchikov equation in quantum field theory.

(2) Weak Hopf algebras. A weak Hopf algebra or a *quantum groupoid* is a unital associative algebra with comultiplication Δ, counit ε and antipode S, satisfying some axioms. The main difference between Hopf algebras and weak Hopf algebras is that in the latter Δ need not map the identity in H to the identity in $H \otimes H$. This relaxation of the axioms of Hopf algebras is very significant. For example, while not every finite (fusion) category is equivalent to $\mathrm{Rep}(H)$ for some finite-dimensional (semisimple) Hopf algebra H, it is known that it is equivalent to $\mathrm{Rep}(H)$ for some finite-dimensional (semisimple) weak Hopf algebra. Thus the theory of weak Hopf algebras is very useful in the study of finite (fusion) categories. See [27,86,87,182].

(3) Multiplier Hopf algebras. A multiplier Hopf algebra is an algebra A with or without identity and a homomorphism Δ from A to the multiplier algebra of $A \otimes A$ satisfying certain axioms (such as a form of coassociativity). If A has an identity then A is a usual Hopf algebra.

Here, as for Hopf algebras, the motivating example arises from groups. Consider the algebra A of all complex valued functions on a group G and define Δ, ε, S as for $\mathrm{F}un(G)$, G a finite group (see the introduction). If G is infinite then $\Delta(f)$ does not necessarily belong to $A \otimes A$ and thus A is not a Hopf algebra. However, if A is the algebra of (continuous) functions with compact support on a discrete group G then $\Delta(f)(g \otimes 1)$, $(1 \otimes f)\Delta(g) \in A \otimes A$ for all $f, g \in A$. This fact implies that A is a multiplier Hopf algebra.

Many results for Hopf algebras can be generalized for multiplier Hopf algebras by using similar methods (see, e.g., [238,70,69]).

References

[1] E. Abe, *Hopf Algebras*, Cambridge Univ. Press (1980) (original Japanese version published by Iwanami Shoten, Tokyo, 1977).

[2] M. Aguiar, *Infinitesimal Hopf algebras*, Contemp. Math. **267** (2000), 1–29.

[3] N. Andruskiewitsch, *Note on extensions of Hopf algebras*, Canad. J. Math. **48** (1996), 3–42.

[4] N. Andruskiewitsch, *About finite dimensional Hopf algebras*, Notes of a course given at the CIMPA School "Quantum Symmetries in Theoretical Physics and Mathematics", Bariloche, 2000, Contemp. Math **294** (2002), 1–57.

[5] N. Andruskiewitsch and J. Devato, *Extensions of Hopf algebras*, St. Petersburg Math. J. **7** (1996), 17–52.

[6] N. Andruskiewitsch and S. Natale, *Counting arguments for low dimensional Hopf algebras*, Tsukuba Math. J. **25** (1) (2001), 187–201.

[7] N. Andruskiewitsch and H.J. Schneider, *Hopf algebras of order p^2 and braided Hopf algebras of order p*, J. Algebra **199** (1998), 430–454.

[8] N. Andruskiewitsch and H.-J. Schneider, *Lifting of quantum linear spaces and pointed Hopf algebras of order p^3*, J. Algebra **209** (2) (1998), 658–691.

[9] N. Andruskiewitsch and H.J. Schneider, *Finite quantum groups and Cartan matrices*, Adv. in Math. **154** (2000), 1–45.

[10] N. Andruskiewitsch and H.J. Schneider, *Finite quantum groups over Abelian groups of prime exponent*, Ann. Sci. École Norm. Sup. (4) **35** (1) (2002), 1–26.

[11] N. Andruskiewitsch and H.J. Schneider, *Pointed Hopf Algebras*, Recent Developments in Hopf Algebras Theory, MSRI Series, Cambridge Univ. Press (2002).

[12] M. Artin, W. Schelter and J. Tate, *Quantum deformations of* GL_n, Comm. Pure Appl. Math. **44** (8–9) (1991), 879–895.

[13] M. Auslander and O. Goldman, *Maximal orders*, Trans. Amer. Math. Soc. **97** (1960), 1–24.

[14] Y. Bahturin, D. Fischman and S. Montgomery, *Bicharacters, twistings, and Scheunert's theorem for Hopf algebras*, J. Algebra **236** (1) (2001), 246–276.

[15] B. Bakalov and A. Kirillov, *Lectures on Tensor Categories and Modular Functors*, University Lecture Series **21**, Amer. Math. Soc. (2001).

[16] M. Beattie, *On the Blattner–Montgomery duality theorem for Hopf algebras*, Contemp. Math. **124** (1992), 23–28.

[17] M. Beattie, S. Dăscălescu and L. Grunenfelder, *On the number of types of finite-dimensional Hopf algebras*, Invent. Math. **136** (1) (1999), 1–7.

[18] M. Beattie, S. Dascalescu, S. Raianu and F. van Oystaeyen, *The categories of Yetter–Drinfeld modules, Doi–Hopf modules and two-sided two-cosided Hopf modules*, Appl. Categ. Structures **6** (2) (1998), 223–237.

[19] A.D. Bell, *Comodule algebras and Galois extensions relative to polynomial algebras, free algebras and enveloping algebras*, Comm. Algebra **28** (2000), 337–362.

[20] I. Bergen and S. Montgomery, *Smash products and outer derivations*, Israel J. Math. **53** (1986), 321–345.

[21] G. Bergman, *Everybody knows what a Hopf algebra is*, Contemp. Math. **43** (1985), 25–48.

[22] Yu.N. Bespalov, *Crossed modules and quantum groups in braided categories*, Appl. Categ. Structures **5** (2) (1997), 155–204.

[23] Y. Bespalov, T. Kerler, V. Lyubashenko and V. Turaev, *Integrals for braided Hopf algebras*, Preprint.

[24] R.J. Blattner, M. Cohen and S. Montgomery, *Crossed products and inner actions of Hopf algebras*, Trans. Amer. Math. Soc. **298** (1986), 671–711.

[25] R.J. Blattner and S. Montgomery, *A duality theorem for Hopf module algebras*, J. Algebra **95** (1985), 153–172.

[26] R.J. Blattner and S. Mongomery, *Crossed products and Galois extensions of Hopf algebras*, Pacific J. Math. **137** (1989), 37–54.

[27] G. Bohm, F. Nill and K. Szlachanyi, *Weak Hopf algebras I: Integral theory and* C^* *structure*, J. Algebra **221** (1999), 385–438.

[28] D. Bulacu, F. Panaite and F. van Oystaeyen, *Quantum traces and quantum dimensions for quasi-Hopf algebras*, Comm. Algebra **27** (12) (1999), 6103–6122.

[29] D. Bulacu, F. Panaite and F. van Oystaeyen, *Quasi-Hopf algebra actions and smash products*, Comm. Algebra **28** (2) (2000), 631–651.

[30] S. Caenepeel and S. Dascalescu, *Pointed Hopf algebras of dimension* p^3, J. Algebra **209** (1998), 622–634.

[31] S. Caenepeel and S. Dascalescu, *On pointed Hopf algebras of dimension* 2^n, Bull. London Math. Soc. **31** (1999), 17–24.

[32] S. Caenepeel, S. Dascalescu and L. le Bruyn, *Forms of pointed Hopf algebras*, Manuscripta Math. **100** (1999), 35–53.

[33] S. Caenepeel, G. Militaru and S. Zhu, *Crossed modules and Doi–Hopf modules*, Israel J. Math. **100** (1997), 221–247.

[34] S. Caenepeel and S. Raianu, *Abelian groups and modules*, Padova, 1994, 73–94.

[35] S. Caenepeel, F. van Oystaeyen and Y.H. Zhang, *Quantum Yang–Baxter module algebras*, K-Theory **8** (1994), 231–255.

[36] P. Cartier, *Groupes algebriques et groupes formels*, Colloq. Theorie des Groups Algebriques, Bruxelles, 1962, Librairie Universitaire, Louvain, Gauthier-Villars, 87–811.

[37] S.U. Chase, D.K. Harrison and A. Rosenberg, *Galois theory and cohomology of commutative rings*, Mem. Amer. Math. Soc. **52** (1965).

[38] W. Chin, *Spectra of smash products*, Israel J. Math. **72** (1990), 84–98.

[39] W. Chin, *Crossed products of semisimple cocommutative Hopf algebras*, Proc. Amer. Math. Soc. **116** (1992), 321–327.

[40] W. Chin and I. Musson, *Multiparameter quantum enveloping algebras*, J. Pure Appl. Algebra **107** (2–3) (1996), 171–191.

[41] M. Cohen, *Smash products, inner actions and quotient rings*, Pacific J. Math. **125** (1986), 45–65.
[42] M. Cohen, *On generalized characters*, Contemp. Math. **267** (2000), 55–65.
[43] M. Cohen and D. Fischman, *Hopf algebra actions*, J. Algebra **100** (1986), 363–379.
[44] M. Cohen, D. Fischman and S. Montgomery, *Hopf Galois extensions, smash products and Morita equivalence*, J. Algebra **133** (1990), 351–372.
[45] M. Cohen, D. Fischman and S. Montgomery, *On Yetter Drinfeld categories and H-commutativity*, Comm. Algebra **27** (1999), 1321–1345.
[46] M. Cohen, D. Fischman and S. Westreich, *Schure double centralizer theorem for triangular Hopf algebras*, Proc. Amer. Math. Soc. **122** (1) (1994), 19–29.
[47] M. Cohen and S. Montgomery, *Group-graded rings, smash products and group actions*, Trans. Amer. Math. Soc. **282** (1984), 237–258.
[48] M. Cohen and L.H. Rowen, *Group graded rings*, Comm. Algebra **11** (1983), 1253–1270.
[49] M. Cohen and S. Westreich, *Central invariance of H-module algebras*, Comm. Algebra **21** (1993), 2859–2883.
[50] M. Cohen and S. Westreich, *From supersymmetry to quantum commutativity*, J. Algebra **168** (1994), 1–27.
[51] M. Cohen, S. Westreich and S. Zhu, *Determinants, integrality and Noether's theorem for quantum commutative algebras*, Israel J. Math. **96** (1996), 185–222.
[52] M. Cohen and S. Zhu, *Invariants of the adjoint coaction and Yetter–Drinfeld categories*, J. Pure Appl. Algebra **159** (2–3) (2001), 149–171.
[53] A.Connes and D. Kreimer, *Hopf algebras, renormalization and noncommutative geometry*, Comm. Math. Phys. **199** (1) (1998), 203–242.
[54] S. Dascalescu, C. Nastasescu and S. Raianu, *Hopf Algebras – An Introduction*, Marcel Dekker (2001).
[55] S. Dascalescu and S. Raianu, *Finite Hopf–Galois coextensions, crossed coproducts and duality*, J. Algebra **178** (1995), 400–413.
[56] P. Deligne, *Categories tannakiennes*, The Grothendieck Festschrift, Vol. II, Prog. Math. **87** (1990), 111–195.
[57] P. Deligne, *Categories tensorielles*, www.math.ias.edu/~phares/deligne/deligne.html (February 2002).
[58] P. Deligne and J.S. Milne, *Tannakien Categories*, Lecture Notes in Math. **900**, Springer (1982).
[59] M. Demazure and A. Grothendieck, *Schémas en groupes*, 1962/64, Lecture Notes in Math. **151**, **152**, **153**, Springer (1970).
[60] Y. Doi, *On the structure of relative Hopf modules*, Comm. Algebra **11** (1983), 243–255.
[61] Y. Doi, *Algebras with total integrals*, Comm. Algebra **13** (1985), 2137–2159.
[62] Y. Doi, *Unifying Hopf modules*, J. Algebra **153** (2) (1992), 373–385.
[63] Y. Doi, *Braided bialgebras and quadratic bialgebras*, Comm. Algebra **21** (1993), 1731–1749.
[64] Y. Doi, *Hopf algebras in Yetter–Drinfeld categories*, Comm. Algebra **26** (1998), 3057–3070.
[65] Y. Doi, *The trace formula for braided Hopf algebras*, Preprint.
[66] Y. Doi and A. Masuoka, *Generalizations of cleft comodule algebras*, Comm. Algebra **20** (12) (1992), 3703–3721.
[67] Y. Doi and M. Takeuchi, *Cleft comodule algebras for a bialgebra*, Comm. Algebra **14** (1986), 801–818.
[68] Y. Doi and M. Takeuchi, *Hopf Galois extensions of algebras, the Miyashita–Ulbrich action and Azumaya algebras*, J. Algebra **121** (1989), 488–516.
[69] B. Drabant and A. van Daele, *Pairing and quantum double of multiplier Hopf algebra*, Alg. Represent. Theory **4** (2001), 109–132.
[70] B. Drabant, A. van Daele and Y.-H. Zhang, *Actions of multiplier Hopf algebras*, Comm. Algebra **27** (9) (1999), 4117–4172.
[71] V.G. Drinfeld, *Quantum groups*, Proc. Int. Cong. Math. Berkeley **1** (1986), 789–820.
[72] V.G. Drinfeld, *On almost cocommutative Hopf algebras*, Leningrad Math. J. **1** (1990), 321–342 (Russian original in Algebra and Analysis, 1989).
[73] V.G. Drinfeld, *Quasi Hopf algebras*, Leningrad Math. J. **1** (1990), 1419–1457 (Russian original in Algebra and Analysis, 1989).
[74] M. Domokos and T. Lenagen, *Weakly multiplicative coactions of quantized function algebras*, J. Pure Appl. Algebra, to appear.
[75] M. Enock and J.M. Schwartz, *Produit croisé d'une algèbre von Neumann per une algebra de Kac II*, Publ. Res. Inst. Math. Sci. **16** (1980), 189–232.

[76] P. Etingof and S. Gelaki, *On finite-dimensional semisimple and cosemisimple Hopf algebras in positive characteristic*, Internat. Math. Res. Notices **16** (1998), 851–864.
[77] P. Etingof and S. Gelaki, *Some properties of finite-dimensional semisimple Hopf algebras*, Math. Res. Lett. **5** (1998), 191–197.
[78] P. Etingof and S. Gelaki, *Semisimple Hopf algebras of dimension pq are trivial*, J. Algebra **210** (1998), 664–669.
[79] P. Etingof and S. Gelaki, *On the exponent of finite-dimensional Hopf algebras*, Math. Res. Lett. **6** (1999), 131–140.
[80] P. Etingof and S. Gelaki, *The classification of triangular semisimple and cosemisimple Hopf algebras over an algebraically closed field*, Internat. Math. Res. Notices **5** (2000), 223–234.
[81] P. Etingof and S. Gelaki, *On cotriangular Hopf algebras*, Amer. J. Math. **123** (4) (2001), 699–713.
[82] P. Etingof and S. Gelaki, *On the quasi-exponent of finite-dimensional Hopf algebras*, Math. Res. Lett. **9** (2002), 277–287.
[83] P. Etingof and S. Gelaki, *On families of triangular Hopf algebras*, Internat. Math. Res. Notices **14** (2002), 757–768.
[84] P. Etingof and S. Gelaki, *The classification of finite-dimensional triangular Hopf algebras over an algebraically closed field of characteristic* 0, Moscow Math. J., to appear, math.QA/0202258.
[85] P. Etingof and S. Gelaki, *On Hopf algebras of dimension pq*, submitted, math.QA/0303359.
[86] P. Etingof, D. Nikshych and V. Ostrik, *On fusion categories*, Preprint, math.QA/0203060.
[87] P. Etingof and V. Ostrik, *Finite tensor categories*, Preprint, math.QA/0301027.
[88] P. Etingof, T. Schedler and A. Soloviev, *Set-theoretic solutions to the quantum Yang–Baxter equation*, Duke Math. J., to appear, math.QA/9801047.
[89] P. Etingof and O. Schiffmann, *Lectures on Quantum Groups*, Lectures in Math. Phys., International Press (1998).
[90] L.D. Faddeev, N.Yu. Reshetikhin and L.A. Takhtajan, *Quantization of Lie groups and Lie algebras*, Algebraic Analysis, Vol. I, Academic Press (1988), 129–139.
[91] L.D. Faddeev, N.Yu. Reshetikhin and L.A. Takhtajan, *Quantum groups. Braid group, knot theory and statistical mechanics*, Adv. Ser. Math. Phys. **9**, World Sci. Publishing (1989), 97–110.
[92] W. Ferrer-Santos, *Finite generation of the invariants of finite-dimensional Hopf algebras*, J. Algebra **165** (3) (1994), 543–549.
[93] D. Fischman and S. Montgomery, *A Schur double centralizer theorem for cotriangular Hopf algebras and generalized Lie algebras*, J. Algebra **168** (2) (1994), 594–614.
[94] I.W. Fisher and S. Montgomery, *Semiprime skew group rings*, J. Algebra **52** (1978), 241–247.
[95] D. Fischman, S. Montgomery and H.J. Schneider, *Frobenius extensions of subalgebras of Hopf algebras*, Trans. Amer. Math. Soc. **349** (1997), 4857–4895.
[96] S. Gelaki, *Pointed Hopf algebras and Kaplansky's* 10*th conjecture*, J. Algebra **209** (2) (1998), 635–657.
[97] S. Gelaki and S. Westreich, *Hopf algebras of types $U_q(sl_n)'$ and $O_q(SL_n)'$ which give rise to certain invariants of knots, links and* 3*-manifolds*, Trans. Amer. Math. Soc. **352** (8) (2000), 3821–3836.
[98] S. Gelaki and S. Westreich, *On semisimple Hopf algebras of dimension pq*, Proc. Amer. Math. Soc. **128** (1) (2000), 39–47.
[99] M. Gerstenhaber and S.D Schack, *Bialgebra cohomology, deformations, and quantum groups*, Proc. Natl. Acad. Sci. USA **87** (1990), 478–481.
[100] M. Grana, *On pointed Hopf algebras of dimension p^5*, Glasgow Math. J. **42** (2000), 405–419.
[101] J.A. Green, *Locally finite representations*, J. Algebra **41** (1975), 137–171.
[102] R. Grossman and R. Larson, *Hopf-algebraic structure of families of trees*, J. Algebra **126** (1) (1989), 184–210.
[103] T. Hayashi, *Quantum groups and quantum determinants*, J. Algebra **152** (1992), 146–165.
[104] T. Hayashi, *Quantum deformations of classical groups*, Publ. RIMS Kyoto Univ. **28** (1992), 57–81.
[105] R.G. Heyneman and D.E. Radford, *Reflexivity and coalgebras of finite type*, J. Algebra (1974), 215–246.
[106] R.G. Heyneman and M.E. Sweedler, *Affine Hopf algebras I*, J. Algebra **13** (1969), 192–241.
[107] G. Hochschild, *Introduction to Affine Algebraic Groups*, Holden-Day (1971), vii+116 pp.
[108] H. Hopf, *Über die Topologie der Gruppen-Mannigfaltigkeiten und ihre Verallgemeinerungen*, Ann. of Math. **42** (1941), 22–52.
[109] R.B. Howlett and I.M. Isaacs, *On groups of central type*, Math. Z. **179** (1982), 555–569.

[110] S.A. Joni and G.C. Rota, *Coalgebras and bialgebras in combinatorics*, Contemp. Math. **6** (1982), 1–47.

[111] A. Joseph and G. Letzter, *Local finiteness of the adjoint action for quantized enveloping algebras*, J. Algebra **153** (1992), 289–318.

[112] A. Joyal and R. Street, *An introduction to Tannaka duality and quantum groups*, Category Theory, Como, 1990, Lecture Notes in Math. **1488**, Springer (1991), 413–492.

[113] A. Joyal and R. Street, *Braided tensor categories*, Adv. Math. **102** (1) (1993), 20–78.

[114] G.I. Kac, *Certain arithmetic properties of ring groups*, Funct. Anal. Appl. **6** (1972), 158–160.

[115] G.I. Kac and V.G. Paljutkin, *Finite ring groups*, Trudy Moskov. Mat. Obshch. **15** (1966), 224–261 (Russian, English translation).

[116] I. Kaplansky, *Bialgebras*, Univ. of Chicago Lecture Notes (1975).

[117] Y. Kashina, *A generalized power map for Hopf algebras*, Hopf Algebras and Quantum Groups, Proc. Colloq. in Brussels, 1998, S. Caenepeel (ed.).

[118] Y. Kashina, *On the order of the antipode of Hopf algebras in* ${}^H_H\mathcal{YD}$, Comm. Algebra **27** (3) (1999), 1261–1273.

[119] Y. Kashina, Y. Sommerhaeuser and Y. Zhu, *Self-dual modules of semisimple Hopf algebras*, Preprint, math.RA/0106254.

[120] C. Kassel, *Quantum Groups*, Springer (1995).

[121] L.H. Kauffman and D.E. Radford, *A necessary and sufficient condition for a finite-dimensional Drinfel'd double to be a ribbon Hopf algebra*, J. Algebra **159** (1993), 98–114.

[122] L.H. Kauffman and D.E. Radford, *Invariants of* 3*-manifolds derived from finite dimensional Hopf algebras*, J. Knots Theory Ramifications **4** (1) (1995), 131–162.

[123] V.K. Kharchenko, *Automorphism and Derivations of Associative Rings*, Kluwer (1991).

[124] V.K. Kharchenko and A.Z. Popov, *Skew derivations of prime rings*, Comm. Algebra **20** (1992), 3321–3345.

[125] D. Kreimer, *On the Hopf algebra structure of perturbative quantum field theories*, Adv. Theor. Math. Phys. **2** (2) (1998).

[126] M. Koppinen, *A Skolem–Noether theorem for coalgebra measurings*, Arch. Math. **57** (1991), 34–40.

[127] M. Koppinen, *A duality theorem for crossed products of Hopf algebras*, J. Algebra **146** (1992), 153–174.

[128] H.F. Kreimer and P.M. Cook, *Galois theories and normal bases*, J. Algebra **43** (1) (1976), 115–121.

[129] H.F. Kreimer and M. Takeuchi, *Hopf algebras and Galois extensions of an algebra*, Indiana Univ. Math. J. **30** (1981), 675–692.

[130] B. Kostant, *Graded manifolds, graded Lie theory, and prequantization*, Differ. Geom. Meth. Math. Phys., Proc. Symp. Bonn, 1975, Lecture Notes in Math. **570** (1977), 177–306.

[131] R.G. Larson, *Characters of Hopf algebras*, J. Algebra **17** (1971), 352–368.

[132] R.G. Larson, *Topological Hopf algebras and braided monoidal categories*, Appl. Categ. Structures **6** (2) (1998), 139–150.

[133] R.G. Larson and D. Radford, *Semisimple cosemisimple Hopf algebras*, Amer. J. Math. **109** (1987), 287–295.

[134] R.G. Larson and D. Radford, *Finite-dimensional cosemisimple Hopf algebras in characteristic* 0 *are semisimple*, J. Algebra **117** (1988), 267–289.

[135] R.G. Larson and M.E. Sweedler, *An associative orthogonal bilinear form for Hopf algebras*, Amer. J. Math. **91** (1969), 75–93.

[136] R.G. Larson and J. Towber, *Two dual classes of bialgebras related to the concepts of "quantum group" and "quantum Lie algebra"*, Comm. Algebra **19** (12) (1991), 3295–3345.

[137] V. Linchenko and S. Montgomery, *A Frobenius–Schur theorem for Hopf algebras*, Alg. Represent. Theory **3** (2000), 347–355.

[138] V. Linchenko, S. Montgomery and L.Y. Small, *Semisimple smash products and the Jacobson radical*, Preprint.

[139] J.L. Loday and M. Ronco, *Hopf algebra of the planar binary trees*, Adv. Math. **139** (1998), 293–309.

[140] J.-H. Lu, *On the Drinfeld double and the Heisenberg double of a Hopf algebra*, Duke Math. J. **74** (1994), 763–776.

[141] J.-H. Lu, M. Yan and Y. Zhu, *On Hopf algebras with positive bases*, J. Algebra **237** (2) (2001), 421–445.

[142] V.V. Lyubashenko, *Hopf algebras and vector symmetries*, Russian Math. Surveys **41** (1986), 153–154.

[143] V.V. Lyubashenko, *Modular transformations for tensor categories*, J. Pure Appl. Algebra **98** (1995), 279–327.

[144] A. Masuoka, *Coalgebra actions on Azumaya algebras*, Tsukuba J. Math. **14** (1990), 107–112.

[145] A. Masuoka, *Existence of a unique maximal subcoalgebra whose action is inner*, Israel J. Math. **72** (1990), 149–157.

[146] A. Masuoka, *Self-dual Hopf algebras of dimension p^3 obtained by extensions*, J. Algebra **178** (1995), 791–806.

[147] A. Masuoka, *The p^n theorem for semisimple Hopf algebras*, Proc. Amer. Math. Soc. **124** (1996), 735–737.

[148] A. Masuoka, *Calculations of some groups of Hopf algebras extensions*, J. Algebra **191** (1997), 568–588.

[149] A. Masuoka, *Defending the negated Kaplansky conjecture*, Proc. Amer. Math. Soc. **129** (11) (2001), 3185–3192.

[150] S. Mac Lane, *Natural associativity and commutativity*, Rice Univ. Studies **49** (1963), 28–46.

[151] S. Mac Lane, *Categories for the Working Mathematician*, Springer (1971).

[152] S. Majid, *Quantum groups and quantum probability*, Quantum Probability and Related Topics VI, Proc. Toronto, 1989, World Scientific (1991).

[153] S. Majid, *Physics for algebraists: non-commutative and non-cocommutative Hopf algebras by a bicrossproduct construction*, J. Algebra **130** (1990), 17–64.

[154] S. Majid, *Quasitriangular Hopf algebras and Yang–Baxter equations*, Internat. J. Modern Phys. A **5** (1) (1990), 1–91.

[155] S. Majid, *Reconstruction theorems and rational conformal field theories*, Internat. J. Modern Phys. A **6** (24) (1991), 4359–4374.

[156] S. Majid, *Doubles of quasitriangular Hopf algebras*, Comm. Algebra **19** (11) (1991), 3061–3073.

[157] S. Majid, *Tannaka–Kreĭ n Theorem for Quasi-Hopf Algebras and Other Results*, Contemp. Math. **134**, Amer. Math. Soc. (1992).

[158] S. Majid, *Cross products by braided groups and bosonization*, J. Algebra **163** (1) (1994), 165–190.

[159] S. Majid, *Algebras and Hopf algebras in braided categories*, Marcel Dekker Lecture Notes in Math. **82** (1994), 55–105.

[160] S. Majid, *Foundations of Quantum Group Theory*, Cambridge Univ. Press (1995).

[161] A. Milinski, *Actions of pointed Hopf algebras on prime algebras*, Comm. Algebra **23** (1995), 313–333.

[162] A. Milinski, *X-Inner objects for Hopf crossed products*, J. Algebra **185** (1996), 390–408.

[163] J. Milnor and J.C. Moore, *On the structure of Hopf algebras*, Ann. Math. **81** (1965), 211–264.

[164] R.K. Molnar, *Semi-direct products of Hopf algebras*, J. Algebra **47** (1977), 29–51.

[165] S. Montgomery, *Fixed Rings of Finite Automorphism Groups of Associative Rings*, Lecture Notes in Math. **818**, Springer (1980).

[166] S. Montgomery, *Hopf Algebras and their Actions on Rings*, CBMS Lectures **82**, Amer. Math. Soc. (1993).

[167] S. Montgomery, *Biinvertible actions of Hopf algebras*, Comm. Algebra **21** (1993), 45–72.

[168] S. Montgomery, *Classifying finite-dimensional semisimple Hopf algebras*, Trends in the Representation Theory of Finite-Dimensional Algebras, Seattle, WA, 1997, Contemp. Math. **229**, Amer. Math. Soc. (1998), 265–279.

[169] S. Montgomery, *Representation theory of semisimple Hopf algebras*, Algebra-Representation Theory, Constanta, 2000, NATO Sci. Ser. II Math. Phys. Chem. **28**, Kluwer Acad. Publ. (2001), 189–218.

[170] S. Montgomery, *Algebra properties invariant under twisting*, Preprint.

[171] S. Montgomery and H.J. Schneider, *Hopf crossed products, rings of quotients, and prime deals*, Adv. Math. **112** (1) (1995), 1–55.

[172] S. Montgomery and H.J. Schneider, *Prime ideals in Hopf Galois extensions*, Israel J. Math. **112** (1999), 187–235.

[173] S. Montgomery and S.J. Witherspoon, *Irreducible representations of crossed products*, J. Pure Appl. Algebra **129** (1998), 315–326.

[174] M. Movshev, *Twisting in group algebras of finite groups*, Funct. Anal. Appl. **27** (1994), 240–244.

[175] Y. Nakagami and M. Takesaki, *Duality for Crossed Products of von Neumann Algebras*, Lecture Notes in Math. **731**, Springer (1979).

[176] C. Nastasescu and F. van Oystaeyen, *Graded Ring Theory*, North-Holland (1982).

[177] W.D. Nichols, *Bialgebras of type one*, Comm. Algebra **6** (1978), 1521–1552.

[178] W.D. Nichols and M.B. Zoeller, *A Hopf algebra freeness theorem*, Amer. J. Math. **111** (1989), 381–385.

[179] W.D. Nichols and B. (Zoeller) Richmond, *The Grothendieck group of a Hopf algebra*, J. Pure Appl. Algebra **106** (1996), 297–306.

[180] D. Nikschych, *K_0-rings and twisting of finite dimensional semisimple Hopf algebras*, Comm. Algebra **26** (1) (1998), 321–342.

[181] D. Nikschych, *Duality for actions of weak Kac algebras and crossed product inclusions of II_1 factors*, Preprint.

[182] D. Nikshych and L. Vainerman, *Finite quantum groupoids and their applications*, New Directions in Hopf Algebras, Math. Sci. Res. Inst. Publ. **43**, Cambridge Univ. Press (2002), 211–262.

[183] U. Oberst and H.J. Schneider, *Über Untergruppen endlicher algebraischer Gruppen*, Manuscripta Math. **8** (1973), 217–241.

[184] F. Panaite, *Relating the Connes–Kreimer and Grossman–Larson Hopf algebras built on rooted trees*, Lett. Math. Phys. **51** (3) (2000), 211–219.

[185] F. Panaite and F. van Oystaeyen, *Quasitriangular structures for some pointed Hopf algebras of dimension 2^n*, Comm. Algebra **27** (10) (1999), 4929–4942.

[186] F. Panaite and F. van Oystaeyen, *Clifford-type algebras as cleft extensions for some pointed Hopf algebras*, Comm. Algebra **28** (2) (2000), 585–600.

[187] B. Pareigis, *A noncommutative noncocommutative Hopf algebra in "nature"*, J. Algebra **70** (2) (1981), 356–374.

[188] D.S. Passman, *Infinite Crossed Products*, Academic Press (1989).

[189] D.E. Radford, *The order of the antipode of a finite-dimensional Hopf algebra is finite*, Bull. Amer. Math. Soc. **81** (1975), 1103–1105.

[190] D.E. Radford, *The structure of Hopf algebras with a projection*, J. Algebra **92** (1985), 322–347.

[191] D.E. Radford, *The group of automorphisms of a semisimple Hopf algebra over a field of* ch. 0 *is finite*, Amer. J. Math. **112** (1990), 331–357.

[192] D.E. Radford, *Minimal quasitriangular Hopf algebras*, J. Algebra **157** (1993), 285–315.

[193] D.E. Radford, *The trace function and Hopf algebras*, J. Algebra **163** (1994), 583–622.

[194] D.E. Radford, *On Kauffman's knot invariants arising from finite-dimensional Hopf algebras*, Advances in Hopf Algebras, Chicago, IL, 1992, Lecture Notes in Pure and Appl. Math. **158**, Marcel Dekker (1994), 205–266.

[195] D. Radford and H.-J. Schneider, *On the even powers of the antipode of a finite-dimensional Hopf algebra*, Preprint.

[196] S. Raianu and M. Saorin, *Finite Hopf–Galois extensions equivalent to crossed products*, Comm. Algebra **29** (11) (2001), 4871–4882.

[197] N.Y. Reshetikhin and M.A. Semenov Tian-Shansky, *Quantum R-matrices and factorization problems*, J. Geom. Phys. **5** (1998), 533–550.

[198] N.Yu. Reshetikhin and V.G. Turaev, *Invariants of* 3*-manifolds via link polynomials and quantum groups*, Invent. Math. **103** (1991), 547–597.

[199] M. Rosso, *Quantum groups and quantum shuffles*, Inventiones Math. **133** (1998), 399–416.

[200] S. Saavedra, *Categories tannakiennes*, Lecture Notes in Math. **265**, Springer (1972).

[201] B. Scharfschwerdt, *The Nichols–Zoeller theorem for Hopf algebras in the category of Yetter–Drinfeld modules*, Comm. Algebra **29** (6) (2001), 2481–2487.

[202] P. Schauenburg, *On coquasitriangular Hopf algebras and the quantum Yang–Baxter equations*, Algebra Berichte **67** (1992), Math. Inst. Uni. München.

[203] P. Schauenburg, *Hopf modules and Yetter–Drinfeld modules*, J. Algebra **169** (3) (1994), 874–890.

[204] P. Schauenburg, *Hopf bi-Galois extensions*, Comm. Algebra **24** (1996), 3797–3825.

[205] M. Scheunert, *Generalized Lie algebras*, J. Math. Phys. **20** (1979), 712–720.

[206] H.-J. Schneider, *Principal homogeneous spaces for arbitrary Hopf algebras*, Israel J. Math. **72** (1990), 167–195.

[207] H.-J. Schneider, *Normal basis and transitivity of crossed products for Hopf algebras*, J. Algebra **151** (1992), 289–312.

[208] H.-J. Schneider, *On inner actions of Hopf algebras and stabilizers of representations*, J. Algebra (1993).

[209] H.-J. Schneider, *Lectures on Hopf Algebras*, Trab. Mat. Mathematical Works 31/95 (1995) (FaMAF).

[210] H.-J. Schneider, *Some properties of factorizable Hopf algebras*, Proc. Amer. Math. Soc. **129** (7) (2001), 1891–1898.

[211] J.-P. Serre, *Local Fields*, Graduate Texts in Math. **67**, Springer.

[212] Y. Sommerhauser, *On Kaplanski's fifth conjecture*, Preprint gk-mp-9702/50 (1997).

[213] Y. Sommerhauser, *On Kaplanski's fifth conjecture*, Interrelations between Ring Theory and Representations of Algebras, Lecture Notes, Pure and Appl. Math., Marcel Dekker (2000), 393–412.
[214] Y. Sommerhauser, *Das Drinfeld-Doppel und die Jones-Konstruction*, Preprint.
[215] J. Stasheft, *Drinfeld's quasi-Hopf algebras and beyond*, Contemp. Math. **134** (1992), 297–307.
[216] D. Stefan, *The set of types of n-dimensional semisimple and cosemisimple Hopf algebras is finite*, J. Algebra **193** (2) (1997), 571–580.
[217] D. Stefan and F. van Oystaeyen, *Hochschild cohomology and coradical filtration of pointed Hopf algebras*, J. Algebra **210** (1998), 535–556.
[218] A. Sudbery, *Consistent multiparameter quantisation of* $\mathrm{GL}(n)$, J. Phys. A **23** (15) (1990), 697–704.
[219] J. Sullivan, *The uniqueness of integrals for Hopf algebras and some existence theorems for commutative Hopf algebras*, J. Algebra **19** (1971), 426–440.
[220] M.E. Sweedler, *Integrals for Hopf algebras*, Ann. of Math. **89** (1969), 323–335.
[221] M.E. Sweedler, *Hopf Algebras*, Benjamin (1969).
[222] E.J. Taft, *The order of the antipode of finite-dimensional Hopf algebra*, Proc. Natl. Acad. Sci. USA **68** (1971), 2631–2633.
[223] E.J. Taft and R.L. Wilson, *On antipodes in pointed Hopf algebras*, J. Algebra **29** (1974), 27–32.
[224] M. Takeuchi, *A correspondence between Hopf ideals and Hopf subalgebras*, Manuscripta Math. **7** (1972), 251–270.
[225] M. Takeuchi, *Matched pairs of groups and bismash products of Hopf algebras*, Comm. Algebra **9** (1981), 841–882.
[226] M. Takeuchi, *The coquasitriangular Hopf algebra associated to a rigid Yang Baxter coalgebra*, Hopf Algebras and Quantum Groups, Proc. Colloq. in Brussels, 1998, S. Caenepeel (ed.).
[227] M. Takeuchi, *Finite Hopf algebras in braided tensor categories*, J. Pure Appl. Algebra **138** (1999), 59–82.
[228] M. Takeuchi, *Survey of braided Hopf algebras*, New Trends in Hopf Algebra Theory, La Falda, 1999, Contemp. Math. **267**, Amer. Math. Soc. (2000), 301–323.
[229] D. Tambara and S. Yamagami, *Tensor categories with fusion rules of self-duality for Abelian groups*, J. Algebra **189** (1997), 23–33.
[230] J. Towber, *Multiparameter quantum forms of the enveloping algebra Ugl_N related to the Faddeev–Reshetikhin–Takhtajan $U(R)$ constructions*, J. Knot Theory Ramifications **4** (2) (1995), 263–317.
[231] Y. Tsang and Y. Zhu, *On the Drinfeld double of a Hopf algebra* (1998).
[232] V. Turaev, *Quantum invariants of knots and 3-manifolds*, Walter de Gruyter (1994).
[233] K.H. Ulbrich, *On Hopf algebras and rigid monoidal categories*, Hopf Algebras, Israel J. Math. **72** (1–2) (1990), 252–256.
[234] K.H. Ulbrich, *Vollgraduierte Algebren*, Abh. Math. Sem. Univ. Hamburg **51** (1981), 136–148.
[235] K.H. Ulbrich, *Galois erweterungen von nicht-kommutativen ringen*, Comm. Algebra **10** (1982), 655–672.
[236] K.H. Ulbrich, *On Hopf algebras and rigid monoidal categories*, Hopf Algebras, Israel J. Math. **72** (1–2) (1990), 252–256.
[237] C. Vafa, *Towards classification of conformal theories*, Phys. Lett. B **206** (1988), 421–426.
[238] A. Van Daele and Y. Zhang, *A survey on multiplier Hopf algebras*, Hopf Algebras and Quantum Groups, Brussels, 1998, Lecture Notes in Pure and Appl. Math. **209**, Marcel Dekker (2000), 269–309.
[239] M. van den Bergh, *A duality theorem for Hopf algebras*, Methods in Ring Theory, NATO ASI Series **129**, Reidel (1984), 517–522.
[240] F. van Oystaeyen and Y.H. Zhang, *Brauer groups of actions*, Rings, Hopf Algebras, and Brauer Groups, Antwerp/Brussels, 1996, Lecture Notes in Pure and Appl. Math. **197**, Marcel Dekker (1998), 299–309.
[241] F. van Oystaeyen and Y.H. Zhang, *Embedding the Hopf automorphism group into the Brauer group*, Canad. Math. Bull. **41** (3) (1998), 359–367 (English. English summary).
[242] E. Verlinde, *Fusion rules and modular transformations in 2D conformal field theory*, Nuclear Phys. B **300** (1988), 360.
[243] W. Waterhouse, *Introduction to Affine Group Schemes*, Springer (1997).
[244] S. Westreich, *A Galois-type correspondence theory for actions of finite dimensional pointed Hopf algebras on prime algebras*, J. Algebra **219** (1999), 606–624.
[245] S. Westreich and T. Yanai, *More about a Galois-type correspondence theory*, J. Algebra **246** (2) (2001), 629–640.

[246] S.L. Woronowicz, *Differential calculus on compact matrix pseudogroups (quantum groups)*, Comm. Math. Phys. **122** (1989), 125–170.

[247] T. Yanai, *Correspondence theory of Kharchenko and X-outer actions of pointed Hopf algebras*, Comm. Algebra **25** (6) (1997), 1713–1740.

[248] D.N. Yetter, *Quantum groups and representations of monoidal categories*, Math. Proc. Cambridge Philos. Soc. **108** (1990), 261–290.

[249] S. Zhu, *On finite-dimensional Hopf algebras*, Comm. Algebra **21** (1993), 3871–3885.

[250] Y. Zhu, *Hopf algebras of prime dimensions*, Internat. Math. Res. Notices **1** (1994), 53–59.

Difference Algebra

Alexander B. Levin

Department of Mathematics, The Catholic University of America, Washington, DC 20064, USA

E-mail: Levin@cua.edu

Contents

HANDBOOK OF ALGEBRA, VOL. 4
Edited by M. Hazewinkel

1. Introduction

Difference algebra as a separate area of mathematics was born in the 1930s when J.F. Ritt (1893–1951) developed the algebraic approach to the study of systems of difference equations over function fields. In a series of papers published during the decade from 1929 to 1939, Ritt worked out the foundations of both differential and difference algebra, the theories of abstract algebraic structures with operators that reflect the algebraic properties of derivatives and shifts of arguments of analytic functions, respectively. One can say that differential and difference algebra grew out of the study of algebraic differential and difference equations with coefficients from function fields in much the same way as the classical algebraic geometry arose from the study of polynomial equations with numerical coefficients.

Ritt's research in differential algebra was continued and extended by H. Raudenbush, H. Levi, A. Seidenberg, A. Rosenfeld, P. Cassidy, J. Johnson, W. Keigher, S. Morrison, W. Sit and many other mathematicians, but the most important role in this area was played by E. Kolchin who recast the whole subject in the style of modern algebraic geometry with the additional presence of derivation operators. In particular, E. Kolchin developed the contemporary theory of differential fields and created differential Galois theory where finite-dimensional algebraic groups played the same role as finite groups play in the theory of algebraic equations. Kolchin's monograph, [86], is the most deep and complete book on the subject, it contains a lot of ideas that determined the main directions of research in differential algebra for the last thirty years.

The rate of development of difference algebra after Ritt's pioneering work and works by F. Herzog, H. Raudenbush and W. Strodt published in the 1930s (see [68,133,134, 136,137,140], and [141]) was much slower than the rate of expansion of its differential counterpart. The situation began to change in the 1950s due to R.M. Cohn whose works, [19–32], not only raised difference algebra to a level comparable with the level of development of differential algebra, but also clarified why many ideas of differential algebra cannot be realized in the difference case and a number of methods and results of difference algebra cannot have differential analogs. R.M. Cohn's book, [29], up to now remains the only fundamental monograph on difference algebra. Since the 60s various problems of difference algebra were studied by A. Babbitt, [2], I. Balaba, [3–5], I. Bentsen, [7], R.M. Cohn, [33–37], P. Evanovich, [49,50], C. Franke, [60–62], B. Greenspan, [63], P. Hendrics, [65,66], M. Kondrateva, [87–91], B. Lando, [96,97], A. Levin, [87,88] and [99–115], A. Mikhalev, [87,88,110–115] and [117–119], E. Pankratev, [87–91,117–119] and [124–127], C. Praagman, [130,131], and some other mathematicians. Difference Galois theory originated in the 60s and 70s in works by C. Franke, [56–59], A. Bialynicki-Birula, [8], H.F. Kreimer, [93–95], R. Infante, [70–75], and E. Pankratev, [124,125], was further actively developed in the last ten years by M. van der Put, M. Singer, and P.A. Hendrix, [64–67,142]. The current state of the theory is fully reflected in the recent monograph by M. van der Put and M. Singer, [142].

Since the 70s difference algebra has been enriched by new methods and ideas from the dimension theory of differential rings (see [41–43,78–82,84,85], and [139]), the theory of Gröbner bases which originated in [13] and computer algebra. A number of deep results were obtained in the model theory of difference fields developed by E. Hrushovski and

Z. Chatzidakis, [15,16] and [69] (see also [46,128] and [129]). Nowadays, difference algebra appears as a rich theory with its own methods and with applications to the study of system of equations in finite differences, functional equations, differential equations with delay, algebraic structures with operators, group and semigroup rings. A number of interesting applications of difference algebra in the theory of discrete-time non-linear systems can be found in the works by M. Fliess, [51–55], E. Aranda-Bricaire, U. Kotta and C. Moog, [1], and some other authors.

In what follows, **Z**, **N**, **Q**, **R**, and **C** denote the sets of integers, non-negative integers, rational numbers, real numbers, and complex numbers respectively. $\mathbf{Q}[t]$ will denote the set of all polynomials in one variable t with rational coefficients. By a ring we always mean an associative ring with a unity. Every ring homomorphism is unitary (maps unity onto unity), every subring of a ring contains the unity of the ring. An ideal I of a ring R is said to be *proper* if $I \neq R$. Unless otherwise indicated, by a module over a ring A we mean a left A-module. Every module over a ring is unitary and every algebra over a commutative ring is also unitary.

2. Basic concepts of difference algebra

2.1. *Difference and inversive difference rings*

A *difference ring* is a commutative ring R together with a finite set $\sigma = \{\alpha_1, \dots, \alpha_n\}$ of mutually commuting injective endomorphisms of R into itself. The set σ is called the *basic set* of the difference ring R, and the endomorphisms $\alpha_1, \dots, \alpha_n$ are called *translations*. In other words, a difference ring R with a basic set $\sigma = \{\alpha_1, \dots, \alpha_n\}$ is a commutative ring possessing n additional unitary operations $\alpha_i : a \mapsto \alpha_i(a)$ such that

$$\alpha_i(a) = 0 \quad \text{if and only if} \quad a = 0,$$

$$\alpha_i(a + b) = \alpha_i(a) + \alpha_i(b),$$

$$\alpha_i(ab) = \alpha_i(a)\alpha_i(b),$$

$$\alpha_i(1) = 1, \quad \text{and}$$

$$\alpha_i\bigl(\alpha_j(a)\bigr) = \alpha_j\bigl(\alpha_i(a)\bigr)$$

for any $a \in R$, $1 \leqslant i, j \leqslant n$. (Formally speaking, a difference ring is an $(n+1)$-tuple $(R, \alpha_1, \dots, \alpha_n)$ where R is a ring and $\alpha_1, \dots, \alpha_n$ are mutually commuting injective endomorphisms of R. Unless the notation is inconvenient or ambiguous, we always write R for $(R, \alpha_1, \dots, \alpha_n)$.)

If $\alpha_1, \dots, \alpha_n$ are automorphisms of R, we say that R is an *inversive difference ring* with the basic set σ.

In what follows, a difference ring R with a basic set $\sigma = \{\alpha_1, \dots, \alpha_n\}$ will be also called a σ-ring. If $n = 1$, R is said to be an *ordinary* difference (σ-) ring, if $\operatorname{Card}\sigma > 1$, the difference ring R is called *partial*.

If R is an inversive difference ring with a basic set $\sigma = \{\alpha_1, \dots, \alpha_n\}$, then the set $\{\alpha_1, \dots, \alpha_n, \alpha_1^{-1}, \dots, \alpha_n^{-1}\}$ is denoted by σ^* and R is also called a σ^*-ring.

If a difference ring with a basic set σ is a field, it is called a difference (or σ-) field. An inversive difference field with a basic set σ is also called a σ^*-field.

Let R be a difference ring with basic set $\sigma = \{\alpha_1, \dots, \alpha_n\}$ and R_0 a subring of R such that $\alpha(R_0) \subseteq R_0$ for any $\alpha \in \sigma$. Then R_0 is called a *difference* (or σ-) *subring* of R and the ring R is said to be a *difference* (or σ-) *overring* of R_0. In this case the restriction of an endomorphism α_i on R_0 is denoted by same symbol α_i. If the σ-ring R is inversive and R_0 a σ-subring of R such that $\alpha^{-1}(R_0) \subseteq R_0$ for any $\alpha \in \sigma$, then R_0 is said to be a σ^*-subring of R. If R is a difference (σ-) field and R_0 a subfield of R such that $\alpha(a) \in R_0$ for any $a \in R_0, \alpha \in \sigma$, then R_0 is said to be a difference (or σ-) subfield of R; R, in turn, is called a difference (or σ-) field extension or a σ-overfield of R_0. In this case we also say that we have a σ-field extension R/R_0. If R is inversive and its subfield R_0 is a σ^*-subring of R, then R_0 is said to be a σ^*-subfield of R while R is called a σ^*-field extension or a σ^*-overfield of R_0. (We also say that we have a σ^*-field extension R/R_0.) If $R_0 \subseteq R_1 \subseteq R$ is a chain of σ- (σ^*-) field extensions, we say that R_1/R_0 is a σ- (respectively, σ^*-) field subextension of R/R_0.

If R is a difference ring with a basic set σ and J is an ideal of the ring R such that $\alpha(J) \subseteq J$ for any $\alpha \in \sigma$, then J is called a *difference* (or σ-) ideal of R. If a prime (maximal) ideal P of R is closed with respect to σ (that is, $\alpha(P) \subseteq P$ for any $\alpha \in \sigma$), it is called a *prime* (respectively, *maximal*) *difference* (or σ-) *ideal* of R.

A σ-ideal J of a σ-ring R is called *reflexive* (or a σ^**-ideal*) if for any translation α, the inclusion $\alpha(a) \in J$ $(a \in R)$ implies $a \in J$. Clearly, if R is an inversive σ-ring, then a σ-ideal J of R is reflexive if and only if it is closed under all automorphisms from the set σ^*. A prime (maximal) reflexive σ-ideal of a σ-ring R is also called a prime (respectively, maximal) σ^*-ideal of R.

EXAMPLES 2.1.1.

(1) Any commutative ring can be treated as a difference (or inversive difference) ring with a basic set σ consisting of one or several identity automorphisms.

(2) Let $z_0 \in \mathbf{C}$ and let U be a region of the complex plane such that $z + z_0 \in U$ whenever $z \in U$ (e.g., $U = \{z \in \mathbf{C} \mid (\operatorname{Re} z)(\operatorname{Re} z_0) \geqslant 0\}$). Furthermore, let M_U denote the field of all functions of one complex variable meromorphic in U. Then M_U can be treated as an ordinary difference field whose basic set consists of one translation α such that $\alpha(f(z)) = f(z + z_0)$ for any function $f(z) \in M_U$. It is clear that this difference field is inversive if and only if $z - z_0 \in U$ for any $z \in U$. (In this case $\alpha^{-1} : f(z) \mapsto f(z - z_0)$ for any $f(z) \in M_U$.)

(3) Let $0 \neq z_0 \in \mathbf{C}$ and let V be a region of the complex plane such that $zz_0 \in U$ whenever $z \in V$ (e.g., $|z_0| \leqslant 1$ and $V = \{z \in \mathbf{C} \mid |z| \leqslant r\}$ for some positive real number r). Then the field of all functions of one complex variable meromorphic in the region V can be considered as an ordinary difference field with a translation β such that $\beta(f(z)) = f(z_0 z)$ for any function $f(z) \in M_V$. Clearly, M_V is inversive if and only if $\frac{z}{z_0} \in V$ for any $z \in V$.

(4) Let A be a ring of functions of n real variables continuous on $\mathbf{R}^n$. Let us fix some real numbers $h_1, \dots, h_n$ and consider the mutually commuting injective endomorphisms

$\alpha_1, \ldots, \alpha_n$ of A given by $(\alpha_i f)(x_1, \ldots, x_n) = f(x_1, \ldots, x_{i-1}, x_i + h_i, x_{i+1}, \ldots, x_n)$ $(i = 1, \ldots, n)$. Then A can be treated as a difference ring with basic set $\sigma = \{\alpha_1, \ldots, \alpha_n\}$. This difference ring is denoted by $A_0(h_1, \ldots, h_n)$.

Similarly, one can introduce the difference structure on the ring $C^p(\mathbf{R}^n)$ of all functions of n real variables that are continuous on $\mathbf{R}^n$ together with all their partial derivatives up to the order p ($p \in \mathbf{N}$ or $p = +\infty$). It is easy to see that $C^p(\mathbf{R}^n)$ can be considered as a difference ring with the basic set $\sigma = \{\alpha_1, \ldots, \alpha_n\}$ described above. This difference ring is denoted by $A_p(h_1, \ldots, h_n)$. It is clear that $A_p(h_1, \ldots, h_n)$ is a σ-subring of the σ-ring $A_q(h_1, \ldots, h_n)$ whenever $p > q$. The difference rings $A_p(h_1, \ldots, h_n)$ often arise in connection with equations in finite differences when the i-th partial finite difference $\Delta_i f(x_1, \ldots, x_n) = f(x_1, \ldots, x_{i-1}, x_i + h_i, x_{i+1}, \ldots, x_n) - f(x_1, \ldots, x_n)$ of a function $f(x_1, \ldots, x_n) \in C^p(\mathbf{R}^n)$ is written as $\Delta_i f = (\alpha_i - 1) f$.

A number of interesting examples of difference rings can be found in [32, Chapter 1], [88, Section 3.3], and [142, Chapter 1].

Let R be a difference ring with a basic set σ. An element $c \in R$ is said to be a *constant* if $\alpha(a) = a$ for any $\alpha \in \sigma$. Clearly, the set of constants of the ring R is a σ-subring of R (it is a σ^*-subring of R, if the difference ring R is inversive). This subring is called the *ring of constants* of R; it is denoted by C_R.

If R is a difference ring with a basic set $\sigma = \{\alpha_1, \ldots, \alpha_n\}$, then T_σ will denote the free commutative semigroup with identity generated by $\alpha_1, \ldots, \alpha_n$. Elements of T_σ will be written in the multiplicative form $\alpha_1^{k_1} \ldots \alpha_n^{k_n}$ ($k_1, \ldots, k_n \in \mathbf{N}$) and considered as endomorphisms of R. If the σ-ring R is inversive, then Γ_σ will denote the free commutative group generated by the set σ. It is clear that elements of the group Γ_σ (written in the multiplicative form $\alpha_1^{i_1} \ldots \alpha_n^{i_n}$ where $i_1, \ldots, i_n \in \mathbf{Z}$) act on R as automorphisms and T_σ is a subsemigroup of Γ_σ.

For any $a \in R$ and for any $\tau \in T_\sigma$, the element $\tau(a)$ is called a *transform* of a. If the σ-ring R is inversive, then an element $\gamma(a)$ ($a \in R, \gamma \in \Gamma_\sigma$) is also called a transform of a.

If J is a σ-ideal of a σ-ring R, then $J^* = \{a \in R \mid \tau(a) \in J$ for some $\tau \in T_\sigma\}$ is a reflexive σ-ideal of R contained in any reflexive σ-ideal of R containing J. The ideal J^* is called a *reflexive closure* of the σ-ideal J.

A difference (σ-) ring R is called *simple* if the only σ-ideals of R are (0) and R. In this case the ring of constants C_R is a field.

Let R be a difference ring with a basic set σ and $S \subseteq R$. Then the intersection of all σ-ideals of R containing S is denoted by $[S]$. Clearly, $[S]$ is the smallest σ-ideal of R containing S; as an ideal, it is generated by the set $T_\sigma S = \{\tau(a) \mid \tau \in T_\sigma, a \in S\}$. If $J = [S]$, we say that the σ-ideal J is generated by the set S called a *set of σ-generators* of J. If S is finite, $S = \{a_1, \ldots, a_k\}$, we write $J = [a_1, \ldots, a_k]$ and say that J is a finitely generated σ-ideal of the σ-ring R. (In this case elements $a_1, \ldots, a_k$ are said to be σ-generators of J.)

If R is an inversive difference (σ-) ring and $S \subseteq R$, then the inverse closure of the σ-ideal $[S]$ is denoted by $[S]^*$. It is easy to see that $[S]^*$ is the smallest σ^*-ideal of R containing S; as an ideal, it is generated by the set $\Gamma_\sigma S = \{\gamma(a) \mid \gamma \in \Gamma_\sigma, a \in S\}$. If S is finite, $S = \{a_1, \ldots, a_k\}$, we write $[a_1, \ldots, a_k]^*$ for $I = [S]^*$ and say that I is a finitely generated σ^*-ideal of R. (In this case, the elements $a_1, \ldots, a_k$ are said to be σ^*-generators of I.)

Let R be a difference ring with a basic set σ, R_0 a σ-subring of R and $B \subseteq R$. The intersection of all σ-subrings of R containing R_0 and B is called the *σ-subring of R generated by the set B over R_0*, it is denoted by $R_0\{B\}$. (As a ring, $R_0\{B\}$ coincides with the ring $R_0[\{\tau(b) \mid b \in B, \tau \in T_\sigma\}]$ obtained by adjoining the set $\{\tau(b) \mid b \in B, \tau \in T_\sigma\}$ to the ring R_0). The set B is said to be the set of *σ-generators* of the σ-ring $R_0\{B\}$ over R_0. If this set is finite, $B = \{b_1, \dots, b_k\}$, we say that $R' = R_0\{B\}$ is a finitely generated difference (or σ-) ring extension (or overring) of R_0 and write $R' = R_0\{b_1, \dots, b_k\}$. If R is a σ-field, R_0 a σ-subfield of R and $B \subseteq R$, then the intersection of all σ-subfields of R containing R_0 and B is denoted by $R_0\langle B\rangle$ (or $R_0\langle b_1, \dots, b_k\rangle$ if $B = \{b_1, \dots, b_k\}$ is a finite set). This is the smallest σ-subfield of R containing R_0 and B; it coincides with the field $R_0(\{\tau(b) \mid b \in B, \tau \in T_\sigma\})$. The set B is called the set of *σ-generators* of the σ-field $R_0\langle B\rangle$ over R_0.

Let R be an inversive difference ring with a basic set σ, R_0 a σ^*-subring of R and $B \subseteq R$. Then the intersection of all σ^*-subrings of R containing R_0 and B is the smallest σ^*-subring of R containing R_0 and B. This ring coincides with the ring $R_0[\{\gamma(b) \mid b \in B, \gamma \in \Gamma_\sigma\}]$; it is denoted by $R_0\{B\}^*$. The set B is said to be a *set of σ^*-generators* of $R_0\{B\}^*$ over R_0. If $B = \{b_1, \dots, b_k\}$ is a finite set, we say that $S = R_0\{B\}^*$ is a finitely generated inversive difference (or σ^*-) ring extension (or overring) of R and write $S = R_0\{b_1, \dots, b_k\}^*$.

If R is a σ^*-field, R_0 a σ^*-subfield of R and $B \subseteq R$, then the intersection of all σ^*-subfields of R containing R_0 and B is denoted by $R_0\langle B\rangle^*$. This is the smallest σ^*-subfield of R containing R_0 and B; it coincides with the field $R_0(\{\gamma(b) \mid b \in B, \gamma \in \Gamma_\sigma\})$. The set B is called the *set of σ^*-generators of the σ^*-field extension $R_0\langle B\rangle^*$ over R_0*. If B is finite, $B = \{b_1, \dots, b_k\}$, we write $R_0\langle b_1, \dots, b_k\rangle^*$ for $R_0\langle B\rangle^*$.

In what follows we shall often consider two or more difference rings $R_1, \dots, R_p$ with the same basic set $\sigma = \{\alpha_1, \dots, \alpha_n\}$. Formally speaking, it means that for every $i = 1, \dots, p$, there is some fixed mapping ν_i from the set σ into the set of all injective endomorphisms of the ring R_i such that any two endomorphisms $\nu_i(\alpha_j)$ and $\nu_i(\alpha_k)$ of R_i commute ($1 \leqslant j, k \leqslant n$). We shall identify elements α_j with their images $\nu_i(\alpha_j)$ and say that elements of the set σ act as mutually commuting injective endomorphisms of the ring R_i ($i = 1, \dots, p$).

Let R_1 and R_2 be difference rings with the same basic set $\sigma = \{\alpha_1, \dots, \alpha_n\}$. A ring homomorphism $\phi : R_1 \to R_2$ is called a *difference* (or σ-) *homomorphism* if $\phi(\alpha(a)) = \alpha(\phi(a))$ for any $\alpha \in \sigma, a \in R_1$. Clearly, if $\phi : R_1 \to R_2$ is a σ-homomorphism of inversive difference rings, then $\phi(\alpha^{-1}(a)) = \alpha^{-1}(\phi(a))$ for any $\alpha \in \sigma$, $a \in R_1$. If a σ-homomorphism is an isomorphism (endomorphism, automorphism, etc.), it is called a difference (or σ-) isomorphism (respectively, difference (or σ-) endomorphism, difference (or σ-) automorphism, etc.). If R_1 and R_2 are two σ-overrings of the same σ-ring R_0 and $\phi : R_1 \to R_2$ is a σ-homomorphism such that $\phi(a) = a$ for any $a \in R_0$, we say that ϕ is a difference (or σ-) homomorphism over R_0 or that ϕ leaves the ring R_0 fixed. It is easy to see that the kernel of any σ-homomorphism of σ-rings $\phi : R \to R'$ is an inversive σ-ideal of R. Conversely, let g be a surjective homomorphism of a σ-ring R onto a ring S such that $\operatorname{Ker} g$ is a σ^*-ideal of R. Then there is a unique structure of a σ-ring on S such that g is a σ-homomorphism. In particular, if I is a σ^*-ideal of a σ-ring R, then the factor ring R/I has a unique structure of a σ-ring such that the canonical surjection $R \to R/I$ is a σ-homomorphism. In this case R/I is said to be the *difference* (or σ-) *factor ring* of R by the σ^*-ideal I.

Since a radical of a difference ideal is a difference ideal, every maximal σ-ideal I of a difference (σ-) ring R is radical and inversive. In this case R/I is a reduced σ-ring (that is, a σ-ring without non-zero nilpotent elements).

EXAMPLE 2.1.2. Let A be the set of all sequences $\mathbf{a} = (a_1, a_2, \ldots)$ of elements of an algebraically closed field C. Consider an equivalence relation on A such that $\mathbf{a} = (a_1, a_2, \ldots)$ is equivalent to $\mathbf{b} = (b_1, b_2, \ldots)$ if and only if $a_n = b_n$ for all sufficiently large $n \in \mathbf{N}$ (that is, there exists $n_0 \in \mathbf{N}$ such that $a_n = b_n$ for all $n > n_0$). Clearly, the corresponding set S of equivalence classes is a ring with respect to coordinatewise addition and multiplication of class representatives. This ring can be treated as an ordinary difference ring with respect to the mapping α sending an equivalence class with a representative $(a_1, a_2, a_3, \ldots)$ to the equivalence class with the representative $(a_2, a_3, \ldots)$. It is easy to see that this mapping is well-defined and it is an automorphism of the ring S. The field C can be naturally identified with the ring of constants of the difference ring S.

Let $\mathbf{C}(z)$ be the field of rational functions in one complex variable z. Then $\mathbf{C}(z)$ can be considered as an ordinary difference field with respect to the automorphism β such that $\beta(z) = z + 1$ and $\beta(a) = a$ for any $a \in \mathbf{C}$. In this case, the mapping $\phi : \mathbf{C}(z) \to S$ that sends a function $f(z)$ to the equivalence class of the element $(f(0), f(1), \ldots)$ is an injective difference ring homomorphism.

Let R be a difference ring with a basic set $\sigma = \{\alpha_1, \ldots, \alpha_n\}$. A σ-overring U of R is called an *inversive closure* of R, if the elements of σ act as mutually commuting automorphisms of the ring U (they are denoted by the same symbols $\alpha_1, \ldots, \alpha_n$) and for any $a \in U$, there exists an automorphism $\tau \in T_\sigma$ of the ring U such that $\tau(a) \in R$.

The ordinary version of the following statement can be found in [32, Chapter 2, Theorem II]; the corresponding theorem for partial difference rings was proved in [7, Theorem 3.1]. (Actually, many results in difference algebra were first proved for ordinary difference rings and then generalized to difference rings with several translations. In such cases we refer to both corresponding publications.)

PROPOSITION 2.1.3 ([32, Chapter 2, Theorem II], [7, Theorem 3.1]).

(i) *Every difference ring has an inversive closure.*

(ii) *If U_1 and U_2 are two inversive closures of a difference ring R, then there exists a difference R-isomorphism of U_1 onto U_2.*

(iii) *If R is a difference ring with a basic set σ and U an inversive σ-ring containing R as a σ-subring, then U contains an inversive closure of R.*

(iv) *If a difference ring R is an integral domain (a field), then its inversive closure is also an integral domain (respectively, a field).*

(v) *Let R_1 and R_2 be two difference rings with the same basic set σ, R_1^* and R_2^* their inversive closures, and $\phi : R_1 \to R_2$ a σ-homomorphism. Then ϕ has a unique extension to a σ-homomorphism $R_1^* \to R_2^*$.*

The inversive closure of an ordinary difference ring R with a basic set $\sigma = \{\alpha\}$ can be constructed as follows. Let $R' = \alpha(R)$ and let R'' be an isomorphic copy of R such that $R \cap R'' = \emptyset$. Let $\beta : R \to R''$ be the corresponding isomorphism and $R''' = \beta(R') = \beta\alpha(R)$. If one replaces all elements of R''' by the corresponding elements of R, then

R'' will be transformed into an overring R_1 of the ring R. The mapping $(\rho\beta)^{-1}$, where ρ denotes the replacement mapping $R'' \to R_1$, is an injective endomorphism of R_1 that extends α. This endomorphism will be also denoted by α and R_1 will be treated as a σ-overring of R. Now, let us consider the sequence of σ-rings $R = R_0, R_1, R_2, \dots$ where every R_n is a σ-overring of R_{n-1} obtained by the forgoing procedure, that is, $R_n = (R_{n-1})_1$ for $n = 1, 2, \dots$. Let us set $R^* = \bigcup_{n\in\mathbf{N}} R_n$ and define the extension of α to R^* as follows (we denote this extension by the same letter α). If $a \in R^*$, then $a \in R_n$ for some n and the extension of α to R_n determines an element $\alpha(a) \in R_n$ which we consider as the image of a under the mapping $\alpha : R^* \to R^*$. It is easy to see that the obtained mapping $\alpha : R^* \to R^*$ is well-defined (the image of an element $a \in R^*$ does not depend on the choice of R_n such that $a \in R_n$) and R^* is an inversive closure of R.

Let R be a partial difference ring with a basic set $\sigma = \{\alpha_1, \dots, \alpha_n\}$. Considering R as an ordinary difference ring whose basic set consists of the endomorphism $\tau = \prod_{i=1}^{n} \alpha_i$, we can construct the inversive closure R^* of this ring. Now we can extend each α_i ($1 \leqslant i \leqslant n$) to R^* as follows. For any $a \in R^*$, let $r(a)$ denote the smallest non-negative integer such that $\tau^{r(a)}(a) \in R$ (we denote the extension of τ to R^* by the same letter τ). Setting $\alpha_i(a) = \tau^{-r(a)}\alpha_i\tau^{r(a)}(a)$ for any $a \in R$ ($1 \leqslant i \leqslant n$), we obtain well-defined extensions of the endomorphisms $\alpha_1, \dots, \alpha_n$ to R^* that make R^* an inversive closure of the σ-ring R.

If H is an inversive difference field with a basic set σ and G a σ-subfield of H, then the set $\{a \in H \mid \tau(a) \in G \text{ for some } \tau \in T_\sigma\}$ is a σ^*-subfield of H denoted by G^*_H (or G^* if one considers subfields of a fixed σ^*-field H). This field is said to be the inversive closure of G in H. Clearly, G^*_H is the intersection of all σ^*-subfields of H containing G.

PROPOSITION 2.1.4 [7, Section 5]. *Let H be a σ^*-field and let $*$ be the operation that assigns to each σ-subfield $F \subseteq H$ its inversive closure in H. Let F and G be two σ-subfields of H and $\langle F, G\rangle$ denote the σ-field $F\langle G\rangle = G\langle F\rangle$ (the "σ-compositum" of F and G). Then*

(i) $F^{**} = F^*$.

(ii) $\langle F, G\rangle^* = \langle F^*, G^*\rangle$.

(iii) *Every σ-isomorphism of F onto G has a unique extension to a σ-isomorphism of F^* onto G^*.*

(iv) *If K is a σ-subfield of H contained in $F \cap G$ and F and G are free (linearly disjoint, quasi-linearly disjoint) over K, then F^* and G^* are free (linearly disjoint, quasi-linearly disjoint) over K^*.*

(The corresponding definitions can be found in [144, Vol. I, Chapter II].)

Let R be a difference ring with a basic set σ. A subset S of the ring R is said to be a *σ-subset* of R if $\sigma(s) \in S$ for any $s \in S$, $\alpha \in \sigma$. If the σ-ring R is inversive and S is a σ-subset of R such that $\alpha^{-1}(s) \in S$ for any $s \in S$, $\alpha \in \sigma$, then S is said to be a *σ^*-subset* of the ring R. By a *multiplicative σ-subset* of a σ-ring R we mean a σ-subset S of R such that $1 \in S$, $0 \notin S$, and $st \in S$ whenever $s \in R$ and $t \in R$. A *multiplicative σ^*-subset* of an inversive σ-ring R is a multiplicative σ-subset of R such that $\alpha^{-1}(s) \in S$ for any $s \in S$, $\alpha \in \sigma$.

The following statement is a natural generalization of [32, Chapter 2, Theorem III].

PROPOSITION 2.1.5. *Let S be a multiplicative σ-subset of a σ-ring R and let $S^{-1}R$ be the ring of fractions of R with denominators in S. Then $S^{-1}R$ has the unique structure of σ-ring such that the natural mapping $\nu : R \to S^{-1}R$ $(a \mapsto \frac{a}{1})$ is a σ-homomorphism. If the σ-ring R is inversive and S is a multiplicative σ^*-subset of R, then $S^{-1}R$ is a σ^*-overring of R.*

If S is a multiplicative σ-subset of a σ-ring R, then the ring $S^{-1}R$ is said to be a *σ-ring of fractions* of R with denominators in S. If the σ-ring R is inversive and S is a multiplicative σ^*-subset of R, then $S^{-1}R$ is called the *σ^*-ring of fractions* of R with denominators in S.

PROPOSITION 2.1.6. *Let R and R' be difference rings with the same basic set σ and let $\phi : R \to R'$ be a σ-homomorphism such that $\phi(s)$ is a unit of R' for any $s \in S$. Then ϕ factors uniquely through the canonical mapping $\nu : R \to S^{-1}R$: there exists a unique σ-homomorphism $\psi : S^{-1}R \to R'$ such that $\psi \circ \nu = \phi$. (This σ-homomorphism is given by $\psi(\frac{a}{s}) = \phi(a)\phi(s)^{-1}$.)*

The last proposition shows that if a difference ring R with a basic set σ is an integral domain, then its quotient field $Q(R)$ can be naturally considered as a σ-overring of R. (We identify an element $a \in R$ with its canonical image $\frac{a}{1}$ in $Q(R)$.) In this case $Q(R)$ is said to be the *quotient difference* (or σ-) *field* of R. Clearly, if the σ-ring R is inversive, then its quotient σ-field $Q(R)$ is also inversive. Furthermore, if a σ-field K contains an integral domain R as its σ-subring, then K contains the quotient σ-field $Q(R)$. Also, if the σ-field K is inversive and R is a σ^*-subring of K, then $Q(R)$ is a σ^*-subfield of K.

2.2. *Rings of difference and inversive difference polynomials. Algebraic difference equations*

Let R be a difference ring with a basic set $\sigma = \{\alpha_1, \dots, \alpha_n\}$, T_σ the free commutative semigroup generated by σ, and $U = \{u_\lambda \mid \lambda \in \Lambda\}$ a family of elements from some σ-overring of R. We say that the family U is *transformally* (or *σ-algebraically*) *dependent* over R, if the family $T_\sigma(U) = \{\tau(u_\lambda) \mid \tau \in T_\sigma, \lambda \in \Lambda\}$ is algebraically dependent over R (that is, there exist elements $v_1, \dots, v_k \in T_\sigma(U)$ and a non-zero polynomial $f(X_1, \dots, X_k)$ with coefficients from R such that $f(v_1, \dots, v_k) = 0$). Otherwise, the family U is said to be *transformally* (or *σ-algebraically*) *independent* over R or a family of *difference* (or σ-) *indeterminates* over R. In the last case, the σ-ring $R\{(u_\lambda)_{\lambda\in\Lambda}\}_\sigma$ is called the *algebra of difference* (or σ-) *polynomials* in the difference (or σ-) indeterminates $\{(u_\lambda)_{\lambda\in\Lambda}\}$ over R. If a family consisting of one element u is σ-algebraically dependent over R, the element u is said to be *transformally algebraic* (or *σ-algebraic*) over the σ-ring R. If the set $\{\tau(u) \mid \tau \in T\}$ is algebraically independent over R, we say that u is *transformally* (or σ-) *transcendental* over the ring R.

Let R be a σ-field, L a σ-overfield of R, and $S \subseteq L$. We say that *the set S is σ-algebraic over R* if every element $a \in S$ is σ-algebraic over R. If every element of L is σ-algebraic over R, we say that L is a *σ-algebraic field extension* of the σ-field R.

PROPOSITION 2.2.1 ([32, Chapter 2, Theorem I], [88, Proposition 3.3.7]). *Let R be a difference ring with a basic set σ and I an arbitrary set. Then there exists an algebra of σ-polynomials over R in a family of σ-indeterminates with indices from the set I. If S and S' are two such algebras, then there exists a σ-isomorphism $S \to S'$ that leaves the ring R fixed. If R is an integral domain, then any algebra of σ-polynomials over R is an integral domain.*

The algebra of σ-polynomials over the σ-ring R can be constructed as follows. Let $T = T_\sigma$ and let S be the polynomial R-algebra in the set of indeterminates $\{y_{i,\tau}\}_{i\in I,\tau\in T}$ with indices from the set $I \times T$. For any $f \in S$ and $\alpha \in \sigma$, let $\alpha(f)$ denote the polynomial from S obtained by replacing every indeterminate $y_{i,\tau}$ that appears in f by $y_{i,\alpha\tau}$ and every coefficient $a \in R$ by $\alpha(a)$. We obtain an injective endomorphism $S \to S$ that extends the original endomorphism α of R to the ring S (this extension is denoted by the same letter α). Setting $y_i = y_{i,1}$ (where 1 denotes the identity of the semigroup T) we obtain a σ-algebraically independent over R set $\{y_i \mid i \in I\}$ such that $S = R\{(y_i)_{i\in I}\}$. Thus, S is an algebra of σ-polynomials over R in a family of σ-indeterminates $\{y_i \mid i \in I\}$.

Let R be an inversive difference ring with a basic set σ, $\Gamma = \Gamma_\sigma$, I a set, and S^* a polynomial ring in the set of indeterminates $\{y_{i,\gamma}\}_{i\in I,\gamma\in\Gamma}$ with indices from the set $I \times \Gamma$. If we extend the automorphisms $\beta \in \sigma^*$ to S^* setting $\beta(y_{i,\gamma}) = y_{i,\beta\gamma}$ for any $y_{i,\gamma}$ and denote $y_{i,1}$ by y_i, then S^* becomes an inversive difference overring of R generated (as a σ^*-overring) by the family $\{(y_i)_{i\in I}\}$. Obviously, this family is σ^*-*algebraically independent* over R, that is, the set $\{\gamma(y_i) \mid \gamma \in \Gamma, i \in I\}$ is algebraically independent over R.

(Note that a set is σ^*-algebraically dependent (independent) over an inversive σ-ring if and only if this set is σ-algebraically dependent (respectively, independent) over this ring.) The ring $S^* = R\{(y_i)_{i\in I}\}^*$ is called the *algebra of inversive difference* (or σ^*-) *polynomials* over R in the set of σ^*-indeterminates $\{(y_i)_{i\in I}\}$. It is easy to see that S^* is the inversive closure of the ring of σ-polynomials $R\{(y_i)_{i\in I}\}$ over R. Furthermore, if a family $\{(u_i)_{i\in I}\}$ from some σ^*-overring of R is σ^*-*algebraically independent* over R, then the inversive difference ring $R\{(u_i)_{i\in I}\}^*$ is naturally σ-isomorphic to S^*. Any such overring $R\{(u_i)_{i\in I}\}^*$ is said to be an algebra of inversive difference (or σ^*-) polynomials over R in the set of σ^*-indeterminates $\{(u_i)_{i\in I}\}$. We obtain the following analog of Proposition 2.2.1.

PROPOSITION 2.2.1* [88, Proposition 3.4.4]. *Let R be an inversive difference ring with a basic set σ and I an arbitrary set. Then there exists an algebra of σ^*-polynomials over R in a family of σ^*-indeterminates with indices from the set I. If S and S' are two such algebras, then there exists a σ^*-isomorphism $S \to S'$ that leaves the ring R fixed. If R is an integral domain, then any algebra of σ^*-polynomials over R is an integral domain.*

Let R be a σ-ring, $R\{(y_i)_{i\in I}\}$ an algebra of difference polynomials in a family of σ-indeterminates $\{(y_i)_{i\in I}\}$, and $\{(\eta_i)_{i\in I}\}$ a set of elements from some σ-overring of R. Since the set $\{\tau(y_i) \mid i \in I, \tau \in T_\sigma\}$ is algebraically independent over R, there exists a unique ring homomorphism $\phi_\eta : R[\tau(y_i)_{i\in I,\tau\in T_\sigma}] \to R[\tau(\eta_i)_{i\in I,\tau\in T_\sigma}]$ that maps every $\tau(y_i)$ onto $\tau(\eta_i)$ and leaves R fixed. Clearly, ϕ_η is a surjective σ-homomorphism of $R\{(y_i)_{i\in I}\}$ onto $R\{(\eta_i)_{i\in I}\}$; it is called the *substitution* of $(\eta_i)_{i\in I}$ for $(y_i)_{i\in I}$. Similarly, if R is an inversive

σ-ring, $R\{(y_i)_{i\in I}\}^*$ an algebra of σ^*-polynomials over R and $(\eta_i)_{i\in I}$ a family of elements from a σ^*-overring of R, one can define a surjective σ-homomorphism $R\{(y_i)_{i\in I}\}^* \to R\{(\eta_i)_{i\in I}\}^*$ that maps every y_i onto η_i and leaves the ring R fixed. This homomorphism is also called the substitution of $(\eta_i)_{i\in I}$ for $(y_i)_{i\in I}$. (It will be always clear whether we talk about substitutions for difference or inversive difference polynomials.) If g is a σ- or σ^*-polynomial, then its image under a substitution of $(\eta_i)_{i\in I}$ for $(y_i)_{i\in I}$ is denoted by $g((\eta_i)_{i\in I})$. The kernel of a substitution ϕ_η is an inversive difference ideal of the σ-ring $R\{(y_i)_{i\in I}\}$ (or the σ^*-ring $R\{(y_i)_{i\in I}\}^*$); it is called the *defining difference* (or σ-) *ideal* of the family $(\eta_i)_{i\in I}$ over R. If R is a σ- (or σ^*-) field and $(\eta_i)_{i\in I}$ is a family of elements from some its σ- (respectively, σ^*-) overfield S, then $R\{(\eta_i)_{i\in I}\}$ (respectively, $R\{(\eta_i)_{i\in I}\}^*$) is an integral domain (it is contained in the field S). It follows that the defining σ-ideal P of the family $(\eta_i)_{i\in I}$ over R is a prime inversive difference ideal of the ring $R\{(y_i)_{i\in I}\}$ (respectively, of the ring of σ^*-polynomials $R\{(y_i)_{i\in I}\}^*$). Therefore, the difference field $R\langle(\eta_i)_{i\in I}\rangle$ can be treated as the quotient σ-field of the σ-ring $R\{(y_i)_{i\in I}\}/P$. (In the case of inversive difference rings, the σ^*-field $R\langle(\eta_i)_{i\in I}\rangle^*$ can be considered as a quotient σ-field of the σ^*-ring $R\{(y_i)_{i\in I}\}^*/P$.)

Let F be a difference field with a basic set σ and s a positive integer. By an *s-tuple over F* we mean an s-dimensional vector $a = (a_1, \dots, a_s)$ whose coordinates belong to some σ-overfield of F. If the σ-field F is inversive, the coordinates of an s-tuple over F are supposed to lie in some σ^*-overfield of F. If each a_i $(1 \leqslant i \leqslant s)$ is σ-algebraic over the σ-field F, we say that the *s-tuple a is σ-algebraic over F.*

DEFINITION 2.2.2. Let F be a difference (inversive difference) field with a basic set σ and let R be the algebra of σ- (respectively, σ^*-) polynomials in finitely many σ- (respectively, σ^*-) indeterminates $y_1, \dots, y_s$ over F. Furthermore, let $\Phi = \{f_j \mid j \in J\}$ be a set of σ- (respectively, σ^*-) polynomials from R. An s-tuple $\eta = (\eta_1, \dots, \eta_s)$ over F is said to be a solution of the set Φ or a solution of the system of difference algebraic equations $f_j(y_1, \dots, y_s) = 0$ $(j \in J)$ if Φ is contained in the kernel of the substitution of $(\eta_1, \dots, \eta_s)$ for $(y_1, \dots, y_s)$. In this case we also say that η annuls Φ. (If Φ is a subset of a ring of inversive difference polynomials, the system is said to be a system of algebraic σ^*-equations.)

As we have seen, if one fixes an s-tuple $\eta = (\eta_1, \dots, \eta_s)$ over a σ-field F, then all σ-polynomials of the ring $F\{y_1, \dots, y_s\}$, for which η is a solution, form a prime inversive difference ideal. It is called the *defining σ-ideal* of η. If η is an s-tuple over a σ^*-field F, then all σ^*-polynomials g of the ring $F\{y_1, \dots, y_s\}^*$ such that $g(\eta_1, \dots, \eta_s) = 0$ form a prime σ^*-ideal of $F\{y_1, \dots, y_s\}^*$. This ideal is called the *defining σ^*-ideal* of η over F.

Let Φ be a subset of the algebra of σ-polynomials $F\{y_1, \dots, y_s\}$ over a σ-field F. An s-tuple $\eta = (\eta_1, \dots, \eta_s)$ over F is called a *generic zero* of Φ if for any σ-polynomial $A \in F\{y_1, \dots, y_s\}$, the inclusion $A \in \Phi$ holds if and only if $A(\eta_1, \dots, \eta_s) = 0$. If the σ-field F is inversive, then the notion of a generic zero of a subset of $F\{y_1, \dots, y_s\}^*$ is defined similarly.

Two s-tuples $\eta = (\eta_1, \dots, \eta_s)$ and $\zeta = (\zeta_1, \dots, \zeta_s)$ over a σ- (or σ^*-) field F are called *equivalent* over F if there is a σ-isomorphism $F\langle\eta_1, \dots, \eta_s\rangle \to F\langle\zeta_1, \dots, \zeta_s\rangle$ (respectively, $F\langle\eta_1,\dots,\eta_s\rangle^* \to F\langle\zeta_1,\dots,\zeta_s\rangle^*$) that maps each η_i onto ζ_i and leaves the field F fixed.

PROPOSITION 2.2.3 ([32, Chapter 2, Theorem VII], [88, Proposition 3.3.20]). *Let S denote the algebra of σ-polynomials $F\{y_1,\dots,y_s\}$ or the algebra of σ^*-polynomials $F\{y_1,\dots,y_s\}^*$ over a difference (respectively, inversive difference) field F with a basic set σ.*

(i) *A set $\Phi \subsetneqq S$ has a generic zero if and only if Φ is a prime σ^*-ideal of S. If $(\eta_1,\dots,\eta_s)$ is a generic zero of Φ, then $F\langle\eta_1,\dots,\eta_s\rangle$ (or $F\langle\eta_1,\dots,\eta_s\rangle^*$ if we consider the algebra of σ^*-polynomials over a σ^*-field F) is σ-isomorphic to the quotient σ-field of S/Φ.*

(ii) *Any s-tuple over F is a generic zero of some prime σ^*-ideal of S.*

(iii) *If two s-tuples over F are generic zeros of the same prime σ^*-ideal of S, then these s-tuples are equivalent.*

EXAMPLE 2.2.4 (see [32, Chapter 2, Example 4]). Let us consider $\mathbf{C}$ as an ordinary difference field whose basic set σ consists of the identity automorphism α. Let $\mathbf{C}\{y\}$ be the algebra of σ-polynomials in one σ-indeterminate y over $\mathbf{C}$ and let ${}^{(k)}y$ denote the k-th transform $\alpha^k y$ ($k = 1, 2, \dots$). Furthermore, let M be the field of functions of one complex variable z meromorphic on the whole complex plane. Then M can be viewed as a σ-overfield of $\mathbf{C}$ if one extends α by setting $\alpha f(z) = f(z+1)$ for any function $f \in M$. It is easy to check that the σ-polynomial $A = ({}^{(1)}y - y)^2 - 2({}^{(1)}y + y) + 1$ is irreducible in $\mathbf{C}\{y\}$ (when this ring is treated as a polynomial ring in the denumerable set of indeterminates y, ${}^{(1)}y$, ${}^{(2)}y, \dots$). Furthermore, if $c(z)$ is a periodic function from M with period 1, then $\xi = (z + c(z))^2$ and $\eta = (c(z)e^{i\pi z} + \frac{1}{2})^2$ are solutions of A. (ξ is a solution of the system of the σ-polynomials A and $A' = {}^{(2)}y - 2\,{}^{(1)}y + y - 2$, while η is the solution of the system of A and $A'' = {}^{(2)}y - y$.) The fact that an irreducible σ-polynomial in one σ-indeterminate may have two distinct sets of solutions, each of which depends on an arbitrary periodic function, does not have an analog in the theory of differential polynomials.

Let F be a difference field with a basic set σ, $R = F\{y_1,\dots,y_s\}$ the algebra of σ-polynomials in a set of s σ-indeterminates $y_1,\dots,y_s$ over F, and $\Phi \subseteq R\{y_1,\dots,y_s\}$. Let $\bar{a} = \{a_{i,\tau} \mid i = 1,\dots,s,\ \tau \in T_\sigma\}$ be a family of elements from some σ-overfield of F. The family $\bar{a}$ (indexed by the set $\{1,\dots,s\} \times T_\sigma$) is said to be an *algebraic solution* of the set of σ-polynomials Φ if $\bar{a}$ is a solution of Φ when this set is treated as a set of polynomials in the polynomial ring $F[\{y_{i,\tau} \mid i = 1,\dots,s,\ \tau \in T_\sigma\}]$. (This polynomial ring in the denumerable family of indeterminates $\{y_{i,\tau} \mid i = 1,\dots,s,\ \tau \in T_\sigma\}$ coincides with R ($y_{i,\tau}$ stands for $\tau(y_i)$), but it is not considered as a difference ring, so the solutions of its subsets are solutions in the sense of classical algebraic geometry.)

It is easy to see that every solution $a = (a_1,\dots,a_s)$ of a set $\Phi \subseteq F\{y_1,\dots,y_s\}$ produces its algebraic solution $\bar{a} = \{\tau(a_i) \mid i = 1,\dots,s,\ \tau \in T_\sigma\}$. On the other hand, not every algebraic solution can be obtained from a solution in this way.

EXAMPLE 2.2.5 (see [32, Chapter 2, Example 6]). Let $\mathbf{C}$ be the field of complex numbers considered as an ordinary difference field whose basic set consists of the complex conjugation (that is, $\sigma = \{\alpha\}$ where $\alpha(a+bi) = a - bi$ for any complex number $a+bi$). Let $\mathbf{C}\{y\}$ be the ring of σ-polynomials in one σ-indeterminate y. If $A = y^2 + 1 \in \mathbf{C}\{y\}$, then the 1-tuples (i) and $(-i)$ are solutions of the σ-polynomial A that produce algebraic solutions $(i, -i, i, -i, \dots)$ and $(-i, i, -i, i, \dots)$ of A. At the same time, the sequence $(-i, i, i, i, \dots)$ is an algebraic solution of A that is not a solution of this σ-polynomial.

If Φ is a subset of an algebra of σ^*-polynomials $F\{y_1, \dots, y_s\}^*$ over an inversive difference field F with a basic set σ, then an algebraic solution of Φ is defined as a family $a^* = \{a_{i,\gamma} \mid i = 1, \dots, s, \gamma \in \Gamma_\sigma\}$ that annuls every polynomial from Φ when Φ is treated as a subset of the polynomial ring $F[\{\gamma(y_i) \mid i = 1, \dots, s, \gamma \in \Gamma_\sigma\}]$. As in the case of σ-polynomials, every solution $a = (a_1, \dots, a_s)$ of a set of σ^*-polynomials generates the algebraic solution $a^* = \{\gamma(a_i) \mid i = 1, \dots, s, \gamma \in \Gamma_\sigma\}$ of this set, but not all algebraic solutions can be obtained in this way.

2.3. *Autoreduced sets of difference and inversive difference polynomials. Characteristic sets*

Let F be a difference field with a basic set $\sigma = \{\alpha_1, \dots, \alpha_n\}$, $T = T_\sigma$, and $R = F\{y_1, \dots, y_s\}$ the algebra of difference polynomials in σ-indeterminates $y_1, \dots, y_s$ over F. Then R can be viewed as a polynomial ring in the set of indeterminates $TY = \{\tau y_i \mid \tau \in T, 1 \leqslant i \leqslant s\}$ over F (here and below we often write τy_i instead of $\tau(y_i)$). Elements of this set are called *terms*. If $\tau = \alpha_1^{k_1} \dots \alpha_n^{k_n} \in T$ $(k_1, \dots, k_n \in \mathbf{N})$, then the number $ord\,\tau = \sum_{\nu=1}^{n} k_\nu$ is called the *order* of τ. The order $ord\,u$ of a term $u = \tau y_i \in TY$ is defined as the order of τ. As usual, if $\tau, \tau' \in T$, we say that τ' divides τ (and write $\tau' | \tau$) if $\tau = \tau'\tau''$ for some $\tau'' \in T$. If $u = \tau y_i$ and $v = \tau' y_j$ are two terms from TY, we say that u divides v (and write $u|v$) if $i = j$ and $\tau | \tau'$.

By a *ranking* of the family of indeterminates $\{y_1, \dots, y_s\}$ we mean a well-ordering of the set of terms from TY that satisfies the following conditions. (We denote the order on TY by the usual symbol $\leqslant$ and write $u < v$ if $u \leqslant v$ and $u \neq v$.)

(i) $u \leqslant \tau u$ for any $u \in TY$, $\tau \in T$.

(ii) If $u, v \in TY$ and $u \leqslant v$, then $\tau u \leqslant \tau v$ for any $\tau \in T$.

A ranking of the family $\{y_1, \dots, y_s\}$ is also referred to as a ranking of the set of terms TY. It is said to be *orderly* if the inequality $ord\,u < ord\,v$ $(u, v \in TY)$ implies that $u < v$. An important example of an orderly ranking is the *standard ranking* defined as follows: $u = \alpha_1^{k_1} \dots \alpha_n^{k_n} y_i \leqslant v = \alpha_1^{l_1} \dots \alpha_n^{l_n} y_j \in TY$ if and only if $(\sum_{\nu=1}^{n} k_\nu, i, k_1, \dots, k_n)$ is less than or equal to $(\sum_{\nu=1}^{n} l_\nu, j, l_1, \dots, l_n)$ with respect to the lexicographic order on $\mathbf{N}^{n+2}$. In what follows, we assume that an orderly ranking of TY has been fixed.

Let $A \in F\{y_1, \dots, y_s\}$. The greatest (with respect to the given ranking) element of TY that appears in the σ-polynomial A is called the *leader* of A; it is denoted by u_A. If A is written as a polynomial in u_A, $A = \sum_{i=0}^{d} I_i u_A^i$ $(d = deg_{u_A} A$ and the σ-polynomials $I_0, \dots, I_d$ do not contain u_A), then I_d is called the *initial* of the σ-polynomial A; it is denoted by I_A.

Let A and B be two σ-polynomials from $F\{y_1, \dots, y_s\}$. We say that A has lower rank than B (or simply A is less than B) and write $A < B$, if either $A \in F$, $B \notin F$ or $u_A < u_B$

or $u_A = u_B = u$, $deg_u A < deg_u B$. If neither $A < B$ nor $B < A$, we say that A and B have the same rank and write $rk\, A = rk\, B$. The σ-polynomial A is said to be *reduced* with respect to B if A does not contain any power of a transform τu_B ($\tau \in T_\sigma$) whose exponent is greater than or equal to $deg_{u_B} B$. If Σ is any subset of $F\{y_1, \dots, y_s\} \setminus F$, then a σ-polynomial $A \in F\{y_1, \dots, y_s\}$ is said to be reduced with respect to Σ if A is reduced with respect to every element of Σ.

A set $\Sigma \subseteq F\{y_1, \dots, y_s\}$ is called an *autoreduced set* if either $\Sigma = \emptyset$ or $\Sigma \cap F = \emptyset$ and every element of Σ is reduced with respect to all other elements of Σ. It is easy to see that distinct elements of an autoreduced set have distinct leaders. It follows from [86, Chapter 0, Lemma 15(a)] (see also [88, Lemma 2.2.1]) that *every autoreduced set is finite*.

The ordinary version of the following reduction theorem was proved in [136, Section 5]. In [36] R.M. Cohn generalized the result to the case of partial difference polynomial rings (actually, to the rings of partial difference-differential polynomials).

THEOREM 2.3.1. *Let $\mathcal{A} = \{A_1, \dots, A_p\}$ be an autoreduced set in a ring of σ-polynomials $F\{y_1, \dots, y_s\}$ over a difference field F with basic set σ. Let $I(\mathcal{A}) = \{B \in F\{y_1, \dots, y_s\} \mid$ either $B = 1$ or B is a product of finitely many σ-polynomials of the form $\tau(I_{A_i})$ ($\tau \in T_\sigma$, $i = 1, \dots, p$)$\}$. Then for any $C \in F\{y_1, \dots, y_s\}$, there exist σ-polynomials $J \in I(\mathcal{A})$ and $C_0 \in F\{y_1, \dots, y_s\}$ such that C_0 is reduced with respect to $\mathcal{A}$ and $JC \equiv C_0 \pmod{[\mathcal{A}]}$ (i.e., $JC - C_0 \in [\mathcal{A}]$).*

With the notation of the theorem, the σ-polynomial C_0 is called the *remainder* of the σ-polynomial C with respect to $\mathcal{A}$. We also say that C *reduces* to C_0 modulo $\mathcal{A}$. (If $\mathcal{A} = \{A\}$, we say that C_0 is a remainder of C with respect to the σ-polynomial A.)

The reduction process, that is, a transition from a given σ-polynomial C to a σ-polynomial C_0 satisfying the conditions of the theorem, can be performed in many ways. Let us describe one of them.

If C is reduced with respect to $\mathcal{A}$, we can take $C_0 = C$ and $J = 1$. If C is not reduced with respect to $\mathcal{A}$, then C contains a power $(\tau u_{A_i})^k$ of some term τu_{A_i} ($\tau \in T_\sigma$, $1 \leqslant i \leqslant p$) whose exponent is greater than or equal to $deg_{u_{A_i}} A_i$. Such a term τu_{A_i} of the highest possible rank is called the *$\mathcal{A}$-leader* of C and denoted by $v_{\mathcal{A},C}$. Obviously, C can be written as $C = Dv^d_{\mathcal{A},C} + E$ where D does not contain $v_{\mathcal{A},C}$ and $deg_{v_{\mathcal{A},C}} Q < d$. Let $v_{\mathcal{A},C} = \tau u_{A_j}$ ($\tau \in T_\sigma$, $1 \leqslant j \leqslant p$). Then $v_{\mathcal{A},C}$ is the leader of τA_j, $I_{\tau A_j} = \tau I_{A_j}$, and $deg_{v_{\mathcal{A},C}}(\tau A_j) = d_j$ where $d_j = deg_{u_{A_j}} A_j$. Consider the σ-polynomial $C' = (\tau I_{A_j})C - v^{d-d_j}_{\mathcal{A},C}(\tau A_j)D$. Clearly, $v_{\mathcal{A},C'} \leqslant v_{\mathcal{A},C}$ and in the case of equality, $deg_{v_{\mathcal{A},C'}} C' < d$. Furthermore, $C' \equiv C \pmod{[\mathcal{A}]}$. Applying the same procedure to C' instead of C and continuing this process, we obtain a σ-polynomial $\overline{C}$ such that $\overline{C} \equiv C \pmod{[\mathcal{A}]}$ and $v_{\mathcal{A},\overline{C}} < v_{\mathcal{A},C}$. Repeating the foregoing procedure we obtain a σ-polynomial C_0 that satisfies the conditions of the last theorem.

In what follows, the elements of an autoreduced set are always written in the order of increasing rank. (Thus, if $\mathcal{A} = \{A_1, \dots, A_p\}$ an autoreduced set in $F\{y_1, \dots, y_s\}$, we assume that $A_1 < \dots < A_p$.)

DEFINITION 2.3.2. Let $\mathcal{A} = \{A_1, \dots, A_p\}$ and $\mathcal{B} = \{B_1, \dots, B_q\}$ be two autoreduced sets in the algebra of difference polynomials $F\{y_1, \dots, y_s\}$. We say that $\mathcal{A}$ has lower rank than $\mathcal{B}$ and write $rk\,\mathcal{A} < rk\,\mathcal{B}$ if one of the following conditions holds:

(i) there exists $k \in \mathbf{N}$, $1 \leqslant k \leqslant \min\{p, q\}$, such that $rk\,A_i = rk\,B_i$ for $i = 1, \dots, k-1$ and $A_k < B_k$;

(ii) $p > q$ and $rk\,A_i = rk\,B_i$ for $i = 1, \dots, q$.

The proof of the following result is similar to the proof of the corresponding statement about autoreduced sets of differential polynomials (see [86, Chapter 1, Proposition 3]).

PROPOSITION 2.3.3. *In every non-empty set of autoreduced subsets of $F\{y_1, \dots, y_s\}$ there exists an autoreduced set of lowest rank.*

If J is a non-empty subset (in particular, an ideal) of the ring $F\{y_1, \dots, y_s\}$, then the family of all autoreduced subsets of J is not empty (if $0 \neq A \in J$, then $\mathcal{A} = \{A\}$ is an autoreduced set). It follows from the last proposition that J contains an autoreduced subset of lowest rank. Such a subset is called a *characteristic set* of J. The following proposition describes some properties of characteristic sets of difference polynomials.

PROPOSITION 2.3.4. *Let F be a difference field with a basic set σ, J a difference ideal of the algebra of σ-polynomials $F\{y_1, \dots, y_s\}$, and Σ a characteristic set of J. Then:*

(i) *The σ-ideal J does not contain non-zero difference polynomials reduced with respect to Σ. In particular, if $A \in \Sigma$, then $I_A \notin J$.*

(ii) *Let $I = \prod_{A \in \Sigma} I_A$. If the ideal J is prime, then $J = [\Sigma] : \Lambda(\Sigma)$ where $\Lambda(\Sigma)$ is the free commutative multiplicative semigroup generated by the set $\{\tau(I) \mid \tau \in T_\sigma\}$.*

Let F be an inversive difference field with a basic set $\sigma = \{\alpha_1, \dots, \alpha_n\}$ and Γ the free commutative group generated by σ. Let $\overline{\mathbf{Z}}_-$ denote the set of all non-positive integers and let $\mathbf{Z}_1, \mathbf{Z}_2, \dots, \mathbf{Z}_{2^n}$ be all distinct Cartesian products of n factors each of which is either $\mathbf{N}$ or $\overline{\mathbf{Z}}_-$ (we assume that $\mathbf{Z}_1 = \mathbf{N}^n$). These sets are called *orthants* of $\mathbf{Z}^n$. For any $j = 1, \dots, 2^n$, we set $\Gamma_j = \{\gamma = \alpha_1^{k_1} \dots \alpha_n^{k_n} \in \Gamma \mid (k_1, \dots, k_n) \in \mathbf{Z}_j\}$. Furthermore, if $\gamma = \alpha_1^{k_1} \dots \alpha_n^{k_n} \in \Gamma$, then the number $ord\,\gamma = \sum_{i=1}^{n} |k_i|$ will be called the *order* of γ.

Let $F\{y_1, \dots, y_s\}^*$ be the algebra of σ^*-polynomials in σ^*-indeterminates $y_1, \dots, y_s$ over F and let Y denote the set $\{\gamma y_i \mid \gamma \in \Gamma, 1 \leqslant i \leqslant s\}$ whose elements are called *terms* (here and below we often write γy_i for $\gamma(y_i)$). By the order of a term $u = \gamma y_j$ we mean the order of the element $\gamma \in \Gamma$. Setting $Y_j = \{\gamma y_i \mid \gamma \in \Gamma_j, 1 \leqslant i \leqslant s\}$ $(j = 1, \dots, 2^n)$ we obtain a representation of the set of terms as a union $Y = \bigcup_{j=1}^{2^n} Y_j$.

DEFINITION 2.3.5. A term $v \in Y$ is called a transform of a term $u \in Y$ if and only if u and v belong to the same Y_j $(1 \leqslant j \leqslant 2^n)$ and $v = \gamma u$ for some $\gamma \in \Gamma_j$. If $\gamma \neq 1$, v is said to be a proper transform of u.

DEFINITION 2.3.6. A well-ordering of the set of terms Y is called a *ranking* of the family of σ^*-indeterminates $y_1, \dots, y_s$ (or a ranking of the set Y) if it satisfies the following conditions. (We use the standard symbol $\leqslant$ for the ranking; it will be always clear what order is denoted by this symbol.)

(i) If $u \in Y_j$ and $\gamma \in \Gamma_j$ ($1 \leqslant j \leqslant 2^n$), then $u \leqslant \gamma u$.

(ii) If $u, v \in Y_j$ ($1 \leqslant j \leqslant 2^n$), $u \leqslant v$ and $\gamma \in \Gamma_j$, then $\gamma u \leqslant \gamma v$.

A ranking of the σ^*-indeterminates $y_1, \dots, y_s$ is called *orderly* if for any $j = 1, \dots, 2^n$ and for any two terms $u, v \in Y_j$, the inequality $ord\, u < ord\, v$ implies that $u < v$ (as usual, $v < w$ means $v \leqslant w$ and $v \neq w$). As an example of an orderly ranking of the σ^*-indeterminates $y_1, \dots, y_s$ one can consider the *standard ranking* defined as follows: $u = \alpha_1^{k_1} \dots \alpha_n^{k_n} y_i \leqslant v = \alpha_1^{l_1} \dots \alpha_n^{l_n} y_j$ if and only if $(\sum_{\nu=1}^{n} |k_\nu|, i, k_1, \dots, k_n)$ is less than or equal to $(\sum_{\nu=1}^{n} |l_\nu|, j, l_1, \dots, l_n)$ with respect to the lexicographic order on $\mathbf{Z}^{n+2}$.

In what follows, we assume that an orderly ranking $\leqslant$ of the set of σ^*-indeterminates $y_1, \dots, y_s$ has been fixed. If $A \in F\{y_1, \dots, y_s\}^*$, then the greatest (with respect to the ranking $\leqslant$) term of Y that appears in A is called the *leader* of A; it is denoted by u_A. If $d = deg_u A$, then the σ^*-polynomial A can be written as $A = I_d u^d + I_{d-1} u^{d-1} + \dots + I_0$ where I_k ($0 \leqslant k \leqslant d$) do not contain u. The σ^*-polynomial I_d is called the *initial* of A; it is denoted by I_A.

The ranking of the set of σ^*-indeterminates $y_1, \dots, y_s$ generates the following relation on $F\{y_1, \dots, y_s\}^*$. If A and B are two σ^*-polynomials, then A is said to have rank less than B (we write $A < B$) if either $A \in F$, $B \notin F$ or $A, B \in F\{y_1, \dots, y_s\}^* \setminus F$ and $u_A < u_B$ or $u_A = u_B = u$, $deg_u A < deg_u B$. If $u_A = u_B = u$ and $deg_u A = deg_u B$, we say that A and B are of the same rank and write $rk\, A = rk\, B$.

Let $A, B \in F\{y_1, \dots, y_s\}^*$. The σ^*-polynomial A is said to be *reduced* with respect to B if A does not contain any power of a transform γu_B ($\gamma \in \Gamma_\sigma$) whose exponent is greater than $deg_{u_B} B$. If $\Sigma \subseteq F\{y_1, \dots, y_s\}^* \setminus F$, then a σ^*-polynomial $A \in F\{y_1, \dots, y_s\}^*$, is said to be reduced with respect to Σ if A is reduced with respect to every element of the set Σ.

A set $\Sigma \subseteq F\{y_1, \dots, y_s\}^*$ is said to be *autoreduced* if either it is empty or $\Sigma \cap F = \emptyset$ and every element of Σ is reduced with respect to all others. As in the case of σ-polynomials, distinct elements of an autoreduced set have distinct leaders and every autoreduced set is finite. The following statement is an analog of Theorem 2.3.1.

THEOREM 2.3.7 [88, Theorem 3.4.27]. *Let $\mathcal{A} = \{A_1, \dots, A_r\}$ be an autoreduced subset of $F\{y_1, \dots, y_s\}^*$ and let $D \in F\{y_1, \dots, y_s\}^*$. Furthermore, let $I(\mathcal{A}) = \{B \in F\{y_1, \dots, y_s\} \mid$ either $B = 1$ or B is a product of finitely many polynomials of the form $\gamma(I_{A_i})$ ($\gamma \in \Gamma_\sigma, i = 1, \dots, r$)$\}$. Then there exist σ-polynomials $J \in I(\mathcal{A})$ and $D_0 \in F\{y_1, \dots, y_s\}$ such that D_0 is reduced with respect to $\mathcal{A}$ and $JD \equiv D_0 \pmod{[\mathcal{A}]}$.*

The transition from a σ^*-polynomial D to a σ^*-polynomial D_0 satisfying the conditions of the theorem can be performed in the same way as in the case of σ-polynomials (see the description of the corresponding reduction process after Theorem 2.3.1). We say that D *reduces to D_0 modulo $\mathcal{A}$*.

As in the case of σ-polynomials, the elements of an autoreduced set in $F\{y_1, \dots, y_s\}^*$ will be always written in the order of increasing rank. If $\mathcal{A} = \{A_1, \dots, A_r\}$ and $\mathcal{B} = \{B_1, \dots, B_s\}$ are two autoreduced sets of σ^*-polynomials, we say that $\mathcal{A}$ has lower rank than $\mathcal{B}$ and write $rk\,\mathcal{A} < rk\,\mathcal{B}$ if either there exists $k \in \mathbf{N}$, $1 \leqslant k \leqslant \min\{r, s\}$, such that $rk\,A_i = rk\,B_i$ for $i = 1, \dots, k-1$ and $A_k < B_k$, or $r > s$ and $rk\,A_i = rk\,B_i$ for $i = 1, \dots, s$.

Repeating the proof of [86, Chapter 1, Proposition 3], one obtains that every family of autoreduced subsets of $F\{y_1, \dots, y_s\}^*$ contains an autoreduced set of lowest rank. In particular, if $\emptyset \neq J \subseteq F\{y_1, \dots, y_s\}^*$, then the set J contains an autoreduced set of lowest rank called a *characteristic set* of J. The following statement is the version of Proposition 2.3.4 for inversive difference polynomials (see [88, Proposition 3.4.32]).

PROPOSITION 2.3.8. *Let F be a difference field with a basic set σ, J a σ^*-ideal of the algebra of σ^*-polynomials $F\{y_1, \dots, y_s\}^*$, and Σ a characteristic set of J. Then*:

(i) *The ideal J does not contain non-zero σ^*-polynomials reduced with respect to Σ. In particular, if $A \in \Sigma$, then $I_A \notin J$.*

(ii) *If J is a prime σ^*-ideal, then $J = [\Sigma] : \Upsilon(\Sigma)$ where $\Upsilon(\Sigma)$ denote the set of all finite products of elements of the form $\gamma(I_A)$ $(\gamma \in \Gamma_\sigma, A \in \Sigma)$.*

Let F be a difference field with a basic set σ and $F\{y_1, \dots, y_s\}$ an algebra of σ-polynomials in σ-indeterminates $y_1, \dots, y_s$ over F. A σ-ideal I of $F\{y_1, \dots, y_s\}$ is called *linear* if it is generated (as a σ-ideal) by linear σ-polynomials (that is, σ-polynomials of the form $\sum_{i=1}^{m} a_i \tau_i y_{k_i}$ where $a_i \in F$, $\tau_i \in T_\sigma$, $1 \leqslant k_i \leqslant s$ for $i = 1, \dots, m$). If the σ-field F is inversive, then a σ^*-ideal of an algebra of σ^*-polynomials $F\{y_1, \dots, y_s\}^*$ is called linear if it is generated (as a σ^*-ideal) by linear σ^*-polynomials, i.e., polynomials of the form $\sum_{i=1}^{m} a_i \gamma_i y_{k_i}$ $(a_i \in F, \gamma_i \in \Gamma_\sigma, 1 \leqslant k_i \leqslant s$ for $i = 1, \dots, m)$. As in the case of linear differential polynomials (see [88, Proposition 3.2.28]), one can show that *if I is a proper linear σ-ideal of $F\{y_1, \dots, y_s\}$ or a proper linear σ^*-ideal of $F\{y_1, \dots, y_s\}^*$ then the ideal I is prime.*

DEFINITION 2.3.9. Let F be a difference field with a basic set σ and $\mathcal{A}$ an autoreduced set in $F\{y_1, \dots, y_s\}$ that consists of linear σ-polynomials (respectively, let F be a σ^*-field and $\mathcal{A}$ an autoreduced set in $F\{y_1, \dots, y_s\}^*$ that consists of linear σ^*-polynomials). The set $\mathcal{A}$ is called *coherent* if the following two conditions hold.

(i) If $A \in \mathcal{A}$ and $\tau \in T_\sigma$ (respectively, $\gamma \in \Gamma_\sigma$), then τA (respectively, γA) reduces to zero modulo $\mathcal{A}$.

(ii) If $A, B \in \mathcal{A}$ and $v = \tau_1 u_A = \tau_2 u_B$ is a common transform of the leaders u_A and u_B $(\tau_1, \tau_2 \in T_\sigma$ or $\tau_1, \tau_2 \in \Gamma_\sigma$ if we consider the case of σ^*-polynomials), then the σ-polynomial $(\tau_2 I_B)(\tau_1 A) - (\tau_1 I_A)(\tau_2 B)$ reduces to zero modulo $\mathcal{A}$.

The following result is proved in [88, Theorem 6.5.3] for autoreduced sets of inversive difference polynomials. The proof for the case of difference polynomials is similar.

THEOREM 2.3.10. *Let F be a difference field with a basic set σ and I a linear σ-ideal of the algebra of σ-polynomials $F\{y_1, \dots, y_s\}$ (respectively, let F be a σ^*-field and I a linear σ^*-ideal of $F\{y_1, \dots, y_s\}^*$). Then any characteristic set of I is a coherent autoreduced set of linear σ- (respectively, σ^*-) polynomials.*

Conversely, if $\mathcal{A} \subseteq F\{y_1, \dots, y_s\}$ (respectively, $\mathcal{A} \subseteq F\{y_1, \dots, y_s\}^*$) is any coherent autoreduced set consisting of linear σ- (respectively, σ^*-) polynomials, then $\mathcal{A}$ is a characteristic set of the linear σ-ideal $[\mathcal{A}]$ (respectively, of the linear σ^*-ideal $[\mathcal{A}]^*$).

COROLLARY 2.3.11. *Let F be an inversive difference field with a basic set σ and let $\preccurlyeq$ be a preorder on $F\{y_1, \dots, y_s\}^*$ such that for any two σ^*-polynomials A_1 and A_2, $A_1 \preccurlyeq A_2$ if and only if u_{A_2} is a transform of u_{A_1}. Let A be a linear σ^*-polynomial from $F\{y_1, \dots, y_s\}^* \setminus F$ and $\Gamma_\sigma A = \{\gamma A \mid \gamma \in \Gamma_\sigma\}$. Then the set of all minimal (with respect to $\preccurlyeq$) elements of $\Gamma_\sigma A$ is a characteristic set of the σ^*-ideal $[A]^*$.*

Theorem 2.3.10 implies the following method of constructing a characteristic set of a proper linear σ^*-ideal I in $F\{y_1, \dots, y_s\}^*$ (a similar method can be used for building a characteristic set of a σ-ideal in $F\{y_1, \dots, y_s\}$). Suppose that $I = [A_1, \dots, A_p]^*$ where $A_1, \dots, A_p$ are linear σ^*-polynomials and $A_1 < \dots < A_p$. It follows from Theorem 2.3.10 that one should find a coherent autoreduced set $\Phi \subseteq F\{y_1, \dots, y_s\}^*$ such that $[\Phi]^* = I$. Such a set can be obtained from the set $\mathcal{A} = \{A_1, \dots, A_p\}$ via the following two-step procedure.

Step 1. Constructing an autoreduced set $\Sigma \subseteq I$ such that $[\Sigma]^* = I$.

If $\mathcal{A}$ is autoreduced, set $\Sigma = \mathcal{A}$. If $\mathcal{A}$ is not autoreduced, choose the smallest i ($1 \leqslant i \leqslant p$) such that some σ^*-polynomial A_j, $1 \leqslant i < j \leqslant p$, is not reduced with respect to A_i. Replace A_j by its remainder with respect to A_i (obtained by the procedure described after Theorem 2.3.1) and arrange the σ^*-polynomials of the new set $\mathcal{A}_1$ in ascending order. Then apply the same procedure to the set $\mathcal{A}_1$ and so on. After each iteration the number of σ^*-polynomials in the set does not increase, one of them is replaced by a σ^*-polynomial of lower or equal rank, and the others do not change. Therefore, the process terminates after a finite number of steps and then we have the desired autoreduced set Σ.

Step 2. Constructing a coherent autoreduced set $\Phi \subseteq I$.

Let $\Sigma_0 = \Sigma$ be an autoreduced subset of I such that $[\Sigma]^* = I$. If Σ is not coherent, we build a new autoreduced set $\Sigma_1 \subseteq I$ by adding to Σ_0 new σ^*-polynomials of the following types.

(a) σ^*-polynomials $(\gamma_1 I_{B_1})\gamma_2 B_2 - (\gamma_2 I_{B_2})\gamma_1 B_1$ constructed for every pair $B_1, B_2 \in \Sigma$ such that the leaders u_{B_1} and u_{B_1} have a common transform $v = \gamma_1 u_{B_1} = \gamma_2 u_{B_2}$ and $(\gamma_1 I_{B_1})\gamma_2 B_2 - (\gamma_2 I_{B_2})\gamma_1 B_1$ is not reducible to zero modulo Σ_0.

(b) σ^*-polynomials of the form γA ($\gamma \in \Gamma_\sigma, A \in \Sigma_0$) that are not reducible to zero modulo Σ_0.

It is clear that $rk\, \Sigma_1 < rk\, \Sigma_0$. Applying the same procedure to Σ_1 and continuing in the same way, we obtain autoreduced subsets $\Sigma_0, \Sigma_1, \dots$ of I such that $rk\, \Sigma_{i+1} < rk\, \Sigma_i$ for $i = 0, 1, \dots$. Obviously, the process terminates after finitely many steps, and so we obtain an autoreduced set $\Phi \subseteq I$ such that $\Phi = \Sigma_k = \Sigma_{k+1} = \cdots$ for some $k \in \mathbf{N}$. It is easy to see that Φ is coherent, so it is a characteristic set of the ideal I.

2.4. *Perfect difference ideals. Ritt difference rings*

DEFINITION 2.4.1. Let R be a difference ring with a basic set σ. A σ-ideal J of the ring R is called perfect if for any $a \in R$, $\tau_1, \dots, \tau_r \in T_\sigma$ and $k_1, \dots, k_r \in \mathbf{N}$, the inclusion $\tau_1(a)^{k_1} \dots \tau_r(a)^{k_r} \in J$ implies $a \in J$.

It is easy to see that every perfect ideal is reflexive and every reflexive prime ideal is perfect. Furthermore, if the σ-ring R is inversive, then a σ-ideal J of R is perfect if and only if any inclusion $\gamma_1(a)^{k_1} \ldots \gamma_r(a)^{k_r} \in J$ $(a \in R, \gamma_1, \ldots, \gamma_r \in \Gamma_\sigma,\ k_1, \ldots, k_r \in \mathbf{N})$ implies $a \in J$.

If B is a subset of a difference ring R with a basic set σ, then the intersection of all perfect σ-ideals of R containing B is the smallest perfect ideal containing B. It is denoted by $\{B\}$ and called the *perfect closure* of the set B. The ideal $\{B\}$ can be obtained from the set B via the following procedure introduced in [137] and called *shuffling*. For any set $M \subseteq R$, let M' denote the set of all $a \in R$ such that $\tau_1(a)^{k_1} \ldots \tau_r(a)^{k_r} \in M$ for some $\tau_1, \ldots, \tau_r \in T_\sigma$ and $k_1, \ldots, k_r \in \mathbf{N}$ $(r \geqslant 1)$. Starting with $B_0 = B$, set $B_1 = [B_0]'$, $B_2 = [B_1]', \ldots$. Then $\{B\} = \bigcup_{i=0}^{\infty} B_i$.

PROPOSITION 2.4.2 [32, Chapter 3, Section 2]. *Let A and B be two subsets of a difference ring R. Then*

(i) $A_k B_k \subseteq (AB)_k$ *for any* $k \in \mathbf{N}$. (*By the product UV of two sets $U, V \subseteq R$ we mean the set* $UV = \{uv \mid u \in U, v \in V\}$.)
(ii) $\{A\}\{B\} \subseteq \{AB\}$.
(iii) $(AB)_k \subseteq A_k \cap B_k$ *for any* $k \in \mathbf{N}, k \geqslant 1$.
(iv) $A_k \cap B_k \subseteq (AB)_{k+1}$ *for any* $k \in \mathbf{N}$.
(v) $\{A\} \cap \{B\} = \{AB\}$.

Let J be a subset of a difference ring R. Then a finite subset A of J is called a *basis* of J if $\{A\} = \{J\}$. If $\{J\} = A_m$ for some $m \in \mathbf{N}$, A is said to be an m-*basis* of J. A difference ring in which every subset has a basis is called a *Ritt difference ring*.

PROPOSITION 2.4.3 [32, Chapter 3, Theorem I]. *A difference ring R is a Ritt difference ring if and only if every perfect difference ideal of R has a basis. If every perfect difference ideal of R has an m-basis, then every set in R has an m-basis.* (*In this and similar statements the number m is not fixed but depends on the set.*)

PROPOSITION 2.4.4 [32, Chapter 3, Theorem II]. *A difference ring R is a Ritt difference ring if and only if it satisfies the ascending chain condition for perfect difference ideals.*

In [35] R.M. Cohn introduced and studied conservative systems of ideals of a commutative ring, that is, sets of ideals closed with respect to unions of linearly ordered subsets and intersections. The set of all perfect difference ideals of a difference ring R is an example of such a system. If R is a Ritt difference ring, then its perfect difference ideals form a Noetherian perfect conservative system, that is, a conservative system where every ideal coincides with its radical and every ascending chain of ideals is finite. A number of results that describe general properties of conservative systems can be also found in [86, Chapter 0, Section 7] and [88, Section 1.4].

The following statement is a version of the J. Ritt and H. Raudenbush theorem, [137], for partial difference rings.

THEOREM 2.4.5 [88, Theorem 3.3.42]. *Let R be a Ritt difference ring with a basic set σ and let $S = R\{\eta_1, \ldots, \eta_s\}$ be a σ-overring of R generated by a finite family of elements $\{\eta_1, \ldots, \eta_s\}$. Then S is a Ritt σ-ring. Moreover, if every set in R has an m-basis, then every* *set in S has an m-basis. In particular, an algebra of difference polynomials $R\{y_1, \dots, y_s\}$ in a finite set of difference indeterminates $y_1, \dots, y_s$ is a Ritt difference ring.*

If R is an inversive Ritt σ-ring and $S^* = R\{\eta_1, \dots, \eta_s\}^*$ a finitely generated σ^*-overring of R, then S^* is a Ritt σ^*-ring. If every set in R has an m-basis, then every set in S^* has an m-basis. In particular, an algebra of σ^*-polynomials in a finite set of σ^*-indeterminates over R is a Ritt σ^*-ring.

It is not known whether the existence of bases of perfect difference ideals implies the existence of m-bases. Another open problem is to find out whether there is a positive integer k such that every set in an algebra of difference polynomials over a difference field has a k-basis. In [32, Chapter 3, Section 13] R.M. Cohn showed that if such an integer k exists, it exceeds 1. More precisely, R.M. Cohn proved that if $\mathbf{Q}$ is considered as an ordinary difference field with the identity automorphism and $S = \mathbf{Q}\{u, v\}$ is an algebra of difference polynomials in two difference indeterminates u, v over $\mathbf{Q}$, then the perfect difference ideal $\{uv\}$ of S has no 1-basis.

The following statement strengthens the result of Theorem 2.4.5 for a wide class of rings of difference polynomials over ordinary difference fields.

THEOREM 2.4.6 [21]. *Let F be an ordinary difference field containing an element t which is distinct from all its non-trivial transforms. Let $S = F\{y_1, \dots, y_s\}$ be a ring of difference polynomials in difference indeterminates $y_1, \dots, y_s$ over F. Then every perfect difference ideal of S has a basis consisting of $n+1$ difference polynomials.*

Let R be a difference ring with a basic set σ and J a proper perfect σ-ideal of R. For every $x \in R \setminus J$, let $\mathcal{P}_x$ denote the set of all perfect σ-ideals I of R such that $J \subseteq I$ and $x \notin I$. By the Zorn lemma, the set $\mathcal{P}_x$ contains, a maximal (relative to inclusion) perfect ideal P_x ($\mathcal{P}_x \neq \emptyset$, since $J \in \mathcal{P}_x$). It follows from Proposition 2.4.2(ii) that the ideal P_x is prime. Since $J = \bigcap_{x \in R \setminus J} P_x$, we obtain that *every proper perfect σ-ideal of a difference ring can be represented as an intersection of prime reflexive σ-ideals.* The following statement, the first version of which appeared in [137], specifies this result for Ritt difference rings. (Recall that a representation of a radical ideal J of a commutative ring R as a finite intersection of prime ideals, $J = \bigcap_{i=1}^{r} P_i$, is called *irredundant* if $P_i \nsubseteq P_j$ for $i \neq j$. Prime ideals P_i from such a representation are called the *essential prime divisors* of J.)

PROPOSITION 2.4.7. *Let R be a difference ring with a basic set σ and J a proper perfect σ-ideal of R.*

(i) *There exists an irredundant representation of J as an intersection of prime σ-ideals, $J = P_1 \cap \dots \cap P_r$.*
(ii) *The σ-ideals $P_1, \dots, P_r$ are reflexive and uniquely determined by the ideal J.*

Let R be a difference ring with a basic set σ and I, J two σ-ideals of R. The ideals I and J are called *separated* if $\{I, J\} = R$ and *strongly separated* if $[I, J] = R$. The σ-ideals $I_1, \dots, I_r$ of R are called (*strongly*) *separated in pairs* if any two of the ideals I_i, I_j ($1 \leqslant i < j \leqslant r$) are (strongly) separated. It is easy to show that if the ideals $I_1, \dots, I_r$ are strongly separated in pairs, then $\bigcap_{i=1}^{r} I_i = \prod_{i=1}^{r} I_i$.

In [32, Chapter 3, Sections 11, 15] R.M. Cohn gives two examples that illustrate the relationships between some characteristics of difference ideals.

Let $S = \mathbf{Q}\{y\}$ be the algebra of σ-polynomials in one σ-indeterminate y over $\mathbf{Q}$ (treated as an ordinary difference field whose basic set consists of the identity isomorphism α). The first example presents two separated σ-ideals that are not strongly separated. Let $A = 1 + y\alpha(y)$, $B = y + \alpha(y) \in S$. Then $\{A, B\} = S$, but $[A, B]$ is a proper ideal of the ring S. (Moreover, even $[\{A\}, \{B\}]$ is a proper ideal of S.)

The second example gives the irreducible representation of the perfect ideal $\{y^2 - 1\}$ of S as an intersection of prime reflexive ideals: $\{y^2 - 1\} = \{y - 1\} \cap \{y + 1\}$. At the same time, the σ-ideal $[y^2 - 1]$ cannot be represented as an intersection (or product) of two σ-ideals whose perfect closures are $\{y - 1\}$ and $\{y + 1\}$. Moreover, $[y^2 - 1]$ cannot be represented as an intersection (or product) of any two proper σ-ideals (such an ideal is called *indecomposable*).

We conclude this section with the description of two more types of difference ideals introduced and studied in [134] (see also [32, Chapter 3]). The corresponding results are formulated for partial difference rings.

Let R be a difference ring with a basic set σ. A σ-ideal I of R is called *complete* if for every element $a \in \{I\}$, there exist $\tau \in T_\sigma$, $k \in \mathbf{N}$ such that $\tau(a)^k \in I$. It is easy to check that a σ-ideal is complete if and only if the presence in I of a product of powers of transforms of an element implies the presence in I of a power of a transform of the element. One can also define a complete σ-ideal as a σ-ideal I such that $\{I\}$ is the reflexive closure of $\sqrt{I}$ (here and below $\sqrt{I}$ denotes the radical of I).

PROPOSITION 2.4.8 [134, Theorem II]. *Let I be a complete difference ideal in a difference ring R with a basic set σ. Suppose that $\{I\}$ is the intersection of s perfect ideals $J_1, \dots, J_s$ which are strongly separated in pairs. Then there exist uniquely determined complete σ-ideals $I_1, \dots, I_s$ such that $I = I_1 \cap \dots \cap I_s$ and $\{I_k\} = J_k$ $(1 \leqslant k \leqslant s)$. In this representation, the ideals $I_1, \dots, I_s$ are strongly separated in pairs. Furthermore, if the ideal I is reflexive, then so are $I_1, \dots, I_s$.*

As we have seen, the ideal $[y^2 - 1]$ of the ring of difference polynomials $\mathbf{Q}\{y\}$ ($\mathbf{Q}$ is treated as an ordinary difference ring with the identity translation) is indecomposable and its essential prime divisors $\{y - 1\}$ and $\{y + 1\}$ are strongly separated. In this case Proposition 2.4.8 shows that the difference ideal $[y^2 - 1]$ is not complete.

Propositions 2.4.7 and 2.4.8 lead to the following decomposition theorem for complete difference ideals.

PROPOSITION 2.4.9 [32, Chapter 3, Theorem IX]. *Let I be a proper complete difference ideal in a Ritt difference ring R with a basic set σ. Then there exist proper complete σ-ideals $I_1, \dots, I_s$ of R such that*

(i) $I = I_1 \cap \dots \cap I_s$,

(ii) *$I_1, \dots, I_s$ are strongly separated in pairs (hence, $I = I_1 I_2 \dots I_s$),*

(iii) *no I_k $(1 \leqslant k \leqslant s)$ is the intersection of two strongly separated proper σ-ideals.*

The ideals $I_1, \dots, I_s$ are uniquely determined. If I is reflexive, so are the I_k $(1 \leqslant k \leqslant s)$.

The ideals $I_1, \dots, I_s$ whose existence is established by Proposition 2.4.9 are called the *essential strongly separated divisors* of the σ-ideal I. The following result is a consequence of Proposition 2.4.9.

PROPOSITION 2.4.10. *Let R be a Ritt difference ring with a basic set σ and let J be a proper perfect σ-ideal of R. Then*

(i) *the essential strongly separated divisors of J are perfect σ-ideals;*

(ii) *the ideal J can be represented as $J = J_1 \cap \dots \cap J_s$ where $J_1, \dots, J_s$ are proper perfect σ-ideals separated in pairs and no J_k $(1 \leqslant k \leqslant s)$ can be represented as an intersection of two separated proper perfect σ-ideals. The ideals $J_1, \dots, J_s$ are uniquely determined by the ideal J. (They are called the essential separated divisors of J.)*

A difference ideal I of a difference (σ-) ring R is called *mixed* if the inclusion $ab \in I$ $(a, b \in R)$ implies that $a\alpha(b) \in I$ for every $\alpha \in \sigma$. It is easy to see that every perfect ideal is mixed and every mixed ideal is complete.

Let $\mathbf{Q}\{y\}$ be the ring of difference polynomials in one difference indeterminate y over $\mathbf{Q}$ (treated as an ordinary difference ring with the identity translation α). In [32, Chapter 3, Section 21] R.M. Cohn showed that the difference ideal $[y\alpha(y)]$ is not complete, while $[y^2]$ is complete, but not mixed. He also proved that if the ideal I in the hypothesis of Proposition 2.4.9 is mixed, then so are the $I_1, \dots, I_s$. Thus, the essential strongly separated divisors of a mixed difference ideal in a Ritt difference ring are mixed difference ideals.

2.5. *Varieties of difference polynomials*

Let F be a difference field with a basic set σ, $F\{y_1, \dots, y_s\}$ an algebra of σ-polynomials in σ-indeterminates $y_1, \dots, y_s$ over F and $\mathcal{E}$ a family of σ-overfields of F. Furthermore, let $\Phi \subseteq F\{y_1, \dots, y_s\}$ and let $\mathcal{M}_{\mathcal{E}}(\Phi)$ denote the set of all s-tuples $a = (a_1, \dots, a_s)$ with coordinates from some field $F_a \in \mathcal{E}$ which are solutions of the set Φ (that is, $f(a_1, \dots, a_s) = 0$ for any $f \in \Phi$). Then $M_{\mathcal{E}}(\Phi)$ is said to be the $\mathcal{E}$-variety defined by the set Φ (it is also called the $\mathcal{E}$-variety of the set Φ over $F\{y_1, \dots, y_s\}$ or over F). The σ-field F is said to be the *ground σ-field* of the $\mathcal{E}$-variety.

Now, let $\mathcal{M}$ be a set of s-tuples such that the coordinates of every point $a \in \mathcal{M}$ belong to some field $F_a \in \mathcal{E}$. If there exists a set $\Phi \subseteq F\{y_1, \dots, y_s\}$ such that $\mathcal{M} = \mathcal{M}_{\mathcal{E}}(\Phi)$, then $\mathcal{M}$ is said to be an $\mathcal{E}$-variety over $F\{y_1, \dots, y_s\}$ (or over F).

DEFINITION 2.5.1. Let F be a difference field with a basic set σ, $G = F(x_1, x_2, \dots)$ the field of rational fractions in a denumerable set of indeterminates $x_1, x_2, \dots$ over F, and $\overline{G}$ the algebraic closure of G. Then the family $\mathcal{U}(F)$ of all σ-overfields of F which are defined on subfields of $\overline{G}$ is called the universal system of σ-overfields of F. If Φ is a subset of the algebra of σ-polynomials $F\{y_1, \dots, y_s\}$ and $\mathcal{U} = \mathcal{U}(F)$, then the $\mathcal{U}$-variety $\mathcal{M}_{\mathcal{U}}(\Phi)$ (also denoted by $\mathcal{M}(\Phi)$) is called the variety defined by the set Φ over F (or over $F\{y_1, \dots, y_s\}$).

A set of s-tuples $\mathcal{M}$ over the field F is said to be a *variety over* $F\{y_1, \dots, y_s\}$ (or *a variety over* F) if there exists a set $\Phi \subseteq F\{y_1, \dots, y_s\}$ such that $\mathcal{M} = \mathcal{M}_{\mathcal{U}(F)}(\Phi)$.

In what follows, we assume that a difference field F with a basic set σ, an algebra of σ-polynomials $F\{y_1, \dots, y_s\}$, and a family $\mathcal{E}$ of σ-overfields of F are fixed. $\mathcal{E}$-varieties and varieties over $F\{y_1, \dots, y_s\}$ will be called simply $\mathcal{E}$-varieties and varieties, respectively.

PROPOSITION 2.5.2 [32, Chapter 4, Section 3]. *Let $\eta = (\eta_1, \dots, \eta_s)$ be an s-tuple over the σ-field F. Then there exists and s-tuple $\zeta = (\zeta_1, \dots, \zeta_s)$ over F such that ζ is equivalent to η and all ζ_i $(1 \leqslant i \leqslant s)$ belong to some field from the universal system $\mathcal{U}(F)$.*

If $\mathcal{A}_1$ and $\mathcal{A}_2$ are two $\mathcal{E}$-varieties and $\mathcal{A}_1 \subseteq \mathcal{A}_2$ ($\mathcal{A}_1 \subsetneqq \mathcal{A}_2$), then $\mathcal{A}_1$ is said to be a $\mathcal{E}$-*subvariety* (respectively, a *proper $\mathcal{E}$-subvariety*) of $\mathcal{A}_2$. $\mathcal{U}(F)$-subvarieties of a variety $\mathcal{A}$ are called subvarieties of $\mathcal{A}$. An $\mathcal{E}$-variety (variety) $\mathcal{A}$ is called *reducible* if it can be represented as a union of two its proper $\mathcal{E}$-subvarieties (subvarieties). If such a representation does not exist, the $\mathcal{E}$-variety (variety) $\mathcal{A}$ is said to be *irreducible*.

Let an $\mathcal{E}$-variety (variety) $\mathcal{A}$ be represented as a union of its $\mathcal{E}$-subvarieties (subvarieties): $\mathcal{A} = \mathcal{A}_1 \cup \dots \cup \mathcal{A}_k$. This representation is called *irredundant* if $\mathcal{A}_i \nsubseteq \mathcal{A}_j$ for $i \neq j$ $(1 \leqslant i, j \leqslant k)$.

The following proposition summarizes basic properties of $\mathcal{E}$-varieties. As before, we assume that a family $\mathcal{E}$ of σ-overfields of F is fixed. Furthermore, if $\mathcal{A}$ is a set of s-tuples a with coordinates from a σ-field $F_a \in \mathcal{E}$ (we say that $\mathcal{A}$ is *a set of s-tuples from $\mathcal{E}$ over F*), then $\Phi_{\mathcal{E}}(\mathcal{A})$ denotes the perfect σ-ideal $\{f \in F\{y_1, \dots, y_s\} \mid f(a_1, \dots, a_s) = 0$ for any $a = (a_1, \dots, a_s) \in \mathcal{A}\}$ of the ring $F\{y_1, \dots, y_s\}$.

PROPOSITION 2.5.3.

(i) *If $\Phi_1 \subseteq \Phi_2 \subseteq F\{y_1, \dots, y_s\}$, then $\mathcal{M}_{\mathcal{E}}(\Phi_2) \subseteq \mathcal{M}_{\mathcal{E}}(\Phi_1)$.*

(ii) *If $\mathcal{A}_1$ and $\mathcal{A}_2$ are two sets of s-tuples from $\mathcal{E}$ over F and $\mathcal{A}_1 \subseteq \mathcal{A}_2$, then $\Phi_{\mathcal{E}}(\mathcal{A}_2) \subseteq \Phi_{\mathcal{E}}(\mathcal{A}_1)$.*

(iii) *If $\mathcal{A}$ is a $\mathcal{E}$-variety, then $\mathcal{A} = \mathcal{M}_{\mathcal{E}}(\Phi_{\mathcal{E}}(\mathcal{A}))$.*

(iv) *If $J_1, \dots, J_k$ are σ-ideals of the ring $F\{y_1, \dots, y_s\}$ and $J = J_1 \cap \dots \cap J_k$, then $\mathcal{M}_{\mathcal{E}}(J) = \mathcal{M}_{\mathcal{E}}(J_1) \cup \dots \cup \mathcal{M}_{\mathcal{E}}(J_k)$.*

(v) *If $\mathcal{A}_1, \dots, \mathcal{A}_k$ are $\mathcal{E}$-varieties over F and $\mathcal{A} = \mathcal{A}_1 \cup \dots \cup \mathcal{A}_k$, then $\mathcal{A}$ is a $\mathcal{E}$-variety over F and $\Phi_{\mathcal{E}}(\mathcal{A}) = \Phi_{\mathcal{E}}(\mathcal{A}_1) \cap \dots \cap \Phi_{\mathcal{E}}(\mathcal{A}_k)$.*

(vi) *The intersection of any family of $\mathcal{E}$-varieties is an $\mathcal{E}$-variety.*

(vii) *An $\mathcal{E}$-variety $\mathcal{A}$ is irreducible if and only if $\Phi_{\mathcal{E}}(\mathcal{A})$ is a prime reflexive ideal of $F\{y_1, \dots, y_s\}$.*

(ix) *Every $\mathcal{E}$-variety $\mathcal{A}$ has a unique irredundant representation as a union of irreducible $\mathcal{E}$-varieties, $\mathcal{A} = \mathcal{A}_1 \cup \dots \cup \mathcal{A}_k$. (The $\mathcal{E}$-varieties $\mathcal{A}_1, \dots, \mathcal{A}_k$ are called* irreducible $\mathcal{E}$-components *of $\mathcal{A}$.) Furthermore, $\mathcal{A}_i \nsubseteq \bigcup_{j \neq i} \mathcal{A}_j$ for $i = 1, \dots, k$.*

(x) *If $\mathcal{A}_1, \dots, \mathcal{A}_k$ are the irreducible $\mathcal{E}$-components of an $\mathcal{E}$-variety $\mathcal{A}$, then the prime σ^*-ideals $\Phi_{\mathcal{E}}(\mathcal{A}_1), \dots, \Phi_{\mathcal{E}}(\mathcal{A}_k)$ are the essential prime divisors of the perfect σ-ideal $\Phi_{\mathcal{E}}(\mathcal{A})$.*

The last proposition implies that $\mathcal{A} \mapsto \Phi_{\mathcal{E}}(\mathcal{A})$ is an injective mapping of the set of all $\mathcal{E}$-varieties over $F\{y_1, \dots, y_s\}$ into a set of all perfect σ-ideals of the ring $F\{y_1, \dots, y_s\}$. If $\mathcal{E}$

is the universal system of σ-overfields of F, then this mapping is bijective. More precisely, we have the following statement about varieties (see [32, Chapter 4, Section 5]).

PROPOSITION 2.5.4.

(i) *If J is a perfect σ-ideal of the ring $F\{y_1,\dots,y_s\}$, then $\Phi(\mathcal{M}(J))=J$.*

(ii) *$\mathcal{M}(J)=\emptyset$ if and only if $J=F\{y_1,\dots,y_s\}$.*

(iii) *The mappings $\mathcal{A}\mapsto\Phi(\mathcal{A})$ and $P\mapsto\mathcal{M}(P)$ are two mutually inverse mappings that establish one-to-one correspondence between the set of all varieties over F and the set of all perfect σ-ideals of the ring $F\{y_1,\dots,y_s\}$.*

(iv) *The correspondence $\mathcal{A}\mapsto\Phi(\mathcal{A})$ maps irreducible components of an arbitrary variety $\mathcal{B}$ onto essential prime divisors of the perfect σ-ideal $\Phi(\mathcal{B})$ in $F\{y_1,\dots,y_s\}$. In particular there is a one-to-one correspondence between irreducible varieties over F and prime σ-ideals of the σ-ring $F\{y_1,\dots,y_s\}$.*

If $\mathcal{A}$ is an irreducible variety over $F\{y_1,\dots,y_s\}$, then a generic zero of the corresponding prime ideal $\Phi(\mathcal{A})$ is called a *generic zero of the variety $\mathcal{A}$*.

The following result is a version of the Hilbert's Nullstellensatz for difference fields.

THEOREM 2.5.5. *Let F be a difference field with a basic set σ and $F\{y_1,\dots,y_s\}$ an algebra of σ-polynomials in σ-indeterminates $y_1,\dots,y_s$ over F. Let $f\in F\{y_1,\dots,y_s\}$, $\Phi\subseteq F\{y_1,\dots,y_s\}$, and $\mathcal{M}(\Phi)$ the variety defined by the set Φ over F. Then the following conditions are equivalent.*

(i) *Every s-tuple from $\mathcal{M}(\Phi)$ is a solution of the σ-polynomial f.*

(ii) *Every s-tuple $\eta=(\eta_1,\dots,\eta_s)\in\mathcal{M}(\Phi)$ which is σ-algebraic over F is a solution of f.*

(iii) $f\in\{\Phi\}$.

Two varieties $\mathcal{A}_1$ and $\mathcal{A}_2$ over the ring of difference polynomials $F\{y_1,\dots,y_s\}$ are called *separated* if $\mathcal{A}_1\cap\mathcal{A}_2=\emptyset$.

If a variety $\mathcal{A}$ is represented as a union of pairwise separated varieties $\mathcal{A}_1,\dots,\mathcal{A}_k$ and no $\mathcal{A}_i$ is the union of two non-empty separated varieties, then $\mathcal{A}_1,\dots,\mathcal{A}_k$ are said to be *essential separated components* of $\mathcal{A}$. (All varieties are considered over the same ring $F\{y_1,\dots,y_s\}$.)

The following statement is due to R.M. Cohn (see [32, Chapter 4, Section 7]).

PROPOSITION 2.5.6. *Let F be a difference field with a basic set σ and $F\{y_1,\dots,y_s\}$ an algebra of σ-polynomials in σ-indeterminates $y_1,\dots,y_s$ over F.*

(i) *Two varieties $\mathcal{A}_1$ and $\mathcal{A}_2$ over $F\{y_1,\dots,y_s\}$ are separated if and only if the perfect σ-ideals $\Phi(\mathcal{A}_1)$ and $\Phi(\mathcal{A}_2)$ are separated.*

(ii) *Two perfect ideals J_1 and J_2 of the ring $F\{y_1,\dots,y_s\}$ are separated if and only if the varieties $\mathcal{M}(J_1)$ and $\mathcal{M}(J_2)$ are separated.*

(iii) *Every non-empty variety $\mathcal{A}$ over $F\{y_1,\dots,y_s\}$ can be represented as a union of a uniquely determined family of its essential separated components. Each of these components is a union of some irreducible components of $\mathcal{A}$.*

(iv) *If* $\mathcal{A}_1, \dots, \mathcal{A}_k$ *are essential separated components of a variety* $\mathcal{A}$, *then* $\Phi(\mathcal{A}_1), \dots, \Phi(\mathcal{A}_k)$ *are essential separated divisors of the perfect* σ*-ideal* $\Phi(\mathcal{A})$ *in* $F\{y_1, \dots, y_s\}$.

Now, let F be an inversive difference field with a basic set σ, $F\{y_1, \dots, y_s\}^*$ an algebra of σ^*-polynomials in σ^*-indeterminates $y_1, \dots, y_s$ over F, and $\mathcal{E}$ a set of σ^*-overfields of F. If $\Phi \subseteq F\{y_1, \dots, y_s\}^*$, then the set $\mathcal{M}_{\mathcal{E}}(\Phi)$ consisting of all s-tuples a which have coordinates in some field $F_a \in \mathcal{E}$ and annul every σ^*-polynomial from Φ is called an *$\mathcal{E}$-variety over $F\{y_1, \dots, y_s\}^*$ determined by the set Φ.*

Let $\mathcal{A}$ be a set of s-tuples over F such that all coordinates of each s-tuple $a \in \mathcal{A}$ belong to some σ^*-field $F_a \in \mathcal{E}$. Then $\mathcal{A}$ is said to be an *$\mathcal{E}$-variety over $F\{y_1, \dots, y_s\}^*$* if there exists a set $\Phi \subseteq F\{y_1, \dots, y_s\}^*$ such that $\mathcal{A} = \mathcal{M}_{\mathcal{E}}(\Phi)$.

Let $L = F(x_1, x_2, \dots)$ be the field of rational fractions in a denumerable set of indeterminates $x_1, x_2, \dots$ over the σ^*-field F and let $\bar{L}$ be the algebraic closure of L. Then the family $\mathcal{U}^*(F)$ consisting of all σ^*-overfields of F defined on subfields of $\bar{L}$ is called the *universal system of σ^*-overfields of F*. As in the case of non-inversive difference fields, one can prove that if $\eta = (\eta_1, \dots, \eta_s)$ is any s-tuple over the σ^*-field F, then there exists an s-tuple $\zeta = (\zeta_1, \dots, \zeta_s)$ such that ζ is equivalent to η and all coordinates of the point ζ lie in some σ^*-field $F_\zeta \in \mathcal{U}^*(F)$ (see [88, Proposition 3.4.34]). A $\mathcal{U}^*(F)$-variety over $F\{y_1, \dots, y_s\}^*$ is called a *variety* over this ring of σ^*-polynomials.

The concepts of $\mathcal{E}$-subvariety, proper $\mathcal{E}$-subvariety, subvariety, and proper subvariety over $F\{y_1, \dots, y_s\}^*$, as well as the notions of reducible and irreducible $\mathcal{E}$-varieties and varieties, are precisely the same as in the case of s-tuples over an algebra of (non-inversive) difference polynomials. If an $\mathcal{E}$-variety (variety) $\mathcal{A}$ over $F\{y_1, \dots, y_s\}^*$ is represented as a union of its $\mathcal{E}$-subvarieties (subvarieties), $\mathcal{A} = \mathcal{A}_1 \cup \dots \cup \mathcal{A}_k$, and $\mathcal{A}_i \subsetneqq \mathcal{A}_j$ for $i \neq j$ $(1 \leqslant i, j \leqslant k)$, then this representation is called *irredundant*.

All properties of $\mathcal{E}$-varieties and varieties over an algebra of difference polynomials listed in Propositions 2.5.3, 2.5.4, 2.5.6 and Theorem 2.5.5 are also valid for $\mathcal{E}$-varieties and varieties over $F\{y_1, \dots, y_s\}^*$. The formulations of the corresponding statements are practically the same. One should just replace the ring $F\{y_1, \dots, y_s\}$ by $F\{y_1, \dots, y_s\}^*$ and treat $\mathcal{M}_{\mathcal{E}}(\Phi)$ $(\Phi \subseteq F\{y_1, \dots, y_s\}^*)$ and $\Phi_{\mathcal{E}}(\mathcal{A})$ ($\mathcal{A}$ is a set of s-tuples a over F whose coordinates belong to some σ^*-field $F_a \in \mathcal{E}$) as the set $\{a = (a_1, \dots, a_s) \mid a_1, \dots, a_s$ belong to some field $F_a \in \mathcal{E}$ and $f(a_1, \dots, a_s) = 0$ for all $f \in \Phi\}$ and the perfect σ-ideal $\{f \in F\{y_1, \dots, y_s\}^* \mid f(a_1, \dots, a_s) = 0$ for any $a = (a_1, \dots, a_s) \in \mathcal{A}\}$ of the ring $F\{y_1, \dots, y_s\}^*$, respectively. (If $\mathcal{E} = \mathcal{U}^*(F)$, then $\Phi_{\mathcal{E}}(\mathcal{A})$ and $\mathcal{M}_{\mathcal{E}}(\Phi)$ are denoted by $\mathcal{M}(\Phi)$ and $\Phi(\mathcal{A})$, respectively.) By a generic zero of an irreducible variety $\mathcal{A}$ over $F\{y_1, \dots, y_s\}^*$ we mean a generic zero of the corresponding perfect σ-ideal $\Phi(\mathcal{A})$ of the ring $F\{y_1, \dots, y_s\}^*$. The concept of separated varieties over $F\{y_1, \dots, y_s\}^*$ is introduced in the same way as in the case of varieties over an algebra of difference polynomials.

A family $\mathcal{E}$ of difference overfields of a difference (σ-) field F is called a *complete system of σ-overfields of F* if distinct perfect σ-ideals of any ring of σ-polynomials over F have distinct $\mathcal{E}$-varieties. Clearly, the universal system of σ-overfields of F is complete.

The proof of the following result can be found in [32, Chapter 8].

PROPOSITION 2.5.7. *Let F be an ordinary difference field with a basic set $\sigma = \{\alpha\}$.*

(i) *There exists a complete system $\mathcal{C}$ of σ-overfields of F where each σ-field is σ-algebraic over F. If F is algebraically closed, $\mathcal{C}$ may be chosen to consist of one σ-field.*

(ii) *Let the σ-field F be aperiodic (that is, there is no $n \in \mathbf{N}$ such that $\alpha^n(a) = a$ for all $a \in F$) and Char $F = 0$. Let $\mathcal{E}$ be a family of difference overfields of F and let $F\{y\}$ be a ring of σ-polynomials in one σ-indeterminate y over F. Then $\mathcal{E}$ is a complete system if and only if given any prime σ^*-ideal P of $F\{y\}$ and any σ-polynomial $A \in F\{y\} \setminus P$, $\mathcal{M}_{\mathcal{E}}(P)$ contains a solution not annulling A.*

We conclude this section with a brief discussion of a realization of an abstract variety of difference polynomials as a set of complex-valued functions. The following results are due to R.M. Cohn, [33], who provided such a realization by means of an existence theorem yielding solutions of difference equations as complex-valued functions defined, except for isolated singularities, on the non-negative real axis.

A complex-valued function $f(x)$ is said to be *permitted function* if it is defined for all real values $x \geqslant 0$ except at a set $S(f)$ which has no limit points, is analytic in each of the intervals into which the non-negative real axis is divided by omission of the points of $S(f)$, and is either identically 0 or is 0 at only finitely many points in any finite interval. A *permitted difference ring* is an ordinary difference ring R whose elements are permitted functions and whose basic set consists of the translation α such that $(\alpha f)(x) = f(x+1)$ for any $f(x) \in R$. (More precisely, the elements of R are equivalence classes of permitted functions, with $f(x)$ equivalent to $g(x)$ if they coincide except possibly on $S(f) \cup S(g)$.) It follows from the definition of permitted functions that α is an isomorphism of R into itself.) A permitted difference ring which is a field is called a *permitted difference field*.

Let K_0 be the ordinary difference field of rational functions $f : \mathbf{R} \to \mathbf{C}$ with complex coefficients whose basic set consists of the translation $\alpha : f(x) \mapsto f(x+1)$ $(f(x) \in K_0)$. Then K_0 may be regarded as a permitted difference field by restricting the domain of its members to $x \geqslant 0$. In what follows, $\mathcal{K}$ denotes the set of permitted difference overfields of K_0 and $\mathcal{K}^* = \{K \in \mathcal{K} \mid$ there exists an infinite set of functions analytic throughout $[0, 1]$ which is algebraically independent over K (regarded as a field of functions over $[0, 1]$)$\}$.

The following existence theorem was proved in [33] where one can also find a discussion of the existence of continuous solutions of reflexive prime ideals in $K_0\{y\}$ and $K\{y\}$ $(K \in \mathcal{K}^*)$.

THEOREM 2.5.8. *With the above notation, let $K \in \mathcal{K}^*$ and let J be a proper reflexive prime ideal of the ring of difference polynomials $K\{y\}$ in one difference indeterminate y. Then*

(i) *$\mathcal{K}$ is a complete system of difference overfields of K.*

(ii) *The ideal J has a generic zero in one of the members of $\mathcal{K}$.*

3. Difference modules

3.1. *Ring of difference operators. Difference modules*

Let R be a difference ring with a basic set $\sigma = \{\alpha_1, \dots, \alpha_n\}$ and T the free commutative semigroup generated by the elements $\alpha_1, \dots, \alpha_n$. Furthermore, for any $r \in \mathbf{N}$, let $T(r) = \{\tau \in T \mid ord\,\tau \leqslant r\}$ (as before, by the order of an element $\tau = \alpha_1^{k_1} \dots \alpha_n^{k_n} \in T$ we mean the number $ord\,\tau = \sum_{i=1}^{n} k_i$).

DEFINITION 3.1.1. An expression of the form $\sum_{\tau \in T} a_\tau \tau$, where $a_\tau \in R$ for any $\tau \in T$ and only finitely many elements a_τ are different from 0, is called a difference (or σ-) operator over the difference ring R. Two σ-operators $\sum_{\tau \in T} a_\tau \tau$ and $\sum_{\tau \in T} b_\tau \tau$ are considered to be equal if and only if $a_\tau = b_\tau$ for any $\tau \in T$.

Let $\mathcal{D}$ denote the set of all σ-operators over the σ-ring R. This set can be equipped with a ring structure if we set $\sum_{\tau \in T} a_\tau \tau + \sum_{\tau \in T} b_\tau \tau = \sum_{\tau \in T} (a_\tau + b_\tau)\tau$, $a \sum_{\tau \in T} a_\tau \tau = \sum_{\tau \in T} (a a_\tau)\tau$, $(\sum_{\tau \in T} a_\tau \tau)\tau_1 = \sum_{\tau \in T} a_\tau (\tau\tau_1)$, $\tau_1 a = \tau_1(a)\tau_1$ for any $\sum_{\tau \in T} a_\tau \tau$, $\sum_{\tau \in T} b_\tau \tau \in \mathcal{D}$, $a \in R$, $\tau_1 \in T$, and extend the multiplication by distributivity. The ring obtained in this way is called *the ring of difference* (or σ-) *operators over* R.

The order of a σ-operator $A = \sum_{\tau \in T} a_\tau \tau \in \mathcal{D}$ is defined as the number $ord\,A = \max\{ord\,\tau \mid a_\tau \neq 0\}$. If for any $q \in \mathbf{N}$ we define $\mathcal{D}^{(q)} = \{\sum_{\tau \in T} a_\tau \tau \in \mathcal{D} \mid ord\,\tau = q$ for every $\tau \in T$ such that $a_\tau \neq 0\}$ and set $\mathcal{D}^{(q)} = 0$ for any $q \in \mathbf{Z}$, $q < 0$, then the ring $\mathcal{D}$ can be considered as a graded ring (with positive grading): $\mathcal{D} = \bigoplus_{q \in \mathbf{Z}} \mathcal{D}^{(q)}$. It can be also treated as a filtered ring with the ascending filtration $(\mathcal{D}_r)_{r \in \mathbf{Z}}$ such that $\mathcal{D}_r = 0$ for any $r < 0$ and $\mathcal{D}_r = \{A \in \mathcal{D} \mid ord\,A \leqslant r\}$ for any $r \in \mathbf{N}$. Below, while considering $\mathcal{D}$ as a graded or filtered ring, we always mean the grading with the homogeneous components $\mathcal{D}^{(q)}$ ($q \in \mathbf{Z}$) or the filtration $(\mathcal{D}_r)_{r \in \mathbf{Z}}$, respectively.

DEFINITION 3.1.2. Let R be a difference ring with a basic set σ and $\mathcal{D}$ the ring of σ-operators over R. Then a left $\mathcal{D}$-module is called a difference R-module or a σ-R-module. In other words, an R-module M is a difference (or σ-) R-module, if the elements of σ act on M in such a way that $\alpha(x + y) = \alpha(x) + \alpha(y)$, $\alpha(\beta x) = \beta(\alpha x)$, and $\alpha(ax) = \alpha(a)\alpha(x)$ for any $x, y \in M$, $\alpha, \beta \in \sigma$, $a \in R$.

If R is a difference (σ-) field, then a σ-R-module M is also called a difference vector space over R or a vector σ-R-space.

We say that a difference R-module M is finitely generated, if it is finitely generated as a left $\mathcal{D}$-module. By a graded difference (σ-) R-module we always mean a graded left module over the ring of σ-operators $\mathcal{D} = \bigoplus_{q \in \mathbf{Z}} \mathcal{D}^{(q)}$. If $M = \bigoplus_{q \in \mathbf{Z}} M^{(q)}$ is a graded σ-R-module and $M^{(q)} = 0$ for all $q < 0$, we say that M is positively graded and write $M = \bigoplus_{q \in \mathbf{N}} M^{(q)}$.

Let R be a difference ring with a basic set σ and $\mathcal{D}$ the ring of σ-operators over R equipped with the ascending filtration $(\mathcal{D}_r)_{r \in \mathbf{Z}}$. In what follows, by a filtered σ-R-module we always mean a left $\mathcal{D}$-module equipped with an exhaustive and separated filtration. Thus, a filtration of a σ-R-module M is an ascending chain $(M_r)_{r \in \mathbf{Z}}$ of R-submodules of

M such that $\mathcal{D}_r M_s \subseteq M_{r+s}$ for all $r, s \in \mathbf{Z}$, $M_r = 0$ for all sufficiently small $r \in \mathbf{Z}$, and $\bigcup_{r\in\mathbf{Z}} M_r = M$.

A filtration $(M_r)_{r\in\mathbf{Z}}$ of a σ-R-module M is called *excellent* if all R-modules M_r $(r \in \mathbf{Z})$ are finitely generated and there exists $r_0 \in \mathbf{Z}$ such that $M_r = \mathcal{D}_{r-r_0} M_{r_0}$ for any $r \in \mathbf{Z}$, $r \geqslant r_0$.

If $(M_r)_{r\in\mathbf{Z}}$ is a filtration of a σ-R-module M, then $gr\, M$ will denote the associate graded $\mathcal{D}$-module with homogeneous components $gr_r M = M_{r+1}/M_r$ $(r \in \mathbf{Z})$. Since the ring $gr\,\mathcal{D} = \bigoplus_{r\in\mathbf{Z}} \mathcal{D}_{r+1}/\mathcal{D}_r$ is naturally isomorphic to $\mathcal{D}$, we identify these two rings.

Let R be a difference ring with basic set σ and let M and N be two σ-R-modules. A homomorphism of R-modules $f: M \to N$ is said to be a *difference* (or σ-) *homomorphism* if $f(\alpha x) = \alpha f(x)$ for any $x \in M$, $\alpha \in \sigma$. A surjective (respectively, injective or bijective) difference homomorphism is called a difference (or a σ-) epimorphism (respectively, a difference monomorphism or a difference isomorphism).

If M and N are equipped with filtrations $(M_r)_{r\in\mathbf{Z}}$ and $(N_r)_{r\in\mathbf{Z}}$, respectively, and a σ-homomorphism $f: M \to N$ has the property that $f(M_r) \subseteq N_r$ for any $r \in \mathbf{Z}$, then f is said to be a homomorphism of filtered σ-R-modules.

The following result generalizes the classical theorem on the Hilbert polynomial. (As usual, by the length of a finitely generated module N over an Artinian ring we mean the length of a composition series of N.)

THEOREM 3.1.3 [88, Theorem 6.1.3]. *Let R be an Artinian difference ring with a basic set $\sigma = \{\alpha_1, \dots, \alpha_n\}$ and $M = \bigoplus_{q\in\mathbf{Z}} M^{(q)}$ a finitely generated positively graded σ-R-module. Then*

(i) *the length $l_R(M^{(q)})$ of every R-module $M^{(q)}$ is finite*;

(ii) *there exists a polynomial $\phi(t) \in \mathbf{Q}[t]$ such that $\phi(q) = l_R(M^{(q)})$ for all sufficiently large $q \in \mathbf{N}$ (i.e., there exists $q_0 \in \mathbf{N}$ such that the equality holds for all $q \geqslant q_0$)*;

(iii) *$\deg \phi(t) \leqslant n-1$ and the polynomial $\phi(t)$ can be written as $\phi(t) = \sum_{i=0}^{n-1} a_i \binom{t+i}{i}$ where $a_0, a_1, \dots, a_{n-1} \in \mathbf{Z}$.*

Let us consider the ring R as a filtered ring with the trivial filtration $(R_r)_{r\in\mathbf{Z}}$ such that $R_r = R$ for all $r \geqslant 0$ and $R_r = 0$ for any $r < 0$. Let P be an R-module and let $(P_r)_{r\in\mathbf{Z}}$ be a non-descending chain of R-submodules of P such that $\bigcup_{r\in\mathbf{Z}} P_r = P$ and $P_r = 0$ for all sufficiently small $r \in \mathbf{Z}$. Then P can be treated as a filtered R-module with the filtration $(P_r)_{r\in\mathbf{Z}}$ and the left $\mathcal{D}$-module $\mathcal{D} \otimes_R P$ can be considered as a filtered σ-R-module with the filtration $((\mathcal{D} \otimes_R P)_r)_{r\in\mathbf{Z}}$ where $(\mathcal{D} \otimes_R P)_r$ is the R-submodule of $\mathcal{D} \otimes_R P$ generated by the set $\{u \otimes x \mid u \in \mathcal{D}_i$ and $x \in P_{r-i}, 0 \leqslant i \leqslant r)\}$. In what follows, while considering $\mathcal{D} \otimes_R P$ as a filtered σ-R-module (P is an exhaustively and separately filtered module over the σ-ring R with the trivial filtration) we shall always mean the filtration $((\mathcal{D} \otimes_R P)_r)_{r\in\mathbf{Z}}$.

THEOREM 3.1.4 [88, Theorem 6.2.5]. *Let R be an Artinian difference ring with a basic set $\sigma = \{\alpha_1, \dots, \alpha_n\}$ and let $(M_r)_{r\in\mathbf{Z}}$ be an excellent filtration of a σ-R-module M. Then there exists a polynomial $\psi(t) \in \mathbf{Q}[t]$ such that $\psi(r) = l_R(M_r)$ for all sufficiently large $r \in \mathbf{Z}$. Furthermore, $\deg \psi(t) \leqslant n$ and the polynomial $\psi(t)$ can be written as $\psi(t) = \sum_{i=0}^{n} c_i \binom{t+i}{i}$ where $c_0, c_1, \dots, c_n \in \mathbf{Z}$.*

The polynomial $\psi(t)$ whose existence is established by theorem 3.1.4 is called the *difference (σ-) dimension polynomial* or *characteristic polynomial* of the module M associated with the excellent filtration $(M_r)_{r\in\mathbf{Z}}$.

EXAMPLE 3.1.5. Let R be a difference field with a basic set $\sigma = \{\alpha_1, \dots, \alpha_n\}$ and $\mathcal{D}$ the ring of σ-operators over R treated as a filtered σ-R-module with the excellent filtration $(\mathcal{D}_r)_{r\in\mathbf{Z}}$. If $r \in \mathbf{N}$, then the elements $\alpha_1^{k_1}, \dots, \alpha_n^{k_n}$, where $k_1, \dots, k_n \in \mathbf{N}$ and $\sum_{i=1}^n k_i \leqslant r$, form a basis of the vector R-space $\mathcal{D}_r$. Therefore, $l_R(\mathcal{D}_r) = \dim_R \mathcal{D}_r = \mathrm{Card}\{(k_1, \dots, k_n) \in \mathbf{N}^n \mid k_1 + \dots + k_n \leqslant r\} = \binom{r+n}{n}$ whence $\psi_{\mathcal{D}}(t) = \binom{t+n}{n}$ is the characteristic polynomial of the ring $\mathcal{D}$ associated with the filtration $(\mathcal{D}_r)_{r\in\mathbf{Z}}$.

Let R be a difference ring with a basic set σ, $\mathcal{D}$ the ring of σ-operators over R, M a filtered σ-R-module with a filtration $(M_r)_{r\in\mathbf{Z}}$, and $R[x]$ the ring of polynomials in one indeterminate x over R. Let $\mathcal{D}'$ denote the subring $\sum_{r\in\mathbf{N}} \mathcal{D}_r \otimes_R Rx^r$ of the ring $\mathcal{D} \otimes_R R[x]$ and M' denote the left $\mathcal{D}'$-module $\sum_{r\in\mathbf{N}} M_r \otimes_R Rx^r$. The proof of the following three results can be found in [88, Section 6.2].

LEMMA 3.1.6. *With the above notation, let all components of the filtration $(M_r)_{r\in\mathbf{Z}}$ be finitely generated R-modules. Then the filtration $(M_r)_{r\in\mathbf{Z}}$ is excellent if and only if M' is a finitely generated $\mathcal{D}'$-module.*

LEMMA 3.1.7. *Let R be a Noetherian inversive difference ring with a basic set σ. Then the ring of σ-operators $\mathcal{D}$ and the ring $\mathcal{D}'$ are left Noetherian.*

THEOREM 3.1.8. *Let R be a Noetherian inversive difference ring with a basic set σ. Let $\rho : N \to M$ be an injective homomorphism of filtered σ-R-modules and let the filtration of the module M be excellent. Then the filtration of N is also excellent.*

Let R be a difference field with a basic set $\sigma = \{\alpha_1, \dots, \alpha_n\}$, $\mathcal{D}$ the ring of σ-operators over R, and M a finitely generated σ-R-module with generators $x_1, \dots, x_m$ (i.e., $M = \sum_{i=1}^m \mathcal{D}x_i$). Then the vector R-spaces $M_r = \sum_{i=1}^m \mathcal{D}_r x_i$ ($r \in \mathbf{Z}$) form an excellent filtration of M. It is easy to see that if $(M'_r)_{r\in\mathbf{Z}}$ is another excellent filtration of M, then there exist $k \in \mathbf{Z}$, $p \in \mathbf{N}$ such that $M_r \subseteq M'_{r+p}$ and $M'_r \subseteq M_{r+p}$ for all $r \in \mathbf{Z}$, $r \geqslant k$. Thus, if $\psi(t)$ and $\psi_1(t)$ are the characteristic polynomials of the σ-R-module M associated with the excellent filtrations $(M_r)_{r\in\mathbf{Z}}$ and $(M'_r)_{r\in\mathbf{Z}}$, respectively, then $\psi(r) \leqslant \psi_1(r+p)$ and $\psi_1(r) \leqslant \psi(r+p)$ for all sufficiently large $r \in \mathbf{Z}$. It follows that $\deg\psi(t) = \deg\psi_1(t)$ and the leading coefficients of the polynomials $\psi(t)$ and $\psi_1(t)$ are equal. Since the degree of a characteristic polynomial of M does not exceed n, $\Delta^n\psi(t) = \Delta^n\psi_1(t) \in \mathbf{Z}$. (The n-th finite difference $\Delta^n f(t)$ of a polynomial $f(t)$ is defined as usual: $\Delta f(t) = f(t+1) - f(t)$, $\Delta^k f(t) = \Delta(\Delta^{k-1} f(t))$ for $k = 1, 2, \dots$.) We arrive at the following result.

THEOREM 3.1.9. *Let R be a difference field with a basic set $\sigma = \{\alpha_1, \dots, \alpha_n\}$, M a finitely generated σ-R-module, and $\psi(t)$ the difference dimension polynomial associated with an excellent filtration of M. Then the integers $\Delta^n\psi(t)$, $d = \deg\psi(t)$, and $\Delta^d\psi(t)$ do not depend on the choice of the excellent filtration.*

With the notation of the last theorem, the numbers $\Delta^n\psi(t)$, $d = deg\,\psi(t)$, and $\Delta^d\psi(t)$ are called the *difference* (or σ-) *dimension*, *difference* (or σ-) *type*, and *typical difference* (or σ-) *dimension of* M, respectively. These characteristics of the σ-R-module M are denoted by $\delta(M)$, $t(M)$, and $t\delta(M)$, respectively.

The following two theorems give some properties of the difference dimension (see [99] or [88, Section 6.2]).

THEOREM 3.1.10. *Let R be an inversive difference field with a basic set $\sigma = \{\alpha_1, \dots, \alpha_n\}$ and let $0 \longrightarrow N \xrightarrow{i} M \xrightarrow{j} P \longrightarrow 0$ be an exact sequence of finitely generated σ-R-modules. Then $\delta(N) + \delta(P) = \delta(M)$.*

THEOREM 3.1.11. *Let R be an inversive difference field with a basic set $\sigma = \{\alpha_1, \dots, \alpha_n\}$, $\mathcal{D}$ the ring of σ-operators over R, and M a finitely generated σ-R-module. Then $\delta(M)$ is equal to the maximal number of elements of M linearly independent over $\mathcal{D}$.*

Type and dimension of difference vector spaces

Let M be a module over a commutative ring R, U a family of R-submodules of M, and $\mathcal{B}_U$ the set of all pairs $(N, N') \in U \times U$ such that $N' \subseteq N$. Furthermore, let $\overline{\mathbf{Z}}$ denote the set $\mathbf{Z} \cup \{\infty\}$ considered as a linearly ordered set with the natural order ($a < \infty$ for any $a \in \mathbf{Z}$). As is shown in [80], there exists a unique map $\mu_U : \mathcal{B}_U \to \overline{\mathbf{Z}}$ such that

(i) $\mu_U(N, N') \geqslant -1$ for every pair $(N, N') \in \mathcal{B}_U$;

(ii) for any $d \in \mathbf{N}$, the inequality $\mu_U(N, N') \geqslant d$ holds if and only if $N \neq N'$ and there exists an infinite chain $N = N_0 \supseteq N_1 \supseteq \cdots \supseteq N'$ such that $N_i \in U$ and $\mu_U(N_{i-1}, N_i) \geqslant d - 1$ for $i = 1, 2, \dots$.

With the above notation, $\sup\{\mu_U(N, N') \mid (N, N') \in \mathcal{B}_U\}$ is called the *type of the R-module M over the family U*; it is denoted by $type_U\, M$. The least upper bound of the lengths p of chains $N_0 \supseteq N_1 \supseteq \cdots \supseteq N_p$ such that $N_i \in U$ $(0 \leqslant i \leqslant p)$ and $\mu_U(N_{i-1}, N_i) = type_U\, M$ for $i = 1, \dots, p$ is called the *dimension of M over U*; it is denoted by $\dim_U M$.

THEOREM 3.1.12 [102]. *Let K be a difference field with a basic set $\sigma = \{\alpha_1, \dots, \alpha_n\}$, M a finitely generated σ-K-module, and U the family of all σ-K-submodules of M.*

(i) *If $\delta(M) > 0$, then $type_U\, M = n$ and* $\dim_U M = \delta(M)$.

(ii) *If $\delta(M) = 0$, then $type_U\, M < n$.*

3.2. *Inversive difference modules. σ^*-dimension polynomials and their invariants*

Let R be an inversive difference ring with a basic set $\sigma = \{\alpha_1, \dots, \alpha_n\}$ and let Γ denote the free commutative group generated by σ. As before, we set $\sigma^* = \{\alpha_1, \dots, \alpha_n, \alpha_1^{-1}, \dots, \alpha_n^{-1}\}$ and call R a σ^*-ring. If $\gamma = \alpha_1^{k_1} \dots \alpha_n^{k_n} \in \Gamma$, then the number $\sum_{i=1}^{n} |k_i|$ is called the *order* of the element γ; it is denoted by $ord\,\gamma$. For any $r \in \mathbf{N}$, the set $\{\gamma \in \Gamma \mid ord\,\gamma \leqslant r\}$ is denoted by $\Gamma(r)$.

DEFINITION 3.2.1. An expression of the form $\sum_{\gamma\in\Gamma} a_\gamma \gamma$, where $a_\gamma \in R$ for any $\gamma \in \Gamma$ and only finitely many elements a_γ are different from 0, is called an inversive difference (or σ^*-) operator over the difference ring R. Two σ^*-operators $\sum_{\gamma\in\Gamma} a_\gamma \gamma$ and $\sum_{\gamma\in\Gamma} b_\gamma \gamma$ are considered to be equal if and only if $a_\gamma = b_\gamma$ for any $\gamma \in \Gamma$.

The set of all σ^*-operators over R can be naturally equipped with a ring structure if one sets $\sum_{\gamma\in\Gamma} a_\gamma \gamma + \sum_{\gamma\in\Gamma} b_\gamma \gamma = \sum_{\gamma\in\Gamma}(a_\gamma + b_\gamma)\gamma$, $a\sum_{\gamma\in\Gamma} a_\gamma \gamma = \sum_{\gamma\in\Gamma}(aa_\gamma)\gamma$, $(\sum_{\gamma\in\Gamma} a_\gamma \gamma)\gamma_1 = \sum_{\gamma\in\Gamma} a_\gamma(\gamma\gamma_1)$, $\gamma_1 a = \gamma_1(a)\gamma_1$ for any σ^*-operators $\sum_{\gamma\in\Gamma} a_\gamma \gamma$, $\sum_{\gamma\in\Gamma} b_\gamma \gamma$ and for any $a \in R$, $\gamma_1 \in \Gamma$, and extends the multiplication by distributivity. The ring obtained in this way is called the *ring of inversive difference* (or σ^*-) *operators over* R; it is denoted by $\mathcal{E}$. Clearly, the ring of difference (σ-) operators $\mathcal{D}$ introduced in the preceding section is a subring of $\mathcal{E}$.

If $A = \sum_{\gamma\in\Gamma} a_\gamma \gamma \in \mathcal{E}$, then the number $ord\, A = \max\{ord\,\gamma \mid a_\gamma \neq 0\}$ is called the *order* of the σ^*-operator A. Setting $\mathcal{E}_r = \{A \in \mathcal{E} \mid ord\, A \leqslant r\}$ for any $r \in \mathbf{N}$ and $\mathcal{E}_r = 0$ for any $r \in \mathbf{Z}, r < 0$, we obtain an ascending filtration $(\mathcal{E}_r)_{r\in\mathbf{Z}}$ of the ring $\mathcal{E}$ called the *standard filtration* of this ring. In what follows, while considering $\mathcal{E}$ as a filtered ring, we always mean this filtration.

THEOREM 3.2.2 [100]. *If R is a Noetherian inversive difference ring with a basic set σ, then the corresponding ring of σ^*-operators $\mathcal{E}$ is left Noetherian. If R is a σ^*-field, then $\mathcal{E}$ is a left Ore ring.*

DEFINITION 3.2.3. Let R be an inversive difference ring with a basic set σ and $\mathcal{E}$ the ring of inversive difference operators over R. Then a left $\mathcal{E}$-module is said to be an inversive difference R-module or a σ^*-R-module. In other words, an R-module M is called a σ^*-R-module if elements of the set σ^* act on M in such a way that $\alpha(x+y) = \alpha x + \alpha y$, $\alpha(\beta x) = \beta(\alpha x)$, $\alpha(ax) = \alpha(a)\alpha(x)$, and $\alpha(\alpha^{-1}x) = x$ for any $\alpha, \beta \in \sigma^*$; $x, y \in M$; $a \in R$.

If R is a σ^*-field, a σ^*-R-module M is said to be a vector σ^*-R-space (or an inversive difference vector space over R).

It is clear that any σ^*-R-module can be naturally treated as a σ-R-module. Also, if M and N are two σ^*-R-modules, then any difference (σ-) homomorphism $f : M \to N$ has the property that $f(\alpha x) = \alpha f(x)$ for any $x \in M$ and $\alpha \in \sigma^*$.

Let R be an inversive difference ring with a basic set σ and $\mathcal{E}$ the ring of σ^*-operators over R. For any σ^*-R-module M, the set $C(M) = \{x \in M \mid \alpha x = x$ for all $\alpha \in \sigma\}$ is called the *set of constants* of M, its elements are called *constants*. Obviously, $C(M)$ is a subgroup of the additive group of M and $C : M \mapsto C(M)$ is a functor from the category of σ^*-R-modules (i.e., the category of all left $\mathcal{E}$-modules) to the category of Abelian groups.

If M and N are two σ^*-R-modules, then each of the R-modules $\mathrm{Hom}_R(M, N)$ and $M \otimes_R N$ can be equipped with a structure of a σ^*-R-module if for any $f \in \mathrm{Hom}_R(M, N)$, $\sum_{i=1}^k x_i \otimes y_i \in M \otimes_R N$ ($x_1, \dots, x_k \in M$; $y_1, \dots, y_k \in N$), and $\alpha \in \sigma^*$, one defines αf by $(\alpha f)x = \alpha f(\alpha^{-1}x)$ and sets $\alpha(\sum_{i=1}^k x_i \otimes y_i) = \sum_{i=1}^k \alpha x_i \otimes \alpha y_i$. It is easy to check that $\alpha f \in \mathrm{Hom}_R(M, N)$ and the actions of elements of σ^* on $\mathrm{Hom}_R(M, N)$ and $M \otimes_R N$ satisfy the conditions of Definition 3.2.3. In what follows, while considering $\mathrm{Hom}_R(M, N)$

and $M \otimes_R N$ as σ^*-R-modules (for some σ^*-R-modules M and N), we always mean these inversive difference structures of these modules.

PROPOSITION 3.2.4 [88, Section 3.4]. *Let R be an inversive difference ring with a basic set σ, and let M, N, and P be σ^*-R-modules.*

(i) *The natural isomorphism of R-modules*

$$\eta : \mathrm{Hom}_R(P \otimes_R M, N) \to \mathrm{Hom}_R(P, \mathrm{Hom}_R(M, N))$$

(defined by $[(\eta f)x](y) = f(x \otimes y)$ for any $f \in \mathrm{Hom}_R(P \otimes_R M, N)$, $x \in P$, $y \in M$) is a σ^*-isomorphism.

(ii) $C(\mathrm{Hom}_R(M, N)) = \mathrm{Hom}_{\mathcal{E}}(M, N)$.

(iii) *The functors C and $\mathrm{Hom}_{\mathcal{E}}(R, \cdot)$ are naturally isomorphic.*

(iv) *The functor C is left exact and for any positive integer p, its p-th right derived functor $\mathcal{R}^p C$ is naturally isomorphic to the functor $\mathrm{Ext}^p_{\mathcal{E}}(R, \cdot)$.*

(v) *The functors $\mathrm{Hom}_{\mathcal{E}}(\cdot \otimes_R M, N)$ and $\mathrm{Hom}_{\mathcal{E}}(\cdot, \mathrm{Hom}_R(M, N))$ are naturally isomorphic and the same is true for the functors $\mathrm{Hom}_{\mathcal{E}}(M \otimes_R \cdot, N)$ and $\mathrm{Hom}_{\mathcal{E}}(M, \mathrm{Hom}_R(\cdot, N))$.*

(vi) *For any positive integers p and q, there exists a spectral sequence converging to $\mathrm{Ext}^{p+q}_{\mathcal{E}}(M, N)$ whose second term is equal to $E_2^{p,q} = (\mathcal{R}^p C)(\mathrm{Ext}^p_R(M, N))$.*

DEFINITION 3.2.5. Let R be an inversive difference ring with a basic set σ, $\mathcal{E}$ the ring of σ^*-operators over R (considered as a filtered ring with the standard filtration $(\mathcal{E}_r)_{r \in \mathbf{Z}}$), and M a σ^*-R-module. An ascending chain $(M_r)_{r \in \mathbf{Z}}$ of R-submodules of M is called a filtration of M if $\mathcal{E}_r M_s \subseteq M_{r+s}$ for all $r, s \in \mathbf{Z}$, $M_r = 0$ for all sufficiently small $r \in \mathbf{Z}$, and $\bigcup_{r \in \mathbf{Z}} M_r = M$. A filtration $(M_r)_{r \in \mathbf{Z}}$ of the σ^*-R-module M is called excellent if all R-modules M_r ($r \in \mathbf{Z}$) are finitely generated and there exists $r_0 \in \mathbf{Z}$ such that $M_r = \mathcal{E}_{r-r_0} M_{r_0}$ for any $r \in \mathbf{Z}, r \geqslant r_0$.

THEOREM 3.2.6. *Let R be an Artinian σ^*-ring with a basic set $\sigma = \{\alpha_1, \dots, \alpha_n\}$ and let $(M_r)_{r \in \mathbf{Z}}$ be an excellent filtration of a σ^*-R-module M. Then there exists a numerical polynomial $\chi(t) \in \mathbf{Q}[t]$ such that*

(i) *$\chi(r) = l_R(M_r)$ for all sufficiently large $r \in \mathbf{Z}$;*

(ii) *$\deg \chi(t) \leqslant n$ and the polynomial $\chi(t)$ can be represented in the form $\chi(t) = \sum_{i=0}^{n} 2^i a_i \binom{t+i}{i}$ where $a_0, \dots, a_n \in \mathbf{Z}$.*

The polynomial $\chi(t)$ whose existence is established by Theorem 3.2.6 is called the σ^*-*dimension polynomial* or the *characteristic polynomial* of the module M associated with the excellent filtration $(M_r)_{r \in \mathbf{Z}}$.

EXAMPLE 3.2.7. Let R be an inversive difference field with a basic set $\sigma = \{\alpha_1, \dots, \alpha_n\}$, $\mathcal{E}$ the ring of σ^*-operators over R, and $\chi_{\mathcal{E}}(t)$ the difference dimension polynomial associated with the standard filtration $(\mathcal{E}_r)_{r \in \mathbf{Z}}$. Then $\chi_{\mathcal{E}}(r) = l_R(\mathcal{E}_r) = \dim_R(\mathcal{E}_r) = \mathrm{Card}\{\gamma = \alpha_1^{k_1} \dots \alpha_n^{k_n} \in \Gamma \mid ord\, \gamma = \sum_{i=1}^{n} |k_i| \leqslant r\}$ for all sufficiently large $r \in \mathbf{Z}$. Applying [88,

Proposition 2.1.9] we obtain three expressions for the last number that lead to the following three forms of the polynomial $\chi_{\mathcal{E}}(t)$:

$$\chi_{\mathcal{E}}(t) = \sum_{i=0}^{n} 2^i \binom{n}{i}\binom{t}{i} = \sum_{i=0}^{n} \binom{n}{i}\binom{t+i}{n} = \sum_{i=0}^{n} (-1)^{n-i} 2^i \binom{n}{i}\binom{t+i}{i}.$$

Let F_m be a free left $\mathcal{E}$-module of rank m ($m \geqslant 1$) with free generators $f_1, \dots, f_m$. Then for every $l \in \mathbf{Z}$, one can consider the excellent filtration $((F_m^l)_r)_{r \in \mathbf{Z}}$ of the module F_m such that $(F_m^l)_r = \sum_{i=1}^{m} \mathcal{E}_{r-l} f_i$ for every $r \in \mathbf{Z}$. We obtain a filtered σ^*-R-module that will be denoted by F_m^l. A finite direct sum of such filtered σ^*-R-modules is called a *free filtered σ^*-R-module*. The representations of $\chi_{\mathcal{E}}(t)$ imply the following expressions for the σ^*-dimension polynomial $\chi(t)$ of F_m^l:

$$\chi(t) = m\chi_{\mathcal{E}}(t-l) = m\sum_{i=0}^{n} 2^i \binom{n}{i}\binom{t-l}{i} = m\sum_{i=0}^{n} \binom{n}{i}\binom{t+i-l}{n}$$

$$= m\sum_{i=0}^{n} (-1)^{n-i} 2^i \binom{n}{i}\binom{t+i-l}{i}.$$

Let R be an inversive difference ring with a basic set $\sigma = \{\alpha_1, \dots, \alpha_n\}$, $\mathcal{E}$ the ring of σ^*-operators over R, M a filtered σ^*-R-module with a filtration $(M_r)_{r \in \mathbf{Z}}$, and $R[x]$ the ring of polynomials in one indeterminate x over R. Let $\mathcal{E}'$ denote the subring $\sum_{r \in \mathbf{N}} \mathcal{E}_r \otimes_R Rx^r$ of the ring $\mathcal{E} \otimes_R R[x]$ and M' denote the left $\mathcal{E}'$-module $\sum_{r \in \mathbf{N}} M_r \otimes_R Rx^r$. The following three results are similar to the corresponding statements for difference modules.

LEMMA 3.2.8. *With the above notation, let all components of the filtration $(M_r)_{r \in \mathbf{Z}}$ be finitely generated R-modules. Then the filtration $(M_r)_{r \in \mathbf{Z}}$ is excellent if and only if M' is a finitely generated $\mathcal{E}'$-module.*

LEMMA 3.2.9. *If R is a Noetherian inversive difference ring, then the ring $\mathcal{E}'$ considered above is left Noetherian.*

Let R be an inversive difference ring with a basic set σ and let M and N be filtered σ^*-R-modules with filtrations $(M_r)_{r \in \mathbf{Z}}$ and $(N_r)_{r \in \mathbf{Z}}$, respectively. Then a σ-homomorphism $f : M \to N$ is said to be a *σ-homomorphism of filtered σ^*-R-modules* if $f(M_r) \subseteq N_r$ for any $r \in \mathbf{Z}$.

THEOREM 3.2.10. *Let R be a Noetherian inversive difference ring with a basic set σ and let $\rho : N \to M$ be an injective homomorphism of filtered σ^*-R-modules. Furthermore, suppose that the filtration of the module M is excellent. Then the filtration of N is also excellent.*

As in the case of difference modules, for any two excellent filtrations $(M_r)_{r \in \mathbf{Z}}$ and $(M'_r)_{r \in \mathbf{Z}}$ of a finitely generated σ^*-module M over an inversive difference (σ^*-) field K, there exist $p \in \mathbf{N}$ such that $M_r \subseteq M'_{r+p}$ and $M'_r \subseteq M_{r+p}$ for all $r \in \mathbf{Z}$. This observation leads to the following statement that gives some invariants of the σ^*-dimension polynomial of M.

THEOREM 3.2.11. *Let K be an inversive difference field with a basic set $\sigma = \{\alpha_1, \dots, \alpha_n\}$, M a finitely generated σ^*-K-module, and $\chi(t)$ the characteristic polynomial associated with an excellent filtration of M. Then the integers $\frac{\Delta^n \chi(t)}{2^n}$, $d = \deg \chi(t)$, and $\frac{\Delta^d \chi(t)}{2^d}$ do not depend on the choice of the excellent filtration of M.*

With the notation of the last theorem, the numbers $\frac{\Delta^n \chi(t)}{2^n}$, $d = \deg \chi(t)$, and $\frac{\Delta^d \chi(t)}{2^d}$ are called the *inversive difference* (or σ^*-) *dimension, inversive difference* (or σ^*-) *type*, and *typical inversive difference* (or *typical* σ^*-) *dimension* of the module M, respectively. These characteristics of the σ^*-K-module M are denoted by $i\delta(M)$, $it(M)$, and $ti\delta(M)$, respectively. (If we want to indicate the basic set with respect to which K is considered as a difference field, we will use the notation $i\delta_\sigma(M)$, $it_\sigma(M)$, and $ti\delta_\sigma(M)$, respectively.)

Let K be an inversive difference field with a basic set $\sigma = \{\alpha_1, \dots, \alpha_n\}$, Γ the free commutative group generated by σ, and M a σ^*-K-module. Elements $z_1, \dots, z_m \in M$ are said to be *σ^*-linearly independent over K* if the set $\{\gamma z_i \mid 1 \leqslant i \leqslant m, \gamma \in \Gamma\}$ is linearly independent over the σ^*-field K. Otherwise, $z_1, \dots, z_m$ are said to be *σ^*-linearly dependent* over K.

THEOREM 3.2.12 [100]. *Let K be an inversive difference field with a basic set $\sigma = \{\alpha_1, \dots, \alpha_n\}$ and let $0 \longrightarrow N \xrightarrow{i} M \xrightarrow{j} P \longrightarrow 0$ be an exact sequence of finitely generated σ^*-K-modules. Then $i\delta(N) + i\delta(P) = i\delta(M)$.*

THEOREM 3.2.13 [100]. *Let K be an inversive difference field with a basic set $\sigma = \{\alpha_1, \dots, \alpha_n\}$, $\mathcal{E}$ the ring of σ^*-operators over K, and M a finitely generated σ^*-K-module. Then $i\delta(M)$ is equal to the maximal number of elements of M that are σ^*-linearly independent over K.*

THEOREM 3.2.14 [102]. *Let K be an inversive difference field with a basic set $\sigma = \{\alpha_1, \dots, \alpha_n\}$, M a finitely generated σ^*-K-module, and U the family of all σ^*-K-submodules of M. Then*

(i) *If $i\delta(M) > 0$, then $\text{type}_U\, M = n$ and $\dim_U M = i\delta(M)$.*
(ii) *If $i\delta(M) = 0$, then $\text{type}_U\, M < n$.*

Let R be an inversive difference ring with a basic set $\sigma = \{\alpha_1, \dots, \alpha_n\}$ and Γ the free commutative group generated by σ. It is easy to see that if $\sigma_1 = \{\tau_1, \dots, \tau_n\}$ is another system of free generators of Γ, then there exists a matrix $K = (k_{ij})_{1 \leqslant i,j \leqslant n} \in GL(n, \mathbf{Z})$ such that $\alpha_i = \tau_1^{k_{i1}} \dots \tau_n^{k_{in}}$ $(1 \leqslant i \leqslant n)$. The σ^*-ring R can be also treated as a σ_1^*-ring and the corresponding ring of σ_1^*-operators coincides with the ring of σ^*-operators $\mathcal{E}$ over R. We say that two finite sets of automorphisms $\sigma = \{\alpha_1, \dots, \alpha_n\}$ and $\sigma_1 = \{\tau_1, \dots, \tau_n\}$ of the same ring R are *equivalent* if there exists a matrix $(k_{ij})_{1 \leqslant i,j \leqslant n} \in GL(n, \mathbf{Z})$ such that $\alpha_i = \tau_1^{k_{i1}} \dots \tau_n^{k_{in}}$ $(1 \leqslant i \leqslant n)$.

THEOREM 3.2.15 [88, Theorem 6.3.19]. *Let K be an inversive difference field with a basic set $\sigma = \{\alpha_1, \dots, \alpha_n\}$, M a finitely generated σ^*-K-module, and $d = it_\sigma(M)$. Then there exists a set $\sigma' = \{\beta_1, \dots, \beta_n\}$ of pairwise commuting automorphisms of K such that*

(i) *The sets σ and σ' are equivalent.*

(ii) *Let $\sigma'' = \{\beta_1, \dots, \beta_d\}$. Then M is a finitely generated σ''^*-K-module and $i\delta_{\sigma''}(M) > 0$.*

There are several publications devoted to methods and algorithm of computation of characteristic polynomials of difference and inversive difference modules (see [87,89–91], [88, Chapters 6, 9], [113,119], and [127]). The corresponding techniques are based either on constructing resolutions of free filtered difference modules (whose characteristic polynomials are given in Example 3.2.7) or by applying the Gröbner basis method to modules over rings of difference and inversive difference operators.

3.3. *Reduction in a free difference vector space. Characteristic sets and multivariable dimension polynomials*

In this section we apply the method of characteristic sets to difference and inversive difference modules whose basic sets of translations are represented as unions of their disjoint subsets. In the case of difference and inversive difference vector spaces, we generalize the results of the two preceding sections and show the existence of characteristic polynomials in several variables associated with partitions of the basic set. We also present invariants of such polynomials.

Let K be a difference field with a basic set $\sigma = \{\alpha_1, \dots, \alpha_n\}$ and let a partition of σ into a union of p its proper disjoint subsets be fixed: $\sigma = \bigcup_{j=1}^{p} \sigma_j$ where $\sigma_1 = \{\alpha_1, \dots, \alpha_{n_1}\}$, $\sigma_2 = \{\alpha_{n_1+1}, \dots, \alpha_{n_1+n_2}\}, \dots, \sigma_p = \{\alpha_{n_1+\dots+n_{p-1}+1}, \dots, \alpha_n\}$ ($p \geqslant 1, n_1, \dots, n_p \in \mathbf{N}$). As before, T and $\mathcal{D}$ denote the free commutative semigroup generated by σ and the ring of σ-operators over K, respectively. If $t = \alpha_1^{k_1} \dots \alpha_n^{k_n} \in T$, then the order of the element t with respect to σ_i ($1 \leqslant i \leqslant p$) is defined as $ord_i\, t = \sum_{\nu=n_1+\dots+n_{i-1}+1}^{n_1+\dots+n_i} k_\nu$ (if $i = 1$, then the lower index in the last sum is 1). The order of t is still defined as $ord\,\theta = k_1 + \dots + k_n$.

Our partition of the set σ induces p orderings $<_1, \dots, <_p$ of the semigroup T defined as follows: $t = \alpha_1^{k_1} \dots \alpha_n^{k_n} <_i t' = \alpha_1^{l_1} \dots \alpha_n^{l_n}$ if and only if the $(n + p + 1)$-tuple $(ord_i\, t, ord\, t, ord_1\, t, \dots, ord_{i-1}\, t, ord_{i+1}\, t, \dots, ord_p\, t, k_{n_1+\dots+n_{i-1}+1}, \dots, k_{n_1+\dots+n_i}, k_1, \dots, k_{n_1+\dots+n_{i-1}}, k_{n_1+\dots+n_i+1}, \dots, k_n)$ is less than the $(n + p + 1)$-tuple $(ord_i\, t', ord\, t', ord_1\, t', \dots, ord_{i-1}\, t', ord_{i+1}\, t', \dots, ord_p\, t', l_{n_1+\dots+n_{i-1}+1}, \dots, l_{n_1+\dots+n_i}, l_1, \dots, l_{n_1+\dots+n_{i-1}}, l_{n_1+\dots+n_i+1}, \dots, l_n)$ with respect to the lexicographic order on $\mathbf{N}^{n+p+1}$.

For any $r_1, \dots, r_p \in \mathbf{N}$, $T(r_1, \dots, r_p)$ will denote the set $\{t \in T \mid ord_1\, t \leqslant r_1, \dots, ord_p\, t \leqslant r_p\}$; the vector K-subspace of $\mathcal{D}$ generated by this set will be denoted by $\mathcal{D}_{r_1 \dots r_p}$. Setting $\mathcal{D}_{r_1 \dots r_p} = 0$ for $(r_1, \dots, r_p) \in \mathbf{Z}^p \setminus \mathbf{N}^p$, we obtain a family $\{\mathcal{D}_{r_1 \dots r_p} \mid (r_1, \dots, r_p) \in \mathbf{Z}^p\}$ called the *standard p-dimensional filtration* of the ring $\mathcal{D}$.

DEFINITION 3.3.1. A family $\{M_{r_1 \dots r_p} \mid (r_1, \dots, r_p) \in \mathbf{Z}^p\}$ of vector K-subspaces of a σ-K-module M is called a p-dimensional filtration of M if

(i) for any fixed integers $r_1, \dots, r_{i-1}, r_{i+1}, \dots, r_p$ $(1 \leqslant i \leqslant p)$, $M_{r_1 \dots r_i \dots r_p} \subseteq M_{r_1 \dots r_{i-1}, r_i+1, r_{i+1} \dots r_p}$ and $M_{r_1 \dots r_p} = 0$ for all sufficiently small $r_i \in \mathbf{Z}$;
(ii) $\bigcup \{M_{r_1 \dots r_p} \mid (r_1, \dots, r_p) \in \mathbf{Z}^p\} = M$;
(iii) $\mathcal{D}_{r_1 \dots r_p} M_{s_1 \dots s_p} \subseteq M_{r_1+s_1, \dots, r_p+s_p}$ for any $(r_1, \dots, r_p) \in \mathbf{N}^p$, $(s_1, \dots, s_p) \in \mathbf{Z}^p$.

If every vector K-space $M_{r_1 \dots r_p}$ is finitely generated and there exists an element $(h_1, \dots, h_p) \in \mathbf{Z}^p$ such that $R_{r_1 \dots r_p} M_{h_1 \dots h_p} = M_{r_1+h_1, \dots, r_p+h_p}$ for any $(r_1, \dots, r_p) \in \mathbf{N}^p$, then the p-dimensional filtration is called *excellent*.

It is easy to see that if $u_1, \dots, u_n$ is a finite system of generators of a left $\mathcal{D}$-module M, then the filtration $\{\sum_{i=1}^{n} R_{r_1 \dots r_p} u_i \mid (r_1, \dots, r_p) \in \mathbf{Z}^p\}$ is excellent.

Let F be a finitely generated free σ-K-module (that is, a free left $\mathcal{D}$-module) and let $f_1, \dots, f_q$ be a fixed basis of F over $\mathcal{D}$. Then elements tf_k $(t \in T, 1 \leqslant k \leqslant q)$ are called *terms*; the set of all terms is denoted by Tf. The order of a term tf_k and the order of this term with respect to σ_i $(1 \leqslant i \leqslant p)$ are defined as the order of the element $t \in T$ and the order of t relative to σ_i, respectively. A term tf_i is said to be a *multiple* of a term $t' f_j$ if $i = j$ and t' divides t in the semigroup T (that is, $t = t'' t'$ for some $t'' \in T$). In this case we also say that the term $t' f_j$ divides tf_i and write $t' f_j \mid tf_i$.

Below we consider p orderings of the set Tf that correspond to the orderings of the set T. These ordering are denoted by the same symbols $<_1, \dots, <_p$ and defined as follows: if $tf_k, t' f_l \in Tf$, then $tf_k <_i t' f_l$ if and only if $t <_i t'$ in T or $t = t'$ and $k < l$.

Since the set Tf is a basis of the vector K-space F, every element $f \in F$ has a unique representation in the form

$$f = a_1 t_1 f_{i_1} + \dots + a_m t_m f_{i_m}, \tag{$*$}$$

where $t_i \in T$, $a_i \in K$, $a_i \neq 0$ $(1 \leqslant i \leqslant m)$, $1 \leqslant i_1, \dots, i_m \leqslant q$ and all terms $t_\nu f_{i_\nu}$ $(1 \leqslant \nu \leqslant m)$ are distinct. For any $j = 1, \dots, p$, the greatest with respect to $<_j$ term of the set $\{t_\nu f_{i_\nu} \mid 1 \leqslant \nu \leqslant m\}$ is called the j-leader of the element f; it is denoted by $u_f^{(j)}$. (Of course, it is possible that $u_f^{(j)} = u_f^{(l)}$ for some distinct numbers j and l.)

In what follows, we say that an element $f \in F$ contains a term tf_j if the term appears in the representation $(*)$ with a non-zero coefficient. The coefficient of the j-leader of an element $f \in F$ will be denoted by $lc_j(f)$ $(1 \leqslant j \leqslant p)$.

DEFINITION 3.3.2. Let f and g be two elements of the free σ-K-module F considered above. The element f is said to be *reduced with respect to* g if f does not contain any multiple $tu_g^{(1)}$ $(t \in T)$ of the 1-leader $u_g^{(1)}$ such that $ord_j(tu_g^{(j)}) \leqslant ord_j u_f^{(j)}$ for $j = 2, \dots, p$. An element $h \in F$ is said to be *reduced with respect to a set* $\mathcal{A} \subseteq F$, if h is reduced with respect to every element of $\mathcal{A}$. A set $\Sigma \subseteq F$ is called autoreduced if every element of Σ is reduced with respect to any other element of this set. An autoreduced set Σ is called normal if $lc_1(g) = 1$ for every element $g \in \Sigma$.

THEOREM 3.3.3. *Let Σ be an autoreduced set in the free σ-K-module F. Then:*
(i) *The set Σ is finite, $\Sigma = \{g_1, \dots, g_r\}$.*
(ii) *For any $f \in F$, there exists an element $g \in F$ such that $f - g = \sum_{i=1}^{r} \lambda_i g_i$ for some $\lambda_1, \dots, \lambda_r \in \mathcal{D}$ and g is reduced with respect to Σ.*

Let f and g be two elements of the free σ-K-module F. We say that f has lower rank than g and write $rk(f) < rk(g)$ if either $u_f^{(1)} <_1 u_g^{(1)}$ or there exists some k, $2 \leqslant k \leqslant p$, such that $u_f^{(\nu)} = u_g^{(\nu)}$ for $\nu = 1, \ldots, k-1$ and $u_f^{(k)} <_k u_g^{(k)}$. If $u_f^{(i)} = u_g^{(i)}$ for $i = 1, \ldots, p$, we say that f and g have the same rank and write $rk(f) = rk(g)$. In what follows, while considering autoreduced sets in F, we always assume that their elements are arranged in order of increasing rank.

DEFINITION 3.3.4. Let $\Sigma = \{h_1, \ldots, h_r\}$ and $\Sigma' = \{h'_1, \ldots, h'_s\}$ be two autoreduced subsets of the free σ-K-module F. An autoreduced set Σ is said to have lower rank than Σ' if one of the following two cases holds:

(1) There exists $k \in \mathbf{N}$ such that $k \leqslant \min\{r, s\}$, $rk(h_i) = rk(h'_i)$ for $i = 1, \ldots, k-1$ and $rk(h_k) < rk(h'_k)$.

(2) $r > s$ and $rk(h_i) = rk(h'_i)$ for $i = 1, \ldots, s$.

If $r = s$ and $rk(h_i) = rk(h'_i)$ for $i = 1, \ldots, r$, then Σ is said to have the same rank as Σ'.

As in the case of autoreduced sets of difference polynomials, one can show that in every non-empty set of autoreduced subsets of the free σ-K-module F there exists an autoreduced subset of lowest rank. If N is a $\mathcal{D}$-submodule of F, then an autoreduced subset of N of lowest rank is called a *characteristic set* of the module N.

THEOREM 3.3.5. *Let N be a $\mathcal{D}$-submodule of the free σ-K-module F and let $\Sigma = \{g_1, \ldots, g_r\}$ be a characteristic set of N.*

(i) *An element $f \in N$ is reduced with respect to Σ if and only if $f = 0$.*

(ii) *If N is a cyclic $\mathcal{D}$-submodule of F generated by an element g, then $\{g\}$ is a characteristic set of N.*

(iii) *The set Σ generates N as a left $\mathcal{D}$-module.*

(iv) *Let the characteristic set Σ be normal and let $\Sigma_1 = \{h_1, \ldots, h_s\}$ be another normal characteristic sets of N. Then $r = s$ and $g_i = h_i$ for $i = 1, \ldots, r$.*

THEOREM 3.3.6. *Let K be a difference field with a basic set σ, $\mathcal{D}$ the ring of σ-operators over K, and M a finitely generated σ-K-module with a system of generators $\{e_1, \ldots, e_q\}$. Furthermore, let F be a free σ-K-module with a basis $f_1, \ldots, f_q$, $\pi : F \to M$ the natural $\mathcal{D}$-epimorphism ($\pi(f_i) = e_i$ for $i = 1, \ldots, q$), and $\Sigma = \{g_1, \ldots, g_d\}$ a characteristic set of the $\mathcal{D}$-module $N = \operatorname{Ker} \pi$. Finally, for any $r_1, \ldots, r_p \in \mathbf{N}$, let $M_{r_1 \ldots r_p} = \sum_{i=1}^{p} \mathcal{D}_{r_1 \ldots r_p} e_i$ and $U_{r_1 \ldots r_p} = \{w \in Tf \mid ord_j\, w \leqslant r_j$ for $j = 1, \ldots, p$, and either w is not a multiple of any $u_{g_i}^{(1)}$ $(1 \leqslant i \leqslant d)$ or for any $t \in T$, $g_i \in \Sigma$ such that $w = tu_{g_i}^{(1)}$, we have $ord_j(tu_{g_j}^{(j)}) > r_j$ for some j, $2 \leqslant j \leqslant p\}$. Then $\pi(U_{r_1 \ldots r_p})$ is a basis of the vector K-space $M_{r_1 \ldots r_p}$.*

THEOREM 3.3.7. *Let K be a difference field with a basic set σ and let $\{M_{r_1 \ldots r_p} \mid (r_1, \ldots, r_p) \in \mathbf{Z}^p\}$ be an excellent p-dimensional filtration of a σ-K-module M. Then there exists a polynomial $\phi(t_1, \ldots, t_p) \in \mathbf{Q}[t_1, \ldots, t_p]$ such that*

(i) *$\phi(r_1, \ldots, r_p) = \dim_K M_{r_1 \ldots r_p}$ for all sufficiently large $(r_1, \ldots, r_p) \in \mathbf{Z}^p$ (i.e., there exists $(r_1^{(0)}, \ldots, r_p^{(0)}) \in \mathbf{Z}^p$ such that the equality holds for all $(r_1, \ldots, r_p)$ that exceed $(r_1^{(0)}, \ldots, r_p^{(0)}) \in \mathbf{Z}^p$ with respect to the product order on $\mathbf{Z}^p$);*

(ii) *$\deg_{t_i} \phi \leqslant n_i$ for $i = 1, \dots, p$ (in particular, the total degree of ϕ does not exceed n) and the polynomial $\phi(t_1, \dots, t_p)$ can be represented as*

$$\phi = \sum_{i_1=0}^{n_1} \cdots \sum_{i_p=0}^{n_p} a_{i_1 \dots i_p} \binom{t_1 + i_1}{i_1} \cdots \binom{t_p + i_p}{i_p},$$

where $a_{i_1 \dots i_p} \in \mathbf{Z}$ for all $i_1, \dots, i_p$.

The numerical polynomial $\phi(t_1, \dots, t_p)$, whose existence is established by Theorem 3.3.7, is called the *difference* (or σ-) *dimension polynomial of the module M associated with the p-dimensional filtration* $\{M_{r_1 \dots r_p} \mid (r_1, \dots, r_p) \in \mathbf{Z}^p\}$.

Any permutation $(j_1, \dots, j_p)$ of the set $\{1, \dots, p\}$, defines a lexicographic order $\leqslant_{j_1, \dots, j_p}$ on $\mathbf{N}^p$ such that $(r_1, \dots, r_p) \leqslant_{j_1, \dots, j_p} (s_1, \dots, s_p)$ if and only if either $r_{j_1} < s_{j_1}$ or there exists $k \in \mathbf{N}$, $1 \leqslant k \leqslant p - 1$, such that $r_{j_\nu} = s_{j_\nu}$ for $\nu = 1, \dots, k$ and $r_{j_{k+1}} < s_{j_{k+1}}$. In what follows, we use these orders to associate with every set $\Sigma \subseteq \mathbf{N}^p$ the set $\Sigma' = \{e \in \Sigma \mid e$ is a maximal element of Σ with respect to one of the $p!$ lexicographic orders $\leqslant_{j_1, \dots, j_p}\}$. For example, if $\Sigma = \{(3, 0, 2), (2, 1, 1), (0, 1, 4), (1, 0, 3), (1, 1, 6), (3, 1, 0), (1, 2, 0)\} \subseteq \mathbf{N}^3$, then $\Sigma' = \{(3, 0, 2), (3, 1, 0), (1, 1, 6), (1, 2, 0)\}$.

THEOREM 3.3.8. *Let K be a difference field with a basic set σ, M be a finitely generated σ-K-module, $\{M_{r_1 \dots r_p} \mid (r_1, \dots, r_p) \in \mathbf{Z}^p\}$ an excellent p-dimensional filtration of M, and*

$$\phi(t_1, \dots, t_p) = \sum_{i_1=0}^{n_1} \cdots \sum_{i_p=0}^{n_p} a_{i_1 \dots i_p} \binom{t_1 + i_1}{i_1} \cdots \binom{t_p + i_p}{i_p}$$

the σ-dimension polynomial associated with this filtration. Let $E = \{(i_1, \dots, i_p) \in \mathbf{N}^p \mid 0 \leqslant i_k \leqslant n_k$ $(k = 1, \dots, p)$ and $a_{i_1 \dots i_p} \neq 0\}$.

Then the total degree d of the polynomial ϕ, $a_{n_1 \dots n_p}$, p-tuples $(j_1, \dots, j_p) \in E'$, the corresponding coefficients $a_{j_1, \dots, j_p}$, and the coefficients of the terms of total degree d do not depend on the choice of the excellent filtration.

Methods of computation of multivariable characteristic polynomials of difference and inversive difference vector spaces are similar to those for multivariable Hilbert and differential dimension polynomials, see [106,107], and [109]. In particular, these papers contain examples showing that multivariable characteristic polynomials can carry essentially more invariants than classical dimension polynomials in one variable.

4. Difference field extensions

4.1. *Transformal dependence. Difference transcendental bases and difference transcendental degree*

Let K be a difference field with a basic set σ, T the free commutative semigroup generated by σ, and L a σ-overfield of K. We say that an element $v \in L$ is *transformally dependent* or

σ-algebraically dependent on a set $A \subseteq L$ over K if v is σ-algebraic over the field $K\langle A\rangle$. Obviously, an element $v \in L$ is σ-algebraically dependent on a set $A \subseteq L$ if and only if there exists a finite family $\{\eta_1, \dots, \eta_s\} \subseteq A$ such that v is σ-algebraic over $K\langle \eta_1, \dots, \eta_s\rangle$.

Let K be an inversive σ-field, L a σ^*-overfield of K and $A \subseteq L$. It is easy to see that an element $v \in L$ is σ^*-algebraically dependent on A over K (that is, v is σ-algebraic over the σ^*-field $K\langle A\rangle^*$) if and only if v is σ-algebraically dependent on A over K.

All the statements in the rest of this section can be proved in the same way as their ordinary versions (see [32, Chapter 5]).

PROPOSITION 4.1.1. *Let K be a difference field with a basic set σ, L a σ-overfield of K and $A \subseteq L$.*

(i) *The set A is σ-algebraically dependent over K if and only if there exists $v \in A$ such that v is σ-algebraically dependent on $A \setminus \{v\}$ over K.*
(ii) *The set A contains a maximal subset σ-algebraically independent over K. In other words, there exists a set $B \subseteq A$ such that B is σ-algebraically independent over K and any subset of A properly containing B is σ-algebraically dependent over K.*

A set B, whose existence is established by Proposition 4.1.1(ii), is called a *basis for transformal transcendence* or a *difference* (or σ-) *transcendence basis of A over K*. If $A = L$, the set B is called a basis for transformal transcendence or a difference (or σ-) transcendence basis of L over K. (We use this terminology when L is a difference overfield of a σ-field K or an inversive difference field extension of a σ^*-field K.)

PROPOSITION 4.1.2. *Let K be a difference (inversive difference) field with a basic set σ and L a σ-overfield of K. Furthermore, let B and B' be two subsets of L and $v, u_1, \dots, u_m \in L$.*

(i) *If v is σ-algebraically dependent on B over K and every element of B is σ-algebraically dependent on B' over K, then v is σ-algebraically dependent on B' over K.*
(ii) *If v is σ-algebraically dependent on $\{u_1, \dots, u_m\}$, but not on $\{u_1, \dots, u_{m-1}\}$ over K, then u_m is σ-algebraically dependent on the set $\{u_1, \dots, u_{m-1}, v\}$ over K.*
(iii) *Suppose that $B' \subseteq B$, $u_1, \dots, u_m \in B$ are σ-algebraically independent over K, and each u_i $(1 \leqslant i \leqslant m)$ is σ-algebraically dependent on B' over K. Then there exist elements $v_1, \dots, v_m \in B'$ such that each v_i is σ-algebraically dependent over K on the set B'' obtained from B' by replacing v_j by u_j $(j = 1, \dots, m)$.*

PROPOSITION 4.1.3. *Let K be a difference field with a basic set σ, L a σ-overfield of K and $A \subseteq L$.*

(i) *Suppose that B is a subset of A which is σ-algebraically independent over K. Then B is a σ-transcendence basis of A over K if and only if every element of A is σ-algebraically dependent on B over K.*
(ii) *All σ-transcendence bases of A over K either contain the same finite number of elements or are infinite.*

DEFINITION 4.1.4. Let K be a difference (in particular, inversive difference) field with a basic set σ, L a σ-overfield of K and $A \subseteq L$. Then the *σ-transcendence degree* of A over K is the number of elements of any σ-transcendence basis of A over K, if this number is finite, or infinity in the contrary case.

The σ-transcendence degree of A over K is denoted by $\sigma\text{-}trdeg_K\, A$. In particular, if $A = L$, then $\sigma\text{-}trdeg_K\, L$ denotes the σ-transcendence degree of the σ- (or σ^*-) field extension L/K.

PROPOSITION 4.1.5. *Let K be a difference field with a basic set σ and L a σ-overfield of K.*

(i) *Any family of σ-generators of L over K contains a σ-transcendence basis of this difference field extension. If the σ-field K is inversive and L a σ^*-overfield of K, then any system of σ^*-generators of L over K contains a σ-transcendence basis of L over K.*

(ii) *Let $\eta_1, \dots, \eta_m \in L$. Then $\sigma\text{-}trdeg_K\, K\langle\eta_1, \dots, \eta_m\rangle \leqslant m$. If K is inversive and L a σ^*-overfield of K, then $\sigma\text{-}trdeg_K\, K\langle\eta_1, \dots, \eta_m\rangle^* \leqslant m$.*

(iii) *Let $\{\eta_1, \dots, \eta_m\}$ and $\{\zeta_1, \dots, \zeta_s\}$ be two finite subsets of L such that $K\langle\eta_1, \dots, \eta_m\rangle = K\langle\zeta_1, \dots, \zeta_s\rangle$ (or $K\langle\eta_1, \dots, \eta_m\rangle^* = K\langle\zeta_1, \dots, \zeta_s\rangle^*$ if K is inversive and L a σ^*-overfield of K). If the set $\{\zeta_1, \dots, \zeta_s\}$ is σ-algebraically independent over K, then $s \leqslant m$.*

(iv) *Let $S = K\{y_1, \dots, y_s\}$ be the ring of σ-polynomials in σ-indeterminates $y_1, \dots, y_s$ over K. If $k \neq s$, then S cannot be a ring of σ-polynomials in k σ-indeterminates over K. Similarly, if the difference field K is inversive, then a ring of σ^*-polynomials $K\{y_1, \dots, y_s\}^*$ cannot be a ring of σ^*-polynomials in k σ^*-indeterminates over K if $k \neq s$.*

PROPOSITION 4.1.6. *Let $H \subseteq K \subseteq L$ be difference (in particular, inversive difference) field extensions with the same basic set σ. Then $\sigma\text{-}trdeg_H\, L = \sigma\text{-}trdeg_K\, L + \sigma\text{-}trdeg_H\, K$.*

4.2. *Dimension polynomials of difference and inversive difference field extensions*

The results of this section first appeared in [99–101,103], and [3–5]. Most of the proofs can be also found in [88, Chapter 6].

Let us consider $\mathbf{N}^n$ as a partially ordered set relative to the *product order* $\leqslant_P$ such that $(a_1, \dots, a_n) \leqslant_P (b_1, \dots, b_n)$ if and only if $a_i \leqslant b_i$ for $i = 1, \dots, n$. For any $A \subseteq \mathbf{N}^n$ and $r \in \mathbf{N}$, let $A(r) = \{(e_1, \dots, e_n) \in A \mid \sum_{i=1}^{n} e_i \leqslant r\}$. Furthermore, let $V_A = \{v = (v_1, \dots, v_n) \in \mathbf{N}^n \mid \text{there is no } a \in A \text{ such that } a \leqslant_P v\}$.

The following result is due to E. Kolchin (see [86, Chapter 0, Lemma 16]).

LEMMA 4.2.1. *With the above notation, there exists a polynomial $\omega_A(t) \in \mathbf{Q}[t]$ with the following properties.*

(i) *$\omega_A(r) = \operatorname{Card} V_A(r)$ for all sufficiently large $r \in \mathbf{N}$.*

(ii) *$deg\, \omega_A \leqslant n$.*

(iii) $\deg \omega_A = n$ *if and only if* $A = \emptyset$. *In this case* $\omega_A(t) = \binom{t+n}{n}$.
(iv) $\omega_A = 0$ *if and only if* $(0, \dots, 0) \in A$.

The polynomial $\omega_A(t)$ is called the *Kolchin polynomial* of the set $A \subseteq \mathbf{N}^n$. Some methods and examples of computation of Kolchin polynomials can be found in [87] and [88, Chapter 2].

In [139] W. Sit proved that the set W of all Kolchin polynomials is well-ordered with respect to the order $\preceq$ on $\mathbf{Q}[t]$ such that $f(t) \preceq g(t)$ if and only if $f(r) \leqslant g(r)$ for all sufficiently large $r \in \mathbf{N}$. Furthermore, this set coincides with the set of all Hilbert polynomials of standard graded algebras over a field, as well as with the set of all differential dimension polynomials of finitely generated differential field extensions (see [108]).

In the rest of this section we assume that all fields have zero characteristic. The following theorem is a difference version of the Kolchin theorem on differential dimension polynomials [84].

THEOREM 4.2.2 [99]. *Let K be a difference field with a basic set $\sigma = \{\alpha_1, \dots, \alpha_n\}$, T the free commutative semigroup generated by σ, and for any $r \in \mathbf{N}$, $T(r) = \{\tau \in T \mid ord\,\tau \leqslant r\}$. Furthermore, let $L = K\langle\eta_1, \dots, \eta_s\rangle$ be a σ-overfield of K generated by a finite family $\eta = \{\eta_1, \dots, \eta_s\}$. Then there exists a polynomial $\phi_{\eta|K}(t) \in \mathbf{Q}[t]$ with the following properties.*

(i) $\phi_{\eta|K}(r) = trdeg_K\, K(\{\tau\eta_j \mid \tau \in T(r), 1 \leqslant j \leqslant s\})$ *for all sufficiently large* $r \in \mathbf{N}$.
(ii) $\deg \phi_{\eta|K}(t) \leqslant n$ *and the polynomial* $\phi_{\eta|K}(t)$ *can be written as* $\phi_{\eta|K}(t) = \sum_{i=0}^{n} a_i \binom{t+i}{i}$ *where* $a_0, \dots, a_n \in \mathbf{Z}$.
(iii) *The integers* a_n, $d = \deg \phi_{\eta|K}(t)$ *and* a_d *are invariants of* $\phi_{\eta|K}(t)$, *that is, they do not depend on the choice of a system of σ-generators η. Furthermore,* $a_n = \sigma\text{-}trdeg_K\, L$.
(iv) *Let P be the defining σ-ideal of $(\eta_1, \dots, \eta_s)$ in the ring of σ-polynomials $K\{y_1, \dots, y_s\}$ and let $\mathcal{A}$ be a characteristic set of P with respect to some orderly ranking of $\{y_1, \dots, y_s\}$. Furthermore, for every $j = 1, \dots, s$, let $E_j = \{(k_1, \dots, k_n) \in \mathbf{N}^n \mid \alpha_1^{k_1} \dots \alpha_n^{k_n} y_j$ is a leader of a σ-polynomial from $\mathcal{A}\}$. Then $\phi_{\eta|K}(t) = \sum_{i=1}^{s} \omega_{E_j}(t)$ where $\omega_{E_j}(t)$ is the Kolchin polynomial of the set E_j.*

The polynomial $\phi_{\eta|K}(t)$ whose existence is established by Theorem 4.2.2 is called the *difference* (or σ-) *dimension polynomial of the difference field extension L of K associated with the system of σ-generators η*. The integers $d = \deg \phi_{\eta|K}(t)$ and a_d are called, respectively, the *difference* (or σ-) *type* and *typical difference* (or σ-) *transcendence degree* of L over K. These invariants of $\phi_{\eta|K}(t)$ are denoted by $\sigma\text{-}type_K\, L$ and $\sigma\text{-}t.trdeg_K\, L$, respectively.

THEOREM 4.2.3. *Let K be a difference field with a basic set $\sigma = \{\alpha_1, \dots, \alpha_n\}$ and let L be a finitely generated σ-field extension of K with a set of σ-generators $\eta = \{\eta_1, \dots, \eta_s\}$ such that $\{\eta_1, \dots, \eta_d\}$ is a σ-transcendence basis of L over K $(1 \leqslant d \leqslant s)$. Then* $\phi_{(\eta_{d+1}, \dots, \eta_s)|K\langle\eta_1, \dots, \eta_d\rangle}(t) \preceq \phi_{\eta|K}(t) - d\binom{t+n}{n}$.

THEOREM 4.2.4. *Let K be a difference field with a basic set $\sigma = \{\alpha_1, \dots, \alpha_n\}$, $L = K\langle\eta_1, \dots, \eta_s\rangle$, and $\phi_{\eta|K}(t)$ the σ-dimension polynomial of the extension L/K associated with the family of σ-generators $\eta = \{\eta_1, \dots, \eta_s\}$. Then $\phi_{\eta|K}(t) = m\binom{t+n}{n}$ ($m \in \mathbf{N}$ if and only if σ-trdeg$_K$ $L = $ trdeg$_K$ $K(\eta_1, \dots, \eta_s) = m$.*

The following theorem gives versions of Theorems 4.2.2–4.2.4 for finitely generated inversive difference field extensions.

THEOREM 4.2.5 ([101], [88, Theorems 6.4.8, 6.4.16 and 6.4.17]). *Let K be an inversive difference field with basic set $\sigma = \{\alpha_1, \dots, \alpha_n\}$, Γ the free commutative group generated by σ and for any $r \in \mathbf{N}$, and $\Gamma(r) = \{\gamma = \alpha_1^{k_1} \dots \alpha_n^{k_n} \in \Gamma \mid \text{ord}\,\gamma = \sum_{i=1}^{n} |k_i| \leqslant r\}$. Furthermore, let $L = K\langle\eta_1, \dots, \eta_s\rangle^*$ be a σ^*-overfield of K generated by a finite family $\eta = \{\eta_1, \dots, \eta_s\}$. Then there exists a polynomial $\psi_{\eta|K}(t) \in \mathbf{Q}[t]$ with the following properties.*

(i) *$\psi_{\eta|K}(r) = $ trdeg$_K$ $K(\{\gamma\eta_j \mid \gamma \in \Gamma(r), 1 \leqslant j \leqslant s\})$ for all sufficiently large $r \in \mathbf{N}$.*

(ii) *deg $\psi_{\eta|K}(t) \leqslant n$ and the polynomial $\psi_{\eta|K}(t)$ can be written as $\psi_{\eta|K}(t) = \sum_{i=0}^{n} 2^i a_i \binom{t+i}{i}$ where $a_0, \dots, a_n \in \mathbf{Z}$.*

(iii) *The integers a_n, $d = \deg \phi_{\eta|K}(t)$ and a_d do not depend on the choice of a system of σ-generators η. Furthermore, $a_n = \sigma$-trdeg$_K$ L.*

(iv) *If $\{\eta_1, \dots, \eta_d\}$ is a σ-transcendence basis of L over K $(1 \leqslant d \leqslant s)$, then $\psi_{(\eta_{d+1},\dots,\eta_s)|K\langle\eta_1,\dots,\eta_d\rangle}(t) \preceq \psi_{\eta|K}(t) - d\sum_{i=0}^{n}(-1)^{n-i}2^i\binom{n}{i}\binom{t+i}{i}$.*

(v) *$\psi_{\eta|K}(t) = m\sum_{i=0}^{n}(-1)^{n-i}2^i\binom{n}{i}\binom{t+i}{i}$ for some $m \in \mathbf{N}$ if and only if*

$$\sigma\text{-}trdeg_K\, L = trdeg_K\, K(\eta_1, \dots, \eta_s) = m.$$

The polynomial $\psi_{\eta|K}(t)$ is called the *σ^*-dimension polynomial* of the σ^*-field extension L/K associated with the system of σ^*-generators η. The integers $d = \deg \psi_{\eta|K}(t)$ and a_d are called, respectively, the *σ^*-type* and *typical σ^*-transcendence degree* of L over K. These invariants of $\psi_{\eta|K}(t)$ are denoted by σ^*-type$_K$ L and σ^*-t.trdeg$_K$ L, respectively.

As in the last part of Theorem 4.2.2, the polynomial $\psi_{\eta|K}(t)$ can be expressed as a sum of certain analogs of Kolchin polynomials. Let $A \subseteq \mathbf{Z}^n$ $(n \geqslant 1)$ and let $\mathbf{Z}_1, \dots, \mathbf{Z}_{2^n}$ be the orthants of $\mathbf{Z}^n$ (introduced just before Definition 2.3.5). Let us consider the following partial order $\trianglelefteq$ on $\mathbf{Z}^n$: $(a_1, \dots, a_n) \trianglelefteq (b_1, \dots, b_n)$ if and only if $(a_1, \dots, a_n)$ and $(b_1, \dots, b_n)$ belong to the same orthant and $|a_i| \leqslant |b_i|$ for $i = 1, \dots, n$. As in the case of subsets of $\mathbf{N}^n$, one can show that if $W_A = \{w \in W \mid a \ntrianglelefteq w$ for any $a \in A\}$ and $W_A(r) = \{(w_1, \dots, w_n) \in W \mid \sum_{i=1}^{n} |w_i| \leqslant r\}$ $(r \in \mathbf{N})$, then there exists a polynomial $\psi_A(t) \in \mathbf{Q}[t]$ such that $\psi_A(r) = \text{Card}\, W_A(r)$ for all sufficiently large $r \in \mathbf{N}$. Furthermore, $\deg \psi_A \leqslant n$, and $\deg \psi_A = n$ if and only if $A = \emptyset$; in the last case $\psi_A(t) = \sum_{i=0}^{n}(-1)^{n-i}2^i\binom{n}{i}\binom{t+i}{i}$. Some properties and methods of computation of polynomials $\psi_A(t)$ $(A \subseteq \mathbf{Z}^n)$ can be found in [87] and [88, Chapter 2].

THEOREM 4.2.6. *With the notation of Theorem 4.2.5, let P be the defining σ^*-ideal of the s-tuple $(\eta_1, \dots, \eta_s)$ in the ring of σ^*-polynomials $K\{y_1, \dots, y_s\}^*$ and let $\mathcal{A}$ be a characteristic set of P with respect to some orderly ranking of $\{y_1, \dots, y_s\}$. Furthermore, for every*

$j = 1, \ldots, s$, *let* $A_j = \{(k_1, \ldots, k_n) \in \mathbf{Z}^n \mid \alpha_1^{k_1} \ldots \alpha_n^{k_n} y_j$ *is a leader of a* σ^**-polynomial from* $\mathcal{A}\}$. *Then* $\psi_{\eta|K}(t) = \sum_{i=1}^{s} \psi_{A_j}(t)$.

As it is shown in [88, Chapter 2] and [108], the set of all polynomials $\psi_A(t)$ $(A \subseteq \mathbf{Z}^n)$, the set of all difference dimension polynomials of finitely generated difference field extensions, the set of all σ^*-dimension polynomials of finitely generated inversive difference (σ^*-) field extensions, and the set W of all Kolchin polynomials coincide. Since the set W is well-ordered, the family of σ-dimension polynomials associated with various finite systems of σ-generators of a given finitely generated difference (σ-) field extension has a minimal element. It is called the *minimal σ-dimension polynomial* of the extension. Similarly, if L is a finitely generated σ^*-overfield of an inversive difference field K with a basic set σ, then the set of all σ^*-dimension polynomials associated with finite systems of σ^*-generators of L/K has a minimal element called the *minimal σ^*-dimension polynomial* of this σ^*-field extension.

Let K be an inversive difference field with a basic set σ and $L = K\langle\eta_1, \ldots, \eta_s\rangle^*$ a σ^*-overfield of K generated by a finite family $\eta = \{\eta_1, \ldots, \eta_s\}$. It is easy to check that the vector L-space $Der_K\, L$ of all K-linear derivations of the field L into itself becomes a σ^*-L-module if one defines the actions of elements of σ on $Der_K\, L$ as follows: $\alpha(D) = \alpha \circ D \circ \alpha^{-1}$ for any $\alpha \in \sigma$, $D \in Der_K\, L$. As in Section 3.2, one can consider a structure of a σ^*-L-module on the dual vector L-space $(Der_K\, L)^* = \mathrm{Hom}_L(Der_K\, L, L)$ and on the vector L-space of differentials $\Omega_K(L)$. (This vector space is generated over L by all elements $d\zeta \in (Der_K\, L)^*$ $(\zeta \in L)$ such that $d\zeta(D) = D(\zeta)$ for any $D \in Der_K\, L$; it is easy to check that $\alpha(d\zeta) = d\alpha(\zeta)$ for any $\zeta \in L$.)

The following theorem clarifies the connection between dimension polynomials of inversive difference field extensions and inversive difference vector spaces.

THEOREM 4.2.7. *With the above notation, let* $\Omega_K(L)_r$ $(r \in \mathbf{N})$ *denote the vector* L*-subspace of* $\Omega_K(L)$ *generated by the set* $\{d\gamma(\eta_i) \mid \gamma \in \Gamma(r), 1 \leqslant i \leqslant s\}$ *and let* $\Omega_K(L)_r = 0$ *for* $r < 0$. *Then*

(i) $((\Omega_K(L))_r)_{r\in\mathbf{Z}}$ *is an excellent filtration of the* σ^**-*L*-module* $\Omega_K(L)$.

(ii) $\dim_K \Omega_K(L)_r = \mathit{trdeg}_K\, K(\{\gamma\eta_j \mid \gamma \in \Gamma(r), 1 \leqslant j \leqslant s\})$ *for all* $r \in \mathbf{N}$.

(iii) *The* σ^**-dimension polynomial* $\phi_{\eta|K}(t)$ *is equal to the* σ^**-dimension polynomial of* $\Omega_K(L)$ *associated with the filtration* $((\Omega_K(L))_r)_{r\in\mathbf{Z}}$.

The last theorem allows one to reduce the computation of a σ^*-dimension polynomial of an inversive (σ^*-) difference field extension L/K to the computation of a σ^*-dimension polynomial of the vector σ^*-L-space $\Omega_K(L)$. The corresponding algorithms can be found in [88, Chapters 6 and 9], [89–91,119], and [127].

REMARK 4.2.8. By analogy with the theory of differential fields, one can expect that if L is a finitely generated σ^*-field extension of an inversive difference field K with a basic set σ and σ-$\mathit{trdeg}_K\, L = 0$, then there is a set $\sigma_1 = \{\beta_1, \ldots, \beta_n\}$ of automorphisms of L such that σ_1 is equivalent to σ and L is a finitely generated as a $\{\beta_1, \ldots, \beta_{n-1}\}^*$-overfield of K. This is not true (see [88, Example 6.4.18]), but we have a weaker version of such a statement proved in [88, Section 6.4].

THEOREM 4.2.9. *Let K be an inversive difference field with a basic set $\sigma = \{\alpha_1, \dots, \alpha_n\}$, L a finitely generated σ^*-field extension of K, and $d = \sigma\text{-}type_K\, L$. Then there exists a set $\sigma_1 = \{\beta_1, \dots, \beta_n\}$ of automorphisms of L and a finite family $\{\zeta_1, \dots, \zeta_q\} \in L$ such that*

(a) *σ_1 is equivalent to σ;*

(b) *if K is treated as an inversive difference field with the basic set $\sigma_2 = \{\beta_1, \dots, \beta_d\}$ and H is the σ_2^*-field extension of K generated by $\zeta_1, \dots, \zeta_q$, then L is an algebraic extension of H.*

We conclude this section with a result on multivariable dimension polynomials that generalizes Theorem 4.2.2.

Let K be a difference field whose basic set $\sigma = \{\alpha_1, \dots, \alpha_n\}$ is a union of p disjoint finite sets ($p \geqslant 1$): $\sigma = \sigma_1 \cup \dots \cup \sigma_p$, where $\sigma_1 = \{\alpha_1, \dots, \alpha_{n_1}\}$, $\sigma_2 = \{\alpha_{n_1+1}, \dots, \alpha_{n_1+n_2}\}, \dots,$ $\sigma_p = \{\alpha_{n_1+\dots+n_{p-1}+1}, \dots, \alpha_n\}$ ($n_1, \dots, n_p \in \mathbf{N}$). Let T be the free commutative semigroup generated by σ, and for any $(r_1, \dots, r_p) \in \mathbf{N}^p$, let $T(r_1, \dots, r_p) = \{\theta \in T \mid ord_i\, \theta \leqslant r_i$ for $i = 1, \dots, p\}$.

THEOREM 4.2.10. *With the above notation, let $L = K\langle \eta_1, \dots, \eta_s\rangle$ be a σ-field extension of K generated by a finite set $\eta = \{\eta_1, \dots, \eta_s\}$. Then there exists a polynomial $\Phi_\eta(t_1, \dots, t_p)$ in p variables $t_1, \dots, t_p$ with rational coefficients such that*

(i) *$\Phi_\eta(r_1, \dots, r_p) = trdeg_K\, K(\bigcup_{j=1}^{n} T(r_1, \dots, r_p)\eta_j)$ for all sufficiently large $(r_1, \dots, r_p) \in \mathbf{N}^p$;*

(ii) *$deg_{t_i}\, \Phi_\eta \leqslant n_i$ ($i = 1, \dots, p$) and the polynomial Φ_η can be written as $\Phi_\eta(t_1, \dots, t_p) = \sum_{i_1=0}^{n_1} \cdots \sum_{i_p=0}^{n_p} a_{i_1 \dots i_p} \binom{t_1+i_1}{i_1} \cdots \binom{t_p+i_p}{i_p}$ where $a_{i_1 \dots i_p} \in \mathbf{Z}$ for all $i_1, \dots, i_p$.*

(iii) *Let $E_\eta = \{(i_1, \dots, i_p) \in \mathbf{N}^p \mid 0 \leqslant i_k \leqslant n_k\ (k = 1, \dots, p)$ and $a_{i_1 \dots i_p} \neq 0\}$. Then the total degree d of the polynomial Φ, $a_{n_1 \dots n_p}$, p-tuples $(j_1, \dots, j_p) \in E'_\eta$ (we use the notation of Theorem 3.3.8), the corresponding coefficients $a_{j_1}, \dots, a_{j_p}$, and the coefficients of the terms of total degree d do not depend on the choice of the system of σ-generators η.*

Theorem 4.2.10, as well as its analog for finitely generated inversive difference field extensions, can be proven in the same way as the corresponding results on multivariable differential and difference-differential dimension polynomials obtained in [107] and [109]. These papers also show that multivariable dimension polynomials can carry essentially more invariants of a difference field extension then the dimension polynomials obtained in Theorems 4.2.2 and 4.2.5.

Dimension polynomials and the strength of a system of difference equations

Let K be an inversive difference field with a basic set $\sigma = \{\alpha_1, \dots, \alpha_n\}$, $\mathcal{E}$ the corresponding ring of σ^*-operators, $K\{y_1, \dots, y_s\}^*$ an algebra of σ^*-polynomials in σ^*-indeterminates $y_1, \dots, y_s$, and P a prime σ^*-ideal of $K\{y_1, \dots, y_s\}^*$. If $\eta = (\eta_1, \dots, \eta_s)$ is a generic zero of P, then the dimension polynomial $\psi_{\eta|K}(t)$ associated with the σ^*-field extension $K\langle \eta_1, \dots, \eta_s\rangle^*/K$ is called the *σ^*-dimension polynomial* of the ideal P; it is denoted by $\psi_P(t)$. (Clearly, if η and ζ are two generic zeros of P, then $\psi_{\eta|K}(t) = \psi_{\zeta|K}(t)$, so that the σ^**-dimension polynomial* of P is well-defined.) It can be shown (see [88, Proposition 6.2.4] that if P_1 and P_2 are prime σ^*-ideals of $K\{y_1, \dots, y_s\}^*$ such that $P_1 \subsetneqq P_2$, then $\psi_{P_2}(t) \prec \psi_{P_2}(t)$ (that is, $\psi_{P_2}(r) < \psi_{P_2}(r)$ for all sufficiently large $r \in \mathbf{N}$).

If $\Phi = \{A_\lambda \mid \lambda \in \Lambda\}$ is a family of σ^*-polynomials in $K\{y_1, \dots, y_s\}^*$, then an s-tuple η that annuls every A_λ is said to be a solution of the *system of algebraic difference* (or σ^*-) *equations* $A_\lambda(y_1, \dots, y_s) = 0\, (\lambda \in \Lambda)$. By Theorem 2.4.5, the last system is equivalent to some its finite subsystem, that is, there is a finite set $\Phi_0 = \{A_1, \dots, A_m\} \subseteq \Phi$ such that the set of solutions of the original system coincides with the set of solutions of the system

$$A_i(y_1, \dots, y_s) = 0 \quad (i = 1, \dots, m). \tag{4.2.1}$$

A system of algebraic σ^*-equations (4.2.1) is called *prime* if the perfect σ^*-ideal $\{A_1, \dots, A_m\}$ of the σ^*-ring $K\{y_1, \dots, y_s\}^*$ is prime. Note that any linear homogeneous system of difference equations (that is, a system of the form $\sum_{j=1}^{s} w_{ij} y_j = 0$ $(i = 1, \dots, m)$ where $w_{ij} \in \mathcal{E}$) is prime.

If (4.2.1) is a prime system of σ^*-equations, then the σ^*-dimension polynomial $\psi_P(t)$ of the prime σ^*-ideal $P = \{A_1, \dots, A_m\}$ is called the σ^**-dimension polynomial of the system.* This polynomial is an algebraic version of the concept of *strength* of a system of equations in finite differences defined as follows (by analogy with the similar notion for a system of differential equations introduced and studied by A. Einstein, [45]). Let us consider a system of equations in finite differences with respect to s unknown grid functions $f_1, \dots, f_s$ of n real variables with coefficients from a function field F. Suppose that the difference grid, whose nodes form the domain of the considered functions, has equal cells of dimension $h_1 \times \cdots \times h_n$ $(h_1, \dots, h_n \in \mathbf{R})$ and fills the whole space $\mathbf{R}^n$. Furthermore, let us fix some node X_0 and say that a node X of the grid is *a node of order* i (with respect to X_0) if the shortest route between X and X_0 passing along the edges of the grid consists of i steps $(i = 0, 1, \dots)$. (By a step we mean a path from a node to a neighboring node along the edge between them.)

Let us consider the values of the grid functions $f_1, \dots, f_s$ at the nodes whose order does not exceed i $(i \in \mathbf{N})$. If the functions f_j $(1 \leqslant j \leqslant s)$ do not satisfy any system of equations, their values at the nodes of any order may be chosen arbitrarily. Because of the system of equations in finite differences (and equations obtained from ones of the system by transformations of the form $f_j(x_1, \dots, x_n) \mapsto f_j(x_1 + r_1 h_1, \dots, x_n + r_n h_n)$ $(r_1, \dots, r_n \in \mathbf{Z})$, the number of independent values of the functions $f_1, \dots, f_s$ at the nodes of order less than or equal to i decreases. This number S_i is a function of i called the *strength* of the system of equations.

Suppose that the mappings $\alpha_j : f(x_1, \dots, x_n) \mapsto f(x_1, \dots, x_j + h_j, \dots, x_n)$ $(j = 1, \dots, n)$ are automorphisms of the field of coefficients F. Then F can be treated as an inversive difference field with a basic set $\sigma = \{\alpha_1, \dots, \alpha_n\}$. If we replace the unknown functions f_k by the σ^*-indeterminates y_k $(1 \leqslant k \leqslant s)$ from the algebra of σ^*-polynomials $F\{y_1, \dots, y_s\}^*$, then the given system of equations in finite differences generates a system of algebraic σ^*-equations of the form (4.2.1). Suppose that the last system is prime (e.g., linear), so that the σ^*-polynomials in its left-hand sides generate a prime σ^*-ideal P in $F\{y_1, \dots, y_s\}^*$. Then the σ^*-dimension polynomial $\psi_P(t)$ of this ideal is said to be the σ^*-dimension polynomial of the original system of equations in finite differences. It is easy to see that $\psi_P(i) = S_i$ for all $i \in \mathbf{N}$, so the strength of the system of equations in finite differences can be determined if one can find the polynomial $\psi_P(t)$ (that is, the dimension polynomial of the σ^*-field extension $F\langle \eta_1, \ldots, \eta_s \rangle^*/F$ where $(\eta_1, \ldots, \eta_s)$ is a generic zero of the ideal P). A number of examples of computation of the strength of systems of difference equations can be found in [88, Chapters 6 and 9], [90,91,119], and [127].

4.3. *Finitely generated difference and inversive difference field extensions. Limit degree. Finitely generated difference algebras*

Let K be an ordinary difference ring with a basic set $\sigma = \{\alpha\}$ and $L = K\langle \eta_1, \ldots, \eta_s \rangle$ a σ-overfield of K generated by a finite set $\eta = \{\eta_1, \ldots, \eta_s\}$. Furthermore, for any $r = 1, 2, \ldots$, let $L_r = K(\{\alpha^i(\eta_j) \mid 1 \leqslant j \leqslant s, 0 \leqslant i \leqslant r\})$ and $d_r = L_r : L_{r-1}$, the dimension of L_r as a vector L_{r-1}-space. It is easy to see that $d_1 \geqslant d_2 \geqslant \cdots$. Moreover, as it is shown in [30], $\min\{d_r \mid r = 1, 2, \ldots\}$ does not depend on the choice of the system of σ-generators η. This minimum value is called the *limit degree* of the σ-field extension L/K, it is denoted by $ld\, L/K$.

If a σ-field extension L of a difference (σ-) field K is not finitely generated, we define its limit degree $ld\, L/K$ as the maximum of limit degrees of all finitely generated σ-field subextensions of L/K, if this maximum exists or ∞ if it does not. Clearly, if $\sigma\text{-}trdeg_K\, L > 0$, then $ld\, L/K = \infty$. If $\sigma\text{-}trdeg_K\, L = 0$ and L is a finitely generated σ-field extension of K, then $ld\, L/K < \infty$.

The concept of limit degree was introduced by R.M. Cohn, [30]. The proofs of the following results on limit degrees of ordinary difference field extensions can be found in [32, Chapter 5].

THEOREM 4.3.1. *Let K be an ordinary difference field with a basic set σ, M a σ-overfield of K and L/K a σ-field subextension of M/K. Then $ld\, M/K = (ld\, M/L)(ld\, L/K)$.*

THEOREM 4.3.2. *Let K be an ordinary difference field and L a difference overfield of K. Furthermore, let K^* and L^* denote the inversive closures of K and L, respectively. Then $ld\, L^*/K = ld\, L^*/K^* = ld\, L/K$.*

REMARK 4.3.3. In the case of ordinary difference fields, there are several concepts related to the notion of limit degree. In the case of fields of characteristic $p > 0$, one can define an analog of limit degree using separable factor of degree in place of degree of L_r over L_{r-1} (we refer to the notation from the beginning of this section). The corresponding invariant of the difference field extension is called the *reduced limit degree* of L/K and denoted by $rld\, L/K$; it has the same properties that are established for $ld\, L/K$ in Theorems 4.3.1 and 4.3.2.

If K is an inversive difference field with a basic set $\sigma = \{\alpha\}$, then K can be also treated as a difference field with the basic set $\sigma' = \{\alpha^{-1}\}$ called the *inverse difference field* of K. It is denoted by K'. Let L be a σ-field extension of K and L' the inverse difference field of L (so that L' is a σ'-field extension of K'). The *inverse limit degree* of L over K is defined to be $ld\, L'/K'$ (it is denoted by $ild\, L/K$), and the *inverse reduced limit degree* of L over K is defined to be $rld\, L'/K'$ (it is denoted by $irld\, L/K$). The analogs of Theorems 4.3.1 and 4.3.2 are valid for these concepts, as well.

THEOREM 4.3.4. *Let K be an ordinary difference field with a basic set σ.*

(i) *If L is a finitely generated σ-field extension of K, then $ld\, L/K = 1$ if and only if $L = K(S)$ for some finite set $S \subseteq L$.*

(ii) *The following two statements are equivalent:*

  (a) *L/K is a finitely generated σ-field extension, L is algebraic over K, and $ld\, L/K = 1$.*

  (b) *$L : K$ is finite.*

THEOREM 4.3.5. *Let K be an ordinary difference field and L a difference field extension of K which is algebraic over K. Then*

(i) *$ld\, L/K = ild\, L/K$ and $rld\, L/K = irld\, L/K$.*

(ii) *If $ld\, L/K = 1$ and K is inversive, then L is inversive.*

The natural generalizations of the concept of limit degree to the case of partial difference fields were obtained in [50].

Let K be a difference field with a basic set $\sigma = \{\alpha_1, \ldots, \alpha_n\}$ and L a finitely generated σ-field extension of K: $L = K\langle S\rangle$ where S is a finite subset of L. In what follows we adopt the following notation. If $\sigma_1 = \{\alpha_{i_1}, \ldots, \alpha_{i_p}\} \subseteq \sigma$, then the set $\{\alpha_{i_1}^{j_1} \ldots \alpha_{i_p}^{j_p}(s) \mid j_1, \ldots, j_p \in \mathbf{N}, s \in S\}$ is denoted by $S^{(\alpha_{i_1}, \ldots, \alpha_{i_p})}$. Furthermore, for any $k = 0, 1, \ldots,$ we set $S_k = \bigcup_{i=0}^{k} \alpha_n^i(S^{(\alpha_{i_1}, \ldots, \alpha_{i_{n-1}})})$ (if $n = 1$, $S^{(\alpha_{i_1}, \ldots, \alpha_{i_{n-1}})} = S$), and for any positive integers $i_1, \ldots, i_t$ $(1 \leqslant t \leqslant n)$, we set

$$S^*(i_n, i_{n-1} \ldots i_t) = \bigcup_{i=0}^{i_n - 1} \alpha_n^i \big(S^{(\alpha_1, \ldots, \alpha_{n-1})}\big) \cup \bigcup_{i=0}^{i_{n-1}-1} \alpha_n^{i_n} \alpha_{n-1}^i \big(S^{(\alpha=_1, \ldots, \alpha_{n-2})}\big) \cup \cdots$$
$$\cup \bigcup_{i=0}^{i_t - 1} \alpha_n^{i_n} \ldots \alpha_{t-1}^{i_{t-1}} \alpha_t^i \big(S^{(\alpha_1, \ldots, \alpha_{t-1})}\big)$$

and $S(i_n, i_{n-1} \ldots i_t) = S^*(i_n, i_{n-1} \ldots i_{t+1}, i_t + 1)$. Finally, if K is treated as a difference field with a basic set σ_1, it is denoted by $(K; \sigma_1)$ or $(K; \alpha_{i_1}, \ldots, \alpha_{i_p})$; the difference transcendence degree of $(K(S_k); \alpha_1, \ldots, \alpha_{n-1})$ over $(K(S_{k-1}); \alpha_1, \ldots, \alpha_{n-1})$ is denoted by δ_k. (If $n = 1$, then $\delta_k = trdeg_{K(S_{k-1})} K(S_k)$.)

With the above notation, the *limit transformal transcendence degree σ_l-$trdeg_K\, L$* of L over K is defined by σ_l-$trdeg_K\, L = \min\{\delta_k \mid k = 1, 2, \ldots\}$.

PROPOSITION 4.3.6 [50]. *Let K be a difference field with a basic set $\sigma = \{\alpha_1, \ldots, \alpha_n\}$ and L a finitely generated σ-field extension of K. Then*

(i) *σ_l-$trdeg_K\, L = \sigma$-$trdeg_K\, L$. Thus, σ_l-$trdeg_K\, L$ is independent of the finite set of σ-generators and the translation from σ chosen as α_n to define σ_l-$trdeg_K\, L$.*

(ii) *If S is a finite set of σ-generators of L over K, then there exists a finite subset $Z \subseteq S$ and positive integers $k_1, \dots, k_n$ with the following properties:*
 (a) *Z is a σ-transcendence basis of L over K.*
 (b) *If $t \in \{1, \dots, n\}$, $i_j \geqslant k_j$ for $j = t, \dots, n$, and $\sigma' = \{\alpha_1, \dots, \alpha_{t-1}\}$, then $\alpha_n^{i_n} \dots \alpha_t^{i_t}(Z)$ is a σ'-transcendence basis of $(K(S(i_n, \dots, i_t); \alpha_1, \dots, \alpha_{t-1})$ over $(K(S^*(i_n, \dots, i_t); \alpha_1, \dots, \alpha_{t-1})$ (and σ-trdeg$_K L$ is the σ'-transcendence degree of this extension).*

With the notation of the last proposition, a set Z that satisfies conditions (a) and (b) for some positive integers $k_1, \dots, k_n$ is called a *limit basis of transformal transcendence* of L over K.

In [50] P. Evanovich introduced an invariant $ld_n(M/K)$ of a partial difference field extension M/K that can be viewed as a generalization of the concept of limit degree. $ld_n(M/K)$ is inductively defined as an element of the set $\mathbf{N} \cup \{\infty\}$ that satisfies the following conditions (ld1)–(ld5). Let K be a difference field with a basic set $\sigma = \{\alpha_1, \dots, \alpha_n\}$, M a σ-overfield of K and L/K is a σ-field subextension of M/K. Then

(ld1) *If there exists a finite set $S \subseteq M$ such that $M = L\langle S\rangle$, then there exists a finitely generated σ-overfield K' of K contained in L such that $ld_n(M/L) = ld_n(K'\langle S\rangle/K')$.*

(ld2) *If $S \subseteq M$, then $ld_n(L\langle S\rangle/K\langle S\rangle) \leqslant ld_n(L/K)$ and $ld_n(L\langle S\rangle/L) \leqslant ld_n(K\langle S\rangle/K)$. Equality will hold in both if S is σ-algebraically independent over L.*

(ld3) *If there is a σ-isomorphism ϕ of L onto a σ-field L' and K' is a σ-subfield of L' such that $\phi(K) = K'$, then $ld_n(L/K) = ld_n(L'/K')$.*

(ld4) *If the σ-field extension L/K is finitely generated, then σ-trdeg$_k L = 0$ if and only if $ld_n(L/K) < \infty$.*

(ld5) *$ld_n(M/K) = ld_n(M/L) \cdot ld_n(L/K)$.*

If $n = 1$, ld_1 is defined to be the limit degree ld for ordinary difference fields. Suppose that ld_{n-1} is defined for difference field extensions whose basic sets consist of $n-1$ translations. Let K be a difference field with a basic set $\sigma = \{\alpha_1, \dots, \alpha_n\}$ and L a σ-field extension of K. Assume first that L/K is finitely generated, say $L = K\langle S\rangle$ for some finite set $S \subseteq L$. For any $m \in \mathbf{N}$, let $L_m = (K(S_m); \alpha_1, \dots, \alpha_{n-1})$. Then L_m is finitely generated extension of L_{m-1} (we set $K(S_{-1}) = K$). Applying (ld3) and (ld2) we obtain that $ld_{n-1}(L_m/L_{m-1}) \geqslant ld_{n-1}(L_{m+1}/L_m)$, whence there exists the limit $\lim_{m\to\infty} ld_{n-1}(L_m/L_{m-1}) = a$ where $a \in \mathbf{N}$ or $a = \infty$. As in the case of limit degree of ordinary difference fields, [30], one can show that a is independent on the choice of σ-generators of L/K. Now we define $ld_n(L/K) = a$.

If L/K is not finitely generated, $ld_n(L/K)$ is defined to be the maximum of $ld_n(K'/K)$ where K'/K is a finitely generated σ-field subextension of L/K, if the maximum exists and ∞ if it does not. As in the case of limit degrees of ordinary difference fields, if L/K itself is finitely generated, then $ld_n(L/K)$ is the maximum of $ld_n(K'/K)$ where K'/K is a finitely generated σ-field subextension of L/K. The proof of the fact that the so defined ld_n satisfies (ld1)–(ld5) can be found in [50, section 3]. In the same paper P. Evanovich used the properties of ld_n to prove the following fundamental result that was first established by R.M. Cohn, [32, Chapter 5], for ordinary difference fields.

THEOREM 4.3.7. *Let K be a difference field with a basic set $\sigma = \{\alpha_1, \dots, \alpha_n\}$, M a finitely generated σ-overfield of K and L/K a σ-field subextension of M/K. Then the σ-field extension L/K is finitely generated.*

We conclude this section with some results on finitely generated difference algebras.

Let K be a difference (inversive difference) ring with a basic set σ. A K-algebra R is said to be a *difference algebra* over K or a σ-K-algebra (respectively, an *inversive difference algebra* over K or a σ^*-K-algebra) if elements of σ (respectively, σ^*) act on R in such a way that R is a σ- (respectively, σ^*-) ring and $\alpha(au) = \alpha(a)\alpha(u)$ for any $a \in K, u \in R$, $\alpha \in \sigma$ (for any $\alpha \in \sigma^*$ if R is a σ^*-K-algebra). A σ-K- (respectively, σ^*-K-) algebra R is said to be finitely generated if there exists a finite family $\{\eta_1, \dots, \eta_s\}$ of elements of R such that $R = K\{\eta_1, \dots, \eta_s\}$ (respectively, $R = K\{\eta_1, \dots, \eta_s\}^*$). If a σ-K- (or σ^*-K-) algebra R is an integral domain, then the σ-transcendence degree of R over K is defined as the σ-transcendence degree of the corresponding σ- (respectively, σ^*-) field of quotients of R over K.

In what follows we consider inversive difference algebras over inversive fields. All results formulated below for such algebras remain valid for difference algebras over difference fields (with replacement of the prefix σ^*- by σ-).

Let R be an inversive difference algebra over an inversive difference field K with a basic set $\sigma = \{\alpha_1, \dots, \alpha_n\}$ and let $\mathcal{U}$ denote the set of all prime σ^*-ideals of R. As at the end of section 3.1, one can consider the set $\mathcal{B}_{\mathcal{U}} = \{(P, Q) \in \mathcal{U} \times \mathcal{U} \mid P \supseteq Q\}$ and the uniquely defined mapping $\mu_{\mathcal{U}} : \mathcal{B}_{\mathcal{U}} \to \overline{\mathbf{Z}}$ such that

(i) $\mu_{\mathcal{U}}(P, Q) \geqslant -1$ for every pair $(P, Q) \in \mathcal{B}_{\mathcal{U}}$;

(ii) for any $d \in \mathbf{N}$, the inequality $\mu_{\mathcal{U}}(P, Q) \geqslant d$ holds if and only if $P \neq Q$ and there exists an infinite chain $P = P_0 \supseteq P_1 \supseteq \cdots \supseteq Q$ such that $P_i \in \mathcal{U}$ and $\mu_{\mathcal{U}}(P_{i-1}, P_i) \geqslant d - 1$ for $i = 1, 2, \dots$.

The *type* and *dimension* of the σ^*-K-algebra R over $\mathcal{U}$ are defined, respectively, as $\sup\{\mu_{\mathcal{U}}(P, Q) \mid (P, Q) \in \mathcal{B}_{\mathcal{U}}\}$ and the least upper bound of the lengths k of chains $P_0 \supseteq P_1 \supseteq \cdots \supseteq P_k$ such that $P_0, \dots, P_k \in \mathcal{U}$ and $\mu_{\mathcal{U}}(P_{i-1}, P_i) = \mathit{type}_{\mathcal{U}}\, R$ for $i = 1, \dots, k$. These characteristics are denoted by $\mathit{type}_{\mathcal{U}}\, R$ and $\dim_{\mathcal{U}} R$, respectively.

THEOREM 4.3.8 [102]. *Let K be an inversive difference field with a basic set $\sigma = \{\alpha_1, \dots, \alpha_n\}$, $R = K\{\eta_1, \dots, \eta_s\}^*$ a σ^*-K-algebra without zero divisors generated by a finite set $\eta = \{\eta_1, \dots, \eta_s\}$, and $\mathcal{U}$ the family of all prime σ^*-ideals of R. Then:*

(i) $\mathit{type}_{\mathcal{U}}\, R \leqslant n$.

(ii) *If σ-$\mathit{trdeg}_K\, R = 0$, then $\mathit{type}_{\mathcal{U}}\, R < n$.*

(iii) *If $\mathit{type}_{\mathcal{U}}\, R = n$, then $\dim_{\mathcal{U}} R \leqslant \sigma$-$\mathit{trdeg}_K\, R$.*

(iv) *If $\eta_1, \dots, \eta_s$ are σ-algebraically independent over K, then $\mathit{type}_{\mathcal{U}}\, R = n$ and $\dim_{\mathcal{U}} R = s$.*

Let K be an inversive difference field with a basic set σ. A σ^*-K-algebra R is said to be a *local σ^*-K-algebra* if R is a local ring (in this case its maximal ideal is a σ^*-ideal). We say that R is a local σ^*-K-algebra *of finitely generated type* if R is a local σ^*-K-algebra and there exist a finite set $\{\eta_1, \dots, \eta_s\} \subseteq R$ such that $R = K\{\eta_1, \dots, \eta_s\}_m$ where m is the maximal ideal of R.

THEOREM 4.3.9 [102]. *Let K be an inversive difference field of zero characteristic with a basic set σ. Let an integral domain R be a local σ^*-K-algebra of finitely generated type, m the maximal ideal of R, and $k = R/m$ the corresponding σ^*-field of residue classes. Then*:

(i) *m/m^2 is a finitely generated vector σ^*-k-space.*

(ii) $\dim_k m/m^2 \geqslant \sigma^*$-$trdeg_K\, R - \sigma^*$-$trdeg_K\, k$.

4.4. *Difference kernels. Realizations*

Let F be an ordinary inversive difference field with a basic set $\sigma = \{\alpha\}$. A *difference kernel of length r* over F is an ordered pair $\mathcal{R} = (F(a_0, \dots, a_r), \tau)$ where each a_i is itself an s-tuple $(a_i^{(1)}, \dots, a_i^{(s)})$ over F (a positive integer s is fixed) and τ is an extension of α to an isomorphism of $F(a_0, \dots, a_{r-1})$ onto $F(a_1, \dots, a_r)$ such that $\tau a_i = a_{i+1}$ for $i = 0, \dots, r-1$. (In other words, $\tau(a_i^{(j)}) = a_{i+1}^{(j)}$ for $0 \leqslant i \leqslant r-1$, $1 \leqslant j \leqslant s$.) If $r = 0$, then $\tau = \alpha$. The *degree of transcendence* of the difference kernel $\mathcal{R}$ is defined to be $trdeg_{F(a_0,\dots,a_{r-1})} F(a_0, \dots, a_r)$; it is denoted by $\delta\mathcal{R}$. Below, while considering difference kernels over a difference field F, we always assume that F is inversive.

A *prolongation* $\mathcal{R}'$ of a difference kernel $\mathcal{R} = (F(a_0, \dots, a_r), \tau)$ is a difference kernel of length $r+1$ consisting of an overfield $F(a_0, \dots, a_r, a_{r+1})$ of $F(a_0, \dots, a_r)$ and an extension τ' of τ to an isomorphism of $F(a_0, \dots, a_r)$ onto $F(a_1, \dots, a_{r+1})$.

In what follows, a set of the form $a = \{a^{(i)} \mid i \in I\}$ will be referred to as an *indexing* of a (with the index set I). If $J \subseteq I$, the set $\{a^{(i)} \mid i \in J\}$ will be called a *subindexing* of a. If $\mathcal{R} = (F(a_0, \dots, a_r), \tau)$ is a difference kernel and $\tilde{a}_0$ is a subindexing of a_0 (that is, $\tilde{a}_0 = (a_0^{(i_1)}, \dots, a_0^{(i_q)})$, $1 \leqslant i_1 < \dots < i_q \leqslant s$), then $\tilde{a}_k$ $(k = 1, \dots)$ will denote the corresponding subindexing of a_k. The proofs of the following results on prolongations of difference kernels over ordinary difference fields can be found in [32, Chapter 6].

THEOREM 4.4.1. *Every difference kernel $\mathcal{R} = (F(a_0, \dots, a_r), \tau)$ over an ordinary difference field F has a prolongation $\mathcal{R}' = (F(a_0, \dots, a_r, a_{r+1}), \tau')$. Moreover, one can chose a prolongation $\mathcal{R}'$ with the following properties.*

(i) *If $\tilde{a}_0$ is a subindexing of a_0 such that $\tilde{a}_r$ is algebraically independent over $F(a_0, \dots, a_{r-1})$, then $\tilde{a}_{r+1}$ is algebraically independent over $F(a_0, \dots, a_r)$ (so that $\delta\mathcal{R} = \delta\mathcal{R}'$).*

(ii) *If $\tilde{a}_0$ is as in* (i), *then $\bigcup_{i=0}^{r+1} \tilde{a}_i$ is algebraically independent over F.*

With the above notation, a prolongation $\mathcal{R}'$ of a difference kernel $\mathcal{R}$ is called *generic* if $\delta\mathcal{R} = \delta\mathcal{R}'$.

A generic prolongation of a difference kernel $\mathcal{R} = (F(a_0, \dots, a_r), \tau)$ over an ordinary difference field can be constructed as follows. Let P be the prime ideal with generic zero a_r of a polynomial ring $F(a_0, \dots, a_{r-1})[X_1, \dots, X_s]$ in s indeterminates $X_1, \dots, X_s$. Let P' be obtained from P by replacing the coefficients of the polynomials of P by their images under τ. Then P' is a prime ideal of $F(a_1, \dots, a_r)[X_1, \dots, X_s]$ and generates an ideal $\widehat{P}$ in $F(a_0, \dots, a_r)[X_1, \dots, X_s]$. Let Q be an essential prime divisor of $\widehat{P}$ in the last ring and let a_{r+1} be a generic zero of Q. Then a_{r+1} is also a generic zero of P' and there is an isomorphism $\tau' : F(a_0, \dots, a_r) \to F(a_1, \dots, a_{r+1})$ that extends τ. We obtain the desired generic prolongation. Conversely, if $\mathcal{R}' = (F(a_0, \dots, a_{r+1}), \tau')$ is any generic prolongation of $\mathcal{R}$, then a_{r+1} is a solution of the ideal $\widehat{P}$ and hence of one of its essential prime divisors Q. It follows that $\dim Q = \dim P = trdeg_{F(a_0,\dots,a_{r-1})} F(a_0, \dots, a_r) = trdeg_{F(a_0,\dots,a_r)} F(a_0, \dots, a_{r+1})$, so that a_{r+1} is a generic zero of Q.

EXAMPLE 4.4.2 [32, Chapter 6, Section 2]. Let F be an ordinary difference field with a basic set $\sigma = \{\alpha\}$, $F\{y\}$ the ring of σ-polynomials in one σ-indeterminate y over F, and A an algebraically irreducible σ-polynomial from $F\{y\}$ (that is, A is irreducible as a polynomial in $y, \alpha(y), \alpha^2(y), \dots$). Assuming that A contains y and $\alpha^m y$, $m > 0$, is the highest transform of y in A, we shall use prolongations of difference kernels to construct a solution of A. First, let us consider an m-tuple $a = (a^{(1)}, \dots, a^{(m)})$ whose coordinates constitute an algebraically independent set over F. Now we define an m-tuple a_1 as follows: we set $a_1^{(i)} = a^{(i+1)}$ for $i = 1, \dots, m-1$, replace y_{i-1} by $a^{(i)}$ in A $(1 \leqslant i \leqslant m)$, find a solution of the resulting polynomial in one unknown y_m and take it as $a^{(m)}$. Since A involves y, $a^{(1)}$ will be algebraically dependent on $a^{(2)}, \dots, a^{(m)}, a_1^{(m)}$ over F. Therefore, $a_1^{(1)}, \dots, a_1^{(m)}$ are algebraically independent over F whence there is a difference kernel $\mathcal{R}_1$ defined over F by the extension of the translation α to an isomorphism $\tau_0 : F(a) \to F(a_1)$. By successive applications of Theorem 4.4.1 we find a sequence $a_0 = a, a_1, \dots$ such that a kernel $\mathcal{R}_{k+1}$ is defined by an isomorphism $\tau_k : F(a_0, \dots, a_k) \to F(a_1, \dots, a_{k+1})$ $(k = 0, 1, \dots)$ and $\mathcal{R}_{k+1}$ is a prolongation of $\mathcal{R}_k$. Then $F(a_0, a_1, \dots)$ becomes a difference overfield of F where the extension of α (denoted by the same letter) is defined by $\alpha(b) = \tau_k(b)$ whenever $b \in F(a_0, \dots, a_k)$. It is clear that this field coincides with $F\langle a^{(1)} \rangle$ and the element $a^{(1)}$ is a solution of A.

PROPOSITION 4.4.3. *Let $\mathcal{R} = (F(a_0, \dots, a_r), \tau)$ be a difference kernel.*

(i) *There are only finitely many distinct (that is, pairwise non-isomorphic) generic prolongations of $\mathcal{R}$.*

(ii) *Let $\mathcal{R}'$ be a generic prolongation of a $\mathcal{R}$ and let $\tilde{a}_0$ be a subindexing of a_0 which is algebraically independent over $F(a_1, \dots, a_r)$. Then $\tilde{a}_0$ is algebraically independent over $F(a_1, \dots, a_{r+1})$.*

Let $\mathcal{R} = (F(a_0, \dots, a_r), \tau)$ be a difference kernel and let $\tilde{a}_0$ be a subindexing of a_0. If $\tilde{a}_r$ is a transcendence basis of a_r over $F(a_0, \dots, a_{r-1})$, then $\tilde{a}$ is called a *special set*. Clearly, such a set consists of $\delta\mathcal{R}$ elements and it is also a special set for any generic prolongation $\mathcal{R}'$ of $\mathcal{R}$. If b denotes a subindexing b_0 of a_0 such that b contains a special set and $trdeg_{F(b,\dots,b_{r-1})} F(b, \dots, b_r) = \delta\mathcal{R}$, then the *order* of $\mathcal{R}$ with respect to b is defined as $ord_b \mathcal{R} = trdeg_{F(b,\dots,b_r)} F(a_0, \dots, a_{r-1})$. (In this case, b is said to be a subindexing of a_0 for which $ord_b \mathcal{R}$ is defined.) Furthermore, if b itself is a special set, we define the *degree* $d_b \mathcal{R}$ and *reduced degree* $rd_b \mathcal{R}$ of $\mathcal{R}$ with respect to b to be $F(a_0, \dots, a_r) : F(a_0, \dots, a_{r-1}; b_r)$ and $[F(a_0, \dots, a_r) : F(a_0, \dots, a_{r-1}; b_r)]_s$, respectively. (As usual, if L is a field extension of a field K, $[L : K]_s$ denotes the separable factor of the degree of L over K.)

PROPOSITION 4.4.4. *With the above notation, let b be a subindexing of a_0 for which $\mathrm{ord}_b\,\mathcal{R}$ is defined.*

(i) *If $\mathcal{R}'$ is a generic prolongation of $\mathcal{R}$, then $\mathrm{ord}_b\,\mathcal{R}'$ is defined and $\mathrm{ord}_b\,\mathcal{R} = \mathrm{ord}_b\,\mathcal{R}'$.*

(ii) *Suppose that b itself is a special set. Then $d_b\,\mathcal{R}$ and $rd_b\,\mathcal{R}$ are finite. Furthermore, if $\mathcal{R}'_1, \dots, \mathcal{R}'_h$ are all distinct (pairwise non-isomorphic) prolongations of $\mathcal{R}$, then $\sum_{i=1}^{h} rd_b\,\mathcal{R}'_i = rd_b\,\mathcal{R}$ and $\sum_{i=1}^{h} d_b\,\mathcal{R}'_i \leqslant d_b\,\mathcal{R}$. In the case of characteristic 0 the last inequality becomes an equality.*

Let F be a difference field with a basic set σ and let $a = \{a^{(i)} \mid i \in I\}$ be an indexing of elements in a σ-overfield of F. A *specialization* of a over F is a σ-homomorphism ϕ of $F\{a\}$ into a σ-overfield of F that leaves F fixed. The image $\phi a = \{\phi a^{(i)} \mid i \in I\}$ is also called a specialization of a over F. A specialization ϕ is called *generic* if it is a σ-homomorphism. Otherwise it is called *proper*.

Let F be an ordinary difference field with a basic set $\sigma = \{\alpha\}$ and $G = F\langle\eta_1, \dots, \eta_s\rangle$ a σ-overfield of F generated by an s-tuple $\eta = (\eta_1, \dots, \eta_s)$. Then the contraction of α to an isomorphism $\tau_r : F(\eta, \alpha\eta, \dots, \alpha^{r-1}\eta) \to F(\alpha\eta, \dots, \alpha^r\eta)$ $(r = 1, 2, \dots)$ defines a difference kernel $\mathcal{R} = (F(a_0, \dots, a_r), \tau_r)$ of length r over F. Conversely, let $\mathcal{R} = (F(a_0, \dots, a_r), \tau)$ be a difference kernel with $a_0 = (a_0^{(1)}, \dots, a_0^{(s)})$. An s-tuple $\eta = (\eta_1, \dots, \eta_s)$ with coordinates from a σ-overfield of F is called a *realization* of $\mathcal{R}$ if $\eta, \alpha\eta, \dots, \alpha^r\eta$ is a specialization of $a_0, \dots, a_r$ over F. If this specialization is generic, the realization is called *regular*. If there exists a sequence $\mathcal{R}^{(0)} = \mathcal{R}, \mathcal{R}^{(1)}, \mathcal{R}^{(2)}, \dots$ of kernels, each a generic prolongation of the preceding, such that η is a regular realization of each $\mathcal{R}^{(i)}$, then η is called a *principal realization* of $\mathcal{R}$.

The proofs of the following two statements can be found in [32, Chapter 6, Section 6].

PROPOSITION 4.4.5. *Let $\mathcal{R} = (F(a_0, \dots, a_r), \tau)$ be a difference kernel over an ordinary difference field F.*

(i) *There exists a principal realization of the kernel $\mathcal{R}$. If η is such a realization, then $\sigma\text{-}\mathrm{trdeg}_F\, F\langle\eta\rangle = \delta\mathcal{R}$.*

(ii) *Let b be a subindexing of a such that $\mathrm{ord}_b\,\mathcal{R}$ is defined and let ζ is the corresponding subindexing of a principal realization η of $\mathcal{R}$. Then $\mathrm{trdeg}_{F\langle\zeta\rangle}\, F\langle\eta\rangle = \mathrm{ord}_b\,\mathcal{R}$. Furthermore, if b is a special set, then ζ is a σ-transcendence basis of $F\langle\eta\rangle$ over F.*

(iii) *The number of distinct principal realizations of the kernel $\mathcal{R}$ is finite. Let us denote them by ${}^{(1)}\eta, \dots, {}^{(h)}\eta$. If b is a special set and ${}^{(i)}\zeta$ is the corresponding subset of the components of ${}^{(i)}\eta$ $(1 \leqslant i \leqslant h)$, then $\sum_{i=1}^{h} rld\, F\langle{}^{(i)}\eta\rangle / F\langle{}^{(i)}\zeta\rangle = rd_b\,\mathcal{R}$ and $\sum_{i=1}^{h} ld\, F\langle{}^{(i)}\eta\rangle / F\langle{}^{(i)}\zeta\rangle \leqslant d_b\,\mathcal{R}$.*

(iv) *If η is a regular realization of the kernel $\mathcal{R}$, but not a principal realization, then $\sigma\text{-}\mathrm{trdeg}_F\, F\langle\eta\rangle < \delta\mathcal{R}$.*

(v) *A realization of the kernel $\mathcal{R}$ which specializes over F to a principal realization is a principal realization, and the specialization is generic.*

PROPOSITION 4.4.6. *Let $\eta = (\eta_1, \dots, \eta_s)$ be an s-tuple over an ordinary difference field F. Then η is the unique principal realization of a kernel $\mathcal{R}$ over F such that every realization of $\mathcal{R}$ is a specialization of η over F.*

Note that every kernel $\mathcal{R} = (F(a_0, \ldots, a_r), \tau)$ over an ordinary difference field F with a basic set $\sigma = \{\alpha\}$ is equivalent to a kernel of length 0 or 1 in the sense that its realizations generate precisely the same extensions. Indeed, if $r > 1$, we can define sr-tuples b_0 and b_1 with components of the s-tuples $a_0, \ldots, a_{r-1}$ and s-tuples $a_1, \ldots, a_r$, respectively. Then the difference kernel $\mathcal{R}^* = (F(b_0, b_1), \tau)$ of length 1 is equivalent to $\mathcal{R}$.

Many concepts and results of this section formulated for the ordinary case can be extended to partial difference fields. In what follows we review such generalizations most of which can be found in [7].

Let F be an inversive difference field with a basic set $\sigma = \{\alpha_1, \ldots, \alpha_n\}$, $\sigma_q = \sigma \setminus \{\alpha_q\}$ for some $\alpha_q \in \sigma$ $(1 \leqslant q \leqslant n)$, and let F^q denote the σ_q-field F (that is, the field F treated as a difference field with the basic set σ_q). A *difference kernel* $\mathcal{R}$ of length r over F $(r \in \mathbf{N}, r > 0)$ is an ordered pair $(F^q\langle a_0, \ldots, a_r\rangle, \tau)$, where $F^q\langle a_0, \ldots, a_r\rangle$ is a σ_q-overfield of F^q generated by a set of s-tuples $a_i = (a_i^{(1)}, \ldots, a_i^{(s)})$, $0 \leqslant i \leqslant r$ (a positive integer s is fixed), and τ is a σ_q-isomorphism of $F^q\langle a_0, \ldots, a_{r-1}\rangle$ onto $F^q\langle a_1, \ldots, a_r\rangle$ such that $\tau a_i^{(k)} = a_{i+1}^{(k)}$ for $i = 0, \ldots, r-1$; $k = 1, \ldots, s$ and the restriction of τ on F^q coincides with α_q. (If $r = 0$, then $\tau = \alpha_q$.) A *prolongation* of $\mathcal{R}$ is a difference kernel $\mathcal{R}'$ consisting of a σ_q-overfield $F^q\langle a_0, \ldots, a_{r+1}\rangle$ of $F^q\langle a_0, \ldots, a_r\rangle$ together with the extension τ' of τ such that $\tau' : a_r^{(k)} \mapsto a_{r+1}^{(k)}$, $1 \leqslant k \leqslant s$. (If a kernel is of length 1 or 0, then $F^q\langle a_1, \ldots, a_r\rangle$ and $F^q\langle a_0, \ldots, a_{r-1}\rangle$ respectively are interpreted as F^q.) A prolongation $\mathcal{R}'$ of $\mathcal{R}$ is called *generic* if $F^q\langle a_0, \ldots, a_{r+1}\rangle^*$ is a free join of $F^q\langle a_0, \ldots, a_r\rangle^*$ and $F^q\langle a_1, \ldots, a_{r+1}\rangle^*$ over $F^q\langle a_1, \ldots, a_r\rangle^*$. It follows from Proposition 2.1.4 that if $\mathcal{R}'$ is a prolongation of a kernel $\mathcal{R} = (F^q\langle a_0, \ldots, a_r\rangle, \tau)$ such that $F^q\langle a_0, \ldots, a_r\rangle$ and $F^q\langle a_1, \ldots, a_{r+1}\rangle$ are free over $F^q\langle a_1, \ldots, a_r\rangle$, then $\mathcal{R}'$ is a generic prolongation of $\mathcal{R}$.

DEFINITION 4.4.7. We say that a kernel $\mathcal{R}$ satisfies property $\mathcal{P}$ if there exists a σ_q-overfield E_1 of $F^q\langle a_0, \ldots, a_r\rangle$, a σ_q-subfield E of E_1 which contains $F^q\langle a_0, \ldots, a_{r-1}\rangle$, and an extension of τ to a σ_q-isomorphism $\tilde{\tau}$ of E into E_1 such that

(a) E_1/E is primary (that is, the algebraic closure of E in E_1 is purely inseparable over E);
(b) E and $F^q\langle a_0, \ldots, a_r\rangle$ are free over $F^q\langle a_0, \ldots, a_{r-1}\rangle$;
(c) if $r = 0$, then $\tilde{\tau}E \subseteq E$. If $r > 0$, then $E_1 = \langle E, \tilde{\tau}E\rangle$ (we use the notation of Proposition 2.1.4).

DEFINITION 4.4.8. A kernel $\mathcal{R}$ is said to satisfy property $\mathcal{P}^*$ if there exists an inversive σ_q-overfield G_1 of $F^q\langle a_0, \ldots, a_r\rangle$, a σ_q^*-subfield G of G_1 which contains $F^q\langle a_0, \ldots, a_{r-1}\rangle$, and an extension of τ to a σ_q-isomorphism $\tilde{\tau}$ of G into G_1 such that

(a) G_1/G is primary;
(b) G and $F^q\langle a_0, \ldots, a_r\rangle^*$ are free over $F^q\langle a_0, \ldots, a_{r-1}\rangle^*$;
(c) if $r = 0$, then $\tilde{\tau}G \subseteq G$. If $r > 0$, then $G_1 = \langle G, \tilde{\tau}G\rangle$.

If the word "free" in (b) is replaced by "quasi-linearly disjoint", the difference kernel $\mathcal{R}$ is said to satisfy property $\mathcal{L}^*$.

PROPOSITION 4.4.9.

(i) *The properties $\mathcal{P}$ and $\mathcal{P}^*$ are equivalent.*

(ii) *With the above notation, a difference kernel $\mathcal{R}$ satisfies $\mathcal{L}^*$ if and only if the field extension $F^q\langle a_0, \dots, a_r\rangle^*/F^q\langle a_0, \dots, a_{r-1}\rangle^*$ is primary.*

(iii) *If a difference kernel $\mathcal{R}$ satisfies $\mathcal{P}^*$ then $\mathcal{R}$ has a generic prolongation $\mathcal{R}'$ which satisfies $\mathcal{P}^*$.*

(iv) *If a generic prolongation $\mathcal{R}'$ of a difference kernel $\mathcal{R}$ satisfies $\mathcal{P}^*$, then there exists a triple $(G, G_1, \tilde{\tau})$ with respect to which $\mathcal{R}$ satisfies $\mathcal{P}^*$ and trough which a generic prolongation $\mathcal{R}''$ of $\mathcal{R}$ can be obtained such that $\mathcal{R}''$ is equivalent to $\mathcal{R}'$ in sense of isomorphism.*

(v) *If a difference kernel satisfies $\mathcal{L}^*$, then all its generic prolongations are equivalent and satisfy $\mathcal{L}^*$.*

Let $\mathcal{R} = (F^q\langle a_0, \dots, a_r\rangle, \tau)$ be a difference kernel over an inversive difference field with a basic set $\sigma = \{\alpha_1, \dots, \alpha_n\}$ (we use the above notation; in particular, $a_0, \dots, a_r$ are s-tuples and $\sigma_q = \sigma \setminus \{\alpha_q\}$, $1 \leqslant q \leqslant n$). Let η denote an s-tuple in a σ^*-overfield H of F and let β be the translation of H which is the extension of α_q. Then with the notation $\eta_0 = \eta, \eta_j = \beta^j\eta$ $(j = 1, 2, \dots)$, we say that η is a *realization* of $\mathcal{R}$ in H over F if $(\eta_0, \dots, \eta_r)$ is a specialization of $(a_0, \dots, a_r)$ over F^q with $a_j^{(k)} \mapsto \eta_j^{(k)}$ $(1 \leqslant k \leqslant s, 0 \leqslant j \leqslant r)$. We also say that β is a realization of τ in H over F. If the specialization is generic, η is called a *regular realization* of $\mathcal{R}$. If there exists a sequence of kernels $\mathcal{R}_0 = \mathcal{R}, \mathcal{R}_1, \dots$ such that for each $h \in \mathbf{N}$, $\mathcal{R}_{h+1}$ is a generic prolongation of $\mathcal{R}_h$ and η is a regular realization of $\mathcal{R}_h$ over F, then η is said to be a *principal realization* of $\mathcal{R}$.

Two realizations η and ζ of a difference kernel $\mathcal{R}$ are said to be *equivalent* if the σ-field extensions $F\langle\eta\rangle/F$ and $F\langle\zeta\rangle/F$ are σ-isomorphic with $\eta^{(k)} \mapsto \zeta^{(k)}$ $(1 \leqslant k \leqslant s)$.

PROPOSITION 4.4.10. *The following statements about a difference kernel $\mathcal{R}$ are equivalent.*

(i) *$\mathcal{R}$ satisfies $\mathcal{P}^*$.*

(ii) *$\mathcal{R}$ has a principal realization.*

(iii) *$\mathcal{R}$ has a regular realization.*

PROPOSITION 4.4.11. *If a difference kernel $\mathcal{R}$ over an inversive difference field F satisfies $\mathcal{L}^*$, then $\mathcal{R}$ has a principal realization over F and all principal realizations of $\mathcal{R}$ are equivalent.*

THEOREM 4.4.12. *If η is a principal realization of a difference kernel $\mathcal{R}$ over an inversive difference field F, then η is not a proper specialization over F of any other realization of $\mathcal{R}$.*

One can also prove an analog of Proposition 4.4.6 for partial difference kernels: with the above notation, if η is an s-tuple over a partial inversive difference field F, there exists a difference kernel $\mathcal{R}$ such that η is the unique principal realization of $\mathcal{R}$ over F and every realization of $\mathcal{R}$ is a specialization of η over F.

As before, if $\mathcal{R} = (F^q\langle a_0, \dots, a_r\rangle, \tau)$ is a difference kernel over an inversive difference (σ^*-) field F (we use the above notation) and $\tilde{a}_0$ a subindexing of a_0, then $\tilde{a}_j$ will denote the corresponding subindexing of a_j $(1 \leqslant j \leqslant r)$. Furthermore, we say that an indexing

$a = \{a^{(i)} \mid i \in I\}$ is σ-algebraically independent over F if $a^{(i)}$ are distinct and form a σ-algebraically independent set over F.

LEMMA 4.4.13. *Let $\mathcal{R}'$ be a generic prolongation of a kernel $\mathcal{R} = (F^q\langle a_0, \dots, a_r\rangle, \tau)$.*

(i) *If $\tilde{a}_0$ is a subindexing of a_0 such that $\tilde{a}_r$ is σ_q-algebraically independent over $F^q\langle a_0, \dots, a_{r-1}\rangle$, then $\tilde{a}_{r+1}$ is σ_q-algebraically independent over $F^q\langle a_0, \dots, a_r\rangle$. Furthermore, the set $\bigcup_{i=1}^{r+1} \tilde{a}_i$ is σ_q-algebraically independent over F^q.*

(ii) *If $\tilde{a}_0$ is a subindexing of a_0 which is σ_q-algebraically independent over $F^q\langle a_1, \dots, a_r\rangle$, then $\tilde{a}_0$ is σ_q-algebraically independent over $F^q\langle a_1, \dots, a_{r+1}\rangle$ and $\bigcup_{i=1}^{r+1} \tilde{a}_i$ is σ_q-algebraically independent over F^q.*

THEOREM 4.4.14. *Let η be a principal realization of a kernel $\mathcal{R} = (F^q\langle a_0, \dots, a_r\rangle, \tau)$.*

(i) *If $\tilde{a}_0$ is a subindexing of a_0 such that $\tilde{a}_r$ is σ_q-algebraically independent over $F^q\langle a_0, \dots, a_{r-1}\rangle$, then the corresponding subindexing of η is σ_q-algebraically independent over F.*

(ii) *If $\tilde{a}_0$ is a subindexing of a_0 which is σ_q-algebraically independent over $F^q\langle a_1, \dots, a_r\rangle$, then the corresponding subindexing of η is σ_q-algebraically independent over F.*

4.5. *Ordinary difference polynomials. Existence theorem*

Let F be an ordinary difference field with basic set $\sigma = \{\alpha\}$ and let $R = F\{y_1, \dots, y_s\}$ be a ring of σ-polynomials in σ-indeterminates $y_1, \dots, y_s$ over F. Throughout this section a k-th transform $\alpha^k g$ of an element of a σ-ring will be also denoted by ${}^{(k)}g$.

Suppose that a σ-polynomial $A \in R$ contains one or more transforms of y_i, $1 \leqslant i \leqslant s$ (that is, one or more transforms of y_i appear in the irreducible representation of A as a linear combination of monomials in $y_1, \dots, y_s$ with coefficients from F; we treat y_i as its transform ${}^{(0)}y_i$). Let ${}^{(p)}y_i$ and ${}^{(q)}y_i$ be the transforms of y_i of lowest and highest order, respectively, contained in A. Then q and $q - p$ are called the *order* and *effective order* of A in y_i; they are denoted by $ord_{y_i} A$ and $Eord_{y_i} A$, respectively. If A does not contain transforms of y_i, both the order and effective order of A in y_i are defined to be 0.

PROPOSITION 4.5.1 [32, Chapter 2, Theorem VIII]. *Let F be an ordinary difference field with a basic set σ and $F\{y\}$ a ring of σ-polynomials in one σ-indeterminate y over F. Let an element η from some σ-overfield of F be σ-algebraic over F and let $r \in \mathbf{N}$ be the smallest integer such that $F\{y\}$ contains a non-zero σ-polynomial of order r with the solution η. Then $trdeg_F F\langle\eta\rangle = r$.*

Let G be a difference overfield of an ordinary difference (σ-) field F. In the theory of ordinary difference fields the transcendence degrees $trdeg_F G$ and $trdeg_{F^*} G^*$ are also called the *order* and *effective order* of the σ-field extension G/F; they are denoted by $ord\, G/F$ and $Eord\, G/F$, respectively. (As usual, K^* denotes the inversive closure of a difference field K.) Clearly, $Eord\, G/F \leqslant ord\, G/F$, $Eord\, G/F = ord\, G/F$ if F is inversive, and $ord\, G/F = \infty$ if σ-$trdeg_F\, G > 0$. Furthermore, the properties of transcendence degree imply that for any chain $F \subseteq G \subseteq H$ of ordinary difference field extensions we have $ord\, H/F = ord\, H/G + ord\, G/F$ and $Eord\, H/F = Eord\, H/G + Eord\, G/F$.

The proofs of the statements in the rest of this section can be found in [32, Chapters 6–10].

PROPOSITION 4.5.2. *Let F be an ordinary difference (σ-) field and ϕ a specialization of an s-tuple $a = (a^{(1)}, \ldots, a^{(s)})$ over F. Let $\{i_1, \ldots, i_q\}$ be a subset of $\{1, \ldots, s\}$, $\tilde{a}$ denote the set $\{a^{(i_1)}, \ldots, a^{(i_q)}\}$, and $\phi\tilde{a} = \{\phi a^{(i_1)}, \ldots, \phi a^{(i_q)}\}$.*

(i) *If $\phi\tilde{a}$ is σ-algebraically independent over F, so is $\tilde{a}$. Thus, $\sigma\text{-}trdeg_F\, F\langle \phi a^{(1)}, \ldots, \phi a^{(s)}\rangle \leqslant \sigma\text{-}trdeg_F\, F\langle a^{(1)}, \ldots, a^{(s)}\rangle$.*

(ii) *If $\phi\tilde{a}$ is a transcendence basis of $\{a^{(1)}, \ldots, a^{(s)}\}$ over F, then $ord\, F\langle \phi a^{(1)}, \ldots, \phi a^{(s)}\rangle / F\langle \phi\tilde{a}\rangle \leqslant ord\, F\langle a^{(1)}, \ldots, a^{(s)}\rangle / F\langle \tilde{a}\rangle$, $Eord\, F\langle \phi a^{(1)}, \ldots, \phi a^{(s)}\rangle / F\langle \phi\tilde{a}\rangle \leqslant Eord\, F\langle a^{(1)}, \ldots, a^{(s)}\rangle / F\langle \tilde{a}\rangle$, and the equality occurs if and only if the specialization ϕ is generic.*

PROPOSITION 4.5.3. *Let F be an ordinary difference field with a basic set σ, $F\{y\}$ a ring of σ-polynomials in one σ-indeterminate y over F, and η an element of some σ-overfield of F which is σ-algebraic over F. Let $r \in \mathbf{N}$ be the smallest non-negative integer such that $F\{y\}$ contains a non-zero σ-polynomial of effective order r with the solution η. Then $Eord\, F\langle \eta\rangle / F = r$.*

Let F be an ordinary difference field with a basic set σ, $F\{y_1, \ldots, y_s\}$ a ring of σ-polynomials in σ-indeterminates $y_1, \ldots, y_s$ over F, and $\mathcal{M}$ a non-empty irreducible variety over $F\{y_1, \ldots, y_s\}$. If $(\eta_1, \ldots, \eta_s)$ is a generic zero of $\mathcal{M}$, then $\sigma\text{-}trdeg_F\, F\langle \eta_1, \ldots, \eta_s\rangle$, $ord\, F\langle \eta_1, \ldots, \eta_s\rangle / F$, $Eord\, F\langle \eta_1, \ldots, \eta_s\rangle / F$, and $ld\, F\langle \eta_1, \ldots, \eta_s\rangle / F$ are called the *dimension*, *order*, *effective order*, and *limit degree* of the variety $\mathcal{M}$, respectively. They are denoted by $\dim \mathcal{M}$, $ord\, \mathcal{M}$, $Eord\, \mathcal{M}$, and $ld\, \mathcal{M}$, respectively. If $\mathcal{M} = \emptyset$, we set $\dim \mathcal{M} = ord\, \mathcal{M} = Eord\, \mathcal{M} = -1$ and $ld\, \mathcal{M} = 0$. Obviously, if $\dim \mathcal{M} > 0$, then $ord\, \mathcal{M} = \infty$.

If P is a prime inversive difference ideal of $F\{y_1, \ldots, y_s\}$ then the *dimension*, *order*, *effective order*, and *limit degree* of P (they are denoted by $\dim P$, $ord\, P$, $Eord\, P$, and $ld\, P$, respectively) are defined as the corresponding values of the variety $\mathcal{M}(P)$. (Thus, these concepts are determined as above through the σ-field extension $F\langle \eta_1, \ldots, \eta_s\rangle / F$ where $(\eta_1, \ldots, \eta_s)$ is a generic zero of P.)

Let $\mathcal{M}$ be a non-empty irreducible variety over $F\{y_1, \ldots, y_s\}$, $(\eta_1, \ldots, \eta_s)$ a generic zero of $\mathcal{M}$, and $\{y_{i_1}, \ldots, y_{i_q}\}$ is a subset of the set of σ-indeterminates $\{y_1, \ldots, y_s\}$. Then the *dimension*, *order*, *effective order*, and *limit degree* of $\mathcal{M}$ *relative to* $y_{i_1}, \ldots, y_{i_q}$ are defined as $\sigma\text{-}trdeg_{F\langle \eta_{i_1}, \ldots, \eta_{i_q}\rangle}\, F\langle \eta_1, \ldots, \eta_s\rangle$, $ord\, F\langle \eta_1, \ldots, \eta_s\rangle / F\langle \eta_{i_1}, \ldots, \eta_{i_q}\rangle$, $Eord\, F\langle \eta_1, \ldots, \eta_s\rangle / F\langle \eta_{i_1}, \ldots, \eta_{i_q}\rangle$, and $ld\, F\langle \eta_1, \ldots, \eta_s\rangle / F\langle \eta_{i_1}, \ldots, \eta_{i_q}\rangle$, respectively. These characteristics of $\mathcal{M}$ are denoted by $\dim(y_{i_1}, \ldots, y_{i_q})\mathcal{M}$, $ord(y_{i_1}, \ldots, y_{i_q})\mathcal{M}$, $Eord(y_{i_1}, \ldots, y_{i_q})\mathcal{M}$, and $ld(y_{i_1}, \ldots, y_{i_q})\mathcal{M}$, respectively. If P is a prime inversive difference ideal of $F\{y_1, \ldots, y_s\}$ then the dimension, order, effective order, and limit degree of P relative to $y_{i_1}, \ldots, y_{i_q}$ are defined as the corresponding characteristics of $\mathcal{M}(P)$ (the notation is the same: $\dim(y_{i_1}, \ldots, y_{i_q})P$, $ord(y_{i_1}, \ldots, y_{i_q})P$, $Eord(y_{i_1}, \ldots, y_{i_q})(P)$, and $ld(y_{i_1}, \ldots, y_{i_q})(P)$, respectively).

A subset $\{y_{i_1}, \ldots, y_{i_q}\}$ of $\{y_1, \ldots, y_s\}$ is called a *set of parameters* of P (or $\mathcal{M}(P)$) if P contains no non-zero σ-polynomial in $F\{y_{i_1}, \ldots, y_{i_q}\}$. A set of parameters of P which is not a proper subset of any set of parameters of P is called *complete*. Clearly, every reflexive prime σ-ideal of $F\{y_1, \ldots, y_s\}$ has at least one complete set of parameters, and every set of parameters can be extended to a complete one.

PROPOSITION 4.5.4. *With the above notation, a set of parameters of a proper inversive difference ideal P of $F\{y_1, \ldots, y_s\}$ is complete if and only if it contains* $\dim P$ *elements.*

PROPOSITION 4.5.5. *Let $\mathcal{M}_1$ and $\mathcal{M}_2$ be two irreducible varieties over $F\{y_1, \ldots, y_s\}$ such that $\mathcal{M}_1 \subseteq \mathcal{M}_2$. Then*

(i) $\dim \mathcal{M}_1 \leqslant \dim \mathcal{M}_2$.

(ii) *If $\{y_{i_1}, \ldots, y_{i_q}\}$ is a complete set of parameters of $\Phi(\mathcal{M}_1)$ (we use the notation of section 2.5), then $ord(y_{i_1}, \ldots, y_{i_q})\mathcal{M}_1 \leqslant ord(y_{i_1}, \ldots, y_{i_q})\mathcal{M}_2$, and the equality occurs if and only if $\mathcal{M}_1 = \mathcal{M}_2$. Furthermore, $Eord(y_{i_1}, \ldots, y_{i_q})\mathcal{M}_1 \leqslant Eord(y_{i_1}, \ldots, y_{i_q})\mathcal{M}_2$, and the equality occurs if and only if $\mathcal{M}_1 = \mathcal{M}_2$.*

Let F be an ordinary difference field with a basic set $\sigma = \{\alpha\}$ and $R = F\{y_1, \ldots, y_s\}$ a ring of σ-polynomials in σ-indeterminates $y_1, \ldots, y_s$ over F. Let $A \in R$ be an algebraically irreducible σ-polynomial, that is, $A \notin F$ and A is irreducible as a polynomial in the indeterminates ${}^{(j)}y_i$ $(1 \leqslant i \leqslant s;\ j = 0, 1, \ldots)$ (in this case R is treated as a polynomial ring in this denumerable set of indeterminates over F; as above, ${}^{(j)}y_i$ denotes $\alpha^j y_i$). An irreducible component $\mathcal{M}$ of $\mathcal{M}(A)$ is called a *principal component* of the variety $\mathcal{M}(A)$ if whenever A contains a transform of y_i $(1 \leqslant i \leqslant s)$, the family $\{y_1, \ldots, y_{i-1}, y_{i+1}, \ldots, y_s\}$ is a complete set of parameters of $\mathcal{M}$ and $Eord(y_1, \ldots, y_{i-1}, y_{i+1}, \ldots, y_s)\mathcal{M}$ is the effective order of A in y_i. (This implies that $\dim \mathcal{M} = s - 1$.) Irreducible components of $\mathcal{M}$ which are not principal are called *singular*.

The following proposition shows that no component of $\mathcal{M}(A)$ is "larger" than a principal component.

PROPOSITION 4.5.6.

(i) *With the above notation, let $\eta = (\eta_1, \ldots, \eta_s)$ be a solution of the non-zero σ-polynomial $A \in R$. Then either σ-$trdeg_F\, F\langle \eta_1, \ldots, \eta_s\rangle < s - 1$ or σ-$trdeg_F\, F\langle \eta_1, \ldots, \eta_s\rangle = s - 1$ and for each k $(1 \leqslant k \leqslant s)$, such that the set $\tilde{\eta}_k = \{\eta_1, \ldots, \eta_{k-1}, \eta_{k+1}, \ldots, \eta_s\}$ is σ-algebraically independent over F, A contains a transform of y_k and $ord\, F\langle \eta_{(1)}, \ldots, \eta_s(s)\rangle / F\langle \hat{\eta}_k\rangle \leqslant ord_{y_k} A$, $Eord\, F\langle \eta_{(1)}, \ldots, \eta_s(s)\rangle / F\langle \hat{\eta}_k\rangle \leqslant Eord_{y_k} A$.*

(ii) *Let a σ-polynomial $A \in R$ be algebraically irreducible, σ-$trdeg_F\, F\langle \eta_1, \ldots, \eta_s\rangle = s - 1$, and for each k $(1 \leqslant k \leqslant s)$ such that A contains a transform of y_k, $Eord\, F\langle \eta_{(1)}, \ldots, \eta_s(s)\rangle / F\langle \hat{\eta}_k\rangle = Eord_{y_k} A$. Then η is a generic zero of a principal component of $\mathcal{M}(A)$.*

The following fundamental result is an abstract form of an existence theorem for ordinary algebraic difference equations.

THEOREM 4.5.7. *Let F be an ordinary difference field with a basic set σ, $R = F\{y_1, \dots, y_s\}$ a ring of σ-polynomials in σ-indeterminates $y_1, \dots, y_s$ over F, and $A \in R$ an algebraically irreducible σ-polynomial. Then*

(i) *The variety $\mathcal{M}(A)$ has principal components.*

(ii) *Let A contain a transform of some y_i $(1 \leqslant i \leqslant s)$. Then*

(a) *If A contains a transform of order 0 of some σ-indeterminate y_j and $\mathcal{M}$ is a principal component of $\mathcal{M}(A)$, then $ord(y_1, \dots, y_{i-1}, y_{i+1}, \dots, y_s)\mathcal{M} = ord_{y_i} A$.*

(b) *If $\mathcal{M}_1, \dots, \mathcal{M}_k$ are the principal components of $\mathcal{M}(A)$ and d and e denote, respectively, the degree and reduced degree of A in the highest transform of y_i which it contains, then $\sum_{j=1}^{k} ld(y_1, \dots, y_{i-1}, y_{i+1}, \dots, y_s)\mathcal{M}_j \leqslant d$ and $\sum_{j=1}^{k} rld(y_1, \dots, y_{i-1}, y_{i+1}, \dots, y_s)\mathcal{M}_j = e$.*

(c) *If d_1 and e_1 denote, respectively, the degree and reduced degree of A in the lowest transform of y_i contained in A, then $\sum_{j=1}^{k} ild(y_1, \dots, y_{i-1}, y_{i+1}, \dots, y_s)\mathcal{M}_j \leqslant d_1$ and $\sum_{j=1}^{k} irld(y_1, \dots, y_{i-1}, y_{i+1}, \dots, y_s)\mathcal{M}_j = e_1$.*

(d) *Let $q = Eord_{y_i} A$ and let $\mathcal{M}$ be a component of $\mathcal{M}(A)$ such that $\{y_1, \dots, y_{i-1}, y_{i+1}, \dots, y_s\}$ is a complete set of parameters of $\mathcal{M}$ and $Eord(y_1, \dots, y_{i-1}, y_{i+1}, \dots, y_s)\mathcal{M} = q$. Then is $\mathcal{M}$ a principal component of $\mathcal{M}(A)$.*

COROLLARY 4.5.8. *Every ordinary difference field F has an algebraic closure G which is a difference overfield of F.*

PROPOSITION 4.5.9. *Let F be an ordinary difference field with a basic set σ, $F\{y\}$ a ring of σ-polynomials in one σ-indeterminate y over F, and A an algebraically irreducible σ-polynomial of order 0. Then, if any component of $\mathcal{M}(A)$ has limit degree 1, so do all components. Similar statements hold for rld, ild, and rild.*

THEOREM 4.5.10. *Let F be an ordinary difference field with a basic set σ and $R = F\{y_1, \dots, y_s\}$ a ring of σ-polynomials in σ-indeterminates $y_1, \dots, y_s$ over F.*

(i) *If $\mathcal{M}$ is an irreducible variety over $F\{y_1, \dots, y_s\}$, then $\dim \mathcal{M} = s - 1$ if and only if $\mathcal{M}$ is a principal component of the variety of an algebraically irreducible σ-polynomial.*

(ii) *Let $A \in R \setminus F$ and let $\mathcal{M}$ be an irreducible component of the variety $\mathcal{M}(A)$ of the σ-polynomial A. Then $\dim \mathcal{M} = s - 1$.*

(iii) *Let $A, B \in R$ be σ-polynomials of order at most 1 in each y_i $(1 \leqslant i \leqslant s)$. If $\mathcal{M}(A, B)$ is not empty, it has a component of dimension not less than $s - 2$.*

R.M. Cohn [32, Chapter 10] conjectured that if η is a realization of a kernel $\mathcal{R}$ over an ordinary difference field F, then η is a specialization over F of a realization ζ such that $\sigma\text{-}trdeg_F\, F\langle\zeta\rangle \geqslant \delta\mathcal{R}$. If this statement is true, it would strengthen the conclusion of Theorem 4.5.10(iii) to the statement that every component of $\mathcal{M}(A, B)$ has dimension at least $s - 2$.

Let F be an ordinary difference field with a basic set $\sigma = \{\alpha\}$ and let $R = F\{y_1, \dots, y_s\}$ be a ring of σ-polynomials in a set of σ-indeterminates $Y = \{y_1, \dots, y_s\}$ over F. Let $\Phi \subseteq$

R, $Y' \subseteq Y$, and let the orders of the σ-polynomials of Φ in each $y_i \in Y \setminus Y'$ be bounded (in particular, Φ may be a finite family). For every $y_i \in Y \setminus Y'$, let r_i denote the maximum of the orders of the σ-polynomials of Φ in y_i. Then the number $\mathcal{R}(Y')\Phi = \sum_i r_i$ (the summation extends over all values of the index i such that $y_i \in Y \setminus Y'$) is called the *Ritt number* of the system Φ associated with the set $Y' \subseteq Y$. If $Y' = \emptyset$, we set $\mathcal{R}(Y')\Phi = \sum_{i=1}^{s} r_i$.

Let $0 \neq A \in \Phi$ and let $h(A)$ be the greatest non-negative integer such that for each i with $y_i \in Y \setminus Y'$, $\alpha^{h(A)}(A)$ is of order at most r_i in y_i. Let $h = \max\{h(A) \mid 0 \neq A \in \Phi\}$, that is, h is the greatest integer such that some polynomial in Φ may be replaced by its h-th transform without altering the Ritt number of Φ. The number $\mathcal{G}(Y')\Phi = \mathcal{R}(Y')\Phi - h$ is called the *Greenspan number* of the system Φ associated with the set $Y' \subseteq Y$. If $Y' = \emptyset$, we write $\mathcal{G}\Phi$ for $\mathcal{G}(Y')\Phi$.

Now, let the system of σ-polynomials Φ be finite, $\Phi = \{A_1, \dots, A_m\}$, and let $r_{ij} = ord_{y_j} A_i$ $(1 \leqslant i \leqslant m, 1 \leqslant j \leqslant s)$. Then the number $\mathcal{J}(\Phi) = \max\{\sum_{i=1}^{s} r_{ij_i} \mid (j_1, \dots, j_s)$ is a permutation of $1, \dots, s\}$ is called the Jacobi number of the system Φ. The following theorem (where we use the above notation) gives some bounds on the effective orders with respect to Y' that involve the Ritt, Greenspan and Jacobi numbers.

THEOREM 4.5.11.

(i) *Suppose that the Ritt number $\mathcal{R}(Y')\Phi$ for a system of σ-polynomials Φ (and a set $Y' \subseteq Y$) is defined. If $\mathcal{M}$ is a component of $\mathcal{M}(\Phi)$ for which Y' contains a complete set of parameters, then $Eord(Y')\mathcal{M} \leqslant \mathcal{R}(Y')\Phi$.*

(ii) *If $\mathcal{G}\Phi$ is defined, $\mathcal{M}$ is a component of $\mathcal{M}(\Phi)$ and Y' contains a complete set of parameters of every component of $\mathcal{M}(\Phi)$, then $Eord(Y')\mathcal{M} \leqslant \mathcal{G}(Y')\Phi$.*

(iii) *Let $\Phi = \{A_1, \dots, A_m\}$ where $A_1, \dots, A_m$ are first-order σ-polynomials (that is $r_{ij} \leqslant 1$ for $i = 1, \dots, m$; $j = 1, \dots, s$ and at least one r_{ij} is equal to 1). If $\mathcal{M}$ is an irreducible component of $\mathcal{M}(\Phi)$ of dimension 0, then $Eord\,\mathcal{M} \leqslant \mathcal{J}(\Phi)$.*

A number of results on the Jacobi bound for systems of algebraic differential equations (see, for example, [41,42] and [88, Section 5.8]) give a hope that the last theorem can be essentially strengthened and generalized to the case of partial difference polynomials.

In [32, Chapter 10, Example 2] R.M. Cohn showed that it is possible for a difference kernel $\mathcal{R}$ to have a realization η such that σ-$trdeg_F F\langle\eta\rangle > \delta\mathcal{R}$. His example also shows that if $\mathcal{R}'$ is a kernel over an ordinary inversive difference field F, ζ is a principal realization of $\mathcal{R}'$, and $\mathcal{R}$ is a kernel over F which specializes to $\mathcal{R}'$, there may be no principal realization of $\mathcal{R}$ which specializes to ζ. The following theorem presents conditions that imply the existence of such a realization. It also implies that *every regular realization of a kernel is the specialization of a principal realization.*

THEOREM 4.5.12 [96]. *Let F be an ordinary inversive difference field. Let $\mathcal{R} = (F(a_0, \dots, a_r), \tau)$ and $\mathcal{R}' = (F(a'_0, \dots, a'_r), \tau)$ $(r \in \mathbf{N})$ be two difference kernels over F such that there is an F-isomorphism of $F(a_0, \dots, a_{r-1})$ onto $F(a'_0, \dots, a'_{r-1})$. Let ζ be a principal realization of $\mathcal{R}'$. If $\mathcal{R}$ specializes to $\mathcal{R}'$, then there exists a principal realization η of $\mathcal{R}$ which specializes to ζ.*

In the rest of this section we use the notation and conventions of Section 4.4. Let F be an ordinary inversive difference field with a basic set $\sigma = \{\alpha\}$ and $\mathcal{R} = (F(a_0, \dots, a_r), \tau)$ a difference kernel ($a_0 = (a_0^{(1)}, \dots, a_0^{(s)})$). A realization $\eta = (\eta_1, \dots, \eta_s)$ of the kernel $\mathcal{R}$ is called *singular* if it is not a specialization of a principal realization of $\mathcal{R}$. A realization of $\mathcal{R}$ is called *multiple* if it is a specialization of two principal realizations which are distinct in the sense of isomorphism. The following two examples (see [32, Chapter 6, Section 21]) illustrate these concepts.

EXAMPLE 4.5.13. Let us consider the algebraically irreducible σ-polynomial $A = {}^{(2)}yy + {}^{(1)}y$ in the ring of σ-polynomials $F\{y\}$ in one difference indeterminate y (as before, ${}^{(k)}y$ stands for $\alpha^k y$). Then $y\alpha(A) - A = {}^{(1)}y({}^{(3)}yy - 1)$. If $\eta \neq 0$ is a generic zero of an irreducible component $\mathcal{M}$ of $\mathcal{M}(A)$, then η must annul ${}^{(3)}yy - 1$, whence ${}^{(3)}yy - 1 \in \Phi(\mathcal{M})$, $0 \notin \mathcal{M}$. Thus, the solution 0 of A itself constitutes an irreducible component of $\mathcal{M}(A)$. Clearly, it is a singular component and furnishes a singular realization of the kernel produced for A by the procedure described in Example 4.4.2. It should be noted (see [19, Theorem V]) that the variety of a first-order σ-polynomial cannot have a singular component.

EXAMPLE 4.5.14. With the notation of the previous example, let $B = ({}^{(1)}y)^2 + y^2 \in F\{y\}$ where $F = \mathbf{Q}$ (and $\alpha = \mathrm{id}_{\mathbf{Q}}$). Then $\alpha(B) - B = ({}^{(2)}y - y)({}^{(2)}y - y)$. It can be shown (see [32, Chapter 6, Theorem IV]) that $\mathcal{M}(B)$ has two principal components $\mathcal{M}_1$ and $\mathcal{M}_2$ which annul ${}^{(2)}y - y$ and ${}^{(2)}y - y$, respectively. Furthermore, $0 \in \mathcal{M}_1 \cap \mathcal{M}_2$, so 0 is a multiple solution of the kernel $\mathcal{R}$ formed for B by the procedure described in Example 4.4.2.

PROPOSITION 4.5.15. *Let $\mathcal{R} = (F(a_0, \dots, a_r), \tau)$ be a difference kernel over an ordinary inversive difference (σ-) field F of zero characteristic. Let S denote the polynomial ring in $s(r+1)$ indeterminates $F[\{{}^{(j)}y_i \mid 1 \leqslant i \leqslant s, 0 \leqslant j \leqslant r\}] \subseteq F\{y_1, \dots, y_s\}$, and let P be the prime ideal of S with the general zero $(a_0, \dots, a_r)$. Then*:

(i) *There exist σ-polynomials $A, B \in S \setminus P$ such that*
  (a) *Every singular realization of $\mathcal{R}$ annuls A.*
  (b) *Every multiple realization of $\mathcal{R}$ annuls B.*

(ii) *If $r = 0$, then $\mathcal{R}$ has no singular realization (even if* Char $F \neq 0$).

Let F be an ordinary difference (σ-) field and $R = F\{y_1, \dots, y_s\}$ a ring of σ-polynomials over F. Suppose that a σ-polynomial $A \in R$ contains one or more transforms of a σ-indeterminate y_i ($1 \leqslant i \leqslant s$) and let ${}^{(p)}y_i$ and ${}^{(q)}y_i$ be such transforms of lowest and highest orders, respectively. Then the formal partial derivatives $\partial A/\partial^{(p)}y_i$ and $\partial A/\partial^{(q)}y_i$ are called the *separants* of A with respect to y_i. If A is written as a polynomial in ${}^{(q)}y_i$, then the coefficient of the highest power of ${}^{(q)}y_i$ is called the *initial of A with respect to y_i*.

PROPOSITION 4.5.16. *With the above notation, let F be a σ-field of zero characteristic and $A \in R$ an algebraically irreducible σ-polynomial. Then*

(i) *Every singular component of $\mathcal{M}(A)$ and every solution common to two principal components of $\mathcal{M}(A)$ annuls the separants of A.*

(ii) *If $\mathcal{M}(A)$ has only one principal component, then every singular component of $\mathcal{M}(A)$ annuls the initials of A.*

PROPOSITION 4.5.17. *Let F be an ordinary difference field and P a reflexive prime difference ideal in a ring of difference polynomials $F\{y_1, \dots, y_s\}$. Unless* $\dim P = 0$, $\operatorname{Eord} P = 0$, *there exist difference overfields of F containing arbitrarily many generic zeros of P.*

An element a of an ordinary difference ring R with a basic set $\sigma = \{\alpha\}$ is called *periodic* if $\alpha^k(a) = a$ for some $k \in \mathbf{N}$. (In particular, every constant $c \in R$ is periodic.) Clearly, the set of all periodic elements of R is a σ-subring of R; it is called the *σ-subring of periodic elements* of R.

A difference ring R with a basic set $\sigma = \{\alpha\}$ is said to be *periodic* if there exists $m \in \mathbf{N}$ such that $\alpha^m(a) = a$ for all $a \in R$. In particular, if $m = 1$, the σ-ring R is said to be *invariant*.

PROPOSITION 4.5.18. *Let F be an ordinary difference field with a basic set σ and let F' and F'' be its σ-subfields of constants and periodic elements, respectively. Then F'' is the algebraic closure of F' in F.*

An ordinary difference field with a basic set $\sigma = \{\alpha\}$ is called *completely aperiodic* if either $\operatorname{Char} F = 0$ or $\operatorname{Char} F = p > 0$ and for any $i, j, k, l \in \mathbf{N}$, $i \neq j, k \neq l$, no element of F satisfies the equation $(\alpha^i(y))^{p^k} = (\alpha^j(y))^{p^l}$. (It is easy to see that if an ordinary difference field F of positive characteristic is aperiodic and contains infinitely many constants, then F is completely aperiodic.)

THEOREM 4.5.19 [25]. *Let F be a completely aperiodic ordinary difference (σ-) field, G a σ-overfield of F, $R = G\{y_1, \dots, y_s\}$ a ring of σ-polynomials in σ-indeterminates $y_1, \dots, y_s$ over G, and $0 \neq A \in R$. Then*

(i) *There exists an s-tuple $(\eta_1, \dots, \eta_s)$ of elements of F which is not a solution of A.*

(ii) *Suppose that G/F is a finitely generated σ-field extension, σ-$\operatorname{trdeg}_F G = 0$, and either* $\operatorname{Char} F = 0$ *or* $\operatorname{Char} F > 0$ *and $\operatorname{rld} G.F = \operatorname{ld} G/F$. Then there exist $\zeta \in G, k \in \mathbf{N}$ such that $\alpha^k(a) \in F\langle\zeta\rangle$ for every element $a \in G$. Moreover, such an element ζ can be chosen as a linear combination of the members of any finite set of σ-generators of G/F with coefficients from any pre-assigned completely aperiodic σ-subfield of F.*

Using the last theorem, R.M. Cohn [32, Chapter 8] showed that the solutions of any irreducible variety $\mathcal{M}$ over an ordinary difference field F can be obtained by rational operations, transforming, and the inverse of transforming from the solutions of a principal component $\mathcal{N}$ of the variety of an algebraically irreducible difference polynomial ($\mathcal{N}$ is the variety over F, but not necessarily over the same ring of difference polynomials, as $\mathcal{M}$). More precisely, we have the following statement.

THEOREM 4.5.20. *Let F be an ordinary difference field with a basic set $\sigma = \{\alpha\}$, $F\{y_1, \dots, y_s\}$ a ring of σ-polynomials in σ-indeterminates $y_1, \dots, y_s$ over F, and $\mathcal{M}$*

an irreducible variety over $F\{y_1, \ldots, y_s\}$. Suppose that F is completely aperiodic or that $\dim \mathcal{M} > 0$, *and also that $\mathcal{M}$ possesses a complete set of parameters $\{y_1, \ldots, y_k\}$ such that $rld(y_1, \ldots, y_k)\mathcal{M} = ld\,(y_1, \ldots, y_k)\mathcal{M}$. (We assume that the σ-indeterminates are so numbered that the first k of them constitute the complete set of parameters. Note also that the last equality always holds if* Char $F = 0$.) *Then there exist*:

(a) *an irreducible variety $\mathcal{N}$ over the ring of σ-polynomials $F\{y_1, \ldots, y_k; w\}$ (w is the $(k+1)$-th σ-indeterminate of this ring)*;

(b) *σ-polynomials $A_{k+1}, \ldots, A_s \in F\{y_1, \ldots, y_k\}$, σ-polynomials $B_{k+1}, \ldots, B_s$, $C \in F\{y_1, \ldots, y_k; w\}$, $C \notin \mathcal{N}$, and an integer $t \geqslant 0$ such that*

  (i) *$\mathcal{N}$ is a principal component of the variety of an algebraically irreducible σ-polynomial of $F\{y_1, \ldots, y_k; w\}$, $y_1, \ldots, y_k$ constitute a complete set of parameters of $\mathcal{N}$, and $Eord(y_1, \ldots, y_k)\mathcal{N} = Eord(y_1, \ldots, y_k)\mathcal{M}$.*

  (ii) *If $(\eta_1, \ldots, \eta_s)$ is any solution in $\mathcal{M}$, there is a solution in $\mathcal{N}$ with $y_i = \eta_i$ for $i = 1, \ldots, k$, and w given by the result of substituting η_j for y_j in $\sum_{i=k+1}^{s} A_i y_i$.*

  (iii) *If $y_i = \zeta_i$, $i = 1, \ldots, k$, $w = \theta$ is a solution in $\mathcal{N}$ which does not annul C, then there is a solution in $\mathcal{M}$ with $y_i = \zeta_i$, $i = 1, \ldots, k$, and y_j, $k+1 \leqslant j \leqslant s$, given by applying α^{-t} to the result of substituting $\zeta_{k+1}, \ldots, \zeta_s, \theta$ for $y_{k+1}, \ldots, y_s, w$, respectively, in B_j/C.*

  (iv) *If $D \in F\{y_1, \ldots, y_s\} \setminus \Phi(\mathcal{M})$, then there exists a σ-polynomial $E \in F\{y_1, \ldots, y_k; w\} \setminus \Phi(\mathcal{N})$ such that any solution in $\mathcal{N}$ not annulling E gives rise by a procedure described in* (iii) *to a solution in $\mathcal{M}$ not annulling D.*

  (v) *The procedures of* (ii) *and* (iii) *carry generic zeros of $\mathcal{M}$ or $\mathcal{N}$ into generic zeros of $\mathcal{N}$ or $\mathcal{M}$. Whenever* (iii) *is defined, these procedures, applied to elements of $\mathcal{M}$ or $\mathcal{N}$, are inverses of each other.*

With the notation of the last theorem, $W = \sum_{i=k+1}^{s} A_i y_i$ is called a *resolvent* for $\mathcal{M}$ or for $\Phi(\mathcal{M})$, and $\Phi(\mathcal{N})$ is called a *resolvent ideal* for $\mathcal{M}$ or for $\Phi(\mathcal{M})$. One can say that $\mathcal{M}$ is obtained from the solutions of its resolvent ideal by the relations $\alpha^t y_i = B_i/C$ $(k+1 \leqslant i \leqslant s)$ and the solutions of the resolvent ideal are obtained from those of $\mathcal{M}$ by the relation $w = W$.

We conclude this section with a summary of the basic properties of a variety of one ordinary difference polynomial (see [32, Chapters 6 and 10]).

Let F be an ordinary difference (σ-) field (we assume Char $F = 0$, except in (i) and (iii)), $R = F\{y_1, \ldots, y_s\}$ a ring of σ-polynomials over F and $A \in R \setminus F$ an algebraically irreducible σ-polynomial. Then

(i) The variety $\mathcal{M}(A)$ consists of one or more principal components and (possibly) of singular components. Each component of $\mathcal{M}(A)$ has dimension $s-1$.

(ii) The relative effective orders and (with certain limitations) the relative orders of principal components of $\mathcal{M}(A)$ are determined by the effective orders and orders of A. Furthermore, it is sufficient for a component to have dimension $s-1$ and one of these effective orders for it to be a principal component.

(iii) Except in the case that $s = 1$ and A is of effective order 0, every principal component of $\mathcal{M}(A)$ contains infinitely many generic zeros.

(iv) The relative effective orders of the singular components of $\mathcal{M}(A)$ are less by at least 2 than those of the principal components. More precisely, if $Eord_{y_i} A = r_i$

$(1 \leqslant i \leqslant s)$ and $\mathcal{M}$ is a singular component of $\mathcal{M}(A)$, then either $y_1, \ldots, y_{i-1}, y_{i+1}, \ldots, y_s$ do not constitute a set of parameters of $\mathcal{M}$ or they do constitute such a set and $Eord(y_1, \ldots, y_{i-1}, y_{i+1}, \ldots, y_s)\mathcal{M} \leqslant r_i - 2$.

(v) Any singular component of $\mathcal{M}(A)$ is itself a principal component of $\mathcal{M}(B)$ for some algebraically irreducible σ-polynomial $B \in R$. It follows from (iv), for each i, $1 \leqslant i \leqslant s$, either B contains no transforms of y_i or $Eord_{y_i} B \leqslant r_i - 2$.

(vi) If $s = 1$ and A is a first-order σ-polynomial in R, then $\mathcal{M}(A)$ has no singular components.

A number of additional important results on varieties of ordinary difference polynomials can be found in [133,19,20,22,23,26,28,33,34], and [32, Chapter 10].

4.6. *Compatibility of difference field extensions. Specializations*

Let F be a difference field with a basic set σ and let G and H be two σ-overfields of F. The difference field extensions G/F and H/F are said to be *compatible* if there exists a σ-field extension E of F such that G/F and H/F have σ-isomorphisms into E/F. (As usual, this means that there exist σ-isomorphisms of G and H into E that leave the σ-field F fixed.) Otherwise, the σ-field extensions G/F and H/F are called *incompatible*.

EXAMPLE 4.6.1 [32, Chapter 1, Example 4]. Let us consider $\mathbf{Q}$ as an ordinary difference field whose basic set σ consists of the identity automorphism α. If one adjoins to $\mathbf{Q}$ an element i such that $i^2 = -1$, then the resulting field $\mathbf{Q}(i)$ has two automorphisms that extend α: one of them is the identity mapping (we denote it by the same letter α) and the other (denoted by β) sends an element $a + bi \in \mathbf{Q}(i)$ $(a, b \in \mathbf{Q})$ to $a - bi$ (complex conjugation). Then $\mathbf{Q}(i)$ can be treated as a difference field with the basic set $\{\alpha\}$, as well as a difference field with the basic set $\{\beta\}$. Denoting these two difference fields by G and H, respectively, we can naturally consider them as σ-field extensions of $\mathbf{Q}$. Let us show that $G/\mathbf{Q}$ and $H/\mathbf{Q}$ are incompatible σ-field extensions. Indeed, suppose that there exists a σ-field extension E of $\mathbf{Q}$ and σ-isomorphisms ϕ and ψ, respectively, of $G/\mathbf{Q}$ and $H/\mathbf{Q}$ into $E/\mathbf{Q}$. Let $j = \phi(i)$, $k = \psi(i)$, and let γ denote the translation of E that extends α and β. Then $j^2 = k^2 = -1$ whence either $j = k$ or $j = -k$. Since $\gamma(j) = j$ and $\gamma(k) = -k$, in both cases we obtain that $j = -j$, that is, $j = 0$. This contradiction implies that the σ-field extensions $G/\mathbf{Q}$ and $H/\mathbf{Q}$ are incompatible.

The existence of incompatible extensions plays an important part in the development of the theory of difference algebra. Of particular concern here is the fact that the presence of incompatible extensions can inhibit the extension of difference isomorphisms for difference field extensions.

EXAMPLE 4.6.2 [32, Chapter 9, Example 1]. As in the preceding example, let us consider $\mathbf{Q}$ as an ordinary difference field with a basic set $\sigma = \{\alpha\}$ where α is the identity automorphism. Let a denote the positive fourth root of 2 (we assume $a \in \mathbf{C}$) and let $i \in \mathbf{C}$ be the square root of -1. Then the field $\mathbf{Q}(a, i)$ can be treated as a σ-field extension of $\mathbf{Q}$ such that $\alpha(i) = -i$ and $\alpha(a) = -a$. Obviously, $\mathbf{Q}(i)/\mathbf{Q}$ can be treated as a σ-field subextension of $\mathbf{Q}(a, i)/\mathbf{Q}$. Let ϕ be a σ-isomorphism of $\mathbf{Q}(a, i)/\mathbf{Q}(i)$ into some σ-field extension $M/\mathbf{Q}(i)$ where M is a σ-overfield of $\mathbf{Q}(a, i)$. Since the field extension $\mathbf{Q}(a, i)/\mathbf{Q}(i)$ is normal, ϕ is an automorphism of $\mathbf{Q}(a, i)$. Clearly, $\phi(a) = i^k a$ for some $k \in \mathbf{N}$. Since $\alpha(a) = -a$, $\alpha(i^k a) = -i^k a$. But $\alpha(i^k a) = (-1)^{k+1}(i^k a)$, so k is even and either $\phi(a) = a$ or $\phi(a) = -a$. It follows that every σ-isomorphism of $\mathbf{Q}(a, i)/\mathbf{Q}(i)$ leaves the elements of $\mathbf{Q}(i)\langle a^2\rangle$ fixed. On the other hand, there exists a σ-automorphism ψ of $\mathbf{Q}(i)\langle a^2\rangle/\mathbf{Q}(i)$ such that $\psi(a^2) = -a^2$. Clearly, such a σ-automorphism has no extension to a σ-isomorphism of $\mathbf{Q}(a, i)/\mathbf{Q}(i)$.

Let F be a difference field with a basic set σ and G/F, H/F two σ-field extensions of F. Then the number of σ-isomorphisms of G/F into H/F is called the *replicability of G/F in H/F*. The *replicability of the σ-field extension G/F* is defined as the maximum of replicabilities of G/F in all σ-field extensions of F, if this maximum exists, or ∞ if it does not. Clearly, if G/F is finitely generated (as a σ-field extension), then it is sufficient to define the replicability of G/F considering only σ-overfields in the universal system over F.

THEOREM 4.6.3 [32, Chapter 7, Theorem II]. *A necessary condition for finite replicability of an ordinary difference field extension G/F is that every element of G have a transform of some order that is algebraic over F. This condition is sufficient if G is finitely generated over F.*

Let F be an ordinary difference field with a basic set σ and a an indexing of elements lying in a σ-overfield of F. We say that *almost every specialization of a over F has property $\mathcal{P}$* if there exists a non-zero element $u \in F\{a\}$ such that every specialization of a over F which does not specialize u to 0 has property $\mathcal{P}$. The statement that almost every specialization which has property $\mathcal{P}$ has property $\mathcal{Q}$ means that almost every specialization has the property "not $\mathcal{P}$ or $\mathcal{Q}$".

In [32, Chapter 7] R.M. Cohn showed that a set of elements and a specialization of the set can generate incompatible difference field extensions over a given difference field. At the same time, the following statement implies that "in most cases" a finite indexing and one of its specializations generate compatible extensions.

THEOREM 4.6.4. *Let F be an ordinary difference field with a basic set σ and a a finite indexing of elements from some σ-overfield of F. Then almost every specialization of a over F generates a σ-field extension of F compatible with $F\langle a\rangle/F$.*

The last theorem implies that if $\mathcal{M}$ is a non-empty irreducible variety over an ordinary difference field F, then almost every solution in $\mathcal{M}$ has the property of compatibility with a generic zero of $\mathcal{M}$. That is, there exists a difference polynomial $A \notin \Phi(\mathcal{M})$ such that if $(\eta_1, \dots, \eta_s)$ is a generic zero of $\mathcal{M}$ and $(\eta'_1, \dots, \eta'_s)$ a solution in $\mathcal{M}$ not annulling A, then $F\langle\eta_1, \dots, \eta_s\rangle/F$ and $F\langle\eta'_1, \dots, \eta'_s\rangle/F$ are compatible.

THEOREM 4.6.5 [32, Chapter 7, Theorem IV]. *Let F be an ordinary difference field with a basic set σ. Let a and b be finite indexings of elements lying in a σ-overfield of F, u a non-*

zero element of $F\{a, b\}$, *and* B *a* σ*-transcendence basis of* b *over* $F\langle a\rangle$. *If* $F\langle a, b\rangle/F\langle a\rangle$ *is primary, then almost every specialization* $a' = \phi a$ *of* a *over* F *can be extended to a specialization* a', b' *of* a, b *in such a way that*

(i) *The specialization of* u *is not* 0.

(ii) *Every* σ*-transcendence basis of* b *over* $F\langle a\rangle$ *specializes to a* σ*-transcendence basis of* b' *over* $F\langle a'\rangle$.

(iii) *If* B' *denote the specialization of* B, *then* $\text{Eord}\, F\langle a', b'\rangle/F\langle a', B'\rangle = \text{Eord}\, F\langle a, b\rangle/F\langle a, B\rangle$.

Let G be a difference overfield of an ordinary difference field F with a basic set σ. The *core* G_F of G over F is defined to be the set of elements $a \in G$ algebraic and separable over F and such that $ld\, F\langle a\rangle/F = 1$. (It follows from Theorem 4.3.1 that G_F is a σ-field and $ld\, G_F/F = 1$.) Example 4.6.1 shows that a core G_F need not to be F. Furthermore, if $\text{Char}\, F = 0$ or G/F is separable, it follows from Theorem 4.3.4 that $G = G_F$ if and only if $G : F$ is finite.

Suppose now that F is an inversive ordinary difference field with a basic set σ and G an algebraic σ-overfield of F (that is, G is algebraic over F in the usual sense). Since it is possible to define a structure of a σ-overfield of G on the algebraic closure of this field (see Corollary 4.5.8), one can define a structure of a σ-overfield of G on a normal closure G' of G over F. In what follows, this difference field will be called a *normal closure of the* σ*-field* G *over* F. Clearly, the normal closures of G over F are generated by the σ-generators of G/F and their conjugates with respect to F. Hence, if G/F is a finitely generated σ-field extension, then any normal closure of G over F is a finitely generated σ-field extension of F. If G is inversive, the normal closures of G over F are inversive.

PROPOSITION 4.6.6. *Let* F *be an inversive ordinary difference field with a basic set* σ, G *an algebraic* σ*-overfield of* F, *and* G' *a normal closure of* G *over* F. *Then* G'_F *contains a normal closure of* G_F *over* F.

Let F be an ordinary difference field with a basic set $\sigma = \{\alpha\}$ and let G be a finitely generated σ-field extension of F which is an algebraic, normal and separable overfield of F. It is easy to see that one can choose a finite set of σ-generators S of G over F such that $F(S)$ is normal over F. Then there exists an element $v \in G$ such that $F(S) = F(v)$ (hence $G = F\langle v\rangle$) and v is normal over F. Furthermore, there exists $k \in \mathbf{N}$ such that $F(v, \dots, \alpha^k v) : F(v, \dots, \alpha^{k-1} v) = ld\, F\langle v\rangle/F$. Let w be such that $F(w) = F(v, \dots, \alpha^{k-1} v)$. Then w is normal over F, $F\langle w\rangle = G$, and $F(w, \alpha w) : F(w) = ld\, F\langle w\rangle/F$. An element with these properties is called the *standard generator of* G *over* F. If w is a standard generator such that $F(w) : F$ is as small as possible, then w is called a *minimal standard generator*.

With the above assumptions, if there exists an element $u \in G$ such that $G = F\langle u\rangle$, u is normal over F, and $F(u) : F = ld\, G/F$, then G/F is said to be a *benign* σ-field extension. (Clearly, a σ-generator of a benign extension has the properties ascribed to u if and only if it is a minimal standard generator.)

PROPOSITION 4.6.7. *Let* G/F *be a benign ordinary difference field extension with a basic set* σ *and let* u *be a minimal standard generator of* G *over* F. *Then*

(i) *G/F is compatible with every σ-field extension of F.*

(ii) *The replicability of G/F is $\operatorname{ld} G/F$.*

(iii) *If K is a σ-subfield of F, η a set of σ-generators of F over K, and $0 \neq v \in G$, then almost every specialization of η over K can be extended to a specialization of η, u, v over K such that the specialization of v is not 0.*

(iv) *Let $F\langle u'\rangle$ be a σ-field extension of F such that the field extensions $F(u')/F$ and $F(u)/F$ are isomorphic with u corresponding to u'. Then $F\langle u'\rangle/F$ and $F\langle u\rangle$ are isomorphic σ-field extensions with u corresponding to u'.*

Let K be an inversive difference field with a basic set σ and let K_1 and K_2 be σ-subfields of this field whose inversive closures in K coincide with K. Let L_1 and L_2 be σ-field extensions of K_1 and K_2, respectively, and let L_1' and L_2' be the inversive closures of L_1 and L_2, respectively. The σ-field extensions L_1/K_1 and L_2/K_2 are called *equivalent if L_1'/K and L_2'/K are σ-isomorphic.*

THEOREM 4.6.8 (Babbitt's decomposition, [2]). *Let F be an ordinary difference field with a basic set $\sigma = \{\alpha\}$ and let G be a finitely generated σ-field extension of F which is an algebraic, normal and separable overfield of F. Then there exist σ-fields $G_1 \subseteq \cdots \subseteq G_r$, with $G_1 = G_F$, such that G_r/F is equivalent to G/F and for every $i = 2, \ldots, r$, the σ-field G_i is inversive and G_i/G_{i-1} is equivalent to a benign σ-field extension of G_{i-1}.*

The following three propositions on compatibility are consequences of the last theorem.

PROPOSITION 4.6.9. *Let F be an ordinary difference field with a basic set σ and let G and H be two σ-field extensions of F. Then the following statements are equivalent.*

(i) *G/F and H/F are incompatible.*

(ii) *There exist finitely generated σ-field extensions G' and H' of F such that $G' \subseteq G$, $H' \subseteq H$, and G'/F and H'/F are incompatible.*

(iii) *G_F/F and H_F/F are incompatible.*

(iv) *G_F/F and H/F are incompatible.*

PROPOSITION 4.6.10. *Let F be an ordinary difference field with a basic set σ and G a σ-overfield of F such that the field extension G/F is primary. Let H and K be σ-overfields of G such that H_G/G is equivalent to $G\langle H_F\rangle/G$, and K_G/G is equivalent to $G\langle K_F\rangle/G$. Then H/G and K/G are compatible if and only if H/F and K/F are compatible.*

PROPOSITION 4.6.11. *Let F be an ordinary difference field with a basic set σ and G a σ-overfield of F such that $G = F\langle S\rangle$ for some σ-algebraically independent over F set $S \subseteq G$.*

(i) *If H is a σ-overfield of G, then the σ-field extensions H_G/G and $G\langle H_F\rangle/G$ are equivalent.*

(ii) *Let L be a σ-overfield of F such that L/F is equivalent to G/F, and let M and N be two σ-overfields of L. Then M/G and N/G are compatible if and only if M/F and N/F are compatible.*

PROPOSITION 4.6.12. *Let F be an ordinary difference field with a basic set σ, and let a and b be finite indexings of elements lying in some σ-overfield of F. Let $G = F\langle a\rangle$, $H = F\langle a, b\rangle$, and let c be a finite set of σ-generators of H_F over G. Let $u \neq 0$ be an element of $F\{a, b\}$ and B a σ-transcendence basis of b over $F\langle a\rangle$. Then almost every specialization a' of a over F which can be extended to a specialization a', c' of a, c over F, can be extended to a specialization a', b' of a, b with the properties* (i), (ii), (iii) *of Theorem* 4.6.5.

THEOREM 4.6.13. *Let F be an ordinary difference field of zero characteristic with a basic set σ. Let $a = (a^{(1)}, \dots, a^{(s)})$ and b be finite indexings of elements from some σ-overfield of F such that $a^{(1)}, \dots, a^{(s)}$ are σ-algebraically independent over F. Furthermore, let B be a σ-transcendence basis of b over $F\langle a\rangle$ and $0 \neq u$ an element of $F\{a, b\}$. Finally, let $R = F\{y_1, \dots, y_s\}$ be the ring of σ-polynomials in σ-indeterminates $y_1, \dots, y_s$ over F. Then there exists a non-zero σ-polynomial $A \in R$ with the following property: if $c = (c^{(1)}, \dots, c^{(s)})$ is an indexing in a σ-overfield of F, c is not a solution of A, and $F\langle c\rangle/F$ is compatible with $F\langle a, b\rangle/F$, then there is a specialization c, b' of a, b with the properties* (i), (ii), (iii) *of Theorem* 4.6.5.

The following three theorems generalize the correspondent statements for ordinary case (see [32, Chapter 7]) to partial difference field extensions.

THEOREM 4.6.14. *Let F be an inversive difference field with a basic set σ and let G/F and H/F be two σ-field extensions of F. Then*

(i) *If the extension G/F is primary, then there exists a σ-field extension L/F such that G/F and H/F have σ-isomorphisms into L/F with the images of G and H quasi-linearly disjoint over F.*

(ii) *Let G' and H' denote the separable parts of G and H, respectively, over F. Then G/F and H/F are compatible if and only if the σ-field extensions G'/F and H'/F are compatible.*

THEOREM 4.6.15 [7, Theorem 3.6]. *Let F be an inversive difference field with a basic set σ, G/F a primary σ-field extension, and τ a σ-isomorphism of F into G. Then*

(i) *τ has an extension to a σ-isomorphism τ_1 of G into a σ-overfield E of G such that $E = \langle G, \tau_1 G\rangle$, G and $\tau_1 G$ are quasi-linearly disjoint over τF, and E/G is a primary extension. Furthermore, if G is inversive, then E is inversive.*

(ii) *Let E' be a σ-overfield of G such that τ has an extension to a σ-isomorphism τ' of G into E' and E' is the free join of G and $\tau' G$ over τF. Then there exists a unique σ-isomorphism ψ of E/G onto E'/G such that $\psi\tau_1(a) = \tau'\psi(a)$ for any $a \in G$.*

Let K be a difference field with a basic set $\sigma = \{\alpha_1, \dots, \alpha_n\}$. We say that K satisfies the *universal compatibility condition* if every two σ-extensions of K are compatible. K is said to satisfy the *stepwise compatibility condition* if there exists a permutation $(i_1, \dots, i_n)$ of $(1, \dots, n)$ such that all difference fields $(K; \alpha_{i_1}, \dots, \alpha_{i_k})$, $1 \leqslant k \leqslant n$, satisfy the universal compatibility condition. (As in Section 4.3, $(K; \alpha_{i_1}, \dots, \alpha_{i_k})$ denotes the difference field K with the basic set $\{\alpha_{i_1}, \dots, \alpha_{i_k}\}$.)

THEOREM 4.6.16 [7]. *Let K be a difference field with a basic set σ, $\alpha \in \sigma$, and let K' denote the field K treated as a difference field with the basic set $\sigma \setminus \{\alpha\}$. If K' satisfies the stepwise compatibility condition, then there exists a σ-overfield L of K such that L is an algebraic closure of K.*

Although any ordinary difference field K has a difference overfield which is an algebraic closure of K (see Corollary 4.5.8), the following example shows that this result cannot be generalized to partial difference fields. Furthermore, the converse of theorem 4.6.16 is not true (see [7, Example 3.9]).

EXAMPLE 4.6.17 [7]. Let us consider $\mathbf{Q}$ as a difference field with a basic set $\sigma = \{\alpha_1, \alpha_2\}$ where α_1 and α_2 are the identity automorphisms. Let b and i denote the positive square root of 2 and the square root of -1, respectively (we assume $b, i \in \mathbf{C}$), and let $F = \mathbf{Q}(b, i)$. Let us extend α_1 and α_2 to automorphisms of the field F by setting $\alpha_1(b) = -b$, $\alpha_1(i) = i$, $\alpha_2(b) = b$, $\alpha_2(i) = -i$, and consider F as a σ-overfield of $\mathbf{Q}$. Then there exists no σ-overfield of F which is an algebraic closure of F. Indeed, if there were one, then there exists a σ-overfield G of F which contains an element a such that $a^2 = b$. Since $(\alpha_1(a))^2 = \alpha_1(a^2) = -b$ and $(\alpha_2(a))^2 = \alpha_2(a^2) = b$, we have $\alpha_1(a) = \lambda a i$ and $\alpha_2(a) = \mu a$ where λ and μ denote plus or minus 1. Then $\alpha_1\alpha_2(a) = \lambda\mu i a$ and $\alpha_2\alpha_1(a) = -\lambda\mu i a$. Thus, α_1 and α_2 do not commute at a, which contradicts the assumption that G is a σ-overfield of F.

REMARK 4.6.18. As we have seen in Section 4.5, every ordinary algebraically irreducible difference polynomial has an abstract solution. One can use the last example to show that this result cannot be extended to partial difference polynomials. Indeed, let F and b be as in the example, and let $F\{y\}$ be a ring of σ-polynomials in one σ-indeterminate y over F. If a is a solution for $A = y^2 - b \in F\{y\}$, then $a^2 = b$ which, as demonstrated in Example 4.6.17, is impossible.

The following statement is the version of the existence theorem for difference fields with two translations.

THEOREM 4.6.19 [7, Theorem 6.1]. *Let F be an inversive difference field with a basic set σ consisting of two translations. Let $R = F\{y_1, \dots, y_s\}$ be the ring of σ-polynomials in σ-indeterminates $y_1, \dots, y_s$ over F. Furthermore, suppose that there exists a translation $\alpha \in \sigma$ such that if F is treated as a difference field with the basic set $\sigma' = \sigma \setminus \{\alpha\}$ (we denote this σ'-field by F'), then every two σ'-field extensions of F' are compatible.*

Then every algebraically irreducible σ-polynomial $A \in R \setminus F$ has a solution $\eta = (\eta_1, \dots, \eta_s)$ with the following properties.

(i) *η is not a proper specialization over F of any solution of A.*

(ii) *If A contains a transform of some y_k ($1 \leqslant k \leqslant s$), then the elements $\eta_1, \dots, \eta_{k-1}, \eta_{k+1}, \dots, \eta_s$ are σ-algebraically independent over F. Furthermore, if a σ-polynomial $B \in R \setminus F$ is annulled by η and contains only those transforms of y_k which are contained in A, then B is a multiple of A.*

COROLLARY 4.6.20.

(i) *If F and A are is in the last theorem, then A has at most a finite number of isomorphically distinct solutions of the type described in the theorem.*

(ii) *The conclusion of Theorem* 4.6.19 *holds if F is an inversive difference field with two translations and F is separably algebraically closed or algebraically closed.*

COROLLARY 4.6.21. *Let F be a difference field whose basic set σ consists of two translations, and let $R = F\{y_1, \dots, y_s\}$ be the ring of σ-polynomials in σ-indeterminates $y_1, \dots, y_s$ over F. Then the following statements are equivalent*:

(a) *F has an algebraic closure which is a σ-field extension of F.*

(b) *Every algebraically irreducible σ-polynomial $A \in R \setminus F$ has a solution η with the properties stated in Theorem* 4.6.19.

(c) *If a is any element that is separably algebraic and normal over F, then F may be extended to the σ-field $F\langle a\rangle$.*

4.7. *Isomorphisms of difference fields. Monadicity*

We have already seen that difference field isomorphisms cannot be extended as freely as field isomorphisms. The following theorem proved in [32, Chapter 9] gives some conditions under which σ-isomorphisms of a given difference (σ-) field can be extended.

THEOREM 4.7.1. *Let F be an ordinary difference field with a basic set σ and G a σ-overfield of F. If G/F is compatible with every σ-field extension of F (in particular, if $G_F = F$), then every σ-isomorphism of F into a σ-overfield H of G extends to a σ-isomorphism of G into a σ-overfield of H.*

COROLLARY 4.7.2. *Let F be an ordinary difference field with a basic set σ, G a σ-overfield of F, and H a σ-field extension of G. Then the replicability of H/F is not less than the replicability of H_G/F.*

Let F be a difference field with a basic set σ. A σ-field extension G/F is called *monadic* if its replicability is 1, that is, G/F has at most one σ-isomorphism into any σ-field extension of F. It follows from Corollary 4.7.2 that a monadic extension has a monadic core.

A monadic extension G/F is called *properly monadic* if $G \neq F$ and there exists an element $a \in G$ such that no transform of a belongs to F.

PROPOSITION 4.7.3 [2]. *Let F be an inversive difference field with a basic set σ and let G and H be σ-overfields of F. Then*

(i) *G/F and H/F are incompatible if and only if the inversive closures G^* and H^* are incompatible σ-field extensions of F.*

(ii) *The σ-field extension G/F is monadic if and only if G^*/F is monadic.*

EXAMPLE 4.7.4 [32, Chapter 1, Example 6]. Let $G = \mathbf{C}(x)$ be the field of rational fractions in one variable x over $\mathbf{C}$. Let us consider G as an ordinary difference field with a basic set $\sigma = \{\alpha\}$ where $\alpha f(x) = f(x^2)$ for any $f(x) \in F$. If $t = x^3$ and $F = \mathbf{C}(t)$, then the field F can be viewed as a σ-subfield of G. Let us show that G/F is a monadic σ-field extension.

Let ϕ and ψ be two σ-isomorphisms of G/F into some σ-field extension H/F. Let $y = \phi(x)$ and $z = \psi(x)$. Then $y^3 = t$, $\alpha(y) = y^2$, $z^3 = t$, and $\alpha(z) = z^2$. Clearly, in order to prove that $\phi = \psi$, one should show that $z = y$. Since $t = y^3 = z^3$, we find $z = \omega y$ where $\omega^3 = 1$, $\omega \in \mathbf{C} \subseteq F$. It follows that $\alpha(\omega) = \omega$ and $\omega\alpha(y) = \omega^2 y^2$, $\alpha(y) = \omega y^2$ (we use the fact that $\alpha(z) = z^2$). Since $\alpha(y) \neq 0$, $\omega = 1$ whence $z = y$ and $\phi = \psi$.

In what follows we show that a finitely generated separable monadic extension coincides with its core (generally speaking, this is not true for extensions of finite replicability). The proofs of the results of the rest of this section can be found in [2] and in [32, Chapter 9].

PROPOSITION 4.7.5. *Let F be an ordinary difference field with a basic set σ, G an algebraic σ-overfield of F, and G' the separable part of G. Then G/F is monadic if and only if G'/F is monadic.*

THEOREM 4.7.6. *Let F be an inversive ordinary difference field with a basic set σ and G/F a finitely generated monadic σ-field extension of F. Then:*

(i) *G is purely inseparable over its core.*

(ii) *If G/F is separable, $G = G_F$.*

(iii) *$\operatorname{ord} G/F = 0$ and $\operatorname{rld} G/F = 1$. (If* Char $F = 0$, *then $\operatorname{ld} G/F = 1$.)*

THEOREM 4.7.7. *Let F be an inversive ordinary difference (σ-) field, B a σ-algebraically independent set over F, and G the inversive closure of $F\langle B\rangle$. If H/G is a finitely generated, separable monadic σ-field extension of G, then there exists a finitely generated monadic σ-field extension K/F such that $H = G\langle K\rangle$. Conversely, if K/F is monadic (whether finitely generated or not), then H/G is monadic, where $H = G\langle K\rangle$.*

THEOREM 4.7.8. *Let F be an ordinary difference field with a basic set σ and G a σ-overfield of F of finite degree (that is $|G:F| < \infty$). Then G/F is compatible with every σ-field extension of F if and only if it is monadic.*

COROLLARY 4.7.9. *A difference subextension of a finitely generated monadic extension of an ordinary difference field is monadic.*

Notice that the restriction to finitely generated difference field extensions made in the last statement is essential (see [32, Chapter 9, Example 4]).

A difference (σ-) field extension G/F is called *pathological* if either it is incompatible with some other σ-field extension of F or G/F is monadic.

EXAMPLE 4.7.10. Let $F = \mathbf{C}(x)$ be the field of rational fractions in one variable x over $\mathbf{C}$ considered as an ordinary difference field with a basic set $\sigma = \{\alpha\}$ where $\alpha f(x) = f(x+1)$ for any $f(x) \in F$. Then F has no finitely generated pathological extensions (see [2, Theorem 2.9]).

EXAMPLE 4.7.11. Let us consider the field $\mathbf{C}(x)$ from the previous example as an ordinary difference field K with a translation $\alpha: f(x) \mapsto f(qx)$ $(f(x) \in \mathbf{C}(x))$, where q is a non-zero complex number such that $q^n \neq 1$ for every $n \in \mathbf{N}$. By [32, Chapter 9, Theorem XX], a σ-field extension G/K is incompatible with some other σ-field extension of K if and only if G contains a k-th root of x for some $k \in \mathbf{N}, k > 1$. Therefore, K has no finitely generated properly monadic extensions.

THEOREM 4.7.12 [2]. *If an ordinary inversive difference field admits a pathological extension, it admits a pathological extension of finite degree of the same type.*

4.8. *Difference valuation rings and extensions of difference specializations*

Let K be a difference field with a basic set σ. A *maximal difference* (or σ-) *specialization* of K is a σ-homomorphism ϕ of a σ-subring R of K onto a difference (σ-) domain Λ such that ϕ cannot be extended to a σ-homomorphism of a larger σ-subring of K onto a domain which is a σ-overring of Λ. (It can be easily shown that the σ-domain Λ is, in fact, a σ-field.) The σ-domain $R \subseteq K$ is called the *maximal difference* (or σ-) *ring* of K. If K is the quotient field of R, we say that R is a *difference* (or σ-) *valuation ring* of K, and ϕ is called a *difference* (or σ-) *place* of K. It is easy to see that if the σ-field K is inversive, then every its maximal σ-ring is also inversive.

A difference ring R with a basic set σ is called a *local difference* (or σ-) *ring* if the non-units of R form a σ-ideal. This ideal will be denoted by $M(R)$. If R_0 is a local σ-subring of R and $M(R) \cap R_0 = M(R_0)$, we say that R *dominates* R_0. Clearly, a localization R_P of a difference (σ-) ring R by a prime σ-ideal P of R is a local σ-ring.

In what follows, we present some results on extensions of difference specializations that are natural generalizations of the corresponding statements proved in [97] for ordinary difference rings and fields.

PROPOSITION 4.8.1. *Let K be a difference field with a basic set σ and R a local σ-subring of K. Then the following statements are equivalent.*

(i) *R is a maximal σ-ring of K.*
(ii) *R is maximal among local σ-subrings of K ordered by domination.*
(iii) *If $x \in K$ and $x \notin R$, then $1 \in \{R\{x\}M(R)\}$, the perfect σ-ideal generated by $M(R)$ in $R\{x\}$.*

Let K be a difference field with a basic set σ and R a difference valuation ring of K. Then the set U of units of R forms a subgroup of $K' = K \setminus \{0\}$ and one may define the natural homomorphism $v: K' \to K'/U$. Let K'/U be denoted by Υ, with the operation written as addition. Then v is called a *difference* (or σ-) *valuation* of K. Let $\Upsilon^+ = v(M(R) \setminus \{0\})$; then for $a \in \Upsilon^+$ we have $-a \notin \Upsilon^+$. For any $a, b \in \Upsilon$ we define $a < b$ if $b - a \in \Upsilon^+$. Then Υ becomes a partially ordered group which is not necessarily linearly ordered. Clearly, $x \in R \setminus \{0\}$ if and only if $v(x) \geqslant 0$, and $x \in M(R) \setminus \{0\}$ if and only if $v(x) > 0$.

Notice that there are difference valuation rings which are not valuation rings (and, thus, difference valuations which are not valuations), see the example in [97, Section 1].

THEOREM 4.8.2. *Let R be a local difference subring of a difference field K with a basic set σ, $M(R)$ the maximal ideal of R, and $x \in K$. Then the natural σ-homomorphism $\phi: R \to R/M(R)$ extends to one sending x to 0 if and only if $1 \notin [x]$ in $R\{x\}$.*

If R is a maximal difference ring of K, then $x \in M(R)$ if and only if $1 \notin [x]$.

COROLLARY 4.8.3. *Let K be a difference field with a basic set σ and let R be a maximal difference ring of K.*

(i) *If S is another maximal difference ring of K such that $R \subseteq S$, then $M(S) \subseteq M(R)$.*
(ii) *Let P be a prime σ^*-ideal of R. Then there is a maximal difference ring R_1 of K such that $R \subseteq R_1$ and $M(R_1) = P$.*
(iii) *The prime σ^*-ideals of R are linearly ordered by inclusion.*
(iv) *Every perfect σ-ideal of R is prime.*
(v) *Let S be a maximal difference ring of K with a specialization $\phi: S \to \Lambda$, and let $R \subseteq S$. Then $\phi(R)$ is a maximal difference ring of Λ.*

Let K be a difference field with a basic set σ and $K\{y\}$ a ring of σ-polynomials in one σ-indeterminate y over K. Let R be a σ-subring of K and g a σ-polynomial in $R\{y\}$ with a constant term $b \in R$. Furthermore, let $\{g\}_R$ and $\{g\}_K$ denote perfect σ-ideals generated by g in the σ-rings $R\{y\}$ and $K\{y\}$, respectively.

PROPOSITION 4.8.4. *With the above notation, let the σ-ideal $\{g\}_K$ be prime, η a generic zero of $\{g\}_K$, and $\phi: R \to \Lambda$ a difference specialization of K with $\phi(b) = 0$. If $\{g\}_K \cap R\{y\} = \{g\}_R$, then ϕ can be extended to $R\{\eta\}$ with $\phi(\eta) = 0$.*

PROPOSITION 4.8.5. *Let K be a difference field with a basic set σ, R_0 a σ-subring of K with prime σ^*-ideals P and Q such that $P \subseteq Q$. Let S be a proper maximal σ-ring of K with $R_0 \subseteq S$ and $M(S) \cap R_0 = P$. Then there exists a proper maximal σ-ring R of K such that $R_0 \subseteq R$ and $M(R) \cap R_0 = Q$. Furthermore, if S is a σ-valuation ring of K then R is also.*

COROLLARY 4.8.6. *Let K be a difference field with a basic set σ and R_0 a local σ-subring of K. Let L be a σ-overfield of K and S a proper maximal σ-ring (valuation ring) of L containing R_0. Then there exists a proper maximal σ-ring (valuation ring) R of L dominating R_0.*

The existence of S in the corollary is equivalent to the condition that L has a subring S_0, $S_0 \subseteq R_0$, which contains a proper non-zero prime σ^*-ideal. This condition does not always hold. For example, if $\mathbf{Q}$ is treated as an ordinary difference (σ-) field with the identity translation α and $\mathbf{Q}\{b\}$ is a σ-overring of $\mathbf{Q}$ such that b is transcendental over $\mathbf{Q}$ and $\alpha(b) = b$, then the σ-specialization $b \to 1$ does not extend to a σ-place $\mathbf{Q}\langle a\rangle$ where $a^2 = b$ and $\alpha(a) = -a$ (see [32, Chapter 7, Example 3]). However, the condition does hold in the situation of the following proposition.

PROPOSITION 4.8.7. *Let R_0 be a local difference ring with a basic set σ and K the difference quotient field of R_0.*

(i) *Suppose that R_0 does not contain minimal non-zero prime σ^*-ideals. If L is a primary finitely generated σ-field extension of K, $L = K\langle\eta_1, \dots, \eta_s\rangle$, then there exists a difference valuation ring R of L dominating R_0.*

(ii) *Let ζ be an element from some σ-overfield of K which is σ-algebraically independent over K. If $N = K\langle\zeta, \eta_1, \dots, \eta_s\rangle$ is a primary σ-field extension of $K\langle\zeta\rangle$, then there exists a difference valuation ring of N dominating R_0.*

5. Difference Galois theory

5.1. *Algebraic difference field extensions. Galois correspondence*

Let K be an inversive difference field with a basic set $\sigma = \{\alpha_1, \dots, \alpha_n\}$ and L a σ^*-overfield of K. As usual, *Gal*(L/K) denotes the corresponding Galois group, that is, the group of all automorphisms (not necessarily σ-automorphisms) that leave the field K fixed. It is easy to see that the mappings $\bar{\alpha}_i : \theta \mapsto \alpha_i^{-1}\theta\alpha_i$ ($\theta \in Gal(L/K)$, $1 \leqslant i \leqslant n$) are automorphisms of the group *Gal*(L/K); they are called the *induced automorphisms of* *Gal*(L/K). A subgroup B of *Gal*(L/K) is called *σ-stable* if $\bar{\alpha}_i(B) = B$ for $i = 1, \dots, n$. If $\bar{\alpha}_i(b) = b$ for every $b \in B$, $\alpha_i \in \sigma$, the subgroup B is called *σ-invariant*. The largest σ-invariant subgroup of *Gal*(L/K) consists of all σ-automorphisms of L that leave the field K fixed. This group is called the *difference* or *σ-Galois group* of L/K; it is denoted by $Gal_\sigma(L/K)$.

If M/K is a σ^*-field subextension of L/K, then $\{\theta \in Gal(L/K) \mid \theta(a) = a$ for every $a \in M\}$ is a σ-stable subgroup of *Gal*(L/K) denoted by M'. Also, if B is an σ-stable subgroup of *Gal*(L/K), then $\{a \in L \mid \theta(a) = a$ for every $\theta \in B\}$ is a σ^*-overfield of K denoted by B'. As usual, we denote the field $(M')'$ by M'' and the group $(B')'$ by B''.

In what follows, the group *Gal*(L/K) is considered as a topological group with the Krull topology. A fundamental system of neighborhoods of the identity in this topology is the set of all groups $Gal(L/M) \subseteq Gal(L/K)$ such that M is a subfield of L which is a Galois extension of K of finite degree. (When L is of finite degree over K, the Krull topology is discrete.) The topological group $G = Gal(L/K)$ is compact, Hausdorff, and has a basis at the identity consisting of the collection of invariant, open (and hence closed and of finite index in G) subgroups of G. If H is a closed invariant stable subgroup of G, then each $\bar{\alpha}_i$ induces a topological automorphism β_i on G/H such that $\beta_i(gH) = \bar{\alpha}_i(g)H$ for any $g \in G$.

THEOREM 5.1.1 [49]. *Let L, K, and $G = Gal(L/K)$ be as above. Then*

(i) *The mapping $M \mapsto M'$ establishes a 1-1 correspondence between the set of σ^*-field subextensions of L/K and the closed stable subgroups of G. If M/K is a σ^*-field subextension of L/K, then $M'' = M$, and if H is a closed stable subgroup of G, then $H'' = H$.*

(ii) *Let M/K be a σ^*-field subextension of L/K such that M is normal over K. Let $\gamma_1, \dots, \gamma_n$ be the automorphisms of the group Gal(M/K) induced by $\alpha_1, \dots, \alpha_n$ (treated as automorphisms of M/K), respectively. Then there is a natural isomorphism of topological groups $\phi : G/M' \to Gal(M/K)$ such that $\phi\beta_i = \gamma_i\phi$ ($1 \leqslant i \leqslant n$) where β_i is the automorphism of G/M' induced by $\bar{\alpha}_i$.*

DEFINITION 5.1.2. Let M be an ordinary inversive difference field with a basic set σ and N a σ^*-overfield of M. The extension N/M is said to be universally compatible if given a σ^*-field extension Q/M, there exist a σ^*-field extension R/M and σ^*-monomorphisms $\phi : N/M \to R/M$ and $\psi : Q/M \to R/M$.

The following two theorems combine the main results on difference Galois groups, universal compatibility and monadicity obtained in [49].

THEOREM 5.1.3. *Let K be an ordinary inversive difference field with a basic set $\sigma = \{\alpha\}$ and L a σ^*-overfield of K such that L/K is a Galois extension. Then:*

(i) *$Gal_\sigma(L/K)$ is a closed subgroup of $Gal(L/K)$.*

(ii) *Let λ be the mapping of $Gal(L/K)$ into itself such that $\lambda(g) = g^{-1}\bar{\alpha}(g)$ (as before, $\bar{\alpha}$ denotes the automorphism of $Gal(L/K)$ induced by α). Then*
   (a) *λ is a continuous function;*
   (b) *L/K is universally compatible if and only if λ maps $Gal(L/K)$ onto itself;*
   (c) *L/K is monadic if and only if λ is one-to-one on $Gal(L/K)$.*

(iii) *If L/K is universally compatible and M/K a σ^*-subextension of L/K, then M/K is universally compatible.*

(iv) *Let M/K be a σ^*-subextension of L/K such that M/K is a Galois extension. If L/M and M/K are universally compatible then L/K is universally compatible. If L/K is monadic and L/M is universally compatible, then M/K is monadic.*

(v) *L/K is universally compatible if and only if every σ^*-subextension of L/K which is finite-dimensional Galois over K is universally compatible.*

THEOREM 5.1.4. *Let K and L be as in Theorem 5.1.3 and let M/K be a σ^*-subextension of L/K such that M is a Galois extension of K. Then*

(i) *$Gal_\sigma(L/M)$ is a closed normal stable subgroup of $Gal_\sigma(L/K)$.*

(ii) *There exists a natural monomorphism $\psi : Gal_\sigma(L/K)/Gal_\sigma(L/M) \to Gal_\sigma(M/K)$. If L/M is universally compatible then ψ is an isomorphism.*

(iii) *If L/K is universally compatible and ψ is an isomorphism, then L/M is universally compatible.*

5.2. *Picard–Vessiot theory of linear homogeneous difference equations*

Throughout this section all fields have characteristic zero and all difference fields are inversive and ordinary. If K is such a difference field with a basic set $\sigma = \{\alpha\}$, then its field of constants $\{c \in K \mid \alpha(c) = c\}$ will be denoted by C_K. All topological statements of this section will refer to the Zariski topology. As before, $K\{y\}$ will denote the ring of σ-polynomials in one σ-variable y over K. Furthermore, for any n-tuple $b = (b_1, \dots, b_n)$, $C^*(b)$ will denote the determinant of the matrix $(\alpha^i b_j)_{0 \leqslant i \leqslant n-1, 1 \leqslant j \leqslant n}$.

Let us consider a linear homogeneous difference equation of order n over K, that is an algebraic difference equation of the form

$$\alpha^n y + a_{n-1}\alpha^{n-1} y + \cdots + a_0 y = 0, \tag{5.2.1}$$

where $a_0, \dots, a_{n-1}, a_n \in K$ $(n > 0)$ and $a_0 \neq 0$. A σ^*-overfield M of K is said to be a *solution field over K for equation* (5.2.1) or *for the σ-polynomial* $f(y) = \alpha^n y + a_{n-1}\alpha^{n-1} y + \cdots + a_0 y$, if $M = K\langle b\rangle^*$ for an n-tuple $b = (b_1, \dots, b_n)$ such that $f(b_j) = 0$ for $j = 1, \dots, n$ and $C^*(b) \neq 0$. Any such n-tuple b is said to be a *basis* of M/K or a *fundamental system of solutions* of (5.2.1). If, in addition, C_K is algebraically closed and $C_M = C_K$, then M is said to be a *Picard–Vessiot extension* (*PVE*) of K (for equation (5.2.1)).

Note that if $C^*(b) \neq 0$, then the elements $b_1, \dots, b_n$ are linearly independent over the constant field of any difference field containing them (see [32, Chapter 8, Lemma II]).

PROPOSITION 5.2.1 [56]. *With the above notation, let M be a σ^*-overfield of K and $R \subseteq C_M$.*

(i) *A subset of R that is linearly (algebraically) dependent over K is linearly (respectively, algebraically) dependent over C_K.*

(ii) *If N is a σ^*-overfield of K with $C_N = C_K$, then N and $K(R)$ are linearly disjoint over K.*

(iii) $C_{K(R)} = C_K(R)$.

If M is a solution field for equation (5.2.1) over K with basis $b = (b_1, \dots, b_n)$ and b' is any solution of (5.2.1) in a σ^*-overfield N of M, then $b' = \sum_{i=1}^{n} c_j b_j$ for some elements $c_1, \dots, c_n \in C_N$ (see [32, Chapter 8, Theorem XII]). It follows that a σ-homomorphism h of $K\{b\}/K$ into a σ^*-overfield N of M determines an $n \times n$-matrix (c_{ij}) over C_N by the equations $h(b_i) = \sum_{i=1}^{n} c_{ij} b_j$. The following theorem and corollary proved in [56] show that the matrices corresponding to σ-homomorphisms satisfy a set of algebraic equations over C_M, and, in the case of a PVE, form an algebraic matrix group.

THEOREM 5.2.2. *If M/K is a solution field with basis $b = (b_1, \dots, b_n)$, then there is a set S_b in the polynomial ring $C_M[x_{ij}]$ $(1 \leqslant i, j \leqslant n)$ so that if N is a σ^*-overfield of M then the following hold.*

(i) *A σ-homomorphism of $K\{b\}/K$ to N/K determines a matrix over C_N that annuls every polynomial of S_b. (In the last case we say "the matrix satisfies S_b".)*

(ii) *A matrix over C_N satisfying S_b defines a σ-homomorphism of $K\{b\}/K$ to N/K.*

(iii) *If $C_M = C_K$ then a σ-homomorphism of $K\{b\}/K$ to N/K determines a σ-isomorphism if and only if its matrix is non-singular.*

COROLLARY 5.2.3. *If M/K is a PVE then the difference Galois group $Gal_\sigma(M/K)$ is an algebraic matrix group over C_K.*

If M is a solution field for equation (5.2.1) and $b = (b_1, \dots, b_n)$ a basis of M/K, then S_b will denote the set of polynomials in Theorem 5.2.2 ($S_b \subseteq C_M[x_{ij}]$, $1 \leqslant i, j \leqslant n$), and T_b will denote the variety of S_b over the algebraic closure of C_M. The following example shows that a matrix in T_b may not correspond to a difference homomorphism of $K\{b\}$.

EXAMPLE 5.2.4 [56]. Let b be a solution of the difference equation $\alpha y + y = 0$ which is transcendental over K. Then the constant field of $K(b)$ contains $C_K(b^2)$, $S_b = \{0\}$, and T_b

contains the algebraic closure of $C_K(b^2)$. Since no σ^*-overfield of $K\langle b\rangle^*$ contains b in its constant field, Theorem 5.2.2 does not apply to the matrix (b). The algebraic isomorphism $h: K\{b\} \to K\{b\}$ defined by $h(b) = b^2$, is not a σ-homomorphism.

Let L be an intermediate σ^*-field of a difference (σ^*-) field extension M/K (M is a solution field for equation (5.2.1) over K). A σ^*-overfield N of M is said to be a *universal extension of M for L* if every σ-isomorphism of L over K can be extended to a σ-isomorphism of M into N. It follows from Theorem 4.7.1 that if L is algebraically closed in M, then universal extensions of M for L exist. At the same time, Example 5.2.4 shows that even if L itself is a solution field over K, M need not be a universal σ^*-field extension for L.

PROPOSITION 5.2.5. *Let M be a solution field for equation* (5.2.1) *over a difference field K and b a basis of M/K. If the field K is algebraically closed in M, then the variety T_b is irreducible and* $\dim T_b = trdeg_K\, M$.

Let M be a σ^*-overfield of a difference field K with a basic set σ. We say that the extension M/K is *σ-normal* if for every $x \in M \setminus K$, there exists a σ-automorphism ϕ of M such that $\phi(x) \neq x$ and $\phi(a) = a$ for every $a \in K$. (Note that C. Franke, [56], called such extensions "normal" while similar differential field extensions are called "weakly normal".) The existence of proper monadic algebraic difference extensions suggests the existence of solution fields that are not σ-normal extensions.

The following result is a version of the fundamental Galois theorem for PVE. As usual, primes indicate the Galois correspondence.

THEOREM 5.2.6 [56]. *Let M/K be a difference (σ^*-) PVE, $G = Gal_\sigma(M/K)$, L an intermediate σ^*-field of M/K, and H an algebraic subgroup of G. Then*

(i) *L' is an algebraic matrix group.*
(ii) $H'' = H$.
(iii) *If L is algebraically closed in M, then M is σ-normal over L and $L'' = L$.*
(iv) *There is a one-to-one correspondence between intermediate σ^*-fields of M/K that are algebraically closed in M and connected algebraic subgroups of G.*
(v) *Let $\overline{K}$ denote the algebraic closure of K in M. If H is a connected normal subgroup of G, then G/H is the full group of H' over K (that is, G/H is isomorphic to $Gal_\sigma(H'/K)$) and H' is σ-normal over $\overline{K}$.*
(vi) *If L is algebraically closed in M and σ-normal over K, then L' is a normal subgroup of G and G/L' is the full group of L over K.*

Let M be a σ^*-overfield of a difference (σ^*-) field K and H a subgroup of $Gal_\sigma(M/K)$. If L is an intermediate σ^*-field of M/K, then L'_H will denote the group $\{h \in H \mid h(a) = a$ for all $a \in L\} \subseteq H$. A subgroup A of H is said to be *Galois closed in H* if $(A')'_H = A$. An intermediate field N of M/K is said to be *Galois closed with respect to H* if $(N'_H)' = N$.

THEOREM 5.2.7 [56,58]. *Let M be a solution field for a difference equation* (5.2.1) *over a difference (σ-) field K. Let b be a basis of M/K and H a subgroup of $Gal_\sigma(M/K)$ which is naturally isomorphic to the set of matrices T_b corresponding to H. Then*:

(i) *Algebraic subgroups of H are Galois closed in H.*
(ii) *Connected subgroups of H correspond to intermediate σ^*-fields of M/K algebraically closed in M.*
(iii) *Let L be an intermediate σ^*-field of M/K which is algebraically closed in M. Let T_b^L be the variety obtained by considering M as a solution field over L with basis b. If L'_H is dense in T_b^L, then $(L'_H)' = L$ and L'_H is connected.*
(iv) *If the algebraic closure of $K(C_M)$ in M coincides with K, then M/K is a σ-normal extension. In this case, there is a one-to-one correspondence between connected algebraic subgroups of $Gal_\sigma(M/K)$ and intermediate σ^*-fields of M/K algebraically closed in M.*

Some generalization of the last theorem was obtained in [59]. Let K and M be as in Theorem 5.2.7, L an intermediate σ^*-field of M/K, and N a σ^*-overfield of M. Furthermore, let I_L denote the set of all σ-isomorphisms of M into N leaving L fixed.

PROPOSITION 5.2.8. *If L is algebraically closed in M, then I_L is a connected algebraic matrix group. Furthermore, $Gal_\sigma(M/L)$ is dense in I_L and I_L is isomorphic to $Gal_\sigma(M(C_N)/L(C_N))$.*

THEOREM 5.2.9. *Let K and M be as in Theorem 5.2.7, H an algebraic subgroup of $Gal_\sigma(M/K)$, and L an intermediate σ^*-field of M/K. Then*
(i) *$H'' = H$.*
(ii) *If the field L is algebraically closed in M, then $L'' = L$ and M is σ-normal over L.*
(iii) *There is a one-to-one correspondence between connected algebraic subgroups of $Gal_\sigma(M/K)$ and intermediate σ^*-fields of M/K algebraically closed in M.*
(iv) *Assume that H is connected and $L = H'$. In this case*
 (a) *H is a normal subgroup of $Gal_\sigma(M/K)$ if and only if L is σ-normal over K.*
 (b) *If H is a normal subgroup of $Gal_\sigma(M/K)$ and N is any universal extension of M for L, then I_L is a normal subgroup of I_K. The homomorphisms defined by restriction and extension determine natural isomorphisms $Gal_\sigma(M/K)/H \to Gal_\sigma(L/K) \to I_K/I_L$ and the image of $Gal_\sigma(M/K)/H$ is dense in I_K/I_L.*

In general a full difference Galois group $G = Gal_\sigma(M/K)$ is not naturally isomorphic to a matrix group (if $g, h \in G$, then the matrix of the composite of g and h is the matrix of g times the matrix obtained by applying g to the entries of the matrix of h). However, if we adjoin C_M to K and consider M as a solution field over $K(C_M)$, we obtain a group D which is naturally isomorphic to a group of matrices contained in an algebraic variety T. Theorem 5.2.2 implies that T consists only of isomorphisms and singular matrices. The Galois correspondence given in Theorem 5.2.7 for D and fields between $K(C_M)$ and M depends in part on whether a subgroup of D is dense in a variety containing it. Examples where this is not the case are not known.

If M is a solution field for (5.2.1) over K with a basis b, then the subsets of T_b and D consisting of non-singular matrices with entries in C_K are automorphism groups. The following two results obtained in [56] deal with these groups.

PROPOSITION 5.2.10. *Let M be a solution field for the difference equation* (5.2.1) *over a difference (σ-) field K. Let b be a basis of M/K, Λ a subfield of C_M, and S_b the subset of the polynomial ring $C_M[x_{ij}]$ $(1 \leqslant i, j \leqslant n)$ whose existence is established by Theorem* 5.2.2. *Then there exists a set $S'_b \subseteq \Lambda[x_{ij}]$ so that the following hold.*

(i) *Every solution of the set S'_b is a solution of S_b.*
(ii) *Every solution of S_b that lies in Λ is a solution of S'_b.*
(iii) *If Λ is algebraically closed and contained in K, then the variety of S'_b over Λ is an algebraic matrix group of automorphisms of M/K plus singular matrices.*

Note that Theorem 5.2.7 can be applied to any group $G_b^{(1)}$ obtained by deleting the singular matrices from a variety $T_b^{(1)}$ determined as in Proposition 5.2.10 by a basis b and a subfield Λ.

PROPOSITION 5.2.11. *Let K, M and b be as in Proposition* 5.2.10, *and let Λ be an algebraically closed field of constants of K. Let $G_b^{(1)}$ be the group determined by b and Λ as in Proposition* 5.2.10 *and let $C_b^{(1)}$ be the component of the identity of $G_b^{(1)}$. Finally, let C_b be the irreducible subvariety of T_b determined by $\overline{K}$ (the algebraic closure of K in M). The following are equivalent and imply that $\overline{K}$ is Galois closed with respect to $C_b^{(1)}$.*

(i) *$C_b^{(1)}$ is dense in C_b.*
(ii) $\dim C_b = \dim C_b^{(1)}$.
(iii) *There is a basis for the ideal of C_b in the polynomial ring $C[x_{ij}]$ $(1 \leqslant i, j \leqslant n)$.*

Let M be a solution field for difference equation (5.2.1) over a difference (σ-) field K. M/K is called a *generalized Picard–Vessiot extension* (GPVE) if there is a basis b of M/K and an algebraically closed subfield Λ of C_K such that $C_b^{(1)}$ is dense in C_b. M is said to be a *generic solution field* for equation (5.2.1) if $trdeg_K M = n^2$ (n is the order of the difference equation).

PROPOSITION 5.2.12. *Every linear homogeneous difference equation over a difference field K has a generic solution field M. Therefore, if C_K contains an algebraically closed subfield, then every linear homogeneous difference equation over K has a solution field which is a GPVE.*

THEOREM 5.2.13. *If $L = K\langle a\rangle^*$ and $M = K\langle b\rangle^*$ are solution fields of equation* (5.2.1) *over a difference (σ-) field K, then $trdeg_{K(C_L)} L = trdeg_{K(C_M)} M$. Furthermore, if L/K and M/K are compatible, then*

(i) *There is a difference (σ-) field M_1 isomorphic to M and a set of constants R such that $L(R) = M_1(R)$.*
(ii) *If L is a PVE, then there is a specialization $b \to b'$ with $L = K\langle b'\rangle^*$.*
(iii) *If L and M are PVE of K, then L and M are σ-isomorphic over K.*

As in the corresponding theory for the differential case, three types of extensions are used in constructing solution fields for linear homogeneous difference equations over a difference field K: solution fields for difference equations $\alpha y = Ay$ or $\alpha y - y = B$ $(A, B \in K)$, and algebraic extensions. The following is a brief account of this approach (the proofs can be found in [56]).

PROPOSITION 5.2.14. *Let K be a difference field with a basic set $\sigma = \{\alpha\}$, $0 \neq B \in K$, and let a be a solution of the difference equation*

$$\alpha y - y = B. \tag{5.2.2}$$

Then $M = K(a)$ is a corresponding solution field over K with basis $b = (a, 1)$.

If there is no solution of equation (5.2.2) *in K, then a is transcendental over K, $K(a)$ has no new constants, and there are no intermediate σ^*-fields different from K and $K(a)$.*

If there is a solution $f \in K$, then $K(a)$ is an extension of K generated by a constant which may be either transcendental or algebraic over K. If a is transcendental, then T_b is the set of matrices $\left(\begin{smallmatrix} 1 & c \\ 0 & 1 \end{smallmatrix}\right)$ where c lies in the algebraic closure of C_M. If $C_M = C_K$, then the full Galois group of M/K is isomorphic to the additive group of C_K.

PROPOSITION 5.2.15. *Let K be as before, a a non-zero solution of the difference equation*

$$\alpha y - Ay = 0 \tag{5.2.3}$$

$(A \in K)$, *and there is no non-zero solution in K of the equation*

$$\alpha y - A^n y = 0 \tag{5.2.4}$$

for $n \in \mathbf{N}, n > 0$. Then a is transcendental over K and $K(a)$ has no new constants.

If L is an intermediate σ^*-field, then $L = K(a^n)$ for some $n \in \mathbf{N}$.

If equation (5.2.4) *has a solution for some $n \in \mathbf{N}$, $n > 0$, then $K(a)$ is obtained from K by an extension by a constant, which may be either transcendental or algebraic over K, followed by an algebraic extension. If a is transcendental over K, the variety T_a is the full set of all constants. If $C_{K(a)} = C_K$, then the full Galois group of $K(a)/K$ is a multiplicative subgroup of C_K.*

DEFINITION 5.2.16. Let K be a difference field with a basic set σ and N a σ^*-overfield of K. N/K is said to be a Liouvillian extension (LE) if there exists a chain

$$K = K_0 \subseteq K_1 \subseteq \cdots \subseteq K_t = N, \quad K_{j+1} = K_j\langle a_j\rangle^* \ (j = 0, \ldots, t-1), \tag{5.2.5}$$

where a_j is one of the following.

(a) A solution of an equation (5.2.2) where $B \in K_j$ and there is no solution of (5.2.2) in the field K_j.

(b) A solution of an equation (5.2.3) where $A \in K_j$ and for any $n \in \mathbf{N}$, $n > 0$, there is no non-zero solution of (5.2.4) in the field K_j.

(c) An algebraic element over K_j.

More generally, N/K is said to be a generalized Liouvillian extension (GLE) if there exists a chain (5.2.5) where a_j is either a solution of (5.2.2) with $B \in K_j$ or a solution of (5.2.3) with $A \in K_j$, or an algebraic element over K_j.

It follows from the definition that if a difference field K has an algebraically closed field of constants and a solution field M for a difference equation (5.2.1) is contained in a Liouvillian extension N of K, then M is a PVE of K.

The following results connect the solvability of a difference equation with the solvability of a matrix group.

THEOREM 5.2.17. *Let M be a solution field for the difference equation* (5.2.1) *over a difference (σ-) field K. Let H be a connected group of automorphisms of M/K with matrix entries with respect to some basis b in an algebraically closed subfield of C_M. (It need not be isomorphic to the set of matrices corresponding to H.)*

(i) *If H is solvable, then M/H' is a GLE.*
(ii) *If H is reducible to diagonal form, then M/H' can be obtained by solving equations of the type* (5.2.3).
(iii) *If H is reducible to special triangular form, then M/H' can be obtained by solving equations of the type* (5.2.2).

THEOREM 5.2.18. *Let K be a difference field with a basic set $\sigma = \{\alpha\}$ and M a σ^*-overfield of K.*

(i) *If M/K is a PVE, then M/K is a GLE if and only if the component of identity of the Galois group is solvable.*
(ii) *If M is a solution field of a difference equation* (5.2.1) *contained in a GLE N/K, then the component of the identity of $Gal(M/K(C_M))$ is solvable. Furthermore, if $M/K(C_M)$ is a GPVE, then M/K is a GLE.*
(iii) *Suppose that M and L are solution fields of a difference equation* (5.2.1) *over K. If L is contained in a GLE N of K and M/K is compatible with N/K, then M is contained in a GLE of K.*
(iv) *If N is a generic solution field for* (5.2.1) *over K and a solution field L for* (5.2.1) *is contained in a GLE of K, then N is contained in a GLE of K.*

The following example indicates that it is not satisfactory to consider equation (5.2.1) to be "solvable by elementary operations" only if its solution field is contained in a GLE.

EXAMPLE 5.2.19. With the notation of Theorem 5.2.18, suppose that K contains an element j with $\alpha(j) \neq j$ and $\alpha^2(j) = j$, and an element u with the following property. If $u^k = v\alpha(v)$ or $u^k = \frac{\alpha(v)}{v}$ for some $v \in K$, $k \in \mathbf{N}$, then $k = 0$.

If η is any non-zero solution of the difference equation $\alpha^2 y - uy = 0$, then $M = K\langle\eta\rangle^*$ is a solution field for this equation with basis $(\eta, \alpha(\eta))$, $trdeg_K M = 2$, $C_M = C_K$ and $Gal_\sigma(M/K)$ is commutative. However, as it is shown in [57, Example 1], M is not a GLE of K.

In what follows we consider some results by C. Franke (see [57] and [62]) that characterize the solvability of a difference equation of the form (5.2.1) "by elementary operations".

Throughout the rest of the section K denotes an inversive difference field with a basic set $\sigma = \{\alpha\}$. If L is a σ^*-overfield of K, then K_L will denote the algebraic closure of $K(C_L)$ in L.

DEFINITION 5.2.20. Let N be a σ^*-overfield of K, and q a positive integer. A q-chain from K to N is a sequence of σ^*-fields $K = K_1 \subseteq K_1 \subseteq \cdots \subseteq K_t = N$, $K_{i+1} = K_i\langle\eta_i\rangle^*$ where η_i is one of the following.

(a) Algebraic over K.
(b) A solution of an equation $\alpha^q y = y + B$ for some $B \in K_i$.
(c) A solution of an equation $\alpha^q y = Ay$ for some $A \in K_i$.

If there is a q-chain from K to N, then N is called a qLE of K.

Let $K^{(q)}$ denote the field K treated as an inversive difference field with basic set $\sigma_q = \{\alpha^q\}$ and let $N^{(q)}$ be a σ^*-overfield N of K treated as a σ_q^*-overfield of $K^{(q)}$. In this case N is a qLE of K if and only if $N^{(q)}$ is a GLE of $K^{(q)}$ (see [57, Proposition 2.1]).

THEOREM 5.2.21. *If M is a σ-normal σ^*-overfield of K such that $K = K_M$ and the group $Gal_\sigma(M/K)$ is solvable, then M is contained in a qLE of K. If, in addition, M/K is a σ^*-field extension generated by a fundamental system of solutions of a difference equation* (5.2.1), *then M is contained in a GLE of K.*

THEOREM 5.2.22. *Let N be a qLE of K and L an intermediate σ^*-field of N/K. Then $Gal_\sigma(L/K_L)$ is solvable.*

Let M be a σ^*-field extension of K. We say that difference equation (5.2.1) is *solvable by elementary operations in M over K* if M is a solution field for (5.2.1) over K and M is contained in a qLE of K. This concept is independent of the solution field M, as follows from the second statement of the next theorem.

THEOREM 5.2.23. *Let a σ^*-field M be a solution field for* (5.2.1) *over K.*

(i) *Equation* (5.2.1) *is solvable by elementary operations in M over K if and only if the group $Gal_\sigma(M/K)$ has a subnormal series whose factors are either finite or commutative.*
(ii) *If* (5.2.1) *is solvable by elementary operations in M over K and N is another solution field for* (5.2.1) *over K, then* (5.2.1) *is solvable by elementary operations in N over K. (This property allows one to say that* (5.2.1) *is solvable by elementary operations over K if it is solvable by elementary operations in some solution field $M \supseteq K$.)*
(iii) *If* (5.2.1) *is solvable by elementary operations over K and L a σ^*-overfield of K, then* (5.2.1) *is solvable by elementary operations over L.*

A number of results that specify the results of this section for the case of second-order difference equations were obtained in [56] and [57]. C. Franke, [56], also showed that the properties of having algebraically closed field of constants and having full sets of solutions of difference equations can be incompatible. Indeed, if K is a difference field with basic set

$\sigma = \{\alpha\}$ (Char $K \neq 2$) such that C_K is algebraically closed, then the difference equation $\alpha y + y = 0$ has no non-zero solution in K. (If b is such a solution, then $\alpha(b^2) = b^2$, so $b^2 \in C_K$. Since C_K is algebraically closed, $b \in C_K$ contradicting the fact that $\alpha(b) = -b$.)

This observation and the fact that one could not associate a Picard–Vessiot-type extension to every difference equation have led to a different approach to the Galois theory of difference equations. This approach, based on the study of simple difference rings rather than difference fields, was realized by M. van der Put and M.F. Singer in their monograph [142]. In the next section we give an outline of the corresponding theory.

We conclude this section with one more theorem on the Galois correspondence for difference fields (see Theorem 5.2.27 below). This result is due to R. Infante who developed the theory of strongly normal difference field extensions (see [70–74]). Under some natural assumptions, the class of such extensions of a difference field K includes, in particular, the class of solution fields of linear homogeneous difference equations over K.

Let K be an ordinary inversive difference field of zero characteristic with basic set $\sigma = \{\alpha\}$. Let M be a finitely generated σ^*-overfield of K such that K is algebraically closed in M and $C_M = C_K = C$. As above, K_M will denote the algebraic closure of $K(C_M)$ in M. Furthermore, for any σ-isomorphism ϕ of M/K into a σ^*-overfield of M, C_ϕ will denote the field of constants of $M\langle\phi M\rangle^*$.

DEFINITION 5.2.24. With the above conventions, M is said to be a strongly normal extension of K if for every σ-isomorphism ϕ of M/K into a σ^*-overfield of M, $M\langle C_\phi\rangle^* = M\langle\phi M\rangle^* = \phi M\langle C_\phi\rangle^*$.

PROPOSITION 5.2.25. *Let M be a solution field of a difference equation of the form* (5.2.1) *over K. Then M is a strongly normal extension of K_M.*

PROPOSITION 5.2.26. *Let M be a strongly normal σ^*-field extension of K. Then*

(i) $ld\, M/K = 1$.

(ii) *If ϕ is any σ-isomorphism of M/K into a σ^*-overfield of M, then C_ϕ is a finitely generated extension of C and $trdeg_M\, M\langle\phi M\rangle^* = trdeg_C\, C_\phi$.*

THEOREM 5.2.27. *If M is a strongly normal σ^*-field extension of K, then there is a connected algebraic group G defined over C_M such that the connected algebraic subgroups of G are in one-to-one correspondence with the intermediate σ^*-fields of M/K algebraically closed in M. Furthermore, there is a field of constants C' such that C'-rational points of G are all the σ-isomorphisms of M/K into $M\langle C'\rangle^*$ and this set is dense in G.*

5.3. *Picard–Vessiot rings and the Galois theory of difference equations*

In this section we discuss some basic results of the Galois theory of difference equations based on the study of simple difference rings associated with such equations. The complete theory is presented in [142] where one can find the proofs of all statements of this section.

All difference rings and fields considered below are supposed to be ordinary and inversive. The basic set of a difference ring will be always denoted by σ and the only element of σ will be denoted by ϕ (we follow the notation of [142]). As usual, $GL_n(R)$ will denote the set of all non-singular $n \times n$-matrices over a ring R.

Let R be a difference ring, $A \in GL_n(R)$, and Y a column vector $(y_1, \dots, y_n)^{\mathrm{T}}$ whose coordinates are σ^*-indeterminates over R. (The corresponding ring of σ^*-polynomials is still denoted by $R\{y_1, \dots, y_n\}^*$.) In what follows, we will study systems of difference equations of the form $\phi Y = AY$ where $\phi Y = (\phi y_1, \dots, \phi y_n)^{\mathrm{T}}$. Notice that an n-th order linear difference equation $\phi^n y + \dots + a_1 \phi y + a_0 y = 0$ ($a_0, a_1, \dots \in R$ and y is a σ^*-indeterminate over R) is equivalent to such a system with $y_i = \phi^{i-1} y$ $(i = 1, \dots, n)$ and

$$A = \begin{pmatrix} 0 & 1 & 0 & \dots & 0 \\ 0 & 0 & 1 & \dots & 0 \\ \dots & \dots & \dots & \dots & \dots \\ -a_0 & -a_1 & -a_2 & \dots & -a_{n-1} \end{pmatrix}.$$

Clearly, $A \in GL_n(R)$ if and only if $a_0 \neq 0$.

With the above notation, a *fundamental matrix* with entries in R for $\phi Y = AY$ is a matrix $U \in GL_n(R)$ such that $\phi U = AU$ (ϕU is the matrix obtained by applying ϕ to every entry of U). If U and V are fundamental matrices for $\phi Y = AY$, then $V = UM$ for some $M \in GL_n(C_R)$ since $U^{-1}V$ is left fixed by ϕ. (As in Section 5.2, C_R denotes the ring of constants of R, that is, $C_R = \{a \in R \mid \phi a = a\}$.)

DEFINITION 5.3.1. Let K be a difference field. A K-algebra R is called a Picard–Vessiot ring (PVR) for an equation

$$\phi Y = AY \quad (A \in GL_n(K)) \tag{5.3.1}$$

if it satisfies the following conditions.

(i) R is a σ^*-K-algebra (as usual, the automorphism of R which extends ϕ is denoted by the same letter).

(ii) R is a simple difference ring, that is, the only difference ideals of R are (0) and R.

(iii) There exists a fundamental matrix for $\phi Y = AY$ with entries in R.

(iv) R is minimal in the sense that no proper subalgebra R satisfies (i)–(iii).

EXAMPLE 5.3.2 (see [142, Examples 1.3 and 1.6]). Let C be an algebraically closed field, $\operatorname{Char} C \neq 2$. Let us define an equivalence relation on the set of all sequences $a = (a_0, a_1, \dots)$ of elements of C by saying that a is equivalent to $b = (b_0, b_1, \dots)$ if there exists $N \in \mathbf{N}$ such that $a_n = b_n$ for all $n > N$. With coordinatewise addition and multiplication, the set of all equivalence classes forms a ring S. This ring can be treated as a difference ring with respect to its automorphism ϕ that maps an equivalent class of $(a_0, a_1, \dots)$ to the equivalent class of $(a_1, a_2, \dots)$. (It is easy to check that ϕ is well-defined.) To simplify notation we shall identify a sequence a with its equivalence class.

Let R be the difference subring of S generated by C and $j = (1, -1, 1, -1, \dots)$, that is, $R = C[j]^*$. The 1×1-matrix whose only entry is j is the fundamental matrix of the equation $\phi y = -y$. This ring is isomorphic to $C[X]/(X^2 - 1)$ ($C[X]$ is the polynomial ring in one indeterminate X over C) whose only non-trivial ideals are generated by the cosets of $X-1$ and $X+1$. Since the ideals generated in R by $j+1$ and $j-1$ are not difference ideals, R is a simple difference ring. Therefore, R is a PVR for $\phi y = -y$ over C. Note that R is reduced but it is not an integral domain.

PROPOSITION 5.3.3. *Let K be a difference (σ^*-) field with an algebraically closed field of constants C_K.*

(i) *If a σ^*-K-algebra R is a simple difference ring finitely generated as a K-algebra, then $C_R = C_K$.*

(ii) *If R_1 and R_2 are two PVR's for a difference equation* (5.3.1), *then there exists a σ-isomorphism between R_1 and R_2 that leaves the field K fixed.*

To form a PVR for a difference equation (5.3.1) one can use the following procedure suggested in [142, Chapter 1]. Let (X_{ij}) denote an $n \times n$-matrix of indeterminates over K and *det* denote the determinant of this matrix. Then one can extend ϕ to an automorphism of the K-algebra $K[X_{ij}, det^{-1}]$ by setting $(\phi X_{ij}) = A(X_{ij})$. If I is a maximal difference ideal of $K[X_{ij}, det^{-1}]$ then $K[X_{ij}, det]/I$ is a PVR for (5.3.1), it satisfies all conditions of Definition 5.3.1. (It is easy to see that I is a radical σ^*-ideal and $K[X_{ij}, det]/I$ is a reduced prime difference ring.) Moreover, any PVR for difference equation (5.3.1) will be of this form.

Let $\overline{K}$ denote the algebraic closure of K and let $D = \overline{K}[X_{ij}, det^{-1}]$. Then the automorphism ϕ extends to an automorphism of $\overline{K}$ which, in turn, extends to an automorphism of D such that $(\phi X_{ij}) = A(X_{ij})$ (the extensions of ϕ are also denoted by ϕ). It is easy to see that every maximal ideal M of D has the form $(X_{11} - b_{11}, \dots, X_{nn} - b_{nn})$ and corresponds to a matrix $B = (b_{ij}) \in GL_n(\overline{K})$. Then $\phi(M)$ is a maximal ideal of D that corresponds to the matrix $A^{-1}\phi(B)$ where $\phi(B) = (\phi(b_{ij}))$. Thus, the action of ϕ on D induces a map τ on $GL_n(\overline{K})$ such that $\tau(B) = A^{-1}\phi(B)$. The elements $f \in D$ are seen as functions on $GL_n(\overline{K})$. For any $f \in D$, $B \in GL_n(\overline{K})$, we have $(\phi f)(\tau(B)) = \phi(f(B))$. Furthermore, if J is an ideal of $K[X_{ij}, det^{-1}]$ such that $\phi(J) \subseteq J$, then $\phi(J) = J$. Also, for reduced algebraic subsets Z of $GL_n(K)$, the condition $\tau(Z) \subseteq Z$ implies $\tau(Z) = Z$.

PROPOSITION 5.3.4 [142, Lemma 1.10]. *The ideal J of a reduced algebraic subset Z of $GL_n(K)$ satisfies $\phi(J) = J$ if and only if $Z(\overline{K})$ satisfies $\tau Z(\overline{K}) = Z(\overline{K})$.*

An ideal I maximal among the ϕ-invariant ideals corresponds to a minimal (reduced) algebraic subset Z of $GL_n(K)$ such that $\tau Z(\overline{K}) = Z(\overline{K})$. Such a set is called a *minimal τ-invariant reduced set.*

Let Z be a minimal τ-invariant reduced subset of $GL_n(K)$ with an ideal $I \subseteq K[X_{ij}, det^{-1}]$ and let $O(Z) = K[X_{ij}, det^{-1}]/I$. Let us denote the image of X_{ij} in $O(Z)$ by x_{ij} and consider the rings

$$K\left[X_{ij}, \frac{1}{det}\right] \subseteq O(Z) \otimes_K K\left[X_{ij}, \frac{1}{det(X_{ij})}\right]$$
$$= O(Z) \otimes_C C\left[Y_{ij}, \frac{1}{det(Y_{ij})}\right] \supseteq C\left[Y_{ij}, \frac{1}{det(Y_{ij})}\right]. \qquad (5.3.2)$$

Let (I) denote the ideal of $O(Z) \otimes_K K[X_{ij}, det^{-1}]$ generated by I and let $J = (I) \cap C[Y_{ij}, det^{-1}]$. The ideal$(I)$ is ϕ-invariant, the set of constants of $O(Z)$ is C, and J generates the ideal (I) in $O(Z) \otimes_K K[X_{ij}, det^{-1}]$. Furthermore, one has natural mappings

$$O(Z) \to O(Z) \otimes_K O(Z) = O(Z) \otimes_C \left(C\left[Y_{ij}, \frac{1}{det(Y_{ij})} \right] / J \right) \leftarrow C\left[Y_{ij}, \frac{1}{det(Y_{ij})} \right] / J. \quad (5.3.3)$$

Suppose that $O(Z)$ is a separable extension of K (for example, $\operatorname{Char} K = 0$ or K is perfect). One can show (see [142, Section 1.2]) that $O(Z) \otimes_K O(Z)$ is reduced. Therefore, $C[Y_{ij}, det(Y_{ij})^{-1}]/J$ is reduced and J is a radical ideal. Furthermore, the following considerations imply that J is the ideal of an algebraic subgroup of $GL_n(C)$.

Let $A \in GL_n(C)$ and let δ_A denote the action on the terms of (5.3.2) defined by $(\delta_A X_{ij}) = (X_{ij})A$ and $(\delta_A Y_{ij}) = (Y_{ij})A$. Then the following eight properties are equivalent:

(1) $ZA = Z$;
(2) $ZA \cap Z \neq \emptyset$;
(3) $\delta_A I = I$;
(4) $I + \delta_A I$ is not the unit ideal of $K[X_{ij}, det^{-1}]$;
(5) $\delta_A(I) = (I)$;
(6) $(I) + \delta_A(I)$ is not the unit ideal of $O(Z) \otimes K[X_{ij}, det^{-1}]$;
(7) $\delta_A J = J$;
(8) $J + \delta_A J$ is not the unit ideal of $O(Z) \otimes C[Y_{ij}, det^{-1}]$.

The set of all matrices $A \in GL_n(C)$ satisfying the equivalent conditions (1)–(8) form a group.

PROPOSITION 5.3.5 (see [142, Lemma 1.12]). *Let $O(Z)$ be a separable extension of K. With the above notation, A satisfies the equivalent conditions* (1)–(8) *if and only if A lies in the reduced subspace V of $GL_n(C)$ defined by J. Therefore, the set of such A is an algebraic group.*

Let G denote the group of all automorphisms of $O(Z)$ over K which commute with the action of ϕ. The group G is called the *difference Galois group* of the equation $\phi(Y) = AY$ over the field K.

If $\delta \in G$, then $(\delta x_{ij}) = (x_{ij})A$ where $A \in GL_n(C)$ is such that δ_A (as defined above) satisfies $\delta_A I = I$. Therefore, one can identify G and the subspace V from the last proposition. Denoting the ring $C[Y_{ij}, det^{-1}]/J$ by $O(G)$ and setting $O(G_k) = O(G) \otimes_C k$, $G_K = spec(O(G_k))$, one can use (5.3.3) to obtain the sequence

$$O(Z) \to O(Z) \otimes_K O(Z) = O(Z) \otimes_C O(G) = O(Z) \otimes_K O(G_K). \quad (5.3.4)$$

The first embedding of rings corresponds to the morphism $Z \times G_K \to Z$ given by $(z, g) \mapsto zg$. The identification $O(Z) \otimes_K O(Z) = O(Z) \otimes_C O(G) = O(Z) \otimes_K O(G_K)$

corresponds to the fact that the morphism $Z \times G_K \to Z \times Z$ given by $(z, g) \mapsto (zg, z)$ is an isomorphism. Thus, Z is a K-homogeneous space for G_K, that is Z/K is a G-*torsor*. The following result (proved in [142, Section 1.2]) shows that a PVR is the coordinate ring of a torsor for its difference Galois group.

THEOREM 5.3.6. *Let R be a separable PVR over K, a difference field with an algebraically closed field of constants C. Let G denote the group of the K-algebra automorphisms of R which commute with ϕ. Then*

(i) *G has a natural structure as reduced linear algebraic group over C and the affine scheme Z over K has the structure of a G-torsor over K.*

(ii) *The set of G-invariant elements of R is K and R has no proper, non-trivial G-invariant ideals.*

(iii) *There exist idempotents $e_0, \ldots, e_{t-1} \in R$ $(t \geqslant 1)$ such that*

 (a) *$R = R_0 \oplus \cdots \oplus R_{t-1}$ where $R_i = e_i R$ for $i = 0, \ldots, t-1$.*

 (b) *$\phi(e_i) = e_{i+1} \pmod t$ and so ϕ maps R_i isomorphically onto $R_{i+1} \pmod t$ and ϕ^t leaves each R_i invariant.*

 (c) *For each i, R_i is a domain and is a Picard–Vessiot extension of $e_i K$ with respect to ϕ^t.*

Let K be a difference field with an algebraically closed field of constants C and R a PVR for an equation $\phi(Y) = AY$ over K. Let $\delta = \delta_A$ and let $R = R_0 \oplus \cdots \oplus R_{t-1}$ $(R_i = e_i R$ for $i = 0, \ldots, t-1)$ be as in the last theorem. Then $\delta : R_i \to R_{i+1}$ is an isomorphism and R_0 is a PVR over K with respect to the automorphism δ^t. Let us define two mappings $\Gamma : Gal(R_0/K) \to Gal(R/K)$ and $\Delta : Gal(R/K) \to \mathbf{Z}/t\mathbf{Z}$ as follows. For any $\psi \in Gal(R_0/K)$, we set $\Gamma(\psi) = \chi$ where for $r = (r_0, \ldots, r_{t-1}) \in R$, $\chi(r_0, \ldots, r_{t-1}) = (\psi(r_0), \delta\psi\delta^{-1}(r_1), \ldots, \delta^{t-1}\psi\delta^{1-t}(r_{t-1}))$. In order to define Δ, notice that if $\chi \in Gal(R/K)$, then χ permutes with each e_i. If $\chi(e_0) = e_j$, we define $\Delta(\chi) = j$.

PROPOSITION 5.3.7. *Let R be a separable PVR over K, a difference field with an algebraically closed field of constants C.*

(i) *With the above notation, we have the exact sequence $0 \to Gal(R_0/K) \xrightarrow{\Gamma} Gal(R/K) \xrightarrow{\Delta} \mathbf{Z}/t\mathbf{Z} \to 0$.*

(ii) *Let G denote the difference Galois group of R over K. If $H^1(Gal(\overline{K}/K), G(\overline{K})) = 0$, then $Z = spec(R)$ is G-isomorphic to the G-torsor G_K and so $R = C[G] \otimes K$.*

In what follows we present a characterization of the difference Galois group of a PVR over the field of rational functions $C(z)$ in one variable z over an algebraically closed field C of zero characteristic (one can assume $C = \mathbf{C}$). We fix $a \in C(z), a \neq 0$, and consider $C(z)$ as an ordinary difference field with the basic automorphism $\phi_a : z \mapsto z + a$ (ϕ_a leaves the field C fixed). This difference field will be denoted by K. Note that ϕ_a does not extend to any proper finite field extension of K (see [142, Lemma 1.19]).

THEOREM 5.3.8. *Let $K = C(z)$ be as above and let G be an algebraic subgroup of $GL_n(C)$. Let $\phi(Y) = AY$ be a system of difference equations with $A \in G(K)$. Then*

(i) *The Galois group of* $\phi(Y) = AY$ *over* K *is a subgroup of* G_C.
(ii) *Any minimal element in the set of* C*-subgroups* H *of* G *for which there exists a* $B \in Gl_n(K)$ *with* $B^{-1}A^{-1}\phi(B) \in H(K)$ *is the difference Galois group of* $\phi(Y) = AY$ *over* K.
(iii) *The difference Galois group of* $\phi(Y) = AY$ *over* K *is* G *if and only if for any* $B \in G(K)$ *and any proper* C*-subgroup* H *of* G, *one has* $B^{-1}A^{-1}\phi(B) \notin H(K)$.

DEFINITION 5.3.9. Let K be an ordinary difference field with a basic set $\sigma = \{\phi\}$ and let $A \in Gl_n(K)$. A difference overring L of K is said to be the total Picard–Vessiot ring (TPVR) of the equation $\phi(Y) = AY$ over K if L is the total ring of fractions of the PVR R of the equation.

As we have seen, a PVR R is a direct sum of domains: $R = R_0 \oplus \cdots \oplus R_{t-1}$ where each R_i is invariant under the action of ϕ^t. The automorphism ϕ of R permutes $R_0, \ldots, R_{t-1}$ in a cyclic way (that is, $\phi(R_i) = R_{i+1}$ for $i = 1, \ldots, t-2$ and $\phi(R_{t-1}) = R_0$). It follows that the TPVR L is the direct sum of fields: $L = L_0 \oplus \cdots \oplus L_{t-1}$ where each L_i is the field of fractions of R_i, and ϕ permutes $L_0, \ldots, L_{t-1}$ in a cyclic way.

PROPOSITION 5.3.10. *With the above notation, let* K *be a perfect difference field with an algebraically closed field of constants* C *and let* $\phi(Y) = BY$ *be a difference equation over* K $(B \in Gl_n(K))$. *Let a difference ring extension* $K' \supseteq K$ *have the following properties*:
(i) K' *has no nilpotent elements and every non-zero divisor of* K' *is invertible*.
(ii) *The set of constants of* K' *is* C.
(iii) *There is a fundamental matrix* F *for the equation with entries in* K'.
(iv) K' *is minimal with respect to* (i), (ii), *and* (iii).
Then K' *is* K*-isomorphic as a difference ring to the TPVR of the equation.*

COROLLARY 5.3.11. *Let* K *be as in the last proposition and let* $\phi(Y) = AY$ *be a difference equation over* K $(A \in Gl_n(K))$. *Let a difference overring* $R \supseteq K$ *have the following properties.*
(i) R *has no nilpotent elements.*
(ii) *The set of constants of the total quotient ring of* R *is* C.
(iii) *There is a fundamental matrix* F *for the equation with entries in* R.
(iv) R *is minimal with respect to* (i), (ii), *and* (iii).
Then R *is a PVR of the equation.*

With the above notation, let $R = R_0 \oplus \cdots \oplus R_{t-1}$ be the PVR of the equation $\phi(Y) = AY$ $(A \in Gl_n(K))$. Let us consider the difference field (K, ϕ^t) (that is, the field K treated as a difference field with the basic set $\sigma_t = \{\phi^t\}$) and the difference equation $\phi^t(Y) = A_tY$ with $A_t = \phi^{t-1}(A) \ldots \phi^2(A)\phi(A)A$.

PROPOSITION 5.3.12.
(i) *Each component* R_i *of* R *is a PVR for the equation* $\phi^t(Y) = A_tY$ *over the difference field* (K, ϕ^t).

(ii) *Let $d \geqslant 1$ be a divisor of t. Using cyclic notation for the indices $\{0, \ldots, t-1\}$, we consider the subrings $\bigoplus_{m=0}^{(t/d)-1} R_{i+md}$ of $R = R_0 \oplus \cdots \oplus R_{t-1}$. Then each of these subrings is a PVR for the equation $\phi^d(Y) = A_d Y$ over the difference field (K, ϕ^d).*

PROPOSITION 5.3.13. *Let L be the TPVR of an equation $\phi(Y) = AY$ $(A \in Gl_n(K))$ over a perfect difference field K whose field of constants $C = C_K$ is algebraically closed. Let G denote the difference Galois group of the equation and let H be an algebraic subgroup of G. Then G acts on L and moreover:*

(i) *L^G, the set of G-invariant elements of L, is equal to K.*

(ii) *If $L^H = K$, then $H = G$.*

The following result describes the Galois correspondence for total Picard–Vessiot rings. As is noticed in [142, Section 1.3], one cannot expect a similar theorem for Picard–Vessiot rings. Indeed, let K be as in the last proposition and $R = K \otimes_C C[G]$ where $C[G]$ is the ring of regular functions on an algebraic group G defined over C. For an algebraic subgroup H of G, the ring of invariants R^H is the ring of regular functions on $(G/H)_K$. In some cases, e.g., $G = Gl_n(C)$ and H a Borel subgroup, the space G/H is a connected projective variety and so the ring of regular functions on $(G/H)_K$ is just K.

THEOREM 5.3.14. *Let K be a difference field of zero characteristic with a basic set $\sigma = \{\phi\}$. Let $A \in GL_n(K)$ and let L be a TVPR of the equation $\phi(Y) = AY$ over K. Let $\mathcal{F}$ denote the set of intermediate difference rings F such that $K \subseteq F \subseteq L$ and every non-zero divisor of F is a unit of F. Furthermore, let $\mathcal{G}$ denote the set of algebraic subgroups of G.*

(i) *For any $F \in \mathcal{F}$, the subgroup $G(L/F) \subseteq G$ of the elements of G which fix F pointwise, is an algebraic subgroup of G.*

(ii) *For any algebraic subgroup H of G, the ring L^H belongs to $\mathcal{F}$.*

(iii) *Let $\alpha : \mathcal{F} \to \mathcal{G}$ and $\beta : \mathcal{G} \to \mathcal{F}$ denote the maps $F \mapsto G(L/F)$ and $H \mapsto L^H$, respectively. Then α and β are each others inverses.*

COROLLARY 5.3.15. *With the notation of Theorem* 5.3.14, *a group $H \in \mathcal{G}$ is a normal subgroup of G if and only if the difference ring $F = L^H$ has the property that for every $z \in F \setminus K$, there is an automorphism δ of F/K which commutes with ϕ and satisfies $\delta z \neq z$. If $H \in \mathcal{G}$ is normal, then the group of all automorphisms δ of F/K which commute with ϕ is isomorphic to G/H.*

COROLLARY 5.3.16. *With the above notation, suppose that an algebraic group $H \subseteq G$ contains G^0, the component of the identity of G. Then the difference ring R^H (R is a PVR for the equation $\phi(Y) = AY$ over K) is a finite dimension vector space over K with dimension equal to $G : H$.*

A number of applications of the above-mentioned results on ring-theoretical difference Galois theory to various types of algebraic difference equations can be found in [64–67] and [142]. Using the technique of difference Galois groups, the monograph [142] also develops the analytic theory of ordinary difference equations over the fields $\mathbf{C}(z)$ and $\mathbf{C}(\{z^{-1}\})$.

We conclude this section with a fundamental result on the inverse problem of ring-theoretical difference Galois theory.

THEOREM 5.3.17 [142, Theorem 3.1]. *Let $K = C(z)$ be the field of fractions of one variable z over an algebraically closed field C of zero characteristic. Consider K as an ordinary difference field with respect to the automorphism ϕ that leaves the field C fixed and maps z to $z + 1$. Then any connected algebraic subgroup G of $Gl_n(C)$ is the difference Galois group of a difference equation $\phi(Y) = AY$, $A \in Gl_n(K)$.*

References

[1] E. Aranda-Bricaire, U. Kotta and C.H. Moog, *Linearization of discrete-time systems*, SIAM J. Control Optim. **34** (1996), 1999–2023.

[2] A.E. Babbitt, *Finitely generated pathological extensions of difference fields*, Trans. Amer. Math. Soc. **102** (1) (1962), 63–81.

[3] I.N. Balaba, *Dimension polynomials of extensions of difference fields*, Vestnik Moskov. Univ., Ser. I, Mat. Mekh., no. 2 (1984), 31–35 (Russian).

[4] I.N. Balaba, *Calculation of the dimension polynomial of a principal difference ideal*, Vestnik Moskov. Univ., Ser. I, Mat. Mekh., no. 2 (1985), 16–20 (Russian).

[5] I.N. Balaba, *Finitely generated extensions of difference fields*, VINITI (Moscow, Russia), no. 6632-87 (1987) (Russian).

[6] T. Becker and V. Weispfenning, *Gröbner Bases. A Computational Approach to Commutative Algebra*, Springer-Verlag (1993).

[7] I. Bentsen, *The existence of solutions of abstract partial difference polynomials*, Trans. Amer. Math. Soc. **158** (2) (1971), 373–397.

[8] A. Bialynicki-Birula, *On Galois theory of fields with operators*, Amer. J. Math. **84** (1962), 89–109.

[9] G.D. Birkhoff, *General theory of linear difference equations*, Trans. Amer. Math. Soc. **12** (1911), 243–284.

[10] G.D. Birkhoff, *The generalized Riemann problem for linear differential equations and the allied problem for difference and q-difference equations*, Proc. Natl. Acad. Sci. **49** (1913), 521–568.

[11] G.D. Birkhoff, *Note on linear difference and differential equations*, Proc. Natl. Acad. Sci. **27** (1941), 65–67.

[12] M. Bronstein, *On solutions of linear ordinary difference equations in their coefficient field*, J. Symbolic Comput. **29** (6) (2000), 841–877.

[13] B. Buchberger, *Ein Algorithmus zum auffinden der Basiselemente des Restklassenringes nach einem nulldimensionalen Polynomideal*, Ph.D. Thesis, University of Innsbruck, Institute for Mathematics (1965).

[14] F. Casorati, *Il calcolo delle differenze finite*, Ann. Mat. Pura Appl., Ser. II **10** (1880–1882), 10–43.

[15] Z. Chatzidakis, *A survey on the model theory of difference fields*, Model Theory, Algebra and Geometry, MSRI Publications **39** (2000), 65–96.

[16] Z. Chatzidakis and E. Hrushovski, *Model theory of difference fields*, Trans. Amer. Math. Soc. **351** (8) (1999), 2997–3071.

[17] P.M. Cohn, *Free Rings and There Relations*, Academic Press (1971).

[18] P.M. Cohn, *Skew Fields. Theory of General Division Rings*, Cambridge Univ. Press (1995).

[19] R.M. Cohn, *Manifolds of difference polynomials*, Trans. Amer. Math. Soc. **64** (1948), 133–172.

[20] R.M. Cohn, *A note on the singular manifolds of a difference polynomial*, Bull. Amer. Math. Soc. **54** (1948), 917–922.

[21] R.M. Cohn, *A theorem on difference polynomials*, Bull. Amer. Math. Soc. **55** (1949), 595–597.

[22] R.M. Cohn, *Inversive difference fields*, Bull. Amer. Math. Soc. **55** (1949), 597–603.

[23] R.M. Cohn, *Singular manifolds of difference polynomials*, Ann. Math. **53** (1951), 445–463.

[24] R.M. Cohn, *Extensions of difference fields*, Amer. J. Math. **74** (1952), 507–530.

[25] R.M. Cohn, *On extensions of difference fields and the resolvents of prime difference ideals*, Proc. Amer. Math. Soc. **3** (1952), 178–182.

[26] R.M. Cohn, *Essential singular manifolds of difference polynomials*, Ann. Math. **57** (1953), 524–530.
[27] R.M. Cohn, *Finitely generated extensions of difference fields*, Proc. Amer. Math. Soc. **6** (1955), 3–5.
[28] R.M. Cohn, *On the intersections of components of a difference polynomial*, Proc. Amer. Math. Soc. **6** (1955), 42–45.
[29] R.M. Cohn, *Specializations over difference fields*, Pacific J. Math. **5** (Suppl. 2) (1955), 887–905.
[30] R.M. Cohn, *An invariant of difference field extensions*, Proc. Amer. Math. Soc. **7** (1956), 656–661.
[31] R.M. Cohn, *An improved result concerning singular manifolds of difference polynomials*, Canad. J. Math. **11** (1959), 222–234.
[32] R.M. Cohn, *Difference Algebra*, Interscience (1965).
[33] R.M. Cohn, *An existence theorem for difference polynomials*, Proc. Amer. Math. Soc. **17** (1966), 254–261.
[34] R.M. Cohn, *Errata to "An existence theorem for difference polynomials"*, Proc. Amer. Math. Soc. **18** (1967), 1142–1143.
[35] R.M. Cohn, *Systems of ideals*, Canad. J. Math. **21** (1969), 783–807.
[36] R.M. Cohn, *A difference-differential basis theorem*, Canad. J. Math. **22** (6) (1970), 1224–1237.
[37] R.M. Cohn, *Types of singularity of components of difference polynomial*, Aequationes Math. **9** (2) (1973), 236–241.
[38] R.M. Cohn, *The general solution of a first order differential polynomial*, Proc. Amer. Math. Soc. **55** (1) (1976), 14–16.
[39] R.M. Cohn, *Solutions in the general solution*, Contribution to Algebra. Collection of Papers Dedicated to Ellis Kolchin, Academic Press (1977), 117–127.
[40] R.M. Cohn, *Specializations of differential kernels and the Ritt problem*, J. Algebra **61** (1) (1979), 256–268.
[41] R.M. Cohn, *The Greenspan bound for the order of differential systems*, Proc. Amer. Math. Soc. **79** (4) (1980), 523–526.
[42] R.M. Cohn, *Order and dimension*, Proc. Amer. Math. Soc. **87** (1) (1983), 1–6.
[43] R.M. Cohn, *Valuations and the Ritt problem*, J. Algebra **101** (1) (1986), 1–15.
[44] R.M. Cohn, *Solutions in the general solution of second order algebraic differential equations*, Amer. J. Math. **108** (3) (1986), 505–523.
[45] A. Einstein, *The Meaning of Relativity*, 4th ed., Princeton Univ. Press (1953), Appendix II (Generalization of gravitation theory), 133–165.
[46] M. Eisen, *Ideal theory and difference algebra*, Math. Japon. **7** (1962), 159–180.
[47] D. Eisenbud, *Commutative Algebra with View towards Algebraic Geometry*, Springer-Verlag (1995).
[48] P.I. Etingof, *Galois groups and connection matrices of q-difference equations*, Electron Res. Announc. Amer. Math. Soc. **1** (1) (1995), 1–9 (electronic).
[49] P. Evanovich, *Algebraic extensions of difference fields*, Trans. Amer. Math. Soc. **179** (1) (1973), 1–22.
[50] P. Evanovich, *Finitely generated extensions of partial difference fields*, Trans. Amer. Math. Soc. **281** (2) (1984), 795–811.
[51] M. Fliess, *Esquisses pour une theorie des systemes non lineaires en temps discret*, Conference on Linear and Nonlinear Mathematical Control Theory, Rend. Sem. Mat. Univ. Politec. Torino, Special Issue (1987), 55–67 (French).
[52] M. Fliess, *Automatique en temps discret et algebre aux differences*, Forum Math. **2** (3) (1990), 213–232 (French).
[53] M. Fliess, *A fundamental result on the invertibility of discrete time dynamics*, Analysis of Controlled Dynamic Systems, Lyon, 1990, Progr. Systems Control Theory **8** (1991), 211–223.
[54] M. Fliess, *Invertibility of causal discrete time dynamical system*, J. Pure Appl. Algebra **86** (2) (1993), 173–179.
[55] M. Fliess, J. Levine, P. Martin and O. Rouchon, *Differential flatness and defect: an overview*, Geometry in Nonlinear Control and Differential Inclusions, Warsaw, 1993, Banach Center Publ. (1995), 209–225.
[56] C. Franke, *Picard–Vessiot theory of linear homogeneous difference equations*, Trans. Amer. Math. Soc. **108** (3) (1963), 491–515.
[57] C. Franke, *Solvability of linear homogeneous difference equations by elementary operations*, Proc. Amer. Math. Soc. **17** (1964), 240–246.
[58] C. Franke, *A note on the Galois theory of linear homogeneous difference equations*, Proc. Amer. Math. Soc. **18** (1967), 548–551.

[59] C. Franke, *The Galois correspondence for linear homogeneous difference equations*, Proc. Amer. Math. Soc. **21** (1969), 397–401.

[60] C. Franke, *Linearly reducible linear difference operators*, Aequationes Math. **6** (1971), 188–194.

[61] C. Franke, *Reducible linear difference operators*, Aequationes Math. **9** (1973), 136–144.

[62] C. Franke, *A characterization of linear difference equations which are solvable by elementary operations*, Aequationes Math. **10** (1974), 97–104.

[63] B. Greenspan, *A bound for the orders of the components of a system of algebraic difference equations*, Pacific J. Math. **9** (1959), 473–486.

[64] P.A. Hendrics, *An algorithm for determining the difference Galois group for second order linear difference equations*, Technical Report, Rijksuniversiteit, Groningen (1996).

[65] P.A. Hendrics, *Algebraic aspects of linear differential and difference equations*, Ph.D. Thesis, Rijksuniversiteit, Groningen (1996).

[66] P.A. Hendrics, *An algorithm for computing a standard form for second-order linear q-difference equations*, Algorithms for Algebra, Eindhoven, 1996, J. Pure Appl. Algebra **117/118** (1997), 331–352.

[67] P.A. Hendrics and M.F. Singer, *Solving difference equations in finite terms*, J. Symbolic Comput. **27** (1999), 239–259.

[68] F. Herzog, *Systems of algebraic mixed difference equations*, Trans. Amer. Math. Soc. **37** (1935), 286–300.

[69] E. Hrushovski, *The Manin–Mamford conjecture and the model theory of difference fields*, Ann. Pure Appl. Logic **112** (1) (2001), 43–115.

[70] R.P. Infante, *Strong normality and normality for difference fields*, Aequationes Math. **20** (1980), 121–122.

[71] R.P. Infante, *Strong normality and normality for difference fields*, Aequationes Math. **20** (1980), 159–165.

[72] R.P. Infante, *The structure of strongly normal difference extensions*, Aequationes Math. **21** (1) (1980), 16–19.

[73] R.P. Infante, *On the Galois theory of difference fields*, Aequationes Math. **22** (1981), 112–113.

[74] R.P. Infante, *On the Galois theory of difference fields*, Aequationes Math. **22** (1981), 194–207.

[75] R.P. Infante, *On the inverse problem in difference Galois theory*, Algebraists' Homage: Papers in the Ring Theory and Related Topics, New Haven, Conn., 1981, Contemp. Math. **13** (1982), 349–352.

[76] K. Iwasawa and T. Tamagawa, *On the group of automorphisms of a functional field*, J. Math. Soc. Japan **3** (1951), 137–147.

[77] C.G.J. Jacobi, *Gesammelte Werke*, Vol. 5, Berlin, Bruck und Verlag von Georg Reimer (1890), 191–216.

[78] J.L. Johnson, *Differential dimension polynomials and a fundamental theorem on differential modules*, Amer. J. Math. **91** (1) (1969), 239–248.

[79] J.L. Johnson, *Kähler differentials and differential algebra*, Ann. of Math. (2) **89** (1969), 92–98.

[80] J.L. Johnson, *A note on Krull dimension for differential rings*, Comment. Math. Helv. **44** (1969), 207–216.

[81] J.L. Johnson, *Kähler differentials and differential algebra in arbitrary characteristic*, Trans. Amer. Math. Soc. **192** (1974), 201–208.

[82] J.L. Johnson and W. Sit, *On the differential transcendence polynomials of finitely generated differential field extensions*, Amer. J. Math. **101** (1979), 1249–1263.

[83] M. Karr, *Summation in finite terms*, J. Assoc. Comput. Mach. **28** (2) (1981), 305–350.

[84] E.R. Kolchin, *The notion of dimension in the theory of algebraic differential equations*, Bull. Amer. Math. Soc. **70** (1964), 570–573.

[85] E.R. Kolchin, *Some problems in differential algebra*, Proc. Int'l Congress of Mathematicians, Moscow, 1966 (1968), 269–276.

[86] E.R. Kolchin, *Differential Algebra and Algebraic Groups*, Academic Press (1973).

[87] M.V. Kondrateva, A.B. Levin, A.V. Mikhalev and E.V. Pankratev, *Computation of dimension polynomials*, Internat. J. Algebra Comput. **2** (2) (1992), 117–137.

[88] M.V. Kondrateva, A.B. Levin, A.V. Mikhalev and E.V. Pankratev, *Differential and Difference Dimension Polynomials*, Kluwer Acad. Publ. (1998).

[89] M.V. Kondrateva and E.V. Pankratev, *A recursive algorithm for the computation of Hilbert polynomial*, Proc. EUROCAL 87, Lecture Notes in Comput. Sci. **378**, Springer-Verlag (1990), 365–375.

[90] M.V. Kondrateva and E.V. Pankratev, *Algorithms of computation of characteristic Hilbert polynomials*, Packets of Applied Programs. Analytic Transformations, Nauka (1988), 129–146 (in Russian).

[91] M.V. Kondrateva, E.V. Pankratev and R.E. Serov, *Computations in differential and difference modules*, Proceedings of the Internat. Conf. on the Analytic Computations and their Applications in Theoret. Physics, Dubna, 1985, 208–213 (in Russian).

[92] P. Kowalski and A. Pillay, *A note on groups definable in difference fields*, Proc. Amer. Math. Soc. **130** (1) (2002), 205–212 (electronic).

[93] H.F. Kreimer, *The foundations for extension of differential algebra*, Trans. Amer. Math. Soc. **111** (1964), 482–492.

[94] H.F. Kreimer, *An extension of differential Galois theory*, Trans. Amer. Math. Soc. **118** (1965), 247–256.

[95] H.F. Kreimer, *On an extension of the Picard–Vessiot theory*, Pacific J. Math. **15** (1965), 191–205.

[96] B. Lando, *Jacobi's bound for first order difference equations*, Proc. Amer. Math. Soc. **32** (1) (1972), 8–12.

[97] B. Lando, *Extensions of difference specializations*, Proc. Amer. Math. Soc. **79** (2) (1980), 197–202.

[98] V.M. Latyshev, A.V. Mikhalev and E.V. Pankratev, *Construction of canonical simplifiers in modules over rings of polynomials*, Visnik Kiev Univ., Ser. Mat. Mekh. **27** (1985), 65–67 (Ukrainian).

[99] A.B. Levin, *Characteristic polynomials of filtered difference modules and difference field extensions*, Uspekhi Mat. Nauk **33** (3) (1978), 177–178 (Russian). English transl.: Russian Math. Surveys **33** (3) (1978), 165–166.

[100] A.B. Levin, *Characteristic polynomials of inversive difference modules and some properties of inversive difference dimension*, Uspekhi Mat. Nauk **35** (1) (1980), 201–202 (Russian). English transl.: Russian Math. Surveys **35** (1) (1980), 217–218.

[101] A.B. Levin, *Characteristic polynomials of difference modules and some properties of difference dimension*, VINITI (Moscow), no. 2175-80 (1980) (Russian).

[102] A.B. Levin, *Type and dimension of inversive difference vector spaces and difference algebras*, VINITI (Moscow), no. 1606-82 (1982) (Russian).

[103] A.B. Levin, *Characteristic polynomials of Δ-modules and finitely generated Δ-field extensions*, VINITI (Moscow), no. 334-85 (1985) (Russian).

[104] A.B. Levin, *Inversive difference modules and problems of solvability of systems of linear difference equations*, VINITI (Moscow), no. 335-85 (1985) (Russian).

[105] A.B. Levin, *Computation of Hilbert polynomials in two variables*, J. Symbolic Comput. **28** (1999), 681–709.

[106] A.B. Levin, *Characteristic polynomials of finitely generated modules over Weyl algebras*, Bull. Austral. Math. Soc. **61** (2000), 387–403.

[107] A.B. Levin, *Reduced Gröbner bases, free difference-differential modules and difference-differential dimension polynomials*, J. Symbolic Comput. **29** (2000), 1–26.

[108] A.B. Levin, *On the set of Hilbert polynomials*, Bull. Austral. Math. Soc. **64** (2001), 291–305.

[109] A.B. Levin, *Multivariable dimension polynomials and new invariants of differential field extensions*, Internat. J. Math. Math. Sci. **27** (4), 201–213.

[110] A.B. Levin and A.V. Mikhalev, *Difference-differential dimension polynomials*, VINITI (Moscow), no. 6848-B88 (1988) (Russian).

[111] A.B. Levin and A.V. Mikhalev, *Dimension polynomials of filtered G-modules and finitely generated G-field extensions*, Algebra (Collection of Papers), Moscow State Univ. (1989), 74–94.

[112] A.B. Levin and A.V. Mikhalev, *Type and dimension of finitely generated vector G-spaces*, Vestnik Moskov. Univ., Ser. I, Mat. Mekh., no. 4 (1991), 72–74 (Russian). English transl.: Moscow Univ. Math. Bull. **46** (4), 51–52.

[113] A.B. Levin and A.V. Mikhalev, *Dimension polynomials of difference-differential modules and of difference-differential field extensions*, Abelian Groups and Modules **10** (1991), 56–82 (Russian).

[114] A.B. Levin and A.V. Mikhalev, *Dimension polynomials of filtered differential G-modules and extensions of differential G-fields*, Contemp. Math. **131** (2) (1992), 469–489.

[115] A.B. Levin and A.V. Mikhalev, *Type and dimension of finitely generated G-algebras*, Contemp. Math. **184** (1995), 275–280.

[116] A. Macintyre, *Generic automorphisms of fields*, Ann. Pure Appl. Logic **88** (2–3) (1997), 165–180.

[117] A.V. Mikhalev and E.V. Pankratev, *Differential dimension polynomial of a system of differential equations*, Algebra (Collection of Papers), Moscow State Univ. (1980), 57–67 (Russian).

[118] A.V. Mikhalev and E.V. Pankratev, *Differential and Difference Algebra*, Algebra, Topology, Geometry **25**, 67–139. Itogi Nauki i Tekhniki, Akad. Nauk SSSR (1987) (Russian). English transl.: J. Soviet Math. **45** (1) (1989), 912–955.
[119] A.V. Mikhalev and E.V. Pankratev, *Computer Algebra. Calculations in Differential and Difference Algebra*, Moscow State Univ. (1989) (Russian).
[120] B. Mishra, *Algorithmic Algebra*, Springer-Verlag (1993).
[121] R. Moosa, *On difference fields with quantifier elimination*, Bull. London Math. Soc. **33** (6) (2001), 641–646.
[122] K. Nishioka, *A note on differentially algebraic solutions of first order linear difference equations*, Aequationes Math. **27** (1984), 32–48.
[123] S. Nonvide, *Corps aux différences finies*, C. R. Acad. Sci. Paris Sér. I Math. **314** (6) (1992), 423–425 (French).
[124] E.V. Pankratev, *The inverse Galois problem for the extensions of difference fields*, Algebra i Logika **11** (1972), 87–118 (Russian). English transl.: Algebra and Logic **11** (1972), 51–69.
[125] E.V. Pankratev, *The inverse Galois problem for extensions of difference fields*, Uspekhi Mat. Nauk **27** (1) (1972), 249–250 (Russian).
[126] E.V. Pankratev, *Fuchsian difference modules*, Uspekhi Mat. Nauk **28** (3) (1973), 193–194 (Russian).
[127] E.V. Pankratev, *Computations in differential and difference modules*, Symmetries of Partial Differential Equations, Part III, Acta Appl. Math. **16** (2) (1989), 167–189.
[128] M. Petkovsek, *Finding closed form solutions of difference equations by symbolic methods*, Thesis, Dept. of Comp. Sci., Carnegie Mellon University (1990).
[129] A. Pillay, *A note on existentially closed difference fields with algebraically closed fixed field*, J. Symbolic Logic **66** (2) (2001), 719–721.
[130] C. Praagman, *The formal classification of linear difference operators*, Proc. Kon. Ned. Ac. Wet., Ser. A **86** (1983), 249–261.
[131] C. Praagman, *Meromorphic linear difference equations*, Thesis, University of Groningen (1985).
[132] J. Riordan, *Combinatorial Identities*, Wiley (1968).
[133] J.F. Ritt, *Algebraic difference equations*, Bull. Amer. Math. Soc. **40** (1934), 303–308.
[134] J.F. Ritt, *Complete difference ideals*, Amer. J. Math. **63** (1941), 681–690.
[135] J.F. Ritt, *Differential Algebra*, Amer. Math. Soc. Coll. Publ. **33**, Amer. Math. Soc. (1973).
[136] J.F. Ritt and J.L. Doob, *Systems of algebraic difference equations*, Amer. J. Math. **55** (1933), 505–514.
[137] J.F. Ritt and H.W. Raudenbush, *Ideal theory and algebraic difference equations*, Trans. Amer. Math. Soc. **46** (1939), 445–453.
[138] T. Scanlon and J.F. Voloch, *Difference algebraic subgroups of commutative algebraic groups over finite fields*, Manuscripta Math. **99** (3) (1999), 329–339.
[139] W. Sit, *Well-ordering of certain numerical polynomials*, Trans. Amer. Math. Soc. **212** (1975), 37–45.
[140] W. Strodt, *Systems of algebraic partial difference equations*, Master essay, Columbia Univ. (1937).
[141] W. Strodt, *Principal solutions of difference equations*, Amer. J. Math. **69** (1947), 717–757.
[142] M. van der Put and M.F. Singer, *Galois Theory of Difference Equations*, Springer (1997).
[143] M. van Hoeij, *Rational solutions of linear difference equations*, Technical Report, Dept. of Mathematics, Florida State University (1998).
[144] O. Zariski and P. Samuel, *Commutative Algebra*, Van Nostrand, Vol. I (1958), Vol. II (1960).

Section 5A
Groups and Semigroups

Reflection Groups

Meinolf Geck

Department of Mathematical Sciences, King's College, Aberdeen University, Aberdeen AB24 3UE, Scotland, UK
E-mail: geck@maths.abdn.ac.uk

Gunter Malle

FB Mathematik, Universität Kaiserslautern, Postfach 3049, D-67653 Kaiserslautern, Germany
E-mail: malle@mathematik.uni-kl.de

Contents

HANDBOOK OF ALGEBRA, VOL. 4
Edited by M. Hazewinkel

This chapter is concerned with the theory of finite reflection groups, that is, finite groups generated by reflections in a real or complex vector space. This is a rich theory, both for intrinsic reasons and as far as applications in other mathematical areas or mathematical physics are concerned. The origin of the theory can be traced back to the ancient study of symmetries of regular polyhedra. Another extremely important impetus comes from the theory of semisimple Lie algebras and Lie groups, where finite reflection groups occur as "Weyl groups". In the last decade, Broué's "Abelian defect group conjecture" (a conjecture concerning the representations of finite groups over fields of positive characteristic) has lead to a vast research program, in which complex reflection groups, corresponding braid groups and Hecke algebras play a prominent role. Thus, the theory of reflection groups is at the same time a well-established classical piece of mathematics and still a very active research area. The aim of this chapter (and a subsequent one on Hecke algebras) is to give an overview of both these aspects.

As far as the study of reflection groups as such is concerned, there are (at least) three reasons why this leads to an interesting and rich theory:

Classification. Given a suitable notion of "irreducible" reflection groups, it is possible to give a complete classification, with typically several infinite families of groups and a certain number of exceptional cases. In fact, this classification can be seen as the simplest possible model for much more complex classification results concerning related algebraic structures, such as complex semisimple Lie algebras, simple algebraic groups and, eventually, finite simple groups. Besides the independent interest of such a classification, we mention that there is a certain number of results on finite reflection groups which can be stated in general terms but whose proof requires a case-by-case analysis according to the classification. (For example, the fact that every element in a finite real reflection group is conjugate to its inverse.)

Presentations. Reflection groups have a highly symmetric "Coxeter type presentation" with generators and defining relations (visualised by "Dynkin diagrams" or generalisations thereof), which makes it possible to study them by purely combinatorial methods (length function, reduced expressions and so on). From this point of view, the associated Hecke algebras can be seen as "deformations" of the group algebras of finite reflection groups, where one or several formal parameters are introduced into the set of defining relations. One of the most important developments in this direction is the discovery of the Kazhdan–Lusztig polynomials and the whole theory coming with them. (This is discussed in more detail in the chapter on Hecke algebras.)

Topology and geometry. The action of a reflection group on the underlying vector space opens the possibility of using geometric methods. First of all, the ring of invariant symmetric functions on that vector space always is a polynomial ring (and this characterises finite reflection groups). Furthermore, we have a corresponding hyperplane arrangement which gives rise to the definition of an associated braid group as the fundamental group of a certain topological space. For the symmetric group, we obtain in this way the classical Artin braid group, with applications in the theory of knots and links.

Furthermore, all these aspects are related to each other which – despite being quite elementary taken individually – eventually leads to a highly sophisticated theory.

We have divided our survey into four major parts. The first part deals with finite complex or real reflection groups in general. The second part deals with finite real reflection groups and the relations with the theory of Coxeter groups. The third part is concerned with the associated braid groups. Finally, in the fourth part, we consider complex irreducible characters of finite reflection groups.

We certainly do not pretend to give a complete picture of all aspects of the theory of reflection groups and Coxeter groups. Our references cover the period up until 2003. Especially, we will not say so much about areas that we do not feel competent in; to our best knowledge, we try to give at least some references for further reading in such cases. This concerns, in particular, all aspects of infinite (affine, hyperbolic, ...) Coxeter groups.

1. Finite groups generated by reflections

1.1. DEFINITIONS. Let V be a finite-dimensional vector space over a field K. A *reflection on* V is a non-trivial element $g \in \mathrm{GL}(V)$ of finite order which fixes a hyperplane in V pointwise. There are two types of reflections, according to whether g is semisimple (hence diagonalisable) or unipotent. Often, the term reflection is reserved for the first type of elements, while the second are called *transvections*. They can only occur in positive characteristic. Here, we will almost exclusively be concerned with ground fields K of characteristic 0, which we may and will then assume to be subfields of the field $\mathbb{C}$ of complex numbers. Then, by our definition, reflections are always semisimple and (thus) diagonalisable. Over fields K contained in the field $\mathbb{R}$ of real numbers, reflections necessarily have order 2, which is the case motivating their name. Some authors reserve the term reflection for this case, and speak of pseudo-reflections in the case of arbitrary (finite) order.

Let $g \in \mathrm{GL}(V)$ be a reflection. The hyperplane $C_V(g)$ fixed point-wise by g is called the *reflecting hyperplane* of g. Then $V = C_V(g) \oplus V_g$ for a unique g-invariant subspace V_g of V of dimension 1. Any non-zero vector $v \in V_g$ is called a *root for* g. Thus, a root for a reflection is an eigenvector with eigenvalue different from 1. Now assume in addition that V is Hermitean. Then conversely, given a vector $v \neq 0$ in V and a natural number $n \geqslant 2$ we may define a reflection in V with root v and of order n by $g.v := \exp(2\pi i/n)v$, and $g|_{V^\perp} = \mathrm{id}$.

A *reflection group on* V is now a finite subgroup $W \leqslant \mathrm{GL}(V)$ generated by reflections. Note that any finite subgroup of $\mathrm{GL}(V)$ leaves invariant a non-degenerate Hermitean form. Thus, there is no loss in assuming that a reflection group W leaves such a form invariant.

1.2. *Invariants*

Let V be a finite-dimensional vector space over $K \leqslant \mathbb{C}$. Let $K[V]$ denote the algebra of symmetric functions on V, i.e., the symmetric algebra $S(V^*)$ of the dual space V^* of V. So $K[V]$ is a commutative algebra over K with a grading $K[V] = \bigoplus_{d \geqslant 0} K[V]^d$, where, for any $d \geqslant 0$, $K[V]^d$ denotes the d-th symmetric power of V^*. If $W \leqslant \mathrm{GL}(V)$ then W

acts naturally on $K[V]$, respecting the grading. Now reflection groups are characterised by the structure of their invariant ring $K[V]^W$:

1.3. THEOREM (Shephard and Todd, [167], Chevalley, [46]). *Let V be a finite-dimensional vector space over a field of characteristic* 0 *and $W \leqslant \mathrm{GL}(V)$ a finite group. Then the following are equivalent*:

(i) *the ring of invariants $K[V]^W$ is a polynomial ring,*
(ii) *W is generated by reflections.*

The implication from (i) to (ii) is an easy consequence of Molien's formula

$$P\big(K[V]^W, x\big) = \frac{1}{|W|} \sum_{g \in W} \frac{1}{\det_V(1 - gx)}$$

for the Hilbert series $P(K[V]^W, x)$ of the ring of invariants $K[V]^W$, see Shephard and Todd, [167, p. 289]. It follows from Auslander's purity of the branch locus that this implication remains true in arbitrary characteristic (see Benson, [7, Theorem 7.2.1], for example). The other direction was proved by Shephard and Todd as an application of their classification of complex reflection groups (see Section 1.13). Chevalley gave a general proof avoiding the classification which uses the combinatorics of differential operators.

Let W be an n-dimensional reflection group. By Theorem 1.3 the ring of invariants is generated by n algebraically independent polynomials (so-called *basic invariants*), which may be taken to be homogeneous. Although these polynomials are not uniquely determined in general, their degrees $d_1 \leqslant \cdots \leqslant d_n$ are. They are called the *degrees of* W. Then $|W| = d_1 \cdots d_n$, and the Molien formula shows that $N(W) := \sum_{i=1}^n m_i$ is the number of reflections in W, where $m_i := d_i - 1$ are the *exponents of* W.

The quotient $K[V]_W$ of $K[V]$ by the ideal generated by the invariants of strictly positive degree is called the *coinvariant algebra* of (V, W). This is again a naturally graded W-module, whose structure is described by:

1.4. THEOREM (Chevalley, [46]). *Let V be a finite-dimensional vector space over a field K of characteristic* 0 *and $W \leqslant \mathrm{GL}(V)$ a reflection group. Then $K[V]_W$ carries a graded version of the regular representation of W. The grading is such that*

$$\sum_{i \geqslant 0} \dim K[V]_W^i \, x^i = \prod_{i=1}^n \frac{x^{d_i} - 1}{x - 1},$$

where $K[V]_W^i$ denotes the homogeneous component of degree i.

(See also Bourbaki, [25, V.5.2, Theorem 2].) The polynomial

$$P_W := \sum_{i \geqslant 0} \dim K[V]_W^i x^i$$

is called the *Poincaré-polynomial of* W.

1.5. *Parabolic subgroups*

Let W be a reflection group on V. The *parabolic subgroups of* W are by definition the pointwise stabilisers

$$W_{V'} := \{g \in W \mid g \cdot v = v \text{ for all } v \in V'\}.$$

of subspaces $V' \leqslant V$. The following result is of big importance in the theory of reflection groups:

1.6. THEOREM (Steinberg, [175]). *Let $W \leqslant \mathrm{GL}(V)$ be a complex reflection group. For any subspace $V' \leqslant V$ the parabolic subgroup $W_{V'}$ is generated by the reflections it contains, that is, by the reflections whose reflecting hyperplane contains V'. In particular, parabolic subgroups are themselves reflection groups.*

For the proof, Steinberg characterises reflection groups via eigenfunctions of differential operators with constant coefficients that are invariant under finite linear groups. Lehrer [129] has recently found an elementary proof. For a generalisation to positive characteristic see Theorem 5.4.

1.7. *Exponents, coexponents and fake degrees*

Let W be a complex reflection group. For $w \in W$ define $k(w) := \dim V^{\langle w \rangle}$, the dimension of the fixed space of w on V. Solomon, [169], proved the following remarkable formula for the generating function of k

$$\sum_{w \in W} x^{k(w)} = \prod_{i=1}^{n} (x + m_i),$$

by showing that the algebra of W-invariant differential forms with polynomial coefficients is an exterior algebra of rank n over the algebra $K[V]^W$, generated by the differentials of a set of basic invariants (see also Flatto, [79], Benson, [7, Theorem 7.3.1]). The formula was first observed by Shephard and Todd, [167, 5.3] using their classification of irreducible complex reflection groups. Dually, Orlik and Solomon, [154], showed

$$\sum_{w \in W} \det_V(w) x^{k(w)} = \prod_{i=1}^{n} (x - m_i^*)$$

for some non-negative integers $m_1^* \leqslant \cdots \leqslant m_n^*$, the *coexponents of* W (see Lehrer and Michel, [130], for a generalisation, and Kusuoka, [123], Orlik and Solomon, [155,156], for versions over finite fields).

Let χ be an irreducible character of W. The *fake degree of* χ is the polynomial

$$R_\chi := \sum_{d \geqslant 0} \langle K[V]^d_W, \chi^* \rangle_W x^d = \frac{1}{|W|} \sum_{w \in W} \frac{\chi(w)}{\det_V(xw - 1)^*} \prod_{i=1}^{n} (x^{d_i} - 1) \in \mathbb{Z}[x],$$

that is, the graded multiplicity of χ^* in the W-module $K[V]_W$. Thus, in particular, R_χ specialises to the degree $\chi(1)$ at $x = 1$. The *exponents* $(e_i(\chi) \mid 1 \leqslant i \leqslant \chi(1))$ *of an irreducible character* χ *of* W are defined by the formula $R_\chi = \sum_{i=1}^{\chi(1)} x^{e_i(\chi)}$. The exponents m_i of W are now just the exponents of the contragradient of the reflection representation $\rho^* := \mathrm{tr}_{V^*}$, that is, $R_{\rho^*} = \sum_{i=1}^n x^{m_i}$. Dually, the coexponents are the exponents of ρ. In particular, for real reflection groups exponents and coexponents coincide. In general $N^*(W) := \sum_{i=1}^n m_i^*$ equals the number of reflecting hyperplanes of W. The $d_i^* := m_i^* - 1$ are sometimes called the *codegrees of* W.

If W is a Weyl group (see Section 2.10), the fake degrees constitute a first approximation to the degrees of principal series unipotent characters of finite groups of Lie type with Weyl group W. See also Section 4.8 for further properties.

1.8. *Reflection data*

In the general theory of finite groups of Lie type (where an algebraic group comes with an action of a Frobenius map) as well as in the study of Levi subgroups it is natural to consider reflection groups together with an automorphism ϕ normalising the reflection representation; see the survey article Broué and Malle, [36]. This leads to the following abstract definition.

A pair $(V, W\phi)$ is called a *reflection datum* if V is a vector space over a subfield $K \subseteq \mathbb{C}$ and $W\phi$ is a coset in $\mathrm{GL}(V)$ of a reflection group $W \subseteq \mathrm{GL}(V)$, where $\phi \in \mathrm{GL}(V)$ normalises W.

A *sub-reflection datum* of a reflection datum $(V, W\phi)$ is a reflection datum of the form $(V', W'(w\phi)|_{V'})$, where V' is a subspace of V, W' is a reflection subgroup of $N_W(V')|_{V'}$ stabilising V' (hence, a reflection subgroup of $N_W(V')/W_{V'}$), and $w\phi$ is an element of $W\phi$ stabilising V' and normalising W'. A *Levi sub-reflection datum of* $(V, W\phi)$ is a sub-reflection datum of the form $(V, W_{V'}(w\phi))$ for some subspace $V' \leqslant V$ (note that, by Theorem 1.6, $W_{V'}$ is indeed a reflection subgroup of W). A *torus of* $\mathbb{G}$ is a sub-reflection datum with trivial reflection group.

Let $\mathbb{G} = (V, W\phi)$ be a reflection datum. Then ϕ acts naturally on the symmetric algebra $K[V]$. It is possible to choose basic invariants $f_1, \dots, f_n \in K[V]^W$, such that $f_i^\phi = \varepsilon_i f_i$ for roots of unity $\varepsilon_1, \dots, \varepsilon_n$. The multiset $\{(d_i, \varepsilon_i)\}$ of *generalised degrees of* $\mathbb{G}$ then only depends on W and ϕ (see, for example, Springer, [170, Lemma 6.1]). The *polynomial order* of the reflection datum $\mathbb{G} = (V, W\phi)$ is by definition the polynomial

$$|\mathbb{G}| := \frac{\varepsilon_{\mathbb{G}} x^{N(W)}}{\frac{1}{|W|}\sum_{w\in W}\frac{1}{\det_V(1-xw\phi)^*}} = x^{N(W)}\prod_{i=1}^n (x^{d_i} - \varepsilon_i),$$

where $\varepsilon_{\mathbb{G}} := (-1)^n \varepsilon_1 \cdots \varepsilon_n$. Let $\Phi(x)$ be a cyclotomic polynomial over K. A torus $\mathbb{T} = (V', (w\phi)|_{V'})$ of $\mathbb{G}$ is called a *Φ-torus* if the polynomial order of $\mathbb{T}$ is a power of Φ.

Reflection data can be thought of as the skeletons of finite reductive groups.

1.9. *Regular elements*

In this section we present results which show that certain subgroups respectively subquotients of reflection groups are again reflection groups. Let $(V, w\phi)$ be a reflection datum over $K = \mathbb{C}$. For $w\phi \in W\phi$ and a root of unity $\zeta \in \mathbb{C}^\times$ write

$$V(w\phi, \zeta) := \{v \in V \mid w\phi \cdot v = \zeta v\}$$

for the ζ-eigenspace of $w\phi$. Note that $(V(w\phi, \zeta), w\phi)$ is an $(x - \zeta)$-torus of $(V, w\phi)$ in the sense defined above. These $(x - \zeta)$-tori for fixed ζ satisfy a kind of Sylow theory. Let $f_1, \ldots, f_n$ be a set of basic invariants for W and H_i the surface defined by $f_i = 0$. Springer, [170], proves that

$$\bigcup_{w\phi \in W\phi} V(w\phi, \zeta) = \bigcap_{i:\varepsilon_i \zeta^{d_i} = 1} H_i,$$

the irreducible components of this algebraic set are just the maximal $V(w\phi, \zeta)$, W acts transitively on these components, and their common dimension is just the number $a(d, \phi)$ of indices i such that $\varepsilon_i \zeta^{d_i} = 1$, where d denotes the order of ζ. (Note that $a(d, \phi)$ only depends on d, not on ζ itself.) From this he obtains:

1.10. THEOREM (Springer, [170, Theorems 3.4 and 6.2]). *Let $(V, W\phi)$ be a reflection datum over $\mathbb{C}$, ζ a primitive d-th root of unity. Then*:

(i) $\max\{\dim V(w\phi, \zeta) \mid w \in W\} = a(d, \phi)$.

(ii) *For any $w \in W$ there exists a $w' \in W$ such that $V(w\phi, \zeta) \subseteq V(w'\phi, \zeta)$ and $V(w'\phi, \zeta)$ has maximal dimension.*

(iii) *If $\dim V(w\phi, \zeta) = \dim V(w'\phi, \zeta) = a(d, \phi)$ then there exists a $u \in W$ with $u \cdot V(w\phi, \zeta) = V(w'\phi, \zeta)$.*

This can be rephrased as follows: Let K be a subfield of $\mathbb{C}$, Φ a cyclotomic polynomial over K. A torus $\mathbb{T}$ of $\mathbb{G}$ is called a *Φ-Sylow torus*, if its order equals the full Φ-part of the order of $\mathbb{G}$. Then Φ-tori of $\mathbb{G}$ satisfy the three statements of Sylow's theorem. (For an analogue of the statement on the number of Sylow subgroups see Broué, Malle and Michel, [37, Theorem 5.1(4)]).

This can in turn be used to deduce a Sylow theory for tori in finite groups of Lie type (see Broué and Malle, [34]).

A vector $v \in V$ is called *regular (for W)* if it is not contained in any reflecting hyperplane, i.e. (by Theorem 1.6), if its stabiliser W_v is trivial. Let $\zeta \in \mathbb{C}$ be a root of unity. An element $w\phi \in W\phi$ is *ζ-regular* if $V(w\phi, \zeta)$ contains a regular vector. By definition, if $w\phi$ is regular for some root of unity, then so is any power of $w\phi$. If $\phi = 1$, theorem 1.6 of Steinberg implies that the orders of w and ζ coincide. An integer d is a *regular number for W* if it is the order of a regular element of W.

1.11. THEOREM (Springer, [170, Theorem 6.4 and Proposition 4.5]). *Let $w\phi \in W\phi$ be ζ-regular of order d. Then*:

(i) $\dim V(w\phi, \zeta) = a(d, \phi)$.
(ii) *The centraliser of* $w\phi$ *in* W *is isomorphic to a reflection group in* $V(w\phi, \zeta)$ *whose degrees are the* d_i *with* $\varepsilon_i \zeta^{d_i} = 1$.
(iii) *The elements of* $W\phi$ *with property* (i) *form a single conjugacy class under* W.
(iv) *Let* $\phi = 1$ *and let* χ *be an irreducible character of* W. *Then the eigenvalues of* w *in a representation with character* χ *are* $(\zeta^{e_i(\chi)} \mid 1 \leqslant i \leqslant \chi(1))$.

In particular, it follows from (iv) that the eigenvalues of a ζ-regular element w on V are $(\zeta^{m_i^*} \mid 1 \leqslant i \leqslant n)$.

Interestingly enough, the normaliser modulo centraliser of arbitrary Sylow tori of reflection data are naturally reflection groups, as the following generalisation of the previous result shows:

1.12. THEOREM (Lehrer and Springer, [131,132]). *Let* $w \in W$ *and* $\widetilde{V} := V(w\phi, \zeta)$ *be such that* $(\widetilde{V}, w\phi)$ *is a* Φ*-Sylow torus. Let* $N := \{w' \in W \mid w' \cdot \widetilde{V} = \widetilde{V}\}$ *be the normaliser,* $C := \{w' \in W \mid w' \cdot v = v \text{ for all } v \in \widetilde{V}\}$ *the centraliser of* $\widetilde{V}$.
(i) *Then* N/C *acts as a reflection group on* $\widetilde{V}$, *with reflecting hyperplanes the intersections with* $\widetilde{V}$ *of those of* W.
(ii) *A set of basic invariants of* N/C *is given by the restrictions to* $\widetilde{V}$ *of those* f_i *with* $\varepsilon_i \zeta^{d_i} = 1$.
(iii) *If* W *is irreducible on* V, *then so is* N/C *on* $\widetilde{V}$.

In the case of regular elements, the second assertion of (i) goes back to Lehrer, [128, 5.8], Denef and Loeser, [64]; see also Broué and Michel, [39, Proposition 3.2].

1.13. *The Shephard–Todd classification*

Let V be a finite-dimensional complex vector space and $W \leqslant \mathrm{GL}(V)$ a reflection group. Since W is finite, the representation on V is completely reducible, and W is the direct product of irreducible reflection subgroups. Thus, in order to determine all reflection groups over $\mathbb{C}$, it is sufficient to classify the irreducible ones. This was achieved by Shephard and Todd, [167].

To describe this classification, first recall that a subgroup $W \leqslant \mathrm{GL}(V)$ is called imprimitive if there exists a direct sum decomposition $V = V_1 \oplus \cdots \oplus V_k$ with $k > 1$ stabilised by W (that is, W permutes the summands). The bulk of irreducible complex reflection groups consists of imprimitive ones. For any $d, e, n \geqslant 1$ let $G(de, e, n)$ denote the group of monomial $n \times n$-matrices (that is, matrices with precisely one non-zero entry in each row and column) with non-zero entries in the set of de-th roots of unity, such that the product over these entries is a d-th root of unity.

Explicit generators may be chosen as follows: $G(d, 1, n)$ is generated on $\mathbb{C}^n$ with standard Hermitean form by the reflection t_1 of order d with root the first standard basis vector b_1 and by the permutation matrices $t_2, \ldots, t_n$ for the transpositions $(1, 2), (2, 3), \ldots, (n-1, n)$. For $d > 1$ this is an irreducible reflection group, isomorphic to the wreath product $C_d \wr \mathfrak{S}_n$ of the cyclic group of order d with the symmetric group $\mathfrak{S}_n$, where the base group is generated by the reflections of order d with roots the standard basis vectors, and a complement consists of all permutation matrices.

Let $\gamma_d : G(d,1,n) \to \mathbb{C}^\times$ be the linear character of $G(d,1,n)$ obtained by tensoring the determinant on V with the sign character on the quotient $\mathfrak{S}_n$. Then for any $e > 1$ we have

$$G(de,e,n) := \ker(\gamma_{de}^d) \leqslant G(de,1,n).$$

This is an irreducible reflection subgroup of $G(de,1,n)$ for all $n \geqslant 2$, $d \geqslant 1$, $e \geqslant 2$, except for $(d,e,n) = (2,2,2)$. It is generated by the reflections

$$t_1^e, t_1^{-1}t_2t_1, t_2, t_3, \ldots, t_n,$$

where the first generator is redundant if $d = 1$. Clearly, $G(de,e,n)$ stabilises the decomposition $V = \mathbb{C}b_1 \oplus \cdots \oplus \mathbb{C}b_n$ of V, so it is imprimitive for $n > 1$. The order of $G(de,e,n)$ is given by $d^n e^{n-1} n!$. Using the wreath product structure it is easy to show that the only isomorphisms among groups in this series are $G(2,1,2) \cong G(4,4,2)$, and $G(de,e,1) \cong G(d,1,1)$ for all d, e, while $G(2,2,2)$ is reducible. All these are isomorphisms of reflection groups.

In its natural action on $\mathbb{Q}^n$ the symmetric group $\mathfrak{S}_n$ stabilises the 1-dimensional subspace consisting of vectors with all coordinates equal and the $(n-1)$-dimensional subspace consisting of those vectors whose coordinates add up to 0. In its action on the latter, $\mathfrak{S}_n$ is an irreducible and primitive reflection group. The classification result may now be stated as follows (see also Cohen, [50]):

1.14. THEOREM (Shephard and Todd, [167]). *The irreducible complex reflection groups are the groups $G(de,e,n)$, for $de \geqslant 2$, $n \geqslant 1$, $(de,e,n) \neq (2,2,2)$, the groups $\mathfrak{S}_n$ $(n \geqslant 2)$ in their $(n-1)$-dimensional natural representation, and 34 further primitive groups.*

Moreover, any irreducible n-dimensional complex reflection group has a generating set of at most $n+1$ reflections.

The primitive groups are usually denoted by $G_4, \ldots, G_{37}$, as in the original article [167] (where the first three indices were reserved for the families of imprimitive groups $G(de,e,n)$ $(de, n \geqslant 2)$, cyclic groups $G(d,1,1)$ and symmetric groups $\mathfrak{S}_{n+1}$). An n-dimensional irreducible reflection groups generated by n of its reflections is called *well-generated*. The groups for which this fails are the imprimitive groups $G(de,e,n)$, $d, e, n > 1$, and the primitive groups

$$G_i \quad \text{with } i \in \{7, 11, 12, 13, 15, 19, 22, 31\}.$$

By construction $G(de,e,n)$ contains the well-generated group $G(de,de,n)$. More generally, the classification implies the following, for which no a priori proof is known:

1.15. COROLLARY. *Any irreducible complex reflection group $W \leqslant \mathrm{GL}(V)$ contains a well-generated reflection subgroup $W' \leqslant W$ which is still irreducible on V.*

The primitive groups $G_4, \ldots, G_{37}$ occur in dimensions 2 up to 8. In Table 1 we collect some data on the irreducible complex reflection groups. (These and many more data for complex reflection groups have been implemented by Jean Michel into the CHEVIE-system, [89].) In the first part, the dimension is always equal to n, in the second it can be read of from the number of degrees. We give the degrees, the codegrees in case they are not described by the following Theorem 1.16 (that is, if W is not well-generated), and the character field K_W of the reflection representation. For the exceptional groups we also give the structure of $W/Z(W)$ and we indicate the regular degrees by boldface (that is, those degrees which are regular numbers for W, see Cohen, [50, p. 395 and p. 412], and Springer, [170, Tables 1–6]). The regular degrees for the infinite series are: $n, n+1$ for $\mathfrak{S}_{n+1}$, dn for $G(de, e, n)$ with $d > 1$, $(n-1)e$ for $G(e, e, n)$ with $n|e$, and $(n-1)e, n$ for $G(e, e, n)$ with $n\nmid e$. Lehrer and Michel, [130, Theorem 3.1], have shown that an integer is a regular number if and only if it divides as many degrees as codegrees.

Fundamental invariants for most types are given in Shephard and Todd, [167], as well as defining relations and further information on parabolic subgroups (see also Coxeter, [56], Shephard, [166], and Broué, Malle and Rouquier, [38], for presentations, and the tables in Cohen, [50], and Broué, Malle and Rouquier, [38, Appendix 2]).

If the irreducible reflection group W has an invariant of degree 2, then it leaves invariant a non-degenerate quadratic form, so the representation may be realised over the reals. Conversely, if W is a real reflection group, then it leaves invariant a quadratic form. Thus the real irreducible reflection groups are precisely those with $d_1 = 2$, that is, the infinite series $G(2, 1, n)$, $G(2, 2, n)$, $G(e, e, 2)$ and $\mathfrak{S}_{n+1}$, and the six exceptional groups $G_{23}, G_{28}, G_{30}, G_{35}, G_{36}, G_{37}$ (see Section 2.5).

The degrees and codegrees of a finite complex reflection group satisfy some remarkable identities. As an example, let us quote the following result, for which at present only a case-by-case proof is known:

1.16. THEOREM (Orlik and Solomon, [154]). *Let W be an irreducible complex reflection group in dimension n. Then the following are equivalent*:

(i) $d_i + d^*_{n-i+1} = d_n$ *for* $i = 1, \ldots, n$,

(ii) $N + N^* = nd_n$,

(iii) $d^*_i < d_n$ *for* $i = 1, \ldots, n$,

(iv) *W is well-generated.*

See also Terao and Yano, [180], for a partial explanation.

From the Shephard–Todd classification, it is straightforward to obtain a classification of reflection data. An easy argument allows to reduce to the case where W acts irreducibly on V. Then either up to scalars ϕ can be chosen to be a reflection, or $W = G_{28}$ is the real reflection group of type F_4, and ϕ induces the graph automorphism on the F_4-diagram (see Broué, Malle and Michel, [37, Proposition 3.13]). An infinite series of examples is obtained from the embedding of $G(de, e, n)$ into $G(de, 1, n)$ (which is the full projective normaliser in all but finitely many cases). Apart from this, there are only six further cases, which we list in Table 2.

Table 1
Irreducible complex reflection groups

Infinite series

W	Degrees	Codegrees	K_W
$G(d,1,n)$ $(d \geqslant 2, n \geqslant 1)$	$d, 2d, \ldots, nd$	$*$	$\mathbb{Q}(\zeta_d)$
$G(de,e,n)$ $(d,e,n \geqslant 2)$	$ed, 2ed, \ldots, (n-1)ed, nd$	$0, ed, \ldots, (n-1)ed$	$\mathbb{Q}(\zeta_{de})$
$G(e,e,n)$ $(e \geqslant 2, n \geqslant 3)$	$e, 2e, \ldots, (n-1)e, n$	$*$	$\mathbb{Q}(\zeta_e)$
$G(e,e,2)$ $(e \geqslant 3)$	$2, e$	$*$	$\mathbb{Q}(\zeta_e + \zeta_e^{-1})$
$\mathfrak{S}_{n+1}$ $(n \geqslant 1)$	$2, 3, \ldots, n+1$	$*$	$\mathbb{Q}$

Exceptional groups

W	Degrees	Codegrees	K_W	$W/Z(W)$
G_4	**4, 6**	$*$	$\mathbb{Q}(\zeta_3)$	$\mathfrak{A}_4$
G_5	6, **12**	$*$	$\mathbb{Q}(\zeta_3)$	$\mathfrak{A}_4$
G_6	4, **12**	$*$	$\mathbb{Q}(\zeta_{12})$	$\mathfrak{A}_4$
G_7	12, **12**	0, 12	$\mathbb{Q}(\zeta_{12})$	$\mathfrak{A}_4$
G_8	**8, 12**	$*$	$\mathbb{Q}(i)$	$\mathfrak{S}_4$
G_9	8, **24**	$*$	$\mathbb{Q}(\zeta_8)$	$\mathfrak{S}_4$
G_{10}	12, **24**	$*$	$\mathbb{Q}(\zeta_{12})$	$\mathfrak{S}_4$
G_{11}	24, **24**	0, 24	$\mathbb{Q}(\zeta_{24})$	$\mathfrak{S}_4$
G_{12}	**6, 8**	0, 10	$\mathbb{Q}(\sqrt{-2})$	$\mathfrak{S}_4$
G_{13}	8, **12**	0, 16	$\mathbb{Q}(\zeta_8)$	$\mathfrak{S}_4$
G_{14}	6, **24**	$*$	$\mathbb{Q}(\zeta_3, \sqrt{-2})$	$\mathfrak{S}_4$
G_{15}	**12**, 24	0, 24	$\mathbb{Q}(\zeta_{24})$	$\mathfrak{S}_4$
G_{16}	**20, 30**	$*$	$\mathbb{Q}(\zeta_5)$	$\mathfrak{A}_5$
G_{17}	20, **60**	$*$	$\mathbb{Q}(\zeta_{20})$	$\mathfrak{A}_5$
G_{18}	30, **60**	$*$	$\mathbb{Q}(\zeta_{15})$	$\mathfrak{A}_5$
G_{19}	60, **60**	0, 60	$\mathbb{Q}(\zeta_{60})$	$\mathfrak{A}_5$
G_{20}	**12, 30**	$*$	$\mathbb{Q}(\zeta_3, \sqrt{5})$	$\mathfrak{A}_5$
G_{21}	12, **60**	$*$	$\mathbb{Q}(\zeta_{12}, \sqrt{5})$	$\mathfrak{A}_5$
G_{22}	**12, 20**	0, 28	$\mathbb{Q}(i, \sqrt{5})$	$\mathfrak{A}_5$
G_{23}	2, **6, 10**	$*$	$\mathbb{Q}(\sqrt{5})$	$\mathfrak{A}_5$
G_{24}	4, **6, 14**	$*$	$\mathbb{Q}(\sqrt{-7})$	$\mathrm{GL}_3(2)$
G_{25}	6, **9, 12**	$*$	$\mathbb{Q}(\zeta_3)$	$3^2 : \mathrm{SL}_2(3)$
G_{26}	6, 12, **18**	$*$	$\mathbb{Q}(\zeta_3)$	$3^2 : \mathrm{SL}_2(3)$
G_{27}	6, 12, **30**	$*$	$\mathbb{Q}(\zeta_3, \sqrt{5})$	$\mathfrak{A}_6$
G_{28}	2, 6, **8, 12**	$*$	$\mathbb{Q}$	$2^4 : (\mathfrak{S}_3 \times \mathfrak{S}_3)$
G_{29}	4, 8, 12, **20**	$*$	$\mathbb{Q}(i)$	$2^4 : \mathfrak{S}_5$
G_{30}	2, **12, 20, 30**	$*$	$\mathbb{Q}(\sqrt{5})$	$\mathfrak{A}_5 \wr 2$
G_{31}	8, 12, **20, 24**	0, 12, 16, 28	$\mathbb{Q}(i)$	$2^4 : \mathfrak{S}_6$
G_{32}	12, 18, **24, 30**	$*$	$\mathbb{Q}(\zeta_3)$	$\mathrm{U}_4(2)$
G_{33}	4, 6, **10**, 12, **18**	$*$	$\mathbb{Q}(\zeta_3)$	$\mathrm{O}_5(3)$
G_{34}	6, 12, 18, 24, 30, **42**	$*$	$\mathbb{Q}(\zeta_3)$	$\mathrm{O}_6^-(3).2$
G_{35}	2, 5, 6, **8, 9, 12**	$*$	$\mathbb{Q}$	$\mathrm{O}_6^-(2)$
G_{36}	2, 6, 8, 10, 12, **14, 18**	$*$	$\mathbb{Q}$	$\mathrm{O}_7(2)$
G_{37}	2, 8, 12, 14, 18, **20, 24, 30**	$*$	$\mathbb{Q}$	$\mathrm{O}_8^+(2).2$

Table 2
Exceptional twisted reflection data

W	d_i	ε_i	Field	Origin of ϕ
$G(4,2,2)$	4, 4	$1, \zeta_3$	$\mathbb{Q}(i)$	$< G_6$
$G(3,3,3)$	3, 6, 3	$1, 1, -1$	$\mathbb{Q}(\zeta_3)$	$< G_{26}$
$G(2,2,4)$	2, 4, 4, 6	$1, \zeta_3, \zeta_3^2, 1$	$\mathbb{Q}$	$< G_{28}$
G_5	6, 12	$1, -1$	$\mathbb{Q}(\zeta_3, \sqrt{-2})$	$< G_{14}$
G_7	12, 12	$1, -1$	$\mathbb{Q}(\zeta_{12})$	$< G_{10}$
G_{28}	2, 6, 8, 12	$1, -1, 1, -1$	$\mathbb{Q}(\sqrt{2})$	graph aut.

2. Real reflection groups

In this section we discuss in more detail the special case where W is a *real* reflection group. This is a well-developed theory, and there are several good places to learn about real reflection groups: the classical Bourbaki volume, [25], the very elementary text by Benson and Grove, [8], the relevant chapters in Curtis and Reiner, [58], Hiller, [99], and Humphreys, [105]. Various pieces of the theory have also been recollected in a concise way in articles by Steinberg, [178]. The exposition here partly follows Geck and Pfeiffer, [95, Chapter 1]. We shall only present the most basic results and refer to the above textbooks and our bibliography for further reading.

2.1. *Coxeter groups*

Let S be a finite non-empty index set and $M = (m_{st})_{s,t \in S}$ be a symmetric matrix such that $m_{ss} = 1$ for all $s \in S$ and $m_{st} \in \{2, 3, 4, \dots\} \cup \{\infty\}$ for all $s \neq t$ in S. Such a matrix is called a *Coxeter matrix*. Now let W be a group containing S as a subset. (W may be finite or infinite.) Then the pair (W, S) is called a *Coxeter system*, and W is called a *Coxeter group*, if W has a presentation with generators S and defining relations of the form

$$(st)^{m_{st}} = 1 \quad \text{for all } s, t \in S \text{ with } m_{st} < \infty;$$

in particular, this means that $s^2 = 1$ for all $s \in S$. Therefore, the above relations (for $s \neq t$) can also be expressed in the form

$$\underbrace{sts\cdots}_{m_{st}\text{ times}} = \underbrace{tst\cdots}_{m_{st}\text{ times}} \quad \text{for all } s, t \in S \text{ with } 2 \leqslant m_{st} < \infty.$$

We say that C is of finite type and that (W, S) is a finite Coxeter system if W is a finite group. The information contained in M can be visualised by a corresponding *Coxeter graph*, which is defined as follows. It has vertices labelled by the elements of S, and two vertices labelled by $s \neq t$ are joined by an edge if $m_{st} \geqslant 3$. Moreover, if $m_{st} \geqslant 4$, we label the edge by m_{st}. The standard example of a finite Coxeter system is the pair $(\mathfrak{S}_n, \{s_1, \dots, s_{n-1}\})$ where $s_i = (i, i+1)$ for $1 \leqslant i \leqslant n-1$. The corresponding graph is

Coxeter groups have a rich combinatorial structure. A basic tool is the *length function* $l: W \to \mathbb{N}_0$, which is defined as follows. Let $w \in W$. Then $l(w)$ is the length of a shortest possible expression $w = s_1 \cdots s_k$ where $s_i \in S$. An expression of w of length $l(w)$ is called a *reduced expression* for w. We have $l(1) = 0$ and $l(s) = 1$ for $s \in S$. Here is a key result about Coxeter groups.

2.2. THEOREM (Matsumoto, [141]; see also Bourbaki, [25]). *Let (W, S) be a Coxeter system and $\mathcal{M}$ be a monoid, with multiplication $\star: \mathcal{M} \times \mathcal{M} \to \mathcal{M}$. Let $f: S \to \mathcal{M}$ be a map such that*

$$\underbrace{f(s) \star f(t) \star f(s) \star \cdots}_{m_{st} \text{ times}} = \underbrace{f(t) \star f(s) \star f(t) \star \cdots}_{m_{st} \text{ times}}$$

for all $s \neq t$ in S such that $m_{st} < \infty$. Then there exists a unique map $F: W \to \mathcal{M}$ such that $F(w) = f(s_1) \star \cdots \star f(s_k)$ whenever $w = s_1 \cdots s_k$ $(s_i \in S)$ is reduced.

Typically, Matsumoto's theorem can be used to show that certain constructions with reduced expressions of elements of W actually do not depend on the choice of the reduced expressions. We give two examples.

(1) *Let $w \in W$ and take a reduced expression $w = s_1 \cdots s_k$ with $s_i \in S$. Then the set $\{s_1, \ldots, s_k\}$ does not depend on the choice of the reduced expression.*

(Indeed, let $\mathcal{M}$ be the monoid whose elements are the subsets of S and product given by $A \star B := A \cup B$. Then the assumptions of Matsumoto's theorem are satisfied for the map $f: S \to \mathcal{M}$, $s \mapsto \{s\}$, and this yields the required assertion.)

(2) *Let $w \in W$ and fix a reduced expression $w = s_1 \cdots s_k$ $(s_i \in S)$. Consider the set of all subexpressions*:

$$\mathcal{S}(w) := \{y \in W \mid y = s_{i_1} \cdots s_{i_l} \text{ where } l \geqslant 0 \text{ and } 1 \leqslant i_1 < \cdots < i_l \leqslant k\}.$$

Then $\mathcal{S}(w)$ does not depend on the choice of the reduced expression for w.

(Indeed, let $\mathcal{M}$ be the monoid whose elements are the subsets of W and product given by $A \star B := \{ab \mid a \in A, b \in B\}$ (for $A, B \subseteq W$). Then the assumptions of Matsumoto's theorem are satisfied for the map $f: S \to \mathcal{M}$, $s \mapsto \{1, s\}$, and this yields the required assertion.)

We also note that the so-called *exchange condition* and the *cancellation law* are further consequences of the above results. The "cancellation law" states that, given $w \in W$ and an expression $w = s_1 \cdots s_k$ $(s_i \in S)$ which is not reduced, one can obtain a reduced expression of w by simply cancelling some of the factors in the given expression. This law together with (2) yields that the relation

$$y \leqslant w \stackrel{\text{def}}{\Longleftrightarrow} y \in \mathcal{S}(w)$$

is a partial order on W, called the *Bruhat–Chevalley order*. This ordering has been extensively studied; see, for example, Verma, [184], Deodhar, [65], Björner, [19], Lascoux and

Schützenberger, [126], and Geck and Kim, [90]. By Chevalley, [48], it is related to the Bruhat decomposition in algebraic groups; we will explain this result in 2.16 below.

2.3. *Cartan matrices*

Let $M = (m_{st})_{s,t\in S}$ be a Coxeter matrix as above. We can also associate with M a group generated by reflections. This is done as follows. Choose a matrix $C = (c_{st})_{s,t\in S}$ with entries in $\mathbb{R}$ such that the following conditions are satisfied:

(C1) For $s \neq t$ we have $c_{st} \leqslant 0$; furthermore, $c_{st} \neq 0$ if and only if $c_{ts} \neq 0$.

(C2) We have $c_{ss} = 2$ and, for $s \neq t$, we have $c_{st}c_{ts} = 4\cos^2(\pi/m_{st})$.

Such a matrix C will be called a *Cartan matrix* associated with M. For example, we could simply take $c_{st} := -2\cos(\pi/m_{st})$ for all $s, t \in S$; this may be called the *standard Cartan matrix* associated with M. We always have $0 \leqslant c_{st}c_{ts} \leqslant 4$. Here are some values for the product $c_{st}c_{ts}$:

m_{st}	2	3	4	5	6	8	∞
$c_{st}c_{ts}$	0	1	2	$(3+\sqrt{5})/2$	3	$2+\sqrt{2}$	4

Now let V be an $\mathbb{R}$-vector space of dimension $|S|$, with a fixed basis $\{\alpha_s \mid s \in S\}$. We define a linear action of the elements in S on V by the rule:

$$s : V \to V, \quad \alpha_t \mapsto \alpha_t - c_{st}\alpha_s \ (t \in S).$$

It is easily checked that $s \in \mathrm{GL}(V)$ has order 2 and precisely one eigenvalue -1 (with eigenvector α_s). Thus, s is a reflection with root α_s. We then define

$$W = W(C) := \langle S \rangle \subset \mathrm{GL}(V);$$

thus, if $|W| < \infty$ is finite, then W will be a real reflection group. Now we can state the following basic result.

2.4. THEOREM (Coxeter, [54,55]). *Let $M = (m_{st})_{s,t\in S}$ be a Coxeter matrix and C be a Cartan matrix associated with M. Let $W(C) = \langle S \rangle \subseteq \mathrm{GL}(V)$ be the group constructed as in 2.3. Then the pair $(W(C), S)$ is a Coxeter system. Furthermore, the group $W(C)$ is finite if and only if*

$$\text{the matrix } \big(-\cos(\pi/m_{st})\big)_{s,t\in S} \text{ is positive-definite,} \qquad (*)$$

i.e., we have $\det(-\cos(\pi/m_{st}))_{s,t\in J} > 0$ for every subset $J \subseteq S$. All finite real reflection groups arise in this way.

The fact that $(W(C), S)$ is a Coxeter system is proved in [95, 1.2.7]. The finiteness condition can be found in Bourbaki, [25, Chapter V, §4, no. 8]. Finally, the fact that all finite real reflection groups arise in this way is established in [25, Chapter V, §3, no. 2]. The "note historique" in [25] contains a detailed account of the history of the above result.

2.5. *Classification of finite Coxeter groups*

(See also Section 1.13.) Let $M = (m_{st})_{s,t \in S}$ be a Coxeter matrix. We say that M is *decomposable* if there is a partition $S = S_1 \amalg S_2$ with $S_1, S_2 \neq \varnothing$ and such that $m_{st} = 2$ whenever $s \in S_1$, $t \in S_2$. If $C = (c_{st})$ is any Cartan matrix associated with M, then this condition translates to: $c_{st} = c_{ts} = 0$ whenever $s \in S_1$, $t \in S_2$. Correspondingly, we also have a direct sum decomposition $V = V_1 \oplus V_2$ where V_1 has basis $\{\alpha_s \mid s \in S_1\}$ and V_2 has basis $\{\alpha_s \mid s \in S_2\}$. Then it easily follows that we have an isomorphism

$$W(C) \xrightarrow{\sim} W(C_1) \times W(C_2), \qquad w \mapsto (w|_{V_1}, w|_{V_2}).$$

In this way, the study of the groups $W(C)$ is reduced to the case where C is *indecomposable* (i.e., there is no partition $S = S_1 \amalg S_1$ as above). If this holds, we call the corresponding Coxeter system (W, S) an *irreducible Coxeter system.*

2.6. THEOREM. *The Coxeter graphs of the indecomposable Coxeter matrices M such that condition* $(*)$ *in Theorem* 2.4 *holds are precisely the graphs in Table* 3.

For the proof of this classification, see [25, Chapter VI, no. 4.1]. The identification with the groups occurring in the Shephard–Todd classification (see Theorem 1.14) is given in the following table.

Coxeter graph	Shephard–Todd
A_{n-1}	$\mathfrak{S}_n$
B_n	$G(2, 1, n)$
D_n	$G(2, 2, n)$
$I_2(m)$	$G(m, m, 2)$
H_3	G_{23}
H_4	G_{30}
F_4	G_{28}
E_6	G_{35}
E_7	G_{36}
E_8	G_{37}

Thus, any finite irreducible real reflection group is the reflection group arising from a Cartan matrix associated with one of the graphs in Table 3.

Now, there are a number of results on finite Coxeter groups which can be formulated in general terms but whose proof requires a case-by-case verification using the above classification. We mention two such results, concerning conjugacy classes.

2.7. THEOREM (Carter, [41]). *Let (W, S) be a finite Coxeter system. Then every element in W is conjugate to its inverse. More precisely, given $w \in W$, there exist $x, y \in W$ such that $w = xy$ and $x^2 = y^2 = 1$.*

Every element $x \in W$ such that $x^2 = 1$ is a product of pairwise commuting reflections in W. Given $w \in W$ and an expression $w = xy$ as above, the geometry of the roots involved in the reflections determining x, y yields a diagram which can be used to label the conjugacy class of w. Complete lists of these diagrams can be found in [41].

Table 3
Coxeter graphs of irreducible finite Coxeter groups

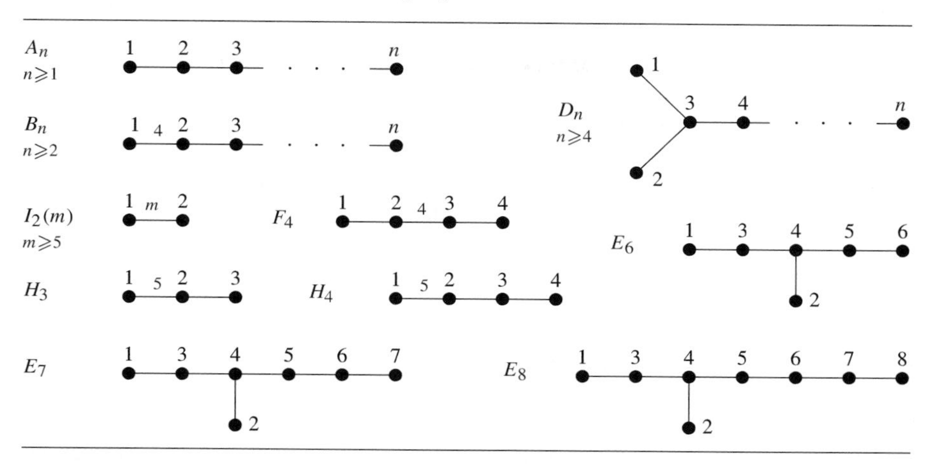

The numbers on the vertices correspond to a chosen labelling of the elements of S.

2.8. *Conjugacy classes and the length function*

Let (W, S) be a Coxeter system and C be a conjugacy class in W. We will be interested in studying how conjugation inside C relates to the length function on W. For this purpose, we introduce two relations, following Geck and Pfeiffer, [94].

Given $x, y \in W$ and $s \in S$, we write $x \xrightarrow{s} y$ if $y = sxs$ and $l(y) \leqslant l(x)$. We shall write $x \to y$ if there are sequences $x_0, x_1, \ldots, x_n \in W$ and $s_1, \ldots, s_n \in S$ (for some $n \geqslant 0$) such that

$$x = x_0 \xrightarrow{s_1} x_1 \xrightarrow{s_2} x_2 \xrightarrow{s_3} \cdots \xrightarrow{s_n} x_n = y.$$

Thus, we have $x \to y$ if we can go from x to y by a chain of conjugations by generators in S such that, at each step, the length of the elements either remains the same or decreases.

In a slightly different direction, let us now consider two elements $x, y \in W$ such that $l(x) = l(y)$. We write $x \overset{w}{\sim} y$ (where $w \in W$) if $wx = yw$ and $l(wx) = l(w) + l(x)$ or $xw = wy$ and $l(wy) = l(w) + l(y)$. We write $x \sim y$ if there are sequences $x_0, x_1, \ldots, x_n \in W$ and $w_1, \ldots, w_n \in W$ (for some $n \geqslant 0$) such that

$$x = x_0 \overset{w_1}{\sim} x_1 \overset{w_2}{\sim} x_2 \overset{w_3}{\sim} \cdots \overset{w_n}{\sim} x_n = y.$$

Thus, we have $x \sim y$ if we can go from x to y by a chain of conjugations with elements of W such that, at each step, the length of the elements remains the same and an additional length condition involving the conjugating elements is satisfied. This additional condition has the following significance. Consider a group $\mathcal{M}$ with multiplication $\star$ and assume that we have a map $f : S \to \mathcal{M}$ which satisfies the requirements in Matsumoto's Theorem 2.2.

Then we have a canonical extension of f to a map $F: W \to \mathcal{M}$ such that $F(ww') = F(w) \star F(w')$ whenever $l(ww') = l(w) + l(w')$. Hence, in this setting, we have

$$x \sim y \Rightarrow F(x), F(y) \text{ are conjugate in } \mathcal{M}.$$

Thus, we can think of the relation "$\sim$" as "universal conjugacy".

2.9. THEOREM (Geck and Pfeiffer, [94,95]). *Let (W, S) be a finite Coxeter system and C be a conjugacy class in W. We set $l_{\min}(C) := \min\{l(w) \mid w \in C\}$ and*

$$C_{\min} := \{w \in C \mid l(w) = l_{\min}(C)\}.$$

Then the following hold:

(a) *For every $x \in C$, there exists some $y \in C_{\min}$ such that $x \to y$.*
(b) *For any two elements $x, y \in C_{\min}$, we have $x \sim y$.*

Precursors of the above result for type A have been found much earlier by Starkey, [173]; see also Ram, [162]. The above result allows to define the *character table* of the Iwahori–Hecke algebra associated with (W, S). This is discussed in more detail in the chapter on Hecke algebras. See Richardson, [165], Geck and Michel, [93], Geck and Pfeiffer, [95, Chapter 3], Geck, Kim and Pfeiffer, [91] and Shi, [168], for further results on conjugacy classes. Krammer, [121], studies the conjugacy problem for arbitrary (infinite) Coxeter groups.

2.10. *The crystallographic condition*

Let $M = (m_{st})_{s,t \in S}$ be a Coxeter matrix such that the connected components of the corresponding Coxeter graph occur in Table 3. We say that M satisfies the *crystallographic condition* if there exists a Cartan matrix C associated with M which has integral coefficients. In this case, the corresponding reflection group $W = W(C)$ is called a *Weyl group*. The significance of this notion is that there exists a corresponding semisimple Lie algebra over $\mathbb{C}$; see 2.15.

Now assume that M is crystallographic. By condition (C2) this implies $m_{st} \in \{2, 3, 4, 6\}$ for all $s \neq t$. Conversely, if m_{st} satisfies this condition, then we have

$$\begin{aligned} c_{st}c_{ts} &= 0 \quad \text{if } m_{st} = 2, \\ c_{st}c_{ts} &= 1 \quad \text{if } m_{st} = 3, \\ c_{st}c_{ts} &= 2 \quad \text{if } m_{st} = 4, \\ c_{st}c_{ts} &= 3 \quad \text{if } m_{st} = 6. \end{aligned}$$

Thus, in each of these cases, we see that there are only two choices for c_{st} and c_{ts}: we must have $c_{st} = -1$ or $c_{ts} = -1$ (and then the other value is determined). We encode this additional information in the Coxeter graph, by putting an arrow on the edge between the nodes labelled by s, t according to the following scheme:

	Edge between $s \neq t$	Values for c_{st}, c_{ts}	
$m_{st} = 3$	no arrow	$c_{st} = c_{ts} = -1$	
$m_{st} = 4$	s t	$c_{st} = -1$	$c_{ts} = -2$
$m_{st} = 6$	s t	$c_{st} = -1$	$c_{ts} = -3$

Table 4
Dynkin diagrams of Cartan matrices of finite type

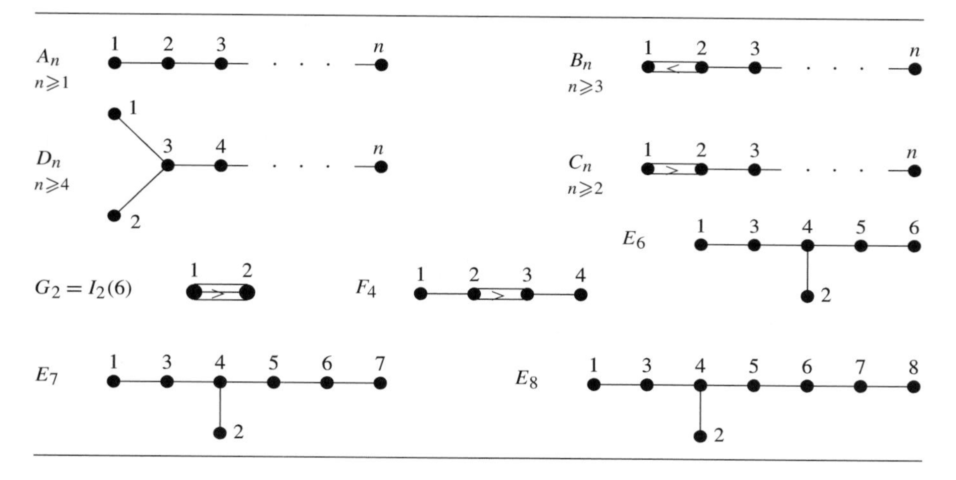

The Coxeter graph of M equipped with this additional information will be called a *Dynkin diagram*; it uniquely determines an integral Cartan matrix C associated with M (if such a Cartan matrix exists). The complete list of connected components of Dynkin diagrams is given in table 4. We see that all irreducible finite Coxeter groups are Weyl groups, except for those of type H_3, H_4 and $I_2(m)$ where $m = 5$ or $m \geqslant 7$. Note that type B_n is the only case where we have two different Dynkin diagrams associated with the same Coxeter graph.

The following discussion of root systems associated with finite reflection groups follows the appendix on finite reflection groups in Steinberg, [176].

2.11. *Root systems*

Let V be a finite-dimensional real vector space and let $(\ ,\ )$ be a positive-definite scalar product on V. Given a non-zero vector $\alpha \in V$, the corresponding reflection $w_\alpha \in \mathrm{GL}(V)$ is defined by

$$w_\alpha(v) = v - 2\frac{(v, \alpha)}{(\alpha, \alpha)}\alpha \quad \text{for all } v \in V.$$

Note that, for any $w \in \mathrm{GL}(V)$, we have $ww_\alpha w^{-1} = w_{w(\alpha)}$. A finite subset $\Phi \subseteq V \setminus \{0\}$ is called a *root system* if the following conditions are satisfied:

(R1) For any $\alpha \in \Phi$, we have $\Phi \cap \mathbb{R}\alpha = \{\pm\alpha\}$;

(R2) For every $\alpha, \beta \in \Phi$, we have $w_\alpha(\beta) \in \Phi$.

Let $W(\Phi) \subseteq \mathrm{GL}(V)$ be the subgroup generated by the reflections w_α ($\alpha \in \Phi$). A subset $\Pi \subseteq \Phi$ is called a *simple system* if Π is linearly independent and if any root in Φ can be written as a linear combination of the elements of Π in which all non-zero coefficients are either all positive or all negative. It is known that simple systems always exist, and that any two simple systems can be transformed into each other by an element of $W(\Phi)$. Let now fix a simple system $\Pi \subseteq \Phi$. Then it is also known that

$$W(\Phi) = \langle w_\alpha \mid \alpha \in \Pi \rangle \quad \text{and} \quad (\alpha, \beta) \geqslant 0 \quad \text{for all } \alpha \neq \beta \text{ in } \Pi.$$

For $\alpha \neq \beta$ in Π, let $m_{\alpha\beta} \geqslant 2$ be the order of $w_\alpha w_\beta$ in $\mathrm{GL}(V)$. Then $w_\alpha w_\beta$ is a rotation in V through the angle $2\pi/m_{\alpha\beta}$ and that we have the relation

$$\frac{(\alpha, \beta)}{(\alpha, \alpha)} \frac{(\beta, \alpha)}{(\beta, \beta)} = \cos^2(\pi/m_{\alpha\beta}) \quad \text{for all } \alpha \neq \beta \text{ in } \Pi.$$

Thus, if we set $M := (m_{\alpha\beta})_{\alpha,\beta\in\Pi}$ (where $m_{\alpha\alpha} = 1$ for all $\alpha \in \Pi$) and define

$$C := (c_{\alpha\beta})_{\alpha,\beta\in\Pi} \quad \text{where } c_{\alpha\beta} := 2\frac{(\alpha, \beta)}{(\alpha, \alpha)} \text{ for } \alpha, \beta \in S,$$

then M is a Coxeter matrix, C is an associated Cartan matrix and $W(\Phi) = W(C)$ is the Coxeter group with Coxeter matrix M; see theorem 2.4. Thus, every root system leads to a finite Coxeter group.

Conversely, let (W, S) be a Coxeter system associated to a Coxeter matrix $M = (m_{st})_{s,t\in S}$. Let C be a corresponding Cartan matrix such that $c_{st} = c_{ts} = -2\cos(\pi/m_{st})$ for every $s \neq t$ in S such that m_{st} is odd. Then one can show that

$$\Phi = \Phi(C) := \{w(\alpha_s) \mid w \in W, s \in S\} \subseteq V$$

is a root system with simple system $\{\alpha_s \mid s \in S\}$ and that $W(\Phi) = W$; see, e.g., Geck and Pfeiffer, [95, Chapter 1]. Thus, every Coxeter group leads to a root system.

Given a root system Φ as above, there is a strong link between the combinatorics of the Coxeter presentation of $W = W(\Phi)$ and the geometry of Φ. To state the following basic result, we set

$$\Phi^+ := \left\{\alpha \in \Pi \,\middle|\, \alpha = \sum_{\beta\in\Pi} x_\beta \beta \text{ where } x_\beta \in \mathbb{R}_{\geqslant 0} \text{ for all } \beta \in \Pi\right\};$$

the roots in Φ^+ will be called *positive roots*. Similarly, $\Phi^- := -\Phi^+$ will be called the set of *negative roots*. By the definition of a simple system, we have $\Phi = \Phi^+ \amalg \Phi^-$.

2.12. PROPOSITION. *Given $\alpha \in \Pi$ and $w \in W$, we have*

$$w^{-1}(\alpha) \in \Phi^+ \Leftrightarrow l(w_\alpha w) = l(w) + 1,$$
$$w^{-1}(\alpha) \in \Phi^- \Leftrightarrow l(w_\alpha w) = l(w) - 1.$$

Furthermore, for any $w \in W$, we have $l(w) = |\{\alpha \in \Phi^+ \mid w(\alpha) \in \Phi^-\}|$.

The root systems associated with finite Weyl groups are explicitly described in Bourbaki, [25, pp. 251–276]. For type H_3 and H_4, see Humphreys, [105, 2.13].

Bremke and Malle, [26,27], have studied suitable generalisations of root systems and length functions for the infinite series $G(d,1,n)$ and $G(e,e,n)$, which have subsequently been extended in weaker form by Rampetas and Shoji, [163], to arbitrary imprimitive reflection groups. For investigations of root systems see also Nebe, [147], and Hughes and Morris, [103]. But there is no general theory of root systems and length functions for complex reflection groups (yet).

2.13. *Torsion primes*

Assume that $\Phi \subseteq V$ is a root system as above, with a set of simple roots $\Pi \subseteq \Phi$. Assume that the corresponding Cartan matrix C is indecomposable and has integral coefficients. Thus, its Dynkin diagram is one of the graphs in Table 4. Following Springer and Steinberg, [172, §I.4], we shall now discuss "bad primes" and "torsion primes" with respect to Φ.

For every $\alpha \in \Phi$, the corresponding coroot is defined by $\alpha^* := 2\alpha/(\alpha,\alpha)$. Then $\Phi^* := \{\alpha^* \mid \alpha \in \Phi\}$ also is a root system, the dual of Φ. The Dynkin diagram of Φ^* is obtained from that of Φ by reversing the arrows. (For example, the dual of a root system of type B_n is of type C_n.)

Let $L(\Phi)$ denote the lattice spanned by Φ in V. A prime number $p > 0$ is called *bad* for Φ if $L(\Phi)/L(\Phi_1)$ has p-torsion for some (integrally) closed subsystem Φ_1 of Φ. The prime p is called a *torsion prime* if $L(\Phi^*)/L(\Phi_1^*)$ has p-torsion for some closed subsystem Φ_1 of Φ. Note that Φ_1^* need not be closed in Φ^*, and so the torsion primes for Φ and the bad primes for Φ^* need not be the same. The bad primes can be characterised as follows. Let $\alpha_0 = \sum_{\alpha \in \Pi} m_\alpha \alpha$ be the unique positive root of maximal height. (The *height* of a root is the sum of the coefficients in the expression of that root as a linear combination of simple roots.) Then we have:

$$p \text{ bad} \Leftrightarrow \begin{array}{c} p = m_\alpha \\ \text{for some } \alpha \end{array} \Leftrightarrow \begin{array}{c} p \text{ divides } m_\alpha \\ \text{for some } \alpha \end{array} \Leftrightarrow \begin{array}{c} p \leqslant m_\alpha \\ \text{for some } \alpha \end{array}.$$

Now let $\alpha_0^* = \sum_{\alpha \in \Pi} m_\alpha^* \alpha^*$. Then p is a torsion prime if and only if p satisfies one of the above conditions, with m_α replaced by m_α^*. For the various roots systems, the bad primes and the torsion primes are given as follows.

Type	A_n	B_n $(n \geqslant 2)$	C_n $(n \geqslant 2)$	D_n $(n \geqslant 4)$	G_2	F_4	E_6	E_7	E_8
Bad	none	2	2	2	2, 3	2, 3	2, 3	2, 3	2, 3, 5
Torsion	none	2	none	2	2	2, 3	2, 3	2, 3	2, 3, 5

The bad primes and torsion primes play a role in various questions related to sub-root systems, centralisers of semisimple elements in algebraic groups, the classification of unipotent classes in simple algebraic groups and so on; see [172] and also the survey in [43, §§1.14–1.15].

2.14. *Affine Weyl groups*

Let $\Phi \subseteq V$ be a root system as above, with Weyl group W. Let $L(\Phi) := \sum_{\alpha \in \Pi} \mathbb{Z}\alpha \subseteq V$ be the lattice spanned by the roots in V. Then W leaves $L(\Phi)$ invariant and we have a natural group homomorphism $W \to \mathrm{Aut}(L(\Phi))$. The semidirect product

$$W_a(\Phi) := L(\Phi) \rtimes W$$

is called the affine Weyl group associated with the root system W; see Bourbaki, [25, Chapter VI, §2]. The group $W_a(\Phi)$ itself is a Coxeter group. The corresponding presentation can also be encoded in a graph, as follows. Let α_0 be the unique positive root of maximal height in Φ. We define an extended Cartan matrix $\widetilde{C}$ by similar rules as before:

$$\tilde{c}_{\alpha\beta} := 2\frac{(\alpha, \beta)}{(\alpha, \alpha)} \quad \text{for } \alpha, \beta \in \Pi \cup \{-\alpha_0\}.$$

The extended Dynkin diagrams encoding these matrices for irreducible W are given in Table 5. They are obtained from the diagrams in Table 4 by adjoining an additional node (corresponding to $-\alpha_0$) and putting edges according to the same rules as before.

Table 5
Extended Dynkin diagrams

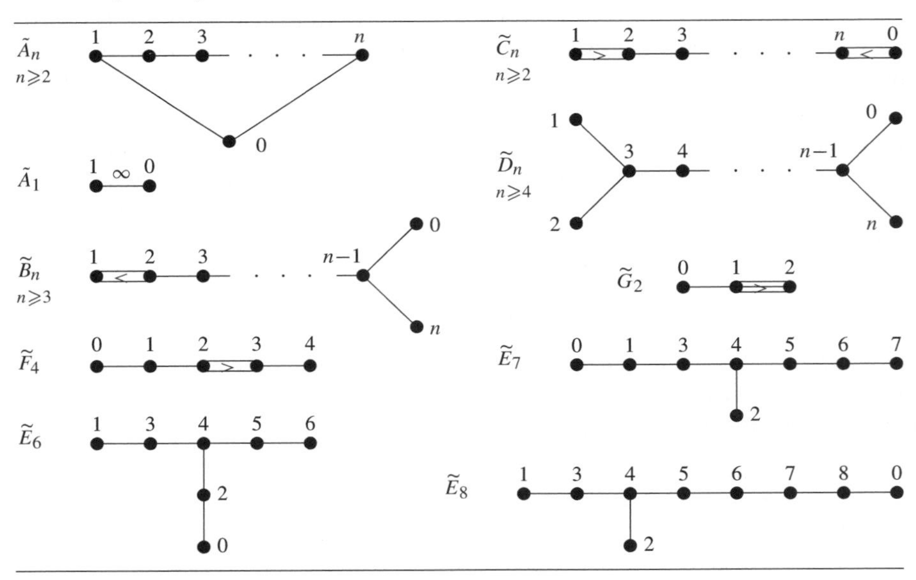

In the following subsections, we describe some situations where Coxeter groups and root systems arise "in nature".

2.15. *Kac–Moody algebras*

Here we briefly discuss how Coxeter groups and root systems arise in the theory of Lie algebras or, more generally, Kac–Moody algebras. We follow the exposition in Kac, [113]. Let $C = (c_{st})_{s,t \in S}$ be a Cartan matrix all of whose coefficients are integers. We also assume that C is symmetrisable, i.e., there exists a diagonal invertible matrix D and a symmetric matrix B such that $C = DB$. A realisation of C is a triple $(\mathfrak{h}, \Pi, \Pi^\vee)$ where $\mathfrak{h}$ is a complex vector space, $\Pi = \{\alpha_s \mid s \in S\} \subseteq \mathfrak{h}^* := \mathrm{Hom}(\mathfrak{h}, \mathbb{C})$ and $\Pi^\vee = \{\alpha_s^\vee \mid s \in S\}$ are subsets of $\mathfrak{h}^*$ and $\mathfrak{h}$, respectively, such that the following conditions hold.

(a) Both sets Π and $\Pi^\vee$ are linearly independent;
(b) we have $\langle \alpha_s^\vee, \alpha_t \rangle := \alpha_t(\alpha_s^\vee) = c_{st}$ for all $s, t \in S$;
(c) $|S| - \mathrm{rank}(C) = \dim \mathfrak{h} - |S|$.

Let $\mathfrak{g}(C)$ be the corresponding Kac–Moody algebra. Then $\mathfrak{g}(C)$ is a Lie algebra which is generated by $\mathfrak{h}$ together with two collections of elements $\{e_s \mid s \in S\}$ and $\{f_s \mid s \in S\}$, where the following relations hold:

$$
\begin{array}{ll}
[e_s, f_t] = \delta_{st}\alpha_s^\vee & (s, t \in S), \\
[h, h'] = 0 & (h, h' \in \mathfrak{h}), \\
[h, e_s] = \langle h, \alpha_s, \rangle e_s & (s \in S, h \in \mathfrak{h}), \\
[h, f_s] = -\langle h, \alpha_s \rangle f_s & (s \in S, h \in \mathfrak{h}), \\
(\mathrm{ad}\, e_s)^{1-c_{st}} e_t = 0 & (s, t \in S, s \neq t), \\
(\mathrm{ad}\, f_s)^{1-c_{st}} f_t = 0 & (s, t \in S, s \neq t).
\end{array}
$$

(By [113, 9.11], this is a set of defining relations for $\mathfrak{g}(C)$.) We have a direct sum decomposition

$$
\mathfrak{g}(C) = \mathfrak{h} \oplus \bigoplus_{0 \neq \alpha \in Q} \mathfrak{g}_\alpha(C), \quad \text{where } Q := \sum_{s \in S} \mathbb{Z}\alpha_s \subseteq \mathfrak{h}^*
$$

and $\mathfrak{g}_\alpha(C) := \{x \in \mathfrak{g}(C) \mid [h, x] = \alpha(h)x \text{ for all } h \in \mathfrak{h}\}$ for all $\alpha \in Q$; here, $\mathfrak{h} = \mathfrak{g}_0$. The set of all $0 \neq \alpha \in Q$ such that $\mathfrak{g}_\alpha(C) \neq \{0\}$ will be denoted by Φ and called the *root system* of $\mathfrak{g}(C)$.

For each $s \in S$, we define a linear map $\sigma_s : \mathfrak{h}^* \to \mathfrak{h}^*$ by the formula

$$
\sigma_s(\lambda) = \lambda - \langle \lambda, \alpha_s^\vee \rangle \alpha_s \quad \text{for } \lambda \in \mathfrak{h}^*.
$$

Then it is easily checked that σ_s is a reflection where $\sigma_s(\alpha_s) = -\alpha_s$. We set

$$
W = W(C) = \langle \sigma_s \mid s \in S \rangle \subseteq \mathrm{GL}(\mathfrak{h}^*).
$$

Now we can state (see [113, 3.7, 3.11 and 3.13]):

(a) The pair $(W, \{\sigma_s \mid s \in S\})$ is a Coxeter system; the corresponding Coxeter matrix is the one associated to C.

(b) The root system Φ is invariant under the action of W and we have $l(\sigma_s w) = l(w)+1$ if and only if $w^{-1}(\alpha_s) \in \Phi^+$, where Φ^+ is defined as in (2.11).

Thus, W and Φ have similar properties as before. Note, however, that here we did not make any assumption on C (except that it is symmetrisable with integer entries) and so W and Φ may be infinite. The finite case is characterised as follows:

$$|W| < \infty \Leftrightarrow |\Phi| < \infty \Leftrightarrow \dim \mathfrak{g}(C) < \infty$$
$$\Leftrightarrow \text{all connected components of } C \text{ occur in table 4.}$$

(This follows from [113, 3.12] and the characterisation of finite Coxeter groups in Theorem 2.4.) In fact, the finite-dimensional Kac–Moody algebras are precisely the "classical" semisimple complex Lie algebras (see, for example, Humphreys, [104]). The Kac–Moody algebras and the root systems associated to so-called Cartan matrices of *affine type* have an extremely rich structure and many applications in other branches of mathematics and mathematical physics; see Kac, [113].

2.16. *Groups with a* BN*-pair*

Let G be an abstract group. We say that G is a *group with a* BN*-pair* or that G admits a *Tits system* if there are subgroups $B, N \subseteq G$ such that the following conditions are satisfied.

(BN1) G is generated by B and N.

(BN2) $T := B \cap N$ is normal in N and the quotient $W := N/T$ is a finite group generated by a set S of elements of order 2.

(BN3) $n_s B n_s \neq B$ if $s \in S$ and n_s is a representative of s in N.

(BN4) $n_s Bn \subseteq Bn_s nB \cup BnB$ for any $s \in S$ and $n \in N$.

The group W is called the *Weyl group* of G. In fact, it is a consequence of the above axioms that the pair (W, S) is a Coxeter system; see [25, Chapter IV, §2, Théorème 2]. The notion of groups with a BN-pair was invented by Tits; see [181]. The standard example of a group with a BN-pair is the general linear group $G = \mathrm{GL}_n(K)$, where K is any field and

$$B := \text{subgroup of all upper triangular matrices in } G,$$
$$N := \text{subgroup of all monomial matrices in } G.$$

(A matrix is called monomial if it has exactly one non-zero entry in each row and each column.) We have

$$T := B \cap N = \text{subgroup of all diagonal matrices in } G$$

and $W = N/T \cong \mathfrak{S}_n$. Thus, $\mathfrak{S}_n$ is the Weyl group of G. More generally, the Chevalley groups (and their twisted analogues) associated with the semisimple complex Lie algebras all have BN-pairs; see Chevalley, [45], Carter, [42], and Steinberg, [176].

The above set of axioms imposes very strong conditions on the structure of a group G with a BN-pair. For example, we have the following *Bruhat decomposition*, which gives the decomposition of G into double cosets with respect to B:

$$G = \coprod_{w \in W} BwB.$$

(More accurately, we should write Bn_wB where n_w is a representative of $w \in W$ in N. But, since any two representatives of w lie in the same coset of $T \subseteq B$, the double coset Bn_wB does not depend on the choice of the representative.)

Furthermore, the proof of the simplicity of the Chevalley groups and their twisted analogues is most economically performed using the simplicity criterion for abstract groups with a BN-pair in Bourbaki, [25, Chapter IV, §2, no. 7].

Groups with a BN-pair play an important rôle in finite group theory. In fact, it is known that every finite simple group possesses a BN-pair, except for the cyclic groups of prime order, the alternating groups of degree $\geqslant 5$, and the 26 sporadic simple groups; see Gorenstein et al., [96]. Given a finite group G with a BN-pair, the irreducible factors of the Weyl group W are of type A_n, B_n, D_n, G_2, F_4, E_6, E_7, E_8 or $I_2(8)$. This follows from the classification by Tits, [181] (rank $\geqslant 3$), Hering, Kantor, Seitz, [98,117] (rank 1) and Fong and Seitz, [81] (rank 2). Note that there is only one case where W is not crystallographic: this is the case where W has a component of type $I_2(8)$ (the dihedral group of order 16), which corresponds to the twisted groups of type F_4 discovered by Ree (see Carter, [42], or Steinberg, [176]).

In another direction, BN-pairs with infinite Weyl groups arise naturally in the theory of p-adic groups; see Iwahori and Matsumoto, [109].

2.17. *Connected reductive algebraic groups*

Here, we assume that the reader has some familiarity with the theory of linear algebraic groups; see Borel, [23], Humphreys, [106], or Springer, [171]. Let G be a connected reductive algebraic group over an algebraically closed field K. Let $B \subseteq G$ be a Borel subgroup. Then we have a semidirect product decomposition $B = U\,T$ where U is the unipotent radical of B and T is a maximal torus. Let $N = \mathrm{N}_G(T)$, the normaliser of T in G. Then the groups B, N form a BN-pair in G; furthermore, W must be a finite Weyl group (and not just a Coxeter group as for general groups with a BN-pair). This is a deep, important result whose proof goes back to Chevalley, [47]; detailed expositions can be found in the monographs by Borel, [23, Chapter IV, 14.15], Humphreys, [106, §29.1], or Springer, [171, Chapter 8].

For example, in $G = \mathrm{GL}_n(K)$, the subgroup B of all upper triangular matrices is a Borel subgroup by the Lie–Kolchin theorem (see, for example, Humphreys, [106, 17.6]). Furthermore, we have a semidirect product decomposition $B = U\,T$ where $U \subseteq B$ is the normal subgroup consisting of all upper triangular matrices with 1 on the diagonal and T is the group of all diagonal matrices in B. Since K is infinite, it is easily checked that $N = \mathrm{N}_G(T)$, the group of all monomial matrices.

Returning to the general case, let us consider the Bruhat cells BwB ($w \in W$). These are locally closed subsets of G since they are orbits of $B \times B$ on G under left and right multiplication. The Zariski closure of BwB is given by

$$\overline{BwB} = \bigcup_{y \in \mathcal{S}(w)} ByB.$$

This yields the promised geometric description of the Bruhat–Chevalley order $\leqslant$ on W (as defined in the remarks following Theorem 2.2.) The proof (see, for example, Springer, [171, §8.5]) relies in an essential way on the fact that G/B is a projective variety.

3. Braid groups

3.1. *The braid group of a complex reflection group*

For a complex reflection group $W \leqslant \mathrm{GL}(V)$, $V = \mathbb{C}^n$, denote by $\mathcal{A}$ the set of its reflecting hyperplanes in V. The topological space

$$V^{\mathrm{reg}} := V \setminus \bigcup_{H \in \mathcal{A}} H$$

is (pathwise) connected in its inherited complex topology. For a fixed base point $x_0 \in V^{\mathrm{reg}}$ we define the *pure braid group of* W as the fundamental group $P(W) := \pi_1(V^{\mathrm{reg}}, x_0)$. Now W acts on V^{reg}, and by the theorem of Steinberg (Theorem 1.6) the covering $\bar{}: V^{\mathrm{reg}} \to V^{\mathrm{reg}}/W$ is Galois, with group W. This induces a short exact sequence

$$1 \to P(W) \to B(W) \to W \to 1 \tag{1}$$

for the *braid group* $B(W) := \pi_1(V^{\mathrm{reg}}/W, \bar{x}_0)$ of W.

If $W = \mathfrak{S}_n$ in its natural permutation representation, the group $B(W)$ is just the classical Artin braid group on n strings, [4].

We next describe some natural generators of $B(W)$. Let $H \in \mathcal{A}$ be a reflecting hyperplane. Let $x_H \in H$ and $r > 0$ such that the open ball $B(x_H, 2r)$ around x_H does not intersect any other reflecting hyperplane and $x_0 \notin B(x_H, 2r)$. Choose a path $\gamma : [0, 1] \to V$ from the base point x_0 to x_H, with $\gamma(t) \in V^{\mathrm{reg}}$ for $t < 1$. Let t_0 be minimal subject to $\gamma(t) \in B(x_H, r)$ for all $t > t_0$. Then $\gamma' := \gamma(t/t_0)$ is a path from x_0 to $\gamma(t_0)$. Then

$$\lambda : [0, 1] \to B(x_H, 2r), \quad t \mapsto \gamma(t_0) \exp(2\pi i t/e_H),$$

where $e_H = |W_H|$ is the order of the fixator of H in W, defines a closed path in the quotient V^{reg}/W. The homotopy class in $B(W)$ of the composition $\gamma' \circ \lambda \circ \gamma'^{-1}$ is then called a *braid reflection* (see Broué, [33]) or *generator of the monodromy around* H. Its image in W is a reflection s_H generating W_H, with non-trivial eigenvalue $\exp(2\pi i/e_H)$. It can be shown that $B(W)$ is generated by all braid reflection, when H varies over the reflecting hyperplanes of W (Broué, Malle and Rouquier, [38, Theorem 2.17]).

Assume from now on that W is irreducible. Recall the definition of N, N^* in Sections 1.2 and 1.7 as the number of reflections respectively of reflecting hyperplanes. The following can be shown without recourse to the classification of irreducible complex reflection groups:

3.2. THEOREM (Bessis, [11]). *Let $W \leqslant \mathrm{GL}_n(\mathbb{C})$ be an irreducible complex reflection group with braid group $B(W)$. Let d be a degree of W which is a regular number for W and let $r := (N + N^*)/d$. Then $r \in \mathbb{N}$, and there exists a subset $\mathbf{S} = \{\mathbf{s}_1, \dots, \mathbf{s}_r\} \subset B(W)$ with:*

(i) *$\mathbf{s}_1, \dots, \mathbf{s}_r$ are braid reflections, so their images $s_1, \dots, s_r \in W$ are reflections.*
(ii) *$\mathbf{S}$ generates $B(W)$, and hence $S := \{s_1, \dots, s_r\}$ generates W.*
(iii) *There exists a finite set $\mathcal{R}$ of relations of the form $w_1 = w_2$, where w_1, w_2 are words of equal length in $\mathbf{s}_1, \dots, \mathbf{s}_r$, such that $\langle \mathbf{s}_1, \dots, \mathbf{s}_r | \mathcal{R} \rangle$ is a presentation for $B(W)$.*
(iv) *Let e_s denote the order of $s \in S$. Then $\langle s_1, \dots, s_r | \mathcal{R}; s^{e_s} = 1\ \forall s \in S \rangle$ is a presentation for W, where now $\mathcal{R}$ is viewed as a set of relations on S.*
(v) *$(\mathbf{s}_1 \cdots \mathbf{s}_r)^d$ is central in $B(W)$ and lies in $P(W)$.*
(vi) *The product $c := s_1 \cdots s_r$ is a $\zeta := \exp(2\pi i/d)$-regular element of W (hence has eigenvalues $\zeta^{-m_1}, \dots, \zeta^{-m_r}$).*

It follows from the classification (see table 1) that there always exists a regular degree. In many cases, for example if W is well-generated, the number $(N + N^*)/r$ is regular, when r is chosen as the minimal number of generating reflections for W (so $n \leqslant r \leqslant n + 1$). Thus, in those cases $B(W)$ is finitely presented on the same minimal number of generators as W. Under the assumptions (i) or (ii) of Theorem 1.16, the largest degree d_n is regular, whence Theorem 1.16(iv) is a consequence of the previous theorem.

At present, presentations of the type described in Theorem 3.2 have been found for all but six irreducible types, by case-by-case considerations, see Bannai, [5], Naruki, [146], Broué, Malle and Rouquier, [38]. For the remaining six groups, conjectural presentations have been found by Bessis and Michel using computer calculations.

For the case of real reflection groups, Brieskorn, [28], and Deligne, [61], determined the structure of $B(W)$ by a nice geometric argument. They show that the generators in Theorem 3.2 (with $r = n$) can be taken as suitable preimages of the Coxeter generators, and the relations $\mathcal{R}$ as the Coxeter relations. For the case of $W(A_n) = \mathfrak{S}_{n+1}$ of the classical braid group, this was first shown by Artin, [4].

A topological space X is called $K(\pi, 1)$ if all homotopy groups $\pi_i(X)$ for $i \neq 1$ vanish. The following is conjectured by Arnol'd to be true for all irreducible complex reflection groups:

3.3. THEOREM. *Assume that W is not of type G_i, $i \in \{24, 27, 29, 31, 33, 34\}$. Then V^{reg} and V^{reg}/W are $K(\pi, 1)$-spaces.*

This was proved by a general argument for Coxeter groups by Deligne, [61], after Fox and Neuwirth, [82], showed it for type A_n and Brieskorn, [29], for those of type different from H_3, H_4, E_6, E_7, E_8. For the non-real Shephard groups (non-real groups with Coxeter braid diagrams), it was proved by Orlik and Solomon, [158]. The case of the infinite series $G(de, e, r)$ has been solved by Nakamura, [145]. In that case, there exists a locally trivial fibration

$$V^{\text{reg}}\big(G(de, e, n)\big) \to V^{\text{reg}}\big(G(de, e, n-1)\big),$$

with fiber isomorphic to $\mathbb{C}$ minus $m(de, e, n)$ points, where

$$m(de, e, n) := \begin{cases} (n-1)de + 1 & \text{for } d \neq 1, \\ (n-1)(e-1) & \text{for } d = 1. \end{cases}$$

This induces a split exact sequence

$$1 \to F_m \to P\big(G(de, e, n)\big) \to P\big(G(de, 1, n-1)\big) \to 1$$

for the pure braid group, with a free group F_m of rank $m = m(de, e, n)$. In particular, the pure braid group has the structure of an iterated semidirect product of free groups (see Broué, Malle and Rouquier, [38, Proposition 3.37]).

3.4. *The centre and regular elements*

Denote by $\boldsymbol{\pi}$ the class in $P(W)$ of the loop

$$[0, 1] \to V^{\text{reg}}, \quad t \mapsto x_0 \exp(2\pi i t).$$

Then $\boldsymbol{\pi}$ lies in the centre $Z(P(W))$ of the pure braid group. Furthermore,

$$[0, 1] \to V^{\text{reg}}, \quad t \mapsto x_0 \exp\big(2\pi i t / |Z(W)|\big),$$

defines a closed path in V^{reg}/W, so an element $\boldsymbol{\beta}$ of $B(W)$, which is again central. Clearly $\boldsymbol{\pi} = \boldsymbol{\beta}^{|Z(W)|}$.

The following was shown independently by Brieskorn and Saito, [30], and Deligne, [61], for Coxeter groups, and by Broué, Malle and Rouquier, [38, Theorem 2.24], for the other groups:

3.5. THEOREM. *Assume that W is not of type G_i, $i \in \{24, 27, 29, 31, 33, 34\}$. Then the centre of $B(W)$ is infinite cyclic generated by $\boldsymbol{\beta}$, the centre of $P(W)$ is infinite cyclic generated by $\boldsymbol{\pi}$, and the exact sequence* (1) *induces an exact sequence*

$$1 \to Z\big(P(W)\big) \to Z\big(B(W)\big) \to Z(W) \to 1.$$

In their papers, Brieskorn and Saito, [30], and Deligne, [61], also solve the word problem and the conjugation problem for braid groups attached to real reflection groups.

For each $H \in \mathcal{A}$ choose a linear form $\alpha_H : V \to \mathbb{C}$ with kernel H. Let $e_H := |W_H|$, the order of the minimal parabolic subgroup fixing H. The *discriminant of* W, defined as

$$\delta := \delta(W) := \prod_{H \in \mathcal{A}} \alpha_H^{e_H},$$

is then a W-invariant element of the symmetric algebra $S(V^*)$ of V^*, well-defined up to non-zero scalars (Cohen, [50, 1.8]). It thus induces a continuous function $\delta : V^{\text{reg}}/W \to \mathbb{C}^\times$, hence by functoriality a group homomorphism $\pi_1(\delta) : B(W) \to \pi_1(\mathbb{C}^\times, 1) \cong \mathbb{Z}$. For $\mathbf{b} \in B(W)$ let $l(\mathbf{b}) := \pi_1(\delta)(\mathbf{b})$ denote the *length of* $\mathbf{b}$. For example, every braid reflection $\mathbf{s}$ has length $l(\mathbf{s}) = 1$, and we have

$$l(\boldsymbol{\beta}) = (N + N^*)/\big|Z(W)\big| \quad \text{and hence} \quad l(\boldsymbol{\pi}) = N + N^*$$

by Broué, Malle and Rouquier, [38, Corollary 2.21].

The elements $\mathbf{b} \in B(W)$ with $l(\mathbf{b}) \geqslant 0$ form the *braid monoid* $B^+(W)$. A *d-th root of* $\boldsymbol{\pi}$ is by definition an element $\mathbf{w} \in B^+$ with $\mathbf{w}^d = \boldsymbol{\pi}$.

Let d be a regular number for W, and $\mathbf{w}$ a d-th root of $\boldsymbol{\pi}$. Assume that the image w of $\mathbf{w}$ in W is ζ-regular for some d-th root of unity ζ (in the sense of 1.9). (This is, for example, the case if W is a Coxeter group by Broué and Michel, [39, Theorem 3.12].) By Theorem 1.11(ii) the centraliser $W(w) := C_W(w)$ is a reflection group on $V(w, \zeta)$, with reflecting hyperplanes the intersections of $V(w, \zeta)$ with the hyperplanes in $\mathcal{A}$ by Theorem 1.12(i). Thus the hyperplane complement of $W(w)$ on $V(w, \zeta)$ is just $V^{\text{reg}}(w) := V^{\text{reg}} \cap V(w, \zeta)$. Assuming that the base point x_0 has been chosen in $V^{\text{reg}}(w)$, this defines natural maps

$$P\big(W(w)\big) \to P(W) \quad \text{and} \quad \psi_w : B\big(W(w)\big) \to B(W).$$

By Broué and Michel, [39, 3.4], the image of $B(W(w))$ in $B(W)$ centralises $\mathbf{w}$. It is conjectured (see Bessis, Digne and Michel, [12, Conjecture 0.1]) that ψ_w defines an isomorphism $B(W(w)) \cong C_{B(W)}(\mathbf{w})$. The following partial answer is known:

3.6. THEOREM (Bessis, Digne and Michel, [12, Theorem 0.2]). *Let W be an irreducible reflection group of type $\mathfrak{S}_n$, $G(d, 1, n)$ or G_i, $i \in \{4, 5, 8, 10, 16, 18, 25, 26, 32\}$, and let $w \in W$ be regular. Then ψ_w induces an isomorphism $B(W(w)) \cong C_{B(W)}(\mathbf{w})$.*

This has also been proved by Michel, [142, Corollary 4.4], in the case that W is a Coxeter group and w acts on W by a diagram automorphism. The injectivity of ψ_w was shown for all but finitely many types of W by Bessis, [10, Theorem 1.3].

The origin of Artin's work on the braid group associated with the symmetric group lies in the theory of knots and links. We shall now briefly discuss this connection and explain the construction of the "HOMFLY-PT" invariant of knots and links (which includes the famous Jones polynomial as a special case). We follow the exposition in Geck and Pfeiffer, [95, §4.5].

3.7. *Knots and links, Alexander and Markov theorem*

If n is a positive integer, an oriented *n-link* is an embedding of n copies of the interval $[0, 1] \subset \mathbb{R}$ into $\mathbb{R}^3$ such that 0 and 1 are mapped to the same point (the orientation is induced by the natural ordering of $[0, 1]$); a 1-link is also called a *knot*. We are only interested in knots and links modulo isotopy, i.e., homeomorphic transformations which preserve the orientation. We refer to Birman, [16], Crowell and Fox, [57], or Burde and Zieschang, [40], for precise versions of the above definitions.

By Artin's classical interpretation of $B(\mathfrak{S}_n)$ as the braid group on n strings, each generator of $B(\mathfrak{S}_n)$ can be represented by oriented diagrams as indicated below; writing any $g \in B(\mathfrak{S}_n)$ as a product of the generators and their inverses, we also obtain a diagram for g, by concatenating the diagrams for the generators. "Closing" such a diagram by joining the end points, we obtain the plane projection of an oriented link in $\mathbb{R}^3$:

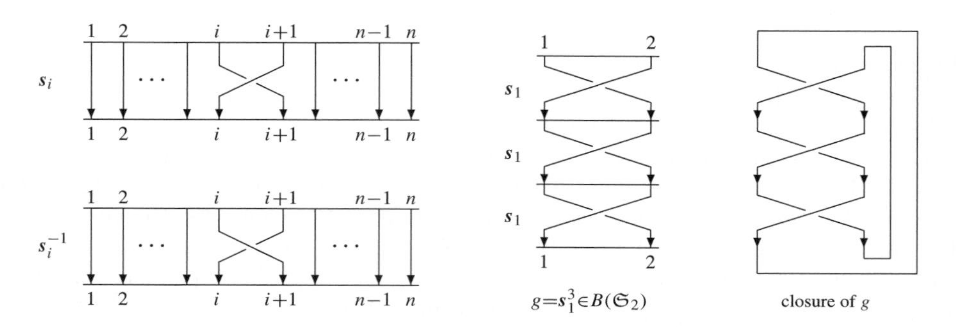

By Alexander's theorem (see Birman, [16], or, for a more recent proof, Vogel, [186]), every oriented link in $\mathbb{R}^3$ is isotopic to the closure of an element in $B(\mathfrak{S}_n)$, for some $n \geqslant 1$. The question of when two links in $\mathbb{R}^3$ are isotopic can also be expressed algebraically. For this purpose, we consider the infinite disjoint union

$$B_\infty := \coprod_{n \geqslant 1} B(\mathfrak{S}_n).$$

Given $g, g' \in B_\infty$, we write $g \sim g'$ if one of the following relations is satisfied:

(I) We have $g, g' \in B(\mathfrak{S}_n)$ and $g' = x^{-1}gx$ for some $x \in B(\mathfrak{S}_n)$.

(II) We have $g \in B(\mathfrak{S}_n)$, $g' \in B(\mathfrak{S}_{n+1})$ and $g' = g\mathbf{s}_n$ or $g' = g\mathbf{s}_n^{-1}$.

The above two relations are called *Markov relations*. By a classical result due to Markov (see Birman, [16], or, for a more recent proof, Traczyk, [183]), two elements of B_∞ are equivalent under the equivalence relation generated by $\sim$ if and only if the corresponding links obtained by closure are isotopic. Thus, to define an invariant of oriented links is the same as to define a map on B_∞ which takes equal values on elements $g, g' \in B_\infty$ satisfying (I) or (II).

We now consider the Iwahori–Hecke algebra $H_\mathbb{C}(\mathfrak{S}_n)$ of the symmetric group $\mathfrak{S}_n$ over $\mathbb{C}$. By definition, $H_\mathbb{C}(\mathfrak{S}_n)$ is a quotient of the group algebra of $B(\mathfrak{S}_n)$, where we factor by an ideal generated by certain quadratic relations depending on two parameters $u, v \in \mathbb{C}$. This is done such that

$$T_{s_i}^2 = uT_1 + vT_{s_i} \quad \text{for } 1 \leqslant i \leqslant n-1,$$

where T_{s_i} denotes the image of the generator $\mathbf{s}_i$ of $B(\mathfrak{S}_n)$ and T_1 denotes the identity element. For each $w \in \mathfrak{S}_n$, we have a well-defined element T_w such that

$$T_w T_{w'} = T_{ww'} \quad \text{whenever } l(ww') = l(w) + l(w').$$

This follows easily from Matsumoto's Theorem 2.2. In fact, one can show that the elements $\{T_w \mid w \in \mathfrak{S}_n\}$ form a $\mathbb{C}$-basis of $H_{\mathbb{C}}(\mathfrak{S}_n)$. (For more details, see the chapter on Hecke algebras.) The map $\boldsymbol{w} \mapsto T_w$ $(w \in \mathfrak{S}_n)$ extends to a well-defined algebra homomorphism from the group algebra of $B(\mathfrak{S}_n)$ over $\mathbb{C}$ onto $H_{\mathbb{C}}(\mathfrak{S}_n)$. Furthermore, the inclusion $\mathfrak{S}_{n-1} \subseteq \mathfrak{S}_n$ also defines an inclusion of algebras $H_{\mathbb{C}}(\mathfrak{S}_{n-1}) \subseteq H_{\mathbb{C}}(\mathfrak{S}_n)$.

3.8. THEOREM (Jones, Ocneanu, [112]). *There is a unique family of $\mathbb{C}$-linear maps $\tau_n : H_{\mathbb{C}}(\mathfrak{S}_n) \to \mathbb{C}$ $(n \geqslant 1)$ such that the following conditions hold*:

(M1) $\tau_1(T_1) = 1$;

(M2) $\tau_{n+1}(hT_{s_n}) = \tau_{n+1}(hT_{s_n}^{-1}) = \tau_n(h)$ *for all $n \geqslant 1$ and $h \in H_{\mathbb{C}}(\mathfrak{S}_n)$*;

(M3) $\tau_n(hh') = \tau_n(h'h)$ *for all $n \geqslant 1$ and $h, h' \in H_{\mathbb{C}}(\mathfrak{S}_n)$*.

Moreover, we have $\tau_{n+1}(h) = v^{-1}(1-u)\tau_n(h)$ for all $n \geqslant 1$ and $h \in H_{\mathbb{C}}(\mathfrak{S}_n)$.

In [112], Jones works with an Iwahori–Hecke algebra of $\mathfrak{S}_n$ where the parameters are related by $v = u - 1$. The different formulation above follows a suggestion by J. Michel. It results in a simplification of the construction of the link invariants below. (The simplification arises from the fact that, due to the presence of two different parameters in the quadratic relations, the "singularities" mentioned in [112, p. 349, notes (1)] simply disappear.) Generalisations of Theorem 3.8 to types $G(d, 1, n)$ and D_n have been found in Geck and Lambropoulou, [92], Lambropoulou, [125], and Geck, [87].

3.9. *The HOMFLY-PT polynomial*

We can now construct a two-variable invariant of oriented knots and links as follows. Consider an oriented link L and assume that it is isotopic to the closure of $g \in B(\mathfrak{S}_n)$ for $n \geqslant 1$. Then we set

$$X_L(u, v) := \tau_n(\bar{g}) \in \mathbb{C} \quad \text{with } \tau_n \text{ as in Theorem 3.8.}$$

Here, $\bar{g}$ denotes the image of g under the natural map $\mathbb{C}[B(\mathfrak{S}_n)] \to H_{\mathbb{C}}(\mathfrak{S}_n)$, $\boldsymbol{w} \mapsto T_w$ $(w \in \mathfrak{S}_n)$. It is easily checked that $X_L(u, v)$ can be expressed as a Laurent polynomial in u and v; the properties (M2) and (M3) make sure that $\tau_n(\bar{g})$ does not depend on the choice of g. If we make the change of variables $u = t^2$ and $v = tx$, we can identify the above invariant with the HOMFLY-PT polynomial $P_L(t, x)$ discovered by Freyd et al., [83], and

Przytycki and Traczyk, [161]; see also Jones, [112, (6.2)]. Furthermore, the *Jones polynomial* $J_L(t)$ is obtained by setting $u = t^2$, $v = \sqrt{t}(t-1)$ (see [112, §11]). Finally, setting $u = 1$ and $v = \sqrt{t} - 1/\sqrt{t}$, we obtain the classical *Alexander polynomial* $A_L(t)$ whose definition can be found in Crowell and Fox, [57].

For a survey about recent developments in the theory of knots and links, especially since the discovery of the Jones polynomial, see Birman, [17].

3.10. *Further aspects of braid groups*

One of the old problems concerning braid groups is the question whether or not they are linear, i.e., whether there exists a faithful linear representation on a finite-dimensional vector space. Significant progress has been made recently on this problem. Krammer, [122], and Bigelow, [15], proved that the classical Artin braid group is linear. Then Digne, [69], and Cohen and Wales, [53], extended this result and showed that all Artin groups of crystallographic type have a faithful representation of dimension equal to the number of reflections of the associated Coxeter group.

On the other hand, there is one particular representation of the braid group associated with $\mathfrak{S}_n$, the so-called Burau representation (see Birman, [16], for which it has been a longstanding problem to determine for which values of n it is faithful. Moody, [143], showed that it is not faithful for $n \geqslant 10$; this bound was improved by Long and Paton, [133], to 6. Recently, Bigelow, [14], showed that the Burau representation is not faithful already for $n = 5$. (It is an old result of Magnus and Peluso that the Burau representation is faithful for $n = 3$.)

In a different direction, Deligne's and Brieskorn–Saito's solution of the word and conjugacy problem in braid groups led to new developments in combinatorial group and monoid theory; see, for example, Dehornoy and Paris, [60], and Dehornoy, [59].

4. Representation theory

In this section we report about the representation theory of finite complex reflection groups.

4.1. *Fields of definition*

Let W be a finite complex reflection group on V. Let K_W denote the character field of the reflection representation of W, that is, the field generated by the traces $\mathrm{tr}_V(w)$, $w \in W$. It is easy to see that the reflection representation can be realised over K_W (see, for example, [7, Proposition 7.1.1]). But we have a much stronger statement:

4.2. THEOREM (Benard, [6], Bessis, [9]). *Let W be a complex reflection group. Then the field K_W is a splitting field for W.*

The only known proof for this result is case-by-case, treating the reflection groups according to the Shephard–Todd classification.

The field K_W has a nice description at least in the case of well-generated groups, that is, irreducible groups generated by $\dim(V)$ reflections. In this case, the largest degree d_n of W is regular, so there exists an element $c := s_1 \cdots s_n$ as in theorem 3.2(vi) of Bessis, called a *Coxeter element* of W, with eigenvalues $\zeta^{-m_1}, \ldots, \zeta^{-m_n}$ in the reflection representation, where $\zeta := \exp(2\pi i/d_n)$ and the m_i are the exponents of W.

If W is a real reflection group, then W is a Coxeter group associated with some Coxeter matrix M (see Theorem 2.4) and we have

$$K_W = \mathbb{Q}\big(\cos(2\pi/m_{st}) \mid s, t \in S\big) \subset \mathbb{R}.$$

In particular, this shows that $K_W = \mathbb{Q}$ if W is a finite Weyl group.

For well-generated irreducible complex reflection groups $W \leqslant \mathrm{GL}(V)$, the field of definition K_W is generated over $\mathbb{Q}$ by the coefficients of the characteristic polynomial on V of a Coxeter element, see Malle, [140, Theorem 7.1]. This characterisation is no longer true for non-well generated reflection groups.

4.3. *Macdonald–Lusztig–Spaltenstein induction*

Let W be a complex reflection group on V. Recall from Section 1.7 the definition of the fake degree R_χ of an irreducible character $\chi \in \mathrm{Irr}(W)$. The *b-invariant* b_χ *of* χ is defined as the order of vanishing of R_χ at $x = 0$, that is, as the minimum of the exponents $e_i(\chi)$ of χ. The coefficient of x^{b_χ} in R_χ is denoted by γ_χ.

4.4. THEOREM (Macdonald, [136], Lusztig and Spaltenstein, [135]). *Let W be a complex reflection group on V, W' a reflection subgroup (on $V_1 := V/C_V(W')$). Let ψ be an irreducible character of W' such that $\gamma_\psi = 1$. Then $\mathrm{Ind}_{W'}^{W}(\psi)$ has a unique irreducible constituent $\chi \in \mathrm{Irr}(W)$ with $b_\chi = b_\psi$. This satisfies $\gamma_\chi = 1$. All other constituents have b-invariant bigger than b_ψ*

The character $\chi \in \mathrm{Irr}(W)$ in Theorem 4.4 is called the *j-induction* $j_{W'}^{W}(\psi)$ of the character $\psi \in \mathrm{Irr}(W')$. Clearly, j-induction is transitive; it is also compatible with direct products.

An important example of characters ψ with $\gamma_\psi = 1$ is given by the determinant character $\det_V : W \to \mathbb{C}^\times$ of a complex reflection group (see Geck and Pfeiffer, [95, Theorem 5.2.10]).

4.5. *Irreducible characters*

There is no general construction of all irreducible representations of a complex reflection group known. Still, we have the following partial result:

4.6. THEOREM (Steinberg). *Let $W \leqslant \mathrm{GL}(V)$ be an irreducible complex reflection group. Then the exterior powers $\Lambda^i(V)$, $1 \leqslant i \leqslant \dim V$, are irreducible, pairwise non-equivalent representations of W.*

A proof in the case of well-generated groups can be found in Bourbaki, [25, V, §2, Example 3(d)], and Kane, [114, Theorem 24-3 A], for example. The general case then follows with corollary 1.15.

We now give some information on the characters of individual reflection groups. The irreducible characters of the symmetric group $\mathfrak{S}_n$ were determined by Frobenius, [84], see also Macdonald, [137], and Fulton, [85]. Here we follow the exposition in Geck and Pfeiffer, [95, 5.4].

Let $\lambda = (\lambda_1, \dots, \lambda_r) \vdash n$ be a partition of n. The corresponding *Young subgroup* $\mathfrak{S}_\lambda$ *of* $\mathfrak{S}_n$ is the common setwise stabiliser $\{1, \dots, \lambda_1\}, \{\lambda_1 + 1, \dots, \lambda_1 + \lambda_2\}, \dots,$ abstractly isomorphic to $\mathfrak{S}_\lambda = \mathfrak{S}_{\lambda_1} \times \cdots \times \mathfrak{S}_{\lambda_r}$. This is a parabolic subgroup of $\mathfrak{S}_n$ in the sense of Section 1.5. For any $m \geqslant 1$, let 1_m, ε_m denote the trivial respectively the sign character of $\mathfrak{S}_m$. For each partition we have the two induced characters

$$\pi_\lambda := \mathrm{Ind}_{\mathfrak{S}_\lambda}^{\mathfrak{S}_n}(1_{\lambda_1} \# \cdots \# 1_{\lambda_r}), \qquad \theta_\lambda := \mathrm{Ind}_{\mathfrak{S}_\lambda}^{\mathfrak{S}_n}(\varepsilon_{\lambda_1} \# \cdots \# \varepsilon_{\lambda_r}).$$

Then π_λ and θ_{λ^*} have a unique irreducible constituent $\chi_\lambda \in \mathrm{Irr}(\mathfrak{S}_n)$ in common, where λ^* denotes the partition dual to λ. This constituent can also be characterised in terms of j-induction as

$$\chi_\lambda = j_{\mathfrak{S}_\lambda}^{\mathfrak{S}_n}(1_{\lambda_1} \# \cdots \# 1_{\lambda_r}).$$

Then the χ_λ are mutually distinct and exhaust the irreducible characters of $\mathfrak{S}_n$, so $\mathrm{Irr}(\mathfrak{S}_n) = \{\chi_\lambda \mid \lambda \vdash n\}$. From the above construction it is easy to see that all χ_λ are afforded by rational representations.

The construction of the irreducible characters of the imprimitive group $G(d, 1, n)$ goes back at least to Osima, [159] (see also Read, [164], Hughes, [102], Bessis, [9]) via their abstract structure as wreath product $C_d \wr \mathfrak{S}_n$. Let us fix $d \geqslant 2$ and write $W_n := G(d, 1, n)$. A d-tuple $\alpha = (\alpha_0, \alpha_1, \dots, \alpha_{d-1})$ of partitions $\alpha_i \vdash n_i$ with $\sum n_i = n$ is called a *d-partition of n*. We denote by W_α the natural subgroup $W_{n_0} \times \cdots \times W_{n_{d-1}}$ of W_n, where $\alpha_j \vdash n_j$, corresponding to the Young subgroup $\mathfrak{S}_{n_0} \times \cdots \times \mathfrak{S}_{n_{d-1}}$ of $\mathfrak{S}_n$. Via the natural projection $W_{n_j} \to \mathfrak{S}_{n_j}$ the characters of $\mathfrak{S}_{n_j}$ may be regarded as characters of W_{n_j}. Thus each α_j defines an irreducible character χ_{α_j} of W_{n_j}. For any m, let $\zeta_d : W_m \to \mathbb{C}^\times$ be the linear character defined by $\zeta_d(t_1) = \exp(2\pi i/d)$, $\zeta_d(t_i) = 1$ for $i > 1$, with the standard generators t_i from 1.13. Then for any d-partition α of n we can define a character χ_α of W_n as the induction of the exterior product

$$\chi_\alpha := \mathrm{Ind}_{W_\alpha}^{W_n}\big(\chi_{\alpha_0} \# (\chi_{\alpha_1} \otimes \zeta_d) \# \cdots \# \big(\chi_{\alpha_{d-1}} \otimes \zeta_d^{d-1}\big)\big).$$

By Clifford-theory χ_α is irreducible, $\chi_\alpha \neq \chi_\beta$ if $\alpha \neq \beta$, and all irreducible characters of W_n arise in this way, so

$$\mathrm{Irr}(W_n) = \big\{\chi_\alpha \mid \alpha = (\alpha_0, \dots, \alpha_{d-1}) \vdash_d n\big\}.$$

We describe the irreducible characters of $G(de, e, n)$ in terms of those of $W_n := G(de, 1, n)$. Recall that the imprimitive reflection group $G(de, e, n)$ is generated by the reflections $t_2, \tilde{t}_2 := t_1^{-1}t_2t_1, t_3, \ldots, t_n$, and t_1^e. Denote by π the cyclic shift on de-partitions of n, i.e.,

$$\pi(\alpha_0, \ldots, \alpha_{de-1}) = (\alpha_1, \ldots, \alpha_{de-1}, \alpha_0).$$

By definition we then have $\chi_{\pi(\alpha)} \otimes \zeta_{de} = \chi_\alpha$. Let $s_e(\alpha)$ denote the order of the stabiliser of α in the cyclic group $\langle \pi^d \rangle$. Then upon restriction to $G(de, e, n)$ the irreducible character χ_α of W_n splits into $s_e(\alpha)$ different irreducible constituents, and this exhausts the set of irreducible characters of $G(de, e, n)$. More precisely, let α be a de-partition of n with $\tilde{e} := s_e(\alpha)$, $W_{\alpha,e} := W_\alpha \cap G(de, e, n)$, and ψ_α the restriction of

$$\chi_{\alpha_0}\#(\chi_{\alpha_1} \otimes \zeta_{de})\#\cdots\#\big(\chi_{\alpha_{de-1}} \otimes \zeta_{de}^{de-1}\big)$$

to $W_{\alpha,e}$. Then ψ_α is invariant under the element $\sigma := (t_2 \cdots t_n)^{n/\tilde{e}}$ (note that $\tilde{e} = s_e(\alpha)$ divides n), and it extends to the semidirect product $W_{\alpha,e}.\langle\sigma\rangle$. The different extensions of ψ_α induced to $G(de, e, n)$ then exhaust the irreducible constituents of the restriction of χ_α to $G(de, e, n)$. Thus, we may parametrise $\mathrm{Irr}(G(de, e, n))$ by de-partitions of n up to cyclic shift by π^d in such a way that any α stands for $s_e(\alpha)$ different characters.

In order to describe the values of the irreducible characters we need the following definitions. We identify partitions with their Young diagrams. A d-partition α is called a *hook* if it has just one non-empty part, which is a hook (i.e., does not contain a 2×2-block). The position of the non-empty part is then denoted by $\tau(\alpha)$. If α, β are d-partitions such that β_i is contained in α_i for all $0 \leqslant i \leqslant d-1$, then $\alpha \setminus \beta$ denotes the d-partition $(\alpha_i \setminus \beta_i \mid 0 \leqslant i \leqslant d-1)$, where $\alpha_i \setminus \beta_i$ is the set theoretic difference of α_i and β_i. If $\alpha \setminus \beta$ is a hook, we denote by l_β^α the number of rows of the hook $(\alpha \setminus \beta)_{\tau(\alpha\setminus\beta)}$ minus 1. With these notations the values of the irreducible characters of $G(d, 1, n)$ can be computed recursively with a generalised Murnaghan–Nakayama rule (see also Stembridge, [179]):

4.7. THEOREM (Osima, [159]). *Let α and γ be d-partitions of n, let m be a part of γ_t for some $1 \leqslant t \leqslant d$, and denote by γ' the d-partition of $n-m$ obtained from γ by deleting the part m from γ_t. Then the value of the irreducible character χ_α on an element of $G(d, 1, n)$ with cycle structure γ is given by*

$$\chi_\alpha(\gamma) = \sum_{\alpha\setminus\beta\vdash_d m} \zeta_d^{st}(-1)^{l_\beta^\alpha}\chi_\beta(\gamma')$$

where the sum ranges over all d-partitions β of $n-m$ such that $\alpha \setminus \beta$ is a hook and where $s = \tau(\alpha \setminus \beta)$.

An overview of the irreducible characters of the exceptional groups (and of their projective characters) is given in Humphreys, [107]. All character tables of irreducible complex reflection groups are also available in the computer algebra system CHEVIE, [89].

4.8. *Fake degrees*

The fake degrees (introduced in Section 1.7) of all complex reflection groups are known. For the symmetric groups, they were first determined by Steinberg, [174], as generic degrees of the unipotent characters of the general linear groups over a finite field. From that result, the fake degrees of arbitrary imprimitive reflection groups can easily be derived (see Malle, [138, Bem. 2.10 and 5.6]).

4.9. THEOREM (Steinberg, [174], Lusztig, [134]). *The fake degrees of the irreducible complex reflection groups $G(de, e, n)$ are given as follows:*

(i) *Let $\chi \in \operatorname{Irr}(G(d,1,n))$ be parameterised by the d-partition $(\alpha_0, \ldots, \alpha_{d-1})$, where $\alpha_i = (\alpha_{i1} \geqslant \cdots \geqslant \alpha_{im_i}) \vdash n_i$, and let $(S_0, \ldots, S_{d-1})$, where $S_i = (\alpha_{i1} + m_i - 1, \ldots, \alpha_{im_i})$, denote the corresponding tuple of β-numbers. Then*

$$R_\chi = \prod_{i=1}^{n}(x^{id} - 1) \prod_{i=0}^{d-1} \frac{\Delta(S_i, x^d) x^{in_i}}{\Theta(S_i, x^d) x^{d\binom{m_i-1}{2} + d\binom{m_i-2}{2} + \cdots}},$$

where, for a finite subset $S \subset \mathbb{N}$,

$$\Delta(S, x) := \prod_{\substack{\lambda, \lambda' \in S \\ \lambda' < \lambda}} (x^\lambda - x^{\lambda'}), \qquad \Theta(S, x) := \prod_{\lambda \in S} \prod_{h=1}^{\lambda} (x^h - 1).$$

(ii) *The fake degree of $\chi \in \operatorname{Irr}(G(de, e, n))$ is obtained from the fake degrees in $G(de, 1, n)$ as*

$$R_\chi = \frac{x^{nd} - 1}{x^{nde} - 1} \sum_{\psi \in \operatorname{Irr}(G(de,1,n))} \langle \chi, \psi|_{G(de,e,n)} \rangle R_\psi.$$

For exceptional complex reflection groups, the fake degrees can easily be computed, for example in the computer algebra system CHEVIE, [89]. In the case of exceptional Weyl groups, they were first studied by Beynon and Lusztig, [13].

The fake degrees of reflection groups satisfy a remarkable palindromicity property:

4.10. THEOREM (Opdam, [151,152], Malle, [140]). *Let W be a complex reflection group. There exists a permutation δ on $\operatorname{Irr}(W)$ such that for every $\chi \in \operatorname{Irr}(W)$ we have*

$$R_\chi(x) = x^c R_{\delta(\bar{\chi})}(x^{-1}),$$

where $c = \sum_r (1 - \chi(r)/\chi(1))$, the sum running over all reflections $r \in W$.

The interest of this result also lies in the fact that the permutation δ is strongly related to the irrationalities of characters of the associated Hecke algebra. Theorem 4.10 was first observed empirically by Beynon and Lusztig, [13], in the case of Weyl groups. Here δ

is non-trivial only for characters such that the corresponding character of the associated Iwahori–Hecke algebra is non-rational. An a priori proof in this case was later given by Opdam, [151]. If W is complex, Theorem 4.10 was verified by Malle, [140, Theorem 6.5], in a case-by-case analysis. Again in all but possibly finitely many cases, δ comes from the irrationalities of characters of the associated Hecke algebra. Opdam, [152, Theorem 4.2 and Corollary 6.8], gives a general argument which proves theorem 4.10 under a suitable assumption on the braid group $B(W)$.

The above discussion is exclusively concerned with representations over a field of characteristic 0 (the "semisimple case"). We close this chapter with some remarks concerning the modular case.

4.11. *Modular representations of* $\mathfrak{S}_n$

Frobenius' theory (as developed further by Specht, James and others) yields a parametrisation of $\mathrm{Irr}(\mathfrak{S}_n)$ and explicit formulas for the degrees and the values of all irreducible characters. As soon as we consider representations over a field of characteristic $p > 0$, the situation changes drastically. James, [110], showed that the irreducible representations of $\mathfrak{S}_n$ still have a natural parametrisation, by so-called p-regular partitions. Furthermore, the decomposition matrix relating representations in characteristic 0 and in characteristic p has a lower triangular shape with 1s on the diagonal. This result shows that, in principle, a knowledge of the irreducible representations of $\mathfrak{S}_n$ in characteristic p is equivalent to the knowledge of the decomposition matrix.

There are a number of results known about certain entries of that decomposition matrix, but a general solution to this problem is completely open; see James, [110], for a survey. Via the classical Schur algebras (see Green, [97]) it is known that the decomposition numbers of $\mathfrak{S}_n$ in characteristic p can be obtained from those of the finite general linear group $\mathrm{GL}_n(\mathbb{F}_q)$ (where q is a power of p). Now, at first sight, the problem of computing the decomposition numbers for $\mathrm{GL}_n(\mathbb{F}_q)$ seems to be much harder than for $\mathfrak{S}_n$. However, Erdmann, [77], has shown that, if one knows the decomposition numbers of $\mathfrak{S}_m$ (for sufficiently large values of m), then one will also know the decomposition numbers of $\mathrm{GL}_n(\mathbb{F}_q)$. Thus, the problem of determining the decomposition numbers for symmetric groups appears to be as difficult as the corresponding problem for general linear groups.

In a completely different direction, Dipper and James, [70], showed that the decomposition numbers of $\mathfrak{S}_n$ can also be obtained from the so-called q-Schur algebra, which is defined in terms of the Hecke algebra of $\mathfrak{S}_n$. Thus, the problem of determining the representations of $\mathfrak{S}_n$ in characteristic p is seen to be a special case of the more general problem of studying the representations of Hecke algebras associated with finite Coxeter groups. This is discussed in more detail in the chapter on Hecke algebras. In this context, we just mention here that James' result on the triangularity of the decomposition matrix of $\mathfrak{S}_n$ is generalised to all finite Weyl groups in Geck, [88].

5. Hints for further reading

Here, we give some hints on topics which were not touched in the previous sections.

5.1. *Crystallographic reflection groups*

Let W be a discrete subgroup of the group of all affine transformations of a finite-dimensional affine space E over $K = \mathbb{C}$ or $K = \mathbb{R}$, generated by affine reflections. If W is finite, it necessarily fixes a point and W is a finite complex reflection group. If W is infinite and $K = \mathbb{R}$, the irreducible examples are precisely the affine Weyl groups (see Section 2.14). In the complex case $K = \mathbb{C}$ there are two essentially different cases. If E/W is compact, the group W is called *crystallographic*. The non-crystallographic groups are now just the complexifications of affine Weyl groups. The crystallographic reflection groups have been classified by Popov, [160] (see also Kaneko, Tokunaga and Yoshida, [182,115], for related results). As in the real case they are extensions of a finite complex reflection group W_0 by an invariant lattice which is generated by roots for W_0.

It turns out that presentations for these groups can be obtained as in the case of affine Weyl groups by adding a further generating reflection corresponding to a highest root in a root system for W_0 (see Malle, [139]). As for the finite complex reflection groups in Section 3.1, the braid group $B(W)$ of W is defined as the fundamental group of the hyperplane complement. For many of the irreducible crystallographic complex reflection groups it is known that a presentation for $B(W)$ can be obtained by omitting the order relations from the presentation of W described above (see Dũng, [72], for affine groups, Malle, [139], for the complex case).

5.2. *Quaternionic reflection groups*

Cohen, [51], has obtained the classification of finite reflection groups over the quaternions. This is closely related to finite linear groups over $\mathbb{C}$ generated by *bireflections*, that is, elements of finite order which fix pointwise a subspace of codimension 2. Indeed, using the identification of the quaternions as a certain ring of 2×2-matrices over the complex numbers, reflections over the quaternions become complex bireflections. The primitive bireflection groups had been classified previously by Huffman and Wales, [101,189]. One of the examples is a 3-dimensional quaternionic representation of the double cover of the sporadic Hall–Janko group J_2, see Wilson, [191].

Presentations for these groups resembling the Coxeter presentations for Weyl groups are given by Cohen, [52].

In recent work on the McKay correspondence, quaternionic reflection groups play an important rôle under the name of *symplectic reflection groups* in the construction of so-called symplectic reflection algebras, see, for example, Etingof and Ginzburg, [78].

5.3. *Reflection groups over finite fields*

Many of the general results for complex reflection groups presented in section 1 are no longer true for reflection groups over fields of positive characteristic. Most importantly, the ring of invariants of such a reflection group is not necessarily a polynomial ring. Nevertheless we have the following criterion due to Serre, [25, V.6, Example 8], and Nakajima, [144], generalising Theorem 1.6:

5.4. THEOREM (Serre, Nakajima, [144]). *Let V be a finite-dimensional vector space over a field K and $W \leqslant \mathrm{GL}(V)$ a finite group such that $K[V]^W$ is polynomial. Then the pointwise stabiliser of any subspace $U \leqslant V$ has polynomial ring of invariants (and thus is generated by reflections).*

The irreducible reflection groups over finite fields were classified by Wagner, [187,188], and Zalesskiĭ and Serežkin, [192], the determination of transvection groups was completed by Kantor, [116]. In addition to the modular reductions of complex reflection groups, there arise the infinite families of classical linear, symplectic, unitary and orthogonal groups, as well as some further exceptional examples. For a complete list see, for example, Kemper and Malle, [118, Section 1].

The results of Wagner, [188], and Kantor, [116], are actually somewhat stronger, giving a classification of all indecomposable reflection groups W over finite fields of characteristic p for which the maximal normal p-subgroup is contained in the intersection $W' \cap Z(W)$ of the centre with the derived group.

Using this classification, the *irreducible* reflection groups over finite fields with polynomial ring of invariants could be determined, leading to the following criterion:

5.5. THEOREM (Kemper and Malle, [118]). *Let V be a finite-dimensional vector space over K, $W \leqslant \mathrm{GL}(V)$ a finite irreducible linear group. Then $K[V]^W$ is a polynomial ring if and only if W is generated by reflections and the pointwise stabiliser in W of any non-trivial subspace of V has a polynomial ring of invariants.*

The list of groups satisfying this criterion can be found in [118, Theorem 7.2]. That paper also contains some information on indecomposable groups.

It is an open question whether at least the field of invariants of a reflection group in positive characteristic is purely transcendental (by Kemper and Malle, [119], the answer is positive in the irreducible case).

For further discussions of modular invariant theory of reflection groups see also Derksen and Kemper, [68, 3.7.4].

5.6. *p-adic reflection groups*

Let R be an integral domain, L an R-lattice of finite rank, i.e., a torsion-free finitely generated R-module, and W a finite subgroup of $\mathrm{GL}(L)$ generated by reflections. Again one can ask under which conditions the invariants of W on the symmetric algebra $R[L]$ of the dual L^* are a graded polynomial ring. In the case of Weyl groups Demazure shows the following extension of Theorem 1.3:

5.7. THEOREM (Demazure, [63]). *Let W be a Weyl group, L the root lattice of W, and R a ring in which all torsion primes of W are invertible. Then the invariants of W on $R[L]$ are a graded polynomial algebra, and $R[L]$ is a free graded module over $R[L]^W$.*

In the case of general lattices for reflection groups, the following example may be instructive: Let $W = \mathfrak{S}_3$ the symmetric group of degree 3. Then the weight lattice L of S_3,

Table 6
p-adic reflection groups

W	Conditions
$G(de,e,n)$ $((d,n)\neq(1,2))$	$p\equiv 1 \pmod{de}$; none when $de=2$
$G(e,e,2)$ $(e\geqslant 3)$	$p\equiv \pm 1 \pmod{e}$; none when $e=3,4,6$
$\mathfrak{S}_{n+1}$	–

W	Conditions	W	Conditions
G_4	$p\equiv 1 \pmod 3$	G_{21}	$p\equiv 1,49 \pmod{60}$
G_5	$p\equiv 1 \pmod 3$	G_{22}	$p\equiv 1,9 \pmod{20}$
G_6	$p\equiv 1 \pmod{12}$	G_{23}	$p\equiv 1,4 \pmod 5$
G_7	$p\equiv 1 \pmod{12}$	G_{24}	$p\equiv 1,2,4 \pmod 7$
G_8	$p\equiv 1 \pmod 4$	G_{25}	$p\equiv 1 \pmod 3$
G_9	$p\equiv 1 \pmod 8$	G_{26}	$p\equiv 1 \pmod 3$
G_{10}	$p\equiv 1 \pmod{12}$	G_{27}	$p\equiv 1,4 \pmod{15}$
G_{11}	$p\equiv 1 \pmod{24}$	G_{28}	–
G_{12}	$p\equiv 1,3 \pmod 8$	G_{29}	$p\equiv 1 \pmod 4$
G_{13}	$p\equiv 1 \pmod 8$	G_{30}	$p\equiv 1,4 \pmod 5$
G_{14}	$p\equiv 1,19 \pmod{24}$	G_{31}	$p\equiv 1 \pmod 4$
G_{15}	$p\equiv 1 \pmod{24}$	G_{32}	$p\equiv 1 \pmod 3$
G_{16}	$p\equiv 1 \pmod 5$	G_{33}	$p\equiv 1 \pmod 3$
G_{17}	$p\equiv 1 \pmod{20}$	G_{34}	$p\equiv 1 \pmod 3$
G_{18}	$p\equiv 1 \pmod{15}$	G_{35}	–
G_{19}	$p\equiv 1 \pmod{60}$	G_{36}	–
G_{20}	$p\equiv 1,4 \pmod{15}$	G_{37}	–

considered as $\mathbb{Z}_3$-lattice, yields a faithful reflection representation of $\mathfrak{S}_3$ with the following property: $\mathbb{Z}_3[L]^{\mathfrak{S}_3}$ is not polynomial, while both the reflection representations over the quotient field $\mathbb{Q}_3$ and over the residue field $\mathbb{F}_3$ have polynomial invariants, the first with generators in degrees 2 and 3, the second with generators in degrees 1 and 6.

The list of all irreducible p-adic reflection groups, that is, reflection groups over the field of p-adic numbers $\mathbb{Q}_p$, was obtained by Clark and Ewing, [49], building on the Shephard–Todd theorem. We reproduce it in Table 6.

Using a case-by-case argument based on the Clark–Ewing classification and his own classification of p-adic lattices for reflection groups, Notbohm, [150], was able to determine all finite reflection groups W over the ring of p-adic integers $\mathbb{Z}_p$, $p>2$, with polynomial ring of invariants. This was subsequently extended by Andersen, Grodal, Møller and Viruel, [3], to include the case $p=2$.

5.8. *p-compact groups*

The p-adic reflection groups play an important rôle in the theory of so-called p-compact groups, which constitute a homotopy theoretic analogue of compact Lie groups. By definition, a *p-compact group* is a p-complete topological space BX such that the homology $H_*(X;\mathbb{F}_p)$ of the loop space $X=\Omega BX$ is finite. Examples for p-compact groups are p-completions of classifying spaces of compact Lie groups. Further examples were constructed by Clark and Ewing, [49], Aguadé, [2], Dwyer and Wilkerson, [73], and Notbohm, [149]. To each p-compact group X Dwyer and Wilkerson, [74], associate a maximal torus (unique up to conjugacy) together with a 'Weyl group', which comes equipped with a representation as a reflection group over the p-adic integers $\mathbb{Z}_p$, which is faithful if X is connected. Conversely, by a theorem of Andersen et al., [3], a connected p-compact group, for $p > 2$, is determined up to isomorphism by its Weyl group data, that is, by its Weyl group in a reflection representation on a $\mathbb{Z}_p$-lattice.

It has been shown that at least for $p > 2$ all p-adic reflection groups (as classified by Clark and Ewing) and all their $\mathbb{Z}_p$-reflection representations arise in that way (see Andersen et al., [3], Notbohm, [149], and also Adams and Wilkerson, [1], Aguadé, [2]).

References

[1] J.F. Adams and C.W. Wilkerson, *Finite H-spaces and algebras over the Steenrod algebra*, Ann. of Math. **111** (1980), 95–143.

[2] J. Aguadé, *Constructing modular classifying spaces*, Israel J. Math. **66** (1989), 23–40.

[3] K. Andersen, J. Grodal, J. Møller and A. Viruel, *The classification of p-compact groups for p odd*, Ann. of Math., to appear.

[4] E. Artin, *Theory of braids*, Ann. of Math. **48** (1947), 101–126.

[5] E. Bannai, *Fundamental groups of the spaces of regular orbits of the finite unitary reflection groups of dimension* 2, J. Math. Soc. Japan **28** (1976), 447–454.

[6] M. Benard, *Schur indices and splitting fields of the unitary reflection groups*, J. Algebra **38** (1976), 318–342.

[7] D. Benson, *Polynomial Invariants of Finite Groups*, London Math. Soc. Lecture Note Series **190**, Cambridge Univ. Press (1993).

[8] C.T. Benson and L.C. Grove, *Finite Reflection Groups*, 2nd ed., Graduate Texts in Math. **99**, Springer-Verlag (1985).

[9] D. Bessis, *Sur le corps de définition d'un groupe de réflexions complexe*, Comm. Algebra **25** (1997), 2703–2716.

[10] D. Bessis, *Groupes des tresses et éléments réguliers*, J. Reine Angew. Math. **518** (2000), 1–40.

[11] D. Bessis, *Zariski theorems and diagrams for braid groups*, Invent. Math. **145** (2001), 487–507.

[12] D. Bessis, F. Digne and J. Michel, *Springer theory in braid groups and the Birman–Ko–Lee monoid*, Pacific J. Math. **205** (2002), 287–310.

[13] W. Beynon and G. Lusztig, *Some numerical results on the characters of exceptional Weyl groups*, Math. Proc. Cambridge Philos. Soc. **84** (1978), 417–426.

[14] S. Bigelow, *The Burau representation is not faithful for* $n = 5$, Geometry and Topology **3** (1999), 397–404.

[15] S. Bigelow, *Braid groups are linear*, J. Amer. Math. Soc. **14** (2001), 471–486.

[16] J.S. Birman, *Braids, Links and Mapping Class Groups*, Annals of Math. Stud. **84**, Princeton Univ. Press (1974).

[17] J.S. Birman, *New points of view in knot theory*, Bull. Amer. Math. Soc. **28** (1993), 253–287.

[18] J. Birman, K.H. Ko and S.J. Lee, *A new approach to the word and conjugacy problems in the braid groups*, Adv. Math. **139** (1998), 322–353.

[19] A. Björner, *Orderings of Coxeter groups*, Combinatorics and Algebra, Contemporary Math. **34**, Amer. Math. Soc. (1984), 175–195.

[20] J. Blair and G.I. Lehrer, *Cohomology actions and centralisers in unitary reflection groups*, Proc. London Math. Soc. **83** (2001), 582–604.

[21] F. Bleher, M. Geck and W. Kimmerle, *Automorphisms of integral group rings of finite Coxeter groups and Iwahori–Hecke algebras*, J. Algebra **197** (1997), 615–655.

[22] R.E. Borcherds, *Coxeter groups, Lorentzian lattices, and K3 surfaces*, Internat. Math. Res. Notices **19** (1998), 1011–1031.

[23] A. Borel, *Linear Algebraic Groups*, 2nd enlarged edition, Graduate Texts in Math. **126**, Springer-Verlag (1991).

[24] A.V. Borovik, I.M. Gelfand and N. White, *Coxeter matroid polytopes*, Ann. Comb. **1** (1997), 123–134.

[25] N. Bourbaki, *Groupes et algèbres de Lie, chap. 4, 5 et 6*, Hermann (1968).

[26] K. Bremke and G. Malle, *Reduced words and a length function for $G(e,1,n)$*, Indag. Math. **8** (1997), 453–469.

[27] K. Bremke and G. Malle, *Root systems and length functions*, Geom. Dedicata **72** (1998), 83–97.

[28] E. Brieskorn, *Die Fundamentalgruppe des Raumes der regulären Orbits einer endlichen komplexen Spiegelungsgruppe*, Invent. Math. **12** (1971), 37–61.

[29] E. Brieskorn, *Sur les groupes de tresses* [d'après V.I. Arnold], Séminaire Bourbaki, 24ème année, 1971/72, Lecture Notes in Math. **317**, Springer (1973).

[30] E. Brieskorn and K. Saito, *Artin-Gruppen und Coxeter-Gruppen*, Invent. Math. **17** (1972), 245–271.

[31] B. Brink and R.B. Howlett, *A finiteness property and an automatic structure for Coxeter groups*, Math. Ann. **296** (1993), 179–190.

[32] B. Brink and R.B. Howlett, *Normalizers of parabolic subgroups in Coxeter groups*, Invent. Math. **136** (1999), 323–351.

[33] M. Broué, *Reflection groups, braid groups, Hecke algebras, finite reductive groups*, Current Developments in Mathematics, Int. Press (2001), 1–107.

[34] M. Broué and G. Malle, *Théorèmes de Sylow génériques pour les groupes réductifs sur les corps finis*, Math. Ann. **292** (1992), 241–262.

[35] M. Broué and G. Malle, *Zyklotomische Heckealgebren*, Astérisque **212** (1993), 119–189.

[36] M. Broué and G. Malle, *Generalized Harish-Chandra theory*, Representations of Reductive Groups, Cambridge Univ. Press (1998), 85–103.

[37] M. Broué, G. Malle and J. Michel, *Towards spetses I*, Transform. Groups **4** (1999), 157–218.

[38] M. Broué, G. Malle and R. Rouquier, *Complex reflection groups, braid groups, Hecke algebras*, J. Reine Angew. Math. **500** (1998), 127–190.

[39] M. Broué and J. Michel, *Sur certains éléments réguliers des groupes de Weyl et les variétés de Deligne–Lusztig associées*, Finite Reductive Groups, Related Structures and Representations, Progress in Math. **141**, Birkhäuser (1997), 73–139.

[40] G. Burde and H. Zieschang, *Knots*, De Gruyter Studies in Math. **5**, Walter de Gruyter & Co. (1985).

[41] R.W. Carter, *Conjugacy classes in the Weyl group*, Compositio Math. **25** (1972), 1–59.

[42] R.W. Carter, *Simple Groups of Lie Type*, Wiley (1972); reprinted 1989 as Wiley Classics Library Edition.

[43] R.W. Carter, *Finite Groups of Lie Type: Conjugacy Classes and Complex Characters*, Wiley, New York (1985); reprinted 1993 as Wiley Classics Library Edition.

[44] W.A. Casselman, *Machine calculations in Weyl groups*, Invent. Math. **116** (1994), 95–108.

[45] C. Chevalley, *Sur certains groupes simples*, Tôhoku Math. J. (2) **7** (1955), 14–66.

[46] C. Chevalley, *Invariants of finite groups generated by reflections*, Amer. J. Math. **77** (1955), 778–782.

[47] C. Chevalley, *Classification des groupes de Lie algébriques*, Séminaire École Normale Supérieure, Mimeographed Notes, Paris (1956–1958).

[48] C. Chevalley, *Sur les décompositions cellulaires des espaces G/B*. With a foreword by A. Borel, Proc. Symp. Pure Math. **56**, Amer. Math. Soc. (1994), 1–23.

[49] A. Clark and J. Ewing, *The realization of polynomial algebras as cohomology rings*, Pacific J. Math. **50** (1974), 425–434.

[50] A.M. Cohen, *Finite complex reflection groups*, Ann. Sci. École Norm. Sup. **9** (1976), 379–436. Erratum: ibid. **11** (1978), 613.

[51] A.M. Cohen, *Finite quaternionic reflection groups*, J. Algebra **64** (1980), 293–324.

[52] A.M. Cohen, *Presentations for certain finite quaternionic reflection groups*, Advances in Finite Geometries and Designs, Oxford Univ. Press (1991), 69–79.

[53] A.M. Cohen and D. Wales, *Linearity of Artin groups of finite type*, Israel J. Math. **131** (2002), 101–123.

[54] H.S.M. Coxeter, *Discrete groups generated by reflections*, Ann. of Math. **35** (1934), 588–621.

[55] H.S.M. Coxeter, *The complete enumeration of finite groups of the form $R_i^2 = (R_i R_j)^{k_{ij}} = 1$*, J. London Math. Soc. **10** (1935), 21–25.

[56] H.S.M. Coxeter, *Groups generated by unitary reflections of period two*, Canad. J. Math. **9** (1957), 243–272.

[57] R.H. Crowell and R.H. Fox, *Introduction to Knot Theory*, Graduate Texts in Math. **57**, Springer-Verlag (1977); reprint of the 1963 original.

[58] C.W. Curtis and I. Reiner, *Methods of Representation Theory*, Vol. II, Wiley (1987); reprinted 1994 as Wiley Classics Library Edition.

[59] P. Dehornoy, *Braids and Self-Distributivity*, Progress in Math. **192**, Birkhäuser (2000).

[60] P. Dehornoy and L. Paris, *Gaussian groups and Garside groups, two generalizations of Artin groups*, Proc. London Math. Soc. **79** (1999), 569–604.

[61] P. Deligne, *Les immeubles des groupes de tresses généralisés*, Invent. Math. **17** (1972), 273–302.

[62] P. Deligne, *Action du groupe de tresses sur une catégorie*, Invent. Math. **128** (1997), 159–175.

[63] M. Demazure, *Invariants symétriques entiers des groupes de Weyl et torsion*, Invent. Math. **21** (1973), 287–301.

[64] J. Denef and F. Loeser, *Regular elements and monodromy of discriminants of finite reflection groups*, Indag. Math. **6** (1995), 129–143.

[65] V.V. Deodhar, *Some characterizations of Bruhat ordering on a Coxeter group and determination of the relative Möbius function*, Invent. Math. **39** (1977), 187–198.

[66] V. V. Deodhar, *On the root system of a Coxeter group*, Comm. Algebra **10** (1982), 611–630.

[67] V.V. Deodhar, *A note on subgroups generated by reflections in Coxeter groups*, Arch. Math. (Basel) **53** (1989), 543–546.

[68] H. Derksen and G. Kemper, *Computational Invariant Theory*, Encyclopaedia of Mathematical Sciences **130**, Springer-Verlag (2002).

[69] F. Digne, *On the linearity of Artin braid groups*, J. Algebra **268** (2003), 39–57.

[70] R. Dipper and G.D. James, *The q-Schur algebra*, Proc. London Math. Soc. **59** (1989), 23–50.

[71] F. du Cloux, *A transducer approach to Coxeter groups*, J. Symbolic Comput. **27** (1999), 1–14.

[72] N.V. Dũng, *The fundamental groups of the spaces of regular orbits of the affine Weyl groups*, Topology **22** (1983), 425–435.

[73] W.G. Dwyer and C.W. Wilkerson, *A new finite loop space at the prime two*, J. Amer. Math. Soc. **6** (1993), 37–64.

[74] W.G. Dwyer and C.W. Wilkerson, *Homotopy fixed-point methods for Lie groups and finite loop spaces*, Ann. of Math. **139** (1994), 395–442.

[75] M. Dyer, *Hecke algebras and reflections in Coxeter groups*, Ph.D. thesis, University of Sydney (1987).

[76] M. Dyer, *Reflection subgroups of Coxeter systems*, J. Algebra **135** (1990), 57–73.

[77] K. Erdmann, *Decomposition numbers for symmetric groups and composition factors of Weyl modules*, J. Algebra **180** (1996), 316–320.

[78] P. Etingof and V. Ginzburg, *Symplectic reflection algebras, Calogero–Moser space, and deformed Harish-Chandra homomorphism*, Invent. Math. **147** (2002), 243–348.

[79] L. Flatto, *Invariants of finite reflection groups*, Enseign. Math. **24** (1978), 237–292.

[80] P. Fleischmann and I. Janiszczak, *Combinatorics and Poincaré polynomials of hyperplane complements for exceptional Weyl groups*, J. Combin. Theory Ser. A **63** (1993), 257–274.

[81] P. Fong and G.M. Seitz, *Groups with a BN-pair of rank 2, I*, Invent. Math. **21** (1973), 1–57; *II*, ibid. **24** (1974), 191–239.

[82] R.H. Fox and L. Neuwirth, *The braid groups*, Math. Scand. **10** (1962), 119–126.

[83] D. Freyd, D. Yetter, J. Hoste, W.B.R. Lickorish, K. Millet and A. Ocneanu, *A new polynomial invariant of knots and links*, Bull. Amer. Math. Soc. **12** (1985), 239–246.

[84] F.G. Frobenius, *Über die Charaktere der symmetrischen Gruppe*, Sitzungsber. Preuss. Akad. Wiss. Berlin (1900), 516–534.

[85] W. Fulton, *Young Tableaux*, London Math. Soc. Student Texts **35**, Cambridge Univ. Press (1997).

[86] F.A. Garside, *The braid group and other groups*, Quart. J. Math. Oxford **20** (1969), 235–254.

[87] M. Geck, *Trace functions on Hecke algebras*, Banach Center Publ. **42**, Polish Acad. Sci. (1998), 87–109.

[88] M. Geck, *Kazhdan–Lusztig cells and decomposition numbers*, Represent. Theory **2** (1998), 264–277 (electronic).

[89] M. Geck, G. Hiss, F. Lübeck, G. Malle and G. Pfeiffer, CHEVIE – *A system for computing and processing generic character tables*, Appl. Algebra Engrg. Comm. Comput. **7** (1996), 175–210.

[90] M. Geck and S. Kim, *Bases for the Bruhat–Chevalley order on all finite Coxeter groups*, J. Algebra **197** (1997), 278–310.

[91] M. Geck, S. Kim and G. Pfeiffer, *Minimal length elements in twisted conjugacy classes of finite Coxeter groups*, J. Algebra **229** (2000), 570–600.

[92] M. Geck and S. Lambropoulou, *Markov traces and knot invariants related to Iwahori–Hecke algebras of type B*, J. Reine Angew. Math. **482** (1997), 191–213.

[93] M. Geck and J. Michel, *"Good" elements of finite Coxeter groups and representations of Iwahori–Hecke algebras*, Proc. London Math. Soc. **74** (1997), 275–305.

[94] M. Geck and G. Pfeiffer, *On the irreducible characters of Hecke algebras*, Adv. Math. **102** (1993), 79–94.

[95] M. Geck and G. Pfeiffer, *Characters of Finite Coxeter Groups and Iwahori–Hecke Algebras*, Oxford Univ. Press (2000).

[96] D. Gorenstein, R. Lyons and R. Solomon, *The Classification of the Finite Simple Groups*, Math. Surveys Monographs **40**, no. 1, Amer. Math. Soc. (1994).

[97] J.A. Green, *Polynomial Representations of* GL_n, Lecture Notes in Math. **830**, Springer-Verlag (1980).

[98] C. Hering, W.M. Kantor and G.M. Seitz, *Finite groups with a split BN-pair of rank* 1, J. Algebra **20** (1972), 435–475.

[99] H. Hiller, *Geometry of Coxeter Groups*, Res. Notes Mathematics **54**, Pitman (1982).

[100] R.B. Howlett, *Normalizers of parabolic subgroups of reflection groups*, J. London Math. Soc. (2) **21** (1980), 62–80.

[101] W. Huffman and D. Wales, *Linear groups containing an involution with two eigenvalues* -1, J. Algebra **45** (1977), 465–515.

[102] M. Hughes, *Representations of complex imprimitive reflection groups*, Math. Proc. Cambridge Philos. Soc. **94** (1983), 425–436.

[103] M. Hughes and A. Morris, *Root systems for two dimensional complex reflection groups*, Sém. Lothar. Combin. **45**, 18 pp. (electronic).

[104] J.E. Humphreys, *Introduction to Lie Algebras and Representation Theory*, Graduate Texts in Math. **9**, Springer-Verlag (1972).

[105] J.E. Humphreys, *Reflection Groups and Coxeter Groups*, Cambridge Stud. Adv. Math. **29**, Cambridge Univ. Press (1990).

[106] J.E. Humphreys, *Linear Algebraic Groups*, 2nd ed., Graduate Texts in Math. **21**, Springer-Verlag (1991).

[107] J.F. Humphreys, *Character tables for the primitive finite unitary reflection groups*, Comm. Algebra **22** (1994), 5777–5802.

[108] N. Iwahori, *On the structure of a Hecke ring of a Chevalley group over a finite field*, J. Fac. Sci. Univ. Tokyo **10** (1964), 215–236.

[109] N. Iwahori and H. Matsumoto, *On some Bruhat decomposition and the structure of the Hecke ring of $\mathfrak{p}$-adic Chevalley groups*, Publ. Math. IHES **25** (1965), 5–48.

[110] G.D. James, *The Representation Theory of the Symmetric Groups*, Lecture Notes in Math. **682**, Springer-Verlag (1978).

[111] G.D. James and A. Kerber, *The Representation Theory of the Symmetric Group*, Encyclopedia of Mathematics **16**, Addison-Wesley (1981).

[112] V.F.R. Jones, *Hecke algebra representations of braid groups and link polynomials*, Ann. of Math. **126** (1987), 335–388.

[113] V.G. Kac, *Infinite Dimensional Lie Algebras*, Cambridge Univ. Press (1985).

[114] R. Kane, *Reflection Groups and Invariant Theory*, CMS Books in Math. **5**, Springer (2001).

[115] J. Kaneko, S. Tokunaga and M. Yoshida, *Complex crystallographic groups. II*, J. Math. Soc. Japan **34** (1982), 595–605.

[116] W.M. Kantor, *Subgroups of classical groups generated by long root elements*, Trans. Amer. Math. Soc. **248** (1979), 347–379.

[117] W.M. Kantor and G.M. Seitz, *Finite groups with a split BN-pair of rank* 1. *II*, J. Algebra **20** (1972), 476–494.

[118] G. Kemper and G. Malle, *The finite irreducible linear groups with polynomial ring of invariants*, Transform. Groups **2** (1997), 57–89.

[119] G. Kemper and G. Malle, *Invariant fields of finite irreducible reflection groups*, Math. Ann. **315** (1999), 569–586.

[120] S. Kim, *Mots de longueur maximale dans une classe de conjugaison d'un groupe symétrique, vu comme groupe de Coxeter*, C. R. Acad. Sci. Paris **327** (1998), 617–622.

[121] D. Krammer, *The conjugacy problem for Coxeter groups*, Ph.D. thesis, University of Utrecht (1994).
[122] D. Krammer, *Braid groups are linear*, Ann. of Math. **155** (2002), 131–156.
[123] S. Kusuoka, *On a conjecture of L. Solomon*, J. Fac. Sci. Univ. Tokyo Sect. IA Math. **24** (1977), 645–655.
[124] S. Lambropoulou, *Solid torus links and Hecke algebras of B-type*, Proceedings of the Conference on Quantum Topology, Manhattan, KS, 1993, World Sci. Publishing (1994), 225–245.
[125] S. Lambropoulou, *Knot theory related to generalized and cyclotomic Hecke algebras of type B*, J. Knot Theory Ramifications **8** (1999), 621–658.
[126] A. Lascoux and M.P. Schützenberger, *Treillis et bases des groupes de Coxeter*, Electron. J. Combin. **3** (1995), 35.
[127] G.I. Lehrer, *On the Poincaré series associated with Coxeter group actions on complements of hyperplanes*, J. London Math. Soc. **36** (1987), 275–294.
[128] G.I. Lehrer, *Poincaré polynomials for unitary reflection groups*, Invent. Math. **120** (1995), 411–425.
[129] G.I. Lehrer, *A new proof of Steinberg's fixed-point theorem*, Internat. Math. Res. Notices **28** (2004), 1407–1411.
[130] G.I. Lehrer and J. Michel, *Invariant theory and eigenspaces for unitary reflection groups*, C. R. Acad. Sci. Paris Sér. I Math. **336** (2003), 795–800.
[131] G.I. Lehrer and T.A. Springer, *Intersection multiplicities and reflection subquotients of unitary reflection groups I*, Geometric Group Theory Down Under, Canberra, 1996, de Gruyter (1999), 181–193.
[132] G.I. Lehrer and T.A. Springer, *Reflection subquotients of unitary reflection groups*, Canad. J. Math. **51** (1999), 1175–1193.
[133] D.D. Long and M. Paton, *The Burau representation is not faithful for $n \geqslant 6$*, Topology **32** (1993), 439–447.
[134] G. Lusztig, *Irreducible representations of finite classical groups*, Invent. Math. **43** (1977), 125–175.
[135] G. Lusztig and N. Spaltenstein, *Induced unipotent classes*, J. London Math. Soc. **19** (1979), 41–52.
[136] I.G. Macdonald, *Some irreducible representations of Weyl groups*, Bull. London Math. Soc. **4** (1972), 148–150.
[137] I.G. Macdonald, *Symmetric Functions and Hall Polynomials*, 2nd ed., Oxford Univ. Press (1995).
[138] G. Malle, *Unipotente Grade imprimitiver komplexer Spiegelungsgruppen*, J. Algebra **177** (1995), 768–826.
[139] G. Malle, *Presentations for crystallographic complex reflection groups*, Transform. Groups **1** (1996), 259–277.
[140] G. Malle, *On the rationality and fake degrees of characters of cyclotomic algebras*, J. Math. Sci. Univ. Tokyo **6** (1999), 647–677.
[141] H. Matsumoto, *Générateurs et relations des groupes de Weyl généralisées*, C. R. Acad. Sci. Paris **258** (1964), 3419–3422.
[142] J. Michel, *A note on words in braid monoids*, J. Algebra **215** (1999), 366–377.
[143] J.A. Moody, *The faithfulness question for the Burau representation*, Proc. Amer. Math. Soc. **119** (1993), 671–679.
[144] H. Nakajima, *Invariants of finite groups generated by pseudoreflections in positive characteristic*, Tsukuba J. Math. **3** (1979), 109–122.
[145] T. Nakamura, *A note on the $K(\pi, 1)$ property of the orbit space of the unitary reflection group $G(m, l, n)$*, Sci. Papers College Arts Sci. Univ. Tokyo **33** (1983), 1–6.
[146] I. Naruki, *The fundamental group of the complement for Klein's arrangement of twenty-one lines*, Topology Appl. **34** (1990), 167–181.
[147] G. Nebe, *The root lattices of the complex reflection groups*, J. Group Theory **2** (1999), 15–38.
[148] D. Notbohm, *p-adic lattices of pseudo reflection groups*, Algebraic Topology: New Trends in Localization and Periodicity, Progr. Math. **136**, Birkhäuser (1996), 337–352.
[149] D. Notbohm, *Topological realization of a family of pseudoreflection groups*, Fund. Math. **155** (1998), 1–31.
[150] D. Notbohm, *For which pseudo-reflection groups are the p-adic polynomial invariants again a polynomial algebra?* J. Algebra **214** (1999), 553–570. Erratum: ibid. **218** (1999), 286–287.
[151] E.M. Opdam, *A remark on the irreducible characters and fake degrees of finite real reflection groups*, Invent. Math. **120** (1995), 447–454.
[152] E.M. Opdam, *Lecture Notes on Dunkl Operators for Real and Complex Reflection Groups*, MSJ Memoirs **8**, Math. Soc. Japan (2000).

[153] P. Orlik and L. Solomon, *Combinatorics and topology of complements of hyperplanes*, Invent. Math. **56** (1980), 167–189.

[154] P. Orlik and L. Solomon, *Unitary reflection groups and cohomology*, Invent. Math. **59** (1980), 77–94.

[155] P. Orlik and L. Solomon, *A character formula for the unitary group over a finite field*, J. Algebra **84** (1983), 136–141.

[156] P. Orlik and L. Solomon, *Arrangements in unitary and orthogonal geometry over finite fields*, J. Combin. Theory Ser. A **38** (1985), 217–229.

[157] P. Orlik and L. Solomon, *Braids and discriminants*, Braids, Santa Cruz, CA, 1986, Contemp. Math. **78**, Amer. Math. Soc. (1988), 605–613.

[158] P. Orlik and L. Solomon, *Discriminants in the invariant theory of reflection groups*, Nagoya Math. J. **109** (1988), 23–45.

[159] M. Osima, *On the representations of the generalized symmetric group*, Math. J. Okayama Univ. **4** (1954), 39–56.

[160] V. Popov, *Discrete Complex Reflection Groups*, Communications of the Mathematical Institute **15**, Rijksuniversiteit Utrecht (1982).

[161] J.H. Przytycki and P. Traczyk, *Invariants of links of Conway type*, Kobe J. Math. **4** (1987), 115–139.

[162] A. Ram, *A Frobenius formula for the characters of the Hecke algebras*, Invent. Math. **106** (1991), 461–488.

[163] K. Rampetas and T. Shoji, *Length functions and Demazure operators for $G(e,1,n)$. I, II*, Indag. Math. **9** (1998), 563–580, 581–594.

[164] E.W. Read, *On the finite imprimitive unitary reflection groups*, J. Algebra **45** (1977), 439–452.

[165] R.W. Richardson, *Conjugacy classes of involutions in Coxeter groups*, Bull. Austral. Math. Soc. **26** (1982), 1–15.

[166] G.C. Shephard, *Abstract definitions for reflection groups*, Canad. J. Math. **9** (1957), 273–276.

[167] G.C. Shephard and J.A. Todd, *Finite unitary reflection groups*, Canad. J. Math. **6** (1954), 274–304.

[168] J.-Y. Shi, *Conjugacy relation on Coxeter elements*, Adv. Math. **161** (2001), 1–19.

[169] L. Solomon, *Invariants of finite reflection groups*, Nagoya Math. J. **22** (1963), 57–64.

[170] T.A. Springer, *Regular elements of finite reflection groups*, Invent. Math. **25** (1974), 159–198.

[171] T.A. Springer, *Linear Algebraic Groups*, 2nd ed., Progress in Math. **9**, Birkhäuser (1998).

[172] T.A. Springer and R. Steinberg, *Conjugacy classes*, Seminar on Algebraic Groups and Related Finite Groups (The Institute for Advanced Studies, Princeton, NJ, 1968/69), Lecture Notes in Math. **131**, Springer (1970), 167–266.

[173] A.J. Starkey, *Characters of the generic Hecke algebra of a system of BN-pairs*, Ph.D. thesis, University of Warwick (July 1975).

[174] R. Steinberg, *A geometric approach to the representations of the full linear group over a Galois field*, Trans. Amer. Math. Soc. **71** (1951), 274–282; see also [178], 1–9.

[175] R. Steinberg, *Differential equations invariant under finite reflection groups*, Trans. Amer. Math. Soc. **112** (1964), 392–400; see also [178], 173–181.

[176] R. Steinberg, *Lectures on Chevalley groups*, Mimeographed notes, Department of Mathematics, Yale University (1967).

[177] R. Steinberg, *Endomorphisms of linear algebraic groups*, Mem. Amer. Math. Soc. **80** (1968), 1–108; see also [178], 229–285.

[178] R. Steinberg, *Collected Papers*, with a foreword by J.-P. Serre, Amer. Math. Soc. (1997).

[179] J. Stembridge, *On the eigenvalues of representations of reflection groups and wreath products*, Pacific J. Math. **140** (1989), 353–396.

[180] H. Terao and T. Yano, *The duality of the exponents of free deformations associated with unitary reflection groups*, Algebraic Groups and Related Topics, Kyoto/Nagoya, 1983, Adv. Stud. Pure Math. **6**, North-Holland (1985), 339–348.

[181] J. Tits, *Buildings of Spherical Types and Finite BN-Pairs*, Lecture Notes in Math. **386**, Springer (1974).

[182] S. Tokunaga and M. Yoshida, *Complex crystallographic groups. I*, J. Math. Soc. Japan **34** (1982), 581–593.

[183] P. Traczyk, *A new proof of Markov's braid theorem*, Banach Center Publ. **42**, Polish Acad. Sci. (1998), 409–419.

[184] D.N. Verma, *Möbius inversion for the Bruhat ordering on a Weyl group*, Ann. Sci. École Norm. Sup. **4** (1971), 393–398.

[185] D.N. Verma, *The rôle of affine Weyl groups in the representation theory of algebraic Chevalley groups and their Lie algebras*, Lie Groups and Their Representations, Halsted, New York (1975), 653–705.

[186] P. Vogel, *Representation of links by braids: a new algorithm*, Comment. Math. Helv. **65** (1990), 104–113.

[187] A. Wagner, *Collineation groups generated by homologies of order greater than* 2, Geom. Dedicata **7** (1978), 387–398.

[188] A. Wagner, *Determination of the finite primitive reflection groups over an arbitrary field of characteristic not* 2. *I*, Geom. Dedicata **9** (1980), 239–253; *II* ibid. **10**, 191–203; *III* ibid. **10**, 475–523.

[189] D. Wales, *Linear groups of degree n containing an involution with two eigenvalues* -1. *II*, J. Algebra **53** (1978), 58–67.

[190] N. White, *The Coxeter matroids of Gelfand et al.*, Contemp. Math. **197**, Amer. Math. Soc. (1996), 401–409.

[191] R.A. Wilson, *The geometry of the Hall–Janko group as a quaternionic reflection group*, Geom. Dedicata **20** (1986), 157–173.

[192] A.E. Zalesskiĭ and V.N. Serežkin, *Finite linear groups generated by reflections*, Izv. Akad. Nauk SSSR Ser. Mat. **44** (1980), 1279–1307 (in Russian). English transl.: Math. USSR-Izv. **17** (1981), 477–503.

Hurwitz Groups and Hurwitz Generation

M.C. Tamburini

Dipartimento di Matematica e Fisica, Universitá Cattolica del Sacro Cuore, Brescia, Italy
E-mail: c.tamburini@dmf.unicatt.it

M. Vsemirnov

St. Petersburg Division of Steklov Institute of Mathematics, St. Petersburg, Russia
E-mail: vsemir@pdmi.ras.ru

Contents

HANDBOOK OF ALGEBRA, VOL. 4
Edited by M. Hazewinkel

1. Introduction

A $(2,3,7)$-generated group is a group generated by two elements of order 2 and 3 respectively such that their product has order 7. Such a group is called Hurwitz if it is finite. In other words, Hurwitz groups are the non-trivial finite homomorphic images of the abstract triangle group $T(2,3,7)$ defined by the presentation

$$T(2,3,7)=\langle X,Y \mid X^2=Y^3=(XY)^7=1\rangle.$$

In particular, the Hurwitz groups form a wide and remarkable class of the so-called $(2,3)$-*generated* groups, i.e. the non-trivial epimorphic images of the free product

$$C_2 * C_3=\langle X,Y \mid X^2=Y^3=1\rangle.$$

It is well known that $C_2 * C_3$ is isomorphic to $\mathrm{PSL}_2(\mathbb{Z})$ (R. Fricke and F. Klein, [32]). The structure of normal subgroups of $\mathrm{PSL}(2,\mathbb{Z})$ and the corresponding factor groups were the subject of intensive study; for instance, see [26,27,77,81,80] and, especially, the remarkable paper of M.W. Liebeck and A. Shalev, [50].

The study of Hurwitz groups goes back to the late XIX century and shows an important connection with the theory of Riemann surfaces. In 1893, A. Hurwitz, [37], proved that the automorphism group of an algebraic curve of genus $g \geqslant 2$ always has order at most $84(g-1)$ and that this upper bound is attained precisely when the group is an image of $T(2,3,7)$. Hurwitz's discovery originated from the example (due to F. Klein, [45]), of $\mathrm{PSL}_2(7)$, the smallest Hurwitz group, acting as the automorphism group of the quartic $x^3y+y^3z+z^3x=0$ of genus 3.

Since then, examples of Hurwitz groups were rather fragmentary until the pioneering paper of A.M. Macbeath, [57], appeared in 1969. In this paper he describes all prime powers q such that the group $\mathrm{PSL}_2(q)$ is Hurwitz. On the other hand, a result of J. Cohen, [6], asserts that the Hurwitz subgroups of $\mathrm{PSL}_3(q)$ are just those which arise from representations of the above groups discovered by A.M. Macbeath. And this fact may have erroneously discouraged, for a long time, the search for (projective) linear groups which are Hurwitz.

The next significant step in the positive direction was done by G. Higman and M.D.E. Conder who developed a very powerful method of building new permutational representations of $T(2,3,7)$ via combinatorial diagrams. As a result, Conder, [8], proved that almost all alternating groups are Hurwitz. Later in the papers of A. Lucchini, M.C. Tamburini and J.S. Wilson, [55,56], these constructive ideas were generalized to a linear context, providing a new bunch of Hurwitz groups, which include most finite classical simple groups of sufficiently large rank. Actually several authors considered the problem of determining which finite simple groups are Hurwitz. Among them, G. Malle, [60,61], gave precise answers for many classes of exceptional simple groups of Lie type. And, by the contributions of A. Woldar, [95], R.A. Wilson, [92–94], and others, it is now known exactly which of the 26 sporadic simple groups are Hurwitz.

It can be shown that there are $2^{\aleph_0}$ non-isomorphic $(2,3,7)$-generated groups, [56]. So any attempt to classify all of them is not realistic. But, as mentioned above, there have been significant achievements in studying specific classes of groups (e.g., finite simple groups) with respect to the property of being Hurwitz. And there have been achievements in classifying the low-dimensional linear and projective representations of $T(2,3,7)$ over an algebraically closed field $\mathbb{F}$ of characteristic $p \geqslant 0$, [82,78].

In this connection there is a crucial formula, due to L.L. Scott, [68]. Given a group $H = \langle a_1, \ldots, a_m \rangle$ and a representation $f : H \to \mathrm{GL}_n(\mathbb{F})$, this formula restricts the similarity invariants of $f(a_1), \ldots, f(a_m)$ and of the product $f(a_1 \cdots a_m)$. Using Scott's result, L. Di Martino, M.C. Tamburini and A.E. Zalesskii, [25], excluded most of the linear classical groups in dimensions up to 19 from being Hurwitz. For small n, further combination of Scott's formula with results of K. Strambach and H. Völklein, [74], on linearly rigid triples allows to classify the irreducible Hurwitz subgroups of $\mathrm{SL}_n(\mathbb{F})$ and $\mathrm{PSL}_n(\mathbb{F})$, for $n \leqslant 5$. We refer to Section 4.3, which is devoted to this classification, and to [82].

There are other aspects of Hurwitz groups which are interesting in themselves, and also shed more light on the understanding of the groups. For example, using number theory, M. Vsemirnov, V. Mysovskikh and M.C. Tamburini, [88], gave an alternative definition of $T(2,3,7)$ as a unitary group over an appropriate ring. This result also has a strict relation to Macbeath's theorem.

The aim of this chapter is to survey the main achievements and ideas in this field as well as bring together recent results widely dispersed in the literature. Some of the results or proofs appear here for the first time. However, we do not touch some specific matters already covered in previous survey articles. For further reading we recommend the survey articles [13,22,23,40], and [91].

2. Triangle groups

DEFINITION 2.1. Let G be a group and k, l, m be integers $\geqslant 2$. If $x, y \in G$ have orders k and l, respectively, and $z = xy$ has order m, we say that the triple (x, y, z) is a (k,l,m)-*triple*. A group G is called (k,l,m)-*generated* if it can be generated by two elements x and y such that (x, y, xy) is a (k,l,m)-triple. In this case we also say that (x, y, xy) is a (k,l,m)-*generating triple*.

In particular, any (k,l,m)-generated group is a homomorphic image of the abstract triangle group $T(k,l,m)$ defined by the presentation

$$T(k,l,m) = \langle X, Y \mid X^k = Y^l = (XY)^m = 1 \rangle.$$

The groups $T(k,l,m)$ have a nice geometric description.[1] Let $\Delta = \Delta(k,l,m)$ be a triangle having angles of size $\frac{\pi}{k}, \frac{\pi}{l}, \frac{\pi}{m}$, that is Δ is a spherical, Euclidean or hyperbolic triangle depending on whether

$$\frac{1}{k} + \frac{1}{l} + \frac{1}{m}$$

[1] This geometric interpretation also explains the terminology 'triangle group'.

is greater than, equal to or less than 1. Then $T(k,l,m)$ can be defined as a group of motions of the two-dimensional space (sphere, Euclidean plane or hyperbolic plane, respectively), namely, as the group generated by rotations of angles $\frac{2\pi}{k}, \frac{2\pi}{l}\ \frac{2\pi}{m}$ around the corresponding vertices of $\triangle$. We just mention that hyperbolic triangle groups are special cases of *Fuchsian* groups, i.e., finitely generated discontinuous groups of orientation-preserving non-Euclidean motions (for instance, see [59]).

Let $T^*(k,l,m)$ be the group of motions, generated by reflections around the sides of $\triangle$. It can be shown that the images of $\triangle$ under $T^*(k,l,m)$ tessellate the corresponding space without overlapping. In addition, $T^*(k,l,m)$ admits the presentation

$$T^*(k,l,m) = \langle X, Y, T \mid X^k = Y^l = (XY)^m = T^2 = (XT)^2 = (YT)^2 = 1\rangle$$

and $T(k,l,m)$ is a subgroup of index 2 in $T^*(k,l,m)$. In particular, $T(k,l,m)$ is finite precisely when

$$\frac{1}{k} + \frac{1}{l} + \frac{1}{m} > 1,$$

see, e.g., [18, Section 6.4], where these groups appear under the name *polyhedral groups*.

The above geometric description of triangle groups $T(k,l,m)$ allows to embed them into $\mathrm{PSL}(2,\mathbb{C})$. We indicate an explicit embedding only when

$$\frac{1}{k} + \frac{1}{l} + \frac{1}{m} < 1.$$

The following construction is taken from [59, Chapter II, Exercises 5, 6]. Set

$$\kappa = \mathrm{e}^{\frac{-i\pi}{k}}, \qquad \lambda = \mathrm{e}^{\frac{i\pi}{l}}, \qquad \mu = \mathrm{e}^{\frac{i\pi}{m}}, \qquad r = \rho^{-1} - \rho,$$

where ρ is the positive root of

$$t^2\big(\mu + \mu^{-1} + \lambda\kappa^{-1} + \kappa\lambda^{-1}\big) = \mu + \mu^{-1} + \lambda\kappa + (\lambda\kappa)^{-1}.$$

Let X, Y be the Möbius transformations with matrices

$$\begin{pmatrix} \kappa & 0 \\ 0 & \kappa^{-1} \end{pmatrix} \quad \text{and} \quad r^{-1}\begin{pmatrix} \rho\lambda^{-1} - \lambda\rho^{-1} & \lambda - \lambda^{-1} \\ \lambda^{-1} - \lambda & \lambda\rho - (\lambda\rho)^{-1} \end{pmatrix},$$

respectively. Then X, Y map the interior of the unit disc $|z| < 1$ into itself and k, l, m are respectively the exact orders of X, Y, XY. In addition, the fixed points of X, Y, XY within the unit disc are the vertices of a non-Euclidean triangle $\triangle$ with angles $\frac{\pi}{k}, \frac{\pi}{l}, \frac{\pi}{m}$ and X, Y actually generate the triangle group $T(k,l,m)$.

The above interpretation relates triangle groups and in particular $T(2,3,7)$ to hyperbolic geometry and the theory of Riemann surfaces. The following theorem explains the importance of $T(2,3,7)$ and Hurwitz groups in this context.

THEOREM 2.2 (Hurwitz, [37]). *Let S be a compact Riemann surface of genus $g \geqslant 2$ and H be its automorphism group. Then $|H| \leqslant 84(g-1)$. Moreover, a finite group H of order $84(g-1)$ is the automorphism group of a compact Riemann surface of genus g if and only if H is $(2,3,7)$-generated.*

The proof of this result is based on the fact that, among all Fuchsian groups, $T(2,3,7)$ has the fundamental domain of the smallest volume. Details can be found in many standard textbooks like [59, Section II.7] or [42, Section 5.11]. For example, the smallest Hurwitz group $\mathrm{PSL}_2(7)$ of order 168 is the group of automorphisms of Klein's quartic $x^3y + y^3z + z^3x = 0$ of genus 3. As we will see later, there are infinitely many non-isomorphic Hurwitz groups. In other words there are infinitely many other values of the genus g for which the Hurwitz upper bound is attained. However, it is not attained when $g = 2$. Moreover, it can be shown that there are infinitely many values of g for which it is not attained. The precise values of g for which the Hurwitz bound is attained are still unknown, see, e.g., [42, Section 5.11] where this problem was posed. M.D.E. Conder, [11,12], showed that in the range $1 < g < 11905$ there are just 32 integers g such that there exists a compact Riemann surface of genus g with the automorphism group of the maximal possible order $84(g-1)$. Moreover, Conder also determined all the 92 normal subgroups of $T(2,3,7)$ of index less than 10^6. In particular, there are exactly 14 simple Hurwitz groups of order less than one million, [12, Table 1].

Finally, the above geometric interpretation also gives some information about indices of subgroups of $T(2,3,7)$. Let G be a subgroup of $T(2,3,7)$ of index n. It has a fundamental domain consisting of n translates of the hyperbolic triangle $\Delta(2,3,7)$. The domain has, say, r (respectively, s, t) elliptic vertices of order 2 (respectively, 3, 7) and the corresponding Riemann surface has genus g. The numbers n, g, r, s, t are not independent: they are related via *the genus formula*

$$n = 84(g-1) + 21r + 28s + 36t. \tag{1}$$

As an easy consequence, we have that n must satisfy

$$\left[\frac{n}{2}\right] + 2\left[\frac{n}{3}\right] + 6\left[\frac{n}{7}\right] \geqslant 2n - 2. \tag{2}$$

W.W. Stothers, [73], showed that, with the exception of $(16,0,0,1,2)$, $(21,1,1,0,0)$ and $(31,1,0,0,1)$, any quintuple (n,g,r,s,t) satisfying (1) corresponds to a subgroup of $T(2,3,7)$.

3. Finite simple and quasi-simple groups which are Hurwitz

Throughout this chapter $\mathbb{F}$ always denotes an algebraically closed field of characteristic $p \geqslant 0$. The first significant and well-known result is the following:

THEOREM 3.1 (Macbeath, [57]). *Let $p > 0$. The group $\mathrm{PSL}_2(\mathbb{F})$ contains exactly one conjugacy class of Hurwitz subgroups. Namely,*

(1) $\mathrm{PSL}_2(p)$, *if* $p \equiv 0, \pm 1 \pmod 7$;
(2) $\mathrm{PSL}_2(p^3)$, *if* $p \equiv \pm 2, \pm 3 \pmod 7$.

We will give a short proof of a more general statement in Theorem 4.9. In fact Macbeath himself showed more, because he classified all the subgroups of $\mathrm{PSL}_2(\mathbb{F})$ which are finite epimorphic images of triangle groups. His original proof is based on a detailed analysis of conjugacy classes and knowledge of the subgroups of $\mathrm{PSL}_2(q)$.

This subject is investigated further in [47] and [49], where criteria are established to determine for which finite fields $\mathrm{GF}(q)$ a given triangle group has $\mathrm{PSL}_2(q)$ or $\mathrm{PGL}_2(q)$ as factor group with torsion free kernel. Finally, permutational representations of the triangle group $T(2,3,7)$, which arise from the action of $\mathrm{PSL}_2(q)$ on the corresponding finite projective line, are studied in [63].

3.1. *Alternating groups*

As noted at the end of the previous section, the above result provides many transitive permutational representations of $T(2,3,7)$. For example those arising from the action of $\mathrm{PSL}_2(q)$, when Hurwitz, on the $q+1$ points of the projective line. Permutational representations of $T(2,3,7)$ of small degrees, and a method of joining them via handles developed by G. Higman, were the starting point for constructive methods, and culminated in the famous theorem that almost all the alternating groups are Hurwitz, which appeared in the paper of M.D.E. Conder, [8]. The same methods, later generalized to a linear context by A. Lucchini, M.C. Tamburini and J.S. Wilson, led to the discovery that most finite classical groups are Hurwitz. In this section we attempt to describe the above results in a uniform way, which is close to the approach used in [55].

Let V be a free module over a ring R, with basis Ω of cardinality n, and let $\mathrm{GL}_n(R)$ act on V. It is natural to identify $\mathrm{Sym}(\Omega)$ with the subgroup of $\mathrm{GL}_n(R)$ consisting of permutation matrices, and $\mathrm{Alt}(\Omega)$ with the subgroup of $\mathrm{SL}_n(R)$ of even permutation matrices. Assume that (X,Y,Z) is a $(2,3,7)$-generating triple of the triangle group $T(2,3,7)$, and that

$$\psi : T(2,3,7) \to \mathrm{GL}_n(R)$$

is a representation.

DEFINITION 3.2. An ordered pair (v_1, v_2) of distinct elements from Ω is called a *handle* for ψ if the following conditions hold:
(1) $\psi(X)$ fixes v_1 and v_2 and leaves invariant the submodule $\langle \Omega \setminus \{v_1, v_2\}\rangle$;
(2) $\psi(Y)$ acts as the permutation (v_1, v_2, v_3) for some $v_3 \in \Omega$, and leaves invariant $\langle \Omega \setminus \{v_1, v_2, v_3\}\rangle$.

The role of handles, in the process of joining representations of the triangle group, is made clear by the following lemma. Here we assume that $\{e_1, \dots, e_n\}$ is the canonical basis of the free R-module R^n and, similarly, that $\{e'_1, \dots, e'_{n'}\}$ is the canonical basis of $R^{n'}$.

LEMMA 3.3. *Given two representations*

$$\psi : T(2,3,7) \to \mathrm{GL}_n(R), \qquad \psi' : T(2,3,7) \to \mathrm{GL}_{n'}(R)$$

assume that ψ *has handles* $\{e_1, e_2\}$ *and* ψ' *has handles* $\{e'_1, e'_2\}$.
Let X_i *to be one of the following involutions of* $\mathrm{GL}_{n+n'}(R)$:

$$X_1 := \left(\begin{array}{cc|cc} & & I_2 & \\ & I_{n-2} & & \\ \hline I_2 & & & \\ & & & I_{n'-2} \end{array}\right), \qquad X_2 := \left(\begin{array}{cc|cc} I_2 & & tI_2 & \\ & I_{n-2} & & \\ \hline & & -I_2 & \\ & & & I_{n'-2} \end{array}\right),$$

where $t \in R$. *Then, for* $i = 1, 2$, *the map*

$$X \mapsto X_i \begin{pmatrix} \psi_1(X) & \\ & \psi_2(X) \end{pmatrix}, \qquad Y \mapsto \begin{pmatrix} \psi_1(Y) & \\ & \psi_2(Y) \end{pmatrix}$$

defines a representation $T(2,3,7) \to \mathrm{GL}_{n+n'}(R)$.

This elementary lemma rests on the following two facts.

(1) For $i = 1, 2$, the involution X_i and the involution

$$\begin{pmatrix} \psi_1(X) & \\ & \psi_2(X) \end{pmatrix}$$

commute, having disjoint supports. Hence their product is again an involution.

(2) The product

$$X_i \begin{pmatrix} \psi_1(XY) & \\ & \psi_2(XY) \end{pmatrix}$$

has order 7, being conjugate to

$$\begin{pmatrix} \psi_1(XY) & \\ & \psi_2(XY) \end{pmatrix}.$$

In particular, if ψ and ψ' are transitive permutational representations of $T(2,3,7)$, of respective degrees n and n', the representation described in Lemma 3.3, relative to X_1, is a transitive permutational representation of degree $n + n'$.

In [8], to define a $(2,3,7)$-generating triple (x, y, z) of $\mathrm{Alt}(\Omega)$ when $n > 167$ and $n \neq 173, 174, 181, 188, 202$, Conder uses $3 + 14$ transitive permutational representations of $T(2,3,7)$, each of which is depicted by a *diagram*, whose vertices are permuted by $T(2,3,7)$. The first three diagrams, denoted A, E and G, have respectively 14, 28 and 42 vertices. (A, for example, corresponds to the action of $\mathrm{PSL}_2(13)$ on the 14 points of the projective line.) The remaining fourteen diagrams can be labeled H_d, $d = 0, \ldots, 13$. Each H_d has d' vertices, where d' is the unique integer determined by the conditions

$$d' \in D := \begin{Bmatrix} 36, & 42, & 57, & 77, & 115, & 135, & 136, \\ 142, & 144, & 165, & 180, & 187, & 195, & 216 \end{Bmatrix} \tag{3}$$

and $d' \equiv d \pmod{14}$. To avoid too many details we do not include these diagrams here. However in the appendix we present an explicit description for Conder's generators, which allows to restore all these diagrams. As the numbers in D give all residues modulo 14, each n big enough can be written in the form

$$n = 42a + 14b + d', \quad a \geqslant 2, \ b \in \{0, 1, 2\}, \ d' \in D.$$

So, if Ω is a set of cardinality n, one can take

$$\Omega := G_1 \cup \cdots \cup G_a \cup H_d \cup \Omega_0,$$

where $d \equiv n \pmod{14}$, each G_i is a copy of G and Ω_0 is empty if $b = 0$, whereas Ω_0 coincides with A if $b = 1$ or with E if $b = 2$. As the diagram G has 3 handles and each of the diagrams A, E and H_d has at least one handle, repeated application of Lemma 3.3 with respect to X_1, gives a transitive permutational representation of $T(2, 3, 7)$ over Ω. In fact it is possible to join the a copies of G into a chain, join G_a with H_d and, if necessary, with A or E. There is a certain degree of flexibility in making the joins, depending on the choice of the handles. But, no matter how the joins are performed, any representation $\psi : T(2, 3, 7) \to \mathrm{Sym}(\Omega)$ obtained in this way has the following properties. Set

$$\psi(X) = x, \qquad \psi(Y) = y. \tag{4}$$

Then we have (see appendix):

(i) $[x, y]^{9 \cdot 11 \cdot 13}$ fixes each vector in $\Omega \setminus (H_d \cup G_a)$;

(ii) there exists a multiple $k = k(d)$ of $9 \cdot 11 \cdot 13$ such that

$$c := [x, y]^k$$

is a cycle of prime length $r \notin \{2, 3, 11, 13\}$, with support $\Gamma \subseteq H_d$;

(iii) Γ contains an orbit of x and two points from an orbit of y;

(iv) $|\Gamma \cup \Gamma y| \geqslant r + 3$.

In the appendix the cycle c is written explicitly in each case.

In particular, $\langle c, c^y \rangle = \mathrm{Alt}(\Gamma \cup \Gamma y)$ and the normal closure of this group under the transitive subgroup $\langle x, y \rangle$ of $\mathrm{Alt}(\Omega)$ is easily seen to be $\mathrm{Alt}(\Omega)$.

Similar considerations, with more specific arguments for some values of n, lead to the following:

THEOREM 3.4 (Conder, [8]). *The alternating group A_n is Hurwitz for all $n > 167$ and for the values of n displayed in the following table.*

	15						21	22					
28	29						35	36	37				
42	43		45				49	50	51	52			
56	57	58					63	64	65	66			
70	71	72	73				77	78	79	80	81		
84	85	86	87	88			91	92	93	94		96	
98	99	100	101	102			105	106	107	108	109		
112	113	114	115	116	117		119	120	121	122	123	124	
126	127	128	129	130		132	133	134	135	136	137	138	
140	141	142	143	144	145		147	148	149	150	151	152	153
154	155	156	157	158	159	160	161	162	163	164	165	166	

Actually, in the paper [8], the previous theorem is obtained as a corollary of the following remarkable result. Whenever $n > 167$, the symmetric group S_n is an epimorphic image of $T^*(2,3,7)$. To prove this it is essential to have a third generator T, which corresponds to a symmetry in the vertical axis of each of the 17 diagrams mentioned above.

There are variations of these results in several directions. For example, in [14], Conder shows that all but finitely many of the alternating groups A_n can be generated by a $(2,3,7)$-generating triple (x, y, xy) satisfying the further relation $[x, y]^{84} = 1$.

Many authors considered other triangle groups. Conder, [9], obtained the following result.

THEOREM 3.5 (Conder, [9]). *For each $k \geqslant 7$, there exists an n_k such that, for all $n \geqslant n_k$, A_n is an epimorphic image of $T(2,3,k)$.*

Actually he proves more than he claims, because a careful analysis of his diagrams leads to the stronger conclusion that A_n is $(2,3,k)$-generated.

Using a similar technique Q. Mushtaq and G.-C. Rota, [64], proved

THEOREM 3.6 (Mushtaq and Rota, [64]). *Let k be even, $k \geqslant 6$ and $l \geqslant 5k-3$. For sufficiently large n, the group A_n is a homomorphic image of $T(2,k,l)$.*

An even more striking generalization of the above results was given by B. Everitt.

THEOREM 3.7 (Everitt, [29]). *Any Fuchsian group surjects almost all of the alternating groups.*

Recently, M.W. Liebeck and A. Shalev, [51], gave another proof of this result. Their proof uses character-theoretic and probabilistic methods and it is totally independent from Higman's and Conder's diagrams.

3.2. *Classical groups*

Already in [77] constructive, permutational methods had been used by the first author of this survey to show that, for all $n \geqslant 25$ and all prime powers q, the special linear group

$SL_n(q)$ can be generated by an element of order 2 and an element of order 3, i.e., is an epimorphic image of the modular group $PSL_2(\mathbb{Z})$. This result actually originated from the following:

THEOREM 3.8 (Tamburini and Wilson, [79]). *Let A and B be finite groups which are non-trivial. If $|A||B| \geqslant 12$, then for all $n \geqslant |A||B| + 12$ the group $PSL_n(q)$ has subgroups $\bar{A} \simeq A$ and $\bar{B} \simeq B$ such that $\langle \bar{A}, \bar{B} \rangle = PSL_n(q)$.*

A key tool in the proof of this theorem is a simple and beautiful idea of H. Wielandt, [89], which requires the hypothesis that at least one of the groups A and B has order $\geqslant 4$. But an appropriate variation of the proof of this theorem was used in [77] to establish the similar result for the smallest possible values of $|A|$ and $|B|$, namely 2 and 3. And, indeed, the $(2, 3)$-generation of the projective special linear groups $PSL_n(q)$, provided $n > 4$ when q is odd and $n > 12$ when q is even, has been established by L. Di Martino and N. Vavilov in [26] and [27], in a constructive way which involves quite a lot of case by case analysis and heavy computation.

A combination of the linear methods in [77] with the $(2, 3, 7)$ generators for $\mathrm{Alt}(\Omega)$ of Theorem 3.4, gives the following:

THEOREM 3.9 (Lucchini, Tamburini and Wilson, [56]). *For all $n \geqslant 287$, the special linear group $SL_n(q)$ is Hurwitz.*

The authors take $n = |\Omega|$ big enough in order to guarantee that the representation of $T(2, 3, 7)$ affording the generators x, y of $\mathrm{Alt}(\Omega)$ in (4) has a couple of handles (e_1, e_2) and (e'_1, e'_2), Thus they can apply Lemma 3.3 to extend the generating triple (x, y, xy) of $\mathrm{Alt}(\Omega)$ to a $(2, 3, 7)$-generating triple (xX_2, y, xX_2y) of $SL_n(q)$. In the definition of X_2 they take $t \neq 2$ to be a generator of $GF(q)$ (as a ring). Their proof consists in showing that

$$\mathrm{Alt}(\Omega) \leqslant \langle xX_2, y \rangle$$

and their claim follows from the fact that for $n \geqslant 6$, $SL_n(q)$ is generated by X_2 and $\mathrm{Alt}(\Omega)$.

An inspection of the proof shows that $SL_n(q)$ (and $SL_n(\mathbb{Z})$) are $(2, 3, 7)$-generated for all n in the set

$$\{14m + d \mid m \geqslant 6,\ d \in D\} \cup \{42 + d \mid d \in D\},$$

where D is as in (3). There are 93 integers less than 286 in this set. Further improvement was made in [86], where 60 new values of n were found. In particular it follows that $SL_n(q)$ is Hurwitz for all $n \geqslant 252$ and for 118 more values of n, the smallest of which is 49.

Duplication of the $(2, 3)$-generators of $SL_n(q)$, according to a well-known embedding of this group into the classical groups of degree $2n$ or $2n + 1$ over $GF(q)$, had already been used in [81] and [80] to show that classical groups of sufficiently large rank are $(2, 3)$-generated. In a similar way, duplication of the $(2, 3, 7)$-generators of $SL_n(q)$, and an application of Lemma 3.3 with appropriate choices of X_i, leads to the following results.

THEOREM 3.10 (Lucchini and Tamburini, [55]). *For each $n \geqslant 371$, the following classical groups are Hurwitz:*

$$\mathrm{Sp}_{2n}(q),\ \mathrm{SU}_{2n}(q),\ \Omega^{+}_{2n}(q),\quad \textit{all } q;$$

$$\mathrm{SU}_{2n+7}(q),\ \Omega_{2n+7}(q),\quad q \textit{ odd}.$$

As above, analysis of the proof shows that this result holds for all $n = 42a + 14b + d$ with $d \in D$ as in (3), and either $a \geqslant 4$ or $a = 3$ and $b = 0$. There are many such integers less than 371, the smallest of which is 162.

As in the permutational case, further generalizations to other triangle groups are possible. For example, A. Lucchini in [54] and J.S. Wilson in [91] independently proved that for any $k \geqslant 7$ there is an integer n_k such that the group $\mathrm{SL}_n(q)$ is $(2,3,k)$-generated provided $n \geqslant n_k$. In fact, their result is a consequence of a more general statement; see Theorem 5.9.

It may be worth noting that the problem of determining which finite classical groups are Hurwitz deserves further investigation, having received only partial answers. In fact there are two classes which are not even considered in the above theorem, namely the orthogonal groups $\Omega^{-}_{2n}(q)$ and the unitary groups $\mathrm{SU}_{2n+1}(2^t)$. Moreover, although the above results are satisfactory if considered asymptotically with respect to the ranks of the groups under consideration, the lower bounds for their ranks are certainly much higher than necessary for the existence of Hurwitz generators. Some evidence for this claim will be given in Section 4.3 dedicated to groups of small rank. In fact it will be shown that the reason why the Hurwitz subgroups of $\mathrm{PSL}_n(\mathbb{F})$, with $n \leqslant 4$, are essentially those discovered by Macbeath is rigidity. On the other hand, already for $n = 5$ and $n = 7$ there are new $(2,3,7)$-generated projective subgroups. In the theorems mentioned above, the assumptions on the lower bounds are forced by the permutational approach, based on the diagrams of Conder, which has the advantage of being constructive and allowing rather uniform proofs.

But the treatment of linear groups of relatively small ranks requires different techniques, which may in any case involve quite a lot of computation and consideration of special cases. In fact the property of being Hurwitz for groups of small rank depends also on the size of the field and its characteristic. Evidence of this is given by the above mentioned result of Macbeath and also by the results in [60,61] and [82,78] which will be mentioned in the following sections.

3.3. *Exceptional groups of Lie type and sporadic simple groups*

A key tool in the approach to these groups has been the use of multiplication constants defined as follows. Let $X_1, \ldots, X_r$ denote the conjugacy classes of a finite group G. For a fixed $z \in X_k$, the number of pairs (x, y) such that $x \in X_i$, $y \in X_j$ and $xy = z$ coincides with the number

$$a_{ijk} := \frac{|X_i||X_j|}{|G|} \sum_{\chi \in \mathrm{Irr}(G)} \frac{\chi(g_i)\chi(g_j)\overline{\chi(g_k)}}{\chi(1)},$$

where $g_\ell \in X_\ell$, $1 \leqslant \ell \leqslant r$. (See [33, Theorem 2.12], for example.) These numbers, also called the multiplication constants, are useful in determining whether G is Hurwitz. One first computes the class constants for each choice of classes of elements of order 2, 3 and 7 in G. Clearly, if they are all 0, then G has no Hurwitz subgroup. But, apart from this trivial case, one can use rather sophisticated techniques, based on additional information about G. Like calculating the class constants in appropriate subgroups of G, in order to evaluate how many solutions generate proper subgroups. A refined version of this technique is due to Philip Hall, [35] (see also [41]).

Using the Green–Deligne–Lusztig parameterizations of characters of groups of Lie type, G. Malle studied the Hurwitz generation of many exceptional simple groups of Lie type. For a more detailed description we refer also to the survey [22] of L. Di Martino, and to the survey [41] of G. Jones.

THEOREM 3.11 (Malle, [60,61]). *As to the exceptional simple groups of Lie type*:

(1) $G_2(p^m)$ *are Hurwitz if and only if* $p^m \geqslant 5$;
(2) ${}^2G_2(3^{2m+1})$ *are Hurwitz if and only if* $m \neq 1$;
(3) ${}^3D_4(p^m)$ *are Hurwitz if and only if* $p \neq 3$, $p^m \neq 4$;
(4) ${}^2F_4(2^{2m+1})'$ *are Hurwitz if and only if* $m \equiv 1 \bmod 3$.

For the Ree groups ${}^2G_2(3^{2m+1})$ see also, [40] and [67]. The results of Theorem 3.11 do not produce explicit (2, 3, 7) generators. With different methods two explicit matrices $x, y \in \mathrm{SL}_7(p)$, $p \geqslant 5$, are constructed in [87] such that $x^2 = y^3 = (xy)^7 = [x, y]^{2p} = I$ and $\langle x, y\rangle$ is isomorphic to $G_2(p)$.

By the contribution of several authors, the problem of determining which of the 26 sporadic simple groups are Hurwitz, has now a complete answer.

THEOREM 3.12. *The sporadic simple groups which are Hurwitz are the following*:

(1) J_1 (Sah, [67]);
(2) J_2 (Finkelstein and Rudvalis, [31]);
(3) Co_3 (Worboys, [97] and Woldar, [95]);
(4) He, Ru, HN, Ly, Fi_{22}, J_4 (Woldar, [95,96]);
(5) Th (Linton, [52]);
(6) Fi_{24}' (Linton and Wilson, [53]);
(7) M (Wilson, [94]).

The orders of M_{11}, M_{12} and J_3 are not divisible by 7. The proof that the remaining simple groups are not Hurwitz comes from the determination of their symmetric genus, [16,44]. The technique is a combination of the above method of multiplication constants and Scott's formula. The latter will be discussed in Section 4.

4. Low-dimensional representations of Hurwitz groups

Clearly, a (2, 3, 7)-generated group G must have order divisible by 2, 3 and 7 and it must be perfect, i.e. with trivial abelianization G/G'. Moreover, for any subgroup S of G, its index n, when finite, must satisfy the classical genus formula (2). But there are other methods which can be used to exclude that a group G is (2, 3, 7)-generated. They will be illustrated in this section.

4.1. *Scott's formula and negative results*

Theorem 4.1 below, which is a special case of a result of L.L. Scott, provides a very efficient tool to show that certain groups are not $(2,3,7)$-generated. We state this result only in the form needed for our purposes. However the original theorem of Scott applies to a more general context and deals with representations of any finitely generated group H. To state the theorem, we need some notation. Given a group H and a representation

$$f : H \to \mathrm{GL}_n(\mathbb{F})$$

let V be the vector space $\mathbb{F}^n$. For any subset A of H, define V_A as the subspace of fixed points of $f(A)$ and denote by d_V^A its dimension over $\mathbb{F}$. In symbols:

$$V_A := \{v \in V \mid f(a)v = v, \text{ for all } a \in A\}, \qquad d_V^A := \dim(V_A). \tag{5}$$

Define $\hat{d}_V^A$ in the same way, with respect to the dual representation, namely set

$$\widehat{V}_A := \{v \in V \mid (f(a))^t v = v, \text{ for all } a \in A\}, \qquad \hat{d}_V^A := \dim(\widehat{V}_A). \tag{6}$$

In the above notations:

THEOREM 4.1 (Scott, [68]). *Assume that H is generated by x and y. Then*

$$d_V^x + d_V^y + d_V^{xy} \leqslant n + d_V^H + \hat{d}_V^H. \tag{7}$$

PROOF. Consider V as an H-module via f. Set $z = (xy)^{-1}$ and let C be the direct sum

$$C := (1-x)V \oplus (1-y)V \oplus (1-z)V.$$

Define the linear transformations $\beta : V \to C$ and $\delta : C \to V$ respectively by

$$v \mapsto \big((1-x)v, (1-y)v, (1-z)v\big),$$

$$(v_1, v_2, v_3) \mapsto v_1 + xv_2 + z^{-1}v_3.$$

We have $\dim(\operatorname{Im}\beta) = n - d_V^H$, since $\operatorname{Ker}\beta$ is the space of fixed points of H. Using the identity $a(1-b) = (1-a)(b-1) + (1-b)$, for all a, b in the group algebra $\mathbb{F}H$, it is easy to deduce that $\operatorname{Im}\delta$ coincides with the subspace $(1-x)V + (1-y)V + (1-z)V$ and, moreover, that $\operatorname{Im}\delta$ is H-invariant. Thus $\operatorname{Im}\delta$ is the smallest H-submodule of V with trivial action on the quotient. Let $\mathcal{B} = \mathcal{B}_0 \cup \mathcal{B}_1$ be a basis of V such that $\mathcal{B}_0$ is a basis of $\operatorname{Im}\delta$. With respect to $\mathcal{B}$, $f(H)$ consists of matrices of the form

$$\begin{pmatrix} * & * \\ 0 & I \end{pmatrix}.$$

This observation easily implies that $|\mathcal{B}_1| = \hat{d}_V^H$, hence $\dim(\operatorname{Im}\delta) = n - \hat{d}_V^H$.

From $\operatorname{Im}\beta \leqslant \operatorname{Ker}\delta \leqslant C$ we deduce

$$\begin{aligned}\dim C &= \dim(\operatorname{Im}\beta) + \dim\frac{\operatorname{Ker}\delta}{\operatorname{Im}\beta} + \dim\frac{C}{\operatorname{Ker}\delta}\\ &\geqslant \dim(\operatorname{Im}\beta) + \dim\frac{C}{\operatorname{Ker}\delta} = \dim(\operatorname{Im}\beta) + \dim(\operatorname{Im}\delta).\end{aligned}$$

We conclude

$$\dim C = \left(n - d_V^x\right) + \left(n - d_V^y\right) + \left(n - d_V^z\right) \geqslant \left(n - d_V^H\right) + \left(n - \hat{d}_V^H\right)$$

whence $d_V^x + d_V^y + d_V^{xy} \leqslant n + d_V^H + \hat{d}_V^H$. □

As noticed by L.L. Scott in [68], the genus formula (2) itself is a consequence of (7). The argument is the following. Assume that a $(2,3,7)$-generated group G has a subgroup S of index n. Let $f : G \to \mathrm{GL}_n(\mathbb{C})$ be the linear representation of G induced by the transitive permutational action on the (left) cosets of S and let $V = \mathbb{C}^n$ be the corresponding G-module. For every $g \in G$ of prime order r whose cyclic structure consists of ℓ non-trivial cycles, we have

$$n - d_V^g = (r-1)\ell \leqslant (r-1)\left[\frac{n}{r}\right].$$

So, if (x, y, xy) is a $(2,3,7)$-generating triple for G, then

$$\left[\frac{n}{2}\right] \geqslant n - d_V^x, \qquad 2\left[\frac{n}{3}\right] \geqslant n - d_V^y, \qquad 6\left[\frac{n}{7}\right] \geqslant n - d_V^{xy},$$

hence

$$\left[\frac{n}{2}\right] + 2\left[\frac{n}{3}\right] + 6\left[\frac{n}{7}\right] \geqslant 3n - \left(d_V^x + d_V^y + d_V^{xy}\right).$$

By the transitivity, the multiplicity of the trivial representation is 1. Hence $d_V^G = \hat{d}_V^G = 1$, and Scott's formula gives

$$d_V^x + d_V^y + d_V^{xy} \leqslant n + 2.$$

We conclude

$$\left[\frac{n}{2}\right] + 2\left[\frac{n}{3}\right] + 6\left[\frac{n}{7}\right] \geqslant 2n - 2.$$

Scott himself observed that his formula could be used for proving that certain linear groups are not $(2,3,7)$-generated. In [68] he considered, as examples, the groups $\mathrm{SL}_6(3)$

and $\mathrm{SL}_9(3)$. But a more systematic application was first made in [25], where this formula was applied essentially to the following representations of an absolutely irreducible subgroup H of $\mathrm{SL}_n(\mathbb{F})$, with $(2,3,7)$-generating triple (x, y, xy).

(1) The conjugation action of H on $M = \mathrm{Mat}_n(\mathbb{F})$. In this case, the fixed-points subspace of M is the centralizer of H and therefore, by Schur's lemma, it consists of scalar matrices. Hence Scott's formula reads

$$d_M^x + d_M^y + d_M^{xy} \leqslant n^2 + 2. \tag{8}$$

The values of the left-hand side of this equation are easily calculated using a well-known formula, due to F.G. Frobenius (e.g., see [39, p. 207, Theorem 3.16]). Namely, let $n_1 \leqslant \cdots \leqslant n_s$ be the degrees of the similarity invariants of $a \in M$. Then

$$d_M^a = \sum_{j=1}^{s}(2s - 2j + 1)n_j = (2s+1)n - 2\sum_{j=1}^{s} jn_j. \tag{9}$$

In particular $d_M^a \geqslant n + s^2 - s$.

(2) The diagonal action of H on the symmetric square S of $\mathbb{F}^n$. Scott's formula takes the shape

$$d_M^x + d_M^y + d_M^{xy} \leqslant \frac{n(n+1)}{2} + 2. \tag{10}$$

Moreover if $\frac{n(n+1)}{2} < d_M^x + d_M^y + d_M^{xy} \leqslant \frac{n(n+1)}{2} + 2$ then H is orthogonal for $p \neq 2$ and H is symplectic for $p = 2$. This claim was first stated in Lemma 4.1 of [25], but the proof in characteristic 2 was inaccurate. For a revised proof see [78, Lemma 2.1] or [84].

The values of the left-hand side of (10), for the relevant elements $g \in \mathrm{GL}_n(\mathbb{F})$, are afforded by the following formulas (see [25]). Assume first that g is semisimple. If ν is an eigenvalue of g, let m_ν denote the multiplicity of ν. Then

$$d_S^g = \frac{m_1(m_1+1) + m_{-1}(m_{-1}+1)}{2} + \sum m_\nu m_{\nu^{-1}},$$

where the summation runs over all pairs ν, ν^{-1} of eigenvalues of g in $\mathbb{F}$ with $\nu \neq \nu^{-1}$. Next assume that g is unipotent, of prime order p. Let k_i be the number of similarity invariants of g of degree i, $1 \leqslant i \leqslant p$. Then, if $p = 2$

$$d_S^g = \frac{k_1^2 + 2k_2^2}{2} + k_1k_2 + \frac{k_1 + 2k_2}{2};$$

otherwise

$$d_S^g = \sum_{i=1}^{p} \frac{ik_i^2}{2} + \sum_{i=1}^{p-1} \sum_{j=i+1}^{p} ik_ik_j + \sum_{i=0}^{\frac{p-1}{2}} \frac{k_{2i+1}}{2}.$$

A detailed analysis of conjugacy classes and comparison of (8) with (10) lead to conclude that many classical linear groups of rank $\leqslant 19$ are not Hurwitz. As an example, we quote some of the results.

THEOREM 4.2 (Di Martino, Tamburini and Zalesski, [25]). *Let H denote an irreducible subgroup of $\mathrm{SL}_n(\mathbb{F})$, with $n \in \{4, 5, 6, 7, 10\}$. Assume that H is not contained in an orthogonal group if $p \neq 2$, and that H is not contained in a symplectic group if $p = 2$. If $n = 6$ and $p = 2$, assume further that $H = \mathrm{SL}_6(q)$ or $\mathrm{SU}_6(q)$. Then H is not $(2,3,7)$-generated. In particular, if $n \in \{4, 5, 6, 7, 10\}$, then*

(1) *the groups $\mathrm{SL}_n(q)$, $\mathrm{Sp}_n(q)$, $\mathrm{SU}_n(q^2)$ are not Hurwitz, with the only possible exception of $\mathrm{Sp}_n(2^t)$, $n \geqslant 6$;*
(2) *every complex irreducible character of degree n of a $(2,3,7)$-generated group is real.*

REMARK 4.3. The case of $\mathrm{Sp}_4(2^t)$ was not excluded in [25]. But the fact that the symplectic groups $\mathrm{Sp}_4(2^t)$ are not Hurwitz can be deduced either from [50], where it is shown that they are not even $(2,3)$-generated, or from our Theorem 4.16.

THEOREM 4.4 (Di Martino, Tamburini and Zalesski, [25]). *Let H denote an absolutely irreducible subgroup of $\mathrm{SL}_n(\mathbb{Q})$, with $n \leqslant 19$ or $n = 22$. If H is not contained in an orthogonal group, then H is not $(2,3,7)$-generated. In particular, for these values of n, the group $\mathrm{SL}_n(\mathbb{Z})$ is not $(2,3,7)$-generated.*

We will not prove Theorems 4.2 and 4.4 here. We just observe that, when $n = 4, 5$, a stronger result holds. In fact there is now a complete classification of the irreducible Hurwitz subgroups of $\mathrm{PSL}_4(\mathbb{F})$ and $\mathrm{PSL}_5(\mathbb{F})$. For details see Section 4.3 and [82]. Recently R. Vincent and A. Zalesskii have extended Theorems 4.2 and 4.4 to other values of n, [84]. Their results depend on the residues of q modulo 42.

4.2. *Rigidity*

The following definition is a special case of a more general one. Among the first who used it we quote G.V. Belyi, [1], and J.G. Thompson, [83]. But for more complete historical information, we refer to [62].

DEFINITION 4.5. Let $a_1, a_2, a_3 \in \mathrm{GL}_n(\mathbb{F})$ be such that $a_1 a_2 = a_3$. The triple (a_1, a_2, a_3) is called *linearly rigid* if, whenever b_1, b_2, b_3 are matrices such that $b_1 b_2 = b_3$ and each b_i is conjugate to a_i, there exists $g \in \mathrm{GL}_n(\mathbb{F})$ such that $g b_i g^{-1} = a_i$, for $i = 1, 2, 3$.

Rigid generators of finite groups have been studied in the inverse Galois problem (see [62] and [85]). The same concept, under the name of physical rigidity (for $\mathbb{F} = \mathbb{C}$) appeared in the totally different context of linear differential equations and local systems on the sphere (see [43]).

In Section 4.3 we will illustrate some applications of linear rigidity to the context of Hurwitz generation, based on a useful criterion for recognizing rigid triples. In order to describe this criterion we recall some notation.

As above we set $M = \mathrm{Mat}_n(\mathbb{F})$ and, for each $a \in M$, $d_M^a = \dim C_M(a)$.

THEOREM 4.6 (Strambach and Völklein, [74]). *Assume that $a_1, a_2 \in \mathrm{GL}_n(\mathbb{F})$ generate an irreducible subgroup. Set $a_3 = a_1 a_2$ and suppose that*

$$\sum_{i=1}^{3} d_M^{a_i} = n^2 + 2. \tag{11}$$

Then the triple a_1, a_2, a_3 is linearly rigid.

PROOF. Let $b_i = a_i^{g_i}$ for $i \leqslant 3$, with $b_3 = b_1 b_2$. Consider the linear transformations σ_i of $M = \mathrm{Mat}_n(\mathbb{F})$ defined by

$$m \mapsto b_i{}^{-1} m a_i.$$

For $c \in M$, let us denote by λ_c and ρ_c the endomorphisms of M given by left and right multiplication by c. Then

$$\sigma_i = \lambda_{b_i^{-1}} \rho_{a_i} = \lambda_{g_i^{-1}} \lambda_{a_i^{-1}} \lambda_{g_i} \rho_{a_i} = \lambda_{g_i}{}^{-1} (\lambda_{a_i^{-1}} \rho_{a_i}) \lambda_{g_i}.$$

Thus σ_i is conjugate in $\mathrm{GL}(M)$ to conjugation by a_i. Therefore $d_M^{\sigma_i} = \dim C_M(a_i)$, hence $d_M^{\sigma_i} = d_M^{a_i}$. Set $H = \langle \sigma_1, \sigma_2 \rangle$. Since $H \leqslant \mathrm{GL}(M)$, we may consider M as an H-module. Thus, applying Theorem 4.1, we obtain

$$\sum_{i=1}^{3} d_M^{\sigma_i} \leqslant n^2 + d_M^H + \hat{d}_M^H.$$

Together with assumption (11) this yields

$$n^2 + 2 \leqslant n^2 + d_M^H + \hat{d}_M^H.$$

It follows that $d_M^H > 0$ or $\hat{d}_M^H > 0$. Thus there exists a non-zero matrix g such that either $b_i{}^{-1} g a_i = g$ or $b_i^{t}{}^{-1} g a_i^t = g$, for $i = 1, 2, 3$. We claim that g is non-singular. Otherwise let W be the eigenspace of g relative to the eigenvalue 0. In the first case W would be invariant under the irreducible subgroup $\langle a_1, a_2 \rangle$. In the second case it would be invariant under its transpose, again a contradiction. We conclude either $b_i = a_i^{g^{-1}}$ or $b_i = a_i^{g^t}$, for $i \leqslant 3$. □

The above proof shows that $d_M^H > 0$ if and only if $\hat{d}_M^H > 0$. Moreover, in this case, both of them must be 1 because, by Schur's lemma, the centralizer in $\mathrm{Mat}_n(\mathbb{F})$ of the irreducible group $\langle a_1, a_2 \rangle$ consists of scalar matrices.

REMARK 4.7. If there are a_1, a_2 and $a_3 = a_1 a_2 \in \mathrm{GL}_n(\mathbb{F})$ that satisfy (11) but generate a reducible subgroup of $\mathrm{GL}_n(\mathbb{F})$, then theorem 4.6 also implies that no other triple with the same set of similarity invariants can generate an irreducible subgroup of $\mathrm{GL}_n(\mathbb{F})$. However there may be more that one conjugacy class of triples generating reducible subgroups.

To conclude this section we quote the following result which, ultimately, depends on a well known theorem of S. Lang and R. Steinberg (see, e.g., [72]).

THEOREM 4.8. *Let (a_1, a_2, a_3) be a linearly rigid triple, with $a_i \in \mathrm{GL}_n(\mathbb{F})$. Let C_i be the conjugacy class of a_i and suppose that $C_i \cap \mathrm{GL}_n(q)$ (respectively, $C_i \cap \mathrm{U}(n, q^2)$) is non-empty, for $i = 1, 2, 3$. Then there exists $g \in \mathrm{GL}_n(\mathbb{F})$ such that $a_i^g \in \mathrm{GL}_n(q)$ (respectively $\in \mathrm{U}_n(q^2)$), for $i = 1, 2, 3$.*

4.3. *Classical groups of small rank which are Hurwitz*

THEOREM 4.9 (Macbeath, [57]). *Let k be a prime number $\geqslant 7$. If $p > 0$ and $p \neq k$ denote by n the order of p modulo k. The group $\mathrm{PSL}_2(\mathbb{F})$ contains exactly one isomorphism type of $(2, 3, k)$-generated subgroups, namely*

(1) $\Delta(2, 3, k)$, *if $p = 0$;*
(2) $\mathrm{PSL}_2(p)$, *if $p = k$;*
(3) $\mathrm{PSL}_2(p^n)$, *if $p \neq k$ and n is odd;*
(4) $\mathrm{PSL}_2(p^{\frac{n}{2}})$, *if $p \neq k$ and n is even.*

Moreover, if $p > 0$, there is just one conjugacy class of such groups.

REMARK 4.10. We recall that $\mathrm{PSL}_2(q)$ has order $q(q+1)(q-1)/(2, q-1)$. So the groups listed in items (2), (3) and (4) in the statement correspond precisely to the smallest power of p such that $\mathrm{PSL}_2(q)$ has order divisible by k.

PROOF. The assumption $k \geqslant 7$ implies that any $(2, 3, k)$-generated group is perfect, hence non-soluble. The latter fact will be used in the following without further mention.

If $p \neq k$, define $\varepsilon \in \mathbb{F}$ to be a primitive k-th root of unity. If $p = k$, put $\varepsilon = 1$.

First, we consider the case when $p = 0$. Note that, for each ℓ such that $(\ell, k) = 1$, the projective image of

$$x = \begin{pmatrix} 0 & -1 \\ 1 & 0 \end{pmatrix}, \qquad y_\ell = \begin{pmatrix} 0 & \varepsilon^{-\ell} \\ -\varepsilon^{\ell} & -1 \end{pmatrix}, \qquad x y_\ell = \begin{pmatrix} \varepsilon^{\ell} & 1 \\ 0 & \varepsilon^{-\ell} \end{pmatrix} \tag{12}$$

is a $(2, 3, k)$-triple. The group $\langle x, y_1 \rangle$ is isomorphic to $\langle x, y_\ell \rangle$, for each ℓ, under the automorphism of $\mathrm{Mat}_2(\mathbb{Z}[\varepsilon])$ induced by the map $\varepsilon \mapsto \varepsilon^\ell$. On the other hand, a slight modification of the arguments given in the proof of theorem 1 in [25] shows that the preimage of a $(2, 3, k)$-generated subgroup of $\mathrm{PSL}_2(\mathbb{F})$ must be conjugate to $\langle x, y_\ell \rangle$, for some ℓ. Our claim follows from the classical embedding of $T(2, 3, k)$ into $\mathrm{PSL}_2(\mathbb{C})$ (see [59, Theorem 2.8]).

Now assume $p > 0$. Let $q = p$ if $p = k$; $q = p^n$ if $p \neq k$ and n is odd; $q = p^{n/2}$ if $p \neq k$ and n is even. So $\theta_\ell = \varepsilon^\ell + \varepsilon^{-\ell}$ is an element of $\mathrm{GF}(q)$. For each ℓ such that $1 \leqslant \ell \leqslant \frac{k-1}{2}$, define

$$x = \begin{pmatrix} 0 & -1 \\ 1 & 0 \end{pmatrix}, \qquad y_\ell = \begin{pmatrix} b & a \\ a - \theta_\ell & -1 - b \end{pmatrix},$$
$$xy_\ell = \begin{pmatrix} \theta_\ell - a & 1 + b \\ b & a \end{pmatrix}, \tag{13}$$

where

$$a(a - \theta_\ell) + b(1 + b) = -1. \tag{14}$$

If $p = 2$, we can take $a = b = (1 + \theta_\ell)^{-1}$. For $p > 2$, equation (14) is equivalent to

$$(2a - \theta_\ell)^2 + (2b + 1)^2 = -3 + \theta_\ell^2,$$

which is always solvable over $\mathrm{GF}(q)$ since every element of a finite field is a sum of two squares. Thus, $\langle x, y_\ell \rangle \leqslant \mathrm{SL}_2(q)$ and the projective image of (x, y_ℓ, xy_ℓ) is a $(2, 3, k)$-triple. As observed in Remark 4.10, n and q are defined so that $\mathrm{SL}(2, q_0)$ does not have elements of projective order k, for any proper divisor q_0 of q. It follows from Dickson's classification of the subgroups of $\mathrm{PSL}_2(q)$ (see, for example, [20, Chapter XII] or [36, 8.27]) that the perfect group $\langle x, y_\ell \rangle$ coincides with $\mathrm{SL}_2(q)$.

On the other hand, every triple $(x', y', x'y')$ in $\mathrm{SL}_2(\mathbb{F})$, whose projective image is a $(2, 3, k)$-triple, is such that $x' \sim x$, $y' \sim y_\ell$ and $x'y' \sim xy_\ell$ for some ℓ, where x and y_ℓ are as in (13). Moreover we can assume that $1 \leqslant \ell \leqslant \frac{k-1}{2}$. Thus our final claim follows from Theorem 4.6. □

Actually the factorizations in (12) and (13), may be viewed as a special case of the following (constructive) factorization theorem for matrices.

THEOREM 4.11 (Sourour, [70]). *Let $a \in \mathrm{GL}_n(\mathbb{F})$ be non-scalar, and let β_i and γ_i ($1 \leqslant i \leqslant n$) be elements of $\mathbb{F}$ such that*

$$\prod_{j=1}^{n} \beta_j \gamma_j = \det a.$$

There exist b and c in $\mathrm{GL}_n(\mathbb{F})$, with respective eigenvalues β_i and γ_i, such that $a = bc$.

In order to classify the Hurwitz subgroups of $\mathrm{PSL}_n(\mathbb{F})$, it is necessary to keep in mind the irreducible representations of $\mathrm{PSL}_2(q)$, when Hurwitz. In the natural characteristic they are described in [3]. But we prefer to give an independent proof of what is relevant for us.

In what follows we consider a field $\mathbb{K}$ of characteristic p, and describe a bunch of absolutely irreducible representations of $\mathrm{SL}_2(\mathbb{K})$ over $\mathbb{K}$. For any automorphism σ of the field $\mathbb{K}$, $\mathrm{SL}_2(\mathbb{K})$ acts on the polynomial ring $\mathbb{K}[t_1, t_2]$ via

$$t_1^i t_2^j \mapsto \left(\sigma(a_{11})t_1 + \sigma(a_{21})t_2\right)^i \left(\sigma(a_{12})t_1 + \sigma(a_{22})t_2\right)^j,$$

where $\left(\begin{smallmatrix} a_{11} & a_{12} \\ a_{21} & a_{22} \end{smallmatrix}\right)$ is in $\mathrm{SL}_2(\mathbb{K})$. For any m and any σ, the space of homogeneous polynomials of degree m is invariant under this action. We will denote this module by V_m^σ.

THEOREM 4.12. *Let $m \leqslant p-1$ if $p > 0$ and $m \geqslant 1$ if $p = 0$. The $\mathrm{SL}_2(\mathbb{K})$-module V_m^σ is absolutely irreducible for each σ.*

PROOF. Let $\overline{\mathbb{K}}$ be the algebraic closure of $\mathbb{K}$ and let U be a non-zero $\mathrm{SL}_2(\mathbb{K})$-invariant subspace of $V_m^\sigma \otimes \overline{\mathbb{K}}$. Let us fix $0 \neq f \in U$. We first show that $t_2^m \in U$. This is clear if $f = \lambda_0 t_2^m$, otherwise write

$$f(t_1, t_2) = \lambda_d t_1^d t_2^{m-d} + \lambda_{d-1} t_1^{d-1} t_2^{m-d+1} + \cdots,$$

where $\lambda_d \neq 0$. Let Δ be the difference operator

$$(\Delta f)(t_1, t_2) = f(t_1, t_2) - f(t_1 - t_2, t_2).$$

Then $\Delta f \in U$ and its d-th iterate $\Delta^{(d)}$ satisfies

$$\left(\Delta^{(d)} f\right)(t_1, t_2) = d!\lambda_d t_2^m.$$

Note that $d! \neq 0$ in $\mathbb{K}$ for any p (for $p > 0$ we use $d \leqslant p-1$). Therefore $t_2^m \in U$. Now, for $j = 0, 1, \ldots, m$, the polynomials $(t_2 + jt_1)^m$ are in U and are linearly independent. We conclude that $U = V_m \otimes \overline{\mathbb{K}}$. □

For each σ and $m \geqslant 1$, the above action has a non-trivial kernel, namely $\langle -I \rangle$, exactly when q is odd and m is even. Otherwise it is faithful. Moreover, this action consists of linear transformations of determinant 1. Thus it gives an embedding of $\mathrm{PSL}_2(\mathbb{K})$ into $\mathrm{SL}_{m+1}(\mathbb{K})$ when m is even, and into $\mathrm{PSL}_{m+1}(\mathbb{K})$ when m is odd. Thus the above result, together with Theorem 4.9, allows to construct projective representations of $T(2,3,7)$ of degree m, for each $m \leqslant p$.

THEOREM 4.13 (Cohen, [6]). *Any Hurwitz subgroup $\overline{H}$ of $\mathrm{PSL}_3(\mathbb{F})$ has a preimage H in $\mathrm{SL}_3(\mathbb{F})$ which is also Hurwitz. If H is reducible, then $p = 2$, $H \simeq \mathrm{PSL}_2(8) = \mathrm{SL}_2(8)$, and there are two conjugacy classes of such groups. If H is irreducible, one of the following holds*:

(1) $p \neq 7$ *and* $H \simeq \mathrm{PSL}_2(7)$;
(2) $p \equiv 0, \pm 1 \pmod 7$ *and* $H = \Omega_3(p) \simeq \mathrm{PSL}_2(p)$, *or* $2 < p \equiv \pm 2, \pm 3 \pmod 7$ *and* $H = \Omega_3(p^3) \simeq \mathrm{PSL}_2(p^3)$.

Moreover, in both cases (1) *and* (2) *there is just one conjugacy class of Hurwitz groups.*

PROOF. Let $(\bar{x}, \bar{y}, \bar{x}\bar{y} = \bar{z})$ be a $(2,3,7)$-generating triple for $\overline{H}$. First, we show that we can choose $x, y \in \mathrm{SL}_3(\mathbb{F})$ such that $x \mapsto \bar{x}$, $y \mapsto \bar{y}$ and $(x, y, xy = z)$ is also a $(2,3,7)$-triple. This is true if the characteristic of $\mathbb{F}$ is 3 since $\mathrm{PSL}_3(\mathbb{F}) = \mathrm{SL}_3(\mathbb{F})$ in that case. So suppose that the characteristic is different from 3. Multiplying x and y by scalar matrices, if necessary, we can assume that $x^2 = I$ and $z^7 = (xy)^7 = I$. In particular, x has eigenvalues -1 (with multiplicity 2) and 1. We claim that y cannot be a matrix of order 9 such that y^3 is scalar. In fact, in this case, y is diagonalizable with eigenvalues η, with multiplicity 2, and η^7, for some primitive 9-th root of unity η. Therefore there exists a non-zero vector v such that $xv = -v$, $yv = \eta v$. It follows $zv = -\eta v$, a contradiction since z has order 7. We conclude that y must have order 3.

Set $H = \langle x, y \rangle$. Assume that H is reducible. Replacing H with its transpose, if necessary, we may assume that H fixes a one-dimensional space U, hence a non-zero vector. In odd characteristic, the action of x modulo U should be scalar. From here it is easy to deduce that H would be soluble, a contradiction. Thus the characteristic is 2 and, by Theorem 4.9, H should induce on the quotient $\mathbb{F}^3/U$ the group $\mathrm{PSL}_2(8)$. In particular, up to conjugation, we can take

$$z = \begin{pmatrix} \varepsilon & 1 & 0 \\ 0 & \varepsilon^{-1} & 0 \\ 0 & 0 & 1 \end{pmatrix}, \qquad x = \begin{pmatrix} 0 & 1 & 0 \\ 1 & 0 & 0 \\ \alpha & \alpha & 1 \end{pmatrix},$$

where $\varepsilon^7 = 1$ and α is either 0 or 1. Direct computation shows that both choices of α give a group isomorphic to $\mathrm{PSL}_2(8)$. Moreover these two groups are not conjugate.

Now suppose that H is irreducible. Then the similarity invariants must be $t+1$, t^2-1 for x, t^3-1 for y and $(t-\varepsilon^{\ell})(t-\varepsilon^{2\ell})(t-\varepsilon^{4\ell})$ with $\ell = 1, 3$ or $(t-1)(t-\varepsilon^{\ell})(t-\varepsilon^{-\ell})$ with $\ell = 1, 2, 3$ for z. Thus, by Theorem 4.6, H belongs to at most five conjugacy classes (just one, if $p = 7$), corresponding to the five possibilities for xy.

(1) If $p \neq 7$, then $\mathrm{PSL}_2(7)$ has two dual irreducible representations of degree 3 over $\mathbb{F}$. This fact can be deduced from the knowledge of its ordinary and Brauer characters. These representations exhaust the first two possibilities for xy.

(2) As to the remaining possibilities for xy, let us consider the embedding of $\mathrm{PSL}_2(q)$ into $\mathrm{SL}_3(q)$ described just before Theorem 4.12, with $\sigma = 1$ and $m = 2$. Under this embedding, the three non-conjugate $(2,3,7)$-triples generating $\mathrm{PSL}_2(q)$ are mapped to three non-conjugate triples in $\mathrm{SL}_3(\mathbb{F})$. Moreover the image of $\mathrm{PSL}_2(q)$ preserves a symmetric (non-zero) bilinear form on the space of homogeneous polynomials of degree 2. This form is non-degenerate precisely when $p > 2$ (see also the proof of Theorem 1 in [25]). Thus, when $p = 2$, the remaining possibilities for xy do not give rise to irreducible subgroups of $\mathrm{SL}_3(\mathbb{F})$, by Remark 4.7. On the other hand, when $p > 2$, this embedding exhausts the remaining possibilities for xy giving rise to an irreducible subgroup of $\Omega_3(q)$. Our claim (2) follows from the isomorphism $\mathrm{PSL}_2(q) \simeq \Omega_3(q)$ and Theorem 4.9. □

The following result deals with classical groups of rank 4. The first statement is related to a fact already proved by Macbeath in [57]. Namely that, when $p \equiv \pm 1 \pmod 7$, there are three normal subgroups of $T(2,3,7)$ with quotient isomorphic to $\mathrm{PSL}_2(p)$.

LEMMA 4.14. *Let $p > 0$ and let $\overline{H}$ be a Hurwitz subgroup of* $\mathrm{PSL}_2(\mathbb{F}) \times \mathrm{PSL}_2(\mathbb{F})$. *Assume that $\overline{H}$ is not isomorphic to the Hurwitz subgroup of* $\mathrm{PSL}_2(\mathbb{F})$. *Then $p \equiv \pm 1 \pmod 7$ and $\overline{H}$ is conjugate to* $\mathrm{PSL}_2(p) \times \mathrm{PSL}_2(p)$.

PROOF. For $i = 1, 2$ let π_i be the projections of $\mathrm{PSL}_2(\mathbb{F}) \times \mathrm{PSL}_2(\mathbb{F})$ onto $\mathrm{PSL}_2(\mathbb{F})$. Since $\mathrm{Ker}\,\pi_i \simeq \mathrm{PSL}_2(\mathbb{F})$, our assumption implies that $1 \neq \pi_1(\overline{H})$ and $1 \neq \pi_2(\overline{H})$. Hence $\pi_1(\overline{H}) \simeq \pi_2(\overline{H}) \simeq \mathrm{PSL}_2(q)$, with $q \in \{p, p^3\}$ as in Theorem 4.9, case $k = 7$. By the same theorem, up to conjugation we may assume that $\overline{H}$ is the image of

$$H = \langle (x, x), (y_\ell, y_m) \rangle \leqslant \mathrm{SL}_2(q) \times \mathrm{SL}_2(q), \tag{15}$$

where x, y_ℓ, y_m are as in (13), $\ell \neq m$ by our assumptions. In particular, $p \neq 7$.

If $q = p^3$ with $p \equiv \pm 2, \pm 3 \pmod 7$, then there exists a field automorphism σ of $\mathrm{PSL}_2(q)$ such that $x = \sigma(x)$ and $y_m = \sigma(y_\ell)$. We conclude that $\overline{H}$ is isomorphic to $\mathrm{PSL}_2(q)$.

Now assume $q = p \equiv \pm 1 \pmod 7$. We note that

$$\pi_i(\mathrm{Ker}\,\pi_j \cap \overline{H}) \trianglelefteq \pi_i(\overline{H}) = \mathrm{PSL}_2(q).$$

If $\pi_1(\mathrm{Ker}\,\pi_2 \cap \overline{H})$ is trivial, it follows that $\mathrm{Ker}\,\pi_1 \cap \overline{H} = 1$. In this case the restriction $\pi_1 : \overline{H} \to \mathrm{PSL}_2(q)$ would be an isomorphism, against our assumption. Thus $\pi_1(\mathrm{Ker}\,\pi_2 \cap \overline{H}) = \mathrm{PSL}_2(q)$ by the simplicity of $\mathrm{PSL}_2(q)$, i.e. $\mathrm{Ker}\,\pi_2 \cap \overline{H} = \mathrm{PSL}_2(q)$. From $\pi_2(\overline{H}) = \mathrm{PSL}_2(q)$ we conclude that $\overline{H} = \mathrm{PSL}_2(q) \times \mathrm{PSL}_2(q)$. □

For any field K, consider the homomorphism $\varphi : \mathrm{SL}_2(K) \times \mathrm{SL}_2(K) \to \mathrm{SL}_4(K)$

$$(a, b) \mapsto a \otimes b. \tag{16}$$

The kernel of φ is $\langle (-I, -I) \rangle$. Moreover the image of φ preserves the bilinear symmetric form defined by the matrix

$$\begin{pmatrix} 0 & 1 \\ -1 & 0 \end{pmatrix} \otimes \begin{pmatrix} 0 & 1 \\ -1 & 0 \end{pmatrix} = \mathrm{antidiag}(1, -1, -1, 1).$$

Hence the image of φ is an orthogonal group in odd characteristic, a symplectic group in characteristic 2.

REMARK 4.15. Assume $K = \mathrm{GF}(q)$, with $q \in \{p, p^3\}$ as in the theorem of Macbeath, $p \neq 7$, and let H be defined as in (15) with $\ell \neq m$. Then $y_\ell \otimes y_m$ does not have the eigenvalue 1. It follows that $\varphi(H)$ does not fix any one-dimensional subspace and this fact easily implies that it is absolutely irreducible. Moreover $\varphi(H)$ is a subgroup of $\Omega_4^+(q)$ for p odd, of $\mathrm{Sp}_4(8)$ for $p = 2$. In particular, when $p \equiv \pm 1 \pmod 7$, by order reasons φ induces an isomorphism from the Hurwitz central product $\mathrm{SL}_2(p) \circ \mathrm{SL}_2(p)$ onto

$\Omega_4^+(p)$. Factorizing this central product by its center we obtain the Hurwitz direct product $\mathrm{PSL}_2(p) \times \mathrm{PSL}_2(p) \simeq \mathrm{P}\Omega_4^+(p)$. The details are left to the reader.

THEOREM 4.16. *If $p > 0$, the irreducible subgroups of* $\mathrm{PSL}_4(\mathbb{F})$ *which are Hurwitz are isomorphic to*:

(1) $\mathrm{PSL}_2(p) \times \mathrm{PSL}_2(p) \simeq \mathrm{P}\Omega_4^+(p)$, *when* $p \equiv \pm 1 \pmod 7$;

(2) $\mathrm{PSL}_2(q)$ *with* $q \in \{p, p^3\}$ *as in Macbeath's theorem*;

(3) $\mathrm{PSL}_2(7)$, *when* $p \neq 2$.

In particular $\mathrm{P}\Omega_4^+(q) \simeq \mathrm{PSL}_2(q) \times \mathrm{PSL}_2(q)$ *is Hurwitz if and only if* $q = p \equiv \pm 1 \pmod 7$, *whereas the following groups are never Hurwitz*: $\mathrm{P}\Omega_4^-(q) \simeq \mathrm{PSL}_2(q^2)$, $\mathrm{PSL}_4(q) \simeq \mathrm{P}\Omega_6^+(q)$, $\mathrm{PSU}_4(q^2) \simeq \mathrm{P}\Omega_6^-(q)$, *and* $\mathrm{PSp}_4(q) \simeq \mathrm{P}\Omega_5(q)$.

PROOF. Let $x, y \in \mathrm{SL}_4(\mathbb{F})$ be such that the projective image of (x, y, xy) is a $(2,3,7)$-generating triple of an irreducible subgroup of $\mathrm{PSL}_4(\mathbb{F})$. Multiplying x and y by scalar matrices of determinant 1, if necessary, we may assume that $y^3 = (xy)^7 = 1$ and

$$x^2 = I, \quad \text{if } p = 2; \qquad x^2 \in \langle iI \rangle, \quad \text{where } i \in \mathbb{F} \text{ has order 4, if } p > 2.$$

By Scott's formula, in the notation of Section 4.1,

$$d_M^x + d_M^y + d_M^{xy} \leqslant 18. \tag{17}$$

Direct calculation based on the formula (9) of Frobenius shows that

$$d_M^x \geqslant 8, \qquad d_M^y \geqslant 6, \qquad d_M^{xy} \geqslant 4. \tag{18}$$

It follows that in (17) and (18) we have all equalities. In particular x must have two equal similarity invariants, namely $t^2 - 1, t^2 - 1$, or $t^2 + 1, t^2 + 1$. In the first case $x^2 = I$, in the second $x^2 = -I$. On the other hand y must have similarity invariants $t - 1$ and $t^3 - 1$; xy must have a unique similarity invariant. Thus, when $p = 7$, xy is conjugate to the Jordan block of order 4. When $p \neq 7$, xy has 4 different eigenvalues and its similarity invariants can only be either

$$(t - \varepsilon^{\ell})(t - \varepsilon^{-\ell})(t - \varepsilon^{2\ell})(t - \varepsilon^{-2\ell}), \quad \text{with } \ell = 1, 2, 3,$$

or

$$(t - 1)(t - \varepsilon^{\ell})(t - \varepsilon^{2\ell})(t - \varepsilon^{4\ell}), \quad \text{with } \ell = 1, 3.$$

Assume first $x^2 = I$ and $p > 2$. By what shown in [25], $\langle x, y \rangle$ is a subgroup of the orthogonal group $\Omega_4(\mathbb{F}, f)$, where f is a non-degenerate quadratic form of Witt index 2. Since $\mathrm{P}\Omega_4(\mathbb{F}, f)$ is isomorphic to $\mathrm{PSL}_2(\mathbb{F}) \times \mathrm{PSL}_2(\mathbb{F})$ (see [21]), by the previous lemma the projective image of $\langle x, y \rangle$ can only be of type (1) or of type (2).

Remark 4.15 tells us that (1) actually occurs, and that we obtain an irreducible subgroup of type (2) whenever $p \neq 7$.

Now assume either $p = 2$ and $x^2 = I = -I$, or p odd and $x^2 = -I$. Equality in relation (17) implies that (x, y, xy) is a linearly rigid triple, by Theorem 4.6. Thus $\langle x, y\rangle$ belongs to at most five conjugacy classes of irreducible subgroups (just one, if $p = 7$) corresponding to the five possibilities for xy. Moreover, if $\mathrm{SL}_4(\mathbb{F})$ contains a reducible subgroup generated by a triple $(x', y', x'y')$ such that $x \sim x'$, $y \sim y'$, $xy \sim x'y'$, then the triple (x, y, xy) cannot generate an irreducible subgroup; see Remark 4.7.

Let $q \in \{p, p^3\}$ be defined as in the theorem of Macbeath. If $p > 2$, we consider the embeddings of $\mathrm{PSL}_2(q)$ into $\mathrm{PSL}_4(\mathbb{F})$ arising from the action of $\mathrm{SL}_2(q)$ on homogeneous polynomials of degree 3. They are irreducible when $p \neq 3$. If $p = 2$, we consider the irreducible embeddings of $\mathrm{PSL}_2(8)$ into $\mathrm{PSL}_4(\mathbb{F})$ corresponding to the projective image of $\varphi(H)$, as in Remark 4.15. These embeddings of $\mathrm{PSL}_2(q)$ exhaust the first three possibilities for xy. Hence they give rise to a unique conjugacy class of irreducible subgroups, whenever $p \neq 3$. On the other hand, when $p = 3$, they are reducible. By what was observed above there is no irreducible subgroup generated by a triple of this kind.

Finally, when $p \neq 2$, $\mathrm{SL}_2(7)$ has faithful irreducible representations of degree 4 over $\mathbb{F}$, the existence of which can be deduced from the knowledge of ordinary, modular and Brauer characters of $\mathrm{SL}_2(7)$. They give rise to projective representations of $\mathrm{PSL}_2(7)$ which exhaust the remaining cases for xy. On the other hand, when $p = 2$, there are copies of $\mathrm{PSL}_2(7) \sim \mathrm{SL}_3(2)$ arising from the embedding of $\mathrm{SL}_3(\mathbb{F})$ into $\mathrm{SL}_4(\mathbb{F}) = \mathrm{PSL}_4(\mathbb{F})$ which exhaust the possibilities for xy. Since this embedding is reducible, there is no irreducible subgroup of this kind when $p = 2$.

For the reader's convenience, we give explicitly these representations of $\mathrm{PSL}_2(7)$ for $p \neq 2$. When $p = 7$, we have the above representation on the homogeneous polynomials of degree 3. So we may assume $p \neq 7$. Let ε be a primitive 7-th root of unity in $\mathbb{F}$ and let $\tau = \varepsilon + \varepsilon^2 + \varepsilon^4$ or $\tau = \varepsilon^3 + \varepsilon^6 + \varepsilon^5$. Define

$$x := \begin{pmatrix} 0 & 0 & -1 & 0 \\ 0 & 0 & 0 & -1 \\ 1 & 0 & 0 & 0 \\ 0 & 1 & 0 & 0 \end{pmatrix}, \qquad y := \begin{pmatrix} 1 & 0 & 0 & \tau \\ 0 & 1 & 0 & 1+\tau \\ 0 & 0 & 0 & -1 \\ 0 & 0 & 1 & -1 \end{pmatrix}.$$

Direct calculation shows that for both values of τ we have

$$x^2 = -I, \qquad y^3 = (xy)^7 = I, \qquad [x, y]^4 = -I,$$

and, moreover, that $\langle x, y\rangle$ fixes only the zero vector. It follows easily that this group is irreducible. Since $\mathrm{PSL}_2(7)$ has the presentation $x^2 = y^3 = (xy)^7 = [x, y]^4 = 1$, the projective image of $\langle x, y\rangle$ is isomorphic to $\mathrm{PSL}_2(7)$.

The remaining claims which do not follow directly from what shown above are consequences of Theorem 4.9 and of the isomorphisms in the statement. These isomorphisms are well known and can be found in [17, page xii]. □

THEOREM 4.17 (Tamburini and Zalesskii, [82]). *Assume that $k \geqslant 7$ is a prime number. If $k = p$, set $q = p$. Otherwise, let n be the order of p modulo k, and suppose $n \not\equiv 0 \pmod 4$. Set $q = p^n$ if n is odd, $q = p^{\frac{n}{2}}$ if n is even.*

(1) *The following groups are* $(2,3,k)$*-generated*:
$\mathrm{PSL}_5(q)$, *if* $p \equiv 1 \pmod 5$;
$\mathrm{PSU}_5(q^2)$, *if* $p \equiv -1 \pmod 5$;
$\mathrm{PSU}_5(q^4)$, *if* $p \equiv \pm 2 \pmod 5$.

(2) *If* $k = 7$, *the only other irreducible Hurwitz subgroups of* $\mathrm{PSL}_5(\mathbb{F})$ *are isomorphic to* $\mathrm{PSL}_2(q)$ *with* q *as above, except* $\mathrm{PSL}_2(8)$ *and* $\mathrm{PSL}_2(27)$.

Note that the assumptions on n are certainly satisfied if $k \equiv 3 \pmod 4$. In particular, for $k = 7$, this theorem gives a new example of a Hurwitz group in each characteristic $p \neq 0, 5$. The proof, which avoids calculations, relies on the following results.

We recall that a pair (b, c) of elements in $\mathrm{Mat}_n(\mathbb{F})$ is said to be spectrally complete with respect to the product if, for every $\lambda_1, \dots, \lambda_n \in \mathbb{F}$ such that $\lambda_1 \cdot \dots \cdot \lambda_n = \det(bc)$, there exist b' conjugate to b and c' conjugate to c under $\mathrm{GL}_n(\mathbb{F})$, such that $b'c'$ has eigenvalues $\lambda_1, \dots, \lambda_n$.

THEOREM 4.18 (Silva, [69]). *Assume* $n \geqslant 3$ *and let* $b, c \in \mathrm{GL}_n(\mathbb{F})$. *Denote by*

$$f_1(t), \dots, f_r(t), \; g_1(t), \dots, g_s(t)$$

the non-trivial similarity invariants of $tI_n - b$ *and of* $tI_n - c$, *respectively. Then* (b, c) *is spectrally complete with respect to the product provided that* $r + s \leqslant n$ *and at least one of the polynomials* $f_i(t)$ *or* $g_j(t)$ *has degree* $\neq 2$.

This theorem, at least in characteristic $p \neq k$, guarantees the existence of a $(2, 3, 5k)$ triple $(x, y, \eta z)$, where $(\eta z)^k$ is scalar, of order 5. It follows immediately that $\langle x, y\rangle$ is an irreducible subgroup of $\mathrm{SL}_5(\mathbb{F})$. Then the proof that the projective image of $\langle x, y\rangle$ coincides with one of the above groups is based on the knowledge of the subgroups of $\mathrm{SL}_5(K)$, when K is a finite field, and Theorem 4.8.

5. Related results

5.1. *Number-theoretic aspects*

In this section we give an alternative description of some triangle groups $T(2, 3, k)$, and in particular $T(2, 3, 7)$, as two-dimensional projective unitary groups over rings of algebraic integers. The corresponding unitary groups can be identified, in turn, with groups of principal units in certain orders of generalized quaternion algebras. This description sheds more light on number-theoretic aspects of Hurwitz generation and provides a new view on some classical results like Macbeath's Theorem 3.1. The treatment in this section mainly follows, [88].

Let $\mathbb{F} = \mathbb{C}$ and define $\varepsilon \in \mathbb{C}$ to be a primitive $2k$-th root of unity if k is even, and a primitive k-th root of unity if k is odd. We also set

$$\eta = \varepsilon - \varepsilon^{-1}, \qquad \theta = \varepsilon + \varepsilon^{-1}.$$

It follows from the proof of Theorem 4.9 that $T(2,3,k)$ is isomorphic to the projective image of the group generated by matrices x and $z = z_1$ (or x and y_1), which were defined in (12). It is immediate to verify that both x and z preserves the Hermitian form defined by the matrix

$$B = \begin{pmatrix} \eta^2 & \eta \\ -\eta & \eta^2 \end{pmatrix},$$

which is non-degenerate provided $k \neq 6$. Thus, $T(2,3,k)$ is isomorphic to the projective image of a subgroup of

$$\mathrm{SU}\big(2, B, \mathbb{Z}[\varepsilon]\big) = \big\{A \in \mathrm{SL}\big(2, \mathbb{Z}[\varepsilon]\big) \mid \bar{A}^t B A = B\big\}.$$

The problem when $\langle x, z\rangle$ coincides with $\mathrm{SU}(2, B, \mathbb{Z}[\varepsilon])$ was studied in [88].

THEOREM 5.1 (Vsemirnov, Mysovskikh and Tamburini, [88]). *The equality*

$$\langle x, z\rangle = \mathrm{SU}\big(2, B, \mathbb{Z}[\varepsilon]\big)$$

holds if and only if $k \in \{2,3,4,5,7,9,11\}$. *In particular,* $T(2,3,k) \cong \mathrm{PSU}(2, B, \mathbb{Z}[\varepsilon])$ *precisely for the values of* k *listed above.*

In fact, the proof in [88] shows more. Namely, for k odd $\geqslant 13$ or k even $\geqslant 6$, the group generated by x and z has infinite index in $\mathrm{SU}(2, B, \mathbb{Z}[\varepsilon])$. As C. Maclahlan noticed in a private correspondence, [58], this 'negative' result can be also deduced from the description of all arithmetic Fuchsian groups given by K. Takeuchi, [75,76]. However, the 'positive' part of Theorem 5.1 is more delicate: it ensures not only that $T(2,3,k)$ is arithmetic for $k = 7, 9$, and 11, or equivalently, that it has a finite index in $\mathrm{PSU}(2, B, \mathbb{Z}[\varepsilon])$, but also proves the coincidence of these two groups.

It is convenient to treat the elements of $\mathrm{SU}(2, B, \mathbb{Z}[\varepsilon])$ as quaternions of norm 1. For this purpose, we note that the matrices

$$\mathbf{1} = \begin{pmatrix} 1 & 0 \\ 0 & 1 \end{pmatrix}, \qquad \mathbf{i} = \begin{pmatrix} 0 & -1 \\ 1 & 0 \end{pmatrix}, \qquad \mathbf{j} = \begin{pmatrix} \eta & 1 \\ 1 & -\eta \end{pmatrix}, \qquad \mathbf{k} = \begin{pmatrix} -1 & \eta \\ \eta & 1 \end{pmatrix}$$

form a standard basis over $\mathbb{Q}(\theta)$ for the generalized quaternion algebra $\big(\frac{-1,\theta^2-3}{\mathbb{Q}(\theta)}\big)$. Moreover, for any element (given as a matrix) in this algebra, its quaternion norm coincides with the determinant. It is then easy to see that $\mathrm{SU}(2, B, \mathbb{Z}[\varepsilon])$ is exactly the set $\mathcal{H}_1^*$ of all quaternions of norm 1 in the subring

$$\mathcal{H} = \left\{ a_0\mathbf{1} + a_1\mathbf{i} + a_2\mathbf{j} + a_3\mathbf{k} \,\middle|\, \begin{array}{l} 2a_i \in \mathbb{Z}[\theta],\ a_0 - a_3 - a_2\theta \in \mathbb{Z}[\theta], \\ a_1 + a_2 - a_3\theta \in \mathbb{Z}[\theta] \end{array} \right\}.$$

Thus, Theorem 5.1 can be restated in the following way.

THEOREM 5.2. *We have* $\langle x, z\rangle = \mathcal{H}_1^*$ *if and only if* $k \in \{2, 3, 4, 5, 7, 9, 11\}$. *In particular,* $T(2,3,k) \cong \mathcal{H}_1^*/\{\pm\mathbf{1}\}$ *precisely for the values of k listed above.*

Note that the norm in the corresponding quaternion algebra is given by

$$a_0^2 + a_1^2 - (\theta^2 - 3)a_2^2 - (\theta^2 - 3)a_3^2.$$

For a given k these norm forms depend on θ, i.e. on the choice of a primitive root of unity, and there is a natural action of $\mathrm{Gal}(\mathbb{Q}(\theta)/\mathbb{Q})$ on their coefficients. If $k = 2, 3, 4$ or 5 all corresponding forms are positively definite, while for each $k = 7, 9$ and 11 there is exactly one indefinite form. The proof of Theorems 5.1 and 5.2 is effective in the following sense. It provides a procedure for representing any element of $\mathcal{H}_1^*$ (or $\mathrm{SU}(2, B, \mathbb{Z}[\varepsilon])$) as a word in the generators x, z. For example, when $k = 7, 9$, or 11 let us choose the quaternion algebra $\mathcal{H}$ which corresponds to the indefinite form above. Then a more detailed analysis of the proof in [88] shows that any $u = a_0\mathbf{1} + a_1\mathbf{i} + a_2\mathbf{j} + a_3\mathbf{k} \in \mathcal{H}_1^*$ can be written as a word in x and z of length $\mathrm{O}(\log(a_2^2 + a_3^2))$.

To conclude this section we indicate some relations with Macbeath's theorem. For the sake of simplicity we deal only with Hurwitz groups, i.e., we assume $k = 7$. However, a similar observation can be applied when $k = 9$ or 11. It is well known that, for $k = 7$, the ring $\mathbb{Z}[\theta]$ is a principal ideal domain (see [66]; the tables from [66] are reproduced in [19, pp. 141–145]; also see [2, Table 7]). In addition, by a special case of a theorem due to E. Kummer (see, e.g., [28, §2.11, Corollary 1]) we have that for any rational prime p the following holds:

- p remains a prime in $\mathbb{Z}[\theta]$ if and only if $p \equiv \pm 2, \pm 3 \pmod 7$;
- p splits into a product of three different primes in $\mathbb{Z}[\theta]$ if and only if $p \equiv \pm 1 \pmod 7$;
- p ramifies in $\mathbb{Z}[\theta]$ if and only if $p = 7$.

In particular, if $\mathfrak{p}$ is a prime in $\mathbb{Z}[\theta]$ lying over a rational prime p, then $\mathbb{Z}[\theta]/\mathfrak{p} = \mathrm{GF}(q)$, where $q = p$ for $p \equiv 0, \pm 1 \pmod 7$ and $q = p^3$ otherwise.

If $\mathfrak{p}$ lies over an odd rational prime, then there is a natural residue homomorphism

$$\psi : \mathcal{H} \to \left(\frac{-1, \theta^2 - 3}{\mathbb{Z}[\theta]/\mathfrak{p}}\right) \cong \mathrm{Mat}_2\big(\mathrm{GF}(q)\big).$$

Now, Theorem 4.9 combined with Theorem 5.2 asserts that the restriction of ψ to $\mathcal{H}_1^*$ is onto $\mathrm{SL}_2(q)$. It is interesting to notice that we can also go in the opposite direction and deduce Macbeath's theorem from the fact that the above restriction is onto. From this point of view it would be very interesting to find a purely number-theoretic proof of Macbeath's theorem not using Dickson's classification of subgroups of $\mathrm{SL}_2(q)$.

We remark that results similar to the above were obtained, independently, in [71].

5.2. *Other groups which are* $(2, 3, 7)$*-generated*

The $(2, 3, 7)$-generation of $\mathrm{SL}_n(q)$, for $n \geqslant 287$, was actually obtained as a special case of a more general result, which has many other applications.

Namely, given a ring R with identity element, let $\mathrm{E}_n(R)$ denote the group generated by the set of elementary matrices:

$$\{1 + re_{ij} \mid r \in R, 1 \leqslant i \neq j \leqslant n\}.$$

Clearly $\mathrm{E}_n(R)$ can only be a finitely generated group if R is a finitely generated ring. On the other hand, applications of the permutational methods of Higman and Conder in a linear context, as explained in Section 3.1, led to the following result.

THEOREM 5.3 (Lucchini, Tamburini and Wilson, [56]). *Let R be a ring which is generated by elements $\alpha_1, \ldots, \alpha_m$, where $2\alpha_1 - \alpha_1^2$ is a unit of R of finite multiplicative order. Then $\mathrm{E}_n(R)$ is $(2, 3, 7)$-generated for all $n \geqslant 287 + 84(m-1)$.*

For $n \geqslant 3$, the groups $\mathrm{E}_n(R)$ and $\mathrm{SL}_n(R)$ coincide, in particular, if R is commutative and either semi-local or a Euclidean domain, see, e.g., [34, 1.2.11 and 4.3.9]. As the hypothesis on R of the above theorem holds with $m = 1$ if R is a finite field or the ring $\mathbb{Z}$ of integers, the same theorem implies that $\mathrm{SL}_n(q)$ and $\mathrm{SL}_n(\mathbb{Z})$ are $(2, 3, 7)$-generated for $n \geqslant 287$. This lower bound for n was improved to 252 in [86].

An easy corollary of Theorem 5.3 also shows that the derived group of the automorphism group $\mathrm{Aut}(F_n)$ of a free group of rank n is $(2, 3, 7)$-generated, provided that $n \geqslant 329$.

But there are other applications of this theorem, which shed further light on the class of $(2, 3, 7)$-generated groups.

Let p be a prime number. For each positive integer l, let N_l be the kernel of the epimorphism $\mathrm{SL}_n(\mathbb{Z}) \to \mathrm{SL}_n(\mathbb{Z}/p^l\mathbb{Z})$. Thus, N_1/N_l is a finite p-group for each l, and $\bigcap_{l \geqslant 1} N_l = 1$. Since N_1 (the group of matrices in $\mathrm{SL}_n(\mathbb{Z})$ congruent to 1 modulo p) is non-soluble, we conclude that there is no bound on the derived lengths of the groups N_1/N_l.

Applying Theorem 5.3 we have the following result:

COROLLARY 5.4 [56]. *Let $n \geqslant 287$ and let p be a prime. There exist Hurwitz groups which are extensions of p-groups of arbitrarily large derived length by the group $\mathrm{SL}_n(p)$.*

Since the direct product of two Hurwitz groups without common composition factors is again a Hurwitz group, one is led to study direct powers of simple groups. To this purpose, consider the polynomial ring $R = \mathrm{GF}(q)[t_1, \ldots, t_l]$. Then R can be generated by $l + 1$ elements, the first of which can be chosen to be a non-zero element α of $\mathrm{GF}(q)$ satisfying $2\alpha - \alpha^2 \neq 0$. Let I be the intersection of the kernels of the homomorphisms from R to $\mathrm{GF}(q)$ which extend the identity map on $\mathrm{GF}(q)$. By the Chinese remainder theorem, R/I is isomorphic to the direct product of q^l copies of $\mathrm{GF}(q)$, and the quotient map induces an epimorphism from $\mathrm{E}_n(R)$ to the direct product of q^l copies of $\mathrm{SL}_n(q)$. We conclude from the theorem that this direct product is a Hurwitz group provided that $n \geqslant 287 + 84l$. Therefore we have the following result, which shows that, for large n, the direct power of many copies of $\mathrm{SL}_n(q)$ is a Hurwitz group.

COROLLARY 5.5 [56]. *Let q be a prime power and let $n \geqslant 287$. Then the direct product of r copies of $\mathrm{SL}_n(q)$ is a Hurwitz group, where $r = q^{[(n-287)/84]}$.*

In [90], J. Wilson constructed a family of simple non-commutative, finitely generated rings S, with the property that, for each $n \geqslant 8$, the central quotient group $\mathrm{PE}_n(S)$ of $\mathrm{E}_n(S)$ is simple. Moreover, it was shown that the groups $\mathrm{PE}_n(S)$ arising from rings S in this family fall into $2^{\aleph_0}$ isomorphism classes. It follows from Theorem 5.3 that each of these groups is $(2, 3, 7)$-generated for n sufficiently large. Therefore we have the following result, which makes irrealistic any attempt of classifying all $(2, 3, 7)$-generated groups.

COROLLARY 5.6 [56]. *There are $2^{\aleph_0}$ isomorphism classes of infinite simple $(2, 3, 7)$-generated groups.*

Of course, there are no more than $2^{\aleph_0}$ isomorphism classes of finitely generated groups.

Another application of Theorem 5.3, recently made by M. Conder, gives a complete answer to the question of what centres are possible in finite quotients of the triangle group $T(2, 3, 7)$. This question was raised in 1965 by John Leech, [48], who later produced two infinite families of Hurwitz groups with centres of order 2 and 4. In [10], M. Conder used similar methods to prove the existence of infinitely many Hurwitz groups with a centre of order 3 and in [13] he constructed a family of central products of 2-dimensional special linear groups to show that the centre of a Hurwitz group could be an elementary Abelian 2-group of arbitrarily large order. Actually the centre of a Hurwitz group can be anything, in virtue of the following:

THEOREM 5.7 (Conder, [15]). *Given any finite Abelian group A, there exist infinitely many Hurwitz groups G such that the centre $Z(G)$ of G is isomorphic to A.*

We give a sketch of the elegant proof, which consists in taking a product of appropriately chosen special linear Hurwitz groups. Indeed, let A be any finite Abelian group and write

$$A = C_{m_1} \times \cdots \times C_{m_s}$$

as a direct product of cyclic groups. Now choose any prime p such that $(p, |A|) = 1$ and let $q = p^e$ with $e = \phi(|A|)$, where ϕ denotes Euler's totient function. Then $q - 1 = \ell_i m_i$ for some integer ℓ_i $(1 \leqslant i \leqslant s)$. Further, if k_i is any positive integer coprime to ℓ_i, then

$$(k_i m_i, q - 1) = (k_i m_i, \ell_i m_i) = m_i.$$

In particular there are infinitely many possibilities for each k_i, and all can be chosen such that $k_i m_i \neq k_j m_j$ for $i \neq j$ and $k_i m_i \geqslant 287$, for $1 \leqslant i \leqslant s$. Next let $H_i = \mathrm{SL}_{k_i m_i}(q)$ and set

$$G = \prod_{1 \leqslant i \leqslant s} H_i.$$

Using Theorem 5.3, it is easily seen that G has the required properties.

We mention another application of the same theorem, to the maximal parabolic subgroups of $E_{n+\bar{n}}(R)$. Namely, let:

$$P_{n,\bar{n}}(R) = \left\{ \begin{pmatrix} A & B \\ 0 & \bar{A} \end{pmatrix} \mid A \in \mathrm{E}_n(R),\ \bar{A} \in \mathrm{E}_{\bar{n}}(R),\ B \in \mathrm{Mat}_{n,\bar{n}}(R) \right\}.$$

THEOREM 5.8 (Di Martino and Tamburini, [24]). *Let R be a ring generated by elements $\alpha_1, \ldots, \alpha_m$, where $2\alpha_1 - \alpha_1^2$ is a unit of R of finite multiplicative order. Then $P_{n,\bar{n}}(R)$ (and therefore also the Levi subgroup L) is $(2,3,7)$-generated for all $n, \bar{n} \geqslant 84(m+1) + 396$.*

Finally, we mention that A. Lucchini in [54] and, independently, J.S. Wilson in [91] have generalized Theorem 5.3 to any $k \geqslant 7$. Lucchini's version reads:

THEOREM 5.9 (Lucchini, [54]). *Let R be a ring which is generated by elements $\alpha_1, \ldots, \alpha_m$, where $2\alpha_1 - \alpha_1^2$ is a unit of R of finite multiplicative order. For any fixed $k \geqslant 7$ there exist two integers n_k and a_k such that $\mathrm{E}_n(R)$ is $(2,3,k)$-generated for all $n \geqslant n_k + a_k m$.*

Acknowledgements

We are very grateful to A. Lucchini and A. Zalesskii for useful remarks and valuable advice on many parts of the manuscript.

Appendix

As mentioned in Section 3, to define a $(2,3,7)$-generating triple (x, y, z) of $\mathrm{Alt}(\Omega)$ when $n > 167$ and $n \neq 173, 174, 181, 188, 202$, Conder uses $3 + 14$ transitive permutational representations of $T(2,3,7)$, each of which is depicted by a diagram (see [8]). Here we describe explicitly each of these representations and give some information about the commutator $[x, y]$, useful to understand the considerations above theorem 3.4. We assume that the joins of diagrams, in order to obtain $[x, y]$, are made as described there: in particular each diagram of type H_d is joined to a diagram of type G.

The following rule, which can easily be checked, has repeated application. Let ψ and ψ' be permutation representations of $T(2,3,7) = \langle X, Y \rangle$ on disjoint sets Δ and Δ', with respective handles (e_1, e_2) and (e_1', e_2'). Let e_3 be the image of e_2 under $\psi(Y)$ and e_3' be the image of e_2' under $\psi'(Y)$. Consider the representation φ on the set $\Delta \cup \Delta'$ defined by

$$\varphi(X) = \psi(X)\psi'(X)(e_1, e_1')(e_2, e_2'), \qquad \varphi(Y) = \psi(Y)\psi'(Y)$$

as described in Lemma 3.3. Assume, for simplicity, that e_1 and e_3 are in the same cycle of the commutator $[\psi(X), \psi(Y)]$ and let Γ be the support of this cycle. Let Γ' be the union of the supports of the cycles of $[\psi'(X), \psi'(Y)]$ (not necessarily distinct) which contain e_1' and e_3'. Then the set $\Gamma \cup \Gamma'$ is invariant under the commutator $[\varphi(X), \varphi(Y)]$. Moreover, setting

$|\Gamma \cup \Gamma'| = 2m + s$, $0 \leqslant s \leqslant 1$, the cycle structure of the restriction $[\varphi(X), \varphi(Y)]_{\Gamma\cup\Gamma'}$ consists of two cycles of length m if $s = 0$ and of one cycle of length $2m+1$ if $s = 1$. Furthermore, the cycle structure of $[\varphi(X), \varphi(Y)]$ on $\Delta \cup \Delta' \setminus \{\Gamma \cup \Gamma'\}$ is the juxtaposition of the cycles of $[\psi(X), \psi(Y)]$ and $[\psi'(X), \psi'(Y)]$.

In view of the application to the linear context, our treatment differs slightly from that of Conder. In particular we avoid the use of the odd involution $t \in \mathrm{Sym}(\Omega)$, which centralizes x and inverts y by conjugation. But the two approaches are essentially equivalent in virtue of the obvious relation $(xyt)^2 = [x, y]^t$.

NOTATION. The cycle structure of a permutation will be denoted by

$$\underbrace{(i, \ldots)}_{j}\, \ell_1^{k_1} \ell_2^{k_2} \ldots.$$

This means a cycle of length j whose support contains i, followed by k_1 cycles of length ℓ_1, followed by k_2 cycles of length ℓ_2, etc.

DIAGRAM G: degree 42. Handles: $(2, 3)$, $(14, 15)$, $(32, 33)$.

$$\begin{aligned} x_G = {} & (1,4)(5,7)(6,10)(8,12)(9,24)(11,29)(13,16)(17,19)(18,25)(20,27)\\ & (21,23)(22,39)(26,30)(28,41)(31,34)(35,37)(36,40)(38,42)(2)(3)\\ & (14)(15)(32)(33). \end{aligned}$$

$$y_G = \prod_{i=0}^{13} (3i+1, 3i+2, 3i+3).$$

$$[x_G, y_G] = \underbrace{(2, \ldots, 1, \ldots)}_{13}\underbrace{(14, \ldots, 13, \ldots)}_{13}\underbrace{(32, \ldots, 31, \ldots)}_{13}\, 1^3.$$

REMARK. This representation is given by the action of $\mathrm{PSL}_2(13)$ on the cosets of a subgroup N of index 42.

DIAGRAM A: degree 14. Handle: $(1, 2)$.

$$x_A = (3,4)(5,9)(6,11)(7,10)(8,13)(12,14)(1)(2).$$

$$y_A = \prod_{i=0}^{3} (3i+1, 3i+2, 3i+3)(13)(14).$$

$$[x_A, y_A] = \underbrace{(1, \ldots, 3, \ldots)}_{13}\, 1^1.$$

REMARK. This representation is given by the action of $PSL_2(13)$ on the points of the projective line.

DIAGRAM E: degree 28. Handle: $(1,2)$.

$$x_E = (3,4)(5,9)(6,11)(7,10)(8,13)(12,24)(14,26)(15,16)(18,19)(21,22)(23,25)(27,28)(1)(2)(17)(20).$$

$$y_E = \prod_{i=0}^{8}(3i+1, 3i+2, 3i+3)(28).$$

$$[x_E, y_E] = \underbrace{(1,\ldots,3,\ldots)}_{9}1^1 9^2.$$

DIAGRAM H_0: degree $42 \equiv 0 \pmod{14}$. Handle: $(1,2)$.

$$x_{H_0} = (3,5)(4,12)(6,7)(8,11)(9,17)(10,32)(13,21)(14,29)(15,16)(18,36)(19,23)(22,27)(24,38)(25,30)(28,33)(31,40)(34,39)(37,42)(1)(2)(20)(26)(35)(41).$$

$$y_{H_0} = \prod_{i=0}^{13}(3i+1, 3i+2, 3i+3).$$

$$[x_{H_0}, y_{H_0}] = \underbrace{(1,\ldots)}_{5}\underbrace{(3,\ldots)}_{5}1^1 3^1 11^1 17^1.$$

$$\begin{aligned} c &= [x,y]^{3\cdot 5\cdot 11\cdot 13\cdot 23} \\ &= (14,31,38,20,24,40,29,19,25,15,34,22,27,39,16,30,23). \end{aligned}$$

DIAGRAM H_1: degree $57 \equiv 1 \pmod{14}$. Handle: $(16,17)$.

$$x_{H_1} = (2,9)(3,5)(4,35)(6,12)(7,41)(8,11)(10,13)(14,15)(18,20)(19,27)(21,22)(23,26)(24,32)(25,47)(28,36)(29,44)(30,31)(33,51)(34,38)(37,42)(39,53)(40,45)(43,48)(46,55)(49,54)(52,57)(1)(16)(17)(50)(56).$$

$$y_{H_1} = \prod_{i=0}^{18}(3i+1, 3i+2, 3i+3).$$

$$[x_{H_1}, y_{H_1}] = \underbrace{(16, \ldots)}_{5}\underbrace{(18, \ldots)}_{5} 1^1 3^1 5^1 7^2 12^2.$$

$$c = [x, y]^{4 \cdot 3 \cdot 7 \cdot 11 \cdot 13 \cdot 23} = (1, 6, 9, 2, 12).$$

DIAGRAM H_2: degree $142 \equiv 2 \pmod{14}$. Handle: $(77, 78)$.

$$\begin{aligned} x_{H_2} = {} & (1,8)(2,20)(3,4)(5,6)(7,14)(9,10)(11,19)(12,22)(13,24)(15,16) \\ & (17,18)(21,26)(23,30)(25,48)(27,29)(28,45)(31,59)(32,60)(33,34) \\ & (35,39)(36,41)(37,40)(38,50)(42,54)(43,47)(44,49)(46,52)(51,56) \\ & (53,55)(57,58)(61,62)(63,67)(64,69)(65,68)(66,71)(70,86)(72,90) \\ & (73,74)(75,81)(76,83)(79,80)(82,84)(85,87)(88,100)(89,98)(91,96) \\ & (92,101)(93,102)(94,95)(97,99)(103,104)(105,109)(106,111) \\ & (107,110)(108,113)(112,130)(114,135)(115,116)(117,120) \\ & (118,125)(121,132)(123,131)(124,126)(127,128)(129,142) \\ & (133,137)(136,138)(139,141)(77)(78)(119)(122)(134)(140). \end{aligned}$$

$$y_{H_2} = \prod_{i=0}^{18}(3i+1, 3i+2, 3i+3)(58)\prod_{i=20}^{47}(3i-1, 3i, 3i+1).$$

$$[x_{H_2}, y_{H_2}] = \underbrace{(77, 79, \ldots)}_{13} 1^5 3^1 11^3 12^4 17^1 23^1.$$

$$\begin{aligned} c &= [x, y]^{4 \cdot 3 \cdot 7 \cdot 11 \cdot 13 \cdot 23} \\ &= (115, 125, 126, 121, 132, 124, 118, 116, 129, 136, 137, 131, 119, 123, 133, \\ &\quad 138, 142). \end{aligned}$$

DIAGRAM H_3: degree $115 \equiv 3 \pmod{14}$. Handle: $(1, 2)$.

$$\begin{aligned} x_{H_3} = {} & (3,4)(5,9)(6,11)(7,10)(8,14)(12,24)(13,17)(15,26)(16,20)(18,38) \\ & (19,22)(21,35)(23,25)(27,28)(30,32)(41,31)(33,44)(34,36)(37,39) \\ & (40,42)(43,47)(45,84)(46,51)(48,50)(49,85)(52,53)(54,56)(55,65) \\ & (57,69)(58,60)(59,107)(61,63)(62,110)(64,66)(67,79)(68,72) \\ & (70,81)(71,77)(73,74)(75,76)(78,80)(82,83)(86,93)(87,105) \\ & (88,89)(90,91)(92,99)(94,95)(96,104)(97,113)(98,115)(100,101) \\ & (102,103)(106,108)(109,111)(112,114)(1)(2)(29). \end{aligned}$$

$$y_{H_3} = \prod_{i=0}^{8}(3i+1, 3i+2, 3i+3)(28)\prod_{i=10}^{38}(3i-1, 3i, 3i+1).$$

$$[x_{H_3}, y_{H_3}] = \underbrace{(1, 3, \ldots)}_{9} 1^1 2^2 5^2 11^4 15^2 17^1.$$

$$[x, y]^{2\cdot 3\cdot 5\cdot 11\cdot 13}$$
$$= (13, 17, 30, 21, 27, 37, 40, 25, 16, 29, 20, 23, 42, 39, 28, 35, 32).$$

DIAGRAM H_4: degree $144 \equiv 4 \pmod{14}$. Handle: (64, 65).

$$x_{H_4} = (1, 10)(2, 7)(3, 4)(5, 76)(6, 77)(8, 13)(9, 11)(12, 28)(14, 37)(15, 17)(16, 20)(18, 32)(19, 25)(21, 22)(23, 24)(26, 31)(27, 29)(30, 40)(33, 34)(35, 48)(36, 55)(38, 42)(39, 43)(41, 60)(44, 62)(45, 46)(47, 53)(49, 50)(51, 52)(54, 56)(57, 58)(59, 72)(61, 70)(63, 68)(66, 67)(69, 71)(73, 82)(74, 85)(75, 80)(78, 79)(81, 83)(84, 102)(86, 103)(87, 89)(88, 95)(90, 91)(92, 111)(93, 118)(94, 100)(96, 97)(98, 126)(99, 133)(101, 141)(104, 143)(105, 106)(107, 109)(110, 116)(112, 113)(114, 115)(117, 119)(120, 121)(122, 142)(123, 124)(125, 131)(127, 128)(129, 130)(132, 134)(135, 137)(138, 139)(140, 144)(64)(65)(108)(136).$$

$$y_{H_4} = \prod_{i=0}^{47}(3i + 1, 3i + 2, 3i + 3).$$

$$[x_{H_4}, y_{H_4}] = \underbrace{(64, 66, \ldots)}_{11}\, 1^3 5^3 8^2 11^2 17^1 30^2.$$

$$c = [x, y]^{8\cdot 3\cdot 5\cdot 11\cdot 13}$$
$$= (108, 129, 115, 120, 128, 110, 135, 113, 125, 131, 112, 137, 116, 127, 121, 114, 130).$$

DIAGRAM H_5: degree $187 \equiv 5 \pmod{14}$. Handle: (1, 2).

$$x_{H_5} = (3, 4)(5, 11)(6, 21)(8, 13)(9, 17)(10, 14)(12, 20)(15, 75)(16, 19)(18, 69)(22, 90)(23, 46)(24, 25)(26, 28)(27, 37)(29, 63)(30, 35)(31, 32)(33, 34)(36, 38)(39, 81)(40, 102)(41, 103)(42, 43)(44, 53)(45, 47)(48, 55)(49, 50)(51, 52)(54, 56)(57, 58)(59, 108)(60, 61)(62, 64)(65, 67)(66, 78)(68, 70)(71, 111)(72, 73)(74, 76)(77, 79)(80, 82)(83, 114)(84, 85)(86, 95)(87, 88)(89, 99)(91, 92)(93, 94)(96, 97)(98, 100)(101, 117)(104, 118)(105, 106)(107, 109)(110, 112)(113, 115)(116, 128)(119, 125)(120, 129)(121, 133)(122, 134)(123, 124)(126, 127)(130, 139)(131, 143)(132, 137)(135, 136)(138, 140)(141, 153)(142, 145)(144, 155)(146, 182)(147, 148)(149, 179)(150, 151)(152, 154)(156, 157)(158, 165)(159, 177)(160, 161)(162, 163)(164, 171)(166, 167)(168, 176)(169, 185)(170, 187)(172, 173)(174, 175)(178, 180)(181, 183)(184, 186)(1)(2)(7).$$

$$y_{H_5} = \prod_{i=0}^{51}(3i+1, 3i+2, 3i+3)(157)\prod_{i=53}^{62}(3i-1, 3i, 3i+1).$$

$$[x_{H_5}, y_{H_5}] = \underbrace{(1, 3, \ldots)}_{17}\, 1^3 4^2 9^2 10^2 12^4 15^2 43^1.$$

$$\begin{aligned} c &= [x, y]^{4\cdot 9\cdot 5\cdot 11\cdot 13} \\ &= (7, 100, 62, 58, 41, 18, 94, 76, 52, 107, 71, 92, 70, 50, 72, 113, 95, 66, 53, \\ &\quad 13, 102, 82, 80, 40, 8, 44, 78, 86, 115, 73, 49, 68, 91, 111, 109, 51, 74, 93, \\ &\quad 69, 103, 57, 64, 98). \end{aligned}$$

DIAGRAM H_6: degree $216 \equiv 6 \pmod{14}$. Handle: (70, 71).

$$\begin{aligned} x_{H_6} = {} & (1, 9)(2, 5)(4, 32)(6, 11)(7, 26)(8, 10)(12, 13)(14, 15)(16, 40)(17, 41) \\ & (18, 19)(20, 23)(21, 36)(22, 25)(24, 35)(27, 28)(29, 31)(33, 34)(37, 46) \\ & (38, 49)(39, 44)(42, 43)(45, 47)(48, 66)(50, 68)(51, 52)(53, 59)(54, 61) \\ & (55, 170)(56, 58)(57, 169)(60, 62)(63, 64)(65, 78)(67, 76)(69, 74)(72, 73) \\ & (72, 73)(75, 77)(79, 83)(80, 121)(81, 92)(82, 88)(84, 85)(86, 87)(89, 91) \\ & (90, 162)(93, 94)(95, 97)(96, 114)(98, 127)(99, 100)(101, 116)(102, 103) \\ & (104, 211)(105, 106)(107, 208)(108, 109)(110, 115)(111, 112)(113, 150) \\ & (117, 118)(119, 141)(120, 142)(122, 182)(123, 124)(125, 134)(126, 128) \\ & (129, 136)(130, 131)(132, 133)(135, 137)(138, 139)(140, 163)(143, 180) \\ & (144, 145)(146, 155)(147, 148)(149, 159)(151, 152)(153, 154)(156, 157) \\ & (158, 160)(161, 185)(164, 181)(165, 166)(167, 173)(168, 175)(171, 172) \\ & (174, 176)(177, 178)(179, 186)(183, 184)(187, 194)(188, 206)(189, 190) \\ & (191, 192)(193, 200)(195, 196)(197, 205)(198, 214)(199, 216)(201, 202) \\ & (203, 204)(207, 209)(210, 212)(213, 215)(3)(30)(70)(71). \end{aligned}$$

$$y_{H_6} = \prod_{i=0}^{71}(3i+1, 3i+2, 3i+3).$$

$$[x_{H_6}, y_{H_6}] = \underbrace{(70, \ldots, 72, \ldots)}_{13}\, 1^5 4^2 5^1 6^2 7^2 11^1 12^2 13^2 17^2 32^2.$$

$$c = [x, y]^{32\cdot 3\cdot 7\cdot 11\cdot 13\cdot 17} = (1, 3, 9, 11, 6).$$

DIAGRAM H_7: degree $77 \equiv 7 \pmod{14}$. Handle: (50, 51).

$$\begin{aligned} x_{H_7} = {} & (1, 22)(2, 23)(3, 4)(5, 8)(6, 21)(7, 11)(9, 20)(10, 13)(14, 16)(18, 19) \\ & (24, 25)(26, 30)(27, 32)(28, 31)(29, 35)(33, 45)(34, 37)(36, 47)(38, 67) \\ & (39, 40)(41, 69)(42, 43)(44, 46)(48, 49)(52, 53)(54, 58)(55, 60)(56, 59) \\ & (57, 63)(61, 73)(62, 66)(64, 75)(65, 68)(70, 71)(72, 74)(76, 77)(12)(15) \\ & (17)(50)(51). \end{aligned}$$

$$y_{H_7} = \prod_{i=0}^{15}(3i+1, 3i+2, 3i+3)(49)\prod_{i=17}^{25}(3i-1, 3i, 3i+1)(77).$$

$$[x_{H_7}, y_{H_7}] = \underbrace{(50, \ldots, 52, \ldots)}_{9}\, 1^3 2^2 4^2 9^4 17^1.$$

$$\begin{aligned} c &= [x, y]^{4\cdot 9\cdot 7\cdot 11\cdot 13} \\ &= (1, 22, 24, 3, 26, 5, 33, 19, 36, 12, 47, 18, 45, 8, 30, 4, 25). \end{aligned}$$

DIAGRAM H_8: degree $36 \equiv 8 \pmod{14}$. Handle: (16, 17).

$$\begin{aligned} x_{H_8} = {} & (1, 9)(2, 5)(4, 32)(6, 11)(7, 27)(8, 10)(12, 13)(14, 15)(18, 19)(20, 23) \\ & (21, 36)(22, 26)(24, 35)(25, 29)(30, 31)(33, 34)(3)(16)(17)(28). \end{aligned}$$

$$y_{H_8} = \prod_{i=0}^{11}(3i+1, 3i+2, 3i+3).$$

$$[x_{H_8}, y_{H_8}] = \underbrace{(16, \ldots, 18, \ldots)}_{11}\, 1^1 4^2 5^1 11^1.$$

$$c = [x, y]^{4\cdot 3\cdot 11\cdot 13} = (1, 11, 3, 6, 9).$$

DIAGRAM H_9: degree $135 \equiv 9 \pmod{14}$. Handle: (124, 125).

$$\begin{aligned} x_{H_9} = {} & (3, 4)(5, 8)(6, 21)(7, 10)(9, 20)(11, 48)(12, 13)(14, 16)(17, 42)(18, 19) \\ & (22, 68)(23, 69)(24, 25)(26, 30)(27, 32)(28, 31)(29, 34)(33, 54)(35, 55) \\ & (36, 37)(38, 40)(39, 51)(41, 43)(44, 60)(45, 46)(47, 49)(50, 52)(53, 62) \\ & (56, 58)(59, 61)(64, 73)(65, 70)(66, 67)(71, 76)(72, 74)(75, 93)(77, 100) \\ & (78, 79)(80, 95)(81, 83)(82, 88)(84, 85)(86, 87)(89, 94)(90, 91)(92, 103) \\ & (96, 97)(98, 111)(99, 118)(101, 105)(102, 106)(104, 123)(107, 131) \\ & (108, 109)(110, 116)(112, 113)(114, 115)(117, 119)(120, 121)(122, 135) \\ & (126, 127)(128, 132)(129, 134)(130, 133)(1)(2)(15)(57)(63)(124)(125). \\ y_{H_9} = {} & \textstyle\prod_{i=0}^{44}(3i+1, 3i+2, 3i+3). \end{aligned}$$

$$[x_{H_9}, y_{H_9}] = \underbrace{(124, 126, \ldots)}_{11}\, 1^4 3^1 4^2 5^2 8^2 11^2 19^1 21^2.$$

$$\begin{aligned} c &= [x, y]^{8\cdot 3\cdot 5\cdot 7\cdot 11\cdot 13} \\ &= (1, 11, 31, 29, 18, 3, 40, 32, 52, 8, 5, 50, 27, 38, 4, 19, 34, 28, 48). \end{aligned}$$

DIAGRAM H_{10}: degree $136 \equiv 10 \pmod{14}$. Handle: $(55, 56)$.

$$x_{H_{10}} = (1,9)(2,5)(4,32)(6,11)(7,26)(8,10)(12,13)(14,15)(16,37)(17,38)(18,19)(20,23)(21,36)(22,25)(24,35)(27,28)(29,31)(33,34)(39,40)(41,45)(42,47)(43,46)(44,49)(48,66)(50,74)(51,52)(53,59)(54,61)(57,58)(60,62)(63,64)(65,78)(67,79)(68,80)(69,70)(71,75)(72,77)(73,76)(81,82)(83,87)(84,89)(85,88)(86,92)(90,102)(91,94)(93,104)(95,131)(96,97)(98,128)(99,100)(101,103)(105,106)(107,114)(108,126)(109,110)(111,112)(113,120)(115,116)(117,125)(118,134)(119,136)(121,122)(123,124)(127,129)(130,132)(133,135)(3)(30)(55)(56).$$

$$y_{H_{10}} = \prod_{i=0}^{34}(3i+1, 3i+2, 3i+3)(106)\prod_{i=36}^{45}(3i-1, 3i, 3i+1).$$

$$[x_{H_{10}}, y_{H_{10}}] = \underbrace{(55, \ldots, 57, \ldots)}_{13} 1^5 4^2 5^1 11^3 12^6.$$

$$c = [x, y]^{4\cdot 3\cdot 11\cdot 13} = (1, 11, 3, 6, 9).$$

DIAGRAM H_{11}: degree $165 \equiv 11 \pmod{14}$. Handle: $(160, 161)$.

$$x_{H_{11}} = (1,26)(2,27)(3,4)(5,8)(6,21)(7,11)(9,20)(10,14)(15,16)(18,19)(22,31)(23,28)(24,25)(29,34)(30,32)(33,51)(35,58)(36,38)(37,41)(39,53)(40,46)(42,43)(44,45)(47,52)(48,49)(50,61)(54,55)(56,69)(57,76)(59,63)(60,64)(62,81)(65,83)(66,67)(68,74)(70,72)(71,73)(75,77)(78,79)(80,93)(82,91)(84,89)(85,98)(86,99)(87,88)(90,92)(94,103)(95,100)(96,97)(101,112)(102,104)(105,123)(106,118)(107,117)(108,109)(110,111)(113,130)(114,115)(116,125)(119,124)(120,121)(122,133)(126,127)(128,141)(129,148)(131,135)(132,136)(134,153)(137,155)(138,139)(140,146)(142,143)(144,145)(147,149)(150,151)(152,159)(154,157)(156,164)(158,165)(162,163)(12)(13)(17)(160)(161).$$

$$y_{H_{11}} = \prod_{i=0}^{54}(3i+1, 3i+2, 3i+3).$$

$$[x_{H_{11}}, y_{H_{11}}] = \underbrace{(160, 162, \ldots)}_{11} 1^5 2^2 4^2 5^4 8^4 11^6 19.$$

$$c = [x, y]^{8\cdot 3\cdot 5\cdot 11\cdot 13}$$
$$= (1, 53, 3, 4, 39, 26, 8, 49, 24, 18, 29, 31, 12, 22, 34, 19, 25, 48, 5).$$

DIAGRAM H_{12}: degree $180 \equiv 12 \pmod{14}$. Handle: (5, 6).

$x_{H_{12}} = (1,10)(2,7)(3,4)(8,13)(9,11)(12,30)(14,37)(15,16)(17,32)(18,19)$
$(20,22)(21,25)(23,24)(26,31)(27,28)(29,40)(33,34)(35,48)(36,55)$
$(38,42)(39,43)(41,60)(44,62)(45,46)(47,53)(49,50)(51,52)(54,56)$
$(57,58)(59,72)(61,70)(63,68)(64,163)(65,164)(66,67)(69,71)$
$(73,82)(74,79)(75,76)(77,78)(80,121)(81,86)(83,85)(84,156)$
$(87,88)(89,91)(90,108)(92,115)(93,94)(95,110)(96,97)(99,100)$
$(102,103)(104,109)(105,106)(107,118)(111,112)(113,138)$
$(114,139)(116,126)(117,133)(119,144)(120,153)(122,176)$
$(123,124)(125,131)(127,128)(129,130)(132,134)(135,136)$
$(137,157)(140,174)(141,142)(143,149)(145,146)(147,148)$
$(150,151)(152,154)(155,179)(158,175)(159,160)(161,167)$
$(162,169)(165,166)(168,170)(171,172)(173,180)(177,178)$
$(5)(6)(98)(101).$

$$y_{H_{12}} = \prod_{i=0}^{59} (3i+1, 3i+2, 3i+3).$$

$$[x_{H_{12}}, y_{H_{12}}] = \underbrace{(4,\ldots,5,\ldots)}_{11} 1^3 5^2 6^2 7^2 8^2 11^2 13^2 19^1 47^1.$$

$$c = [x,y]^{8\cdot 3\cdot 5\cdot 7\cdot 11\cdot 13\cdot 19}$$
$$= (36,155,113,66,157,101,137,67,138,179,55,124,111,163,171,100,$$
$$142,63,174,178,45,120,104,166,167,96,117,59,175,158,72,133,97,$$
$$161,165,109,153,46,177,140,68,141,99,172,64,112,123).$$

DIAGRAM H_{13}: degree $195 \equiv 13 \pmod{14}$. Handle: (180, 178).

$x_{H_{13}} = (1,8)(2,20)(3,4)(5,6)(7,11)(9,16)(10,30)(12,13)(14,15)(17,19)$
$(18,28)(21,23)(22,37)(24,26)(25,40)(27,29)(32,34)(33,43)(35,46)$
$(36,38)(39,41)(42,44)(45,51)(47,86)(48,53)(49,52)(50,87)(54,55)$
$(56,58)(57,61)(59,77)(60,65)(62,69)(63,81)(64,116)(66,68)$
$(67,113)(70,79)(71,76)(72,73)(74,75)(78,83)(80,82)(84,85)$
$(88,97)(89,94)(90,91)(92,93)(95,138)(96,101)(98,100)(99,170)$
$(102,103)(104,106)(105,123)(107,130)(108,109)(110,125)$
$(111,112)(114,115)(117,118)(119,124)(120,121)(122,135)$
$(126,127)(128,152)(129,156)(131,140)(132,149)(133,158)$
$(134,168)(136,190)(137,141)(139,145)(142,144)(143,147)$
$(146,150)(148,153)(151,174)(154,188)(155,159)(157,163)$
$(160,162)(161,165)(164,166)(167,171)(169,193)(172,192)$
$(173,177)(175,181)(176,186)(179,180)(182,184)(185,189)$
$(187,194)(191,195)(31)(178)(180).$

$$y_{H_{13}} = \prod_{i=0}^{64} (3i+1, 3i+2, 3i+3).$$

$$[x_{H_{13}}, y_{H_{13}}] = \underbrace{(179, \ldots, 180, \ldots)}_{51} 1^1 2^2 5^2 6^2 7^2 13^4 14^2 23^1.$$

$$\begin{aligned} c &= [x, y]^{32 \cdot 3 \cdot 5 \cdot 7 \cdot 11 \cdot 13} \\ &= (2, 19, 37, 32, 42, 24, 9, 18, 7, 27, 39, 31, 41, 29, 11, 28, 16, 26, 44, 34, \\ &\quad 22, 17, 20). \end{aligned}$$

References

[1] G.V. Belyi, *On Galois extensions of a maximal cyclotomic field*, Izv. Akad. Nauk SSSR Ser. Mat. **43** (2) (1979), 267–276.

[2] Z.I. Borevich and I.R. Shafarevich, *Number Theory*, Academic Press (1966).

[3] R. Burkhardt, *Die Zerlegungmatrizen der Gruppen* PSL(2, p^f), J. Algebra **40** (1976), 75–96.

[4] J.W.S. Cassels and A. Fröhlich (eds), *Algebraic Number Theory*, Academic Press (1967).

[5] J. Cohen, *On Hurwitz extensions by* $\mathrm{PSL}_2(7)$, Math. Proc. Cambridge Philos. Soc. **86** (1979), 395–400.

[6] J. Cohen, *On non-Hurwitz groups and non-congruence subgroups of the modular group*, Glasgow Math. J. **22** (1981), 1–7.

[7] J. Cohen, *Homomorphisms of cocompact Fuchsian groups on* $\mathrm{PSL}_2(\mathbb{Z}_{p^n}[x]/f(x))$, Trans. Amer. Math. Soc. **281** (2) (1984), 571–585.

[8] M.D.E. Conder, *Generators for alternating and symmetric groups*, J. London Math. Soc. (2) **22** (1980), 75–86.

[9] M.D.E. Conder, *More on generators for alternating and symmetric groups*, Quart. J. Math. Oxford (2) **32** (1981), 137–163.

[10] M.D.E. Conder, *A family of Hurwitz groups with non-trivial centre*, Bull. Austral. Math. Soc. **33** (1986), 123–130.

[11] M.D.E. Conder, *A note on maximal automorphism groups of compact Riemann surfaces*, Indian J. Pure Appl. Math. **17** (1) (1986), 58–60.

[12] M.D.E. Conder, *The genus of compact Riemann surfaces with maximal automorphism group*, J. Algebra **108** (1) (1987), 204–247.

[13] M.D.E. Conder, *Hurwitz groups: a brief survey*, Bull. Amer. Math. Soc. **23** (1990), 359–370.

[14] M.D.E. Conder, *A question by Graham Higman concerning quotients of the* (2, 3, 7) *triangle group*, J. Algebra **141** (1991), 275–286.

[15] M.D.E. Conder, *Hurwitz groups with given centre*, Bull. London Math. Soc. **34** (6) (2002), 725–728.

[16] M.D.E. Conder, R.A. Wilson and A.J. Woldar, *The symmetric genus of sporadic groups*, Proc. Amer. Math. Soc. **116** (3) (1992), 653–663.

[17] J.H. Conway, R.T. Curtis, S.P. Norton, R.A. Parker and R.A. Wilson, *Atlas of Finite Groups*, Clarendon Press (1985).

[18] H.S.M. Coxeter and W.O.J. Moser, *Generators and Relations for Discrete Groups*, Springer-Verlag (1980).

[19] B.N. Delone and D.K. Faddeev, *The Theory of Irrationalities of the Third Degree*, Transl. Math. Monographs **10**, Amer. Math. Soc. (1964).

[20] L.E. Dickson, *Linear Groups with an Exposition of the Galois Field Theory*, Teubner (1901).

[21] J.A. Dieudonné, *La geometrie des groupes classiques*, troisiéme ed., Springer-Verlag (1971).

[22] L. Di Martino, *The normal structure of the classical modular group*, Atti del Convegno in onore di F. Brioschi, Istit. Lombardo Accad. Sci. Lett. Rend. (1999).

[23] L. Di Martino and M.C. Tamburini, *2-generation of finite simple groups and some related topics*, Generators and Relations in Groups and Geometries, A. Barlotti et al. (eds) (1991), 195–233.

[24] L. Di Martino and M.C. Tamburini, *On the* (2, 3, 7)*-generation of maximal parabolic subgroups*, J. Austral. Math. Soc. **71** (2) (2001), 187–199.

[25] L. Di Martino, M.C. Tamburini and A.E. Zalesski, *On Hurwitz groups of low rank*, Comm. Algebra **28** (11) (2000), 5383–5404.

[26] L. Di Martino and N. Vavilov, *$(2,3)$-generation of $\mathrm{SL}_n(q)$. I: Cases $n=5,6,7$*, Comm. Algebra **22** (4) (1994), 1321–1347.

[27] L. Di Martino and N. Vavilov, *$(2,3)$-generation of $\mathrm{SL}_n(q)$. II: Cases $n \geqslant 8$*, Comm. Algebra **24** (1996), 487–515.

[28] H.M. Edwards, *Divisor Theory*, Birkhäuser (1990).

[29] B. Everitt, *Alternating quotients of Fuchsian groups*, J. Algebra **223** (2000), 457–476.

[30] R.D. Feuer, *Torsion-free subgroups of triangle groups*, Proc. Amer. Math. Soc. **30** (2) (1971), 235–240.

[31] L. Finkelstein and A. Rudvalis, *Maximal subgroups of the Hall–Janko–Wales group*, J. Algebra **24** (1973), 486–493.

[32] R. Fricke and F. Klein, *Vorlesungen uber die Theorie der Elliptischen Modul Funktionen*, Vols. 1, 2, Teubner (1890).

[33] D. Gorenstein, *Finite Groups*, Harper and Row (1968).

[34] A.J. Hahn and O.T. O'Meara, *The Classical Groups and K-Theory*, Springer-Verlag (1989).

[35] P. Hall, *The Eulerian functions of a group*, Quart. J. Math. (Oxford) **7** (1936), 134–151.

[36] B. Huppert, *Endliche Gruppen I*, Springer-Verlag (1967).

[37] A. Hurwitz, *Über algebraische Gebilde mit eindeutigen Transformationen in sich*, Math. Ann. **41** (1893), 408–442.

[38] I.M. Isaacs, *Character Theory of Finite Groups*, Academic Press (1976).

[39] N. Jacobson, *Basic Algebra I*, 2nd ed., W.H. Freeman and Company (1985).

[40] G.A. Jones, *Ree groups and Riemann surfaces*, J. Algebra **165** (1994), 41–62.

[41] G.A. Jones, *Characters and surfaces: a survey*, The Atlas of Finite Groups: Ten Years on, Birmingham, 1995, London Math. Soc. Lecture Note Ser. **249** (1998), 90–118.

[42] G.A. Jones and D. Singerman, *Complex Functions: An Algebraic and Geometric Viewpoint*, Cambridge Univ. Press (1987).

[43] N. Katz, *Rigid Local Systems*, Ann. of Math. Stud. **139**, Princeton Univ. Press (1996).

[44] P.B. Kleidman, R.A. Parker and R.A. Wilson, *The maximal subgroups of the Fisher group Fi_{23}*, J. London Math. Soc. (2) **39** (1989), 89–101.

[45] F. Klein, *Über die Transformation siebenter Ordnung der elliptischen Functionen*, Math. Ann. **14** (1879), 428–471.

[46] C. Kurtz, *On the product of diagonal conjugacy classes*, Comm. Algebra **29** (2) (2001), 769–779.

[47] U. Langer and G. Rosenberger, *Erzeugende endlicher projectiver linearer Gruppen*, Results Math. **15** (1–2) (1989), 119–148.

[48] J. Leech, *Generators for certain normal subgroups of $(2,3,7)$*, Math. Proc. Cambridge Philos. Soc. **61** (1965), 321–332.

[49] F. Levin and G. Rosenberger, *Generators of finite projective linear groups: Part 2*, Results Math. **17** (1–2) (1990), 120–127.

[50] M.W. Liebeck and A. Shalev, *Classical groups, probabilistic methods, and the $(2,3)$-generation problem*, Ann. Math. **144** (1996), 77–125.

[51] M.W. Liebeck and A. Shalev, *Fuchsian groups, coverings of Riemann surfaces, subgroup growth, random quotients and random walks*, J. Algebra **276** (2) (2004), 552–601.

[52] S.A. Linton, *The maximal subgroups of the Thompson group*, J. London Math. Soc. (2) **39** (1989), 79–88.

[53] S.A. Linton and R.A. Wilson, *The maximal subgroups of the Fischer groups Fi'_{24} and Fi_{24}*, Proc. London Math. Soc. **63** (1991), 113–164.

[54] A. Lucchini, *$(2,3,k)$-generated groups of large rank*, Arch. Math. **73** (4) (1999), 241–248.

[55] A. Lucchini and M.C. Tamburini, *Classical groups of large rank as Hurwitz groups*, J. Algebra **219** (1999), 531–546.

[56] A. Lucchini, M.C. Tamburini and J.S. Wilson, *Hurwitz groups of large rank*, J. London Math. Soc. (2) **61** (2000), 81–92.

[57] A.M. Macbeath, *Generators of the linear fractional groups*, Proc. Symp. Pure Math. **12** (1969), 14–32.

[58] C. Maclahlan, private correspondence.

[59] W. Magnus, *Noneuclidean Tessellations and Their Groups*, Academic Press (1974).

[60] G. Malle, *Hurwitz groups and $G_2(q)$*, Canad. Math. Bull. **33** (1990), 349–357.

[61] G. Malle, *Small rank exceptional Hurwitz groups*, Groups of Lie Type and their Geometries, London Math. Soc. Lecture Note Ser. **207** (1995), 173–183.

[62] G. Malle, B.H. Matzat and B. Heinrich, *Inverse Galois Theory*, Springer Monographs in Math., Springer-Verlag (1999).
[63] Q. Mushtaq, *Coset diagrams for Hurwitz groups*, Comm. Algebra **18** (11) (1990), 3857–3888.
[64] Q. Mushtaq and G.-C. Rota, *Alternating groups as quotients of two generator groups*, Adv. Math. **96** (1) (1992), 113–121.
[65] W. Plesken and D. Robertz, *Representations, commutative algebra, and Hurwitz groups*, J. Algebra, Computational Section, to appear.
[66] L.W. Reid, *Tafel der Klassenzahlen für kubische Zahlkörper*, Inaugural Dissertation, Göttingen (1899).
[67] C.H. Sah, *Groups related to compact Riemann surfaces*, Acta Math. **123** (1969), 13–42.
[68] L.L. Scott, *Matrices and cohomology*, Ann. Math. **105** (1977), 473–492.
[69] F.C. Silva, *The eigenvalues of the product of matrices with prescribed similarity classes*, Linear and Multilinear Algebra **34** (1993), 269–277.
[70] A.R. Sourour, *A factorization theorem for matrices*, Linear and Multilinear Algebra **19** (2) (1986), 141–147.
[71] M. Stefanutti, *Superficie di Hurwitz*, Tesi di Laurea, Relatore: P. Corvaja, Universitá degli Studi di Udine (2000–2001).
[72] R. Steinberg, *Endomorphisms of linear algebraic groups*, Mem. Amer. Math. Soc. **80** (1968).
[73] W.W. Stothers, *Subgroups of the* $(2,3,7)$*-triangle group*, Manuscripta Math. **20** (4) (1977), 323–334.
[74] K. Strambach and H. Völklein, *On linearly rigid tuples*, J. Reine Angew. Math. **510** (1999), 57–62.
[75] K. Takeuchi, *A characterization of arithmetic Fuchsian groups*, J. Math. Soc. Japan **27** (1975), 600–612.
[76] K. Takeuchi, *Arithmetic triangle groups*, J. Math. Soc. Japan **29** (1977), 91–106.
[77] M.C. Tamburini, *Generation of certain simple groups by elements of small order*, Istit. Lombardo Accad. Sci. Lett. Rend. A **121** (1987), 21–27.
[78] M.C. Tamburini and M. Vsemirnov, *Irreducible* $(2,3,7)$*-subgroups of* $\mathrm{PGL}_n(\mathbb{F})$, $n \leqslant 7$, J. Algebra, submitted.
[79] M.C. Tamburini and J.S. Wilson, *On the generation of finite simple groups by pairs of subgroups*, J. Algebra **116** (2) (1988), 316–333.
[80] M.C. Tamburini and J.S. Wilson, *On the* $(2,3)$*-generation of some classical groups.* II, J. Algebra **176** (1995), 667–680.
[81] M.C. Tamburini, J.S. Wilson and N. Gavioli, *On the* $(2,3)$*-generation of some classical groups.* I, J. Algebra **168** (1994), 353–370.
[82] M.C. Tamburini and A. Zalesskii, *Classical groups in dimension* 5 *which are Hurwitz*, Proceedings of the Gainesville Conference on Finite Groups, 2003, C.Y. Ho, P. Sin, P.H. Tiep and A. Turull (eds), Walter de Gruyter (2004), 363–371.
[83] J.G. Thompson, *Some finite groups which appear as* $\mathrm{Gal}(L/K)$*, where* $K \subseteq \mathbb{Q}(\mu_n)$, J. Algebra **89** (1984), 437–499.
[84] R. Vincent and A.E. Zalesskii, *More on non-Hurwitz classical groups*, Preprint.
[85] H. Völklein, *Groups as Galois Groups – An Introduction*, Cambr. Stud. Adv. Math. **53**, Cambridge Univ. Press (1996).
[86] M. Vsemirnov, *Hurwitz groups of intermediate rank*, LMS J. Comput. Math. **7** (2004), 300–336.
[87] M. Vsemirnov, *Groups* $G_2(p)$, $p \geqslant 5$*, as quotients of* $(2,3,7;2p)$, Transformation Groups, to appear.
[88] M. Vsemirnov, V. Mysovskikh and M.C. Tamburini, *Triangle groups as subgroups of unitary groups*, J. Algebra **245** (2001), 562–583.
[89] H. Wielandt, *Einbettung zweier Gruppen in eine einfache Gruppe*, Math. Z. **73** (1960), 20–21.
[90] J.S. Wilson, *On characteristically simple groups*, Math. Proc. Cambridge Philos. Soc. **80** (1976), 19–35.
[91] J.S. Wilson, *Simple images of triangle groups*, Quart. J. Math. Oxford (2) **50** (200) (1999), 523–531.
[92] R.A. Wilson, *The symmetric genus of the Baby Monster*, Quart. J. Math. Oxford (2) **44** (1993), 513–516.
[93] R.A. Wilson, *The symmetric genus of the Fischer group* Fi_{23}, Topology **36** (2) (1997), 379–380.
[94] R.A. Wilson, *The Monster is a Hurwitz group*, J. Group Theory **4** (4) (2001), 367–374.
[95] A.J. Woldar, *On Hurwitz generation and genus action of sporadic groups*, Illinois Math. J. **33** (1989), 416–437.
[96] A.J. Woldar, *Sporadic simple groups which are Hurwitz*, J. Algebra **144** (1991), 443–450.
[97] M.F. Worboys, *Generators for the sporadic group* Co_3 *as a* $(2,3,7)$*-group*, Proc. Edinburgh Math. Soc. (2) **25** (1982), 65–68.

Braids, their Properties and Generalizations

V.V. Vershinin*

Département des Sciences Mathématiques, Université Montpellier II, Place Eugéne Bataillon, 34095 Montpellier cedex 5, France
E-mail: *vershini@math.univ-montp2.fr*

Sobolev Institute of Mathematics, Novosibirsk, 630090, Russia
E-mail: *versh@math.nsc.ru*

Contents

*The author was supported in part by the by CNRS-NSF grant No 17149, INTAS grant No 03-5-3251 and the ACI project ACI-NIM-2004-243 “Braids and Knots”.

HANDBOOK OF ALGEBRA, VOL. 4
Edited by M. Hazewinkel

Abstract

In the chapter we give a survey on braid groups and subjects connected with them. We start with the initial definition, then we give several interpretations as well as several presentations of these groups. Burau presentation for the pure braid group and the Markov normal form are given next. Garside normal form and his solution of the conjugacy problem are presented as well as more recent results on the ordering and on the linearity of braid groups. Next topics are the generalizations of braids, their homological properties and connections with the other mathematical fields, like knot theory (via Alexander and Markov theorems) and homotopy groups of spheres.

1. Introduction

Braid groups describe intuitive concept of classes of continuous deformations of braids, which are collections of intertwining strands whose endpoints are fixed. Mathematically they can be considered from various points of view. The first intuitive approach is formalized naturally as isotopy classes of a collection of n connected curves (strings) in 3-dimensional space. This point of view is connected with the definition of a braid group as the fundamental group of a configuration space of n points on a plane. Also braids can be interpreted as a mapping class group of a punctured disc and as a subgroup of the automorphism group of a free group (Section 3.3).

The present survey is organized as follows. In Section 2 we make some historical remarks. Definition and general properties are considered in Section 3. Configuration spaces appear in Section 3.2. Connections with groups of automorphisms of free groups are given in Section 3.3. Presentations of the braid group which appeared quite recently are observed in Section 3.4. Section 4 is devoted to F.A. Garside's classical work, [91], and Section 5 to that of P. Dehornoy on ordering for braids. Representations and in particular linearity are discussed in Section 6. In Section 7 various generalization of braids are presented. Homological properties are observed in Section 8. In the last Section 9 we discuss connection with the knot theory given by the Alexander and Markov theorems and with the homotopy groups of spheres.

2. Historical remarks

Braids were rigorously defined by E. Artin, [7], in 1925, although the roots of this natural concept are seen in the works of A. Hurwitz ([117], 1891), R. Fricke and F. Klein ([89], 1897) and even in the notebooks of C.-F. Gauss. E. Artin, [7], gave the presentation of the braid group (see formulas (3.2) in Section 3) which is common now. Already in the book of Felix Klein, [126], published in 1926 there appeared a chapter about braids. Essential topics about braids were also presented in the Reidemeister's Knotentheorie, [175], published in 1932.

In the 30ies there appeared a series of papers of Werner Burau, [48–50], where he in particular gave the presentation of the pure braid group (see Section 3.5) and introduced the Burau representation (Section 6.1). Wilhelm Magnus in his work [138] published in 1934 established relations between braid groups and the mapping class groups. At the same time there appeared the work of A.A. Markov, [149], which together with the Alexander theorem, [3], builds a bijection between links and equivalence classes of braids. It became an essential ingredient in study of links and knots (in the work of V.F.R. Jones, [121], for example). In 1936–37 were published papers of O. Zariski, [200,201], where he discovered connections between braid groups and the fundamental group of the complement of the discriminant of the general polynomial

$$f_n(t) = a_0 t^n + a_1 t^{n-1} + \cdots + a_{n-1} t + a_n,$$

a point of view later rediscovered by V.I. Arnold, [5]. Zariski also understood connections between braids and configuration spaces, gave the presentation of the braid group of the sphere, and studied the braid groups of Riemann surfaces. Amazingly and unfortunately these works of Zariski were not noticed by the specialists on braids and are not mentioned even in books and papers where the presentations of braids of surfaces are discussed.

In the beginning of 60ies R. Fox and L. Neuwirth, [86], and E. Fadell and L. Neuwirth, [80], studied configuration spaces which turned out to be $K(\pi, 1)$-spaces and so give a natural geometrical model of the classifying spaces for the braid groups. Later, V.I. Arnold, [5], in this direction proved the first results on the cohomology of braids. The motivation for his study was a connection (which he discovered) with the problem of representing algebraic functions in several variables by superposing algebraic functions in fewer variables. Also, in 1969 V.I. Arnold completely described the cohomology of pure braid groups, [4].

In 1969 there appeared the publication of F.A. Garside's work [91] where he suggests a new normal form of elements in the braid group and with its help gives a new solution of the word problem and also solves the conjugacy problem. In 1968 was published a two-page note of G.S. Makanin, [142], where he sketches his algorithm for the solution of the conjugacy problem. The complete publication of Makanin's work didn't appear (as far as the author is aware).

In the 70ies the study of cohomology of braids was continued independently and by different methods by D.B. Fuks, [90], who determined the mod 2 cohomology, and F.R. Cohen, [54–56], who described the homology with coefficients in $\mathbb{Z}$ and in $\mathbb{Z}/p$ as modules over the Steenrod algebra.

In 1984–85, independently, N.V. Ivanov, [118], and J. McCarthy, [151], proved the "Tits alternative" for the mapping class groups of surfaces and as a consequence it is true for the braid groups. Namely, they proved that every subgroup of the mapping class group either contains an Abelian subgroup of finite index, or contains a non-Abelian free group.

The question of whether braid groups are linear attracted significant attention. It was realized that the Burau representation is faithful for Br_3, [92,141]. Then, after a long break, in 1991 J.A. Moody, [153], proved that Burau representation is unfaithful for $n \geqslant 9$. This bound was improved to $n \geqslant 6$ by D.D. Long and M. Paton, [137], and to $n = 5$ by S. Bigelow, [28]. In 1999–2000 there appeared preprints of papers of D. Krammer, [128,129], and S. Bigelow, [29], who proved that Br_n is linear for all n (using the other representation).

At the beginning of nineties P. Dehornoy, [68–70], proved that there exists a left order in braid groups.

Interesting generalizations of braids were introduced in the work of E. Brieskorn, [42]. The configuration space can also be considered as the orbit space of the complement of the complexification of the arrangement of hyperplanes corresponding to the Coxeter group $A_{n-1} = \Sigma_n$. Generalizing this approach to any finite Coxeter group, E. Brieskorn defined the so-called generalized braid groups which are also called Artin groups.

Another way of generalization is to consider braid groups in 3-manifolds, possibly with a boundary. The simplest examples are braid groups in handlebodies. A.B. Sossinsky, [182], was the first who studied them. Such a group can be interpreted as the fundamental group of the configuration space of a plane without g points where g is the genus of the handlebody. The generalized braid group of type C is isomorphic to the braid group in the solid torus.

In the context of the influence of the theory of Vassiliev–Goussarov (finite-type) invariants singular braids were introduced. The corresponding algebraic structures are the Baez–Birman monoid, [10,33], and the braid-permutation group by R. Fenn, R. Rimányi and C. Rourke, [83,84]. Various properties of these objects were studied in [85,202,93,94, 66,119,62,104].

3. Definitions and general properties

3.1. *Systems of n curves in three-dimensional space and braid groups*

First of all, as was already mentioned, braids naturally arise as objects in 3-space. Let us consider two parallel planes P_0 and P_1 in $\mathbb{R}^3$, which contain two ordered sets of points $A_1, \ldots, A_n \in P_0$ and $B_1, \ldots, B_n \in P_1$. These points are lying on parallel lines L_A and L_B respectively. The space between the planes P_0 and P_1 we denote by Π. Suppose that the point B_i is lying under the point A_i, as a result of the orthogonal projection of the plane P_0 onto the plane P_1. Let us connect the set of points $A_1, \ldots, A_n$ with the set of points $B_1, \ldots, B_n$ by simple nonintersecting curves $C_1, \ldots, C_n$ lying in the space Π and such that each curve meets only once each parallel plane P_t lying in the space Π (see Figure 1). This object is called a *braid* and the curves are called the *strings* of a braid. Usually braids are depicted by projections on the plane passing through the lines L_A and L_B. This projection is supposed to be in general position so that there is only finite number of double points of intersection which are lying on pairwise different levels and intersections are transversal. The simplest braid σ_i (Figure 2) corresponds to the transposition $(i, i+1)$.

Let us introduce the following equivalence relation on the set of all braids with n strings and with fixed P_0, P_1, A_i and B_i. It is defined by homeomorphisms $h : \Pi \to \Pi$, which are the identity on $P_0 \cup P_1$ and such that $h(P_t) = P_t$. Braids β and β' are equivalent if there exists a homeomorphism h such that $h(\beta) = \beta'$. On the set Br_n of equivalence classes under the considered relation the structure of a group is introduced as follows. We put a copy Π' of the domain Π under the Π in such a way that P'_0 coincides with P_1 and each A_i coincides with B_i and we glue the braids β and β'. This gluing gives a composition of braids $\beta\beta'$ (Figure 3). The unit element is the equivalence class containing a braid of n parallel intervals, the braid β^{-1} inverse to β is defined by reflection of β with respect to

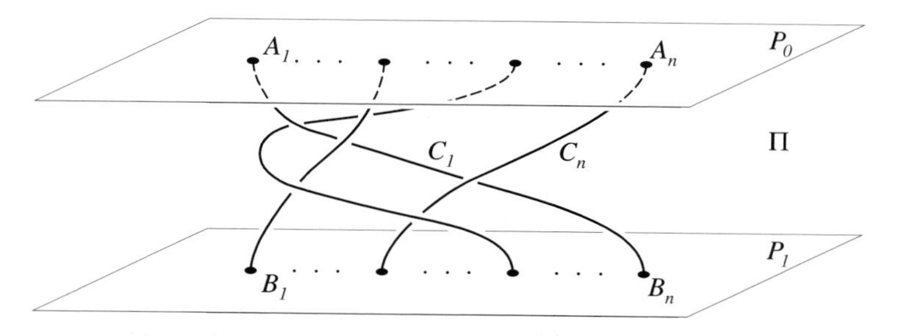

Fig. 1.

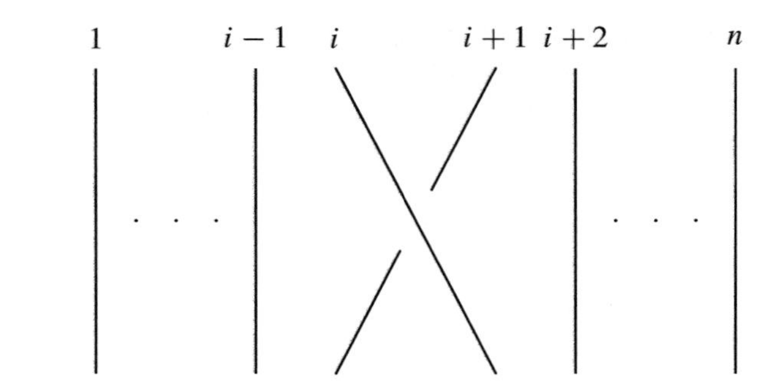

Fig. 2.

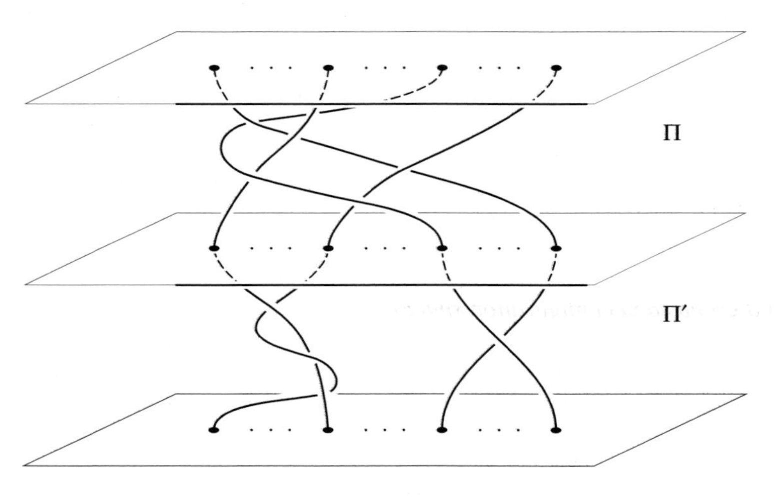

Fig. 3.

the plane $P_{1/2}$. A string C_i of a braid β connects the point A_i with the point B_{k_i} defining a permutation S^β. If this permutation is the identity then the braid β is called *pure*. The map $\beta \to S^\beta$ defines an epimorphism τ_n of the braid group Br_n on the permutation group Σ_n with the kernel consisting of all pure braids:

$$1 \to P_n \to Br_n \xrightarrow{\tau_n} \Sigma_n \to 1. \tag{3.1}$$

The following presentation of the braid group Br_n with generators σ_i, $i = 1, \ldots, n-1$, and two types of relations:

$$\begin{cases} \sigma_i \sigma_j = \sigma_j \sigma_i, & \text{if } |i-j| > 1, \\ \sigma_i \sigma_{i+1} \sigma_i = \sigma_{i+1} \sigma_i \sigma_{i+1} \end{cases} \tag{3.2}$$

is the algebraic expression of the fact that any isotopy of braids can be broken down into "elementary moves" of two types that correspond to the two types of relations.

If we add a vertical interval to the system of curves on Figure 1 we can get a canonical inclusion j_n of the group Br_n into the group Br_{n+1}

$$j_n : Br_n \to Br_{n+1}.$$

If the symmetric group Σ_n is given by its canonical presentation with generators s_i, $i = 1, \dots, n-1$, and relations:

$$\begin{cases} s_i s_j = s_j s_i, & \text{if } |i-j| > 1, \\ s_i s_{i+1} s_i = s_{i+1} s_i s_{i+1}, \\ s_i^2 = 1, \end{cases} \tag{3.3}$$

then the homomorphism τ_n is given by the formula

$$\tau_n(\sigma_i) = s_i, \quad i = 1, \dots, n-1.$$

It is possible to consider braids as classes of equivalence of *braid diagrams* which are generic projections of three-dimensional braids on a plane. The classes of equivalence are defined by the *Reidemeister moves* depicted in Figure 4.

3.2. *Braid groups and configuration spaces*

If we look at Figure 1, then this picture can be interpreted as a graph of a loop in the *configuration space* of n points on a plane, that is the space of unordered sets of n points on a plane, see Figure 5. So, it is possible to interpret the braid group as the fundamental

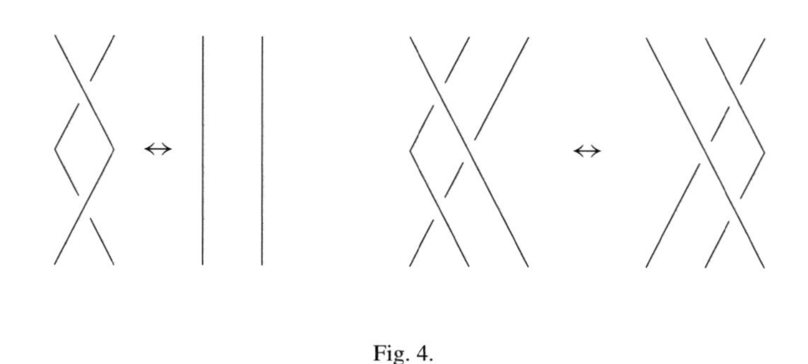

Fig. 4.

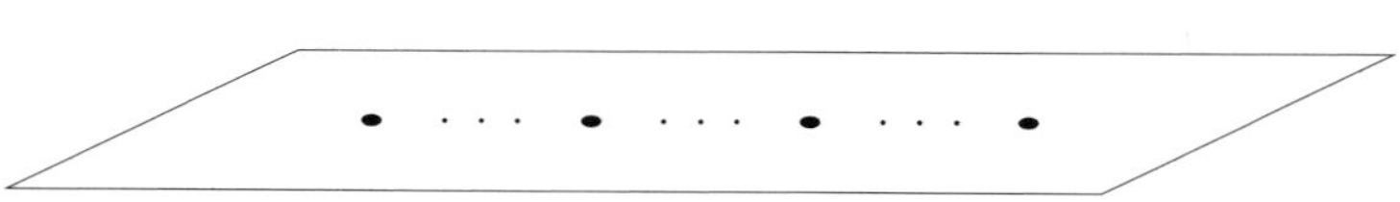

Fig. 5.

group of the configuration space. Formally it is done as follows. The symmetric group Σ_m acts on the Cartesian power $(\mathbb{R}^2)^m$ of the space $\mathbb{R}^2$:

$$w(y_1, \dots, y_m) = (y_{w^{-1}(1)}, \dots, y_{w^{-1}(m)}), \quad w \in \Sigma_m. \tag{3.4}$$

Denote by $F(\mathbb{R}^2, m)$ the space of m-tuples of pairwise different points in $\mathbb{R}^2$:

$$F(\mathbb{R}^2, m) = \{(p_1, \dots, p_m) \in (\mathbb{R}^2)^m \colon p_i \neq p_j \text{ for } i \neq j\}.$$

This is the space of regular points of our action. We call the orbit space of this action $B(\mathbb{R}^2, m) = F(\mathbb{R}^2, m)/\Sigma_m$ the *configuration space of n points on a plane*. The braid group Br_m is the fundamental group of this configuration space

$$Br_m = \pi_1(B(\mathbb{R}^2, m)).$$

The pure braid group P_m is the fundamental group of the space $F(\mathbb{R}^2, m)$. The covering

$$p \colon F(\mathbb{R}^2, m) \to B(\mathbb{R}^2, m)$$

defines an exact sequence:

$$1 \to \pi_1(F(\mathbb{R}^2, m)) \xrightarrow{p_*} \pi_1(B(\mathbb{R}^2, m)) \to \Sigma_n \to 1, \tag{3.5}$$

which is equivalent to the sequence (3.1).

It can be used for proving the canonical presentation of the braid group (3.2) as is done, for example, in the book of J. Birman, [32].

Such considerations were done by R. Fox and L. Neuwirth, [86].

3.3. *Braid groups as automorphism groups of free groups and the word problem*

Another important approach to the braid group is based on the fact that this group may be considered as a subgroup of the automorphism group of a free group.

Let F_n be the free group of rank n with the set of generators $\{x_1, \dots, x_n\}$. Denote by $\operatorname{Aut} F_n$ the automorphism group of F_n.

We have the standard inclusions of the symmetric group Σ_n and the braid group Br_n into $\operatorname{Aut} F_n$. For the braid group it may be described as follows. Let $\bar{\sigma}_i \in \operatorname{Aut} F_n$, $i = 1, 2, \dots, n-1$, be given by the following formula, which describes its action on the generators:

$$\begin{cases} x_i \mapsto x_{i+1}, \\ x_{i+1} \mapsto x_{i+1}^{-1} x_i x_{i+1}, \\ x_j \mapsto x_j, \quad j \neq i, i+1. \end{cases} \tag{3.6}$$

Let us define a map ν of the generators σ_i, $i = 1, \dots, n-1$, of the braid group Br_n to these automorphisms:

$$\nu(\sigma_i) = \bar{\sigma}_i. \tag{3.7}$$

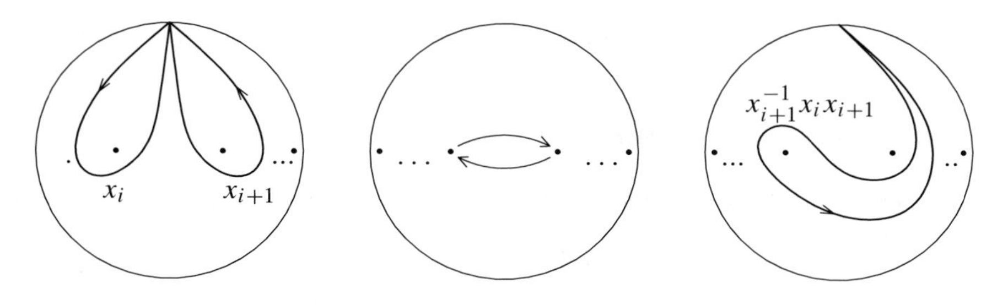

Fig. 6.

THEOREM 3.1. *Formulas* (3.7) *define correctly a homomorphism*

$$\nu : Br_n \to \operatorname{Aut} F_n$$

which is a monomorphism.

Theorem 3.1 gives a solution of the word problem for the braid groups. This was done first by E. Artin, [7].

The free group F_n is a fundamental group of a disc D_n without n points and the generator x_i corresponds to a loop going around the i-th point. The braid group Br_n is the mapping class group of a disc D_n with its boundary fixed, [32], and so it acts on the fundamental group of D_n. This action is described by the formulas (3.6) where x_i corresponds to the canonical loops on D_n which form the generators of the fundamental group. Geometrically this action is depicted in Figure 6.

3.4. *Commutator subgroup and other presentations*

Let us define a homomorphism from the braid group to the integers by taking the sum of exponents of the entries of the generators σ_i in the expression of any element of the group through these canonical generators:

$$\deg : Br_n \to \mathbb{Z}, \quad \deg(b) = \sum_j m_j, \text{ where } b = (\sigma_{i_1})^{m_1} \cdots (\sigma_{i_k})^{m_k}.$$

PROPOSITION 3.1. *The homomorphism*

$$\deg : Br_n \to \mathbb{Z}$$

gives the Abelianization of the braid group and the commutator subgroup Br'_n is characterized by the condition

$$b \in Br'_n \quad \textit{if and only if} \quad \deg(b) = 0.$$

PROOF. Let $a: Br_n \to A$ be a homomorphism to any other Abelian group A, then from the relations (3.2) we have:

$$a(\sigma_i)a(\sigma_{i+1})a(\sigma_i) = a(\sigma_{i+1})a(\sigma_i)a(\sigma_{i+1}).$$

The commutativity of A gives that $a(\sigma_{i+1}) = a(\sigma_i)$. This means that the homomorphism deg is universal. $\square$

Of course, there exist another presentations of the braid group. Let

$$\sigma = \sigma_1\sigma_2\cdots\sigma_{n-1},$$

then the group Br_n is generated by σ_1 and σ because

$$\sigma_{i+1} = \sigma^i\sigma_1\sigma^{-i}, \quad i = 1, \dots, n-2.$$

The relations for the generators σ_1 and σ are the following

$$\begin{cases} \sigma_1\sigma^i\sigma_1\sigma^{-i} = \sigma^i\sigma_1\sigma^{-i}\sigma_1 & \text{for } 2 \leqslant i \leqslant n/2, \\ \sigma^n = (\sigma\sigma_1)^{n-1}. \end{cases} \tag{3.8}$$

This was observed by Artin in the initial paper [7].

An interesting series of presentations was given by V. Sergiescu, [181]. For every planar graph he constructed a presentation of the group Br_n, where n is the number of vertices of the graph, with generators corresponding to edges and relations reflecting the geometry of the graph. Artin's presentation in this context corresponds to the graph consisting of the interval from 1 to n with the natural numbers (from 1 to n) as vertices and with segments between them as edges. For generalizations of braids graph presentations of these type were considered by P. Bellingeri and V. Vershinin, [17,21].

J.S. Birman, K.H. Ko and S.J. Lee, [35], introduced the presentation with the generators a_{ts} with $1 \leqslant s < t \leqslant n$ and relations

$$\begin{cases} a_{ts}a_{rq} = a_{rq}a_{ts} & \text{for } (t-r)(t-q)(s-r)(s-q) > 0, \\ a_{ts}a_{sr} = a_{tr}a_{ts} = a_{sr}a_{tr} & \text{for } 1 \leqslant r < s < t \leqslant n. \end{cases} \tag{3.9}$$

The generators a_{ts} are expressed by the canonical generators σ_i in the following form:

$$a_{ts} = (\sigma_{t-1}\sigma_{t-2}\cdots\sigma_{s+1})\sigma_s\left(\sigma_{s+1}^{-1}\cdots\sigma_{t-2}^{-1}\sigma_{t-1}^{-1}\right) \quad \text{for } 1 \leqslant s < t \leqslant n. \tag{3.10}$$

Geometrically the generators $a_{s,t}$ are depicted in Figure 7.

The set of generators for braid groups were even enlarged in the work of Jean Michel, [152], as follows. Let $|\;|: \Sigma_n \to \mathbb{Z}$ be the length function on the symmetric group with respect to the generators s_i: for $x \in \Sigma_n$, $|x|$ is the smallest natural number k such that x is a product of k elements of the set $\{s_1, \dots, s_{n-1}\}$. It is known ([41], Section 1, Example 13(b)) that two minimal expressions for an element of Σ_n are equivalent by using only the relations (3.2). This implies that the canonical projection $\tau_n: Br_n \to \Sigma_n$ has

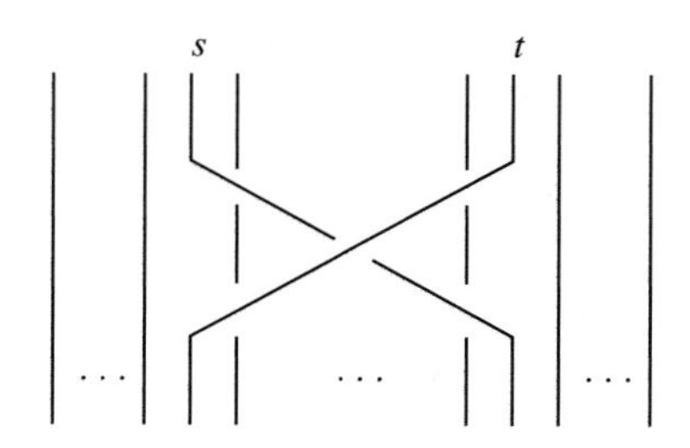

Fig. 7.

a unique set-theoretic section $r: \Sigma_n \to Br_n$ such that $r(s_i) = \sigma_i$ for $i = 1, \ldots, n-1$ and $r(xy) = r(x)r(y)$ whenever $|xy| = |x| + |y|$. Then the group Br_n admits a presentation by generators $\{r(x) \mid x \in \Sigma_n\}$ and relations $r(xy) = r(x)r(y)$ for all $x, y \in \Sigma_n$ such that $|xy| = |x| + |y|$.

3.5. *Presentation of the pure braid group and Markov normal form*

Let $f(y_1, \ldots, y_m)$ be a word with (possibly empty) entries of y_i^ε, where the y_i are some letters and ε may be ± 1. If y_i are elements of a group G then $f(y_1, \ldots, y_m)$ will be considered as the corresponding element of G.

Let us define the elements $s_{i,j}$, $1 \leqslant i < j \leqslant m$, of the braid group Br_m by the formula:

$$s_{i,j} = \sigma_{j-1} \cdots \sigma_{i+1} \sigma_i^2 \sigma_{i+1}^{-1} \cdots \sigma_{j-1}^{-1}.$$

These elements satisfy the following Burau relations ([48,150], see also [134]):

$$\begin{cases} s_{i,j} s_{k,l} = s_{k,l} s_{i,j} & \text{for } i < j < k < l \text{ and } i < k < l < j, \\ s_{i,j} s_{i,k} s_{j,k} = s_{i,k} s_{j,k} s_{i,j} & \text{for } i < j < k, \\ s_{i,k} s_{j,k} s_{i,j} = s_{j,k} s_{i,j} s_{i,k} & \text{for } i < j < k, \\ s_{i,k} s_{j,k} s_{j,l} s_{j,k}^{-1} = s_{j,k} s_{j,l} s_{j,k}^{-1} s_{i,k} & \text{for } i < j < k < l. \end{cases} \tag{3.11}$$

W. Burau and later A.A. Markov proved that the elements $s_{i,j}$ with the relations (3.11) give a presentation of the pure braid group P_m, [150]. The following formula is a consequence of the Burau relations and is also due to A.A. Markov:

$$[s_{i,l}, s_{j,k}^\varepsilon] = f(s_{1,l}, \ldots, s_{l-1,l}), \quad \varepsilon = \pm 1, \ k < l. \tag{3.12}$$

Let us define the elements $\sigma_{k,l}$, $1 \leqslant k \leqslant l \leqslant m$, by the formulas

$$\sigma_{k,k} = e,$$

$$\sigma_{k,l} = \sigma_k^{-1} \cdots \sigma_{l-1}^{-1}.$$

Let P_m^k be the subgroup of P_m generated by the elements $s_{i,j}$ with $k < j$.

THEOREM 3.2 (A.A. Markov).

(i) *Every element of the group Br_m can be uniquely written in the form*

$$f_m(s_{1,m}, \dots, s_{m-1,m}) \cdots f_j(s_{1,j}, \dots, s_{j-1,j}) \\ \cdots f_2(s_{1,2})\sigma_{i_m,m} \cdots \sigma_{i_j,j} \cdots \sigma_{i_2,2}. \tag{3.13}$$

(ii) *The factor group P_m^k / P_m^{k+1} is the free group on free generators $s_{i,k+1}$, $1 \leqslant i \leqslant k$.*

The form (3.13) is called the *Markov normal form*, it also gives the solution of the word problem for the braid groups.

4. Garside normal form, center and conjugacy problem

An essential role in Garside's work [91] is played by the monoid of *positive* braids Br_n^+, that is the monoid which has a presentation with generators σ_i, $i = 1, \dots, n$, and relations (3.2). In other words each element of this monoid can be represented as a word on the elements σ_i, $i = 1, \dots, n$, with no entrances of the σ_i^{-1}. Two positive words A and B in the alphabet $\{\sigma_i,\ i = 1, \dots, n-1\}$ will be said to be *positively equal* if they are equal as elements of Br_n^+. In this case we shall write $A \doteq B$.

First of all Garside proves the following statement.

PROPOSITION 4.1. *In Br_n^+ for $i, k = 1, \dots, n-1$, given $\sigma_i A \doteq \sigma_k B$, it follows that*
if $k = i$, then $A \doteq B$,
if $|k-i| = 1$, then $A \doteq \sigma_k \sigma_i Z$, $B \doteq \sigma_i \sigma_k Z$ for some Z,
if $|k-i| \geqslant 2$, then $A \doteq \sigma_k Z$, $B \doteq \sigma_i Z$ for some Z.
The same is true for the right multiples of σ_i.

COROLLARY 4.1. *If $A \doteq P$, $B \doteq Q$, $AXB \doteq PYQ$, ($L(A) \geqslant 0$, $L(B) \geqslant 0$), then $X \doteq Y$. That is, monoid Br_n^+ is left and right cancellative.*

Garside's *fundamental word* Δ in the braid group Br_{n+1} is defined by the formula:

$$\Delta = \sigma_1 \cdots \sigma_n \sigma_1 \cdots \sigma_{n-1} \cdots \sigma_1 \sigma_2 \sigma_1.$$

If we use Garside's notation $\Pi_t \equiv \sigma_1 \cdots \sigma_t$, then $\Delta \equiv \Pi_{n-1} \cdots \Pi_1$.

For a positive word W in σ_i, $i = 1, \dots, n$, we say that Δ is a *factor* of W or simply W contains Δ, if $W \doteq A\Delta B$ with A and B being arbitrary positive words, probably empty. If W does not contain Δ we shall say W is *prime to* Δ.

Garside's transformation of words $\mathcal{R}$ is defined by the formula

$$\mathcal{R}(\sigma_i) \equiv \sigma_{n-i}.$$

This gives the automorphism of Br_n and the positive braid monoid Br_n^+.

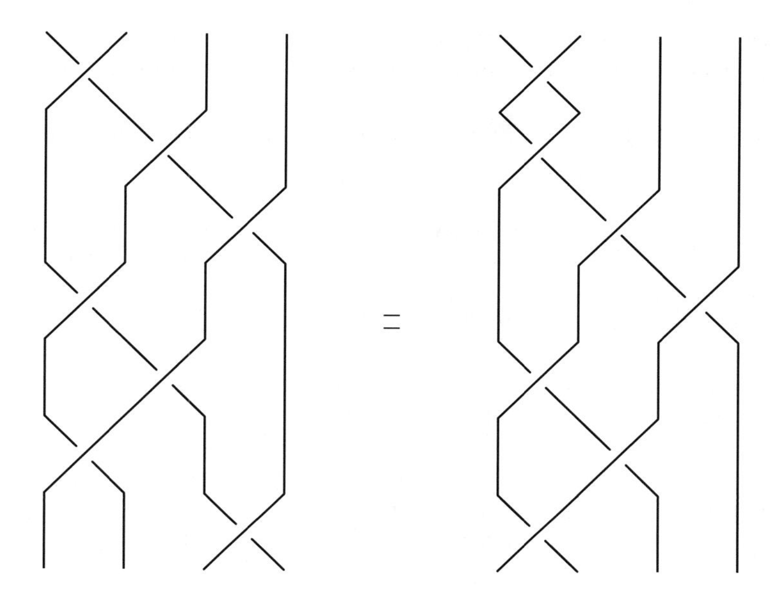

Fig. 8.

PROPOSITION 4.2. *In Br_n*

$$\sigma_i \Delta \doteq \Delta \mathcal{R}(\sigma_i).$$

Geometrically this commutation is shown on Figure 8 ($\Delta\sigma_3 = \sigma_1\Delta$).

PROPOSITION 4.3. *If W is an arbitrary positive word in Br_n^+ such that either*

$$W \doteq \sigma_1 A_1 \doteq \sigma_2 A_2 \doteq \cdots \doteq \sigma_{n-1} A_{n-1},$$

or

$$W \doteq B_1\sigma_1 \doteq B_2\sigma_2 \doteq \cdots \doteq B_{n-1}\sigma_{n-1},$$

then $W \doteq \Delta Z$ for some Z.

PROPOSITION 4.4. *The canonical homomorphism*

$$Br_n^+ \to Br_n$$

is a monomorphism.

Among positive words on the alphabet $\{\sigma_1 \cdots \sigma_n\}$ let us introduce a lexicographical ordering with the condition that $\sigma_1 < \sigma_2 < \cdots < \sigma_n$. For a positive word W the *base* of W is the smallest positive word which is positively equal to W. The base is uniquely determined. If a positive word A is prime to Δ, then for the base of A the notation $\bar{A}$ will be used.

THEOREM 4.1 (F.A. Garside). *Every word W in Br_{n+1} can be uniquely written in the form $\Delta^m \bar{A}$, where m is an integer.*

The form of a word W established in this theorem we call the *Garside left normal form* and the index m we call the *power* of W. The same way the *Garside right normal form* is defined and the corresponding variant of Theorem 4.1 is true. The Garside normal form also gives a solution to the word problem in the braid group.

THEOREM 4.2 (F.A. Garside). *The necessary and sufficient condition that two words in Br_{n+1} are equal is that their Garside normal forms are identical.*

Garside normal form for the braid groups was precised in the subsequent works of S.I. Adyan, [1], W. Thurston, [78], E. El-Rifai and H.R. Morton, [76]. Namely, there was introduced the *left-greedy form* (in the terminology of W. Thurston, [78])

$$\Delta^t A_1 \cdots A_k,$$

where A_i are the successive possible longest *fragments of the word* Δ (in the terminology of S.I. Adyan, [1]) or *positive permutation braids* (in the terminology of E. El-Rifai and H.R. Morton, [76]). Certainly, the same way the *right-greedy form* is defined. With the help of this form it was proved that the braid group is biautomatic.

The center of the braid group was fist found by W.-L. Chow, [53]. Namely, as follows from the presentation of braid groups with two generators σ_1 and σ and relations (3.8) given in Section 3.1 the element σ^n commutes with $\sigma\sigma_1$ and so with σ_1. Chow proved that it generates the center. Garside normal form gives an elegant proof of the following theorem.

THEOREM 4.3.
(i) *When $n = 1$, the center of the group Br_{n+1} is generated by Δ.*
(ii) *When $n > 1$ the center of the group Br_{n+1} is generated by Δ^2.*

Let α be a positive word such that $\Delta \doteq \alpha X$, where X is a positive word, possibly empty. For any word W in Br_{n+1}, the word $\alpha^{-1} W \alpha$, reduced to Garside normal form is called an *α-transformation of W*.

For any word W in Br_{n+1} with the Garside normal form $\Delta^m \bar{A} \equiv W_1$ consider the following chains of α-transformations: take all the α-transformations of W_1 and let those which are of power $\geqslant m$ and which are distinct from each other be $W_2, W_3, \ldots, W_t$. Now repeat the process for each of the words $W_2, W_3, \ldots, W_t$ in turn, denoting successively by W_{t+1}, $W_{t+2}, \ldots,$ any new words occurring, the condition being always that each new word must be of power $\geqslant m$. Continue to repeat the process for every new distinct word arising, as the sequence $W_1, W_2, W_{t+2}, \ldots,$ expands.

PROPOSITION 4.5. *The set* $W_1, W_2, W_{t+2}, \ldots,$ *is finite.*

Suppose that in the set $W_1, W_2, W_{t+2}, \ldots,$ the highest power reached is s and that the words of power s form the subset $V_1, V_2, \ldots$. Then this set $V_1, V_2, \ldots$ is called the *summit set* of W.

THEOREM 4.4 (F.A. Garside). *Two elements A and B of the group* Br_{n+1} *are conjugate if and only if their summit sets are identical.*

J.S. Birman, K.H. Ko, and S.J. Lee considered the word $\delta = a_{n(n-1)} \cdots a_{32}a_{21} = \sigma_{n-1} \cdots \sigma_2\sigma_1$, as a fundamental in their system of generators and proved that every element in Br_n has a representative $W = \delta^j A_1 A_2 \cdots A_k$ with positive A_i in a unique way in some sense. Based on this form they gave an algorithm for the word problem in B_n which runs in time (nm^2) for a given word of length m.

5. Ordering of braids

A group G is said *totally* (or *linearly*) *left* (correspondingly *right*) *ordered* if it has a total order $<$ invariant by left (right) multiplication, i.e. if $a < b$, then $ca < cb$ for any $c \in G$. If this order is also invariant by right (left) multiplication, then the group G is called *ordered*.

For any left ordered group G denote by P the set of positive elements $\{x \in G\colon x > 1\}$, then the set of negative elements is defined by the formula: $P^{-1} = \{x \in G\colon x \in P\}$. The total character of an order on G is expressed by the partition

$$G = P \amalg \{1\} \amalg P^{-1}.$$

The invariance of multiplication is expressed by the inclusion $P^2 \subset P$, where P^2 is formed by products of couples of elements of P. Conversely, if there exists a subset P of a group G with the properties:

$$G = P \amalg \{1\} \amalg P^{-1}, \quad P^2 \subset P,$$

then G is left ordered by the order defined by: $x < y$ if and only if $x^{-1}y \in P$. A group G then is ordered if and only if $xPx^{-1} \subset P$ for all $x \in G$.

Let $i \in \{1, \ldots, n\}$ and considered a word w on the alphabet $\{\sigma_1, \ldots, \sigma_n\}$ expressed in the form

$$w_0\sigma_i w_1\sigma_i \cdots \sigma_i w_r,$$

where the subwords $w_0, \ldots, w_r$ are the words on the letters $\sigma_j^{\pm 1}$ with $j > i$. Then such a word is called σ_i*-positive*. This means that all entries of $\sigma_i^{\pm 1}$ in the word w with i minimal must be positive. If all such entries are negative then a word w is called σ_i*-negative*. A braid of Br_{n+1} is called σ_i*-positive* (σ_i*-negative*) if there exists its expression as a word on the standard generators which is σ_i*-positive* (σ_i*-negative*). A braid is called σ*-positive* (σ*-negative*) if there exists a number i, such that it is σ_i*-positive* (σ_i*-negative*).

THEOREM 5.1 (P. Dehornoy). *Every braid in Br_{n+1} different from* 1 *is either* σ*-positive or* σ*-negative.*

COROLLARY 5.1. *For all n the braid group Br_{n+1} is left ordered.*

6. Representations

6.1. *Burau representation*

Let us map the generators of the braid group Br_n to the following elements of the group $GL_n\mathbb{Z}[t, t^{-1}]$

$$\sigma_i \mapsto \begin{pmatrix} E_{i-1} & 0 & 0 & 0 \\ 0 & 1-t & t & 0 \\ 0 & 1 & 0 & 0 \\ 0 & 0 & 0 & E_{n-i-1} \end{pmatrix}, \tag{6.1}$$

where E_i is the unit $i \times i$ matrix. The formula (6.1) gives a well-defined representation of the braid group in $GL_n\mathbb{Z}[t, t^{-1}]$:

$$r : Br_n \to GL_n\mathbb{Z}[t, t^{-1}],$$

which is called *Burau representation*, [50].

THEOREM 6.1. *Burau representation is faithful for $n = 3$.*

THEOREM 6.2 (J.A. Moody, D.D. Long and M. Paton, S. Bigelow). *The Burau representation is not faithful for $n \geqslant 5$.*

The case $n = 4$ remains open.

6.2. *Lawrence–Krammer representation*

Consider the ring $K = \mathbb{Z}[q^{\pm 1}, t^{\pm 1}]$ of Laurent polynomials in two variables q, t, and the free K-module

$$V = \bigoplus_{1 \leqslant i < j \leqslant n} K x_{i,j}.$$

For $k \in \{1, 2, \ldots, n-1\}$, define the action of the braid generators σ_k on the specified basis of V by the formula:

$$\sigma_k(x_{i,j}) = \begin{cases} x_{i,j}, & k < i-1 \text{ or } j < k; \\ x_{i-1,j} + (1-q)x_{i,j}, & k = i-1; \\ tq(q-1)x_{i,i+1} + qx_{i+1,j}, & k = i < j-1; \\ tq^2 x_{i,j}, & k = i = j-1; \\ x_{i,j} + tq^{k-i}(q-1)^2 x_{k,k+1}, & i < k < j-1; \\ x_{i,j-1} + tq^{j-i}(q-1)x_{j-1,j}, & k = j-1; \\ (1-q)x_{i,j} + qx_{i,j+1}, & k = j. \end{cases} \tag{6.2}$$

Direct computation shows that this defines a representation

$$\rho_n : Br_n \to GL(V),$$

which was firstly defined by R. Lawrence, [133], in topological terms and in the explicit form (6.2) by D. Krammer, [129].

THEOREM 6.3 (S. Bigelow, [29], D. Krammer, [129]). *The representation*

$$\rho_n : Br_n \to GL(V)$$

is faithful for all $n \geqslant 1$.

REMARK 6.1. Actually, S. Bigelow, [29], proved this theorem for the representation ρ_n characterized in homological terms and D. Krammer, [129], proved the following. Let $K = \mathbb{R}[t^{\pm 1}]$, $q \in \mathbb{R}$, and $0 < q < 1$. Then the representation ρ_n defined by (6.2) is faithful for all $n \geqslant 1$. This result implies Theorem 6.3: if a representation over $\mathbb{Z}[q^{\pm 1}, t^{\pm 1}]$ becomes faithful after assigning a real value to q, then it is faithful itself.

M.G. Zinno, [203], established a connection between the Birman–Murakami–Wenzl algebra, [40,155], and the Lawrence–Krammer representation. Namely, he proved that the Lawrence–Krammer representation is identical to the irreducible representations of the Birman–Murakami–Wenzl algebra parametrized by Young diagrams of shapes $(n-2)$ and (1^{n-2}). This means that the Young diagram in the case considered consists of one row (respectively of one column) only, with $n-2$ boxes. It follows that Lawrence–Krammer representation is irreducible.

7. Generalizations of braids

7.1. *Configuration spaces of manifolds*

The notion of a configuration space as in Section 3.2 can be naturally generalized for a configuration space of a manifold as follows. Let Y be a connected topological manifold and let W be a finite group acting on Y. A point $y \in Y$ is called *regular* if its stabilizer $\{w \in W\colon\ wy = y\}$ is trivial, i.e. consists only of the unit of the group W. The set $\widetilde{Y}$

of all regular points is open. Suppose that it is connected and nonempty. The subspace $ORB(Y, W)$ of the space of all orbits $Orb(Y, W)$ consisting of the orbits of all regular points is called the *space of regular orbits*. There is a free action of W on $\widetilde{Y}$ and the projection $p : \widetilde{Y} \to \widetilde{Y}/W = ORB(Y, W)$ defines a covering. Let us consider the initial segment of the long exact sequence of this covering:

$$1 \to \pi_1(\widetilde{Y}, y_0) \xrightarrow{p_*} \pi_1(ORB(Y, W), p(y_0)) \to W \to 1. \tag{7.1}$$

The fundamental group $\pi_1(ORB(Y, W), p(y_0))$ of the space of regular orbits is called the *braid group of the action of W on Y* and is denoted by $Br(Y, W)$. The fundamental group $\pi_1(\widetilde{Y}, y_0)$ is called the *pure braid group of the action of W on Y* and is denoted by $P(Y, W)$. The spaces $\widetilde{Y}$ and $ORB(Y, W)$ are path connected, so the pair of these groups is defined uniquely up to isomorphism and we may omit mentioning the base point y_0 in the notations.

For any space Y the symmetric group Σ_m acts on the Cartesian power Y^m of the space Y by the formulas (3.4). We denote by $F(Y, m)$ the space of m-tuples of pairwise different points in Y:

$$F(Y, m) = \{(p_1, \dots, p_m) \in Y^m \colon\ p_i \neq p_j \text{ for } i \neq j\}.$$

This is the space of regular points of this action. In the case when Y is a connected topological manifold M without boundary and $\dim M \geqslant 2$, the space of regular orbits $ORB(M^m, \Sigma_m)$ is open, connected and nonempty. We call $ORB(M^m, \Sigma_m)$ the *configuration space of the manifold M* and denote by $B(M, m)$. The braid group $Br(M^m, \Sigma_m)$ is called the *braid group on m strings of the manifold M* and is denoted by $Br(m, M)$. Analogously, we call the group $P(M^m, \Sigma_m)$ the *pure braid group on m strings of the manifold M* and denote it by $P(m, M)$. These definitions of braid groups were given by R. Fox and L. Neuwirth, [86].

7.2. *Artin–Brieskorn braid groups*

The braid groups are included in the series of so-called generalized braid groups (this was their name in the work of E. Brieskorn of 1971, [42]), or Artin groups (as they were called by E. Brieskorn and K. Saito in the paper of 1972, [45]). They were defined by E. Brieskorn, [42], so we call them Artin–Brieskorn groups.

Let V be a finite-dimensional real vector space ($\dim V = n$) with Euclidean structure. Let W be a finite subgroup of $GL(V)$ generated by reflections. Let $\mathcal{M}$ be the set of hyperplanes such that W is generated by orthogonal reflections with respect to $M \in \mathcal{M}$. We suppose that for every $w \in W$ and every hyperplane $M \in \mathcal{M}$ the hyperplane $w(M)$ belongs to $\mathcal{M}$.

The group W is generated by the reflections $w_i = w_i(M_i)$, $i \in I$, satisfying only the following relations

$$(w_i w_j)^{m_{i,j}} = e, \quad i, j \in I,$$

where the natural numbers $m_{i,j} = m_{j,i}$ form the *Coxeter matrix* of W from which the *Coxeter graph* $\Gamma(W)$ of W is constructed, [41]. We use the following notation of P. Deligne, [74]: $\text{prod}(m; x, y)$ denotes the product $xyxy\cdots$ (m factors). The *generalized braid group* (or *Artin–Brieskorn group*) $Br(W)$ *of* W, [42,74], is defined as the group with generators $\{s_i, i \in I\}$ and relations:

$$\text{prod}(m_{i,j}; s_i, s_j) = \text{prod}(m_{j,i}; s_j, s_i).$$

From this we obtain the presentation of the group W by adding the relations:

$$s_i^2 = e; \quad i \in I.$$

We will see in Theorem 7.1 that this definition of the generalized braid group agrees with our general definition of a braid group of an action of a group W (Section 7.1). We denote by τ_W the canonical homomorphism from $Br(W)$ to W. The classical braids on k strings Br_k are obtained by this construction if W is the symmetric group on $k+1$ symbols. In this case $m_{i,i+1} = 3$, and $m_{i,j} = 2$ if $j \neq i, i+1$.

The classification of *irreducible* (with connected Coxeter graph) Coxeter groups is well known (see, for example, Theorem 1, Chapter VI, §4 of [41]). It consists of the three infinite series: A, C (which is also denoted by B because in the corresponding classification of simple Lie algebras two different series B and C have this group as their Weyl group) and D as well as the exceptional groups E_6, E_7, E_8, F_4, G_2, H_3, H_4 and $I_2(p)$.

Now let us consider the complexification V_C of the space V and the complexification M_C of $M \in \mathcal{M}$. Let $Y_W = V_C - \bigcup_{M\in\mathcal{M}} M_C$. The group W acts freely on Y_W. Let $X_W = Y_W/W$ then Y_W is a covering over X_W corresponding to the group W. Let $y_0 \in A_0$ be a point in some chamber A_0 and let x_0 stand for its image in X_W. We are in the situation described in Section 7.1 in the definition of the braid group of the action of the group W. This braid group is defined as the fundamental group of the space of regular orbits of the action of W. In our case $ORB(V_C, W) = X_W$. So, the generalized braid group is $\pi_1(X_W, x_0)$. For each $j \in I$, let ℓ'_j be the homotopy class of paths in Y_W starting from y_0 and ending in $w_j(y_0)$ which contains a polygon line with successive vertices: $y_0, y_0 + iy_0, w_j(y_0) + iy_0, w_j(y_0)$. The image ℓ_j of the class ℓ'_j in X_W is a loop with base point x_0.

THEOREM 7.1. *The fundamental group* $\pi_1(X_W, x_0)$ *is generated by the elements* ℓ_j *satisfying the following relations*:

$$\text{prod}(m_{j,k}; \ell_j, \ell_k) = \text{prod}(m_{k,j}; \ell_k, \ell_j).$$

This theorem was proved by E. Brieskorn, [43].

The word problem and the conjugacy problem for Artin–Brieskorn groups were solved by E. Brieskorn and K. Saito, [45], and P. Deligne, [74]. The biautomatic structure of these groups was established by R. Charney, [51].

In the case when V is complex finite-dimensional space and W is a finite subgroup of $GL(V)$ generated by *pseudo-reflections* the corresponding braid groups were studied by M. Broué, G. Malle and R. Rouquier, [47], and also by D. Bessis and J. Michel, [27].

7.3. *Braid groups of surfaces*

The braid groups of a sphere $Br_n(S^2)$ also have simple geometric interpretation as a group of isotopy classes of braids lying in a layer between two concentric spheres. It has the presentation with generators $\delta_i,\ i = 1, \dots, n-1$, and relations:

$$\begin{cases} \delta_i\delta_j = \delta_j\delta_i, & \text{if } |i-j| > 1, \\ \delta_i\delta_{i+1}\delta_i = \delta_{i+1}\delta_i\delta_{i+1}, \\ \delta_1\delta_2\cdots\delta_{n-2}\delta_{n-1}^2\delta_{n-2}\cdots\delta_2\delta_1 = 1. \end{cases} \tag{7.2}$$

This presentation was found by O. Zariski, [200], in 1936 and then rediscovered by E. Fadell and J. Van Buskirk, [81], in 1961.

Presentations of braid groups of all closed surfaces were obtained by G.P. Scott, [180], and others.

7.4. *Braid groups in handlebodies*

The subgroup $Br_{1,n+1}$ of the braid group Br_{n+1} consisting of braids with the first string fixed can be interpreted also as the braid group in a solid torus. Here we study braids in a handlebody of the arbitrary genus g.

Let H_g be a handlebody of genus g. The braid group Br_n^g on n strings in H_g was first considered by A.B. Sossinsky, [182]. Let Q_g denote a subset of the complex plain $\mathbb{C}$, consisting of g different points, $Q_g = \{z_1^0, \dots, z_g^0\}$, say, $z_j^0 = j$. The interior of the handlebody H_g may be interpreted as the direct product of the complex plane $\mathbb{C}$ without g points: $\mathbb{C} \setminus Q_g$, and an open interval, for example, $(-1, 1)$:

$$\dot{H}_g = (\mathbb{C} \setminus Q_g) \times (-1, 1).$$

The space $F(\mathbb{C} \setminus Q_g, n)$ can be interpreted as the complement of the arrangement of hyperplanes in $\mathbb{C}^{g+n}$ given by the formulas:

$$H_{j,k}\colon\ z_j - z_k = 0 \quad \text{for all } j, k;$$

$$H_j^i\colon\ z_j = z_i^0 \quad \text{for } i = 1, \dots, g;\ j = 1, \dots, n.$$

The braids in Br_n^g are considered as lying between the planes with coordinates $z = 0$ and $z = 1$ and connecting the points $((g+1, 0), \dots, (g+n, 0))$. So Br_n^g can be considered as a subgroup of the classical braid group Br_{g+n} on $g+n$ strings such that the braids from Br_n^g leave the first g strings unbraided. In this subsection we denote by $\bar{\sigma}_j$ the standard generators of the group Br_{g+n}. Let τ_k, $k = 1, 2, \dots, g$, be the following braids:

$$\tau_k = \bar{\sigma}_g\bar{\sigma}_{g-1}\cdots\bar{\sigma}_{k+1}\bar{\sigma}_k^2\bar{\sigma}_{k+1}^{-1}\cdots\bar{\sigma}_{g-1}^{-1}\bar{\sigma}_g^{-1}.$$

Such a braid is depicted in Figure 9. The elements τ_k, $k = 1, 2, \dots, g$, generate a free subgroup F_g in the braid group Br_{g+n}. It follows for example from the Markov normal form that the elements τ_k, $k = 1, 2, \dots, g$, together with the standard generators

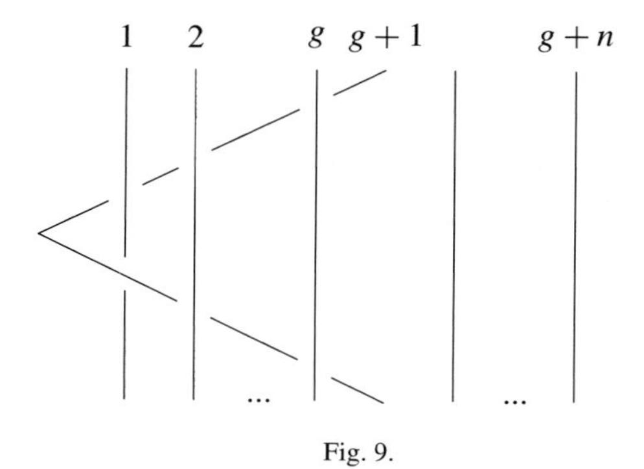

Fig. 9.

$\bar{\sigma}_{g+1}, \dots, \bar{\sigma}_{g+n-1}$ generate the group Br_n^g. So, the braid group in the handlebody Br_n^g can be considered as a subgroup of Br_{g+n}, generated by two subgroups: F_g and Br_n. Denote by $\sigma_1, \dots, \sigma_{n-1}$ the standard generators of Br_n considered as the elements of Br_n^g, $\sigma_i = \bar{\sigma}_{g+i}$, $i = 1, \dots, n-1$. So we have the presentation of Br_n^g with the generators τ_k and σ_i and relations, [182,189,192]:

$$\begin{cases} \sigma_i \sigma_j = \sigma_j \sigma_i & \text{if } |i-j| > 1, \\ \sigma_i \sigma_{i+1} \sigma_i = \sigma_{i+1} \sigma_i \sigma_{i+1}, & \\ \tau_k \sigma_i = \sigma_i \tau_k & \text{if } k \geqslant 1,\ i \geqslant 2, \\ \tau_k \sigma_1 \tau_k \sigma_1 = \sigma_1 \tau_k \sigma_1 \tau_k, & k = 1, 2, \dots, g, \\ \tau_k \sigma_1^{-1} \tau_{k+l} \sigma_1 = \sigma_1^{-1} \tau_{k+l} \sigma_1 \tau_k, & k = 1, 2, \dots, g-1;\ l = 1, 2, \dots, g-k. \end{cases} \tag{7.3}$$

The relation of the fourth type in (7.3) is the relation of the braid group of type B (C). The relations of the fifth type in (7.3) describe the interaction between the generators of the free group and their closest neighbor σ_1. Geometrically this is seen in Figure 10. If we introduce new generators θ_k, $k = 1, 2, \dots, g-1$; by the formulas:

$$\theta_k = \sigma_1^{-1} \tau_k \sigma_1$$

we obtain the "positive" presentation of the group B_n^g with generators of the types σ_i, τ_k, θ_k and relations:

$$\begin{cases} \sigma_i \sigma_j = \sigma_j \sigma_i & \text{if } |i-j| > 1, \\ \sigma_i \sigma_{i+1} \sigma_i = \sigma_{i+1} \sigma_i \sigma_{i+1}, & \\ \tau_k \sigma_i = \sigma_i \tau_k & \text{if } k \geqslant 1,\ i \geqslant 2, \\ \tau_k \sigma_1 \tau_k \sigma_1 = \sigma_1 \tau_k \sigma_1 \tau_k, & k = 1, 2, \dots, g, \\ \tau_k \theta_{k+l} = \theta_{k+l} \tau_k, & k = 1, 2, \dots, g-1;\ l = 1, 2, \dots, g-k, \\ \sigma_1 \theta_k = \tau_k \sigma_1, & k = 1, 2, \dots, g-1. \end{cases} \tag{7.4}$$

There is an analog of Markov Theorem 3.2 for the group Br_n^g, [189,192].

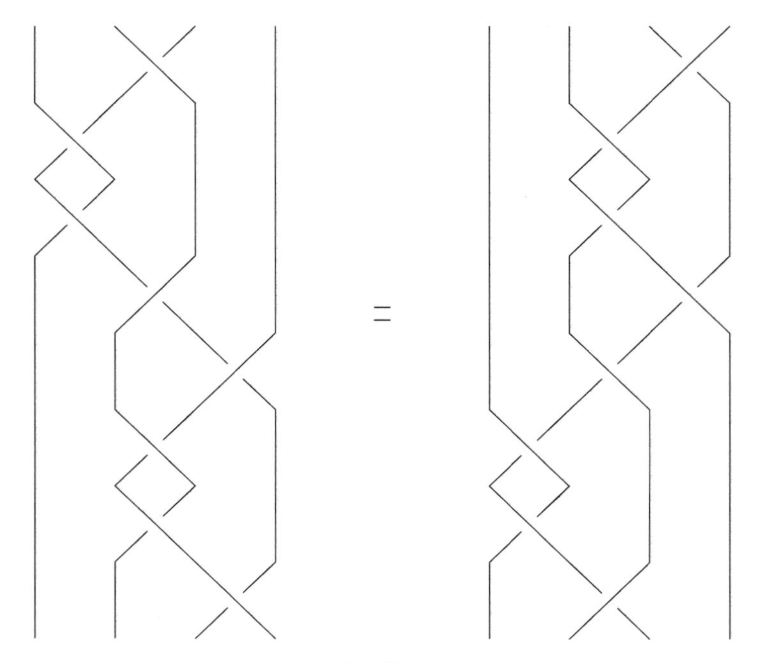

Fig. 10.

7.5. *Braids with singularities*

Let BP_n be the subgroup of Aut F_n, generated by both sets of the automorphisms σ_i of (3.6) and ξ_i of the following form:

$$\begin{cases} x_i \mapsto x_{i+1}, \\ x_{i+1} \mapsto x_i, \\ x_j \mapsto x_j, \quad j \neq i, i+1. \end{cases}$$

This is the *braid-permutation group*. R. Fenn, R. Rimányi and C. Rourke proved, [83,84], that this group is given by the set of generators: $\{\xi_i, \sigma_i, \ i = 1, 2, \dots, n-1\}$ and relations:

$$\begin{cases} \xi_i^2 = 1, \\ \xi_i \xi_j = \xi_j \xi_i, \quad \text{if } |i-j| > 1, \\ \xi_i \xi_{i+1} \xi_i = \xi_{i+1} \xi_i \xi_{i+1}. \end{cases}$$

The symmetric group relations

$$\begin{cases} \sigma_i \sigma_j = \sigma_j \sigma_i, \quad \text{if } |i-j| > 1, \\ \sigma_i \sigma_{i+1} \sigma_i = \sigma_{i+1} \sigma_i \sigma_{i+1}. \end{cases}$$

The braid group relations

$$\begin{cases} \sigma_i \xi_j = \xi_j \sigma_i, & \text{if } |i-j| > 1, \\ \xi_i \xi_{i+1} \sigma_i = \sigma_{i+1} \xi_i \xi_{i+1}, \\ \sigma_i \sigma_{i+1} \xi_i = \xi_{i+1} \sigma_i \sigma_{i+1}. \end{cases}$$

The mixed relations

R. Fenn, R. Rimányi and C. Rourke also gave a geometric interpretation of BP_n as a group of *welded braids*. First they define a *welded braid diagram* on n strings as a collection of n monotone arcs starting from n points at a horizontal line of a plane (the top of the diagram) and going down to n points at another horizontal line (the bottom of the diagram). The diagrams may have crossings of two types: (1) the same as usual braids as for example on Figure 2 or (2) welds as depicted in Figure 11.

Composition of welded braid diagrams on n strings is defined by stacking one diagram under the other. The diagram with no crossings or welds is the identity with respect to composition. So the set of welded braid diagrams on n strings forms a semigroup which is denoted by WD_n.

R. Fenn, R. Rimányi and C. Rourke defined the allowable moves on welded braid diagrams. They consist of the usual Reidemeister moves (Figure 4) and the specific moves depicted in Figures 12, 13, 14. The automorphisms of F_n which lie in BP_n can be characterized as follows. Let $\pi \in \Sigma_n$ be a permutation and w_i, $i = 1, 2 \ldots, n$, be words in F_n. Then the mapping

$$x_i \mapsto w_i^{-1} x_{\pi(i)} w_i$$

determines an injective endomorphism of F_n. If it is also surjective, we call it an automorphism of *permutation-conjugacy type*. The automorphisms of this type comprise a subgroup of $\operatorname{Aut} F_n$ which is precisely BP_n.

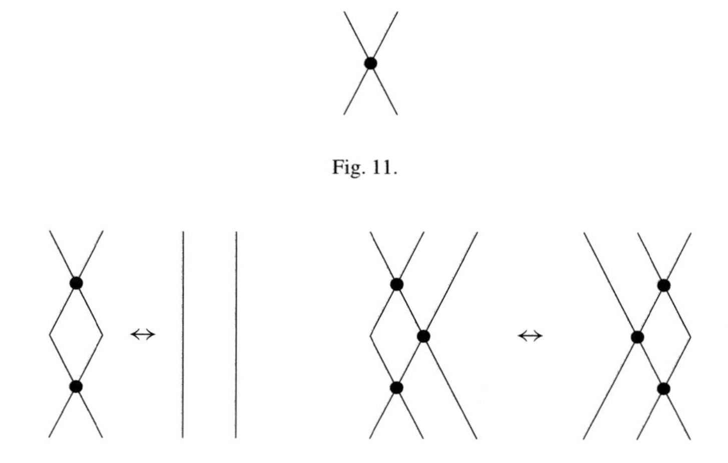

Fig. 11.

Fig. 12.

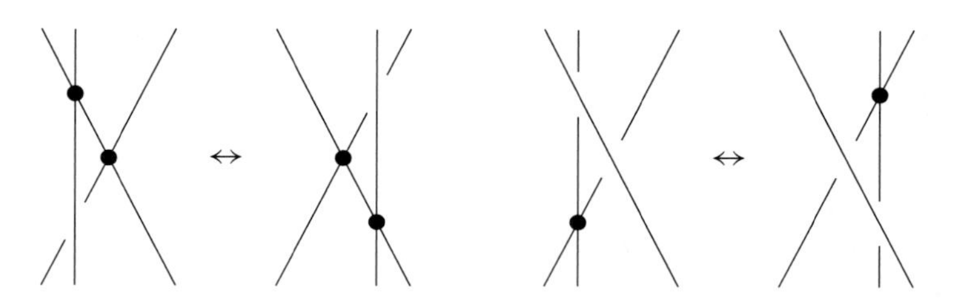

Fig. 13.

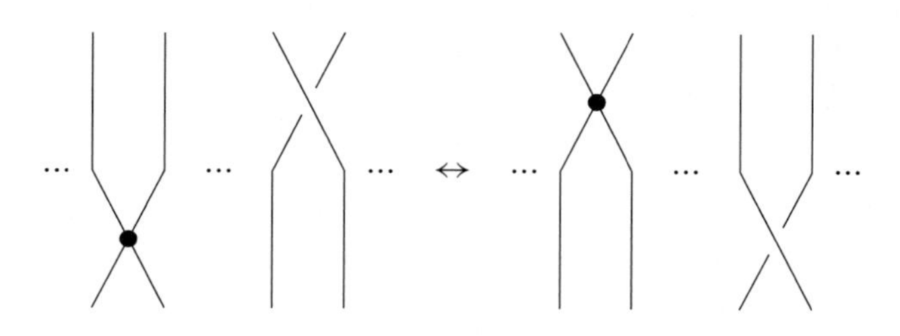

Fig. 14.

The *Baez–Birman monoid* SB_n or *singular braid monoid*, [10,33], is defined as the monoid with generators $g_i, g_i^{-1}, a_i, i = 1, \dots, n-1$, and relations

$$\begin{aligned}
&g_i g_j = g_j g_i \quad \text{if } |i-j| > 1, \\
&a_i a_j = a_j a_i \quad \text{if } |i-j| > 1, \\
&a_i g_j = g_j a_i \quad \text{if } |i-j| \neq 1, \\
&g_i g_{i+1} g_i = g_{i+1} g_i g_{i+1}, \\
&g_i g_{i+1} a_i = a_{i+1} g_i g_{i+1}, \\
&g_{i+1} g_i a_{i+1} = a_i g_{i+1} g_i, \\
&g_i g_i^{-1} = g_i^{-1} g_i = 1.
\end{aligned}$$

In these pictures g_i corresponds to canonical generator of the braid group and a_i represents an intersection of the i-th and $(i+1)$-st strand as in Figure 15. A more detailed geometric interpretation of the Baez–Birman monoid can be found in the article of J. Birman, [33]. R. Fenn, E. Keyman and C. Rourke proved, [82], that the Baez–Birman monoid embeds in a group SG_n which they called the *singular braid group*:

$$SB_n \to SG_n.$$

So, in SG_n the elements a_i become invertible and all relations of SB_n remain true.

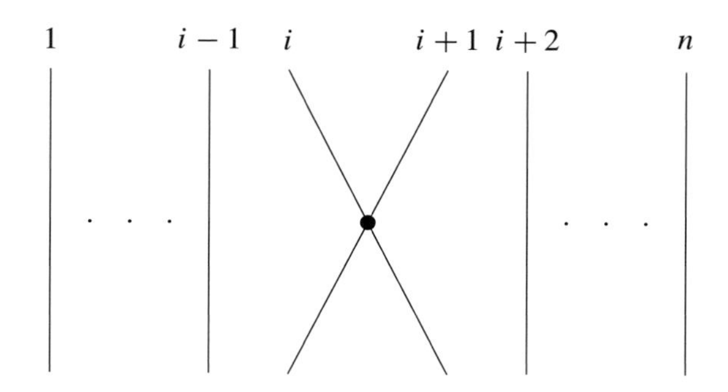

Fig. 15.

The analogue of the Birman–Ko–Lee presentation for the singular braid monoid was obtained in [198]. Namely, it was proved that the monoid SB_n has a presentation with generators a_{ts}, a_{ts}^{-1} for $1 \leqslant s < t \leqslant n$ and b_{qp} for $1 \leqslant p < q \leqslant n$ and relations

$$\begin{cases} a_{ts}a_{rq} = a_{rq}a_{ts} & \text{for } (t-r)(t-q)(s-r)(s-q) > 0, \\ a_{ts}a_{sr} = a_{tr}a_{ts} = a_{sr}a_{tr} & \text{for } 1 \leqslant r < s < t \leqslant n, \\ a_{ts}a_{ts}^{-1} = a_{ts}^{-1}a_{ts} = 1 & \text{for } 1 \leqslant s < t \leqslant n, \\ a_{ts}b_{rq} = b_{rq}a_{ts} & \text{for } (t-r)(t-q)(s-r)(s-q) > 0, \\ a_{ts}b_{ts} = b_{ts}a_{ts} & \text{for } 1 \leqslant s < t \leqslant n, \\ a_{ts}b_{sr} = b_{tr}a_{ts} & \text{for } 1 \leqslant r < s < t \leqslant n, \\ a_{sr}b_{tr} = b_{ts}a_{sr} & \text{for } 1 \leqslant r < s < t \leqslant n, \\ a_{tr}b_{ts} = b_{sr}a_{tr} & \text{for } 1 \leqslant r < s < t \leqslant n, \\ b_{ts}b_{rq} = b_{rq}b_{ts} & \text{for } (t-r)(t-q)(s-r)(s-q) > 0. \end{cases} \tag{7.5}$$

The elements a_{ts} are defined the same way as in (3.10) and the elements b_{qp} for $1 \leqslant p < q \leqslant n$ are defined by

$$b_{qp} = (\sigma_{q-1}\sigma_{q-2}\cdots\sigma_{p+1})x_p\left(\sigma_{p+1}^{-1}\cdots\sigma_{q-2}^{-1}\sigma_{q-1}^{-1}\right) \quad \text{for } 1 \leqslant p < q \leqslant n. \tag{7.6}$$

Geometrically the generators $b_{s,t}$ are depicted in Figure 16.

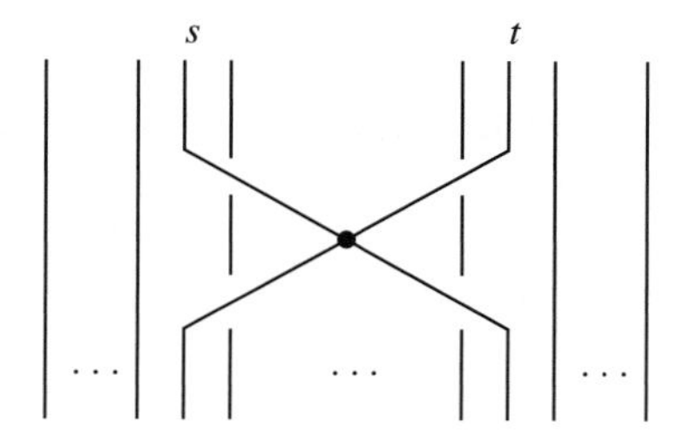

Fig. 16.

8. Homological properties

8.1. *Configuration spaces and $K(\pi, 1)$-spaces*

Let $(q_i)_{i\in\mathbb{N}}$ be a fixed sequence of distinct points in the manifold M and put $Q_m = \{q_1, \ldots, q_m\}$. We use

$$Q_{m,l} = (q_{l+1}, \ldots, q_{l+m}) \in F(M \setminus Q_l, m)$$

as the standard base point of the space $F(M \setminus Q_l, m)$. If $k < m$ we define the projection

$$\mathrm{proj}: F(M \setminus Q_l, m) \to F(M \setminus Q_l, k)$$

by the formula: $\mathrm{proj}(p_1, \ldots, p_m) = (p_1, \ldots, p_k)$. The following theorems were proved by E. Fadell and L. Neuwirth, [80].

THEOREM 8.1. *The triple* $\mathrm{proj}: F(M \setminus Q_l, m) \to F(M \setminus Q_l, k)$ *is a locally trivial fiber bundle with fiber* $\mathrm{proj}^{-1} Q_{k,l}$ *homeomorphic to* $F(M \setminus Q_{k+l}, m-k)$.

Consideration of the sequence of fibrations

$$\begin{aligned}
&F(M \setminus Q_{m-1}, 1) \to F(M \setminus Q_{m-2}, 2) \to M \setminus Q_{m-2},\\
&F(M \setminus Q_{m-2}, 2) \to F(M \setminus Q_{m-3}, 3) \to M \setminus Q_{m-3},\\
&\qquad\qquad\qquad\qquad\ldots\\
&F(M \setminus Q_1, m-1) \to F(M, m) \to M
\end{aligned}$$

leads to the following theorem.

THEOREM 8.2. *For any manifold M*

$$\pi_i\big(F(M \setminus Q_1, m-1)\big) = \bigoplus_{k=1}^{m-1} \pi_i(M \setminus Q_k)$$

for $i \geqslant 2$. If $\mathrm{proj}: F(M, m) \to M$ *admits a section then*

$$\mathrm{proj}_i\, \pi_i\big(F(M, m)\big) = \bigoplus_{k=0}^{m-1} \pi_i(M \setminus Q_k), \quad i \geqslant 2.$$

COROLLARY 8.1. *If M is Euclidean r-space, then*

$$\pi_i\big(F(M, m)\big) = \bigoplus_{k=0}^{m-1} \pi_i\big(\underbrace{S^{r-1} \vee \cdots \vee S^{r-1}}_{k}\big), \quad i \geqslant 2.$$

COROLLARY 8.2. *If M is Euclidean 2-space, then $F(\mathbb{R}^2, m)$ is the $K(P_m, 1)$-space and $B(\mathbb{R}^2, m)$ is the $K(Br_m, 1)$-space.*

Let X_W be the space defined in Section 7.2.

THEOREM 8.3. *The universal covering of X_W is contractible, and so X_W is a $K(\pi; 1)$-space.*

This theorem for the groups of types C_n, G_2 and $I_2(p)$, was proved by E. Brieskorn, [42], in much the same way as E. Fadell and L. Neuwirth, [80], proved Theorems 8.1, 8.2 and Corollary 8.2. For the groups of types D_n and F_4 E. Brieskorn used this method with minor modifications. In the general case Theorem 8.3 was proved by P. Deligne, [74].

It follows from Theorem 8.2 that $F(\mathbb{C} \setminus Q_g, n)$ and $B(\mathbb{C} \setminus Q_g)$ are $K(\pi, 1)$-spaces, that $\pi_1 B(\mathbb{C} \setminus Q_g) = Br_n^g$, and so, $B(\mathbb{C} \setminus Q_g)$ can be considered as the classifying space of Br_n^g.

8.2. *Cohomology of pure braid groups*

The cohomology of pure braid groups was first calculated by V.I. Arnold, [4]. The map

$$\phi : S^{n-1} \to F(\mathbb{R}^n, 2),$$

described by the formula $\phi(x) = (x, -x)$, is a Σ_2-equivariant homotopy equivalence. Denote by A the generator of $H^{n-1}(F(\mathbb{R}^n, 2), \mathbb{Z})$ that is mapped by ϕ^* to the standard generator of $H^{n-1}(S^{n-1}, \mathbb{Z})$. For i and j, such that $1 \leqslant i, j \leqslant m$, $i \neq j$, specify $\pi_{i,j} : F(\mathbb{R}^n, m) \to F(\mathbb{R}^n, 2)$ by the formula $\pi_{i,j}(p_1, \dots, p_m) = (p_i, p_j)$. Put

$$A_{i,j} = \pi_{i,j}^*(A) \in H^{n-1}(F(\mathbb{R}^n, m), \mathbb{Z}).$$

It follows that $A_{i,j} = (-1)^n A_{j,i}$ and $A_{i,j}^2 = 0$. For $w \in \Sigma_m$ there is an action $w(A_{i,j}) = A_{w^{-1}(i), w^{-1}(j)}$, since $\pi_{i,j} w = \pi_{w^{-1}(i), w^{-1}(j)}$. Note also that under restriction to

$$F(\mathbb{R}^n \setminus Q_k, m-k) \cong \pi^{-1}(Q_k) \subset F(\mathbb{R}^n, m),$$

the classes $A_{i,j}$ with $1 \leqslant i, j \leqslant k$ go to zero since in this case the map $\pi_{i,j}$ is constant on $\pi^{-1}(Q_k)$.

THEOREM 8.4. *The cohomology group $H^*(F(\mathbb{R}^n \setminus Q_k, m-k), \mathbb{Z})$ is the free Abelian group with generators*

$$A_{i_1, j_1} A_{i_2, j_2} \cdots A_{i_s, j_s},$$

where $k < j_1 < j_2 < \cdots < j_s \leqslant m$ and $i_r < j_r$ for $r = 1, \dots, s$.

The multiplicative structure and the Σ_m-algebra structure of $H^*(F(\mathbb{R}^n, m), \mathbb{Z})$ are given by the following theorem which is proved using the Σ_3-action on $H^*(F(\mathbb{R}^n, 3), \mathbb{Z})$.

THEOREM 8.5. *The cohomology ring $H^*(F(\mathbb{R}^n, m), \mathbb{Z})$ is multiplicatively generated by the square-zero elements*

$$A_{i,j} \in H^{n-1}\big(F\big(\mathbb{R}^n, m\big), \mathbb{Z}\big), \quad 1 \leqslant i < j \leqslant m,$$

subject only to the relations

$$A_{i,k}A_{j,k} = A_{i,j}A_{j,k} - A_{i,j}A_{i,k} \quad \text{for } i < j < k. \tag{8.1}$$

The Poincaré series for $F(\mathbb{R}^n, m)$ is the product $\prod_{j=1}^{m-1}(1 + jt^{n-1})$.

REMARK 8.1. In the case of $\mathbb{R}^2 = \mathbb{C}$ the cohomology classes $A_{j,k}$ can be interpreted as the classes of cohomology of the differential forms

$$\omega_{j,k} = \frac{1}{2\pi i}\frac{\mathrm{d}z_j - \mathrm{d}z_k}{z_j - z_k}.$$

E. Brieskorn calculated the cohomology of pure generalized braid groups, [42], using ideas of V.I. Arnold for the classical case. Let $\mathcal{V}$ be a finite-dimensional complex vector space and $H_j \in \mathcal{V}$, $j \in I$, be the finite family of complex affine hyperplanes given by linear forms l_j. E. Brieskorn proved the following fact.

THEOREM 8.6. *The cohomology classes, corresponding to the holomorphic differential forms*

$$\omega_j = \frac{1}{2\pi i}\frac{\mathrm{d}l_j}{l_j},$$

generate the cohomology ring $H^*(\mathcal{V} \setminus \bigcup_{j\in I} H_j, \mathbb{Z})$. Moreover, this ring is isomorphic to the $\mathbb{Z}$-subalgebra generated by the forms ω_j in the algebra of meromorphic forms on $\mathcal{V}$.

The cohomology of pure generalized braid groups is described as follows.

THEOREM 8.7.

(i) *The cohomology group $H^k(P(W), \mathbb{Z})$ of the pure braid group $P(W)$ with integer coefficients is a free Abelian group, and its rank is equal to the number of elements $w \in W$ of length $l(w) = k$, where l is the length considered with respect to the system of generators consisting of all reflections of W.*

(ii) *The Poincaré series for $H^*(P(W), \mathbb{Z})$ is the product $\prod_{j=1}^{n}(1 + m_j t)$, where the m_j are the exponents of the group W.*

(iii) *The multiplicative structure of $H^*(P(W), \mathbb{Z})$ coincides with the structure of the algebra generated by the 1-forms described in the previous theorem.*

8.3. *Homology of braid groups*

To study the cohomology of the classical braid groups $H^*(Br_n, \mathbb{Z})$, V.I. Arnold, [5], interpreted the space $K(Br_n, 1) \cong B(\mathbb{R}^2, n)$ as the space of monic complex polynomials of degree n without multiple roots

$$P_n(t) = t^n + z_1 t^{n-1} + \cdots + z_{n-1} t + z_n.$$

Using this idea he proved theorems of finiteness, of recurrence and of stabilization. Homology with coefficients in $\mathbb{Z}/2$ was calculated by D.B. Fuks in the following theorems, [90].

THEOREM 8.8. *The homology of the braid group on an infinite number of strings with coefficients in $\mathbb{Z}/2$ as a Hopf algebra is isomorphic to the polynomial algebra on infinitely many generators a_i, $i = 1, 2, \ldots$; $\deg a_i = 2^i - 1$:*

$$H_*(Br_\infty, \mathbb{Z}/2) \cong \mathbb{Z}/2[a_1, a_2, \ldots, a_i, \ldots]$$

with the coproduct given by the formula:

$$\Delta(a_i) = 1 \otimes a_i + a_i \otimes 1.$$

THEOREM 8.9. *The canonical inclusion $Br_n \to Br_\infty$ induces a monomorphism in homology with coefficients in $\mathbb{Z}/2$. Its image is the subcoalgebra of the polynomial algebra $\mathbb{Z}/2[a_1, a_2, \ldots, a_i, \ldots]$ with $\mathbb{Z}/2$-basis consisting of the monomials*

$$a_1^{k_1} \cdots a_l^{k_l} \quad \text{such that} \quad \sum_i k_i 2^i \leqslant n.$$

THEOREM 8.10. *The canonical homomorphism $Br_n \to BO_n$, $1 \leqslant n \leqslant \infty$, induces a monomorphism (of Hopf algebras if $n = \infty$)*

$$H_*(Br_n, \mathbb{Z}/2) \to H_*(BO_n, \mathbb{Z}/2).$$

F.R. Cohen calculated the homology of braid groups with coefficients $\mathbb{Z}/p$, $p > 2$, also as modules over the Steenrod algebra, [54–56].

Later V.V. Goryunov, [108,109], applied the methods of Fuks and expressed the cohomology of the generalized braid groups of types C and D in terms of the cohomology of the classical braid groups.

9. Connections with the other domains

9.1. *Markov theorem*

Suppose a braid depicted in Figure 1 is placed in a cube. On the boundary of the cube join the point A_i to the point B_i by a simple arc D_i, such that D_i and D_j are mutually disjoint

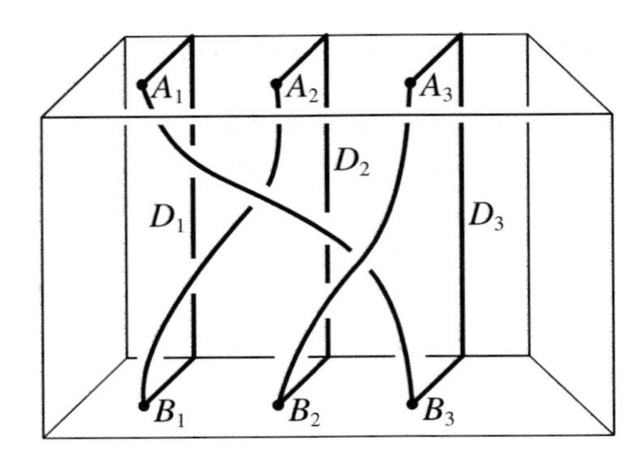

Fig. 17.

if $i \neq j$. Since our initial braid does not intersect the boundary of the cube except at the points $A_1, \ldots, A_n$ and $B_1, \ldots, B_n$ we obtain a link (or, in particular, a knot), i.e. a system of simple closed curves in $\mathbb{R}^3$. A link obtained in such a manner is called the *closure* of the braid, see Figure 17.

THEOREM 9.1 (J.W. Alexander). *Any link can be represented by a closed braid.*

The next step is to understand equivalence classes of braids which correspond to links. The following Markov theorem gives an answer to this question. At first we define two types of Markov moves for braids.

Type 1 *Markov move* replaces a braid β on n strings by its conjugate $\gamma\beta\gamma^{-1}$.

Type 2 *Markov move* replaces a braid β on n strings by the braid $j_n(\beta)\sigma_n$ on $n+1$ strings or by $j_n(\beta)\sigma_n^{-1}$ where j_n is the canonical inclusion of the group Br_n into the group Br_{n+1} (see Section 3.1)

$$j_n : Br_n \to Br_{n+1}.$$

THEOREM 9.2 (A.A. Markov). *Suppose that β and β' are two braids (not necessary with the same number of strings). Then, the closures of β and β' represent the same link if and only if β can be transformed into β' by means of a finite number of type* 1 *and type* 2 *Markov moves. Namely there exists the following sequence,*

$$\beta = \beta_0 \to \beta_1 \to \cdots \to \beta_m = \beta',$$

such that, for $i = 0, 1 \ldots, m-1$, β_{i+1} is obtained from β_i by the application of a type 1 *or* 2 *Markov moves or their inverses.*

In other words, if we consider the disjoint union of all braid groups

$$\coprod_{n=1}^{n} Br_n,$$

then the Markov moves of types 1 and 2 define the equivalence relation $\sim$ on this set such that the quotient set

$$\coprod_{n=1}^{n} Br_n/\sim$$

is in one-to-one correspondence with isotopy classes of links.

There exist a lot of proofs of Markov theorem, see, for example, the work of P. Traczyk, [185].

9.2. *Homotopy groups of spheres and Makanin braids*

Consider the coordinate projections for the spaces $F(M,m)$ where M is a manifold (see Section 7.1)

$$d_i : F(M, n+1) \to F(M, n), \quad i = 0, \dots, n,$$

defined by the formula

$$d_i(p_1, \dots, p_{i+1}, \dots, p_{n+1}) = (p_1, \dots, p_i, p_{i+2}, \dots, p_{n+1}).$$

By taking the fundamental group the maps d_i induces group homomorphisms

$$d_{i*} : P_{m+1}(M) \to P_m(M), \quad i = 0, \dots, n.$$

A braid $\beta \in Br_{n+1}$ is called *Makanin* (*smooth* in the terminology of D.L. Johnson, [120], *Brunnian* in the terminology of J.A. Berrick, F.R. Cohen, Y.L. Wong and J. Wu, [22]) if $d_i(\beta) = 1$ for all $0 \leqslant i \leqslant n$. We call them Makanin, because up to our knowledge it was G.S. Makanin who first mentioned them, [127, page 78, Question 6.23]. In other words the group of Makanin braids $Mak_{n+1}(M)$ is given by the formula

$$Mak_{n+1}(M) = \bigcap_{i=0}^{n} \operatorname{Ker}\bigl(d_{i*} : P_{m+1}(M) \to P_m(M)\bigr).$$

The canonical embedding of the open disc D^2 into the sphere S^2

$$f : D^2 \to S^2$$

induces a group homomorphism

$$f_* : Mak_n\bigl(D^2\bigr) \to Mak_n\bigl(S^2\bigr),$$

where $Mak_n(D^2)$ is the Makanin subgroup Mak_n of the classical braid group Br_n. The group Mak_n is free, [112,120]. The following theorem is proved in [22].

THEOREM 9.3. *The is an exact sequence of groups*

$$1 \to Mak_{n+1}(S^2) \to Mak_n(D^2) \to \pi_{n-1}(S^2) \to 1$$

for $n \geqslant 5$.

Here as usual $\pi_k(S^2)$ denote the k-th homotopy group of the sphere S^2.

For instance, $Mak_5(S^2)$ modulo Mak_5 is $\pi_4(S^2) = \mathbb{Z}/2$. The other homotopy groups of S^2 are as follows

$$\pi_5(S^2) = \mathbb{Z}/2,\ \pi_6(S^2) = \mathbb{Z}/12,\ \pi_7(S^2) = \mathbb{Z}/2,\ \pi_8(S^2) = \mathbb{Z}/2, \ldots.$$

Thus, up to certain range, $Mak_n(S^2)$ modulo Mak_n are known by nontrivial calculation of $\pi_*(S^2)$.

Acknowledgements

The author is grateful to F.R. Cohen and Wu Jie for fruitful discussions on theory of braids and their applications. The author is also thankful to Vik.S. Kulikov who attracted his attention to the works of O. Zariski, [200,201], and to E.P. Volokitin for the help and advice on the presentation of the chapter.

References

[1] S.I. Adyan, *Fragments of the word Δ in the braid group*, Mat. Zametki **36** (1) (1984), 25–34 (Russian).

[2] A. Adem, D. Cohen and F.R. Cohen, *On representations and K-theory of the braid groups*, Math. Ann. **326** (3) (2003), 515–542.

[3] J.W. Alexander, *A lemma on systems of knotted curves*, Proc. Nat. Acad. Sci. USA **9**, 93–95.

[4] V.I. Arnold, *The cohomology ring of colored braids*, Mat. Zametki **5** (1969), 227–231 (Russian). English transl. in Math. Notes **5** (1969), 138–140.

[5] V.I. Arnold, *On some topological invariants of algebraic functions*, Trudy Moskov. Mat. Obshch. **21** (1970), 27–46 (Russian). English transl. in Trans. Moscow Math. Soc. **21** (1970), 30–52.

[6] V.I. Arnold, *Topological invariants of algebraic functions II*, Functional Anal. Appl. **4** (2) (1970), 1–9 (Russian).

[7] E. Artin, *Theorie der Zöpfe*, Abh. Math. Semin. Univ. Hamburg **4** (1925), 47–72.

[8] E. Artin, *Theory of braids*, Ann. Math. **48** (1) (1947), 101–126.

[9] E. Artin, *Braids and permutations*, Ann. of Math. (2) **48** (1947), 643–649.

[10] J.C. Baez, *Link invariants of finite type and perturbation theory*, Lett. Math. Phys. **26** (1) (1992), 43–51.

[11] V.G. Bardakov, *On the theory of braid groups*, Mat. Sb. **183** (6) (1992), 3–42 (Russian). English transl. in Russian Acad. Sci. Sb. Math. **76** (1) (1993), 123–153.

[12] V.G. Bardakov, *The width of verbal subgroups of some Artin groups*, Group and Metric Properties of Mappings, Novosibirsk. Gos. Univ. (1995), 8–18 (Russian).

[13] V.G. Bardakov, *The structure of a group of conjugating automorphisms*, Algebra i Logika **42** (5) (2003), 515–541, 636 (Russian). English transl. in Algebra and Logic **42** (5) (2003), 287–303.

[14] V.G. Bardakov, *The virtual and universal braids*, Fund. Math. **184** (2004), 1–18.

[15] V.G. Bardakov, *Linear representations of the braid groups of some manifolds*, Acta Appl. Math. **85** (1–3) (2005), 41–48.

[16] V.G. Bardakov, *Linear representations of the group of conjugating automorphisms and the braid groups of some manifolds*, Sibirsk. Mat. Zh. **46** (1) (2005), 17–31 (Russian). English transl. in Siberian Math. J. **46** (1) (2005), 13–23.
[17] P. Bellingeri, *Tresses sur surface et invariants d'entrelacs*, Thèse, Univ. Grenoble I (2003).
[18] P. Bellingeri, *Centralisers in surface braid groups*, Comm. Algebra **32** (10) (2004), 4099–4115.
[19] P. Bellingeri, *On presentations of surface braid groups*, J. Algebra **274** (2) (2004), 543–563.
[20] P. Bellingeri and L. Funar, *Braids on surfaces and finite type invariants*, C. R. Acad. Sci. Paris Sér. Math. **338** (2) (2004), 157–162.
[21] P. Bellingeri and V. Vershinin, *Presentations of surface braid groups by graphs*, Fund. Math., to appear.
[22] J.A. Berrick, F.R. Cohen, Y.L. Wong and J. Wu, *Configurations, braids, and homotopy groups*, J. Amer. Math. Soc., to appear.
[23] D. Bessis, *Groupes des tresses et éléments réguliers* (French) [Braid groups and regular elements], J. Reine Angew. Math. **518** (2000), 1–40.
[24] D. Bessis, *Zariski theorems and diagrams for braid groups*, Invent. Math. **145** (3) (2001), 487–507.
[25] D. Bessis, *The dual braid monoid*, Ann. Sci. École Norm. Sup. (4) **36** (5) (2003), 647–683.
[26] D. Bessis, F. Digne and J. Michel, *Springer theory in braid groups and the Birman–Ko–Lee monoid*, Pacific J. Math. **205** (2) (2002), 287–309.
[27] D. Bessis and J. Michel, *Explicit presentations for exceptional braid groups*, Experiment. Math. **13** (3) (2004), 257–266.
[28] S. Bigelow, *The Burau representation is not faithful for* $n = 5$, Geom. Topol. **3** (1999), 397–404 (electronic).
[29] S. Bigelow, *Braid groups are linear*, J. Amer. Math. Soc. **14** (2) (2001), 471–486.
[30] J.S. Birman, *On braid groups*, Comm. Pure Appl. Math. **22** (1968), 41–72.
[31] J.S. Birman, *Mapping class groups and their relationship to braid groups*, Comm. Pure Appl. Math. **22** (1969), 213–238.
[32] J.S. Birman, *Braids, Links, and Mapping Class Groups*, Ann. Math. Stud. **82** (1974).
[33] J.S. Birman, *New points of view in knot theory*, Bull. Amer. Math. Soc. **28** (2) (1993), 253–387.
[34] J.S. Birman and T.E. Brendle, *Braids: a survey*, Arxiv.math.GT/0409205.
[35] J.S. Birman, K.H. Ko and S.J. Lee, *A new approach to the word and conjugacy problems in the braid groups*, Adv. Math. **139** (2) (1998), 322–353.
[36] J.S. Birman, K.H. Ko and S.J. Lee, *The infimum, supremum, and geodesic length of a braid conjugacy class*, Adv. Math. **164** (1) (2001), 41–56.
[37] J.S. Birman, D.D. Long and J.A. Moody, *Finite-dimensional representations of Artin's braid group*, The Mathematical Legacy of Wilhelm Magnus: Groups, Geometry and Special Functions, Brooklyn, NY, 1992, Contemp. Math. **169**, Amer. Math. Soc. (1994), 123–132.
[38] J.S. Birman and W.W. Menasco, *On Markov's theorem*, Knots 2000, Korea, vol. 1 (Yongpyong), J. Knot Theory Ramifications **11** (3) (2002), 295–310.
[39] J.S. Birman and B. Wajnryb, *Markov classes in certain finite quotients of Artin's braid group*, Israel J. Math. **56** (2) (1986), 160–178.
[40] J.S. Birman and H. Wenzl, *Braids, link polynomials and a new algebra*, Trans. Amer. Math. Soc. **313** (1) (1989), 249–273.
[41] N. Bourbaki, *Groupes et algèbres de Lie, Chaps. 4–6*, Masson (1981).
[42] E. Brieskorn, *Sur les groupes de tresses* [d'après V.I. Arnol'd], Séminaire Bourbaki, 24ème année (1971/1972), Exp. No. 401, Lecture Notes in Math. **317**, Springer (1973), 21–44 (French).
[43] E. Brieskorn, *Die Fundamentalgruppe des Raumes der regulären Orbits einer endlichen komplexen Spiegelungsgruppe*, Invent. Math. **12** (1971), 57–61 (German).
[44] E. Brieskorn, *Automorphic sets and braids and singularities*, Braids, Santa Cruz, CA, 1986, Contemp. Math. **78**, Amer. Math. Soc. (1988), 45–115.
[45] E. Brieskorn and K. Saito, *Artin-Gruppen und Coxeter-Gruppen*, Invent. Math. **17** (1972), 245–271 (German).
[46] C. Broto and V.V. Vershinin, *On the generalized homology of Artin groups*, J. Math. Sci. (NY) **113** (4) (2003), 545–547. English transl. from Russian: Zap. Nauchn. Sem. S.-Peterburg. Otdel. Mat. Inst. Steklov. (POMI) **266** (2000), Teor. Predst. Din. Sist. Komb. i Algoritm. Metody **5**, 7–12, 336.

[47] M. Broué, G. Malle and R. Rouquier, *Complex reflection groups, braid groups, Hecke algebras*, J. Reine Angew. Math. **500** (1998), 127–190.

[48] W. Burau, *Über Zopfinvarianten*, Abh. Math. Semin. Univ. Hamburg **9** (1932), 117–124 (German).

[49] W. Burau, *Über Verkettungsgruppen*, Abh. Math. Semin. Univ. Hamburg **11** (1935), 171–178 (German).

[50] W. Burau, *Über Zopfgruppen und gleichsinnig verdrillte Verkettungen*, Abh. Math. Semin. Univ. Hamburg **11** (1935), 179–186 (German).

[51] R. Charney, *Artin groups of finite type are biautomatic*, Math. Ann. **292** (4) (1992), 671–683.

[52] R. Charney, *Geodesic automation and growth functions for Artin groups of finite type*, Math. Ann. **301** (2) (1995), 307–324.

[53] W.-L. Chow, *On the algebraical braid group*, Ann. Math. **49** (3) (1948), 654–658.

[54] F. Cohen, *Cohomology of braid spaces*, Bull. Amer. Math. Soc. **79** (4) (1973), 763–766.

[55] F. Cohen, *Homology of* $\Omega^{n+1}\Sigma^{n+1}X$ *and* $C_{n+1}X$, $n > 0$, Bull. Amer. Math. Soc. **79** (6) (1973), 1236–1241.

[56] F. Cohen, *Braid orientations and bundles with flat connections*, Invent. Math. **46** (1978), 99–110.

[57] F. Cohen, *Artin's braid groups, classical homotopy theory, and other curiosities*, Braids, Contemp. Math. **78** (1988), 167–206.

[58] F. Cohen, *On braid groups, homotopy groups, and modular forms*, Advances in Topological Quantum Field Theory, NATO Sci. Ser. II Math. Phys. Chem. **179**, Kluwer Acad. Publ. (2004), 275–288.

[59] F. Cohen, T. Lada and J.P. May, *The Homology of Iterated Loop Spaces*, Lecture Notes in Math. **533**, Springer-Verlag (1976), vii+490 pp.

[60] F. Cohen and V.V. Vershinin, *Thom spectra which are wedges of Eilenberg–MacLane spectra*, Stable and Unstable Homotopy, Toronto, ON, 1996, Fields Inst. Commun. **19**, Amer. Math. Soc. (1998), 43–65.

[61] F. Cohen and J. Wu, *On braid groups, free groups, and the loop space of the 2-sphere*, Categorical Decomposition Techniques in Algebraic Topology, Isle of Skye, 2001, Progr. Math. **215**, Birkhäuser (2004), 93–105.

[62] R. Corran, *A normal form for a class of monoids including the singular braid monoids*, J. Algebra **223** (1) (2000), 256–282.

[63] H.S.M. Coxeter, *Factor groups of the braid group*, Proc. Fourth Canad. Math. Congr., Banff, 1957 (1959), 95–122.

[64] H.S.M. Coxeter and W.O.J. Moser, *Generators and Relations for Discrete Groups*, 3rd ed. Ergeb. Math. Grenzgeb. **IX 14**, Springer-Verlag (1972), 161 p.

[65] J. Crisp and L. Paris, *The solution to a conjecture of Tits on the subgroup generated by the squares of the generators of an Artin group*, Invent. Math. **145** (1) (2001), 19–36.

[66] O. Dasbach and B. Gemein, *A faithful representation of the singular braid monoid on three strands*, Knots in Hellas '98 (Delphi), Ser. Knots Everything **24**, World Sci. Publishing (2000), 48–58.

[67] J. Daz-Cantos, J. Gonzlez-Meneses and J.M. Tornero, *On the singular braid monoid of an orientable surface*, Proc. Amer. Math. Soc. **132** (10) (2004), 2867–2873.

[68] P. Dehornoy, *Deux propriétés des groupes de tresses* (French) [Two properties of braid groups], C. R. Acad. Sci. Paris Sér. I Math. **315** (6) (1992), 633–638.

[69] P. Dehornoy, *Braid groups and left distributive operations*, Trans. Amer. Math. Soc. **345** (1) (1994), 115–150.

[70] P. Dehornoy, *Braids and Self-distributivity*, Progr. Math. **192**, Birkhäuser (2000), xx+623 pp.

[71] P. Dehornoy, I. Dynnikov, D. Rolfsen and B. Wiest, *Why Are Braids Orderable?* Panoramas et Synthèses [Panoramas and Syntheses] **14**, Société Mathématique de France (2002), xiv+190 pp.

[72] P. Dehornoy and Y. Lafont, *Homology of Gaussian groups*, Ann. Inst. Fourier (Grenoble) **53** (2) (2003), 489–540.

[73] P. Dehornoy and L. Paris, *Gaussian groups and Garside groups, two generalisations of Artin groups*, Proc. London Math. Soc. (3) **79** (3) (1999), 569–604.

[74] P. Deligne, *Les immeubles des groupes de tresses généralisés*, Invent. Math. **17** (1972), 273–302 (French).

[75] G. Dethloff, S. Orevkov and M. Zaidenberg, *Plane curves with a big fundamental group of the complement*, Voronezh Winter Mathematical Schools, Amer. Math. Soc. Transl. Ser. 2 **184**, Amer. Math. Soc. (1998) 63–84.

[76] E. El-Rifai and H.R. Morton, *Algorithms for positive braids*, Quart. J. Math. Oxford Ser. (2) **45** (180) (1994), 479–497.

[77] M. Epple, *Orbits of asteroids, a braid, and the first link invariant*, Math. Intell. **20** (1) (1998), 45–52.

[78] D.B.A. Epstein, J.W. Cannon, D.E. Holt, S.V.F. Levy, M.S. Paterson and W.P. Thurston, *Word Processing in Groups*, Jones and Bartlett Publishers (1992), xii+330 pp.

[79] E. Fadell and S.Y. Husseini, *Geometry and Topology of Configuration Spaces*, Springer Monographs in Math., Springer-Verlag (2001), xvi+313 pp.

[80] E. Fadell and L. Neuwirth, *Configuration spaces*, Math. Scand. **10** (Fasc. I) (1962), 111–118.

[81] E. Fadell and J. Van Buskirk, *The braid groups of E^2 and S^2*, Duke Math. J. **29** (1962), 243–257.

[82] R. Fenn, E. Keyman and C. Rourke, *The singular braid monoid embeds in a group*, J. Knot Theory Ramifications **7** (7) (1998), 881–892.

[83] R. Fenn, R. Rimányi and C. Rourke, *Some remarks on the braid-permutation group*, Topics in Knot Theory, Kluwer Acad. Publ. (1993), 57–68.

[84] R. Fenn, R. Rimányi and C. Rourke, *The braid-permutation group*, Topology **36** (1) (1997), 123–135.

[85] R. Fenn, D. Rolfsen and J. Zhu, *Centralisers in the braid group and singular braid monoid*, Enseign. Math. (2) **42** (1–2) (1996), 75–96.

[86] R. Fox and L. Neuwirth, *The braid groups*, Math. Scand. **10** (Fasc. I) (1962), 119–126.

[87] N. Franco and J. Gonzlez-Meneses, *Computation of centralizers in braid groups and Garside groups*, Proceedings of the International Conference on Algebraic Geometry and Singularities, Sevilla, 2001, Rev. Mat. Iberoamericana **19** (2) (2003), 367–384 (Spanish).

[88] N. Franco and J. Gonzlez-Meneses, *Conjugacy problem for braid groups and Garside groups*, J. Algebra **266** (1) (2003), 112–132.

[89] R. Fricke and F. Klein, *Vorlesungen über die Theorie der Automorphen Funktionen. Bd. I. Gruppentheoretischen Grundlagen*, Teubner (1897) (Johnson Repr. Corp., NY, 1965).

[90] D.B. Fuks, *Cohomology of the braid group* mod 2, Funktsional. Anal. i Prilozhen. **4** (2) (1970), 62–75 (Russian). English transl. in Funct. Anal. Appl. **4** (1970), 143–151.

[91] F.A. Garside, *The braid group and other groups*, Quart. J. Math. Oxford **20** (1969), 235–254.

[92] B.J. Gassner, *On braid groups*, Abh. Math. Sem. Univ. Hamburg **25** (1961), 10–22.

[93] B. Gemein, *Singular braids and Markov's theorem*, J. Knot Theory Ramifications **6** (4) (1997), 441–454.

[94] B. Gemein, *Representations of the singular braid monoid and group invariants of singular knots*, Topology Appl. **114** (2) (2001), 117–140.

[95] E. Godelle and L. Paris, *On singular Artin monoids*, Geometric Methods in Group Theory, Contemp. Math. **372**, Amer. Math. Soc. (2005), 43–57.

[96] V.A. Golubeva and V.P. Leksin, *Quadratic relations in the cohomology of generalized colored braids, and Dunkl identities*, Funktsional. Anal. i Prilozhen. **30** (2) (1996), 73–76 (Russian). English transl. in Funct. Anal. Appl. **30** (2) (1996), 131–134.

[97] V.A. Golubeva and V.P. Leksin, *On two types of representations of the braid group associated with the Knizhnik–Zamolodchikov equation of the B_n type*, J. Dynam. Control Systems **5** (4) (1999), 565–596.

[98] D.L. Gonalves and J. Guaschi, *On the structure of surface pure braid groups*, J. Pure Appl. Algebra **182** (1) (2003), 33–64.

[99] D.L. Gonalves and J. Guaschi, *The roots of the full twist for surface braid groups*, Math. Proc. Cambridge Philos. Soc. **137** (2) (2004), 307–320.

[100] D.L. Gonalves and J. Guaschi, *The braid groups of the projective plane*, Algebr. Geom. Topol. **4** (2004), 757–780 (electronic).

[101] D.L. Gonalves and J. Guaschi, *The braid group $B_{n,m}(\mathbb{S}^2)$ and a generalisation of the Fadell–Neuwirth short exact sequence*, J. Knot Theory Ramifications **14** (30 (2005), 375–403.

[102] J. González-Meneses, *New presentations of surface braid groups*, J. Knot Theory Ramifications **10** (3) (2001), 431–451.

[103] J. González-Meneses, *Ordering pure braid groups on compact, connected surfaces*, Pacific J. Math. **203** (2) (2002), 369–378.

[104] J. González-Meneses, *Presentations for the monoids of singular braids on closed surfaces*, Comm. Algebra **30** (6) (2002), 2829–2836.

[105] J. González-Meneses, *The nth root of a braid is unique up to conjugacy*, Algebr. Geom. Topol. **3** (2003), 1103–1118.

[106] J. González-Meneses, *Improving an algorithm to solve multiple simultaneous conjugacy problems in braid groups*, Geometric Methods in Group Theory, Contemp. Math. **372**, Amer. Math. Soc. (2005), 35–42.

[107] J. González-Meneses and L. Paris, *Vassiliev invariants for braids on surfaces*, Trans. Amer. Math. Soc. **356** (1) (2004), 219–243.
[108] V.V. Goryunov, *The cohomology of braid groups of series C and D and certain stratifications*, Funktsional. Anal. i Prilozhen. **12** (2) (1978), 76–77 (Russian).
[109] V.V. Goryunov, *Cohomology of braid groups of series C and D*, Trudy Moskov. Mat. Obshch. **42** (1981), 234–242 (Russian).
[110] J. Guaschi, *Representations of Artin's braid groups and linking numbers of periodic orbits*, J. Knot Theory Ramifications **4** (2) (1995), 197–212.
[111] J. Guaschi, *Nielsen theory, braids and fixed points of surface homeomorphisms*, Topology Appl. **117** (2) (2002), 199–230.
[112] G.G. Gurzo, *On the group of smooth braids*, 16th All-Union Algebra Conference, Abstracts II (1981), 39–40 (Russian).
[113] G.G. Gurzo, *Systems of generators for normalizers of certain elements of a braid group*, Izv. Akad. Nauk SSSR Ser. Mat. **48** (3) (1984), 476–519 (Russian).
[114] G.G. Gurzo, *Centralizers of finite sets of elements of the braid group* $\mathfrak{B}_{n+1}$, Mat. Zametki **37** (1) (1985), 3–6, 137 (Russian).
[115] G.G. Gurzo, *Systems of generators for the centralizers of the rigid elements of a braid group*, Izv. Akad. Nauk SSSR Ser. Mat. **51** (5) (1987), 915–935 (Russian).
[116] V. Hansen, *Braids and Coverings: Selected Topics*, London Math. Soc. Student Text **18**, Cambridge University Press (1989), 191 p.
[117] A. Hurwitz, *Über Riemannische Flächen mit gegebenen Verzweigungspunkten*, Math. Ann. **39** (1891), 1–61.
[118] N.V. Ivanov, *Algebraic properties of the Teichmüller modular group*, Dokl. Akad. Nauk SSSR **275** (4) (1984), 786–789 (Russian).
[119] A. Járai, Jr., *On the monoid of singular braids*, Topology Appl. **96** (2) (1999), 109–119.
[120] D.L. Johnson, *Towards a characterization of smooth braids*, Math. Proc. Cambridge Philos. Soc. **92** (3) (1982), 425–427.
[121] V.F.R. Jones, *Hecke algebra representations of braid groups and link polynomials*, Ann. Math. **126** (1987), 335–388.
[122] A. Joyal and R. Street, *Braided tensor categories*, Adv. Math. **102** (1) (1993), 20–78.
[123] S. Kamada, *Braid presentation of virtual knots and welded knots*, Arxiv.math.GT/0008092.
[124] S. Kamada and Y. Matsumoto, *Certain racks associated with the braid groups*, Knots in Hellas '98 (Delphi), Ser. Knots Everything **24**, World Sci. Publishing (2000), 118–130.
[125] L.H. Kauffman and S. Lambropoulou, *Virtual braids*, Fund. Math. **184** (2004), 159–186.
[126] F. Klein, *Vorlesungen über höhere Geometrie*, 3. Aufl., bearbeitet und herausgegeben von W. Blaschke. (German) VIII + 405 S. J. Springer (Die Grundlehren der mathematischen Wissenschaften in Einzeldarstellungen **22**) (1926).
[127] *Kourovka Notebook: Unsolved Problems in Group Theory*, 7th ed. Akad. Nauk SSSR Sibirsk. Otdel., Inst. Mat. (1980), 115 pp. (Russian).
[128] D. Krammer, *The braid group* B_4 *is linear*, Invent. Math. **142** (3) (2000), 451–486.
[129] D. Krammer, *Braid groups are linear*, Ann. of Math. (2) **155** (1) (2002), 131–156.
[130] Vik.S. Kulikov, *A full-twist factorization formula with a double number of strings*, Izv. Ross. Akad. Nauk Ser. Mat. **68** (1) (2004), 123–158 (Russian). English transl. in Izv. Math. **68** (1) (2004), 125–158.
[131] Vik.S. Kulikov and V.M. Kharlamov, *On braid monodromy factorizations*, Izv. Ross. Akad. Nauk Ser. Mat. **67** (3) (2003), 79–118 (Russian). English transl. in Izv. Math. **67** (3) (2003), 499–534.
[132] S. Lambropoulou and C.P. Rourke, *Markov's theorem in* 3*-manifolds*, Special issue on Braid Groups and Related Topics, Jerusalem, 1995, Topology Appl. **78** (1–2) (1997), 95–122.
[133] R.J. Lawrence, *Homological representations of the Hecke algebra*, Comm. Math. Phys. **135** (1) (1990), 141–191.
[134] V.Ya. Lin, *Artinian braids and groups and spaces connected with them*, Itogi Nauki i Tekhniki (Algebra, Topologiya, Geometriya) **17** (1979), 159–227 (Russian). English transl. in J. Soviet Math. **18** (1982), 736–788.
[135] V.Ya. Lin, *Configuration spaces of* $\mathbb{C}$ *and* $\mathbb{C}P^1$*: some analytic properties*, Arxiv: math.AG/0403120.
[136] V.Ya. Lin, *Braids and permutations*, Arxiv: math.GR/0404528.

[137] D.D. Long and M. Paton, *The Burau representation is not faithful for $n \geqslant 6$*, Topology **32** (2) (1993), 439–447.

[138] W. Magnus, *Über Automorphismen von Fundamentalgruppen berandeter Flächen*, Math. Ann. **109** (1934), 617–646 (German).

[139] W. Magnus, *Braid groups: a survey*, Proc. 2nd Internat. Conf. Theory of Groups, Canberra, 1973, Lecture Notes Math. **372** (1974), 463–487.

[140] W. Magnus, A. Karrass and D. Solitar, *Combinatorial Group Theory. Presentations of Groups in Terms of Generators and Relations*, 2nd rev. ed. Dover Books on Advanced Mathematics **XII**, Dover Publications, Inc. (1976), 444 p.

[141] W. Magnus and A. Peluso, *On knot groups*, Comm. Pure Appl. Math. **20** (1967), 749–770.

[142] G.S. Makanin, *The conjugacy problem in the braid group*, Dokl. Akad. Nauk SSSR **182** (1968), 495–496 (Russian).

[143] G.S. Makanin, *The normalizers of a braid group*, Mat. Sb. (N.S.) **86** (128) (1971), 171–179 (Russian).

[144] G.S. Makanin, *Separable closed braids*, Mat. Sb. (N.S.) **132** (174) (4) (1987), 531–540 (Russian). English transl. in Math. USSR-Sb. **60** (2) (1988), 521–531.

[145] G.S. Makanin, *A representation of an oriented knot*, Dokl. Akad. Nauk SSSR **299** (5) (1988), 1060–1063 (Russian). English transl. in Soviet Math. Dokl. **37** (2) (1988), 522–525.

[146] G.S. Makanin, *An analogue of the Alexander–Markov theorem*, Izv. Akad. Nauk SSSR Ser. Mat. **53** (1) (1989), 200–210 (Russian). English transl. in Math. USSR-Izv. **34** (1) (1990), 201–211.

[147] G.S. Makanin and A.G. Savushkina, *An equation in a free group that defines colored braids*, Mat. Zametki **70** (4) (2001), 591–602 (Russian). English transl. in Math. Notes **70** (3–4) (2001), 535–544.

[148] S. Manfredini, *Some subgroups of Artin's braid group*, Special issue on Braid Groups and Related Topics, Jerusalem, 1995, Topology Appl. **78** (1–2) (1997), 123–142.

[149] A.A. Markoff, *Über die freie Äquivalenz der geschlossenen Zöpfe*, Math. Sb. **16** (1936), 73–78.

[150] A.A. Markoff, *Foundations of the Algebraic Theory of Tresses*, Trudy Mat. Inst. Steklova **16** (1945) (Russian, English summary).

[151] J. McCarthy, *A "Tits-alternative" for subgroups of surface mapping class groups*, Trans. Amer. Math. Soc. **291** (2) (1985), 583–612.

[152] J. Michel, *A note on words in braid monoids*, J. Algebra **215** (1) (1999), 366–377.

[153] J.A. Moody, *The Burau representation of the braid group B_n is unfaithful for large n*, Bull. Amer. Math. Soc. (N.S.) **25** (20 (1991), 379–384.

[154] S. Moran, *The Mathematical Theory of Knots and Braids. An Introduction*, North-Holland Mathematics Studies **82**, North-Holland Publishing Co. (1983), xii+295 pp.

[155] J. Murakami, *The Kauffman polynomial of links and representation theory*, Osaka J. Math. **24** (4) (1987), 745–758.

[156] K. Murasugi and B. Kurpita, *A Study of Braids*, Math. Appl. **484**, Kluwer Acad. Publ. (1999), x+272 pp.

[157] F. Napolitano, *Configuration spaces on surfaces*, C. R. Acad. Sci. Paris Sér. I Math. **327** (10) (1998), 887–892.

[158] F. Napolitano, *On the cohomology of configuration spaces on surfaces*, J. London Math. Soc. (2) **68** (2) (2003), 477–492.

[159] F. Napolitano, *Cohomology rings of spaces of generic bipolynomials and extended affine Weyl groups of series A*, Ann. Inst. Fourier (Grenoble) **53** (3) (2003), 927–940.

[160] T. Ohtsuki, *Quantum Invariants. A Study of Knots, 3-manifolds, and their Sets*, Series on Knots and Everything **29**, World Scientific Publishing Co., Inc. (2002), xiv+489 pp.

[161] S.Yu. Orevkov, *Strong positivity in the right-invariant order on a braid group and quasipositivity*, Mat. Zametki **68** (5) (2000), 692–698 (Russian). English translation in Math. Notes **68** (5–6) (2000), 588–593.

[162] S.Yu. Orevkov, *Markov moves for quasipositive braids*, C. R. Acad. Sci. Paris Sér. I Math. **331** (7) (2000), 557–562.

[163] S.Yu. Orevkov, *Quasipositivity test via unitary representations of braid groups and its applications to real algebraic curves*, J. Knot Theory Ramifications **10** (7) (2001), 1005–1023.

[164] S.Yu. Orevkov, *Quasipositivity problem for 3-braids*, Turkish J. Math. **28** (1) (2004), 89–93.

[165] S.Yu. Orevkov, *Solution of the word problem in the singular braid group*, Turkish J. Math. **28** (1) (2004), 95–100.

[166] S.Yu. Orevkov and V.V. Shevchishin, *Markov theorem for transversal links*, J. Knot Theory Ramifications **12** (7) (2003), 905–913.

[167] E. Ossa, *On the cohomology of configuration spaces*, Algebraic Topology: New Trends in Localization and Periodicity, Sant Feliu de Guxols, 1994, Progr. Math. **136**, Birkhäuser (1996), 353–361.

[168] E. Ossa and V.V. Vershinin, *Thom spectra of commutants of generalized braid groups*, Funct. Anal. Appl. **32** (4) (1998), 219–226 (1999). Translated from Russian: Funktsional. Anal. i Prilozhen. **32** (4) (1998), 1–9.

[169] L. Paris, *Parabolic subgroups of Artin groups*, J. Algebra **196** (2) (1997), 369–399.

[170] L. Paris, *Centralizers of parabolic subgroups of Artin groups of type* A_l, B_l, *and* D_l, J. Algebra **196** (2) (1997), 400–435.

[171] L. Paris, *Artin monoids inject in their groups*, Comment. Math. Helv. **77** (3) (2002), 609–637.

[172] L. Paris, *The proof of Birman's conjecture on singular braid monoids*, Geom. Topol. **8** (2004), 1281–1300 (electronic).

[173] L. Paris and D. Rolfsen, *Geometric subgroups of surface braid groups*, Ann. Inst. Fourier (Grenoble) **49** (2) (1999), 417–472.

[174] V.V. Prasolov and A.B. Sossinsky, *Knots, Links, Braids and 3-manifolds. An Introduction to the New Invariants in Low-dimensional Topology*, Transl. Math. Monographs **154**, Amer. Math. Soc. (1997), viii+239 pp.

[175] K. Reidemeister, *Knotentheorie*, Ergeb. Math. Grenzgeb. **1**, No. 1, Julius Springer (1932). VI, 74 S., 114 Fig. English transl.: *Knot Theory*, BCS Associates (1983), xv+143 pp.

[176] A.G. Savushkina, *On the commutator subgroup of the braid group*, Vestnik Moskov. Univ. Ser. I Mat. Mekh. no. 6 (1993), 11–14, 118 (1994) (Russian). English transl. in Moscow Univ. Math. Bull. **48** (6) (1993), 9–11.

[177] A.G. Savushkina, *On a group of conjugating automorphisms of a free group*, Mat. Zametki **60** (1) (1996), 92–108, 159 (Russian). English transl. in Math. Notes **60** (1–2) (1996), 68–80 (1997).

[178] A.G. Savushkina, *The center of a colored braid group*, Vestnik Moskov. Univ. Ser. I Mat. Mekh. no. 1 (1996), 32–36, 103 (Russian). English transl. in Moscow Univ. Math. Bull. **51** (1) (1996), 27–30.

[179] A.G. Savushkina, *Basis-conjugating automorphisms of a free group*, Vestnik Moskov. Univ. Ser. I Mat. Mekh. no. 4 (1996), 17–21, 110 (Russian). English transl. in Moscow Univ. Math. Bull. **51** (4) (1996), 14–17.

[180] G.P. Scott, *Braid groups and the group of homeomorphisms of a surface*, Proc. Cambridge Philos. Soc. **68** (1970), 605–617.

[181] V. Sergiescu, *Graphes planaires et présentations des groupes de tresses*, Math. Z. **214** (1993), 477–490.

[182] A.B. Sossinsky, *Preparation theorems for isotopy invariants of links in 3-manifolds*, Quantum Groups, Leningrad, 1990, Lecture Notes in Math. **1510**, Springer (1992), 354–362.

[183] R. Street, *Braids among the groups*, Seminarberichte aus dem Fachbereich Mathematik **63** (5) (1998), 699–703.

[184] I. Sysoeva, *Dimension n representations of the braid group on n strings*, J. Algebra **243** (2) (2001), 518–538.

[185] P. Traczyk, *A new proof of Markov's braid theorem*, Knot Theory, Warsaw, 1995, Banach Center Publ. **42** (1998), 409–419.

[186] V. Turaev, *Faithful linear representations of the braid groups*, Séminaire Bourbaki, vol. 1999/2000, Astérisque **276** (2002), 389–409.

[187] V.A. Vassiliev, *Complements of Discriminants of Smooth Maps: Topology and Applications*, Transl. Math. Monographs **98**, Amer. Math. Soc. (1992), vi+208 pp.

[188] A.M. Vershik, S. Nechaev and R. Bikbov, *Statistical properties of locally free groups with applications to braid groups and growth of random heaps*, Comm. Math. Phys. **212** (2) (2000), 469–501.

[189] V.V. Vershinin, *On braid groups in handlebodies*, Siberian Math. J. **39** (4) (1998), 645–654. Translation from Sibirsk. Mat. Zh. **39** (4) (1998), 755–764.

[190] V.V. Vershinin, *On homological properties of singular braids*, Trans. Amer. Math. Soc. **350** (6) (1998), 2431–2455.

[191] V.V. Vershinin, *Homology of braid groups and their generalizations*, Knot Theory, Warsaw, 1995, Banach Center Publ. **42** (1998), 421–446.

[192] V.V. Vershinin, *Generalizations of braids from a homological point of view*, Sib. Adv. Math. **9** (2) (1999), 109–139.

[193] V.V. Vershinin, *Mapping class groups and braid groups*, St. Petersburg Math. J. **10** (6) (1999), 997–1003. Translation from Russian: Algebra i Analiz **10** (6) (1998), 135–143.

[194] V.V. Vershinin, *Braid groups and loop spaces*, Russian Math. Surveys **54** (2) (1999), 273–350. Translation from Uspekhi Mat. Nauk **54** (2) (1999), 3–84.

[195] V.V. Vershinin, *Thom spectra of generalized braid groups*, J. London Math. Soc. (2) **61** (1) (2000), 245–258.

[196] V.V. Vershinin, *On homology of virtual braids and Burau representation*, Knots in Hellas '98 (Delphi), vol. 3, J. Knot Theory Ramifications **10** (5) (2001), 795–812.

[197] V.V. Vershinin, *On presentations of generalizations of braids with few generators*, Fundam. Prikl. Mat., to appear.

[198] V.V. Vershinin, *On the singular braid monoid*, ArXiv:mathGR/0309339.

[199] J. Wu, *A braided simplicial group*, Proc. London Math. Soc. (3) **84** (3) (2002), 645–662.

[200] O. Zariski, *On the Poincaré group of rational plane curves*, Amer. J. Math. **58** (1936), 607–619.

[201] O. Zariski, *The topological discriminant group of a Riemann surface of genus p*, Amer. J. Math. **59** (1937), 335–358.

[202] J. Zhu, *On singular braids*, J. Knot Theory Ramifications **6** (3) (1997), 427–440.

[203] M.G. Zinno, *On Krammer's representation of the braid group*, Math. Ann. **321** (1) (2001), 197–211.

Groups with Finiteness Conditions

V.I. Senashov*

Institute of Computational Modelling of Siberian Division of Russian Academy of Sciences, Krasnoyarsk, Russia
E-mail: *sen@icm.krasn.ru*

Contents

*The work is supported by the Russian Fund of Fundamental Researches (grant 05-01-00576).

HANDBOOK OF ALGEBRA, VOL. 4
Edited by M. Hazewinkel

1. Finiteness conditions, examples

Groups with conditions of finiteness were traditionally studied by V.P. Shunkov's school. Weak conditions imposed on subgroups, on normalizers of finite subgroups, suddenly yield unexpected effects and extend over the whole group or give it some interesting properties.

Examples of infinite groups with different conditions of finiteness always played important role in the theory of infinite groups. This section cites definitions of different finiteness conditions and constructs several examples, illustrating non-coincidence and differences of the properties of some classes of infinite groups.

The conditions of biprimitive finiteness, conjugate biprimitive finiteness are consecutive weakening of the conditions of local finiteness and binary finiteness. They appear in the work of V.P. Shunkov in the 70ies when he was studying periodic groups. Here we consistently state these conditions of finiteness and give examples of biprimitively finite groups, which are not binary finite, constructed by M.Yu. Bakhova, [3], and A.A. Cherep, [4].

We shall need the following definitions:

A group G is *locally finite*, if every finite set of elements in it generates a finite subgroup.

A group G is called *s-finite*, if any s elements in it generate a finite subgroup.

An s-finite group with $s = 2$ is called *binary finite*.

A group is called *conjugately n-finite*, if any its n conjugate elements generate a finite subgroup.

A group G is called *biprimitively finite*, if for every finite subgroup K from G every two elements of prime order in $N_G(K)/K$ generate a finite subgroup.

A group G is called *p-biprimitively finite*, if for every finite subgroup K from G every two elements of the prime order p in $N_G(K)/K$ generate a finite subgroup.

A group G is called *conjugately biprimitively finite* or a *Shunkov group*, if for every finite subgroup K from G in $N_G(K)/K$ every two conjugate elements of prime order generate a finite subgroup.

This class of groups receive the title *Shunkov group* in the articles of L. Hammoudi, A.V. Rojkov, V.I. Senashov, A.I. Sozutov, A.K. Shlepkin.

These classes of groups have been introduced by V.P. Shunkov. These conditions have been successfully used in the proof of many theorems in for already more then thirty years.

Examples of periodic non-locally finite groups are not a sensation already: there are examples of S.P. Novikov, S.I. Adian, [1], examples of A.Yu. Ol'shanskii, [21], examples of E.S. Golod, [11]).

It is not difficult to see, that the class of finite groups belongs to the class of locally finite groups; the class of locally finite groups belongs to the class of binary finite groups; the class of binary finite groups belongs to the class of biprimitively finite groups.

From the Golod examples, [11], ensues, that there exist binary finite, but not locally finite groups.

There exist biprimitively finite groups, which are not binary finite. Examples of such groups have been constructed by M.Yu. Bakhova, [3], and A.A. Cherep, [4]. We shall cite both constructions.

Here is a construction of M.Yu. Bakhova, [3].

Let n be a composite number, F_n be a free group with n generators $x_1, \dots, x_n$. Denote by c an element from $\operatorname{Aut} F_n$, for which $x_i^c = x_{i+1}$ $(i = 1, \dots, n-1)$ and $x_n^c = x_1$. Take in Hol F_n the subgroup $W = F_n\lambda\langle c\rangle$. Let φ be the homomorphism $F_n \to P$, where P is a Golod p-group with n generators. V is the kernel of this homomorphism. By the homomorphism theorem, [16], $F_n/V \cong P$. Obviously, V^{c^i} $(i = 1, 2, \ldots, n)$ and $D = \bigcap_{i=1}^{n} V^{c^i}$ are normal subgroups in F_n. It is not difficult to show, that D is normal in W. Consider now the quotient group $G = W/D$. The subgroup F_n/D from G will be a sub-Cartesian product of groups isomorphic to P, and therefore the given subgroup is $(n-1)$-finite.

Let's introduce notations:

$$B = F_n/D, \qquad a = cD, \qquad b_i = x_i D \quad (i = 1, \ldots, n).$$

Then $G = B\lambda\langle a\rangle$ and $B = \langle b_1, \ldots, b_n\rangle$. But $b_i^a = b_{i+1}$ $(i = 1, \ldots, n-1)$ and $b_n^a = b_1$. Hence, $G = \langle b_1, a\rangle$, and this means, that the group G is not binary finite, with $\pi(G) = \pi(n) \cup \{p\}$.

Further, as in the group G every subgroup and every quotient group are the extensions of an $(n-1)$-finite group by means of a cyclic group, and n is a composite number, then the group G is biprimitively finite. So, the following assertion is valid: there exists a biprimitively finite group G, which is not binary finite.

As it ensues from the structure of the above given group G, a finite extension of the binary finite group may be not a binary finite group.

Here is the example of A.A. Cherep, [4].

Consider a direct product $A = \prod_{i \in Z}\langle a_i\rangle$ of cyclic groups $\langle a_i\rangle$ of order 2 and the group $G = (A\lambda\langle h\rangle)\lambda\langle t\rangle$, where the element h of infinite order acts on the generators from A by the rule $h^{-1}a_i h = a_{i+2}$ $(i \in Z)$, and the action of the element t of order 4 is determined by the equalities:

$$t^{-1}a_i t = a_{-i} \quad (i \in Z), \qquad t^{-1}ht = h^{-1}a_0a_1.$$

It is obvious that for each $i \in Z$, $t^{-4}a_i t^4 = a_i$. It is also easy to check that

$$t^{-2}ht^2 = a_0a_1ha_0a_{-1} = ha_{-1}a_0a_2a_3,$$
$$t^{-3}ht^3 = h^{-1}a_0a_1a_0a_1a_{-2}a_{-3} = h^{-1}a_{-2}a_{-3},$$
$$t^{-4}ht^4 = a_0a_1ha_2a_3 = h.$$

Hence, the group G has been determined correctly. Let $B = \langle A, t^2\rangle$, $L = \langle B, h\rangle$. As $t^2 \in C_G(A)$ and $t^2A \in Z(G/A)$, B is an elementary Abelian 2-subgroup, and the periodic elements from L lie in B. Further, the relation in the quotient-group G/B

$$(th^k B)^2 = t^2h^{-k}h^k B = B$$

shows that the elements from the set $G \setminus L$ have order 4.

In [4] is shown that the group G is biprimitively finite, but not binary finite.

There is a V.P. Shunkov problem: are the classes of biprimitively finite and conjugately biprimitively finite groups different.

The following theorem solves this problem in the class of soluble groups.

THEOREM 1.1 (A.A. Cherep, [5]). *A soluble conjugately biprimitively finite group is biprimitively finite.*

The proof of this theorem contains more than it states; in fact, it shows, that if in a soluble group a certain periodic element generates with each of its conjugates a finite subgroup, then it lies in the locally finite normal divisor.

A group G is called *group with unmixed factors* if it has an ordered normal series

$$1 = G_0 \leqslant G_1 \leqslant \cdots \leqslant G_\alpha = G,$$

whose factors are either locally finite, or torsion-free.

It can be directly checked, that this property can be extended to the subgroups and quotient groups with respect to a periodic normal divisor. In addition, the periodic subgroups of a group with unmixed factors, are locally finite.

THEOREM 1.2 (A.A. Cherep, [5]). *If G is a group with unmixed factors, then conditions* (1)–(3) *are equivalent*:

(1) *The group G is conjugately biprimitively finite.*
(2) *The group G is biprimitively finite.*
(3) *For every finite subgroup H in the quotient group $N_G(H)/H$ the elements of prime order generate a locally finite subgroup.*

The conditions of conjugate biprimitive finiteness and biprimitive finiteness coincide in a more general case.

THEOREM 1.3 (A.A. Cherep, [5]). *If in the group G every two elements generate a subgroup with unmixed factors, then the conjugate biprimitive finiteness of G is equivalent to its biprimitive finiteness.*

The next theorem gives infinitely many examples of infinite groups which separated classes of n-finite and $(n+1)$-finite p-group for an arbitrary large enough number n.

THEOREM 1.4 (A.V. Rojkov, [26]). *Let p be a prime number, $1 \leqslant n \leqslant k$ be a natural number. Then there exists a finitely generated finitely approximate conjugately k-finite n-finite, but not a $(n+1)$-finite p-group. In particular, a p-group can be non-binary finite, but conjugately k-finite, where k is any large enough number.*

2. Layer-finite groups

Another kind of finiteness conditions of groups is a finiteness of its layers (a *layer* is a set of elements of given order).

Layer-finite groups appeared for the first time in a paper by S.N. Chernikov written in 1945, [6]. This particular kind of group was mentioned, but without any name. Chernikov used the name of "layer-finite" in some of his later works.

A group is *layer-finite* if any set of its elements of any given order is finite.

Investigations of properties of layer-finite groups were carried out by S.N. Chernikov, R. Baer, Kh.Kh. Mukhamedjan in 1945–1960. The basic properties were described in different journals and remained in this form until 1980, when S.N. Chernikov's book, [9], was published. In that book, S.N. Chernikov collected all of his results in one paragraph. It is possible to find practically all the properties of layer-finite groups and the nearest related questions in the monograph [30].

When all the properties of layer-finite groups were described, the question about the place of this class among other groups arose. The first of such characterizations of layer-finite groups was the establishment of the interconnection with the nearest class of groups: locally normal groups. The idea of this connection appeared in articles of S.N. Chernikov, R. Baer, Kh.Kh. Mukhamedjan, [2,6,7,19,20]. But more complete is a theorem by Chernikov:

THEOREM 2.1 (S.N. Chernikov, [9]). *A class of layer-finite groups coincides with a class of locally normal groups if all their Sylow subgroups satisfy the minimality condition.*

It should be noted, that a group G is called a *Chernikov group* if it is a finite extension of a direct product of a finite number of quasi-cyclic groups.

The Shmidt theorem on the closure of locally finite groups with respect to extensions by locally finite groups is valid for locally finite groups. There is no similar theorem for the class of layer-finite groups. Even the finite extensions of layer-finite groups lead us beyond the limits of this class. But nevertheless, some hereditary properties for layer-finite groups do exist:

THEOREM 2.2 (S.N. Chernikov, [9]). *The thin layer-finite groups are precisely locally normal groups, all of whose Sylow subgroups are finite.*

THEOREM 2.3 (S.N. Chernikov, [9]). *A group G, which is an extension of the layer-finite group by the layer-finite group, is layer-finite if and only if it is locally normal.*

THEOREM 2.4 (S.N. Chernikov, [9]). *If the group G can be represented in the form of a product of two layer-finite normal divisors, then the group G is layer-finite.*

THEOREM 2.5 (V.I. Senashov, [28,29]). *A periodic group is layer-finite if and only if it is conjugately biprimitively finite and every one of its locally finite subgroups is layer-finite.*

COROLLARY 2.1. *If in a binary finite group any locally soluble subgroup is layer-finite, then the group is layer-finite too.*

The statement of the corollary immediately follows from Theorem 2.5.

The condition of conjugate biprimitive finiteness in the theorem is necessary because there are examples such as the Novikov–Adian group, [1], and the Ol'shanskii group, [21], in which all the conditions of the theorem are valid except for the condition of finiteness, but they are not layer-finite.

It is impossible to make this condition weaker by replacing it by the F^*-condition (for the definition of the F^*-condition see below), because the Ol'shanskii group from [21] is an F^*-group and any of its locally soluble subgroups is layer-finite, but the group is not layer-finite. So the condition of conjugate biprimitive finiteness is limiting in the theorem.

The next theorem gives one more characterization of layer-finite groups in the class of periodic binary soluble groups: here the condition of layer-finiteness is not global, but only for subgroups with the element a.

THEOREM 2.6 (V.O. Gomer, [12]). *Let G be a periodic binary soluble group, and a an element of prime order p such that*:

(1) *in $C_G(a)$ every locally finite subgroup is layer-finite and has finite Sylow q-subgroups for all primes q*;

(2) *every locally finite subgroup with the element a is layer-finite.*

Then G is a locally soluble (locally finite) layer-finite group.

The next results characterized layer-finite groups in the class of periodic almost locally soluble groups with the condition: the centralizer of any non-identity element from some elementary Abelian subgroup of order p^2 is layer-finite.

THEOREM 2.7 (M.N. Ivko, [13]). *Let G be a periodic almost locally soluble group, possessing an elementary Abelian subgroup V of order p^2. If the centralizer in G of any non-identity element from V is layer-finite, then the group G is layer-finite.*

THEOREM 2.8 (M.N. Ivko, [13]). *The 2-biprimitively-finite group G of the form $G = H\lambda L$, where H is the subgroup without involutions, and L is the Klein four group, is layer-finite if and only if the centralizer in G of any involution from L is layer-finite.*

Recall that an element of the order 2 is called an *involution*.

The next corollary characterizes layer-finite groups in the class of periodic groups without involutions.

COROLLARY 2.2. *The periodic group H without involutions whose holomorph contains a Klein four subgroup L, is layer-finite if and only if the centralizer in H of any involution from L is layer-finite.*

A group G satisfies the *p-minimality condition* (min-p condition), if every descending chain of subgroups of it

$$H_1 > H_2 > \cdots > H_n > \cdots$$

is such that if $H_n \setminus H_{n+1}$ contains p-elements for all n it terminates at a finite number.

By $\pi(G)$ we shall denote a set of prime divisors of orders of elements of the group G.

A group G satisfies the *primary minimality condition*, if it satisfies the p-minimality condition for every prime number $p \in \pi(G)$.

A group G is called *finitely approximate* if for every set of different elements of G there is a homomorphism of G on a finite group such that the images of these elements are different.

The group G is called an *F_q-group* ($q \in \pi(G)$), if for every finite subgroup H and any elements $a, b \in T = N_G(H)/H$ of order q there exists an element $c \in T$, such that $\langle a, c^{-1}bc\rangle$ is finite.

If every subgroup from G is an F_q-group, then G is called an *F_q^*-group*. If G is a F_q- (respectively, F_q^*-) group for any number $q \in \pi(G)$, then G is called simply an *F- (respectively, F^*-) group*.

The next theorem gives the feature of layer-finiteness of periodic finitely approximate F^*-group, in which locally finite subgroups are layer-finite.

THEOREM 2.9 (E.I. Sedova, [27]). *A periodic finitely approximate F^*-group, in which every locally finite subgroup is layer-finite, is a layer-finite group.*

On the base of this result, descriptions of locally solvable layer-finite groups and groups with primary minimality condition in the class of binary solvable groups are given by the next theorems.

THEOREM 2.10 (E.I. Sedova, [27]). *A periodic group is a locally soluble layer-finite group if and only if it is binary soluble and every one of its locally soluble subgroup is layer-finite.*

THEOREM 2.11 (E.I. Sedova, [27]). *A periodic group is a locally soluble group with min-p condition if and only if it is binary soluble and every one of its locally soluble subgroups satisfies min-p condition.*

Recall that a group G satisfies the *p-minimality condition* (min-p condition), if every descending chain of its subgroups

$$H_1 > H_2 > \cdots > H_n > \cdots$$

is such, that if $H_n \setminus H_{n+1}$ contains p-elements for all n, it terminates at a finite number.

3. Groups with layer-finite periodic part

Remind that the *periodic part* of groups is a set of all of its elements that have finite order if they form a subgroup.

Here we adduce some criteria of layer-finiteness for the periodic part of a group.

THEOREM 3.1 (V.I. Senashov, [32]). *A group has a layer-finite periodic part if and only if it is conjugately biprimitively finite and every one of its locally solvable subgroup is layer-finite.*

THEOREM 3.2 (M.N. Ivko, V.P. Shunkov, [15]). *A group G without involutions has a layer-finite periodic part if and only if in it for some element a of prime order p the following conditions hold*:

(1) *the normalizer of any non-trivial $\langle a\rangle$-invariant finite elementary Abelian subgroup of the group G has a layer-finite periodic part*;

(2) *almost all subgroups of the form $\langle a, a^g\rangle$ are finite.*

COROLLARY 3.1. *A group G without involutions is layer-finite if and only if in it for some element a of prime order p the following conditions hold*:

(1) *the normalizer of any non-trivial $\langle a\rangle$-invariant finite elementary Abelian subgroup from G is layer-finite*;

(2) *almost all subgroups of the form $\langle a, a^g\rangle$ are finite.*

To obtain the characterization of groups having a layer-finite periodic part, in one of the group classes not containing involutions, it would be reasonable to obtain such a characterization in the general case. This problem with some additional limitations is solved by:

THEOREM 3.3 (M.N. Ivko, V.P. Shunkov, [15]). *A group G, containing an element a of prime order $p \neq 2$ has a layer-finite periodic part if and only if the following conditions hold*:

(1) *the normalizer of any non-trivial $\langle a\rangle$-invariant finite subgroup of the group G has a layer-finite periodic part*;

(2) *any locally finite subgroup, containing the element a is almost locally soluble*;

(3) *all subgroups of the form $\langle a, a^g\rangle$, where $g \in G$, are finite and almost all of them are soluble.*

COROLLARY 3.2. *An infinite group G is layer-finite if and only if in it for some element a of prime order $p \neq 2$ the following conditions are fulfilled*:

(1) *the normalizer of any non-trivial $\langle a\rangle$-invariant finite subgroup of the group G is layer-finite*;

(2) *any locally finite subgroup, containing the element a, is almost locally soluble*;

(3) *all subgroups of the form $\langle a, a^g\rangle$, where $g \in G$, are finite.*

Removing in Theorem 3.3 condition (2) and replacing the condition (1) with a stronger restriction, we can obtain another characterization of groups, having a layer-finite periodic part. Namely, the following theorem is valid.

THEOREM 3.4 (M.N. Ivko, V.P. Shunkov, [15]). *An infinite group G has a layer-finite periodic part if and only if in it for some element a of prime order $p \neq 2$ the following conditions are fulfilled*:

(1) *the normalizer of any non-trivial $\langle a\rangle$-invariant locally finite subgroup of the group G has a layer-finite periodic part*;

(2) *all subgroups of the form $\langle a, a^g\rangle$, where $g \in G$, are finite and almost all are soluble.*

COROLLARY 3.3. *An infinite group G is layer-finite if and only if in it for some element a of prime order $p \neq 2$ the following conditions are valid:*

(1) *the normalizer of any non-trivial $\langle a \rangle$-invariant locally finite subgroup of the group G is layer-finite;*

(2) *all subgroups of the form $\langle a, a^g \rangle$, where $g \in G$, are finite and almost all are soluble.*

We have so characterized groups with layer-finite periodic part with Φ-group accuracy (see the definition in the section "Φ-groups").

THEOREM 3.5 (M.N. Ivko, V.I. Senashov, [14]). *Let G be a group, a an involution of it, that satisfies the following conditions:*

(1) *all the subgroups of the form $\langle a, a^g \rangle$, $g \in G$, are finite;*

(2) *the normalizer of every non-trivial $\langle a \rangle$-invariant finite subgroup has a layer-finite periodic part.*

Then either the set of all finite order elements generates a layer-finite group or G is a Φ-group.

4. Generalizations of layer-finite groups

Now, let's discuss a few different generalizations of layer-finite groups as described by L.A. Kurdachenko in [17,18]. These generalizations appeared under different considerations of layer-finite groups.

The first generalization appeared on the base of the definition of layer-finite group: from the definition remove the demand of finiteness from a finite number of layers. Such groups are named QLF-groups. For such groups there holds

THEOREM 4.1 (L.A. Kurdachenko, [17]). *For every locally finite QLF-group is either an almost layer-finite extension of finite group by a QLF-group, or all non-trivial layers of it are infinite (so, the group has a finite number of layers).*

COROLLARY 4.1. *The orders of the elements of a locally normal group, which has at least one infinite layer, bounded globally.*

Layer-finite groups can be consider as groups, in which every primary layer is finite. On this basis one more generalization of layer-finiteness is constructed.

A group G is named an *LB-group*, if the number of all its infinite primary layers is finite.

THEOREM 4.2 (L.A. Kurdachenko, [17]). *Let G be a locally finite LB-group. Then it is an extension of locally normal LB-group by a group with a global bound on the orders of its elements. If, in particular, all Sylow subgroups of a group G are Chernikov, then it is almost layer-finite.*

Earlier, layer-finite groups were characterized as locally normal groups with Chernikov Sylow subgroups. Hence appears a second generalization of layer-finite group: groups in which the Sylow p-subgroups are non-Chernikov only for a finite set of prime numbers p (QSE-group). More precisely:

A periodic group G is called a *QSE-group*, if only for a finite (and particularly an empty set) of prime numbers the p Sylow p-subgroups of the group G are non-Chernikov.

The next theorem shows the relation between QSE-groups and LB-groups under the condition of locally normality.

THEOREM 4.3 (L.A. Kurdachenko, [17]). *A locally normal QSE-group is LB-group if and only if when the orders of all its elements are globally bounded.*

Further developments of investigations on these generalizations of layer-finite groups can be find in the paper [17] under additional conditions of finiteness of classes of conjugate elements.

5. Layer-Chernikov groups

Here we consider layer-Chernikov groups, i.e. the groups, in which every set of elements of the same order generates a Chernikov group. This class of groups was called *layer-extreme groups* in the work of Ya.D. Polovitskii, [24], in 1960, when the name "Chernikov groups" for the class of finite extensions of the Abelian groups with minimality condition had not yet been settled down and such groups were called extreme or *ch*-groups.

The layer-Chernikov groups are close to layer-finite groups not in the sense of the definitions only, but also in their properties. In Section 2, in particular, the layer-finite groups have been specified as locally normal groups with Chernikov Sylow p-subgroups. The same role with respect to the layer-Chernikov groups is played by the locally Chernikov groups. A group is called *locally Chernikov* if every element is contained in its Chernikov normal divisor. Further we examine the subclass of the locally Chernikov groups: the locally finite groups, whose quotient groups by every p-center are Chernikov (the *p-center* of an arbitrary group is the name of the intersection of the centralizers of all its p-elements). This allows to define the class of layer-Chernikov groups in a new fashion.

A direct product of a number of periodic groups is called a *primary thin direct product*, if whatever is the prime number p, the p-elements are contained in not more, than in a finite number of direct factors.

THEOREM 5.1 (Ya.D. Polovitskii, [25]). *The layer-Chernikov groups and only they are the subgroups of the primary thin direct products of Chernikov groups.*

This theorem can be strengthened, taking into account Theorem 8.1 of Chernikov from [8] as follows:

THEOREM 5.2. *A layer-Chernikov group embeds into a primary thin direct product of such Chernikov groups, that the maximum complete subgroup of every one of them is a p-group, with the orders of the elements of every two maximum complete subgroups of different multipliers of this direct product being mutually prime.*

COROLLARY 5.1. *The quotient group of the layer-Chernikov group with respect to the arbitrary p-center is a Chernikov group.*

A description of layer-Chernikov groups is given by the next theorem:

THEOREM 5.3 (Ya.D. Polovitskii, [25]). *The group G is layer-Chernikov if and only if it decomposes into a product $G = AB$ where A is an layer-finite complete Abelian group that is invariant in G and B is a thin layer-finite group, and for every number $p \in \pi(G)$ all Sylow q-subgroups of the group A (q is a prime number), except for, maybe, a finite number, are in the p-center of the group G.*

A layer-Chernikov group does not necessarily decompose into the semi-direct product of a complete Abelian group and a thin layer-finite group, as such an assertion does not take place even for the layer-finite groups.

The following theorems describe relations between the locally Chernikov and layer-Chernikov groups.

THEOREM 5.4 (Ya.D. Polovitskii, [25]). *The class of layer-Chernikov groups coincides with the class of locally Chernikov groups which have Chernikov Sylow p-subgroups (with respect to all p).*

THEOREM 5.5 (Ya.D. Polovitskii, [25]). *A central extension of a periodic group by means of a layer-Chernikov group is a locally Chernikov group.*

The next theorem gives a property of π-minimality of a locally Chernikov group.

THEOREM 5.6 (Ya.D. Polovitskii, [25]). *A locally Chernikov group satisfies the π-minimality condition if and only if all its Sylow π-subgroups are Chernikov ones.*

The locally Chernikov groups can also be determined in a different way as seen from

THEOREM 5.7 (Ya.D. Polovitskii, [25]). *A group G is locally Chernikov if and only if every Chernikov subgroup of it is contained in some Chernikov normal divisor of the group G.*

COROLLARY 5.2. *A central extension of a layer-Chernikov group by means of a layer-Chernikov group is a layer-Chernikov group.*

The assertion, analogous to Theorem 5.5 for central extensions by means of layer-finite groups has been proved by I.I. Yeremin, [10].

Let's add some more results for layer-Chernikov groups.

THEOREM 5.8 (Ya.D. Polovitskii, [25]). *The quotient groups of a locally finite group G by each p-center are Chernikov if and only if the group G is a central extension of a periodic group by means of a layer-Chernikov group.*

THEOREM 5.9 (Ya.D. Polovitskii, [44]). *A group G is a central extension of a periodic group by means of a thin layer-finite group if and only if its quotient groups are finite over each p-center.*

COROLLARY 5.3. *A locally finite group such that the quotient groups over each p-center are Chernikov is locally Chernikov.*

COROLLARY 5.4. *A group G is layer-Chernikov if and only if all its Sylow p-subgroups are Chernikov and the quotient groups by each p-center are Chernikov.*

It is of interest to note that by virtue of Theorem 5.3, a layer-Chernikov group is an extension of a complete Abelian group with Chernikov Sylow subgroups by means of the layer-finite group.

6. Generalized Chernikov groups

It is very natural to use the theory of layer-finite groups when one studies generalized Chernikov groups. Because such groups with conditions of a periodic type and almost locally solvability are extensions of layer-finite groups by layer-finite groups.

The imposition of restrictions on chains of subgroups has often been used in investigations concerning the structure of infinite groups. Many authors have in particular considered groups satisfying the minimality condition on all subgroups or the minimality condition on all Abelian subgroups. On the other hand, generalizing the concept of a locally finite group, S.P. Strunkov introduced in [59] binary finite groups, and later V.P. Shunkov, [46], considered the wider class of conjugately biprimitively finite groups. One of the main results relating to these finiteness conditions is due to A.N. Ostylovskii and V.P. Shunkov, [22], and states that a conjugate biprimitive finite group without involutions and satisfying the minimality condition on subgroups is a Chernikov group. Examples of P.S. Novikov, S.I. Adian, [1], and A.Yu. Ol'shanskii, [21], show that this result cannot be generalized to arbitrary periodic groups without involutions.

We shall say that a group G satisfies the *primary minimality condition* if for each prime p every chain

$$G_1 > G_2 > \cdots > G_n > \cdots$$

of subgroups of G, such that each set $G_n \setminus G_{n+1}$ contains an element g_n with $g_n^{p^{k_n}} \in G_{n+1}$ for some k_n, stops after finitely many steps.

An almost locally soluble group satisfying the primary minimality condition will be called a *generalized Chernikov group*. This name is motivated by the following result: every almost locally soluble group G satisfying the primary minimality condition contains a complete part $\widetilde{G}$, the quotient group $G/\widetilde{G}$ is locally normal, and every element of G centralizes all but finitely many Sylow subgroups of $\widetilde{G}$ (see, for instance, [23]).

Recall that a group G has a *complete part* A, if A is an Abelian group generated by all complete Abelian subgroups from G and G/A does not have complete Abelian subgroups.

The next results characterize generalized Chernikov groups in the class of periodic groups without involutions and in the class of mixed groups.

THEOREM 6.1 (V.I. Senashov, [31]). *Let G be a periodic group without involutions. Then G is a generalized Chernikov group if and only if it is conjugately biprimitively finite and the normalizers of all its finite non-trivial subgroups are generalized Chernikov groups.*

The earlier mentioned examples by Novikov, Adian, [1], and Ol'shanskii, [21], prove that in this theorem the condition that G is conjugately biprimitively finite cannot be removed.

It is possible to consider class of groups with generalized Chernikov periodic part of the normalizer of any finite non-trivial subgroup. Among these groups there are the examples of the Novikov–Adian and Ol'shanskii groups.

The term "generalized Chernikov groups" was first used in [57]. Its use can be justified by the fact that according to the theorem of Ya.D. Polovitskii a generalized Chernikov group G is an extension of the direct product A of quasi-cyclic p-groups with a finite number of multipliers for any prime number p by a locally normal group B, and each of the elements from B is element-wise non-permutable with only a finite number of Sylow primary subgroups from A. For comparison a Chernikov group is a finite extension of a direct product of finite number of quasi-cyclic groups.

Here some properties of generalized Chernikov groups.

THEOREM 6.2 (V.I. Senashov, [33]). *In a generalized Chernikov group primary subgroups are Chernikov.*

THEOREM 6.3 (V.I. Senashov, [33]). *If a generalized Chernikov group G does not have complete subgroups, it is a thin layer-finite group.*

THEOREM 6.4 (V.I. Senashov, [33]). *In a non-Chernikov generalized Chernikov group any element has infinite centralizer.*

Let's introduce the definition of a T-group.

DEFINITION. Let G be a group with involutions. Each involution i from G is associated with a subgroup V_i from G defined as follows. If the Sylow 2-subgroups from the order eighth dihedral group and i are contained in a Klein four-subgroup R_i such that $C_G(i) < N_G(R_i)$ and $C_G(i)$ has an infinite torsion subgroup, we take $V_i = N_G(R_i)$. In all other cases V_i is taken to be $V_i = C_G(i)$.

A group G with involutions will be said to be a T-group if it satisfies the conditions:

(1) any two involutions from G generate a finite subgroup;
(2) the normalizer of any locally finite subgroup from G containing involutions has locally finite periodic part;
(3) the set $G \setminus V_i$ possesses involutions and V_i is an infinite subgroup for every involution i from G;
(4) for every element c from $G \setminus V_i$ strictly real with respect to i, for which $c^i = c^{-1}$, there exists in $C_G(i)$ such an element s_c, that the subgroup $\langle c, c^{s_c} \rangle$ is infinite.

The class of T-groups has been introduced by V.P. Shunkov.

Recall that a group, generated by two involutions is called a *dihedral group*. An element a of a group G is called *strictly real* with respect to an involution $i \in G$, if $iai^{-1} = a^{-1}$.

In 1987 there appeared a conjecture by V.P. Shunkov according to which his Theorem 3.1 from [51] could possibly generalize to more wide classes of groups. And, indeed, in Theorem 6.5 below the second condition of the Shunkov theorem is replaced by a weaker condition: a normalizer with involutions of every finite non-trivial subgroup has generalized Chernikov periodic part. We proved that under this condition and if every subgroup generated by two involutions is finite, then it has a generalized Chernikov periodic part or it is T-group. So, the conjecture is proved.

THEOREM 6.5 (V.I. Senashov, [33,34]). *Let G be a group with involutions satisfying the conditions*:

(1) *any two involutions from G generate a finite subgroup*;
(2) *the normalizer of any finite non-trivial subgroup containing involutions has generalized Chernikov periodic part.*

Then either G has a generalized Chernikov periodic part or G is a T-group.

The next theorem characterizes groups with generalized Chernikov periodic part.

THEOREM 6.6 (V.I. Senashov, [33,34]). *A group has a generalized Chernikov periodic part if and only if it is conjugate biprimitive finite and the normalizer of any finite non-trivial subgroup containing involutions has generalized Chernikov periodic part.*

7. T_0-groups

At the beginning of the 90ies the concept of a T_0-group appeared. This class is defined by finiteness conditions. Here is the definition of the class of T_0-groups (V.P. Shunkov).

DEFINITION. Let G be a group with involutions, and let i be one of its involutions. We shall call the group G a T_0-group, if it satisfies the following conditions (for the given involution i):

(1) all subgroups of the form $\langle i, i^g \rangle$, $g \in G$, are finite;
(2) Sylow 2-subgroups from G are cyclic or generalized groups of quaternions;
(3) the centralizer $C_G(i)$ is infinite and has a finite periodic part;
(4) the normalizer of any non-trivial $\langle i \rangle$-invariant finite subgroup from G is either contained in $C_G(i)$, or has a periodic part being a Frobenius group with Abelian kernel and a finite complement of even order;
(5) $C_G(i) \neq G$ and for any element c from $G \setminus C_G(i)$, strictly real relating i (i.e. such that $c^i = c^{-1}$), there is an element s_c in $C_G(i)$, such that the subgroup $\langle c, c^{s_c} \rangle$ is infinite.

Let's consider the construction of Shunkov's example of a T_0-group from [52] based on the well-known example of S.P. Novikov, S.I. Adian, [1].

EXAMPLE OF A T_0-GROUP [52]. Let $A = A(m,n)$ be a torsion-free group $A(m,n)$, which is a central extension of cyclic group with help of group $B(m,n)$, $m > 1$, $n > 664$ an odd number, [1]. The group $A(m,n)$ has non-trivial center $Z(A) = \langle d \rangle$ and $A/\langle d \rangle$ is isomorphic to $B(m,n)$, [1]. Let's consider a group $B = A \wr \langle x \rangle$, where x is an involution.

Let's take an element $u = d \cdot d^{-x}$ from $A \times A^x$. It is obvious, that $u \in Z(A \times A^x)$, and $u^x = u^{-1}$. As is shown in [52], the group $G = B/\langle u \rangle$, and it's involution $i = x \cdot \langle u \rangle$ satisfy conditions (1)–(5) from the definition of a T_0-group, and $G = V\lambda\langle i \rangle$, $C_G(i)$ is an infinite group with periodic part $\langle i \rangle$, all subgroups $\langle i, i^g \rangle$ in G are finite and every maximal finite subgroup from G with involution i is a dihedral group of the order $2n$ and G is a T_0-group.

EXAMPLE OF A T_0-GROUP [53]. Let $V = O(p)$ (see the definition of groups of the type $O(p)$, $C(\infty)$ in the introduction). The group V has a non-trivial center $Z(V) = (t)$ and $V/Z(V) = V/(t) \simeq C(\infty)$, [16].

Consider the group $T = V \wr (k) = (V \times V)\lambda(k)$, where k is an involution. Let us take from $V \times V$ the element $b = (t, t^{-1})$. Obviously, $b \in Z(V \times V)$ and $b^k = b^{-1}$. Let us take quotient group $M = T/(b)$, and in it an involution $j = k(b)$. Further, using abstract properties of the groups $V = O(p)$, $C(\infty)$, [16], it is easy to show that the group M and its involution j satisfy the conditions (1)–(5) of the definition. Hence, $M = T/(b)$ is a T_0-group (with respect to the involution $j = k(b)$). Let us also note that in M any maximum periodic subgroup containing the involution j is a dihedral group of order $2p$.

Here are some results on T_0-groups. Details can be found in [55].

THEOREM 7.1 (V.P. Shunkov, [53,56]). *Let G be a group and a be an element of prime order p, satisfying the following conditions*:

(1) *the subgroups of the form $\langle a, a^g \rangle$, $g \in G$, are finite and almost all are solvable*;
(2) *in the centralizer $C_G(a)$ the set of elements of finite order is finite*;
(3) *in the group G the normalizer of any non-trivial $\langle a \rangle$-invariant finite subgroup has periodic part*;
(4) *for $p \neq 2$ and for $q \in \pi(G)$, $q \neq p$, any $\langle a \rangle$-invariant elementary Abelian q-subgroup of G is finite.*

Then either G has an almost nilpotent periodic part, or G is a T_0-group and $p = 2$.

COROLLARY 7.1. *Let G be a (periodic) group and let a be an element of prime order $p \neq 2$, satisfying conditions* (1)–(4) *of Theorem* 7.1.

Then G has an almost nilpotent periodic part.

The following statement is equivalent to Theorem 7.1 and gives an abstract characterization of T_0-groups in the class of all groups.

COROLLARY 7.2. *Let G be a group, a be an element of prime order p. The group G is a T_0-group and $p = 2$ if and only if for the pair (G, a) the conditions* (1)–(4) *of theorem are satisfied and the subgroup $\langle a^g \mid g \in G \rangle$ is not periodic almost nilpotent.*

The particular case when $p = 2$ requires special consideration, since in this case condition (4) of Theorem 7.1 is superfluous, i.e. the following statements are true.

COROLLARY 7.3. *Let G be a group with involutions and i be one of its involutions, satisfying the following conditions*:

(1) *the subgroups of the form $\langle i, i^g \rangle$, $g \in G$, are finite*;
(2) *in the centralizer $C_G(i)$ the set of elements of finite order is finite*;
(3) *in the group G the normalizer of any non-trivial $\langle i \rangle$-invariant finite subgroup has periodic part.*

Then either G has almost nilpotent periodic part, or G is a T_0-group.

The conditions (1)–(3) of Corollary 7.3 are independent, i.e. none of them follows from the other two.

COROLLARY 7.4. *Let G be a group with involutions, and let i be one of its involutions. The group G is a T_0-group if and only if for the pair (G, i) conditions* (1)–(3) *of Corollary* 7.3 *are satisfied and the subgroup $\langle i^g \mid g \in G \rangle$ is not periodic almost nilpotent.*

THEOREM 7.2 [53]. *Let G be a group with involutions and i be an involution, satisfying the following conditions*:

(1) *the subgroups of the form $\langle i, i^g \rangle$, $g \in G$, are finite*;
(2) *in the centralizer $C_G(i)$ the set of elements of finite order is finite*;
(3) *in the group G the normalizer of any non-trivial $\langle i \rangle$-invariant finite subgroup has periodic part.*

Then either G has almost nilpotent periodic part, or G is a T_0-group.

THEOREM 7.3 [53]. *Let G be a group and a be an element of prime order p, satisfying the following conditions*:

(1) *subgroups of the form $\langle a, a^g \rangle$, $g \in G$, are finite and almost all are solvable*;
(2) *the centralizer $C_G(a)$ is finite*;
(3) *for $p \neq 2$ and for $q \in \pi(G)$, $q \neq p$, any $\langle a \rangle$-invariant elementary Abelian q-subgroup of G is finite.*

Then G is a periodic almost nilpotent group.

THEOREM 7.4 [53]. *A non-trivial finitely generated group G is finite if and only if in it there exists an element a of prime order p satisfying the following conditions*:

(1) *the subgroups of the form $\langle a, a^g \rangle$, $g \in G$, are finite and almost all solvable*;
(2) *the centralizer $C_G(a)$ is finite*;
(3) *when $p \neq 2$ and for $q \in \pi(G)$, $q \neq p$, any (a)-invariant elementary Abelian q-subgroup is finite.*

Theory of T_0-groups was created by V.P. Shunkov in [55].

8. Φ-groups

This section investigates properties of the new class of Φ-groups. Such groups are very close to T_0-groups, but in this sections we also point out the difference.

This group class is rather broad: among them are groups of Burnside type, [1], Ol'shanskii monsters, [21]. It is very closely connected with the groups of Burnside type of odd period $n \geqslant 665$.

DEFINITION. Let G be a group, let i be an involution of G, satisfying the following conditions:

(1) all subgroups of the form $\langle i, i^g \rangle$, $g \in G$, are finite;
(2) $C_G(i)$ is infinite and has a layer-finite periodic part;
(3) $C_G(i) \neq G$ and $C_G(i)$ is not contained in other subgroups from G with a periodic part;
(4) if K is a finite subgroup from G, which is not inside $C_G(i)$, and $V = K \cap C_G(i) \neq 1$, then K is a Frobenius group with complement V.

The group G with a specified involution i satisfying these conditions (1)–(4) is called a *Φ-group*.

This class of groups has been introduced by V.P. Shunkov.

EXAMPLE OF A Φ-GROUP (V.P. Shunkov). Let $A = \langle b, c \rangle$ (where $b^n = c^n = d$ and n is a positive integer) be a torsion free group and let $A/\langle d \rangle$ be the free Burnside group with period n, [1]. Consider the group $B = A \wr \langle x \rangle = (A \times A)\lambda \langle x \rangle$, where x is an involution. Let us take from $A \times A$ the element $v = (d, d^{-1})$. Obviously $v \in Z(A \times A)$ and $v^x = v^{-1}$. Further, the group $G = B/\langle v \rangle$ and its involution $i = x\langle v \rangle$ (which is easy to see from the abstract properties of the group $A = \langle b, c \rangle$, [1]) satisfy all the conditions from the definition of a Φ-group. Hence, $G = B/\langle v \rangle$ is a Φ-group.

THEOREM 8.1 (V.I. Senashov, [14]). *A Φ-group G satisfies the properties*:

(1) *all involutions are conjugate*;
(2) *Sylow 2-subgroups are locally cyclic or finite generalized quaternion groups*;
(3) *there are infinitely many elements of finite order in G, which are strictly real with respect to the involution i and for every such element c of this set there exists an element s_c from the centralizer of i such that $\langle c, c^{s_c} \rangle$ is an infinite group.*

V.P. Shunkov posed the problem of studying groups with some additional limitations in the form that for the given finite subgroup B, the next condition is valid: the normalizer of any non-trivial B-invariant finite subgroup has a layer-finite periodic part.

This problem is partly solved in the class of locally soluble groups and for the case $|B| = 2$ under more general limitations, it is solved with Φ-groups accuracy.

THEOREM 8.2 (M.N. Ivko, V.I. Senashov, [14]). *A periodic locally soluble group is layer-finite if and only if for some finite subgroup B of it the next condition is valid*: *the normalizer of any non-trivial B-invariant finite subgroup is layer-finite.*

THEOREM 8.3 (M.N. Ivko, V.I. Senashov, [14]). *Let G be a group, let a be an involution of G, satisfying the conditions*:

(1) *all subgroups of the form $\langle a, a^g \rangle$, $g \in G$, are finite*;

(2) *the normalizer of every non-trivial* $\langle a\rangle$*-invariant finite subgroup has a layer-finite periodic part.*

Then either the set of all elements of finite orders forms a layer-finite group or G *is an* Φ*-group.*

COROLLARY 8.1 (M.N. Ivko, V.I. Senashov, [14]). *Let* G *be a group with involutions and let* i *be some involution from* G *satisfying the conditions*:

(1) G *is generated by the involutions which are conjugate with* i;

(2) *almost all groups* $\langle i, i^g\rangle$ *are finite*, $g \in G$;

(3) *the normalizer of every* $\langle i\rangle$*-invariant finite subgroup has a layer-finite periodic part.*

Then G *is either a finite or an* Φ*-group.*

In a Φ-group G all involutions are conjugate; the Sylow 2-subgroups are locally cyclic or finite generalized quaternion groups; there are infinitely many elements of finite order in G, which are strictly real with respect to the involution i and for every such element c of this set there exists an element s_c from the centralizer of i such that $\langle c, c^{s_c}\rangle$ is an infinite group.

Layer-finite groups are characterized in the class of locally solvable groups and groups with a layer-finite periodic part in more general case with Φ-groups accuracy.

In the article [52], V.P. Shunkov bring up next question for discussion:

Do the classes of Φ_0-groups and T_0-groups coincide or not?

In the same article V.P. Shunkov specially emphasized that the most difficult part of the problem is the establishing of satisfiability for a Φ_0-group of conditions (4) and (5) from the definition of a T_0-group.

In [35] V.I. Senashov proved, that a Φ_0-group satisfies all conditions from the definition of a T_0-group except for the fourth condition. In the same article an example of a Φ_0-group which is not a T_0-group was constructed, i.e. it was shown that the fourth condition does not hold in every Φ_0-group.

EXAMPLE OF A Φ_0-GROUP (V.I. Senashov, [35]). Let's take isomorphic copies of the T_0-groups $G = V\lambda(i)$ from [52]:

$$G_1 = V_1\lambda(i_1),\ G_2 = V_1\lambda(i_1),\ \ldots,\ G_n = V_n\lambda(i_n),\ \ldots.$$

In the Cartesian product of the groups G_n, $n = 1, 2, \ldots$, consider the subgroup $U = W\lambda(j)$, where W is the direct product of the subgroups V_n, $n = 1, 2, \ldots$, and $j = i_1 \cdot i_2 \cdots$ is an involution from the Cartesian product of the G_n, $n = 1, 2, \ldots$. Such a group U is a Φ_0-group. It is easy to see that fourth condition from the definition of a T_0-group does not hold for the group U.

9. Almost layer-finite groups

A group is said to be the *almost layer-finite* if it is a finite extension of layer-finite group.

To start with here are some theorems which describe almost layer-finite groups in the class of locally finite groups.

THEOREM 9.1 (V.P. Shunkov, [40]). *A locally finite group G is almost layer-finite if and only if in G the following condition is valid: the normalizer of any non-trivial finite subgroup from G is almost layer-finite.*

Next, here is the theorem which characterizes almost layer-finite groups in the class of periodic groups without involutions.

THEOREM 9.2 (V.I. Senashov, [37]). *Let G be a conjugately biprimitively finite group without involutions. If in G the normalizer of any non-trivial finite subgroup has an almost layer-finite periodic part, the group G has an almost layer-finite periodic part.*

The condition of conjugate biprimitive finiteness in this theorem should be taken into account in view of examples of the Novikov–Adian, [1], and Ol'shanskii groups, [21].

The next theorem describes almost layer-finite groups in the class of periodic groups with involutions.

THEOREM 9.3 (V.I. Senashov, [36,38]). *Let G be a periodic group of Shunkov type with strongly embedded subgroup. If in G the normalizer of any non-trivial finite subgroup from G is almost layer-finite, then the group G is almost layer-finite.*

Let's recall that a subgroup H of a group G is called *strongly embedded* in G, if H is a proper subgroup of G and $H \cap x^{-1}Hx$ has odd order for all $x \in G \setminus H$.

We know the structure of the infinite Sylow 2-subgroups of the periodic non-almost layer-finite group of Shunkov:

THEOREM 9.4 (V.I. Senashov, [38]). *Let G be the periodic non-almost layer-finite group of Shunkov with almost layer-finite normalizers of any non-trivial finite subgroups. If the Sylow 2-subgroup of group G is infinite, then it is a quasi-dihedral 2-subgroup.*

Recall, that a quasi-dihedral group is an extension of a quasi-cyclic 2-group with the help of an inverting automorphism (this group received this name because it is an union of infinite number of finite dihedral 2-groups).

Using the known results about locally finite groups with Chernikov primary Sylow subgroups, we now obtain the following characterization of almost layer-finite groups in the class of locally soluble groups, which is an analog of one of the main results of [41].

THEOREM 9.5 (M.N. Ivko, [13]). *Let G be a periodic almost locally soluble group, possessing a Klein four-subgroup L. If the centralizer in G of any involution from L is layer-finite, then the group G is almost layer-finite.*

10. Periodic groups with minimality condition

A group G satisfies the *minimality condition for subgroups* (*Abelian subgroups*), if in G every decreasing chain of subgroups (Abelian subgroups) $H_1 > H_2 > \cdots$ stops at a finite number, i.e. $H_n = H_{n+1} = \cdots$ for some n.

Let any decreasing chain of subgroups in the infinite locally finite group stop at a finite number. Will such a group be a finite extension of the direct product of the finite number of quasi-cyclic groups? This is the problem of minimality in the class of locally finite groups. In 1965 it became possible to reduce that problem to the case when the Sylow 2-subgroups in the group are finite. Those results were published in [42,43]. The problem of minimality in the class of locally finite groups was positively solved by V.P. Shunkov in [45]. Locally finite groups with the condition of minimality for Abelian subgroups were described in [47].

In 1968 P.S. Novikov and S.I. Adian published the solution of the famous Burnside problem. Moreover, the following theorem was proved: the free Burnside group $B(m,n)$, $m \geqslant 2$, of odd period $n \geqslant 4381$, is infinite, the centralizer of any non-trivial element is finite and is contained in a cyclic subgroup of order n from $B(m,n)$. In particular, in such a group all Abelian subgroups are finite, the group also satisfies the condition of minimality for Abelian subgroups. Moreover, for any odd prime number p and natural number s with $p^s > 4381$ the free Burnside p-group $B(m,p^s)$ is infinite and every elementary Abelian p-subgroup is finite. At once the following question emerged: what can be said about the 2-groups in which some maximal elementary Abelian subgroup is finite? In 1970, V.P. Shunkov obtained the answer to that question. In fact, he proved the following

THEOREM 10.1 (V.P. Shunkov, [46]). *If some maximal elementary Abelian subgroup is finite in an infinite 2-group, then the group itself is a finite extension of the direct product of the finite number of quasi-cyclic groups.*

Recall that a group G satisfies the *p-minimality condition* (min-p condition), if every descending chain of subgroups

$$H_1 > H_2 > \cdots > H_n > \cdots$$

that is such, that $H_n \setminus H_{n+1}$ contains p-elements, terminates at a finite number.

Earlier we already pay attention to the fact that the extension of a locally finite group by a locally finite group is locally finite, but at the same time this is incorrect for layer-finite groups. The periodic almost locally soluble groups with min-p condition are the extensions of layer-finite groups by layer-finite groups. In fact, in this case the group G has a complete part A, and G/A is locally normal, and every element of G is unpermutational element-wise only with finite number of Sylow p-subgroups of A. The quotient group G/A is locally normal and obviously has finite Sylow p-subgroups, hence it is layer-finite group. The group A is an Abelian group constructed from a quasi-cyclic group, and by the min-p condition every $p \in \pi(A)$ in A has only a finite number of quasi-cyclic groups. Hence A is a layer-finite group too.

Let's discuss some results for groups with minimality condition.

A positive solution of the minimality problem for locally finite groups is given by the next theorem:

THEOREM 10.2 (V.P. Shunkov, [45]). *A locally finite group with minimality condition for subgroups is either finite or it is a finite extension of a direct product of a finite number of quasi-cyclic groups.*

THEOREM 10.3 (Ya.D. Polovitskii, [24,25]). *Let G be an almost locally soluble group with min-p condition. Then it has a complete part A; moreover G/A is locally normal, and every element of G is unpermutational element-wise only with finite number of Sylow p-subgroups of A.*

THEOREM 10.4 (V.P. Shunkov, [49]). *Let G be a conjugately biprimitively finite group without involution with min-p condition. Then G has a complete part R and the quotient-group G/R is conjugate biprimitive finite with min-p condition and with finite Sylow p-subgroups for every $p \in \pi(G)$.*

THEOREM 10.5 (V.P. Shunkov, A.K. Shlepkin, [49]). *Every periodic conjugately biprimitively finite group without involutions with min-p condition is locally finite.*

COROLLARY 10.1. *Every periodic conjugately biprimitively finite group without involutions with finite Sylow subgroups, satisfying the min-p condition is layer-finite.*

THEOREM 10.6 (E.I. Sedova, [27]). *A periodic group is a locally soluble group with min-p condition if and only if it is binary soluble and every one of its locally soluble subgroups satisfies the min-p condition.*

THEOREM 10.7 (A.N. Ostylovskii, V.P. Shunkov, [22]). *A conjugately biprimitively finite group without involutions with minimality condition is locally finite and is a solvable Chernikov group.*

THEOREM 10.8 (N.G. Suchkova, V.P. Shunkov, [60]). *Every conjugately biprimitively finite group with minimality condition for Abelian subgroups is a Chernikov group.*

Results on groups with minimality condition can be found in the monograph [39].

11. Groups with finitely embedded involution

We now go on to introduce the next concept introduced by V.P. Shunkov at the end of the 80ies.

Let G be a group, i one of its involutions and let $L_i = \{i^g \mid g \in G\}$ be the set of conjugated involutions from G with i. We shall call the involution i *finitely embedded in G*, if for any element g from G the intersection $(L_i L_i) \cap gC_G(i)$ is finite, where $L_i L_i = \{i^{g_1} = i^{g_2} \mid g_1, g_2 \in G\}$.

Let's give the most simple examples of the groups with a finitely embedded involution.

(1) If in the group G there exists an involution i with finite centralizer $C_G(i)$, then i is a finitely embedded involution in G.
(2) If in some group G the involution i is contained in finite normal subgroup from G, then i is a finitely embedded involution in G.
(3) Let G be a Frobenius group with a periodic kernel and infinite complement H, containing an involution i. Then i is a finitely embedded involution in G.

(4) Let

$$B_1, B_2, \dots, B_n, \dots$$

be an infinite sequence of finite groups, in which there is only a finite number of the groups of even order, and let $B_n\lambda(i_n)$ be a subgroup from the holomorph $\mathrm{Hol}(B_n)$, where i_n is an involution, inducing in B_n an automorphism of order two ($n = 1, 2, \dots$). Let's consider the group $G = B\lambda(i)$, where

$$B = B_1 \times B_2 \times \dots \times B_n \times \cdots,$$

and i is the involution

$$i = (i_1, i_2, \dots, i_n, \dots).$$

It is easy to show, that i is a finitely embedded involution.

An involution of a group is called a *finite* involution, if it generates a finite subgroup with every involution, that is conjugate to it.

Now let us formulate some results, the main of which is the following.

THEOREM 11.1 (V.P. Shunkov, [50]). *Let G be a group, and let i be a finite and finitely embedded involution in it, $L_i = \{i^g \mid g \in G\}$, $B = \langle L_i \rangle$, $R = \langle L_i L_i \rangle$, and Z let be the subgroup generated by all 2-elements from R.*

Then B, R, Z are normal subgroups in G and one of the next statements holds:

(1) *B is a finite subgroup*;

(2) *the subgroup B is locally finite, $B = R\lambda(i)$ and Z is a finite extension of a complete Abelian 2-subgroup A_2 with the condition of minimality, and $ici = c^{-1}$ ($c \in A_2$).*

A number of corollaries follow from this theorem.

COROLLARY 11.1. *If a group has a finite involution with a finite centralizer, then it is locally finite.*

COROLLARY 11.2. *If a periodic group has an involution with a finite centralizer, then it is locally finite.*

COROLLARY 11.3. *If a finite finitely embedded involution exists in a group, then its closure is a periodic subgroup.*

COROLLARY 11.4. *A simple group with involutions is finite if and only if one of its involutions is finite and is finitely embedded.*

As in a periodic group any involution is finite, the next results follow from Corollary 11.4.

COROLLARY 11.5. *A periodic simple group with involutions is finite if and only if some involution in it is finitely embedded.*

COROLLARY 11.6. *Let G be a group, let H be a subgroup of it containing a finite involution, and let (G, H) be a Frobenius pair. The group G is a Frobenius group with a periodic Abelian kernel and with complement $H = C_G(i)$ if and only if i is a finitely embedded involution in G.*

Corollary 11.6 does not hold, even in a periodic group, if the involution i is not finitely embedded.

12. Frobenius groups and finiteness conditions

When one study groups with finiteness conditions it is very helpful to use Frobenius group type properties.

A group of the form $G = F\lambda H$ is called a *Frobenius group* with kernel F and complement H, if $H \cap H^g = 1$ for every $g \in G \setminus H$ and $F \setminus (1) = G \setminus \bigcup_{g\in G} H^g$, where H is a proper subgroup.

Let G be a group, H subgroup. G and H are said to form a *Frobenius pair*, if $H \cap H^x = 1$ for every element $x \in G \setminus H$.

Let G be a group, H a subgroup of it, satisfying the following condition: for any $g \in G \setminus H$ the intersection $H \cap g^{-1}H_g = 1$. In this case we shall call the pair (G, H) a Frobenius pair. If G is a finite group, $G = F\lambda H$. This is a famous Frobenius theorem which plays a fundamental role in the theory of finite groups.

Let a Frobenius pair (G, H) be given and $G = F\lambda H$. If $F \setminus (1) = G \setminus \bigcup_{g\in G} H^g$, then G is called a Frobenius group with kernel F and with complement H. According to the Adian theorem from the book "Burnside Problem and Identities in Groups" (Moscow: Science, 1975), in the group $B(m,n)$, $m \geqslant 2$, n is an odd number and $n \geqslant 665$, each finite subgroup is contained in a cyclic subgroup of order n, making with the group $B(m,n)$ a Frobenius pair, and $B(m,n)$ is an infinite group. If we take a prime number $p > 665$ as n, then according to the well-known Kostrikin theorem, $B(m, p)$ has a finitely generated subgroup $H(p)$ of finite index in $B(m, p)$, not having subgroup of the finite index of its own.

The group $H(p)$ with its each cyclic subgroup of the prime order p makes a Frobenius pair. However, it is not a Frobenius group with a non-invariant cyclic multiplier of prime order p. Let us give another interesting example.

Let $V = B(m, p)$, where p is a prime number and $p > 665$. As is well known, V has an automorphism ϕ of order 2 which takes all free generators into their inverse. In the holomorph $\mathrm{Hol}(V)$ take the subgroup $G = V\lambda(i)$, where i is an involution, inducing the automorphism ϕ in V. If the centralizer $C_G(i)$ were finite, then according to statement 4, formulated above, the group G would be finite, it couldn't be possible. Hence, $H = C_G(i)$ is an infinite group. Further, according to the Adian theorem, formulated above, (G, H) is a Frobenius pair, but G is not a Frobenius group with the non-invariant infinite multiplier $H = C_G(i)$.

That is how the situation in periodic groups having Frobenius pairs stands. But, probably, the following V.P. Shunkov theorem can lead to further progress.

THEOREM 12.1 (V.P. Shunkov, [48,50]). *Let G be a group, H a subgroup of it, a an element of prime order $p \neq 2$ from H, satisfying the following condition: for every $g \in G \setminus H$, $\langle a, g^{-1}ag \rangle$ is a Frobenius group with the complement* (a). *Then*

(1) *$H = T\lambda N_G((a))$ and $K = T\lambda(a)$ is either a Frobenius group with a complement (a) and a kernel T, or $K = (a)$;*
(2) *$F_a = T \cup N$ is a subgroup in G and $G = F_a N_G((a))$, where N is the set of all p-real elements from $G \setminus H$ relating the element a;*
(3) *$E = T \setminus L$ is an invariant set in G, where L is the set of all such elements from T, which is p-real relating some element from*

$$L = \bigcup_{x \in G} [(x^{-1}ax) \setminus \{1\}].$$

Finally we note a feature of unsimplicity for infinite groups.

THEOREM 12.2 (A.I. Sozutov, V.P. Shunkov, [58]). *Suppose G is a group, H a proper subgroup, a an element of order $p \neq 2$ in G such that*

(∗) *for almost all (i.e. except for perhaps a finite number) of elements of the form $g^{-1}ag$, where $g \in G \setminus H$, the subgroups $L_g = \langle a, g^{-1}ag \rangle$ are Frobenius groups with complement $\langle a \rangle$.*

Then either $G = F\lambda N_G(\langle a \rangle)$ and $F\lambda\langle a \rangle$ is a Frobenius group with complement $\langle a \rangle$ and kernel F, or the index of $C_G(a)$ in G is finite.

This feature of unsimplicity plays a very important role in the research of infinite groups with finiteness conditions.

References

[1] S.I. Adian, *Burnside Problem and Identities in Groups*, Nauka (1975) (in Russian).
[2] R. Baer, *Finiteness properties of groups*, Duke Math. J. **15** (1948), 1021–1032.
[3] M.Yu. Bakhova, *Example of a biprimitively finite group without involutions*, Abstracts of Reports at the 17th All-Union Algebra Conference, Minsk, 1983, 17.
[4] A.A. Cherep, *On the set of elements of finite order in a biprimitively finite group*, Algebra i Logika **26** (4) (1987), 518–521.
[5] A.A. Cherep, *Groups with (a, b)-conditions of finiteness*, Ph.D. Thesis, Krasnoyarsk (1993).
[6] S.N. Chernikov, *On the theory of infinite p-groups*, Dokl. Akad. Nauk SSSR **50** (1945), 71–74.
[7] S.N. Chernikov, *On the centralizer of a complete Abelian normal divisor in an infinite periodic group*, Dokl. Akad. Nauk SSSR **72** (2) (1950), 243–246.
[8] S.N. Chernikov, *Conditions of finiteness in a common group theory*, Uspekhi Mat. Nauk **14** (5) (1959), 45–96.
[9] S.N. Chernikov, *Groups with Fixed Properties of a Subgroups System*, Nauka (1980) (in Russian).
[10] I.I. Eremin, *On central extensions with thin layer-finite groups*, Izv. Vyssh. Uchebn. Zaved. Mat. **2** (1960), 93–95.

[11] E.S. Golod, *On some problems of Burnside type*, Transactions of International Congress of Mathematicians, Moscow, 1966, 284–289.
[12] V.O. Gomer, *Groups with elements of finite rank*, Ukrainian Math. J. **44** (6) (1992), 836–839.
[13] M.N. Ivko, *Characterization of groups with layer-finite periodic part*, Abstracts of cand. phys.-math. sciences, Ekaterinburg (1993).
[14] M.N. Ivko and V.I. Senashov, *On a new class of infinite groups*, Ukrainian Math. J. **47** (6) (1995), 760–770.
[15] M.N. Ivko and V.P. Shunkov, *On a characterization of groups which have layer-finite periodic parts*, Proceedings of Mathematics Institute of Academy of Science of Ukraine, 1993, Vol. 1: Infinite Groups and Some Algebraic Structures, 87–103.
[16] M.I. Kargapolov and Yu.I. Merzlyakov, *The Foundations of Group Theory*, 3rd ed., Nauka (1982) (in Russian).
[17] L.A. Kurdachenko, *Some generalizations of layer-finite groups*, Groups with Fixed Properties of Subgroups, Institute of Mathematics of AN of the Ukraine, 1973, 270–308.
[18] L.A. Kurdachenko, *Unperiodic groups with limitations for layers of elements*, Ukrainian Math. J. **26** (3) (1974), 386–389.
[19] Kh.Kh. Mukhamedjan, *On groups with ascending central chain*, Mat. Sbornik **28** (1951), 185–196.
[20] Kh.Kh. Mukhamedjan, *On groups with ascending invariant chains*, Mat. Sbornik **39** (2) (1956), 201–218.
[21] A.Yu. Ol'shanskii, *Geometry of Definite Relations in a Group*, Nauka (1989) (in Russian).
[22] A.N. Ostylovskii and V.P. Shunkov, *Local finiteness of a class of groups with minimum condition*, Studies in Group Theory, Krasnoyarsk, 1975, 32–48.
[23] I.I. Parvluk, A.A. Shafiro and V.P. Shunkov, *On locally finite groups with min-p condition for subgroups*, Algebra i Logika **13** (3) (1974), 324–336.
[24] Ya.D. Polovitskii, *Layer-extremal groups*, Dokl. Akad. Nauk SSSR **134** (3) (1960), 533–535.
[25] Ya.D. Polovitskii, *On locally extremal and layer-extremal groups*, Mat. Sbornik **58** (2) (1962), 685–694.
[26] A.V. Rojkov, *Conditions of finiteness in groups of automorphisms of trees*, Algebra i Logika **37** (5) (1998), 568–605.
[27] E.I. Sedova, *Periodic F^*-groups with additional finiteness conditions*, Abstracts of cand. phys. mat. nauk, Krasnoyarsk (1985).
[28] V.I. Senashov, *Characterization of layer-finite groups in the class of periodic groups*, Algebra i Logika **24** (5) (1985), 608–617.
[29] V.I. Senashov, *Characterization of layer-finite groups*, Algebra i Logika **28** (6) (1989), 687–704.
[30] V.I. Senashov, *Layer-Finite Groups*, Nauka (1993) (in Russian).
[31] V.I. Senashov, *Groups with minimality condition*, Proceedings of the International Conference "Infinite Groups 1994", Berlin, Walter de Gruyter (1995), 229–234.
[32] V.I. Senashov, *Groups with layer-finite periodic part*, Siberian Math. J. **38** (6) (1997), 1374–1386.
[33] V.I. Senashov, *Characterization of generalized Chernikov groups*, Dokl. Akad. Nauk **352** (6) (1997), 309–310.
[34] V.I. Senashov, *Characterization of generalized Chernikov groups in groups with involutions*, Math. Notes **62** (4) (1997), 577–588.
[35] V.I. Senashov, *On a question of V.P. Shunkov*, Siberian Math. J. **39** (5) (1998), 1154–1156.
[36] V.I. Senashov, *Sufficient conditions of almost layer-finiteness of a group*, Ukrainian Math. J. **51** (4) (1999), 472–485.
[37] V.I. Senashov, *Almost layer-finiteness of periodic group without involutions*, Ukrainian Math. J. **51** (11) (1999), 1529–1533.
[38] V.I. Senashov, *Structure of an infinite Sylow subgroup in some periodic Shunkov's groups*, Discrete Math. **14** (4) (2002), 133–152.
[39] V.I. Senashov and V.P. Shunkov, *Groups with Finiteness Conditions*, Publishing house of Siberian Division of Russian Academy of Science (2001) (in Russian).
[40] V.I. Senashov and V.P. Shunkov, *Almost layer-finiteness of the periodic part of groups without involutions*, Discrete Math. **13** (4) (2003), 391–404.
[41] A.A. Shafiro and V.P. Shunkov, *On locally finite groups with Chernikov centralizers of involutions*, Research in Group Theory, Institute of Physic of Siberian Division of Academy of Sciences SSSR, Krasnoyarsk (1975), 128–146.

[42] V.P. Shunkov, *On the theory of generally soluble groups*, Proceedings of Academy of Sciences of USSR **160** (6) (1965), 1279–1282.
[43] V.P. Shunkov, *On the theory of locally finite groups*, Proceedings of Academy of Sciences of USSR **168** (6) (1966), 1272–1274.
[44] V.P. Shunkov, *A generalization of the Frobenius theorem to periodic groups*, Algebra i Logika **6** (3) (1967), 113–124.
[45] V.P. Shunkov, *On the problem of minimality for locally finite groups*, Algebra i Logika **9** (2) (1970), 220–248.
[46] V.P. Shunkov, *On a class of p-groups*, Algebra i Logika **9** (4) (1970), 484–496.
[47] V.P. Shunkov, *On locally finite groups with minimality condition for Abelian subgroups*, Algebra i Logika **9** (5) (1970), 579–615.
[48] V.P. Shunkov, *On one indication of non-simplicity of groups*, Algebra i Logika **14** (5) (1975), 576–603.
[49] V.P. Shunkov, *On sufficient conditions for existence in a group of infinite locally finite subgroups*, Algebra i Logika **15** (6) (1976), 716–737.
[50] V.P. Shunkov, *A group with finitely embedding involution*, Algebra i Logika **19** (1) (1990), 102–123.
[51] V.P. Shunkov, *On the Embedding of Prime Order Elements in a Group*, Novosibirsk, Nauka (1992) (in Russian).
[52] V.P. Shunkov, *On placement of involutions in a group*, Siberian Math. J. **34** (2) (1993), 208–219.
[53] V.P. Shunkov, *T_0-groups*, Math. Transactions **1** (1) (1998), 139–202.
[54] V.P. Shunkov, *T_0-groups and their place in group theory*, Ukrainian Math. J. **51** (4) (1999), 572–576.
[55] V.P. Shunkov, *T_0-Groups*, Novosibirsk, Nauka (2000) (in Russian).
[56] V.P. Shunkov, *On the placement of prime order elements in a group*, Ukrainian Math. J. **54** (6) (2002), 881–884.
[57] V.P. Shunkov and A.A. Shafiro, *On a characterization of general Chernikov groups*, Materials of the 15th All Union Algebraic Conf., Krasnoyarsk State University, Krasnoyarsk (1979), 185.
[58] A.I. Sozutov and V.P. Shunkov, *On infinite groups 2*, Algebra i Logika **18** (2) (1979), 206–223.
[59] S.P. Strunkov, *Subgroups of periodic groups*, Dokl. Akad. Nauk SSSR **170** (1966), 279–281.
[60] N.G. Suchkova and V.P. Shunkov, *On the groups with minimality condition for Abelian subgroups*, Algebra i Logika **25** (4) (1986), 445–469.

Subject Index